Applied Mathematical Sciences
Volume 149

For further volumes:
http://www.springer.com/series/34

Kiyohiro Ikeda • Kazuo Murota

Imperfect Bifurcation in Structures and Materials

Engineering Use of Group-Theoretic Bifurcation Theory

Second Edition

Springer

Dr. Kiyohiro Ikeda
Department of Civil Engineering
Tohoku University
Aramaki-aza-Aoba
980-8579 Sendai
Aoba-ku
Japan
ikeda@civil.tohoku.ac.jp

Dr. Kazuo Murota
Department of Mathematical Informatics
Graduate School of Information Science & Technology
University of Tokyo
Hongo 7-3-1
113-8656 Tokyo
Bunkyo-ku
Japan
murota@mist.i.u-tokyo.ac.jp

ISSN 0066-5452
ISBN 978-1-4614-2665-3 ISBN 978-1-4419-7296-5 (eBook)
DOI 10.1007/978-1-4419-7296-5
Springer New York Dordrecht Heidelberg London

Mathematics Subject Classification (2010): 74-X, 74Gxx, 74G60, 58-XX, 58D19, 58E09, 58Kxx, 58K70, 20-xx, 20Cxx, 34-XX, 34Fxx

Printed on acid-free paper

Springer is part of Springer Science+Business Media (www.springer.com)

Contents

Preface

The first edition of this book was published in 2002 for an audience of applied mathematicians and engineers. The response to the first edition, represented by several book reviews, has been most heartening. Accordingly, the second edition of this book maintains its major framework and serves as an expanded revision of our previous work.

In the second edition, the theoretical backgrounds of group representation theory are strengthened and made self-contained, in response to a request of a book reviewer and students of the authors. Based on these strengthened backgrounds, a theory and a numerical procedure on block-diagonalization are presented. Among a number of methodologies, block-diagonalization analysis has come to be acknowledged as a systematic and rigorous procedure for symmetry exploitation for the following two purposes:

- Gain better insight into bifurcation behaviors via blockwise singularity detection.
- Enhance the computational efficiency and accuracy of the numerical analysis.

Included in the second edition are up-to-date topics of bifurcation analysis of diverse materials from rectangular parallelepiped sand specimens to honeycomb cellular solids. Theory, experimentation, and numerical analysis have been used in a synthetic manner to untangle the mechanism of shear band formation of these sand specimens. For the honeycomb cellular solids, the mechanism to engender flower patterns has been elucidated by the analysis of a group defined as the semidirect product of D_6 and $\mathbb{Z}_2 \times \mathbb{Z}_2$. This group is highlighted in view of its suitability for the description of spatial symmetries of a uniform planar domain. This is in line with the tendency that the interest in engineering is shifted from structures to materials.

The title of the book *Imperfect Bifurcation in Structures and Materials* remains unchanged. Structures do not mean mathematical structures but structures in the real world, such as buildings, domes, dams, and towers. Because all physical systems are "imperfect," in that they inevitably involve some imperfections, the study of imperfect bifurcation (bifurcation of imperfect systems) is of great mathematical interest.

Many physical systems lose or gain stability and pattern through bifurcation. Extensive research of such bifurcation behavior is carried out in science and engineering. The study of dynamic bifurcation behavior, for example, has clarified the mechanism of dynamic instability and chaos. Group-theoretic bifurcation theory is an established means to deal with the formation and selection of patterns in association with symmetry-breaking bifurcations.

In structural mechanics, bifurcation behavior has been studied to model the buckling and failure of structural systems. The sharp reduction of the strength of structural systems by initial imperfections is formulated as imperfection sensitivity laws. A series of statistical studies has been conducted to clarify the dependence of the strength of structures on the statistical variation of imperfections. A difficulty in these studies arises from the presence of a large number of imperfections. At this state, most of these studies are conducted based on the Monte Carlo simulation for several imperfections, or, on an imperfection sensitivity law against a single imperfection.

In geomechanics, the predominant role of bifurcation behavior in strengths and deformation patterns of the geomaterials, sand and soil, has come to be acknowledged. Yet the experimental behavior of geomaterials is obscured to a great degree by the presence of imperfections; moreover, observed curves of force versus displacement can differ qualitatively from the bifurcation diagrams predicted by mathematics. Although defects in geomaterials are known to form geometrical patterns, the underlying mechanism of these patterns remains open.

To sum up, notwithstanding extensive studies of bifurcation behavior in many fields of research, a gap exists between mathematical theory and engineering practice. In an attempt to fill this gap, this book presents a modern view of static imperfect bifurcation.

The major objective of this book is to present:

- Theories on the strength variation of (structural) systems due to imperfections,
- A systematic technique for addressing bifurcation diagrams to be observed in experiments as opposed to conceptual and schematic diagrams in mathematics,
- A method for revealing the mechanism of pattern formation of uniform materials,

as extensions of the basic tools:

- Asymptotic bifurcation theory,
- Statistical approach to random imperfections,
- Group-theoretic bifurcation theory.

This book consequently offers a wider and deeper insight into imperfect bifurcation. Our approach to imperfections is pragmatic, rather than mathematically rigorous; it is intended to be an introduction for students in engineering by minimizing the mathematical formalism and by including details of derivations. This book will also be of assistance to mathematicians, showing how bifurcation theory is applied to actual problems. A proper modeling of symmetries of systems, for example, leads to a proper understanding of their bifurcation behavior with the aid of group-theoretic bifurcation theory. This book offers a number of strategies, based on up-to-date mathematics, to untangle the mechanism of actual physical and structural problems undergoing bifurcation, such as soil, sand, kaolin, steel, concrete, and regular-polygonal domes. The present approach to bifurcation is successfully applied to the experimental behaviors of materials. In particular, the symmetry-breaking bifurcation behaviors of uniform materials are introduced as an essential source of the emergence of patterns on the surface of materials. The horizon of static bifurcation has thus been extended.

Theoretically, bifurcation is associated with an instability induced by a singular Jacobian matrix of a system, the linearized eigenvalue problem. A critical (singular) point is the one at which one or more eigenvalues of this matrix vanish. It is such a point (with some additional conditions) at which bifurcation occurs. According to whether the number of zero eigenvalue(s) is equal to or greater than one, the critical point is classified into two types:

- A simple critical point,
- A multiple critical point.

The bifurcation behavior at the multiple critical point is far more complex than that at the simple critical point in that more paths can potentially branch.

This book is divided into three parts. In Part I we aim at a fundamental understanding of the concepts and theories of imperfections related to simple examples, focusing on simple critical points. In Part II we extend them to systems with geometrical symmetries, for which multiple criticality appears generically. In Part III we tackle the bifurcation behavior of realistic systems of various kinds and, in turn, address the issue of modeling symmetries of these systems. The contents of this book are outlined below.

Chapter 1: Overview of Book. This first chapter offers an overview of the book to highlight important results.

Part I: Imperfect Behavior at Simple Critical Points. This first part is devoted to the study of imperfect behavior in the vicinity of a simple critical point. With the help of the simplicity due to simple criticality, imperfect behaviors are investigated in an asymptotic sense. Here the word "asymptotic" means that the results are valid in a sufficiently close neighborhood of the critical point under consideration for a sufficiently small value of imperfection(s). Various important aspects of imperfect behavior, such as the bifurcation equation, imperfection sensitivity, the worst imperfection, probabilistic variation, and observability are introduced in Chapters 2 to 6, respectively. It is to be remarked that the term "critical imperfection" in the first edition is replaced by the "worst imperfection" in the second edition, as the latter is used more commonly in structural mechanics. Emphasis is placed on the case of multiple imperfection parameters, although it is customary in the literature to deal with one or two parameters.

Part II: Imperfect Bifurcation of Symmetric Systems. In this second part, we extend the results of Part I to multiple critical points of symmetric systems. To avoid sophisticated mathematical concepts, we focus on the simplest groups, the dihedral and cyclic groups. Nonetheless, the basic strategy presented is essentially general and is extensible to other groups. In Chapter 7 group-theoretic bifurcation theory is introduced as a mathematical tool to deal with group-theoretic degeneracy. Chapter 8 presents the theory of perfect and imperfect bifurcation behaviors in the vicinity of a critical point of a system with dihedral or cyclic group symmetry. This theory is applied to spherical domes. The worst imperfection and probabilistic variation of imperfections are studied, respectively, in Chapters 9 and 10. In Chapter 11 perfect and imperfect behaviors of domes and soil specimens are investigated by a synthetic application of the procedures presented in Part II. In Chapter 12 the concrete computational procedure for block-diagonalization in bifurcation analysis is explained as a new ingredient of the second edition.

Part III: Modeling of Bifurcation Phenomena. In this third part, we study the bifurcation behaviors of various physical and structural systems by modeling their symmetries appropriately. In Chapter 13 the recursive change of the shapes of cylindrical sand specimens undergoing bifurcation is investigated. In Chapter 14 the mechanism of echelon-mode formation on sand, kaolin and steel specimens is revealed by investigating the bifurcation of an $O(2)\times O(2)$-equivariant system. In Chapter 15 the recursive bifurcation of rectangular parallelepiped steel specimens is studied. Chapter 16 offers, as an up-to-date topic added in the second edition, an analysis of flower patterns of honeycomb cellular solids, which have the symmetry expressed by the semidirect product of D_6 and $\mathbb{Z}_2 \times \mathbb{Z}_2$.

We would like to express our gratitude to those who have contributed to the realization of the first edition of the book. Critical reading of the text by R. Tanaka was invaluable. We thank H. Okamoto and M. Sugihara for helpful comments. We are grateful to M. Nakano, T. Nishimaki, T. Noda, N. Oguma, M. Osada, I. Saiki, and I. Sano for offering important photographs and figures. The assistance of T. Ichimura, Y. Sudo, and Y. Yamakawa in preparing the manuscript is most appreciated. The enthusiastic support of M. Peters and the help of A. Dosanjh, B. Howe, A. D. Orrantia, and R. Putter were indispensable for the publication of the first edition.

In the second edition, the discussion with J. Desrues was vital in the image simulation of shear bands on sands. We are grateful to F. Fujii for offering important figures. Comments of T. Yaguchi were helpful in the revision.

Sendai and Tokyo, *Kiyohiro Ikeda*
March 2010 *Kazuo Murota*

Notation

$\approx$	= asymptotically equal	
$\simeq$	= isomorphic (group)	§7.3
$\langle\cdot\rangle$	= group generated by elements in parentheses	
$\lvert\cdot\rvert$	= size of a set, order of a group	
$(\cdot)^0$	= variable associated with a perfect system	(2.6)
$(\cdot)_\mathrm{c}$	= variable associated with a critical point	(2.2)
$(\cdot)^\top$	= transpose of a vector or a matrix	
$(\cdot)^{-\top}$	= transpose of the inverse of a matrix	
$(\cdot)^\perp$	= orthogonal complement of a subspace	§7.4.1
$(\cdot)^*$	= complex conjugate of a vector or a matrix	
$\overline{\cdot}$	= complex conjugate of a complex number	(8.35)
$\setminus$	= difference of sets	§7.4.3
$\lfloor\cdot\rfloor$	= largest integer not larger than a real number	§8.2.2
$\oplus$	= direct sum	
$\otimes$	= tensor product	
$\times$	= direct product	
$\dot{+}$	= semidirect product of groups	(16.4)
α	= imperfection influence factor	(4.23)
γ, γ^*	= constants for asymptotic sensitivity laws	(6.32), (6.33)
$\Gamma(\cdot)$	= gamma function	§5.2.2
δ_{ij}	= Kronecker's delta	§2.2.3
ε	= magnitude of imperfection	(2.2)
ε_a	= axial strain	§6.5.2
ζ	= normalized incremental critical load	(5.8)
ζ_K	= minimum of ζ for K random imperfections	§5.4
η_i	= ith right eigenvector of J_c^0	§2.2.2
κ	= index of partitioning of displacements	(12.30)
λ_i	= ith eigenvalue of J	§2.2.3
μ	= irreducible representation	§7.4.2
ν	= Poisson's ratio	§14.6.1

ξ	= type of an orbit	(12.37)
$\boldsymbol{\xi}_i$	= ith left eigenvector of J_{c}^0	§2.2.2
Ξ	= permutation matrix	(12.33)
Π	= permutation matrix	(7.117)
Π^μ	= permutation matrix for μ	(7.116)
ρ	= exponent in imperfection sensitivity law	(3.7)
$\widetilde{\sigma}$	= variable related to the variance of f_{c}	(5.6)
σ	= reflection	(7.3)
$\sigma_{\mathrm{h}}, \sigma_{\mathrm{v}}$	= reflections with respect to horizontal and vertical planes	§ 13.2.1
$\sigma_x, \sigma_y, \sigma_z$	= reflections with respect to the yz-, zx-, and xy-planes	§8.2.1
σ_{a}	= axial stress	§6.5.2
$\Sigma(\cdot)$	= isotropy subgroup of a vector or a subspace	(7.86)
$\phi_\zeta(\cdot)$	= probability density function of ζ	(5.11)
$\phi_{\mathrm{N}}(\cdot)$	= density of the standard normal distribution N(0, 1)	(5.7)
$\Phi_\zeta(\cdot)$	= cumulative distribution function of ζ	(5.12)
$\Phi_{\mathrm{N}}(\cdot)$	= cumulative distribution function of N(0, 1)	(5.14)
ω	= complex variable expressing a rotation	(8.50), (12.64)
a^μ	= multiplicity of irreducible representation μ	(7.23)
a	= imperfection coefficient	(4.21), (8.46)
$\arg(\cdot)$	= argument of a complex number	(8.69)
A	= cross-sectional area of a truss member	(2.14)
B	= imperfection sensitivity matrix	(2.3)
$c(\varphi)$	= counterclockwise rotation about the z-axis at angle φ	(13.2)
$\mathbb{C}$	= set of complex numbers	
$C(\boldsymbol{d})$	= coefficient for the imperfection sensitivity	(3.6)
C_∞	= symmetry group of a circular domain	(13.8)
$\mathrm{C}_{\infty\mathrm{v}}$	= symmetry group of a barrel domain	(13.8)
C_{i}	= group of inversion	(13.9)
C_n	= cyclic group of degree n	§8.2.1
$\mathrm{C}_{n\mathrm{v}}, \mathrm{C}_{n\mathrm{h}}$	= groups in the Schoenflies notation	(13.8)
$\boldsymbol{d}$	= imperfection pattern vector	(2.2)
$\boldsymbol{d}^*$	= worst imperfection pattern vector	(4.24)
$\det(\cdot)$	= determinant of a matrix	(2.5)
$\mathrm{diag}(\cdot)$	= diagonal matrix with diagonal entries	
$\dim(\cdot)$	= dimension of a linear space	
$\mathrm{D}_{\infty\infty}$	= symmetry group of a cylindrical domain with twisting	(13.8)
$\mathrm{D}_{\infty\mathrm{h}}$	= symmetry group of a cylindrical domain	(13.8)
D_n	= dihedral group of degree n	(8.1)
$\mathrm{D}_{n\mathrm{h}}, \mathrm{D}_{n\mathrm{d}}, \mathrm{D}_n$	= groups in the Schoenflies notation	(13.8)
$\mathrm{DI}_{n\tilde{n}}$	= symmetry group of a diamond pattern	(14.15)
e	= base of natural logarithm	
e	= identity element of a group	§7.3

E	= Young's modulus	(2.14)
	or set of nodal coordinates	§12.3.1
E_m	= a subset of nodal coordinates	(12.28)
$\mathrm{E}[\cdot]$	= mean of a random variable	§5.2.2
$\mathrm{E}_{\mathrm{sample}}[\cdot]$	= sample mean of a random variable	§5.3
$\mathrm{EC}^{+}_{n\tilde{n}kl}$, $\mathrm{EC}^{-}_{n\tilde{n}kl}$	= symmetry groups of echelon mode	(14.18), (14.20)
f	= bifurcation parameter, loading parameter	(2.1)
$\widetilde{f}$	= increment of bifurcation parameter	(2.33)
F	= $\mathbb{R}$ or $\mathbb{C}$	§7.4.1
$\boldsymbol{F}$	= governing or equilibrium equation	(2.1)
$\widetilde{F}$, $\widehat{F}$, $\widetilde{\boldsymbol{F}}$	= bifurcation equations	(2.39), (2.65), (7.66)
g	= element of a group	§7.3
$\mathrm{gcd}(\cdot,\cdot)$	= greatest common divisor	(8.11)
G	= group	§7.3
G^{μ}	= subgroup of G associated with μ	(7.22)
$\mathrm{GL}(V)$	= group of all nonsingular linear transformations of V onto itself	§7.4.1
$\mathrm{GL}(N,F)$	= group of all nonsingular matrices over F of order N	§7.4.1
H	= transformation matrix for block-diagonalization	(7.120)
H^{μ}	= block of matrix H associated with μ	(7.122)
i	= imaginary unit	
I_n	= $n\times n$ identity matrix	
$\mathrm{Im}(\cdot)$	= imaginary part of a complex number	(8.47)
J	= Jacobian matrix	(2.4)
$\overline{J}$	= block-diagonal form of J	§2.4.2, (7.105)
$\ker(\cdot)$	= kernel space of a matrix	(2.7)
M	= multiplicity of a critical point	(2.7)
$\widehat{n}$	= index associated with a two-dimensional irreducible representation of D_n	(8.11)
N	= dimension of unknown vector $\boldsymbol{u}$	§2.2.1
N^{μ}	= dimension of irreducible representation μ	§7.4.2
$\mathrm{N}(0,1)$	= standard normal distribution	§5.2.1
$\mathrm{N}(\boldsymbol{0},W)$	= normal distribution with mean $\boldsymbol{0}$ and variance–covariance W	§5.2.1
$\mathrm{o}(\cdot)$	= quantity of a smaller order than the term in parentheses	
$\mathrm{O}(\cdot)$	= quantity of a smaller order or the same order	
$\mathrm{O}(2)$	= two-dimensional orthogonal group	(13.5)
$\mathrm{OB}^{+}_{n\tilde{n}}$, $\mathrm{OB}^{-}_{n\tilde{n}}$	= symmetry groups of oblique stripe pattern	(14.16), (14.17)
$\mathrm{orb}(\cdot)$	= orbit of a vector	(7.98)
p	= dimension of imperfection parameter vector $\boldsymbol{v}$	(7.62)
P	= projection matrix	(7.62)
	or set of nodes	(12.27)
P_l	= an orbit of nodes	(12.27)

Q	= transformation matrix for irreducible decomposition	§7.4.2
Q^{μ}	= block of matrix Q associated with μ	(7.102)
$\hat{Q}_{\kappa}$	= local transformation matrix for κ	§12.3.2
$\hat{Q}_{\xi}$	= local transformation matrix for type ξ	(12.40)
r, s	= generators of a dihedral group	§7.3, (8.2)
r, θ	= polar coordinates	§8.6.1
range$(\cdot)$	= range space of a matrix	§7.4.3
rank$(\cdot)$	= rank of a matrix	(2.7)
$\mathbb{R}$	= set of real numbers	
$R(\cdot)$	= family of irreducible representations of a group	§7.4.2
$R_{\mathrm{a}}(\cdot)$	= family of absolutely irreducible representations of a group	§7.4.3
R_n	= 2×2 matrix expressing rotation at a angle of $2\pi/n$	(12.25)
$R_{\zeta}(\cdot)$	= reliability function of ζ	(5.13)
Re$(\cdot)$	= real part of a complex number	(8.39)
S	= 2×2 matrix expressing reflection	(12.25)
$S(g)$	= representation matrix for $\boldsymbol{v}$	(7.38)
S_n	= group in the Schoenflies notation	(13.8)
SO(2)	= two-dimensional rotation group (special orthogonal group)	
sign$(\cdot)$	= sign of a variable	
$t(l)$	= z-directional translation at a length of l	(14.6)
$T(g)$	= representation matrix for $\boldsymbol{u}$	(7.12)
$\widetilde{T}(g)$	= representation matrix for $\boldsymbol{w}$	(7.54)
$T^{\mu}(g)$	= representation matrix of an irreducible representation μ	§7.4.2
$\hat{T}$	= linear mapping $G \to \mathrm{GL}(V)$	§7.4.1
trace$(\cdot)$	= trace of a matrix	
$\boldsymbol{u}$	= state variable vector, displacement vector	(2.1)
$\hat{\boldsymbol{u}}$	= rearranged state variable vector	(12.32)
$\boldsymbol{u}_i$	= nodal displacement vector	(12.18)
U	= total potential energy function	(2.11)
	or a complementary subspace of $\ker(J_{\mathrm{c}}^{0})$	(7.59)
$\boldsymbol{v}$	= imperfection parameter vector	(2.1)
V	= representation space	§7.4.1
	or a complementary subspace of range(J_{c}^{0})	(7.60)
V^{μ}	= isotypic subspace associated with μ	(7.25)
Var$[\cdot]$	= variance of a random variable	§5.2.2
$\mathrm{Var}_{\mathrm{sample}}[\cdot]$	= sample variance of a random variable	§5.3
$w, \boldsymbol{w}$	= incremental transformed state variable (vector)	(2.32), (7.61)
W	= weight matrix for $\boldsymbol{d}$	(4.18)
	or variance–covariance matrix of $\boldsymbol{d}$	§5.2.1
x, y, z	= Cartesian coordinates	
x_i, y_i, z_i	= nodal displacements	(12.18)
z	= complex coordinate	(8.35)
$\mathbb{Z}$	= set of integers	
$\mathbb{Z}_2$	= two-element group or cyclic group of order two	(13.6)

Chapter 1
Overview of Book

1.1 Introduction

This book offers several systematic methods based on up-to-date mathematics to untangle the mechanism of bifurcation of structures and materials, such as soil, sand, kaolin, and concrete. Throughout this book, we place more emphasis on engineering pragmatism than on mathematical rigor. We offer, in this first chapter, an overview of the book using a series of illustrative examples and photographs.

First, an overview of Parts I and II is given. We present in a logical sequence the fundamental issues of a static problem:

- Governing equation with imperfections,
- Simple examples of bifurcation behavior,
- Imperfection sensitivity law.

Moreover, we present theoretical tools and concepts:

- Worst imperfections[1] of structural systems,
- Random imperfections of structures and materials,
- Experimentally observed bifurcation diagrams,
- Group-theoretic bifurcation theory.

The first three of the above deal with asymptotic and probabilistic issues around imperfection sensitivity, and group-theoretic bifurcation theory is a standard tool to describe bifurcation under symmetry.

Next, we present an overview of Part III that gives the application of a series of tools and concepts to analyze experimental results with reference to their symmetry. We cover the following topics:

- Recursive (secondary, tertiary, ...) bifurcation and mode switching of sands,
- Echelon mode formation on uniform materials,
- Recursive bifurcation of steel specimens,

[1] The term "critical imperfection" in the first edition is replaced by the term "worst imperfection" in the second edition, as the latter is used more commonly in structural mechanics.

K. Ikeda and K. Murota, *Imperfect Bifurcation in Structures and Materials*,
Applied Mathematical Sciences 149, DOI 10.1007/978-1-4419-7296-5_1,

- Flower patterns on honeycomb structures.

Group-theoretic bifurcation theory is employed throughout Part III to obtain an exhaustive list of possible symmetries of bifurcated solutions.

This chapter is organized as follows.

- Major theoretical tools and concepts that are introduced in this book are overviewed in §1.2.
- Theoretical description of the bifurcation behaviors of several symmetric systems are overviewed in §1.3.

1.2 Overview of Theoretical Tools and Concepts

We overview fundamental issues of a static problem and major theoretical tools and concepts.

1.2.1 Governing Equation with Imperfections

In this book, we study a static bifurcation problem of a system expressed in terms of a nonlinear governing (equilibrium) equation. Chapter 2 presents the governing equation of the form

$$\boldsymbol{F}(\boldsymbol{u},f,\boldsymbol{v})=\boldsymbol{0}. \tag{1.1}$$

Therein, $\boldsymbol{v}$ denotes a p-dimensional *imperfection parameter vector*; $\boldsymbol{u}$ signifies an N-dimensional *state variable vector*, and f denotes a *bifurcation parameter*; in structural engineering, $\boldsymbol{u}$ signifies a *displacement vector* and f denotes a *loading parameter*. We assume $\boldsymbol{F}$ to be a sufficiently smooth nonlinear function. It is emphasized that we have a distinguished single parameter f, which, mathematically, plays the role of bifurcation parameter.

The system may or may not have the *potential function* $U(\boldsymbol{u},f,\boldsymbol{v})$. For a *potential system*, we have a condition[2]

$$\boldsymbol{F}(\boldsymbol{u},f,\boldsymbol{v})=\left(\frac{\partial U(\boldsymbol{u},f,\boldsymbol{v})}{\partial \boldsymbol{u}}\right)^{\top} \tag{1.2}$$

by the principle of stationary potential energy.[3]

The nominal state of the system is designated as the *perfect system*, which is assumed to correspond to $\boldsymbol{v}=\boldsymbol{v}^0$. A system with an imperfection[4] (i.e., a system deviated from the nominal state) is called an *imperfect system*. In dealing with imperfect systems that are close to the perfect system, we often express the imperfection

---

[2] It is assumed that $\boldsymbol{F}$ is a column vector and $\partial U/\partial \boldsymbol{u}$ is a row vector.

[3] See, for example, Theorem VIII on page 305 of Oden and Ripperger, 1981 [152].

[4] Imperfection is called *initial imperfection* in structural engineering.

parameter vector $\boldsymbol{v}$ as

$$\boldsymbol{v} = \boldsymbol{v}^0 + \varepsilon \boldsymbol{d}, \tag{1.3}$$

where $\boldsymbol{d}$ is called the *imperfection pattern vector* and ε denotes the magnitude of imperfection that represents the amount of deviation from the perfect case. It is noted that the pattern $\boldsymbol{d}$ and the magnitude ε of imperfection parameters are addressed separately.

We are concerned with the loci of $(\boldsymbol{u}, f) = (\boldsymbol{u}(\boldsymbol{v}), f(\boldsymbol{v}))$ that satisfies (1.1) for an imperfect system described by $\boldsymbol{v}$. Specifically, we examine the local behavior in a neighborhood of a *critical point*, at which the Jacobian matrix

$$J(\boldsymbol{u}, f, \boldsymbol{v}) = \frac{\partial \boldsymbol{F}(\boldsymbol{u}, f, \boldsymbol{v})}{\partial \boldsymbol{u}}$$

of $\boldsymbol{F}$ is singular (by definition). Consequently, bifurcation can occur. A critical point is identified as a point where the Jacobian matrix J has zero eigenvalue(s); that is,

$$J\boldsymbol{\eta}_i = \boldsymbol{0}, \qquad i = 1, \ldots, M$$

for M linearly independent vectors $\boldsymbol{\eta}_i$ $(i = 1, \ldots, M)$. Here M denotes the *multiplicity* of the critical point.

Remark 1.1. In this book we restrict ourselves to equations in finite-dimensional unknown vectors, which may be obtained through discretization of differential equations. This enables us to examine central issues of engineering interest specifically, without entering into mathematical complications. The methodology presented in this book can be extended to the case of continuous independent variables, as is expounded in the literature.[5]

With the static bifurcation problem in (1.1), we may associate a dynamical system

$$\frac{d\boldsymbol{u}}{dt} + \boldsymbol{F}(\boldsymbol{u}, f, \boldsymbol{v}) = \boldsymbol{0}$$

and consider dynamic bifurcation problems. The study of dynamic bifurcation is beyond the scope of this book, although it is an important issue. For dynamical problems the readers are referred to the literature.[6] □

1.2.2 Simple Examples of Bifurcation Behavior

Simple examples of systems undergoing bifurcation are investigated.

---

[5] See, for example, Keller and Antman, 1969 [114]; Thompson and Hunt, 1973 [191]; Golubitsky and Schaeffer, 1985 [62]; Bažant and Cedolin, 1991 [12]; and Antman, 1995 [3].

[6] See, for example, Marsden and Hughes, 1983 [133]; Thompson and Hunt, 1984 [192]; Wiggins, 1988 [210]; Iooss and Joseph, 1990 [108]; Hale and Koçak, 1991 [67]; Seydel, 1994 [184]; and Kuznetsov, 1995 [128].

One-Degree-of-Freedom System

The cantilever supported by a linear spring shown in Fig. 1.1(a) is employed as an example of a one-degree-of-freedom system. The loci of $(\boldsymbol{u}, f)$ or, in other words, the equilibrium paths of this system, are obtained to show the mechanism of bifurcation.

The cantilever is rigid and of length L, and the linear spring has a spring constant of k. The vertical force kLf and the horizontal disturbance force $kL\varepsilon$ are applied at the top of the cantilever, where f is a loading parameter, and is chosen as the bifurcation parameter for this case; ε is an imperfection magnitude. For the perfect system ($\varepsilon = 0$), the disturbance load vanishes. The deflected angle of the cantilever is denoted by u ($|u| < \pi/2$) (cf., Fig. 1.1(b)).

The total potential energy[7] of this system is given as

$$\begin{aligned} U(u, f, \varepsilon) &\\ = &\int (\text{spring force})\, d(\text{spring displacement}) - \sum (\text{force}) \times (\text{force displacement}) \\ = &\frac{1}{2}(kL\sin u)(L\sin u) - kL^2 f(1-\cos u) - kL^2 \varepsilon \sin u. \end{aligned} \tag{1.4}$$

By (1.2), differentiating U in (1.4) with respect to u gives the governing equation

$$F(u, f, \varepsilon) = \frac{\partial U}{\partial u} = kL^2 \left(\frac{1}{2} \sin 2u - f \sin u - \varepsilon \cos u \right) = 0. \tag{1.5}$$

We first consider the perfect system with $\varepsilon = 0$. Then (1.5) has the following solutions

$$\begin{cases} u = 0, & \text{fundamental path,} \\ f = \cos u, & \text{bifurcated path,} \end{cases}$$

which are depicted as solid lines in Fig. 1.2. The fundamental and bifurcated paths[8] are connected at a bifurcation point[9] $(u, f) = (0, 1)$, shown as ($\circ$). On the fundamental path, the cantilever remains upright, whereas it tilts on the bifurcated path.

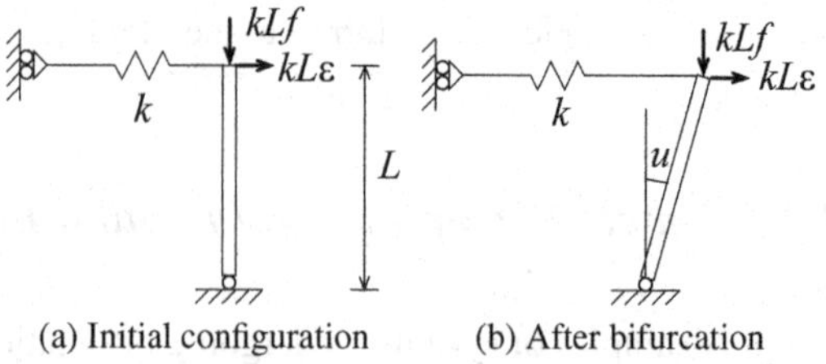

Fig. 1.1 Cantilever supported by a linear spring.

---

[7] It is often simpler to derive the potential energy function, and then to derive the governing equation.

[8] The *fundamental path* is often called the *primary* or *trivial path*, whereas the bifurcated path is called the *secondary* or *postbifurcation path*.

[9] This point is usually called the *pitchfork bifurcation point* in that the shape of these paths resembles a pitchfork.

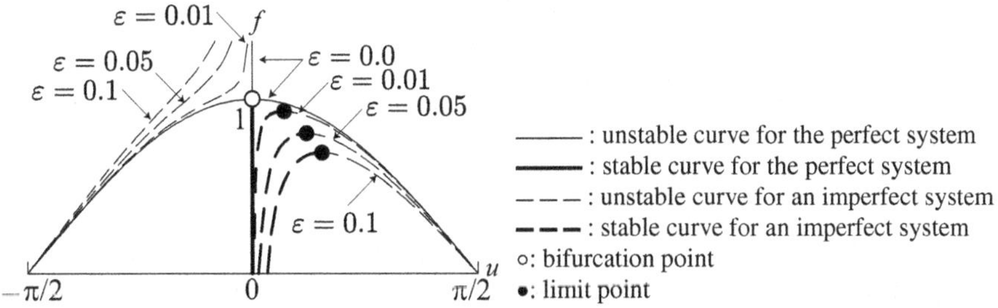

Fig. 1.2 Equilibrium paths of the cantilever supported by a linear spring. Imperfect paths are plotted for $\varepsilon > 0$.

The perfect system, in general, corresponds to idealistic modeling or, in other words, first-order modeling. A realistic system inevitably involves some imperfections; therefore, we proceed to implement the imperfection ε to achieve a more realistic modeling or, second-order modeling. For imperfect systems with $\varepsilon \neq 0$, (1.5) can be solved for f as

$$f = \cos u - \varepsilon \cot u. \tag{1.6}$$

The f versus u curves expressed by this equation are depicted as dashed lines in Fig. 1.2 for $\varepsilon = 0.01$, 0.05, and 0.1. The points portrayed as (•) are the limit points of f of those imperfect systems. The distance between the curves for the perfect system and the imperfect system becomes greater in direct relation with the increase in ε.

The bifurcation point (∘) for the perfect system and the limit point (•) for an imperfect system are instances of a *critical point*, which is defined to be a solution point such that the partial derivative of F with respect to u vanishes. It follows from (1.5) that

$$\frac{\partial F}{\partial u}(u, f, \varepsilon) = kL^2(\cos 2u - f \cos u + \varepsilon \sin u). \tag{1.7}$$

On the fundamental path $u = 0$ of the perfect system ($\varepsilon = 0$), we have

$$\frac{\partial F}{\partial u}(0, f, 0) = kL^2(1 - f). \tag{1.8}$$

This vanishes at $f = 1$, corresponding to the bifurcation point (∘). For the bifurcated path of an imperfect system, substituting (1.6) into (1.7) produces

$$\frac{\partial F}{\partial u}(u, \cos u - \varepsilon \cot u, \varepsilon) = -kL^2\left(\sin^2 u - \frac{\varepsilon}{\sin u}\right). \tag{1.9}$$

This vanishes at $u = \arcsin(\varepsilon^{1/3})$, corresponding to the limit point (•).

Stability (in the physical sense) of an equilibrium state is represented by the sign of $\partial F/\partial u$, which is equal to the second-order derivative of the potential U. For the perfect system, the stability of points on the fundamental path is categorized from (1.8) as

$$\begin{cases} \text{stable} & \text{if } f < 1, \\ \text{unstable} & \text{if } f > 1. \end{cases}$$

The points on the bifurcated path $f = \cos u$ (excluding the bifurcation point at $u = 0$) are unstable, since

$$\frac{\partial F}{\partial u}(u, \cos u, 0) = -kL^2 \sin^2 u < 0$$

for u with $0 < |u| < \pi/2$. On the bifurcated path of an imperfect system with $\varepsilon > 0$, on the other hand, the stability of points is categorized from (1.9) as

$$\begin{cases} \text{stable} & \text{if } 0 < u < \arcsin(\varepsilon^{1/3}), \\ \text{unstable} & \text{if } -\pi/2 < u < 0 \text{ or } \arcsin(\varepsilon^{1/3}) < u < \pi/2. \end{cases}$$

Therefore, the paths portrayed by thick curves in Fig. 1.2 are composed of stable points and those by thin lines are of unstable points.

Multiple-Degree-of-Freedom System

As a simple example of a system with multiple degrees of freedom, we investigate the bifurcation of the two-degree-of-freedom structural system shown in Fig. 1.3(a) that consists of two rigid bars supported by two springs. The horizontal force kLf and a pair of vertical disturbance forces $kL\varepsilon d_1$ and $kL\varepsilon d_2$ are applied.

The total potential energy of this system reads as

$$U(\boldsymbol{u}, f, \varepsilon, d_1, d_2) = kL^2\Big[\frac{3}{2}\sin^2 u_1 + (\sin u_1 + \sin u_2)^2 - (2 - \cos u_1 - \cos u_2)f \\ - (d_1 \sin u_1 + d_2(\sin u_1 + \sin u_2))\varepsilon\Big], \quad (1.10)$$

where $\boldsymbol{u} = (u_1, u_2)^\top$ denotes the independent variable vector, consisting of the deflection angles at nodes 1 and 2 (cf., Fig. 1.3(b)). The imperfection pattern vector is given by $\boldsymbol{v} = (\varepsilon d_1, \varepsilon d_2)^\top$ in the notation of (1.1).

Differentiation of (1.10) leads to the equilibrium equation

$$\boldsymbol{F}(\boldsymbol{u}, f, \varepsilon, d_1, d_2) = \boldsymbol{0} \quad (1.11)$$

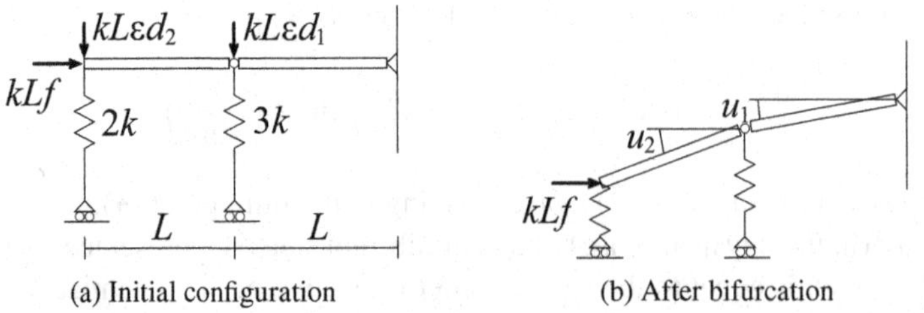

Fig. 1.3 Two rigid bars supported by springs.

with

$$F = \begin{pmatrix} \partial U/\partial u_1 \\ \partial U/\partial u_2 \end{pmatrix} = kL^2 \begin{pmatrix} \frac{5}{2}\sin 2u_1 + 2\cos u_1 \sin u_2 - f\sin u_1 - \varepsilon(d_1+d_2)\cos u_1 \\ 2\sin u_1 \cos u_2 + \sin 2u_2 - f\sin u_2 - \varepsilon d_2 \cos u_2 \end{pmatrix},$$

and the Jacobian matrix is given as

$$J = J(\boldsymbol{u}, f, \varepsilon) = \frac{\partial \boldsymbol{F}}{\partial \boldsymbol{u}} = \begin{pmatrix} J_{11} & J_{12} \\ J_{21} & J_{22} \end{pmatrix} = \begin{pmatrix} \partial^2 U/\partial {u_1}^2 & \partial^2 U/\partial u_1 \partial u_2 \\ \partial^2 U/\partial u_2 \partial u_1 & \partial^2 U/\partial {u_2}^2 \end{pmatrix}, \tag{1.12}$$

where

$$\begin{aligned} J_{11} &= kL^2[5\cos 2u_1 - 2\sin u_1 \sin u_2 - f\cos u_1 + \varepsilon(d_1+d_2)\sin u_1], \\ J_{12} &= J_{21} = 2kL^2 \cos u_1 \cos u_2, \\ J_{22} &= kL^2[-2\sin u_1 \sin u_2 + 2\cos 2u_2 - f\cos u_2 + \varepsilon d_2 \sin u_2]. \end{aligned} \tag{1.13}$$

For the perfect system with $\varepsilon = 0$, the fundamental path is associated with the trivial solution $u_1 = u_2 = 0$. On the fundamental path, the Jacobian matrix is evaluated as

$$J(\boldsymbol{0}, f, 0) = kL^2 \begin{pmatrix} 5-f & 2 \\ 2 & 2-f \end{pmatrix}.$$

This matrix becomes singular at the loads $f^{(1)} = 1$ and $f^{(2)} = 6$, which are called the critical loads (cf., Remark 1.2 below). The critical eigenvectors at these points—the eigenvectors of the Jacobian matrix associated with zero eigenvalues—are given by

$$\boldsymbol{\eta}^{(1)} = \frac{1}{\sqrt{5}} \begin{pmatrix} -1 \\ 2 \end{pmatrix}, \qquad \boldsymbol{\eta}^{(2)} = \frac{1}{\sqrt{5}} \begin{pmatrix} 2 \\ 1 \end{pmatrix}. \tag{1.14}$$

These eigenvectors are depicted in Fig. 1.4. At these critical points, which are simple bifurcation points, bifurcated paths shown in Fig. 1.5 are found in the directions of the eigenvectors in (1.14).

The imperfect behavior of this system is dependent both on the magnitude ε and the pattern $\boldsymbol{d}$ of the imperfection, and its equilibrium paths satisfying (1.11) differ, in general, for different patterns $\boldsymbol{d}$ even for the same magnitude ε. Consider, for example, a particular imperfection magnitude $\varepsilon = 0.05$ and a particular imperfection pattern

$$\boldsymbol{d} = \begin{pmatrix} d_1 \\ d_2 \end{pmatrix} = \begin{pmatrix} 1 \\ 2 \end{pmatrix}. \tag{1.15}$$

Then the solutions of (1.11) yield the equilibrium paths shown by the dashed lines in Fig. 1.5. The maximum loads to be attained by imperfect systems are governed by the limit points that appear in the vicinity of the bifurcation point. Maximum loads of imperfect systems are treated in Chapter 3.

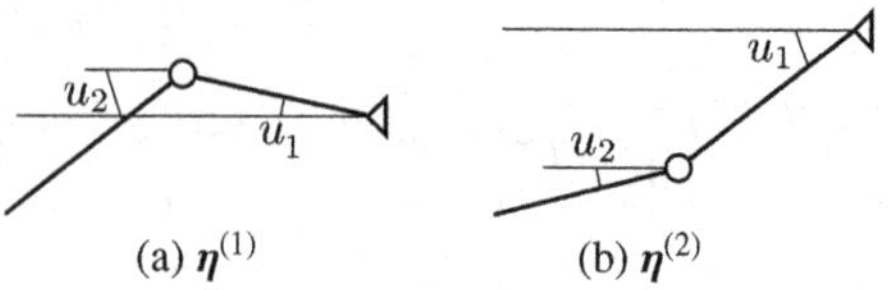

Fig. 1.4 Bifurcation modes of the two rigid bars.

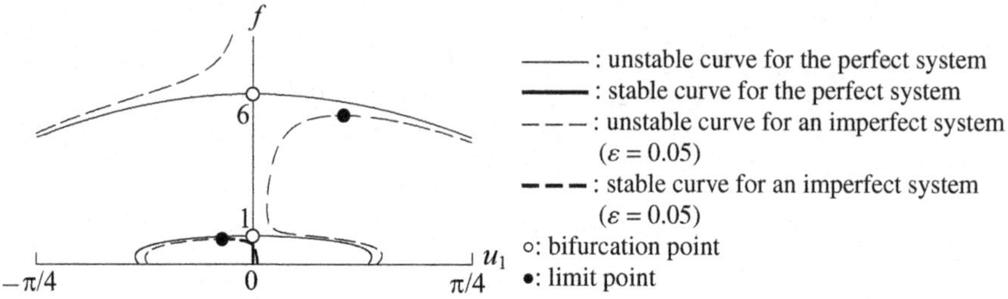

Fig. 1.5 Equilibrium paths of the two rigid bars.

Stability of the equilibrium points can be investigated by the eigenanalysis[10] of the Jacobian matrix (1.12) with (1.13) evaluated at the solution $(\boldsymbol{u}, f)$ of the governing equation (1.11). The paths denoted by thick curves in Fig. 1.5 consist of stable points and those by thin lines of unstable points.

Remark 1.2. In structural engineering, f is increased from zero until reaching the first critical load, $f^{(1)} = 1$ for this case, to evaluate the strength of the structural system under consideration. This load, accordingly, is of engineering importance. The deformation mode expressed by $\boldsymbol{\eta}^{(1)}$ is important because it is related to the collapse mode. □

1.2.3 Imperfection Sensitivity Law

All physical systems are "imperfect" in that they inevitably involve imperfections. Bifurcation behavior is ill-posed in that it is highly sensitive to the variation in imperfections. The strength of shell structures, for example, is well known to deteriorate sharply owing to the presence of small imperfections, such as uneven wall thickness and deformed initial shapes. The discrepancy between their theoretical strength and experimental strength had long been an annoying problem in the design of shell structures until the mechanism of the sensitivity of the strength to imperfections was formulated as imperfection sensitivity laws by Koiter, 1945 [119].

The imperfection sensitivity is demonstrated here using the cantilever in §1.2.2. To investigate the asymptotic properties of the solution in the neighborhood of the

[10] A system is stable if every eigenvalue of the Jacobian matrix has a positive real part, and is unstable if at least one eigenvalue has a negative real part (cf., §2.2.4).

bifurcation point at $(u, f, \varepsilon) = (0, 1, 0)$, we define an increment $\widetilde{f}$ by

$$f = 1 + \widetilde{f}, \tag{1.16}$$

and expand the equilibrium equation (1.5) in this neighborhood to arrive at an incremental form of the equation

$$\begin{aligned}\widehat{F}(u, f, \varepsilon) &\equiv F(u, f, \varepsilon) - F(0, 1, 0)\\ &= F(u, 1 + \widetilde{f}, \varepsilon) - F(0, 1, 0)\\ &= kL^2\left[\frac{1}{2}\sin 2u - (1 + \widetilde{f})\sin u - \varepsilon\cos u\right]\\ &= -kL^2(u^3/2 + \widetilde{f}u + \varepsilon) + \text{h.o.t.} = 0,\end{aligned} \tag{1.17}$$

where h.o.t. means higher-order terms and $\sin u = u - u^3/3! + \cdots$ is used. This equation starts with a cubic term in u. Accordingly, one solution or three solutions exist for a particular value of f: the bifurcating solutions emerge when three solutions exist. Consequently, bifurcation can be understood as the emergence of multiple solutions due to the vanishing of lower-order terms in the equation (1.17). Figure 1.6(b) depicts the force versus displacement curves expressed by the leading terms in the equation (1.17). These curves are almost identical to the actual cantilever behavior presented in Fig. 1.6(a), or in Fig. 1.2(b), especially in the vicinity of the bifurcation point. Consequently, the approximation by higher-order terms inherits the essential bifurcation behavioral characteristics of the original equilibrium equation (1.5).

The first-order derivative (Jacobian matrix) of (1.17) vanishes; that is,

$$\frac{\partial \widehat{F}}{\partial u} = -kL^2\left(\frac{3}{2}u^2 + \widetilde{f}\right) + \text{h.o.t.} = 0 \tag{1.18}$$

at a critical point of an imperfect system. The simultaneous solution of the bifurcation equation (1.17) and condition (1.18) for criticality yields the location of the critical point for an imperfect system

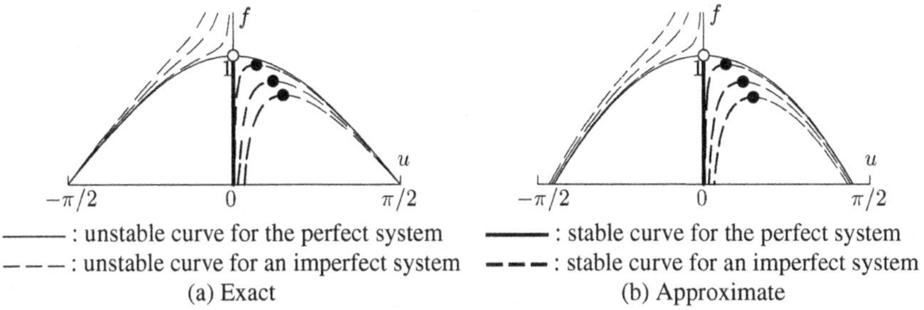

Fig. 1.6 Exact and approximate equilibrium paths of the cantilever. ○: bifurcation point; •: limit point.

$$\widetilde{f}_{\rm c} = -\frac{3}{2}\varepsilon^{2/3} + \text{h.o.t.}, \qquad u_{\rm c} = \varepsilon^{1/3} + \text{h.o.t.}, \tag{1.19}$$

where $(\cdot)_{\rm c}$ denotes a variable related to the critical point. These relations show the sensitivity of the location of the critical point to imperfection magnitude ε.

For a more general case, where the imperfection pattern is given as a vector $\boldsymbol{d}$ (e.g., (1.15)), the reduction of the critical load $\widetilde{f}_{\rm c}$, when ε is small, is proportional to the two-thirds power of ε. That is,

$$\widetilde{f}_{\rm c} \approx C(\boldsymbol{d})\varepsilon^{2/3} \tag{1.20}$$

with a coefficient $C(\boldsymbol{d})$ depending on $\boldsymbol{d}$, where $\approx$ means that the relevant terms are equal in an asymptotic sense for sufficiently small ε. The relation (1.20), found by Koiter, 1945 [119], is known as the *two-thirds power law* for an unstable pitchfork bifurcation point, and has been used to explain the sharp reduction in shell strength resulting from the presence of imperfections.

As demonstrated in Chapter 2, even for multiple-degree-of-freedom systems, a single equation in a single variable, called a *bifurcation equation*, can be derived by eliminating the so-called *passive coordinates* if the critical point is simple. The imperfection sensitivity laws for simple critical points are derived in Chapter 3 using this equation. The form of the bifurcation equation and such laws, however, are dependent on the type of critical points and are therefore to be derived for critical points of each type. In Chapter 8 a two-dimensional bifurcation equation and sensitivity laws are derived for double bifurcation points of systems with the symmetry of dihedral and cyclic groups. Furthermore, in Chapter 14, a four-dimensional bifurcation equation is derived for a quadruple bifurcation point of a system with the symmetry expressed by the group O(2)×O(2).

1.2.4 Worst Imperfection of Structural Systems

By virtue of the imperfection sensitivity law in (1.20), it is possible to grasp the quantitative influence of the magnitude ε of imperfection on the critical load $f_{\rm c}$ of a system undergoing bifurcation for a given pattern $\boldsymbol{d}$ of imperfection. It is nevertheless far more problematic to identify pattern vector $\boldsymbol{d}$ involving numerous variables. Historically, Monte Carlo simulation has been employed for structural systems using randomly chosen imperfections to determine the lower bound of the critical load $f_{\rm c}$. This method is indeed a robust strategy, but it is an awkward means to tackle that problem.

A refined method to address this awkwardness is to determine the worst direction of the imperfection vector that causes the maximum change (decrease) of the critical load. It amounts to minimizing the coefficient $C(\boldsymbol{d})$ in (1.20) (or maximizing $|C(\boldsymbol{d})|$). The theory of the worst imperfection for simple critical points is developed in Chapter 4; and that for double critical points of systems with dihedral symmetry is addressed in Chapter 9. The essence of the results is summarized below.

Problem Formulation

We consider a system of governing nonlinear equations $\boldsymbol{F}(\boldsymbol{u}, f, \boldsymbol{v}) = \boldsymbol{0}$ in (1.1). Recall that $\boldsymbol{u}$ indicates an N-dimensional state variable vector, that f denotes a loading parameter, and that $\boldsymbol{v}$ is a p-dimensional imperfection parameter vector. The vector $\boldsymbol{d}$ is normalized as

$$\boldsymbol{d}^\top W^{-1} \boldsymbol{d} = 1 \tag{1.21}$$

with respect to a *weight matrix* W, which is a $p \times p$ positive-definite symmetric matrix. Then our problem of determining the worst imperfection pattern vector is to find such a vector $\boldsymbol{d} = \boldsymbol{d}^*$ that maximizes $|C(\boldsymbol{d})|$ under the constraint (1.21). Vector $\boldsymbol{d}^*$ is called the *worst imperfection* in that it reduces the critical load most rapidly in an asymptotic sense.

We outline below the computation of the worst imperfection at a simple bifurcation point and at a double bifurcation point for simple structural systems.

Simple Pitchfork Bifurcation Point

It is proved in §4.3 that, for a (simple) pitchfork bifurcation point $(\boldsymbol{u}_\mathrm{c}^0, f_\mathrm{c}^0)$ of the perfect system described by $\boldsymbol{v}^0$, $C(\boldsymbol{d})$ in (1.20) takes the form of

$$C(\boldsymbol{d}) = C_0(\boldsymbol{\xi}_1^\top B_\mathrm{c}^0 \boldsymbol{d})^{2/3}. \tag{1.22}$$

Here C_0 is a constant that is independent of $\boldsymbol{d}$, $\boldsymbol{\xi}_1$ is a vector satisfying

$$\boldsymbol{\xi}_1^\top J_\mathrm{c}^0 = \boldsymbol{0}^\top \tag{1.23}$$

for the $N \times N$ Jacobian matrix J_c^0 at the critical point, and B_c^0 is the $N \times p$ *imperfection sensitivity matrix*, which is defined as

$$B_\mathrm{c}^0 = \frac{\partial \boldsymbol{F}}{\partial \boldsymbol{v}}(\boldsymbol{u}_\mathrm{c}^0, f_\mathrm{c}^0, \boldsymbol{v}^0). \tag{1.24}$$

We adopt the convention that the superscript $(\cdot)^0$ denotes a variable related to the perfect system and $(\cdot)_\mathrm{c}$ denotes a variable related to the critical point.

Consequently, the maximum of $|C(\boldsymbol{d})|$ in (1.22) with respect to $\boldsymbol{d}$ is achieved by $\boldsymbol{d}^*$ that maximizes $|\boldsymbol{\xi}_1^\top B_\mathrm{c}^0 \boldsymbol{d}|$ under the constraint (1.21). We see that such $\boldsymbol{d}^*$ is parallel to $W B_\mathrm{c}^{0\top} \boldsymbol{\xi}_1$, that is,

$$\boldsymbol{d}^* = \frac{1}{\alpha} W B_\mathrm{c}^{0\top} \boldsymbol{\xi}_1 \tag{1.25}$$

or its negative, where α is a positive scalar for the normalization (1.21).

As an example, the structural system of Fig. 1.3, consisting of two rigid bars supported by springs, is recalled. The perfect system is described by $\boldsymbol{v}^0 = (0,0)^\top$ and the imperfection pattern vector is $\boldsymbol{d} = (d_1, d_2)^\top$. The vector $\boldsymbol{\xi}_1$ satisfying (1.23) is equal to the critical eigenvector $\boldsymbol{\eta}^{(1)}$ in (1.14); that is,

$$\xi_1 = \frac{1}{\sqrt{5}} \begin{pmatrix} -1 \\ 2 \end{pmatrix}. \tag{1.26}$$

The weight matrix in (1.21) is chosen as

$$W = \begin{pmatrix} 1 & 0 \\ 0 & 1 \end{pmatrix}. \tag{1.27}$$

The imperfection sensitivity matrix $B_{\rm c}^0$ in (1.24) can be computed from (1.11) as

$$B_{\rm c}^0 = -kL^2 \begin{pmatrix} 1 & 1 \\ 0 & 1 \end{pmatrix}. \tag{1.28}$$

Substituting (1.26)–(1.28) into (1.25) yields the worst imperfection pattern

$$\boldsymbol{d}^* = \frac{1}{\sqrt{2}} \begin{pmatrix} 1 \\ -1 \end{pmatrix}, \tag{1.29}$$

which is distinct from the critical eigenvector $\boldsymbol{\eta}^{(1)} = (1/\sqrt{5})(-1,2)^\top$ in (1.14).

Double Bifurcation Point

As an example of the worst imperfection at a double bifurcation point, we refer to the regular-hexagonal truss tent in Fig 1.7(a), consisting of six truss members connecting one free crown node (node 0) to six fixed nodes (nodes 1 to 6). The system of equilibrium equations under the vertical load $EA\,f$ is described by

$$\boldsymbol{F} = \begin{pmatrix} F_1 \\ F_2 \\ F_3 \end{pmatrix} = \sum_{i=1}^{6} EA_i \left(\frac{1}{L_i} - \frac{1}{\hat{L}_i} \right) \begin{pmatrix} x - x_i \\ y - y_i \\ z - z_i \end{pmatrix} - \begin{pmatrix} 0 \\ 0 \\ EA\,f \end{pmatrix} = \begin{pmatrix} 0 \\ 0 \\ 0 \end{pmatrix}, \tag{1.30}$$

where E expresses Young's modulus; A_i denotes the cross-sectional area of member i ($i = 1,\ldots,6$); (x_i, y_i, z_i) is the initial location of node i ($i = 0,1,\ldots,6$); (x,y,z) is the independent variable denoting the location of node 0 after displacement; and

$$\begin{aligned} L_i &= ((x_0 - x_i)^2 + (y_0 - y_i)^2 + (z_0 - z_i)^2)^{1/2}, & i &= 1,\ldots,6; \\ \hat{L}_i &= ((x - x_i)^2 + (y - y_i)^2 + (z - z_i)^2)^{1/2}, & i &= 1,\ldots,6. \end{aligned}$$

We introduce the imperfection parameter vector

$$\boldsymbol{v} = (A_1, A_2, A_3, A_4, A_5, A_6)^\top,$$

which is equal to

$$\boldsymbol{v}^0 = (A, A, A, A, A, A)^\top$$

for the perfect (or nominal) system: $\boldsymbol{v} = \boldsymbol{v}^0 + \varepsilon \boldsymbol{d}$ in (1.3).

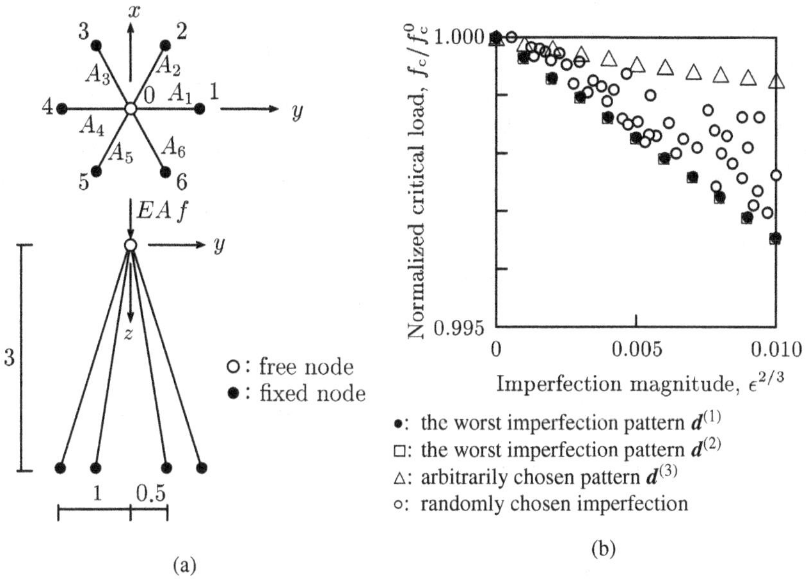

Fig. 1.7 (a) Regular-hexagonal, six-bar truss tent and (b) its critical load versus imperfection magnitude ($f_c/f_c^0 - \varepsilon^{2/3}$) relation.

By numerically tracing the solution path (x,y,z,f) of (1.30) of the perfect system for increasing values of f (starting from zero), it is apparent that the critical load of this tent is governed by a double critical point (bifurcation point) with $(x_c^0, y_c^0, z_c^0, f_c^0) = (0, 0, 0.1877, 0.3173)$.

The theoretical asymptotic formula

$$\widetilde{f_c} = f_c - f_c^0 \approx C(\boldsymbol{d})\varepsilon^{2/3}$$

in (1.20) can be numerically verified as follows. Consider three (arbitrarily chosen) imperfection pattern vectors

$$\begin{aligned}
\boldsymbol{d}^{(1)} &= \frac{A}{\sqrt{12}}(2,1,-1,-2,-1,1)^\top, \\
\boldsymbol{d}^{(2)} &= \frac{A}{2}(0,1,1,0,-1,-1)^\top, \\
\boldsymbol{d}^{(3)} &= A(0.6772,-0.2130,0.1704,0.4978,-0.2444,0.3992)^\top,
\end{aligned}$$

that are normalized as (1.21) with respect to a diagonal matrix $W = A^2 I_6$, where I_6 is the 6×6 identity matrix.

The critical loads for $\boldsymbol{d}^{(1)}$, $\boldsymbol{d}^{(2)}$, and $\boldsymbol{d}^{(3)}$ are plotted in Fig 1.7(b), respectively, by (•), (□), and (△). This plot suggests linear relations between $\widetilde{f_c}$ and $\varepsilon^{2/3}$ with different coefficients $C(\boldsymbol{d})$, consistent with the theoretical formula (1.20). The pattern $\boldsymbol{d}^{(3)}$ is the least influential on $\widetilde{f_c}$ among these three patterns: $\boldsymbol{d}^{(1)}$ and $\boldsymbol{d}^{(2)}$ have similar influence on $\widetilde{f}_{\mathrm{c}}$. The points denoted by (∘) indicate the values of $\widetilde{f}_{\mathrm{c}}$ computed for randomly chosen imperfections with varying $\boldsymbol{d}$ and ε. Since the lower envelope of these points (∘) is given by the points (•) for $\boldsymbol{d}^{(1)}$ and (□) for $\boldsymbol{d}^{(2)}$, the imperfection patterns $\boldsymbol{d}^{(1)}$ and $\boldsymbol{d}^{(2)}$ are candidates of the worst imperfections.

Indeed, the theory developed in Chapter 9 shows that $\boldsymbol{d}^{(1)}$ and $\boldsymbol{d}^{(2)}$ are the worst imperfections, as follows. The critical point is a double point, at which the Jacobian matrix

$$J_{\mathrm{c}}^{0} = EA \begin{pmatrix} 0 & 0 & 0 \\ 0 & 0 & 0 \\ 0 & 0 & 1.672 \end{pmatrix}$$

is singular (of rank 1), having

$$\xi_1 = (1,0,0)^{\top}, \qquad \xi_2 = (0,1,0)^{\top}$$

as the critical eigenvectors. According to the theory, the worst imperfections are given by

$$\begin{aligned} \frac{1}{\alpha} W B_{\mathrm{c}}^{0\,\top} \xi_1 &= \frac{A}{\sqrt{12}} (2,1,-1,-2,-1,1)^{\top}, \\ \frac{1}{\alpha} W B_{\mathrm{c}}^{0\,\top} \xi_2 &= \frac{A}{2} (0,1,1,0,-1,-1)^{\top}, \end{aligned}$$

where

$$\alpha = (\xi_1^{\top} B_{\mathrm{c}}^{0} W B_{\mathrm{c}}^{0\,\top} \xi_1)^{1/2} = (\xi_2^{\top} B_{\mathrm{c}}^{0} W B_{\mathrm{c}}^{0\,\top} \xi_2)^{1/2} = 0.03257 \cdot EA.$$

The worst imperfections thus derived coincide with $\boldsymbol{d}^{(1)}$ and $\boldsymbol{d}^{(2)}$ given above. The imperfection patterns $\boldsymbol{d}^{(1)}$ and $\boldsymbol{d}^{(2)}$ are more influential than arbitrarily chosen patterns, as shown earlier in Fig. 1.7(b).

It is intuitively obvious from the symmetry that the critical load remains unchanged in an asymptotic sense when the components of the imperfection pattern $\boldsymbol{d}$ are permuted appropriately. In particular, we can obtain many different worst imperfections by permuting the components of $\boldsymbol{d}^{(1)}$ and $\boldsymbol{d}^{(2)}$. Furthermore, it turns out that any linear combination of the worst imperfections is also the worst imperfection. That is, any vector represented as

$$\boldsymbol{d}^{*}(\varphi) = \cos\varphi \cdot \boldsymbol{d}^{(1)} + \sin\varphi \cdot \boldsymbol{d}^{(2)} \tag{1.31}$$

for some φ serves as the worst imperfection of this example. The worst imperfections form a two-dimensional subspace. Figure 1.8 shows the patterns of the worst imperfections for $\varphi = 0°, 30°, 45°, 60°$, and $90°$. It is noteworthy that

$$\boldsymbol{d}^{*}(30°) = \frac{A}{2}(1,1,0,-1,-1,0)^{\top}, \qquad \boldsymbol{d}^{*}(60°) = \frac{A}{\sqrt{12}}(1,2,1,-1,-2,-1)^{\top}$$

are equal, respectively, to $\boldsymbol{d}^{(2)}$ and $\boldsymbol{d}^{(1)}$ with components cyclically permuted.

In the analysis of the present example we have encountered degeneracy of two kinds: the multiple criticality of the Jacobian matrix and the degeneracy of the

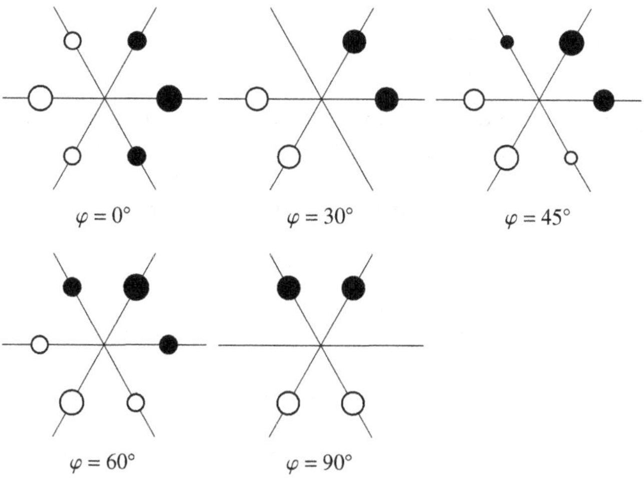

Fig. 1.8 The worst imperfection patterns for the hexagonal truss tent. •: positive component; ◦: negative component; area of • or ◦: magnitude of a component.

worst imperfection pattern, as given in (1.31). Both of these degeneracies are consequences of the inherent symmetry of the system, expressed mathematically as the invariance to a finite group (dihedral group). Group representation theory is the major mathematical tool for dealing with such generic degeneracy.

1.2.5 Random Variation of Imperfections

The worst imperfection patterns introduced in the previous subsection can offer us a lower bound of the critical load. However, this lower bound is often too conservative because the possibility of the occurrence of such a pattern is negligibly small for realistic systems with numerous imperfection parameters. All physical systems, which inevitably involve some imperfection, are probabilistic systems, as a consequence of the fact that the imperfections are inevitably subject to random variations. In particular, the probabilistic variation of the critical loads f_c of these systems is of great engineering interest.

The imperfection sensitivity law in (1.20) lays the foundation on the theoretical formulation of this variation. For a given probabilistic distribution of imperfection vector $\boldsymbol{v}$, the probabilistic variation of the critical load f_c in (1.20) can be obtained. This point is explained briefly here as an overview of Chapters 5 and 10.

Assuming that the imperfection vector $\boldsymbol{v}$ is subject to a multivariate normal distribution with a mean $\boldsymbol{v}^0$ and the variance proportional to ε^2, we derive the probability density function $\phi(f_c)$ of the critical load f_c for a (simple) unstable pitchfork bifurcation point. It turns out that $\phi(f_c)$ takes the form of

$$\phi(f_c) = \frac{3|f_c - f_c^0|^{1/2}}{\sqrt{2\pi}\widehat{C}^{3/2}} \exp\left(\frac{-1}{2}\left|\frac{f_c - f_c^0}{\widehat{C}}\right|^3\right), \qquad f_c \le f_c^0, \tag{1.32}$$

where f_c^0 is the load at the bifurcation point and $\widehat{C}$ is a constant of the order $\varepsilon^{2/3}$. The mean $\mathrm{E}[f_c]$ and variance $\mathrm{Var}[f_c]$ of f_c are expressed, respectively, as

$$\mathrm{E}[f_c] = f_c^0 - 0.802\widehat{C}, \qquad \mathrm{Var}[f_c] = (0.432\widehat{C})^2. \tag{1.33}$$

By respectively equating the sample mean and the sample variance with $\mathrm{E}[f_c]$ and $\mathrm{Var}[f_c]$ in (1.33), it is possible to estimate the values of f_c^0 and $\widehat{C}$.

The statistical approach presented above is useful in the description of the probabilistic variation of the strength of structures and materials. We refer here to a set of 32 experimental curves of stress σ_a versus strain ε_a of cylindrical sand specimens,[11] examples of which are shown in Fig. 1.9(a). Their strength variation is expressed in terms of the histogram of the maximum stress in Fig. 1.9(b). The values of the sample mean $\mathrm{E}[(\sigma_a)_c]$ and the sample variance $\mathrm{Var}[(\sigma_a)_c]$ of the maximum stress, respectively, are

$$\mathrm{E}[(\sigma_a)_c] = 4.49, \qquad \mathrm{Var}[(\sigma_a)_c] = 0.183^2$$

(unit in kgf/cm^2 = 98 kPa). Using these values in the theoretical formula (1.33) yields the estimated values of $(\sigma_a)_c^0$ and $\widehat{C}$ as

$$(\sigma_a)_c^0 = 4.83, \qquad \widehat{C} = 0.424.$$

Substitution of these values into the theoretical formula (1.32) yields an estimate of the probability density function of $(\sigma_a)_c$ portrayed by the solid curve in Fig. 1.9(b).

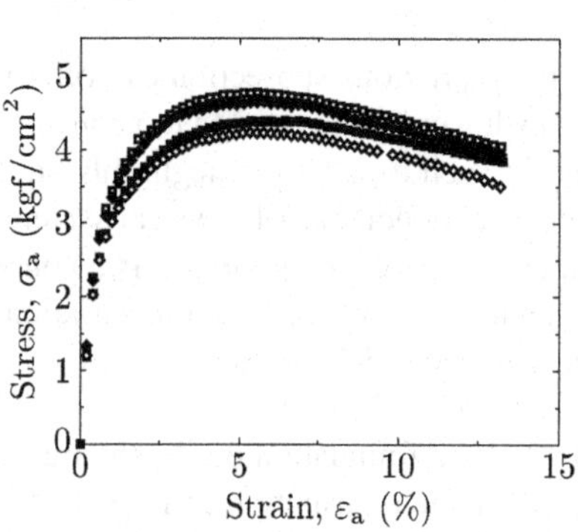

(a) Stress versus strain curves

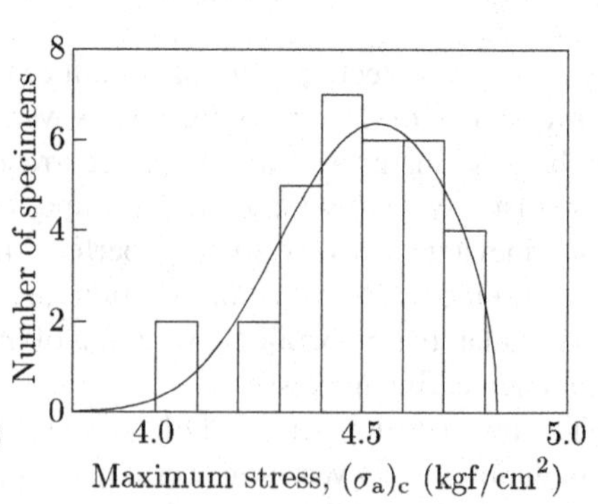

(b) Histogram and a curve (1.32) of probability density function of the maximum stress

Fig. 1.9 Probabilistic scatter of maximum stresses of 32 sand specimens. 1 kgf/cm^2 = 98 kPa.

[11] Details on (triaxial compression) tests on sand and soil specimens are given in Chapter 13.

This curve, which is consistent with the experimental histogram, is pertinent to the probabilistic design of the strength of the sand specimens.

1.2.6 Experimentally Observed Bifurcation Diagrams

There is a gap separating bifurcation diagrams in mathematical theory and those in engineering practice in the experimentation of materials undergoing bifurcation. In mathematical theory, a canonical coordinate for mathematical convenience is chosen as the abscissa of a bifurcation diagram. In contrast, a physically meaningful variable is a natural choice of an abscissa in the bifurcation diagram obtained by an analysis or experiment. Bifurcation diagrams observed in experiments might differ qualitatively from those in mathematics, as illustrated in Fig. 1.10 for a pitchfork bifurcation point (possible observed bifurcation diagrams in Fig. 1.10(a) and Fig. 1.10(b) in comparison with a mathematical bifurcation diagram in Fig. 1.10(c)).

Such a qualitative difference may arise for the following reason. A bifurcation diagram is obtained as the projection of the solution path in a higher-dimensional space to a two-dimensional plane. The resulting picture depends naturally on the

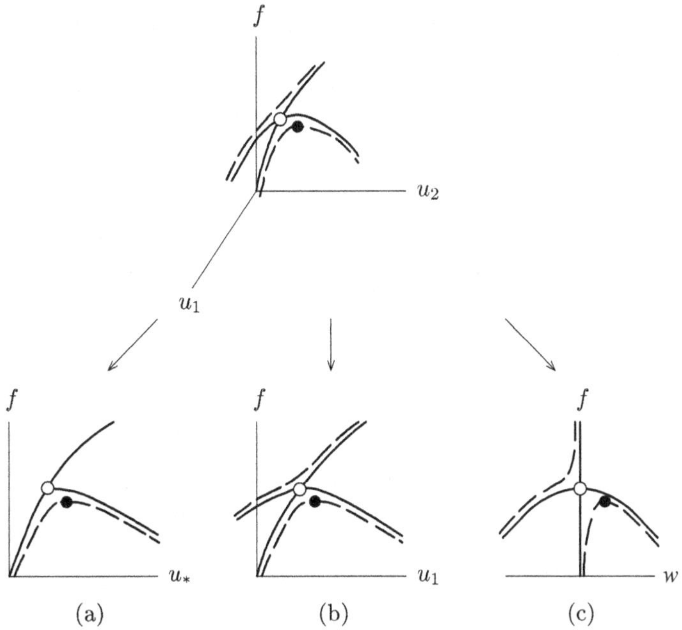

Fig. 1.10 Choice of projections in drawing bifurcation diagrams. (a) and (b) experimentally observed bifurcation diagrams and (c) a mathematical diagram at an unstable pitchfork bifurcation point. ——: curve for the perfect system; – – –: curve for an imperfect system; ∘: bifurcation point; •: limit point.

chosen projection. A canonical choice of the projection yields the mathematical bifurcation diagram (see Fig. 1.10(c)), whereas an arbitrary choice would result in a diagram similar to that shown in Fig. 1.10(b), which is qualitatively similar to the mathematical diagram. If the direction of the projection is so special that it is perpendicular to the bifurcated path, the resulting diagram resembles that shown in Fig. 1.10(a), which differs qualitatively from the mathematical diagram. Such an exceptional situation occurs quite often in experiments as a natural result of geometrical symmetry. To fill the gap resulting from this difference, the theory on imperfections is tailored to be applicable to experimentally observed diagrams in Chapters 6 and 8. Its major results are highlighted below.

For an experimentally observed displacement with geometrical symmetry, the bifurcation diagram is described, for example, by an equation

$$\sqrt{\tilde{u} - \tilde{f}/E}\,[\tilde{f} + p(\tilde{u} - \tilde{f}/E)] \pm q\varepsilon + \text{h.o.t.} = 0 \tag{1.34}$$

expressed as a relation among an imperfection ε, an incremental displacement $\tilde{u}$, and an incremental force $\tilde{f}$. Here E denotes the slope of the fundamental path for the perfect system ($\varepsilon = 0$), and p and q are parameters. Figure 1.10(a) corresponds to the bifurcation diagram expressed by the equation (1.34). It is noted here that the form of (1.34) differs from the form of the mathematical bifurcation equation, such as (1.17).

The analysis of the observed bifurcation diagrams presented above is applied to the data of 32 sand specimens introduced in the previous subsection. Figure 1.11 presents results of the simulation of the curves of stress versus strain for the two specimens by formula (1.34), using the values of the parameters at the right of

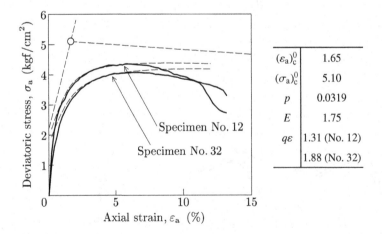

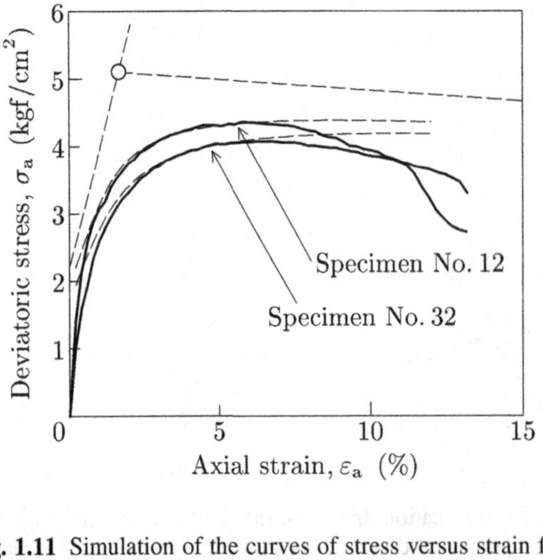

$(\varepsilon_a)^0_c$	1.65
$(\sigma_a)^0_c$	5.10
p	0.0319
E	1.75
$q\varepsilon$	1.31 (No. 12)
	1.88 (No. 32)

Fig. 1.11 Simulation of the curves of stress versus strain for the sand specimens. ——: experimental (imperfect) curve; — — —: theoretically computed curve; o: bifurcation point; 1 kgf/cm^2 = 98 kPa.

Fig. 1.11. The theoretical curves (– – –) correlate fairly well with the experimental curves (——) to underscore the validity of this formula. This ensures the applicability of the experimentally observed bifurcation diagram.

1.2.7 Group-Theoretic Bifurcation Theory

Bifurcation behavior of a physical problem involves various aspects. Its analytical aspects, such as the emergence of multiple solutions, can be characterized by the singularity of the Jacobian matrix. In contrast, its geometrical aspects can be characterized by the formation of patterns. As an example of this, Fig. 1.12 presents the periodic bumps and dents on a cylindrical shell undergoing bifurcation obtained through a computational analysis.

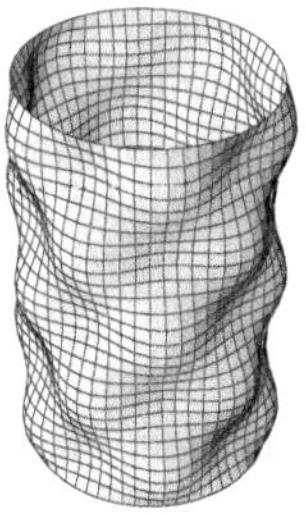

Fig. 1.12 Bifurcation pattern on a cylindrical shell computed by I. Saiki.

One may ask how the mechanism to generate the periodically symmetric bumps and dents is embedded into the cylindrical shell and how it has been inherited by the governing equation. The answer to this question had remained somewhat heuristic until its mechanism was fully untangled by the development of group-theoretic bifurcation theory. The fundamental concept of this theory is summarized in Chapter 7, and abundant examples of symmetric systems undergoing bifurcation are presented in Chapter 8 and Chapters 13 to 16. Major results of this theory on the bifurcation of a symmetric system are:

- The symmetry of the system is reduced at the onset of bifurcation.
- The symmetry of the system on each equilibrium path is labeled using a group until branching into a bifurcated path.
- The recursive bifurcation is associated with a *hierarchy of subgroups*

$$G_1 \to G_2 \to \cdots \to G_i \to G_{i+1} \to \cdots,$$

  where G_{i+1} is a proper subgroup of G_i $(i = 1, 2, \ldots)$.
- Symmetry often generates multiple bifurcation points where the rank deficiency of the Jacobian matrix is greater than one.

As an example of symmetric systems, we consider the elastic regular-triangular truss dome of Fig. 1.13 subjected to a z-directional load f applied to each of the

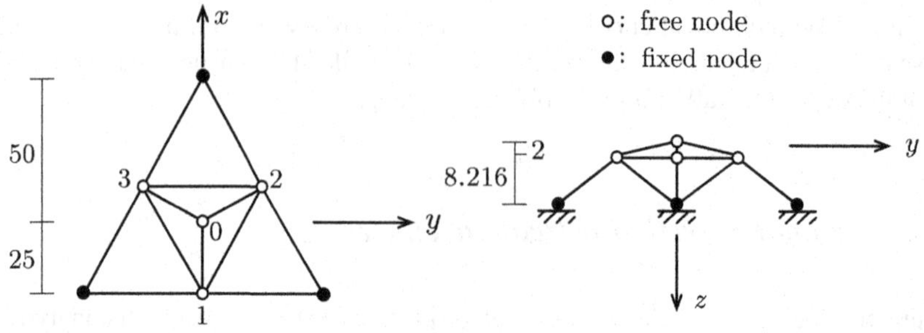

Fig. 1.13 Regular-triangular truss dome (D_3-symmetric).

nodes 1, 2, and 3. All members are assumed to have the same material and sectional properties EA. This dome, which is symmetric in geometrical configuration, in stiffness distribution, and in loading, remains invariant under geometrical transformations of two kinds: the counterclockwise *rotation* $c(2\pi/3)$ about the z-axis at an angle of $2\pi/3$ and the *reflection* $\sigma_y : y \mapsto -y$ with respect to the xz-plane. This geometrical invariance is expressed mathematically as the invariance (or equivariance) with respect to the *dihedral group* of degree three

$$D_3 = \{e, c(2\pi/3), c(4\pi/3), \sigma_y, \sigma_y \cdot c(2\pi/3), \sigma_y \cdot c(4\pi/3)\},$$

where the element e stands for the identity transformation and $\{\cdot\}$ indicates the group with the elements therein.

Deformation (bifurcation) patterns of this dome are often less symmetric than D_3 but may retain partial symmetry, which is represented by its invariance to subgroups of D_3. The subgroups of D_3 are enumerated as

$$C_3 = \{e, c(2\pi/3), c(4\pi/3)\}, \qquad C_1 = \{e\}.$$
$$D_1^{k,3} = \{e, \sigma_y \cdot c(2\pi(k-1)/3)\}, \qquad k = 1, 2, 3.$$

Figure 1.14 shows plane views of deformation patterns of the regular-triangular free nodes 1, 2, and 3 of this dome, where the solid–dashed lines denote the lines of *reflection symmetry*.

As explained in §8.4, the dome displays an interesting bifurcation phenomenon that should be regarded as a process of symmetry breaking associated with a hierarchy of subgroups

$$D_3 \to D_1^{k,3} \to C_1.$$

We may realize that symmetry is the underlying mechanism controlling the bifurcation behavior of this dome.

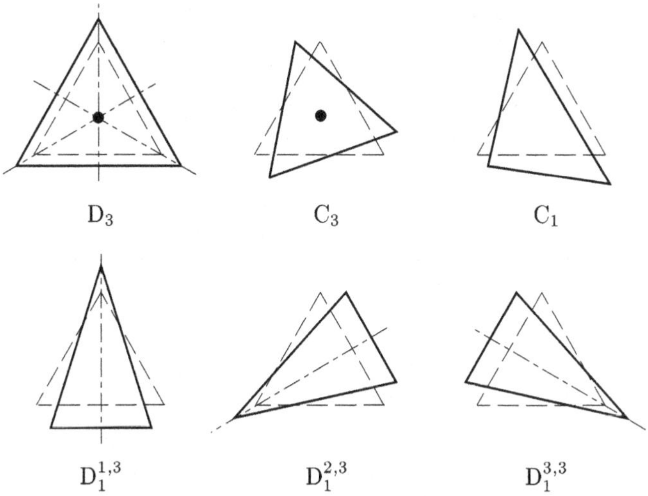

Fig. 1.14 Plane views of deformation patterns of regular-triangular free nodes of the regular-triangular dome. ——: displaced position; — — —: initial position; — - —: line of reflection symmetry; •: center of rotation symmetry.

1.3 Overview of Bifurcation of Symmetric Systems

We overview the bifurcation of symmetric systems. The group-theoretic bifurcation theory introduced in §1.2.7 serves as a standard tool to describe the symmetries of possible bifurcated solutions.

1.3.1 Recursive Bifurcation and Mode Switching of Sands

In the numerical analysis of a system undergoing bifurcation, the occurrence of bifurcation can be identified clearly by the singularity of the Jacobian matrix of the system under consideration (cf., §1.2.2). In the experiment on a physical system that might be undergoing bifurcation, it is usually impossible to identify this matrix; therefore, the occurrence of bifurcation must be identified without resort to the matrix.

The loss of symmetry is the key phenomenon to be observed to sort out the occurrence of bifurcation in experiments. As explained in the previous subsection, the successive loss of symmetry caused by recursive bifurcation[12] (cf., Fig. 1.15(a)) is to be observed and classified using the subgroups.

In most studies of bifurcation behavior, the predominant bifurcation mode is assumed to be unique. Such uniqueness, however, can be jeopardized by mode switch-

[12] *Recursive bifurcation* means repeated occurrence of symmetry-breaking bifurcations.

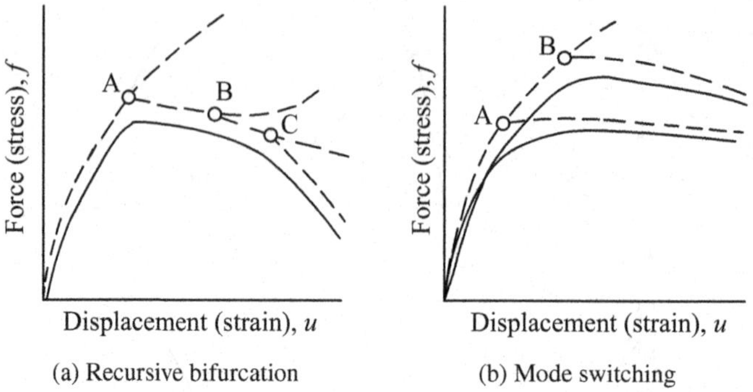

Fig. 1.15 General views of recursive bifurcation and mode switching. ——: experimental (imperfect) curve; — — —: perfect curve; ○: bifurcation point.

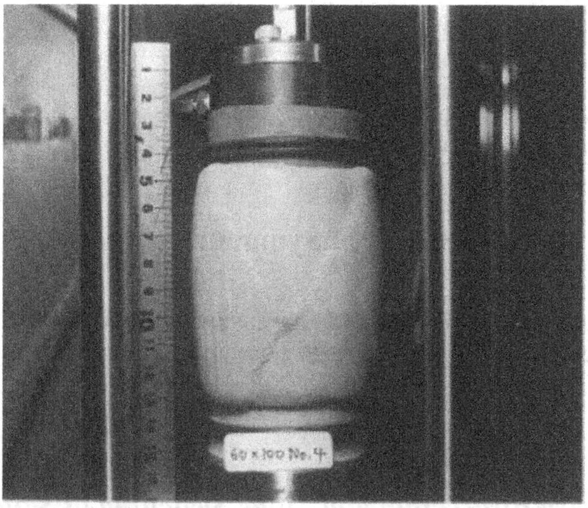

Fig. 1.16 Photograph of a sand specimen after deformation.

ing.[13] When the same experiment is performed repeatedly, each experimental specimen suffers from its particular imperfections. Especially when two or more bifurcation points are located near each other, a number of bifurcation modes may be activated depending on imperfections (cf., Fig. 1.15(b)).

In Chapter 13, recursive bifurcation and mode switching are observed in the triaxial compression test of sand specimens. A cylindrical sand specimen contained in a transparent rubber is loaded axially, and is subjected to uniform water pressure in the circumferential direction. Although this specimen retains its cylindrical shape up to a certain level of loading, it loses its shape at the onset of symmetry-breaking bifurcation, as presented in Fig. 1.16.

[13] *Mode switching* means the change of the bifurcated path that the actual behavior follows due to the difference in imperfections.

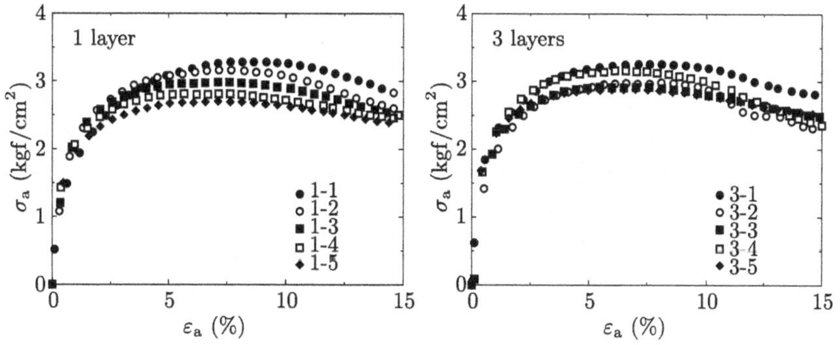

(a) Classification by the number of layers (imperfections)

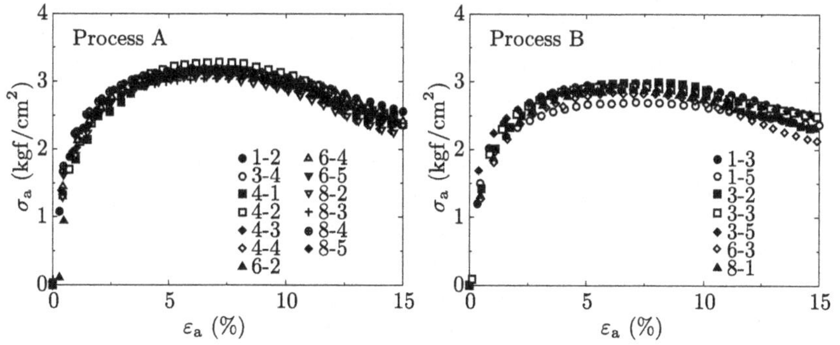

(b) Classification by bifurcation modes

Fig. 1.17 Curves of stress versus strain of sand specimens ($1\,\text{kgf/cm}^2 = 98\,\text{kPa}$).

A series of (triaxial compression) tests is performed on sand specimens, which are made up of a few horizontal layers of sand. We employ specimens of five types with the number of layers of 1, 3, 4, 6, and 8 each with five cases. These layers may be interpreted as artificial imperfections that trigger diverse bifurcation modes. Figure 1.17(a) portrays the curves of stress versus strain categorized based on the number of layers.

During the experiment, the progress of the deformation patterns of the specimens is carefully observed and categorized into five sets of symmetry reduction processes, using groups representing spatial symmetries. One might focus, for example, on two of those symmetry-reduction processes expressed by two hierarchies of subgroups:

$$\text{Process A}: \mathrm{D}_{\infty\mathrm{h}} \to \mathrm{D}_{2\mathrm{h}} \to \mathrm{D}_{1\mathrm{h}} \to \mathrm{C}_{1\mathrm{v}},$$
$$\text{Process B}: \mathrm{D}_{\infty\mathrm{h}} \to \mathrm{C}_{\infty\mathrm{v}} \to \mathrm{C}_{2\mathrm{v}} \to \mathrm{C}_{1\mathrm{v}}$$

with respect to the symmetry of the deformation patterns; the groups representing spatial symmetries (such as $\mathrm{D}_{\infty\mathrm{h}}$ and $\mathrm{D}_{2\mathrm{h}}$) are defined in §13.2. Figure 1.18, for

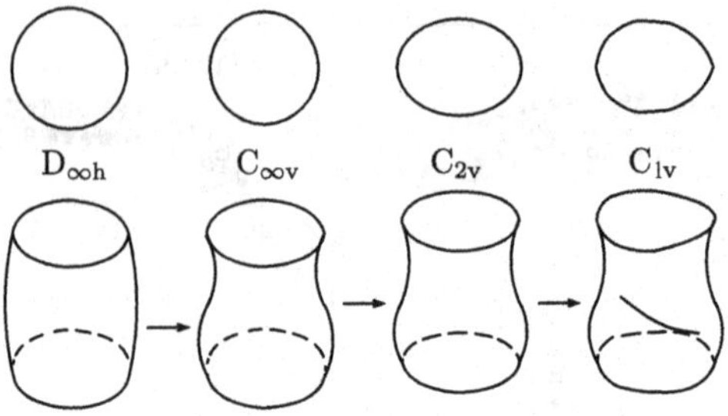

Fig. 1.18 Typical deformation pattern change of a cylindrical sand specimen (Process B).

example, portrays the typical recursive pattern change for Process B. The specimen loses the upside-down symmetry at the onset of the bifurcation $D_{\infty h} \rightarrow C_{\infty v}$; its cross-section becomes elliptic at $C_{\infty v} \rightarrow C_{2v}$; and the elliptic cross-section further deforms and an oblique shear band is formed at $C_{2v} \rightarrow C_{1v}$. It is noteworthy that diverse symmetry-reduction processes—diverse bifurcation processes—exist even in specimens with the same number of layers.

The stress versus strain curves categorized in this manner are presented in Fig. 1.17(b). The variation among the curves for Process A, which are categorized based on the deformation patterns, is significantly less than that portrayed in Fig. 1.17(a) categorized based on the number of layers. Such is also the case for Process B. The curves for Process A have markedly higher peaks (strength) than those of B. This fact underscores the importance of identifying the type of bifurcation process in the proper understanding of the bifurcation behavior of sands; the type of bifurcation process more strongly influences the strength variation than the number of layers.

Figure 1.19(a) depicts the simulation of the recursive bifurcation of a sand specimen belonging to Process B, in association with the loss of symmetry in Fig. 1.18.

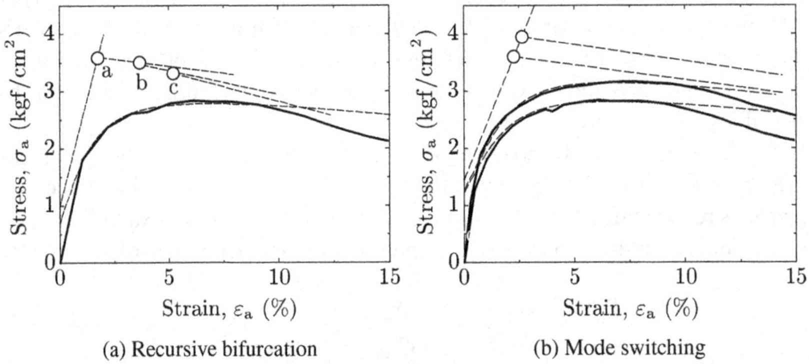

Fig. 1.19 Simulation of stress versus strain curves. ——: experimental curve; – – –: simulated curve; ∘: bifurcation point; $1\,\mathrm{kgf/cm^2} = 98\,\mathrm{kPa}$.

The observation and classification of these patterns based on the groups is a systematic means to sort out the recursive bifurcation for cylindrical sand and soil specimens during experiments. The mode switching and recursive bifurcation of sand specimens are investigated in §13.3.

Figure 1.19(b) shows the simulation of a pair of curves of sand specimens belonging, respectively, to Processes A and B using the asymptotic formula (1.34). The theoretical curves correlate fairly well with the experimental ones. Furthermore, the presence of two bifurcation points on the fundamental path that triggers mode switching is clearly demonstrated.

1.3.2 Echelon Modes on Uniform Materials

Various patterns, such as the echelon, the Riedel, and the self-similar anastomosing patterns, emerge on the surface of uniform or almost uniform geotechnical materials, including soil, rock, and sand. Figure 1.20 shows a *cross-checker pattern* on the surface of kaolin observed in a uniaxial compression test; a similar pattern in Fig. 1.21 has been observed on a hollow cylindrical sand specimen. Figure 1.22 portrays a photograph and a sketch of a soil specimen after shearing; there is an echelonlike series of wrinkles, which is called an *echelon mode* in structural geology.

In Chapter 14, the underlying mechanism of those patterns is explained by group-theoretic bifurcation theory. For successful explanation of this mechanism, it is necessary to exploit the translational symmetry due to the local uniformity of materials. For a cylindrical domain the periodic boundaries are used on the top and bottom surfaces to make the domain infinite in the axial direction and thereby to exploit the translational symmetry in the axial direction. To be more specific, the symmetry of this domain is to be modeled as $\mathrm{O}(2)\times\mathrm{O}(2)$-symmetry, which represents the symmetry of a torus, rather than as the geometrically more natural $\mathrm{O}(2)\times\mathbb{Z}_2$-symmetry. Here the group $\mathrm{O}(2)\times\mathbb{Z}_2$ acts as rotations around the longitudinal axis and the reflections with respect to planes containing or perpendicular to the axis, and the group $\mathrm{O}(2)\times\mathrm{O}(2)$ contains an additional action of translation along the axis.

For the $\mathrm{O}(2)\times\mathrm{O}(2)$-symmetry chosen above, bifurcation rules can be obtained by group-theoretic bifurcation theory. In particular, it can be found that the classical diamond pattern (checkerboard) solution and a pair of oblique stripe pattern solutions branch simultaneously from the bifurcation point of multiplicity four. Figure 1.23 portrays a general view of the formation of an echelon mode through recursive bifurcation. The cross-checker pattern (shown as $\mathrm{EC}^{+}_{n\tilde{n}kl}$) does not appear directly from the $\mathrm{O}(2)\times\mathrm{O}(2)$-symmetry, but it can appear as a consequence of recursive bifurcation from the bifurcated path for the oblique stripe pattern (presented as $\mathrm{OB}^{+}_{n\tilde{n}}$). The shear bands (presented as D_1 and C_1) are made up of numerous parallel wrinkles that are formed at the emergence of the oblique stripe and meshlike patterns. A shear band is depicted in Fig. 1.23(c); it is noteworthy that the shear band and the wrinkles are in completely different directions.

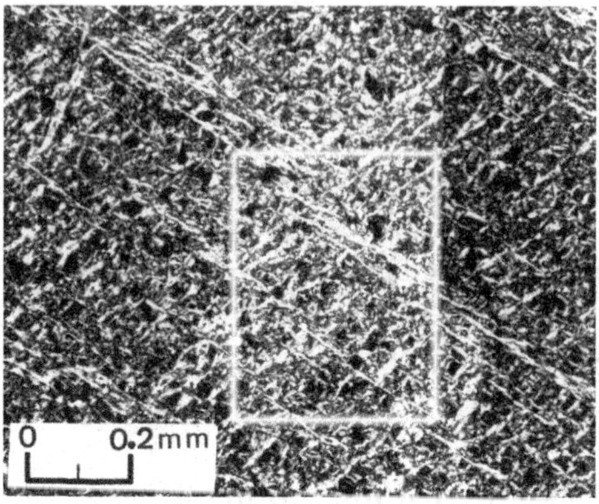

Fig. 1.20 Cross-checker pattern on kaolin. Photograph by I. Sano (Ikeda et al., 2001 [84]).

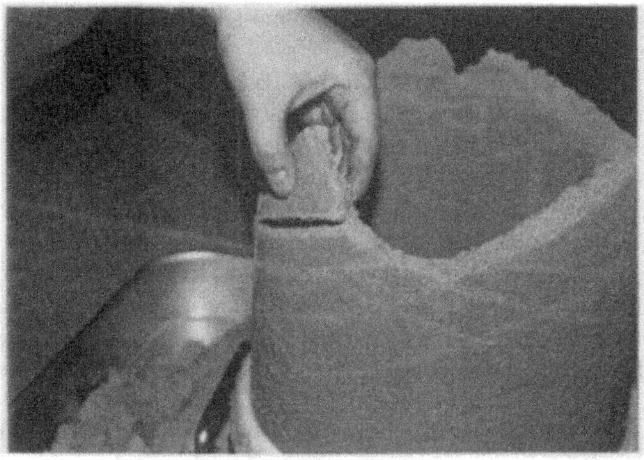

Fig. 1.21 Diamond pattern on a hollow cylindrical sand specimen. Photograph by T. Nishimaki.

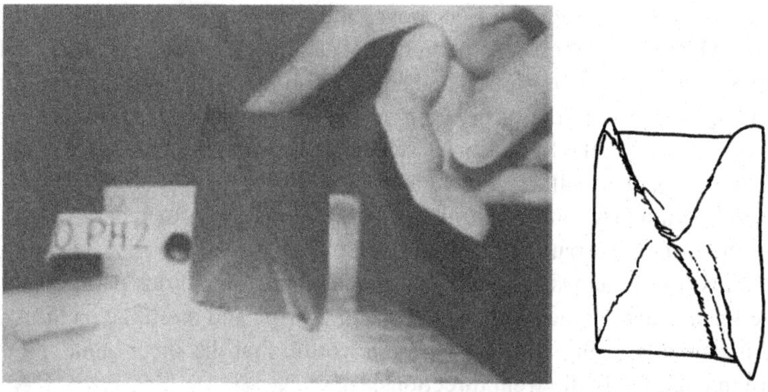

Fig. 1.22 Photograph and a sketch of an echelon mode on a soil specimen by Nakano, 1993 [147].

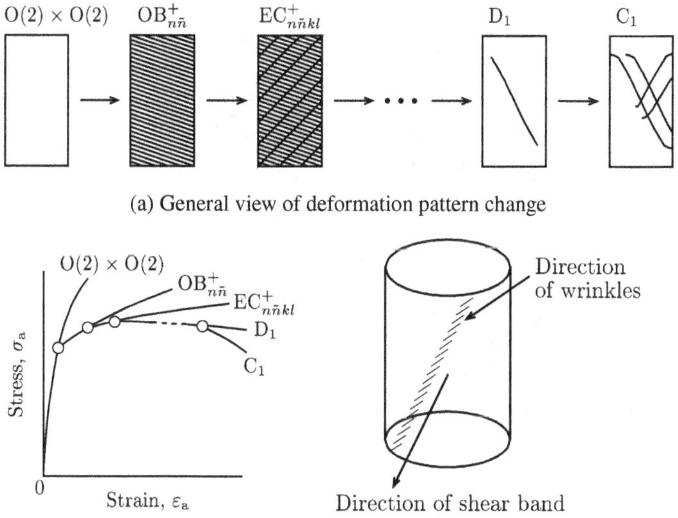

(a) General view of deformation pattern change

(b) Stress versus strain curve (c) Echelon mode on a cylindrical domain

Fig. 1.23 General view of deformation pattern change by recursive bifurcation leading to the stripe pattern labeled by the group $OB^{+}_{n\tilde{n}}$ and the cross-checker pattern labeled by the group $EC^{+}_{n\tilde{n}kl}$, and so on. ∘: bifurcation point.

It is emphasized that the seemingly more "natural" $O(2)\times\mathbb{Z}_2$-symmetry does not yield a spatial pattern that can be interpreted as the echelon mode. The expanded symmetry, accordingly, is more appropriate for the understanding of the echelon-mode formation and, in turn, to the success in the numerical simulation of patterns with high spatial frequencies that can be understood as a consequence of the local uniformity of the materials. The mechanism of the formation of the echelon mode is presented in Chapter 14 by the investigation of the rules of bifurcation of an $O(2)\times O(2)$-symmetric system.

Figure 1.24 presents an example of image simulation of recursive bifurcation. The rectangular domain of the kaolin is cut from the whole domain shown in Fig. 1.20. The observed density u is expanded into the double Fourier series. The deformation history of the kaolin is reconstructed by observing the magnitudes of the Fourier coefficients with the theoretical knowledge of the recursive bifurcation of an $O(2)\times O(2)$-symmetric system. This is portrayed in Fig. 1.24. The deformation history starts from a uniform initial state and ends with the final state labeled C_1 in Fig. 1.24(h). More details on the image simulation of kaolin are given in §14.7.

As another example of image simulation of recursive bifurcation, we refer to an experiment on a sand specimen subject to plane strain compression in §14.8. Photographs taken during the experiment are numbered 1–8. The progression of localization of incremental strain fields between two neighboring photographs is obtained as portrayed in Fig. 1.25(a). Two parallel oblique shear bands are observed during increments 3–5. During increments 5–8, some shear bands diminish gradu-

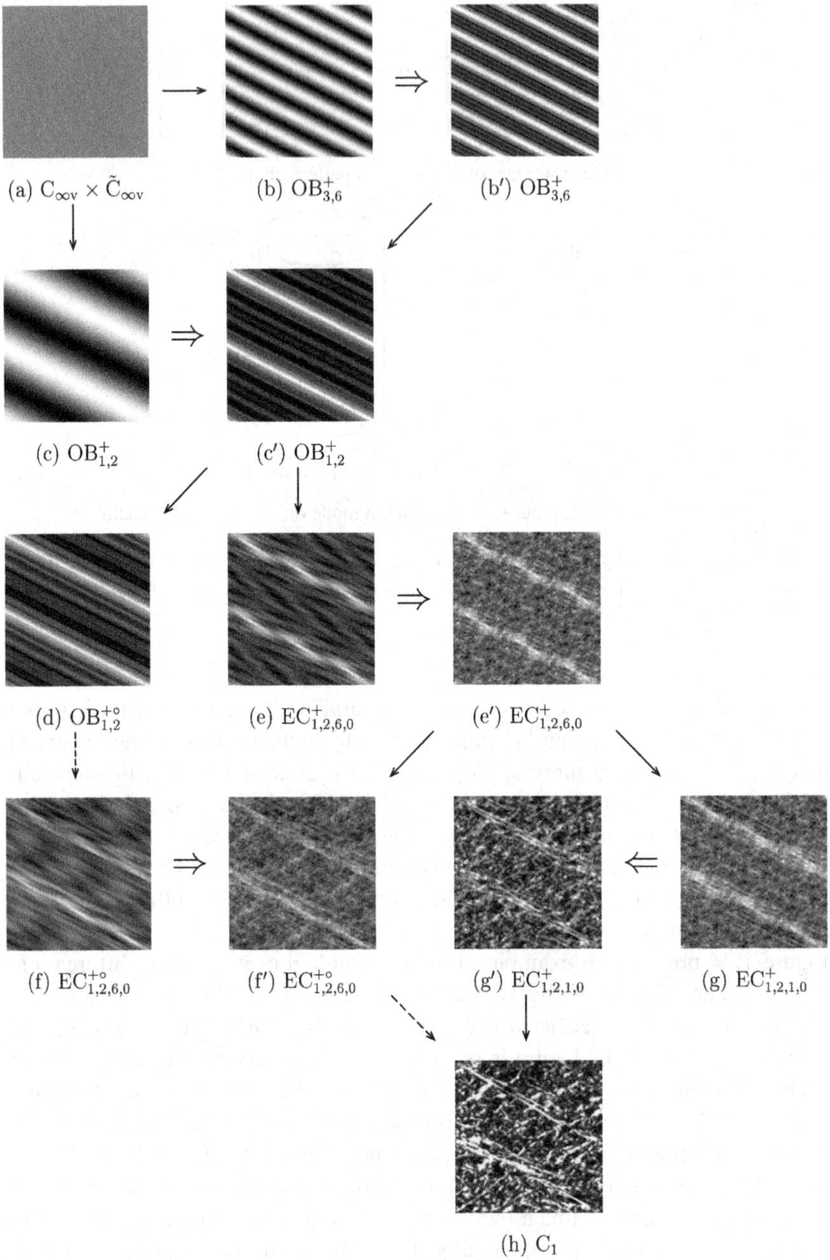

Fig. 1.24 Image simulation of the progress of deformations for the kaolin specimen expressed in terms of a hierarchy of images.

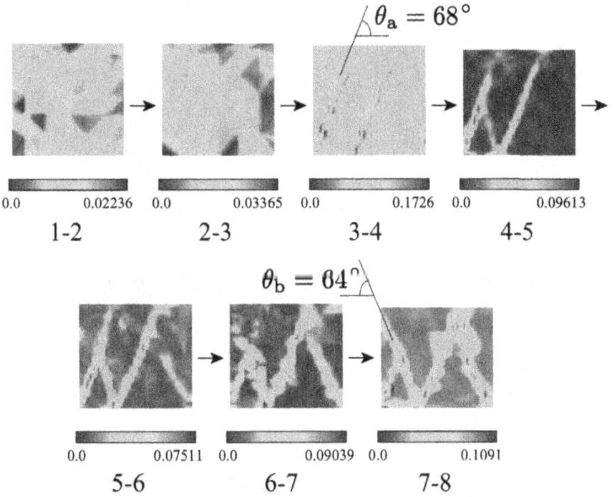

(a) Contour views of stereophotogrammetry-based incremental strain fields

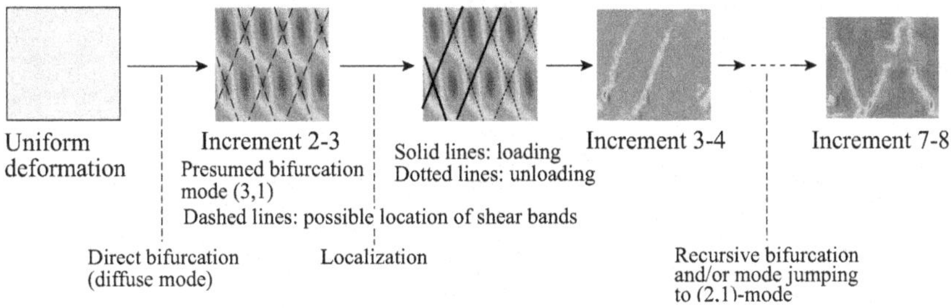

(b) Progress of diffuse-mode bifurcation, followed by localization, further bifurcation, and/or mode jumping

Fig. 1.25 Image simulation of incremental strain fields of patterned shear bands observed on Hostun sand rectangular parallelepiped specimens.

ally in favor of the emergence of two oblique shear bands in a different direction. As depicted in Fig. 1.25(b), the pattern formation takes the course of the evolution of bifurcation with a diamondlike diffuse mode breaking uniformity, followed by further bifurcation, mode jumping, and the formation and disappearance of shear bands through localization. A chaotic explosive increase of possible postbifurcation states is pointed out as a mechanism to diversify geometrical patterns.

1.3.3 Recursive Bifurcation of Steel Specimens

The recursive bifurcation behavior of rectangular parallelepiped steel specimens is studied in Chapter 15 with the aid of group-theoretic bifurcation theory. As an ex-

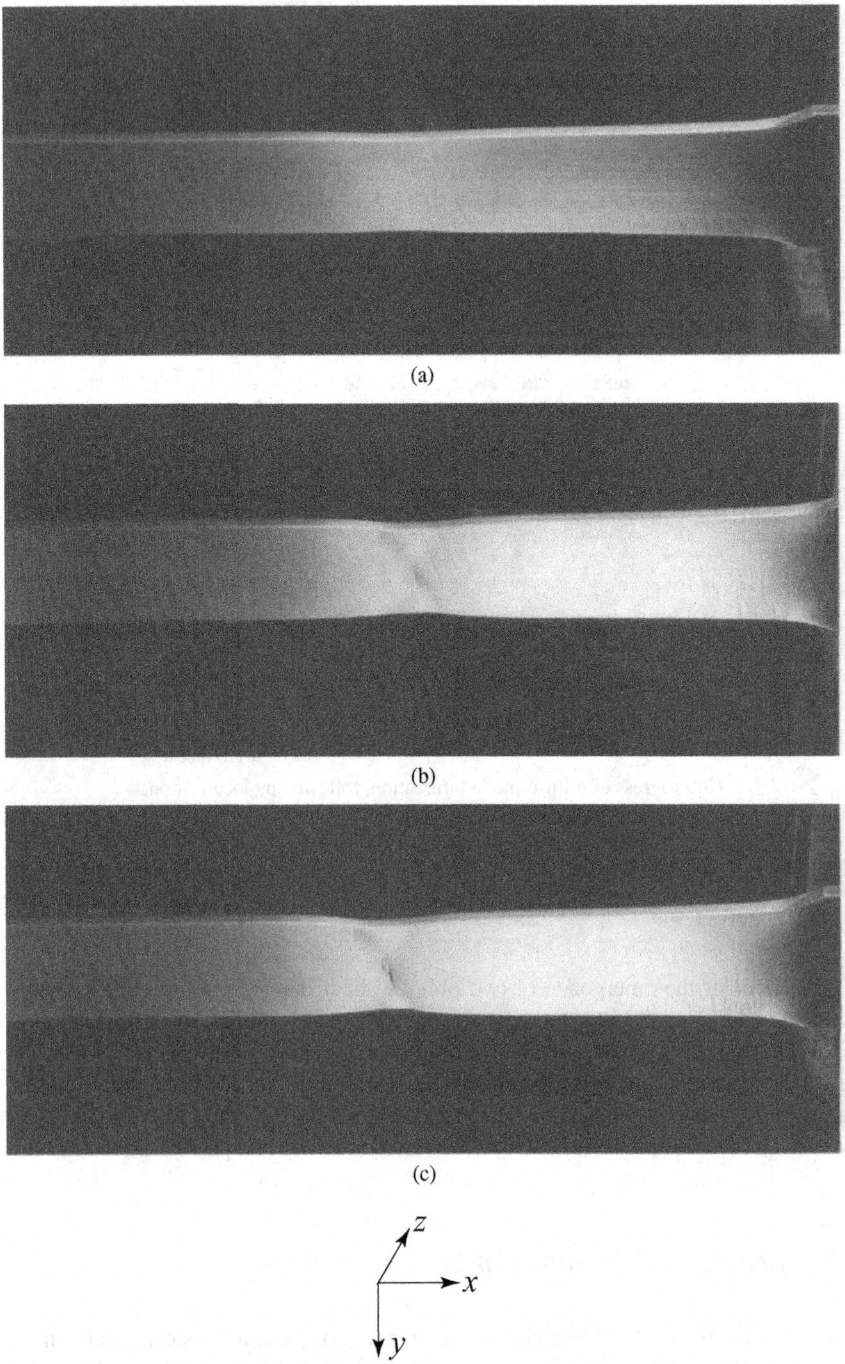

Fig. 1.26 Deformation pattern change of a thin steel specimen (Ikeda et al., 2001 [101]).

ample of this, we refer here to a thin steel specimen; the deformation change of this specimen during the experiment is portrayed in Fig. 1.26.

The seemingly complex experimental behavior of this specimen can be explained by the recursive bifurcation expressed by a hierarchy of subgroups

$$G = \langle \sigma_x, \sigma_y, \sigma_z, t_x \rangle \rightarrow \mathrm{D}_{2\mathrm{h}} = \langle \sigma_x, \sigma_y, \sigma_z \rangle \rightarrow \langle \sigma_y, \sigma_z \rangle, \tag{1.35}$$

where σ_x, σ_y, and σ_z, respectively, denote reflections with respect to the yz-, zx-, and xy-planes; t_x is the x-directional translation at some length; and the angle brackets $\langle \cdot \rangle$ denote the group generated by the elements therein.

The formation of the necking depicted in Fig. 1.26(a) from the uniform state results from the direct bifurcation associated with

$$G \rightarrow \mathrm{D}_{2\mathrm{h}},$$

in which the symmetry of the uniform state is labeled by G and the necking is labeled by the group $\mathrm{D}_{2\mathrm{h}}$. As depicted in Fig. 1.26(b), the diagonal shear bands are formed after the diffuse necking arises from the intense localized straining. Such a formation is characteristic from a physical standpoint, but is not associated with bifurcation in that both the state of necking and that of the diagonal shear bands share the same symmetry labeled by the group $\mathrm{D}_{2\mathrm{h}}$. The secondary bifurcation takes place at the onset of the formation of a single distinct shear band. This secondary bifurcation is associated with a further reduction of symmetry described by

$$\mathrm{D}_{2\mathrm{h}} \rightarrow \langle \sigma_z, \sigma_x \sigma_y \rangle,$$

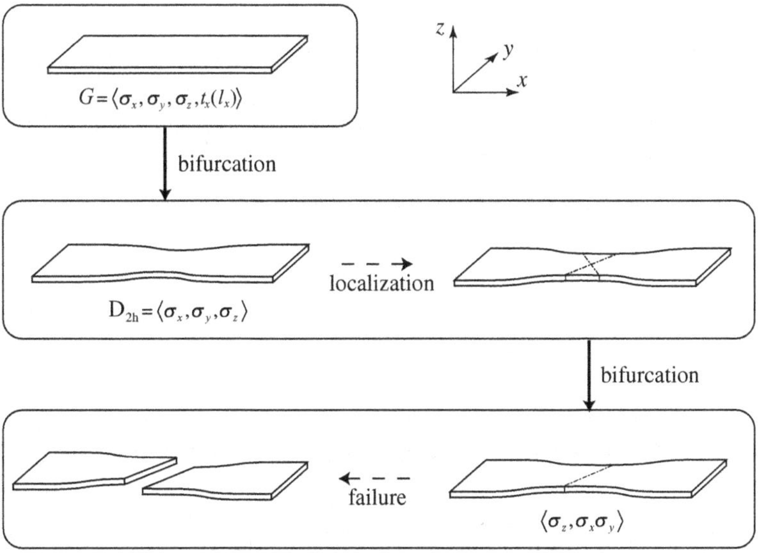

Fig. 1.27 Hierarchical deformation pattern change by recursive bifurcation for the steel specimen.

in which D_{2h} denotes the symmetry of the diagonal shear bands and $\langle \sigma_z, \sigma_x\sigma_y \rangle$ represents that of the single distinct shear band. The final failure state in Fig. 1.26(c), the symmetry of which is labeled by the same group $\langle \sigma_z, \sigma_x\sigma_y \rangle$, is no longer caused directly by bifurcation. The deformation pattern change and the loss of symmetry associated with the direct and secondary bifurcations are presented in Fig. 1.27.

1.3.4 Flower Patterns on a Honeycomb Structure

Beautiful wallpapers of deformation patterns of a honeycomb structure, as depicted in Fig. 1.28, are obtained in Chapter 16. The bifurcation mechanism and an exhaustive list of possible bifurcating deformation patterns are obtained for a honeycomb structure consisting of 2×2 hexagonal cells that has the symmetry of the semidirect product of D_6 and $\mathbb{Z}_2 \times \mathbb{Z}_2$. New deformation patterns of a honeycomb structure have been found and classified. Knowledge of the symmetries of the bifurcating solutions has turned out to be vital for the successful numerical tracing of the bifurcated paths.

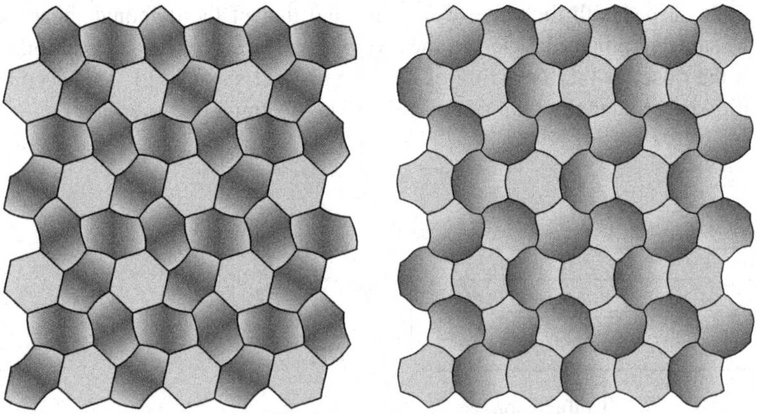

Fig. 1.28 Wallpapers expressed by planforms of flowerlike modes.

Summary

- Major theoretical tools and concepts that are introduced in this book have been overviewed.
- Theoretical descriptions of the bifurcation behaviors of several symmetric systems that are presented in the book have been overviewed.

Part I
Imperfect Behavior at Simple Critical Points

This part, Part I, is devoted to the study of imperfect behavior in the neighborhood of a simple critical point; the study of a double critical point is given in Part II using group theory. Emphasis is placed on the case of multiple imperfection parameters, although it is customary to address a single imperfection parameter. Major viewpoints of this part are:

- Theories of the strength[1] of structural systems,
- A systematic method to address observed bifurcation diagrams.

A nonlinear governing equation in general involves a number of independent variables and nonlinear terms and is highly complex. The nonlinear governing equation is simplified based on the following two procedures.

- Thc governing equation is reduced to the *bifurcation equation* with only a few active independent variables using the Liapunov–Schmidt reduction.
- Higher-order terms of the bifurcation equation are truncated by an asymptotic assumption.

From the bifurcation equation, several pertinent formulas expressing the influence of imperfections are obtained.

In the description of imperfections, the mathematical rigor related to universal unfoldings is not emphasized, in favor of the engineering pragmatism to capture the most important imperfections and their influences in an asymptotic sense.

Part I is organized as follows.

In Chapter 2 bifurcation equations are derived for simple critical points of three types, including: limit point, transcritical point of bifurcation, and pitchfork point of bifurcation. The asymptotic influence of an imperfection on the loci of equilibrium of the bifurcation equation is illustrated.

In Chapter 3, to formulate the dependence of the critical load on imperfections, imperfection sensitivity laws for simple critical points are derived from the bifurcation equation.

In Chapter 4 the worst imperfection pattern that reduces the critical load most rapidly is formulated. An explicit formula for the worst imperfection is derived based on the imperfection sensitivity laws.

In Chapter 5 a variation of the critical load, attributable to the probabilistic variation of imperfections is formulated. Emphasis is placed on imperfections subject to a multivariate normal distribution.

In Chapter 6 an experimentally observed bifurcation diagram is obtained by transforming an independent variable in the bifurcation equation to a variable observed in an experiment. Then imperfection sensitivity laws for this diagram are derived, and their usefulness is assessed based on experimental results.

[1] The strength is given as the value of a bifurcation parameter at a critical point.

Chapter 2
Critical Points and Local Behavior

2.1 Introduction

Bifurcation, which means the emergence of multiple solutions for the same value of parameter f, is induced by the criticality of the Jacobian matrix of the system, as demonstrated using examples in the previous chapter (cf., §1.2.2). The "bifurcation equation" is a standard means to describe bifurcation behavior. In a neighborhood of a simple critical point, for example, a set of equilibrium equations is reduced to a single bifurcation equation, by condensing the influence of a number of independent variables into a single scalar variable by the implicit function theorem. This reduction achieves a drastic simplification, but retains essential properties of the bifurcation. A similar reduction can be conducted on a system with a number of imperfection parameters to arrive at the bifurcation equation for an imperfect system. The process of deriving this equation is called the Liapunov–Schmidt reduction.[1]

The bifurcation equation is useful in that essential asymptotic characteristics of the system around a critical point are expressed by a few leading terms. The word "asymptotic" means that all results are local, valid only for sufficiently small absolute values of imperfection parameters, and in a sufficiently close neighborhood of the critical point. It is quite pertinent to simplify the bifurcation equation by truncating the equation using the leading terms. Investigation of this form leads to the classification of simple critical points.

Imperfections perturb a bifurcation equation and, in turn, perturb the associated bifurcation diagram both quantitatively and qualitatively. Several theories related to

[1] This term, "Liapunov–Schmidt reduction," is widely accepted in nonlinear mathematics (e.g., Sattinger, 1979 [176]; Chow and Hale, 1982 [28]; and Golubitsky and Schaeffer, 1985 [62]). In structural mechanics, it is called the "Liapunov–Schmidt–Koiter reduction" (e.g., Peek and Kheyrkhahan, 1993 [163]) or the "elimination of passive coordinates" (e.g., Thompson and Hunt, 1973 [191]; Thompson, 1982 [190]; El Naschie, 1990 [49]; and Godoy, 2000 [61]).

K. Ikeda and K. Murota, *Imperfect Bifurcation in Structures and Materials*,
Applied Mathematical Sciences 149, DOI 10.1007/978-1-4419-7296-5_2,

imperfect bifurcation, such as imperfection sensitivities and universal unfoldings, have been developed.[2]

In this chapter the influence of imperfections is investigated by conducting the Liapunov–Schmidt reduction on a system with a number of imperfection parameters to arrive at the bifurcation equation for an imperfect system. In our description of imperfections, mathematical rigor is much less emphasized in favor of engineering pragmatism to capture the most important imperfections and their influences in an asymptotic sense.

This chapter is organized as follows.

- A general mathematical framework is presented in §2.2.
- A bifurcation equation of a simple example is derived to illustrate the mechanism of the Liapunov–Schmidt reduction in §2.3.
- This reduction is explained in general terms in §2.4.
- Simple critical points are classified, and the perfect and imperfect bifurcation behaviors at simple critical points are investigated in view of the leading terms of the bifurcation equation in §2.5.
- The bifurcation behavior of an example is investigated in §2.6.

2.2 General Framework

The general mathematical framework is presented.

2.2.1 Governing Equation with Imperfection

We consider a system of nonlinear *governing* or *equilibrium equations*

$$\boldsymbol{F}(\boldsymbol{u}, f, \boldsymbol{v}) = \boldsymbol{0}, \tag{2.1}$$

where $\boldsymbol{u} \in \mathbb{R}^N$ indicates an N-dimensional unknown vector, the *state variable vector* ($\mathbb{R}$ is the set of real numbers); $f \in \mathbb{R}$ denotes an auxiliary parameter, the *bifurcation parameter*; and $\boldsymbol{v} \in \mathbb{R}^p$ denotes a p-dimensional *imperfection parameter vector*. We assume that $\boldsymbol{F}$: $\mathbb{R}^N \times \mathbb{R} \times \mathbb{R}^p \to \mathbb{R}^N$ is a sufficiently smooth nonlinear function in $\boldsymbol{u}$, f, and $\boldsymbol{v}$. It is emphasized that we have a distinguished single parameter f, which, mathematically, plays the role of bifurcation parameter. In structural mechanics, $\boldsymbol{u}$ indicates a *displacement vector* and f denotes a *loading parameter*.

The imperfection parameter vector $\boldsymbol{v}$ is expressed deliberately as

$$\boldsymbol{v} = \boldsymbol{v}^0 + \varepsilon \boldsymbol{d}, \tag{2.2}$$

[2] See, for example, Koiter, 1945 [119]; Keener and Keller, 1973 [113]; Keener, 1974 [111]; Chow, Hale, and Mallet-Panet, 1975 [29], 1976 [30]; Matkowsky and Reiss, 1977 [135]; and Reiss, 1977 [171].

where $\boldsymbol{v}^0$ denotes the value of the imperfection parameter vector $\boldsymbol{v}$ for the perfect system ($(\cdot)^0$ denotes a variable associated with the perfect system); $\boldsymbol{d}$ is called the *imperfection pattern vector* (normalized appropriately); and ε denotes the *magnitude of imperfection* that represents the amount of deviation from the perfect case (ε can be negative).

We define the *imperfection sensitivity matrix* by

$$B(\boldsymbol{u}, f, \boldsymbol{v}) = (B_{ij}) = \left(\frac{\partial F_i}{\partial v_j}\right), \tag{2.3}$$

which is an $N \times p$ matrix. This matrix plays the major role in the description of the influence of imperfection in this book.

Remark 2.1. Equation (2.1) can represent physical phenomena of various kinds. For a structure subjected to an external load, for example, the load is selected to be the bifurcation parameter f, the displacement components to be the state (independent) variable vector $\boldsymbol{u}$, and the member lengths to be the imperfection parameter vector $\boldsymbol{v}$. The temperature is the bifurcation parameter for the combustion problem; the solar radiation is the bifurcation parameter for the problem of climate changes, and so on. □

2.2.2 Critical Point

For a fixed $\boldsymbol{v}$, solutions $(\boldsymbol{u}, f) = (\boldsymbol{u}(\boldsymbol{v}), f(\boldsymbol{v}))$ of the above system of equations (2.1) make up solution curves called *equilibrium paths* or *loci of equilibria*. The solution points are divided into two types, *ordinary* or *critical* (*singular*) points, according to whether the *Jacobian matrix*[3]

$$J = J(\boldsymbol{u}, f, \boldsymbol{v}) = (J_{ij}) = \left(\frac{\partial F_i}{\partial u_j}\right), \tag{2.4}$$

which is an $N \times N$ matrix, is nonsingular or singular. That is,

$$\det(J) = \begin{cases} \text{nonzero} & \text{at the ordinary point,} \\ 0 & \text{at the critical (singular) point,} \end{cases} \tag{2.5}$$

where $\det(\cdot)$ denotes the determinant of the matrix therein.

In a sufficiently small neighborhood of an ordinary point, the implicit function theorem applies. For each f there exists a unique $\boldsymbol{u} = \boldsymbol{u}(f)$ such that $(\boldsymbol{u}(f), f)$ is a solution to (2.1). Here the imperfection parameter $\boldsymbol{v}$ is kept fixed. Consequently, it is suppressed in the notation $\boldsymbol{u} = \boldsymbol{u}(f)$, which should be written more precisely as $\boldsymbol{u} = \boldsymbol{u}(f, \boldsymbol{v})$.

In the vicinity of a critical point, say $(\boldsymbol{u}_\mathrm{c}, f_\mathrm{c}) = (\boldsymbol{u}_\mathrm{c}(\boldsymbol{v}), f_\mathrm{c}(\boldsymbol{v}))$, an interesting phenomenon can occur, where $(\cdot)_\mathrm{c}$ denotes a variable related to the critical point. A

[3] The Jacobian matrix is called the *tangent stiffness matrix* in structural mechanics.

typical interesting phenomenon is *bifurcation*: the emergence of multiple solution paths. The Jacobian matrix $J_{\rm c} = J(\boldsymbol{u}_{\rm c}, f_{\rm c}, \boldsymbol{v})$ at $(\boldsymbol{u}_{\rm c}, f_{\rm c}, \boldsymbol{v})$ is singular by the definition of a critical point; that is,

$$\det[J(\boldsymbol{u}_{\rm c}, f_{\rm c}, \boldsymbol{v})] = 0, \tag{2.6}$$

and the behavior of $\boldsymbol{u} = \boldsymbol{u}(f)$ around $(\boldsymbol{u}_{\rm c}, f_{\rm c})$ is not governed by the implicit function theorem.

The *multiplicity* M of a critical point $(\boldsymbol{u}_{\rm c}, f_{\rm c})$ is defined as the *rank deficiency* of the Jacobian matrix; that is,

$$M = \dim[\ker(J_{\rm c})] = N - \mathrm{rank}(J_{\rm c}), \tag{2.7}$$

where $\ker(\cdot)$ denotes the *kernel space* of the matrix in parentheses and $\mathrm{rank}(\cdot)$ denotes the *rank* of the matrix. The critical point $(\boldsymbol{u}_{\rm c}, f_{\rm c})$ is a *simple* critical point if $M = 1$ and a *multiple* critical point if $M \geq 2$.

Let $\{\boldsymbol{\xi}_i \mid i = 1, \ldots, M\}$ and $\{\boldsymbol{\eta}_i \mid i = 1, \ldots, M\}$ be two families of independent vectors of $\mathbb{R}^N$ such that

$$\boldsymbol{\xi}_i^\top J_{\rm c} = \boldsymbol{0}^\top, \qquad J_{\rm c}\boldsymbol{\eta}_i = \boldsymbol{0}, \qquad i = 1, \ldots, M. \tag{2.8}$$

Such vectors $\boldsymbol{\xi}_i$ $(i = 1, \ldots, M)$ are called the *left critical (eigen)vectors*, and $\boldsymbol{\eta}_i$ $(i = 1, \ldots, M)$ the *right critical (eigen)vectors*. Here $\{\boldsymbol{\xi}_i \mid i = 1, \ldots, M\}$ and $\{\boldsymbol{\eta}_i \mid i = 1, \ldots, M\}$, respectively, span the kernel of $J_{\rm c}{}^\top$ and $J_{\rm c}$. Note that orthogonality is not imposed in general, although in some cases it is a natural and convenient requirement (cf., §2.2.3). Critical eigenvectors play a crucial role in deriving a reduced system of equations, the bifurcation equation, in §2.4. See Remark 2.3 below.

We denote by $(\boldsymbol{u}_{\rm c}^0, f_{\rm c}^0)$ a critical point for the *perfect system* described by $\boldsymbol{v} = \boldsymbol{v}^0$. The Jacobian matrix at the critical point for the perfect system is denoted as

$$J_{\rm c}^0 = J(\boldsymbol{u}_{\rm c}^0, f_{\rm c}^0, \boldsymbol{v}^0). \tag{2.9}$$

The imperfection sensitivity matrix (2.3) at this point is denoted as

$$B_{\rm c}^0 = B(\boldsymbol{u}_{\rm c}^0, f_{\rm c}^0, \boldsymbol{v}^0). \tag{2.10}$$

Remark 2.2. For some problems, a natural problem formulation yields a system of equations of the form (2.1) with more unknown variables than equations. In such cases, a solution is called an *ordinary point* if the rank of the Jacobian matrix J is equal to the number of equations; it is called a *critical point* otherwise. □

Remark 2.3. The term of *critical eigenvectors* introduced in (2.8) might be less than adequate mathematically, but it is conventional in engineering. In (2.8), the matrix $J_{\rm c}$ is regarded as a linear map from one vector space to another. As such, it is meaningless to talk about eigenvectors. As described earlier, $\{\boldsymbol{\xi}_i \mid i = 1, \ldots, M\}$ and $\{\boldsymbol{\eta}_i \mid i = 1, \ldots, M\}$, respectively, represent the bases of the kernel spaces of $J_{\rm c}{}^\top$ and $J_{\rm c}$. Consider, for example, a 2×2 matrix

$$J_c = \begin{pmatrix} 0 & 1 \\ 0 & 0 \end{pmatrix},$$

for which $M = 1$. We may take $\boldsymbol{\xi}_1 = (0,1)^\top$ and $\boldsymbol{\eta}_1 = (1,0)^\top$ in (2.8). For reciprocal systems (see §2.2.3), in which the Jacobian matrix is symmetric, it is natural to refer to the eigenvectors of J_c, and M is equal to the algebraic multiplicity of the characteristic polynomial $\det(\lambda I - J_c)$ at $\lambda = 0$. □

Remark 2.4. Although we restrict ourselves to the discretized system (2.1) in Parts I and II, all the ideas can be extended to a continuous case, that is, to a system of governing equations

$$F(u, f, v) = 0,$$

in which $u = u(x)$ is a function in x (the coordinate of a point in a domain), v is an imperfection parameter, and F is a nonlinear operator (defined on a certain function space and satisfying relevant regularity conditions). Then the Jacobian matrix is to be replaced with the derivative (or the *Fréchet derivative*) of F with respect to u. We refer to the description of continuous problems in Part III. □

2.2.3 Reciprocity

Consider a system that has a *total potential energy* designated by $U(\boldsymbol{u}, f, \boldsymbol{v})$. Then, we can derive, by the *principle of stationary potential energy*,[4] the governing equation $\boldsymbol{F}$ in (2.1) as[5]

$$\boldsymbol{F}(\boldsymbol{u}, f, \boldsymbol{v}) = \left(\frac{\partial U(\boldsymbol{u}, f, \boldsymbol{v})}{\partial \boldsymbol{u}} \right)^\top. \tag{2.11}$$

A system is called a *potential system* or a *gradient system*, if $\boldsymbol{F}$ is given as (2.11) for some scalar function U.

The governing equation $\boldsymbol{F}$ of a potential system has a symmetry of the form

$$\frac{\partial F_i}{\partial u_j} = \frac{\partial F_j}{\partial u_i}, \qquad i, j = 1, \ldots, N, \tag{2.12}$$

as a consequence of the basic fact in calculus:

$$\frac{\partial}{\partial u_j}\left(\frac{\partial U}{\partial u_i} \right) = \frac{\partial}{\partial u_i}\left(\frac{\partial U}{\partial u_j} \right), \qquad i, j = 1, \ldots, N.$$

The symmetry in (2.12), as a property of a system, is referred to as *reciprocity*; and a system equipped with this property is called a *reciprocal system*.

---

4 See, for example, Theorem VIII on page 305 of Oden and Ripperger, 1981 [152].

5 It is assumed in (2.11) that $\boldsymbol{F}$ is a column vector and $\partial U / \partial \boldsymbol{u}$ is a row vector.

Conversely, the existence of a potential function is guaranteed by the reciprocity (2.12), which is an important fundamental fact in calculus. Thus a reciprocal system is a synonym of a potential system.

The reciprocity (2.12) is equivalent to the symmetry ($J_{ij} = J_{ji}$) of the Jacobian matrix J, and then the eigenvalues $\lambda_1, \ldots, \lambda_N$ of J are all real. In this case, it is natural in (2.8) to assume $\boldsymbol{\xi}_i = \boldsymbol{\eta}_i$ $(i = 1, \ldots, M)$ and to impose orthogonality $\boldsymbol{\eta}_i^\top \boldsymbol{\eta}_j = \delta_{ij}$ $(i, j = 1, \ldots, M)$, where δ_{ij} denotes *Kronecker's delta*, which is equal to 1 for $i = j$ and 0 for $i \neq j$.

2.2.4 Stability

For reciprocal systems, the *stability* of a solution can be defined naturally. Let $\boldsymbol{F}$ be given as the gradient of $U(\boldsymbol{u}) = U(\boldsymbol{u}, f, \boldsymbol{v})$ as in (2.11). Then a solution $\boldsymbol{u}$ to $\boldsymbol{F}(\boldsymbol{u}, f, \boldsymbol{v}) = \boldsymbol{0}$ is said to be *stable* if $U(\boldsymbol{u} + \Delta\boldsymbol{u}) \geq U(\boldsymbol{u})$ for any sufficiently small perturbation $\Delta\boldsymbol{u}$. It is called *unstable* if it is not stable. A solution $\boldsymbol{u}$ is called *linearly stable* if the Jacobian matrix $J(\boldsymbol{u}, f, \boldsymbol{v})$, which is symmetric, is positive-definite or, equivalently, if every *eigenvalue* of $J(\boldsymbol{u}, f, \boldsymbol{v})$ is positive. The solution $\boldsymbol{u}$ is said to be *linearly unstable* if $J(\boldsymbol{u}, f, \boldsymbol{v})$ has at least one negative eigenvalue. It is readily apparent that a solution is stable if it is linearly stable, and unstable if it is linearly unstable.

For *nonreciprocal systems*, the stability of a solution $\boldsymbol{u}$ to $\boldsymbol{F}(\boldsymbol{u}, f, \boldsymbol{v}) = \boldsymbol{0}$ can be defined in relation to the associated *dynamical system*

$$\frac{d\widehat{\boldsymbol{u}}}{dt} + \boldsymbol{F}(\widehat{\boldsymbol{u}}, f, \boldsymbol{v}) = \boldsymbol{0} \tag{2.13}$$

as the *asymptotic stability* of the solution $\widehat{\boldsymbol{u}}(t)$ as $t \to +\infty$. A solution $\boldsymbol{u}$ to $\boldsymbol{F}(\boldsymbol{u}, f, \boldsymbol{v}) = \boldsymbol{0}$ is said to be *stable* if every solution $\widehat{\boldsymbol{u}}(t)$ to (2.13) that is initially close to $\boldsymbol{u}$ decays to $\boldsymbol{u}$ as $t \to +\infty$. A solution $\boldsymbol{u}$ is called *unstable* if it is not stable. We designate $\boldsymbol{u}$ as *linearly stable* if every eigenvalue of $J(\boldsymbol{u}, f, \boldsymbol{v})$ has a positive real part, and *linearly unstable* if at least one eigenvalue has a negative real part.

In this book we restrict ourselves to linear stability/instability via the eigenanalysis of the Jacobian matrix and designate it as stability/instability for simplicity.

2.3 Illustrative Example

The general framework introduced in §2.2 is presented based on concrete calculations for a simple example, the propped cantilever shown in Fig. 2.1. This serves as an introduction to the concept of the bifurcation equation, which is treated formally in §2.4.

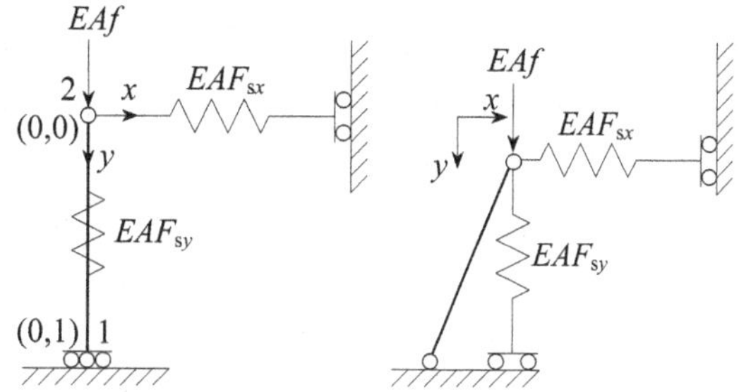

(a) Perfect system (b) Imperfect system

Fig. 2.1 Propped cantilever.

Governing Equation

The propped cantilever comprises a truss member that is simply supported at a rigid foundation and supported by horizontal and vertical springs. The equilibrium equation for this cantilever is

$$\boldsymbol{F}(\boldsymbol{u},f,\boldsymbol{v}) = \begin{pmatrix} F_x \\ F_y \end{pmatrix} = EA \begin{pmatrix} \left(\frac{1}{L} - \frac{1}{\widehat{L}}\right)(x - x_1) + F_{sx} \\ \left(\frac{1}{L} - \frac{1}{\widehat{L}}\right)(y - y_1) + F_{sy} - f \end{pmatrix} = \begin{pmatrix} 0 \\ 0 \end{pmatrix}, \tag{2.14}$$

where f is the nondimensional vertical load normalized with respect to the cross-sectional rigidity EA of the truss member (E signifies Young's modulus and A denotes the cross-sectional area); $\boldsymbol{u} = (x,y)^\top$ is the location of node 2 after displacement; and (x_1,y_1) and (x_2,y_2), respectively, represent the locations of nodes 1 and 2 before displacement; L and $\widehat{L}$ denote the length of the member before and after displacement, respectively, and are given by

$$L = [(x_2 - x_1)^2 + (y_2 - y_1)^2]^{1/2}, \qquad \widehat{L} = [(x - x_1)^2 + (y - y_1)^2]^{1/2}; \tag{2.15}$$

F_{sx} and F_{sy} are the horizontal and vertical normalized nondimensional forces exerted by the springs, respectively, which are assumed to be

$$F_{sx} = \beta_1 + \beta_2 \frac{x - x_2}{L} + \beta_3 \left(\frac{x - x_2}{L}\right)^2, \qquad F_{sy} = \beta_4 \frac{y - y_2}{L}. \tag{2.16}$$

For the perfect system, we have $(x_1,y_1) = (0,1)$ and $(x_2,y_2) = (0,0)$, and the initial member length is given by $L^0 = [(0-0)^2 + (1-0)^2]^{1/2} = 1$.

We set the imperfection parameter vector in (2.14) as

$$\boldsymbol{v} = \left(\frac{x_1}{L^0}, \frac{y_1}{L^0}, \frac{x_2}{L^0}, \frac{y_2}{L^0}, \beta_1, \beta_2, \beta_3, \beta_4\right)^\top; \tag{2.17}$$

and express it in the form[6] of $\boldsymbol{v} = \boldsymbol{v}^0 + \varepsilon \boldsymbol{d}$ with

$$\boldsymbol{v}^0 = (0,1,0,0,0,1,1,1)^\top, \qquad \boldsymbol{d} = (-1,0,1,0,-1,0,0,0)^\top. \tag{2.18}$$

That is,

$$\boldsymbol{v} = (0,1,0,0,0,1,1,1)^\top + \varepsilon(-1,0,1,0,-1,0,0,0)^\top. \tag{2.19}$$

For this imperfection vector $\boldsymbol{v}$, the variables in (2.15) and (2.16) are evaluated as

$$\left.\begin{aligned} &L = L^0(1+4\varepsilon^2)^{1/2}, \qquad \widehat{L} = [(x+L^0\varepsilon)^2 + (y-L^0)^2]^{1/2}, \\ &F_{\mathrm{s}x} = -\varepsilon + \frac{x - L^0\varepsilon}{L} + \left(\frac{x-L^0\varepsilon}{L}\right)^2, \qquad F_{\mathrm{s}y} = \frac{y}{L}. \end{aligned}\right\} \tag{2.20}$$

In the following, "exact" and "asymptotic" analyses for the perfect and imperfect systems are compared. In the exact analysis the reduction to a single equation—the bifurcation equation—is achieved exactly, whereas in the asymptotic analysis only the leading terms of this equation are considered and are shown to represent important local bifurcation behaviors.

Exact Analysis

First, we refer to the bifurcation behavior of the perfect system ($\varepsilon = 0$) of the propped cantilever. Setting $\varepsilon = 0$ in (2.20) yields $L = L^0 = 1$ and the spring characteristic in the x-direction

$$F_{\mathrm{s}x} = \frac{x}{L^0} + \left(\frac{x}{L^0}\right)^2 = x + x^2, \tag{2.21}$$

which exerts greater force for positive x than for negative x. By setting $\varepsilon = 0$ in (2.14) with (2.19), it is possible to reduce the equilibrium equation for the perfect system to

$$\boldsymbol{F}(\boldsymbol{u}, f, \boldsymbol{v}^0) = EA \begin{pmatrix} \left(1 - \dfrac{1}{[x^2+(y-1)^2]^{1/2}}\right)x + x + x^2 \\ \left(1 - \dfrac{1}{[x^2+(y-1)^2]^{1/2}}\right)(y-1) + y - f \end{pmatrix} = \begin{pmatrix} 0 \\ 0 \end{pmatrix}. \tag{2.22}$$

The solution path (x, y, f) of (2.22) is expressed as

$$\left.\begin{aligned} &x = 0, \quad f = 2y, \quad y < 1, && \text{fundamental path,} \\ &y = 1 - [(x+2)^{-2} - x^2]^{1/2}, \quad f = 1 + x[(x+2)^{-2} - x^2]^{1/2}, && -2 < x \leq \sqrt{2} - 1, \\ & && \text{bifurcated path.} \end{aligned}\right\} \tag{2.23}$$

[6] In this form, it is mandatory to let ε be nondimensional. It is preferred to let $\boldsymbol{v}$ and $\boldsymbol{d}$ be nondimensional.

This solution is plotted as solid lines in Fig. 2.2(a). Two paths intersect at the bifurcation point $(x_c^0, y_c^0, f_c^0) = (0, 1/2, 1)$. A critical point of this type is called a *transcritical bifurcation point*.

The location of this point on the fundamental path can be determined as a point where the Jacobian matrix $J(x, y, f, v^0)$ with $x = 0$ becomes singular. Note that

$$J(0, y, f, v^0) = \begin{pmatrix} F_{x,x} & F_{x,y} \\ F_{y,x} & F_{y,y} \end{pmatrix}\Bigg|_{(x,v)=(0,v^0)} = EA \begin{pmatrix} 2 - \dfrac{1}{|y-1|} & 0 \\ 0 & 2 \end{pmatrix},$$

where $F_{x,y}$, for example, denotes the derivative of F_x with respect to y. The Jacobian matrix at the bifurcation point $(x_c^0, y_c^0, f_c^0) = (0, 1/2, 1)$ is given by

$$J_c^0 = EA \begin{pmatrix} 0 & 0 \\ 0 & 2 \end{pmatrix}, \tag{2.24}$$

of which $\eta_1 = (1, 0)^\top$ is the critical eigenvector; physically, η_1 represents the tumbling of node 2 in the x-direction.

Next, the imperfect bifurcation behavior of the cantilever is investigated. Substituting the variables (2.20) into the equilibrium equation (2.14) leads to a nonlinear equation, whose solution is given by

$$y = 1 - |x + \varepsilon| \left[\left(\frac{x + \varepsilon}{L} + F_{sx} \right)^{-2} - 1 \right]^{1/2},$$

$$f = \frac{1}{L} + \left(F_{sx} \operatorname{sign}(x + \varepsilon) - \frac{|x + \varepsilon|}{L} \right) \left[\left(\frac{x + \varepsilon}{L} + F_{sx} \right)^{-2} - 1 \right]^{1/2},$$

where sign(·) denotes the sign of the variable in parentheses (equal to 0 if the variable is equal to 0) and L and F_{sx} are given by (2.20).

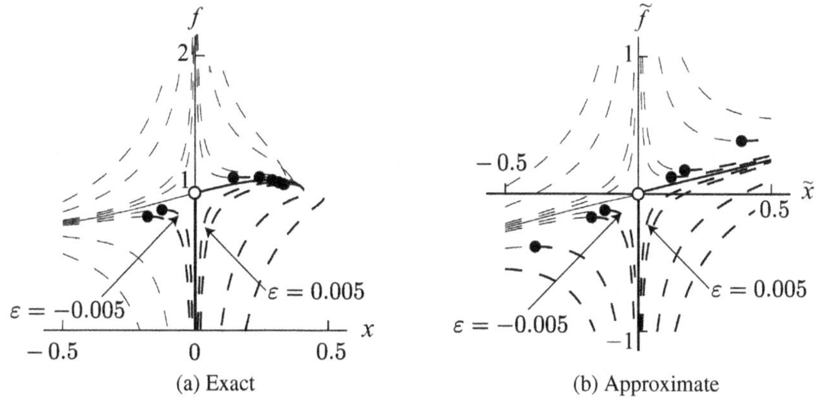

Fig. 2.2 Exact and approximate equilibrium paths for the propped cantilever in the neighborhood of the bifurcation point. ——: path for the perfect system; — — —: path for an imperfect system; thick line: stable; thin line: unstable; ○: bifurcation point; ●: limit point.

This solution is presented in Fig. 2.2(a) for the imperfection magnitude of various values, $\varepsilon = 0, \pm 0.005, \pm 0.01, \pm 0.05$, and ± 0.1. The distance between the curves for the perfect and those for imperfect systems increases in association with the increase in $|\varepsilon|$. Realistic systems, which inevitably involve some imperfections, will behave similarly to these imperfect paths. The behaviors of the imperfect systems with ε and $-\varepsilon$ are not symmetric; see, for example, the imperfect paths with $\varepsilon = \pm 0.005$. Of a pair of imperfect paths ascending from the origin ($f = 0$) for the same value of $|\varepsilon|$, f increases stably beyond the bifurcation load ($f = 1$) for one path; for the other path, f reaches its maximum at the limit point shown by (•).

Asymptotic Analysis

The actual bifurcation analysis is complex even for a system with a few degrees of freedom. Such complexity can be resolved using asymptotic analysis as described below.

We investigate the properties of the solution, locally around the point $(x_c^0, y_c^0, f_c^0) = (0, 1/2, 1)$, which corresponds to the bifurcation point of the perfect system. Define the increment $(\widetilde{x}, \widetilde{y}, \widetilde{f})$ as

$$(\widetilde{x}, \widetilde{y}, \widetilde{f}) = (x, y, f) - (0, 1/2, 1). \tag{2.25}$$

Expanding the equilibrium equation (2.14) around $(x_c^0, y_c^0, f_c^0) = (0, 1/2, 1)$ yields a set of incremental equations

$$\begin{aligned}
&F_x(x,y,f,\varepsilon) - F_x(0,1/2,1,0) \\
&\quad = F_x(\widetilde{x}, 1/2+\widetilde{y}, 1+\widetilde{f}, \varepsilon) - F_x(0,1/2,1,0) \\
&\quad = (F_{x,x})_c^0\,\widetilde{x} + \left(F_{x,y}\right)_c^0\widetilde{y} + \left(F_{x,f}\right)_c^0\widetilde{f} + (F_{x,\varepsilon})_c^0\,\varepsilon \\
&\qquad + \left[\frac{1}{2}(F_{x,xx})_c^0\,\widetilde{x}^2 + \left(F_{x,xy}\right)_c^0\widetilde{x}\,\widetilde{y} + \frac{1}{2}\left(F_{x,yy}\right)_c^0\widetilde{y}^2\right. \\
&\qquad \left. + \left(F_{x,xf}\right)_c^0\widetilde{x}\widetilde{f} + \left(F_{x,yf}\right)_c^0\widetilde{y}\widetilde{f} + \frac{1}{2}\left(F_{x,ff}\right)_c^0\widetilde{f}^2\right] + \text{h.o.t.} \\
&\quad = EA(\widetilde{x}^2 - 4\widetilde{x}\,\widetilde{y} - 3\varepsilon) + \text{h.o.t.} = 0,
\end{aligned} \tag{2.26}$$

$$F_y(x,y,f,\varepsilon) - F_y(0,1/2,1,0) = EA(2\widetilde{y} - \widetilde{f}) + \text{h.o.t.} = 0, \tag{2.27}$$

where $(\cdot)_c^0$ denotes a value at point (x_c^0, y_c^0, f_c^0) and h.o.t. denotes higher-order terms.

Equation (2.27) can be solved for $\widetilde{y}$ as

$$\widetilde{y} = \widetilde{f}/2 + \text{h.o.t.} \tag{2.28}$$

Substitution of this equation into (2.26) yields

$$\widehat{F}(\widetilde{x}, \widetilde{f}, \varepsilon) = \widetilde{x}^2 - 2\widetilde{x}\widetilde{f} - 3\varepsilon + \text{h.o.t.} = 0, \tag{2.29}$$

which denotes an incremental relation between $\widetilde{x}$ (associated with the bifurcation mode) and $\widetilde{f}$. As (2.29) begins with a quadratic term in $\widetilde{x}$ for given $\widetilde{f}$ and ε, it implies the presence of two solutions in the neighborhood of the bifurcation point. Thus, bifurcation can be interpreted as the emergence of the multiplicity of solutions due to the vanishing of lower-order terms.

Equation (2.29) is called the *bifurcation equation*. It has been derived from the full system of equations by eliminating the variable $\widetilde{y}$, which is not associated with the critical eigenvector of the Jacobian matrix J_c^0 in (2.24) at the critical point. Such a reduction process to the bifurcation equation is called the *Liapunov–Schmidt reduction*, the general treatment of which is the topic of the next section.

Figure 2.2(b) shows the force versus displacement curves expressed by the leading terms of the bifurcation equation (2.29) with $\varepsilon = 0, \pm 0.005, \pm 0.01, \pm 0.05$, and ± 0.1. These curves simulate the actual behavior of the cantilever in Fig. 2.2(a) quite well in the neighborhood of the bifurcation point, but less accurately away from it due to the omission of higher-order terms in the bifurcation equation (2.29).

The bifurcation equation (2.29), despite its simplicity, retains important information about bifurcation behavior. It is a fundamental strategy of bifurcation theory to extract information about bifurcation behavior by examining the asymptotic form of the bifurcation equation, which is much simpler and easier to handle than the original equation.

2.4 Liapunov–Schmidt Reduction

This section explains a standard procedure in local bifurcation analysis—the *Liapunov–Schmidt reduction*—that reduces the whole system of equations to a single equation locally in a neighborhood of a simple critical point.

2.4.1 Reduction Procedure

We consider a nonlinear equilibrium, or governing equation, in (2.1):

$$\boldsymbol{F}(\boldsymbol{u}, f, \boldsymbol{v}) = \boldsymbol{0}, \tag{2.30}$$

where $\boldsymbol{u} \in \mathbb{R}^N$ represents a state (independent) variable vector; $f \in \mathbb{R}$ denotes a bifurcation (loading) parameter; $\boldsymbol{v} \in \mathbb{R}^p$ is an imperfection parameter vector; and $\boldsymbol{F}$ is a sufficiently smooth nonlinear function. The reciprocity (2.12) is not assumed.

Let $(\boldsymbol{u}_c^0, f_c^0)$ be a simple critical point of the perfect system with $\boldsymbol{v} = \boldsymbol{v}_0$,

$$J_c^0 = J(\boldsymbol{u}_c^0, f_c^0, \boldsymbol{v}^0)$$

be the Jacobian matrix at the critical point, and $\{\boldsymbol{\xi}_j \mid j = 1,\ldots,N\}$ and $\{\boldsymbol{\eta}_j \mid j = 1,\ldots,N\}$ be bases[7] of $\mathbb{R}^N$ satisfying

$$\boldsymbol{\xi}_1^\top J_{\mathrm{c}}^0 = \boldsymbol{0}^\top, \qquad J_{\mathrm{c}}^0 \boldsymbol{\eta}_1 = \boldsymbol{0}, \tag{2.31}$$

where $\boldsymbol{\xi}_1$ and $\boldsymbol{\eta}_1$ are the left and right critical eigenvectors at $(\boldsymbol{u}_{\mathrm{c}}^0, f_{\mathrm{c}}^0)$.

We express the state variable $\boldsymbol{u}$ as

$$\boldsymbol{u} = \boldsymbol{u}_{\mathrm{c}}^0 + \sum_{j=1}^{N} w_j \boldsymbol{\eta}_j \tag{2.32}$$

in terms of incremental variables $(w_j \mid j = 1,\ldots,N)$, and express the bifurcation parameter f as

$$f = f_{\mathrm{c}}^0 + \widetilde{f}, \tag{2.33}$$

where $\widetilde{f}$ represents the increment of f.

Using these incremental variables in the original system (2.30) gives

$$\boldsymbol{\xi}_1^\top \boldsymbol{F}\Big(\boldsymbol{u}_{\mathrm{c}}^0 + \sum_{j=1}^{N} w_j \boldsymbol{\eta}_j, f_{\mathrm{c}}^0 + \widetilde{f}, \boldsymbol{v}\Big) = 0, \tag{2.34}$$

$$\boldsymbol{\xi}_i^\top \boldsymbol{F}\Big(\boldsymbol{u}_{\mathrm{c}}^0 + \sum_{j=1}^{N} w_j \boldsymbol{\eta}_j, f_{\mathrm{c}}^0 + \widetilde{f}, \boldsymbol{v}\Big) = 0, \qquad i = 2,\ldots,N. \tag{2.35}$$

Equations (2.34) and (2.35) provide a decomposition of the original system (2.30) into two parts; the latter component (2.35) is for the range space of J_{c}^0 and the former component (2.34) is for its complement.

The Jacobian matrix of the left-hand side of (2.35) with respect to w_j $(j = 2,\ldots,N)$, evaluated at $w_j = 0$ $(j = 1,2,\ldots,N)$, $\widetilde{f} = 0$, and $\boldsymbol{v} = \boldsymbol{v}^0$, is nonsingular (cf., (2.48) in §2.4.2). Therefore, by the implicit function theorem, (2.35) can be solved locally for w_j $(j = 2,\ldots,N)$ as

$$w_j = \varphi_j(w, \widetilde{f}, \boldsymbol{v}), \qquad j = 2,\ldots,N, \tag{2.36}$$

where $w \equiv w_1$ and

$$\varphi_j(0, 0, \boldsymbol{v}^0) = 0, \qquad j = 2,\ldots,N. \tag{2.37}$$

Substituting this into (2.34) yields a reduced equation

$$\widetilde{F}(w, \widetilde{f}, \boldsymbol{v}) = 0 \tag{2.38}$$

in w, where

[7] It is intended that $\{\boldsymbol{\eta}_j\}$ be a basis of the space of the vectors $\boldsymbol{u}$, and that $\{\boldsymbol{\xi}_j\}$ be a basis of the (dual) space of the values of $\boldsymbol{F}$.

$$\widetilde{F}(w,\widetilde{f},\boldsymbol{v}) = \boldsymbol{\xi}_1^\top \boldsymbol{F}\Big(\boldsymbol{u}_c^0 + w\boldsymbol{\eta}_1 + \sum_{j=2}^{N} \varphi_j(w,\widetilde{f},\boldsymbol{v})\boldsymbol{\eta}_j, f_c^0 + \widetilde{f}, \boldsymbol{v}\Big). \tag{2.39}$$

This reduced equation (2.38) is called the *bifurcation equation*. It must be emphasized that the reduction to (2.38) is valid locally in a neighborhood of $(\boldsymbol{u}_c^0, f_c^0, \boldsymbol{v}^0)$.

The solutions $(w,\widetilde{f},\boldsymbol{v})$ to the bifurcation equation (2.38) are in one-to-one correspondence through (2.36) with the solutions $(\boldsymbol{u}, f, \boldsymbol{v})$ of the original system (2.30); that is,

$$\boldsymbol{u} = \boldsymbol{u}(w,\widetilde{f},\boldsymbol{v}) = \boldsymbol{u}_c^0 + w\boldsymbol{\eta}_1 + \sum_{j=2}^{N} \varphi_j(w,\widetilde{f},\boldsymbol{v})\boldsymbol{\eta}_j \tag{2.40}$$

in the neighborhood of $(\boldsymbol{u}_c^0, f_c^0, \boldsymbol{v}^0)$. Consequently, the qualitative picture of the solution set of the original system (2.30) is isomorphic to that of the bifurcation equation (2.38).

For later reference we display the key identity

$$\boldsymbol{F}\Big(\boldsymbol{u}_c^0 + w\boldsymbol{\eta}_1 + \sum_{j=2}^{N} \varphi_j(w,\widetilde{f},\boldsymbol{v})\boldsymbol{\eta}_j, f_c^0 + \widetilde{f}, \boldsymbol{v}\Big) = \boldsymbol{0}. \tag{2.41}$$

Remark 2.5. The bifurcation equation obtained by the Liapunov–Schmidt reduction depends on the choice of the bases $\{\boldsymbol{\xi}_j \mid j = 1,\dots,N\}$ and $\{\boldsymbol{\eta}_j \mid j = 1,\dots,N\}$. It is known, however, that different choices yield qualitatively equivalent bifurcation equations. See Chapter I, Appendix 2, Golubitsky and Schaeffer, 1985 [62] for the precise formulation of this equivalence and its proof. □

Remark 2.6. In the case of a reciprocal system, which has a symmetric Jacobian matrix, it is natural (and indeed possible) to choose orthonormal eigenvectors of J_c^0 as the basis vectors $\{\boldsymbol{\xi}_j \mid j = 1,\dots,N\}$ and $\{\boldsymbol{\eta}_j \mid j = 1,\dots,N\}$. We can further assume that $\boldsymbol{\xi}_j = \boldsymbol{\eta}_j$ $(j = 1,\dots,N)$. □

Remark 2.7. The key to the reduction to the single equation (2.38) is the elimination of w_j $(j = 2,\dots,N)$ based on the nonsingularity of the Jacobian matrix of (2.35). This means, in particular, that the reduction to the single equation is also possible at an ordinary point. □

2.4.2 Criticality Condition

The criticality condition (2.6) for the original system (2.30) is equivalent to the criticality condition

$$\frac{\partial \widetilde{F}}{\partial w} = 0 \tag{2.42}$$

for the bifurcation equation, as described below. Recall the notation $J = J(\boldsymbol{u}, f, \boldsymbol{v})$ in (2.4) for the Jacobian matrix of (2.30).

Let $\overline{J} = \overline{J}(\boldsymbol{u}, f, \boldsymbol{v}) = (\overline{J}_{ij} \mid i, j = 1, \ldots, N)$ be an $N \times N$ matrix defined as

$$\overline{J}_{ij} = \boldsymbol{\xi}_i^{\top} J \boldsymbol{\eta}_j, \qquad i, j = 1, \ldots, N \tag{2.43}$$

and partition $\overline{J}$ as

$$\overline{J} = \overline{J}(\boldsymbol{u}, f, \boldsymbol{v}) = \begin{pmatrix} \overline{J}_{[1,1]} & \overline{J}_{[1,2]} \\ \overline{J}_{[2,1]} & \overline{J}_{[2,2]} \end{pmatrix}, \tag{2.44}$$

where $\overline{J}_{[1,1]}$ is a 1×1 matrix, $\overline{J}_{[1,2]}$ is a $1 \times (N-1)$ matrix, $\overline{J}_{[2,1]}$ is an $(N-1) \times 1$ matrix, and $\overline{J}_{[2,2]}$ is an $(N-1) \times (N-1)$ matrix given, respectively, by

$$\begin{aligned} \overline{J}_{[1,1]} &= (\overline{J}_{11}), & \overline{J}_{[1,2]} &= (\overline{J}_{1j} \mid j = 2, \ldots, N), \\ \overline{J}_{[2,1]} &= (\overline{J}_{i1} \mid i = 2, \ldots, N), & \overline{J}_{[2,2]} &= (\overline{J}_{ij} \mid i, j = 2, \ldots, N). \end{aligned} \tag{2.45}$$

Since $\boldsymbol{\xi}_1$ and $\boldsymbol{\eta}_1$ are critical eigenvectors, we have

$$\overline{J}_{[1,1]}(\boldsymbol{u}_{\mathrm{c}}^0, f_{\mathrm{c}}^0, \boldsymbol{v}^0) = 0, \qquad \overline{J}_{[1,2]}(\boldsymbol{u}_{\mathrm{c}}^0, f_{\mathrm{c}}^0, \boldsymbol{v}^0) = \boldsymbol{0}^{\top}, \qquad \overline{J}_{[2,1]}(\boldsymbol{u}_{\mathrm{c}}^0, f_{\mathrm{c}}^0, \boldsymbol{v}^0) = \boldsymbol{0}. \tag{2.46}$$

That is,

$$\overline{J}(\boldsymbol{u}_{\mathrm{c}}^0, f_{\mathrm{c}}^0, \boldsymbol{v}^0) = \begin{pmatrix} 0 & \boldsymbol{0}^{\top} \\ \boldsymbol{0} & \overline{J}_{[2,2]}(\boldsymbol{u}_{\mathrm{c}}^0, f_{\mathrm{c}}^0, \boldsymbol{v}^0) \end{pmatrix}. \tag{2.47}$$

This implies, in particular, that

$$\det[\overline{J}_{[2,2]}(\boldsymbol{u}_{\mathrm{c}}^0, f_{\mathrm{c}}^0, \boldsymbol{v}^0)] \neq 0, \tag{2.48}$$

since $\mathrm{rank}[\overline{J}(\boldsymbol{u}_{\mathrm{c}}^0, f_{\mathrm{c}}^0, \boldsymbol{v}^0)] = \mathrm{rank}[J(\boldsymbol{u}_{\mathrm{c}}^0, f_{\mathrm{c}}^0, \boldsymbol{v}^0)] = N - 1$. Then, by continuity of the determinant, we have $\det(\overline{J}_{[2,2]}(\boldsymbol{u}, f, \boldsymbol{v})) \neq 0$ for $(\boldsymbol{u}, f, \boldsymbol{v})$ sufficiently close to $(\boldsymbol{u}_{\mathrm{c}}^0, f_{\mathrm{c}}^0, \boldsymbol{v}^0)$. Consequently, matrix $\overline{J}_{[2,2]}(\boldsymbol{u}, f, \boldsymbol{v})$ is nonsingular in a neighborhood of $(\boldsymbol{u}_{\mathrm{c}}^0, f_{\mathrm{c}}^0, \boldsymbol{v}^0)$.

The following lemma gives an expression of $\partial \widetilde{F} / \partial w$ in terms of $\overline{J}$.

Lemma 2.1. *In a neighborhood of* $(\boldsymbol{u}_{\mathrm{c}}^0, f_{\mathrm{c}}^0, \boldsymbol{v}^0)$, *we have*

$$\frac{\partial \widetilde{F}}{\partial w}(w, \widetilde{f}, \boldsymbol{v}) = \overline{J}_{[1,1]} - \overline{J}_{[1,2]}(\overline{J}_{[2,2]})^{-1}\overline{J}_{[2,1]} = \frac{\det(\overline{J})}{\det(\overline{J}_{[2,2]})} = \frac{\alpha \cdot \det(J)}{\det(\overline{J}_{[2,2]})}, \tag{2.49}$$

where $J = J(\boldsymbol{u}(w, \widetilde{f}, \boldsymbol{v}), f_{\mathrm{c}}^0 + \widetilde{f}, \boldsymbol{v})$ *and* $\overline{J} = \overline{J}(\boldsymbol{u}(w, \widetilde{f}, \boldsymbol{v}), f_{\mathrm{c}}^0 + \widetilde{f}, \boldsymbol{v})$ *with* $\boldsymbol{u}(w, \widetilde{f}, \boldsymbol{v})$ *defined in* (2.40), *and*

$$\alpha = \det(\boldsymbol{\xi}_1, \ldots, \boldsymbol{\xi}_N) \cdot \det(\boldsymbol{\eta}_1, \ldots, \boldsymbol{\eta}_N) \tag{2.50}$$

is a nonzero real number. In particular,

$$\det[J(\boldsymbol{u}(w, \widetilde{f}, \boldsymbol{v}), f_{\mathrm{c}}^0 + \widetilde{f}, \boldsymbol{v})] = 0 \quad \Longleftrightarrow \quad \frac{\partial \widetilde{F}}{\partial w}(w, \widetilde{f}, \boldsymbol{v}) = 0. \tag{2.51}$$

Proof. As noted already, matrix $\overline{J}_{[2,2]}(\boldsymbol{u}, f, \boldsymbol{v})$ is nonsingular in a neighborhood of $(\boldsymbol{u}_{\mathrm{c}}^0, f_{\mathrm{c}}^0, \boldsymbol{v}^0)$. Differentiation of (2.35) with respect to w, using $w_j = \varphi_j(w, \widetilde{f}, \boldsymbol{v})$ ($j =$

$2,\dots,N$), yields

$$\boldsymbol{\xi}_i^\top J\Big(\boldsymbol{\eta}_1 + \sum_{j=2}^{N} \frac{\partial \varphi_j}{\partial w}\boldsymbol{\eta}_j\Big) = 0, \qquad i = 2,\dots,N; \tag{2.52}$$

that is,

$$\sum_{j=2}^{N} (\boldsymbol{\xi}_i^\top J\boldsymbol{\eta}_j)\frac{\partial \varphi_j}{\partial w} = -\boldsymbol{\xi}_i^\top J\boldsymbol{\eta}_1, \quad i = 2,\dots,N.$$

This is a system of equations in $\partial\varphi_j/\partial w$ ($j = 2,\dots,N$), in which the coefficient matrix is

$$(\boldsymbol{\xi}_i^\top J\boldsymbol{\eta}_j \mid i,j = 2,\dots,N) = (\overline{J}_{ij} \mid i,j = 2,\dots,N) = \overline{J}_{[2,2]}$$

(cf., (2.43) and (2.45)), which is nonsingular, and the right-hand side vector is equal to

$$(-\boldsymbol{\xi}_i^\top J\boldsymbol{\eta}_1 \mid i = 2,\dots,N) = (-\overline{J}_{i1} \mid i = 2,\dots,N) = -\overline{J}_{[2,1]}$$

(cf., (2.43) and (2.45)). Consequently, we obtain

$$\left(\frac{\partial \varphi_j}{\partial w} \;\middle|\; j = 2,\dots,N\right) = -(\overline{J}_{[2,2]})^{-1}\overline{J}_{[2,1]}. \tag{2.53}$$

Differentiation of (2.39) with respect to w, followed by the substitution of this expression, yields

$$\frac{\partial \widetilde{F}}{\partial w} = \boldsymbol{\xi}_1^\top J\Big(\boldsymbol{\eta}_1 + \sum_{j=2}^{N} \frac{\partial \varphi_j}{\partial w}\boldsymbol{\eta}_j\Big) = \overline{J}_{[1,1]} - \overline{J}_{[1,2]}(\overline{J}_{[2,2]})^{-1}\overline{J}_{[2,1]}. \tag{2.54}$$

On the other hand, a well-known formula in matrix algebra

$$\det\begin{pmatrix} A & B \\ C & D \end{pmatrix} = \det(A - BD^{-1}C)\cdot\det D$$

(where D is nonsingular) is applied to $\overline{J}$ in (2.44) to yield

$$\det(\overline{J}) = (\overline{J}_{[1,1]} - \overline{J}_{[1,2]}(\overline{J}_{[2,2]})^{-1}\overline{J}_{[2,1]})\cdot\det(\overline{J}_{[2,2]}), \tag{2.55}$$

whereas by (2.50), we have

$$\det(\overline{J}) = \det(\boldsymbol{\xi}_1,\dots,\boldsymbol{\xi}_N)\cdot\det(J)\cdot\det(\boldsymbol{\eta}_1,\dots,\boldsymbol{\eta}_N) = \alpha\cdot\det(J). \tag{2.56}$$

Combination of (2.54)–(2.56) yields (2.49). □

2.4.3 Direction of Bifurcated Paths

The direction of bifurcated paths can be analyzed as follows. We assume that the bifurcated path of our interest can be described in the form[8] of

$$\widetilde{f} = \psi(w, \boldsymbol{v}^0) \tag{2.57}$$

with a differentiable function ψ. Substituting this into (2.40) yields

$$\boldsymbol{u} = \boldsymbol{u}(w, \psi(w, \boldsymbol{v}^0), \boldsymbol{v}^0) = \boldsymbol{u}_\mathrm{c}^0 + w\boldsymbol{\eta}_1 + \sum_{j=2}^{N} \varphi_j(w, \psi(w, \boldsymbol{v}^0), \boldsymbol{v}^0)\boldsymbol{\eta}_j$$

for the bifurcated path. The direction of bifurcation at $(\boldsymbol{u}_\mathrm{c}^0, f_\mathrm{c}^0, \boldsymbol{v}^0)$ is then obtained as the derivative of $\boldsymbol{u}$ with respect to w, evaluated at $w = 0$:

$$\begin{aligned}
\left(\frac{\partial \boldsymbol{u}}{\partial w}\right)_\mathrm{c}^0 &= \boldsymbol{\eta}_1 + \sum_{j=2}^{N} \left(\frac{\partial \varphi_j}{\partial w}(0,0,\boldsymbol{v}^0) + \frac{\partial \varphi_j}{\partial \widetilde{f}}(0,0,\boldsymbol{v}^0) \cdot \frac{\partial \psi}{\partial w}(0,\boldsymbol{v}^0)\right)\boldsymbol{\eta}_j \\
&= \boldsymbol{\eta}_1 + \frac{\partial \psi}{\partial w}(0,\boldsymbol{v}^0) \sum_{j=2}^{N} \frac{\partial \varphi_j}{\partial \widetilde{f}}(0,0,\boldsymbol{v}^0)\boldsymbol{\eta}_j,
\end{aligned} \tag{2.58}$$

where the second equality follows from Lemma 2.2 below.

Lemma 2.2.

$$\frac{\partial \varphi_j}{\partial w}(0,0,\boldsymbol{v}^0) = 0, \qquad j = 2, \ldots, N.$$

Proof. The expression above is immediate from the evaluation of (2.53) at $(w, \widetilde{f}, \boldsymbol{v}) = (0, 0, \boldsymbol{v}^0)$ with the equality $\widetilde{J}_{[2,1]}(\boldsymbol{u}_\mathrm{c}^0, f_\mathrm{c}^0, \boldsymbol{v}^0) = \boldsymbol{0}$ in (2.46). □

The expression (2.58) can be rewritten as

$$\left(\frac{\partial \boldsymbol{u}}{\partial w}\right)_\mathrm{c}^0 = \boldsymbol{\eta}_1 + C\boldsymbol{\eta}_* \tag{2.59}$$

with

$$C = \frac{\partial \psi}{\partial w}(0,\boldsymbol{v}^0), \qquad \boldsymbol{\eta}_* = \sum_{j=2}^{N} \frac{\partial \varphi_j}{\partial \widetilde{f}}(0,0,\boldsymbol{v}^0)\boldsymbol{\eta}_j. \tag{2.60}$$

This shows that the direction of the bifurcation path, in the space of $(\boldsymbol{u}, f) \in \mathbb{R}^N \times \mathbb{R}$, is given by[9]

---

[8] Such is almost always the case with practical examples. Mathematically, however, this represents a restrictive assumption. For example, the bifurcation equation $w^2 - \widetilde{f}^4 = 0$ gives $\widetilde{f} = \pm\sqrt{|w|}$.

[9] Stated precisely, it is necessary to distinguish column and row vectors in the expression (2.61). Nevertheless it seems more comprehensive as it is.

$$\left(\frac{\partial \boldsymbol{u}}{\partial w}, \frac{\partial \psi}{\partial w}\right)_{\mathrm{c}}^{0} = (\boldsymbol{\eta}_1 + C\boldsymbol{\eta}_*, C) = (\boldsymbol{\eta}_1, 0) + C(\boldsymbol{\eta}_*, 1). \tag{2.61}$$

In the space of $\boldsymbol{u}$, the direction of bifurcation is not necessarily the same as the critical eigenvector $\boldsymbol{\eta}_1$, but is given by $\boldsymbol{\eta}_1 + C\boldsymbol{\eta}_*$ involving an extra direction $\boldsymbol{\eta}_*$ if the coefficient C is nonzero.

Although the coefficient C can be determined only from the (nonlinear) analysis of the bifurcation equation, the extra direction $\boldsymbol{\eta}_*$ can be determined from the linear part of the full system of equations, as follows. Differentiation of (2.41) with respect to $\widetilde{f}$, with subsequent evaluation at $(w, \widetilde{f}) = (0,0)$, yields

$$J_{\mathrm{c}}^{0}\boldsymbol{\eta}_* + \left(\frac{\partial \boldsymbol{F}}{\partial f}\right)_{\mathrm{c}}^{0} = \boldsymbol{0}. \tag{2.62}$$

Although the matrix J_{c}^{0} is singular with rank $N-1$, this system of equations admits a unique solution.[10] More specific analysis based on (2.58) and (2.61) is described in §2.5.2 and §2.5.3 for bifurcation points of various types.

2.4.4 Stability

With a suitable choice of bases $\{\boldsymbol{\xi}_j \mid j = 1, \ldots, N\}$ and $\{\boldsymbol{\eta}_j \mid j = 1, \ldots, N\}$, the Liapunov–Schmidt reduction can be done consistently with the stability property in the sense that a solution to the bifurcation equation is (linearly) stable or unstable according to whether the corresponding solution to the original system is (linearly) stable or unstable (see Chapter I, §4, of Golubitsky and Schaeffer, 1985 [62]).

Here we investigate the linear stability in a reciprocal system with reference to Lemma 2.1 in §2.4.2. Assume that $\{\boldsymbol{\xi}_j = \boldsymbol{\eta}_j \mid j = 1, \ldots, N\}$ are orthonormal eigenvectors of J_{c}^{0}, as in Remark 2.6 in §2.4.1. Then

$$\overline{J}(\boldsymbol{u}_{\mathrm{c}}^{0}, f_{\mathrm{c}}^{0}, \boldsymbol{v}_{\mathrm{c}}^{0}) = \mathrm{diag}(\lambda_1^0, \lambda_2^0, \ldots, \lambda_N^0).$$

We further assume that $\lambda_1^0 = 0$ and $\lambda_i^0 > 0$ for $i = 2, \ldots, N$, expressing that the solution on the fundamental path with $f < f_{\mathrm{c}}^{0}$ is stable. In (2.49) we have $\alpha = 1$ and $\det[\overline{J}_{[2,2]}(\boldsymbol{u}, f, \boldsymbol{v})] > 0$, since the eigenvectors are orthonormal, $\det[\overline{J}_{[2,2]}(\boldsymbol{u}_{\mathrm{c}}^{0}, f_{\mathrm{c}}^{0}, \boldsymbol{v}^{0})] = \prod_{i=2}^{N} \lambda_i^0 > 0$ and $(\boldsymbol{u}, f, \boldsymbol{v})$ is assumed to lie in a sufficiently close neighborhood of $(\boldsymbol{u}_{\mathrm{c}}^{0}, f_{\mathrm{c}}^{0}, \boldsymbol{v}^{0})$. Therefore,

$$\mathrm{sign}\left(\frac{\partial \widetilde{F}}{\partial w}\right) = \mathrm{sign}[\det(J)]. \tag{2.63}$$

Denote by $\lambda_1, \lambda_2, \ldots, \lambda_N$ the eigenvalues of $J(\boldsymbol{u}, f, \boldsymbol{v})$ that are respectively close to $\lambda_1^0, \lambda_2^0, \ldots, \lambda_N^0$. Then we have $\lambda_i > 0$ for $i = 2, \ldots, N$ and, therefore,

---

[10] At a bifurcation point we have $\boldsymbol{\xi}_1^{\top}(\partial \boldsymbol{F}/\partial f)_{\mathrm{c}}^{0} = 0$ (cf., Remark 2.8 in §2.5). Since $\boldsymbol{\xi}_1$ is the left critical eigenvector of J_{c}^{0}, $(\partial \boldsymbol{F}/\partial f)_{\mathrm{c}}^{0}$ lies in the range space of J_{c}^{0}.

$$\mathrm{sign}[\det(J)] = \mathrm{sign}(\lambda_1). \tag{2.64}$$

A solution to the bifurcation equation is linearly stable or unstable if $\partial\widetilde{F}/\partial w > 0$ or < 0, whereas the corresponding solution to the original system is linearly stable or unstable if $\lambda_1 > 0$ or < 0. The expressions (2.63) and (2.64) show the consistency of the linear stability/instability in the bifurcation equation and in the original system.

2.4.5 Power Series Expansion of Bifurcation Equation

The direct use of the bifurcation equation (2.38) in the investigation of the bifurcation behavior is difficult in general. It is much simpler and pertinent to investigate its asymptotic behavior by expanding the bifurcation equation into a power series and examining the leading terms.

Referring to $\boldsymbol{v} = \boldsymbol{v}^0 + \varepsilon\boldsymbol{d}$ in (2.2), we consider

$$\widehat{F}(w,\widetilde{f},\varepsilon) = \widetilde{F}(w,\widetilde{f},\boldsymbol{v}^0 + \varepsilon\boldsymbol{d}) \tag{2.65}$$

to regard ε as an independent variable for imperfection thereby regarding $\boldsymbol{d}$ as a constant vector. Then the bifurcation equation (2.38) can be expressed, alternatively, as

$$\widehat{F}(w,\widetilde{f},\varepsilon) = 0. \tag{2.66}$$

The nature of (2.66) can be grasped by expanding $\widehat{F}$ into a power series involving an appropriate number of terms

$$\widehat{F}(w,\widetilde{f},\varepsilon) \approx \sum_{i=0}\sum_{j=0}\sum_{k=0} A_{ijk} w^i \widetilde{f}^j \varepsilon^k, \tag{2.67}$$

where

$$A_{ijk} = \frac{1}{i!\,j!\,k!}\frac{\partial^{i+j+k}\widehat{F}}{\partial w^i\,\partial\widetilde{f}^j\,\partial\varepsilon^k}(0,0,0).$$

Since $(w,\widetilde{f}) = (0,0)$ is a simple critical point for the perfect system, both the condition for equilibrium

$$\widehat{F}(0,0,0) = A_{000} = 0, \tag{2.68}$$

and the condition for criticality

$$\frac{\partial\widehat{F}}{\partial w}(0,0,0) = A_{100} = 0 \tag{2.69}$$

(cf., (2.42)) hold.

The coefficient A_{010} plays a major role in the classification of the critical point in §2.5. It can be shown (cf., Problem 2-8) that

$$A_{010} = \boldsymbol{\xi}_1^\top \left(\frac{\partial \boldsymbol{F}}{\partial f}\right)_\mathrm{c}^0, \tag{2.70}$$

where $(\cdot)_\mathrm{c}^0$ denotes the evaluation at $(\boldsymbol{u}_\mathrm{c}^0, f_\mathrm{c}^0, \boldsymbol{v}^0)$.

The coefficient A_{001} in (2.67), which is termed the *imperfection coefficient*, represents the influence of the imperfection on the bifurcation equation. With the imperfection sensitivity matrix B_c^0 in (2.10), the coefficient A_{001} can be represented as (2.71) below. This expression plays a pivotal role in this book.

Lemma 2.3.

$$A_{001} = \boldsymbol{\xi}_1^\top B_\mathrm{c}^0 \boldsymbol{d}. \tag{2.71}$$

Proof. We use the notations in Lemma 2.1 in §2.4.2. Define an N-dimensional vector $\overline{\boldsymbol{b}} = (\overline{b}_1, \dots, \overline{b}_N)^\top$ as $\overline{b}_i = \boldsymbol{\xi}_i^\top B\boldsymbol{d}$ $(i = 1, \dots, N)$, and partition $\overline{\boldsymbol{b}}$ as

$$\overline{\boldsymbol{b}} = \begin{pmatrix} \overline{b}_{[1]} \\ \overline{\boldsymbol{b}}_{[2]} \end{pmatrix},$$

where $\overline{b}_{[1]} = (\overline{b}_1)$ and $\overline{\boldsymbol{b}}_{[2]} = (\overline{b}_2, \dots, \overline{b}_N)^\top$. Just as in (2.54), we obtain

$$\frac{\partial \widehat{F}}{\partial \varepsilon}(w, \widetilde{f}, \varepsilon) = \overline{b}_{[1]} - \overline{J}_{[1,2]}(\overline{J}_{[2,2]})^{-1} \overline{\boldsymbol{b}}_{[2]}.$$

Evaluation of this at $(w, \widetilde{f}, \varepsilon) = (0, 0, 0)$ yields

$$A_{001} = (\overline{b}_{[1]})_\mathrm{c}^0 = \boldsymbol{\xi}_1^\top B_\mathrm{c}^0 \boldsymbol{d},$$

since $\overline{J}_{[1,2]}(\boldsymbol{u}_\mathrm{c}^0, f_\mathrm{c}^0, \boldsymbol{v}^0) = \boldsymbol{0}^\top$ as shown in (2.46). Note also that

$$\frac{\partial \widehat{F}}{\partial \varepsilon}(w, \widetilde{f}, \varepsilon) \neq \boldsymbol{\xi}_1^\top B(\boldsymbol{u}, f, \boldsymbol{v})\boldsymbol{d}$$

in general. □

2.5 Classification of Simple Critical Points

It is possible to classify equilibrium points satisfying the bifurcation equation (2.66), based on the vanishing or nonvanishing of the coefficients A_{ijk} in the expansion (2.67):

$$\widehat{F}(w, \widetilde{f}, \varepsilon) \approx \sum_{i=0} \sum_{j=0} \sum_{k=0} A_{ijk} w^i \widetilde{f}^j \varepsilon^k. \tag{2.72}$$

First, by the vanishing or nonvanishing of A_{100}, equilibrium points are classified into

$$\begin{cases} \textit{ordinary point} & \text{if } A_{100} \neq 0, \\ \textit{simple critical point} & \text{if } A_{100} = 0. \end{cases}$$

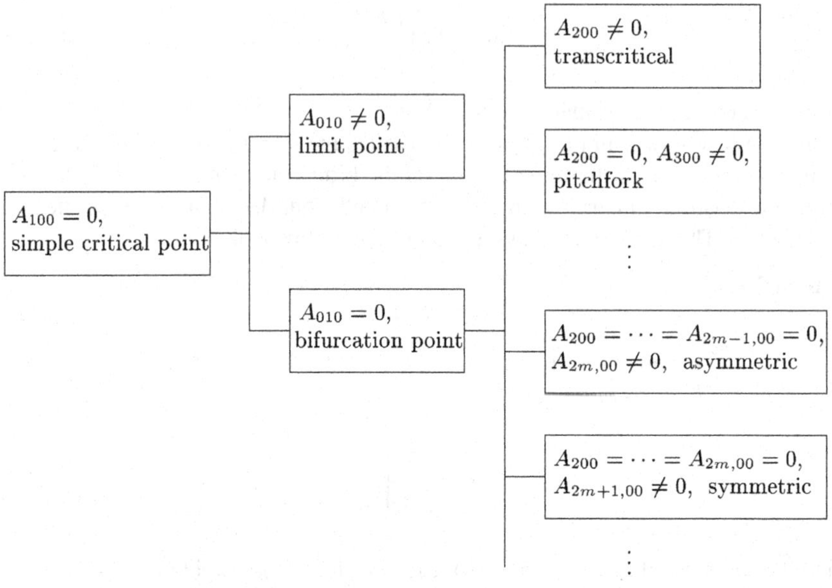

Fig. 2.3 Classification of simple critical points.

Next, the simple critical point ($A_{100} = 0$) is classified further into

$$\begin{cases} \textit{limit point} & \text{if } A_{010} \neq 0 \ (A_{200} \neq 0), \\ \textit{bifurcation point} & \text{if } A_{010} = 0. \end{cases} \tag{2.73}$$

Last, the bifurcation point is classified further into

$$\begin{cases} \textit{transcritical} & \text{if } A_{200} \neq 0, \\ \textit{pitchfork} & \text{if } A_{200} = 0, A_{300} \neq 0, \\ m\text{th order } (m \geq 4) & \text{if } A_{200} = \cdots = A_{(m-1)00} = 0, \ \ A_{m00} \neq 0. \end{cases} \tag{2.74}$$

This classification is summarized in Fig. 2.3.

The literature offers other names for critical points: limit points are also called *turning points*, *saddle nodes*, and *fold bifurcations* (e.g., Seydel, 1994 [184]); transcritical and pitchfork bifurcation points are called *asymmetric* and *symmetric* bifurcation points, respectively, in Thompson and Hunt, 1973 [191]. Our terminology follows that of Marsden and Hughes, 1983 [133] (Table 7.1.1, in particular).

The remainder of this section presents the perfect and imperfect behaviors of $\widetilde{f}$ versus w curves in the neighborhood of simple critical points using the bifurcation equation (2.66). Since $A_{000} = A_{100} = 0$ by (2.68) and (2.69), w cannot be uniquely determined as a function of $\widetilde{f}$ from the bifurcation equation (2.66). This equation, in general, has a number of nonlinear terms and is therefore quite complex. Nonetheless, such complexity can be reduced greatly in asymptotic analysis, where it suffices to consider a small number of leading terms of the bifurcation equation. Emphasis is placed on investigating the stability of the perfect system.

For the investigation of the asymptotic influence of imperfection ε, we assume

$$A_{001} \neq 0, \tag{2.75}$$

that is, that the effect of imperfections is of the first order. This assumption is customary in the study[11] of imperfections, and more issues on this account are discussed in Chapter 8 in connection with group symmetry. Discussion of stability assumes the consistency of the bifurcation equation with respect to the stability explained in §2.4.4.

Remark 2.8. The coefficient A_{010} used in (2.73) for the classification of critical points admits an explicit expression $A_{010} = \boldsymbol{\xi}_1^\top (\partial \boldsymbol{F}/\partial f)_\mathrm{c}^0$ in (2.70). In practice, it is more convenient to resort to the classification

$$\begin{cases} \text{limit point} & \text{if } \boldsymbol{\xi}_1^\top (\partial \boldsymbol{F}/\partial f)_\mathrm{c}^0 \neq 0, \\ \text{bifurcation point} & \text{if } \boldsymbol{\xi}_1^\top (\partial \boldsymbol{F}/\partial f)_\mathrm{c}^0 = 0, \end{cases} \tag{2.76}$$

because it is often possible to compute both $\boldsymbol{\xi}_1$ and $(\partial \boldsymbol{F}/\partial f)_\mathrm{c}^0$. □

Remark 2.9. For an ordinary point ($A_{100} \neq 0$), the bifurcation equation with (2.72) becomes

$$\widehat{F}(w, \widetilde{f}, \varepsilon) = A_{100} w + A_{010} \widetilde{f} + A_{001} \varepsilon + \text{h.o.t.} = 0 \tag{2.77}$$

(see Remark 2.7 in §2.4.1). Therefore, w can be determined uniquely as

$$w = -\frac{A_{010}}{A_{100}} \widetilde{f} - \frac{A_{001}}{A_{100}} \varepsilon + \text{h.o.t.}$$

A set of ordinary points therefore forms an equilibrium path on which the bifurcation parameter $\widetilde{f}$ monotonically increases or decreases and from which no solution path branches, as shown, respectively, by the solid line for a perfect system ($\varepsilon = 0$) and by the dashed lines for imperfect ones ($\varepsilon \neq 0$) in Fig. 2.4. The $\widetilde{f}$ versus w curve shifts in proportion with ε. □

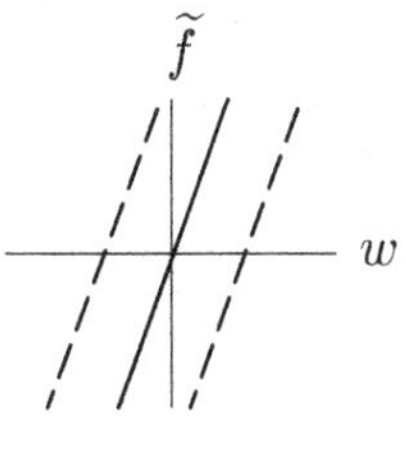

Fig. 2.4 Solution curves in the neighborhood of an ordinary point expressed by the leading terms of the bifurcation equation (2.77). ——: path for the perfect system; – – –: path for an imperfect system.

[11] See, for example, Thompson and Hunt, 1973 [191]; Budiansky, 1974 [20]; and El Naschie, 1990 [49].

2.5.1 Limit Point

At a limit point, where $A_{100} = 0$ and $A_{010} \neq 0$ (cf., (2.73)), the bifurcation equation with (2.72) becomes

$$\widehat{F}(w,\widetilde{f},\varepsilon) = A_{200}w^2 + A_{010}\widetilde{f} + A_{001}\varepsilon + A_{101}w\varepsilon + A_{110}w\widetilde{f} + \text{h.o.t.} = 0. \tag{2.78}$$

Perfect Behavior

The perfect system ($\varepsilon = 0$) satisfying the nondegeneracy condition[12]

$$A_{200} \neq 0 \tag{2.79}$$

is considered. Then the leading terms of (2.78) yield an asymptotic curve

$$A_{200}w^2 + A_{010}\widetilde{f} = 0, \tag{2.80}$$

which has a limit point at $(w,\widetilde{f},\varepsilon) = (0,0,0)$, as depicted by the solid curves in Fig. 2.5.

The limit point of the perfect system ($\varepsilon = 0$) is classified according to the sign of A_{200}/A_{010} as (cf., Fig. 2.5)

$$\begin{cases} \text{limit point (maximum)} & \text{for } A_{200}/A_{010} > 0, \\ \text{limit point (minimum)} & \text{for } A_{200}/A_{010} < 0. \end{cases}$$

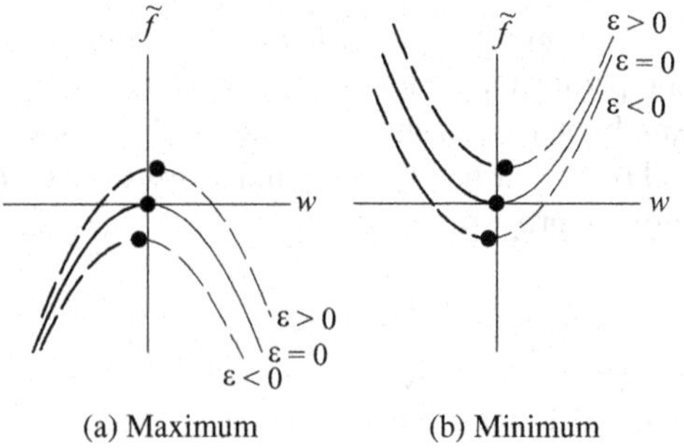

Fig. 2.5 Solution curves in the neighborhood of limit (maximum and minimum) points expressed by the leading terms of the bifurcation equation (2.78) for $A_{001}/A_{010} < 0$ and $A_{200} < 0$. ——: path for the perfect system; — — —: path for an imperfect system; thick line: stable; thin line: unstable; ●: limit point.

The Jacobian on the asymptotic curve (2.80) becomes

$$\frac{\partial \widehat{F}}{\partial w}(w, -w^2(A_{200}/A_{010}), 0) = 2A_{200}w + \text{h.o.t.}$$

[12] The point is degenerate if $A_{200} = 0$; in particular, it is a stationary point if $A_{200} = 0$ and $A_{300} \neq 0$.

At a limit point, the stability of the system under consideration alters from unstable to stable or vice versa according to the sign of A_{200}, as presented in Fig. 2.5 for $A_{200} < 0$.

Imperfect Behavior

For an imperfect system ($\varepsilon \neq 0$), the leading terms of the bifurcation equation (2.78) express the $\widetilde{f}$ versus w curves depicted as the dashed curves in Fig. 2.5. The curves shift due to the presence of ε.

2.5.2 Transcritical Bifurcation Point

At a transcritical bifurcation point, where $A_{100} = A_{010} = 0$ and $A_{200} \neq 0$ (cf., (2.74)), the bifurcation equation with (2.72) becomes

$$\widehat{F}(w, \widetilde{f}, \varepsilon) = A_{200}w^2 + A_{110}w\widetilde{f} + A_{020}\widetilde{f}^2 + A_{001}\varepsilon + \text{h.o.t.} = 0. \tag{2.81}$$

Perfect Behavior

For the perfect system with $\varepsilon = 0$, the bifurcation equation (2.81) has two real-valued solutions

$$w = \frac{-A_{110} \pm (A_{110}^2 - 4A_{020}A_{200})^{1/2}}{2A_{200}}\widetilde{f} + \text{h.o.t.} \tag{2.82}$$

provided the condition of nondegeneracy[13]

$$A_{110}^2 - 4A_{200}A_{020} > 0 \tag{2.83}$$

holds. Consequently, this is a bifurcation point at which two solution curves intersect, as shown in Fig. 2.6.

The directions of the fundamental path and the bifurcated path are given by formula (2.58) with $(\partial\psi/\partial w)(0, v^0)$ being equal to the reciprocal of the coefficients in (2.82); that is,

$$\frac{\partial\psi}{\partial w}(0, v^0) = \frac{2A_{200}}{-A_{110} \pm (A_{110}^2 - 4A_{020}A_{200})^{1/2}}.$$

In particular, the direction of the bifurcated path does not necessarily coincide with that of the critical eigenvector $\boldsymbol{\eta}_1$. As a concrete example, we refer to the propped

[13] When the condition of nondegeneracy (2.83) is not satisfied, the bifurcation equation (2.81) has $(w, \widetilde{f}, \varepsilon) = (0, 0, 0)$ as the only real-valued solution in a neighborhood of the bifurcation point $(0, 0, 0)$. This point is called an *isola center* (e.g., Seydel, 1994 [184]).

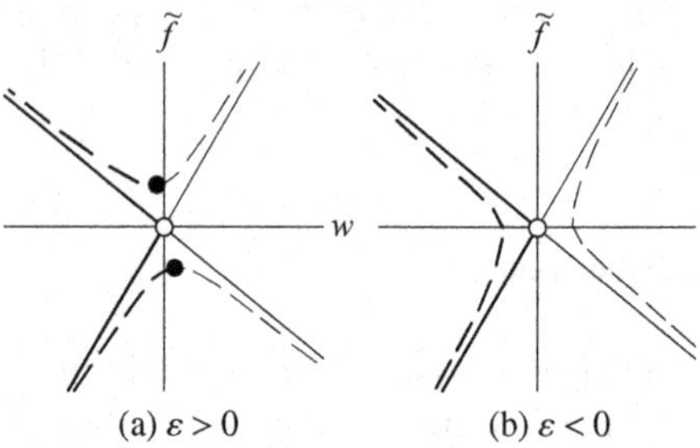

Fig. 2.6 Solution curves in the neighborhood of a transcritical bifurcation point expressed by the leading terms of the bifurcation equation (2.81) for $A_{200} > 0$, $A_{020} < 0$, and $A_{001} > 0$. ——: path for the perfect system; – – –: path for an imperfect system; thick line: stable; thin line: unstable; ○: bifurcation point; ●: limit point.

cantilever in §2.3. The solution path represented by (2.23) branches in the direction of $(x,y) = (1,1/4)$, which is distinct from $\boldsymbol{\eta}_1 = (1,0)^\top$. See also Example 2.1 below.

For stability, we note that

$$\frac{\partial \widehat{F}}{\partial w} = 2A_{200}w + A_{110}\widetilde{f} + \text{h.o.t.} = \pm({A_{110}}^2 - 4A_{200}A_{020})^{1/2}\widetilde{f} + \text{h.o.t.}$$

for the solution (2.82) and $\varepsilon = 0$. Hence, one of the two curves is stable for $\widetilde{f} < 0$ and unstable for $\widetilde{f} > 0$, and vice versa for the other curve, which is a famous phenomenon called *Poincaré's exchange of stability*.

Example 2.1. Consider a system of equations

$$\begin{cases} F_1 = 3x^2 - 2xf = 0, \\ F_2 = -f + 2y = 0. \end{cases}$$

This system has two solution curves

$$\begin{cases} x = 0, \ f = 2y, & \text{trivial solution,} \\ f = 3x/2, \ f = 2y, & \text{bifurcated solution,} \end{cases}$$

which intersect at the transcritical bifurcation point $(x,y,f) = (0,0,0)$. The Jacobian matrix is given by

$$J(x,y,f) = \begin{pmatrix} 6x - 2f & 0 \\ 0 & 2 \end{pmatrix}, \qquad J(0,0,0) = \begin{pmatrix} 0 & 0 \\ 0 & 2 \end{pmatrix}, \tag{2.84}$$

and the critical eigenvector of $J(0,0,0)$ is $\boldsymbol{\eta}_1 = (1,0)^\top$. The bifurcated solution emanates in the directions

$$\pm(1,3/4,3/2) = \pm[(1,0,0) + (3/2)(0,1/2,1)],$$

which is of the form of (2.61) with $\boldsymbol{\eta}_1 = (1,0)^\top$, $\boldsymbol{\eta}_* = (0,1/2)^\top$, and $C = 3/2$. Note that the vector $\boldsymbol{\eta}_* = (0,1/2)^\top$ is determined as the solution of (2.62) with $J_c^0 = J(0,0,0)$ in (2.84) and $(\partial \boldsymbol{F}/\partial f)(0,0,0) = (0,-1)^\top$. □

Imperfect Behavior

As depicted in Fig. 2.6, an imperfect system with $\varepsilon \neq 0$ has two separate $\widetilde{f}$ versus w curves (shown by the dashed lines) in the neighborhood of the transcritical bifurcation point (shown by (∘)). The sign of the imperfection ε controls imperfect behaviors in a pivotal manner. For $\varepsilon > 0$ (and $A_{200}A_{001} > 0$), the imperfect path approaching the bifurcation point from downward ($\widetilde{f} < 0$) becomes unstable at the maximum point of $\widetilde{f}$ shown by (•), for which $\widetilde{f}$ cannot attain $\widetilde{f} = 0$ (i.e., $f = f_c^0$), whereas such a path for $\varepsilon < 0$ is stable, for which $\widetilde{f}$ increases stably beyond $\widetilde{f} = 0$.

2.5.3 Pitchfork Bifurcation Point

At a pitchfork bifurcation point, where $A_{100} = A_{010} = A_{200} = 0$ and $A_{300} \neq 0$ (cf., (2.74)), the bifurcation equation with (2.72) becomes

$$\widehat{F}(w, \widetilde{f}, \varepsilon) = A_{300}w^3 + A_{110}w\widetilde{f} + A_{020}\widetilde{f}^2 + A_{001}\varepsilon + \text{h.o.t.} = 0. \tag{2.85}$$

Perfect Behavior

For the perfect system ($\varepsilon = 0$), this equation (2.85) has the solution (see Remark 2.10 below) of

$$\begin{cases} \widetilde{f} = -\dfrac{A_{110}}{A_{020}}w + \text{h.o.t.}, & \text{fundamental path}, \\ \widetilde{f} = -\dfrac{A_{300}}{A_{110}}w^2 + \text{h.o.t.}, & \text{bifurcated path}. \end{cases} \tag{2.86}$$

The set of $\widetilde{f}$ versus w curves expressed by (2.86) is shown in Fig. 2.7 by the solid lines. This kind of bifurcation point is called a pitchfork, as this set of curves looks like a pitchfork.

The solution path branches in the direction of the critical eigenvector $\boldsymbol{\eta}_1$ in the space of $\boldsymbol{u}$. This follows from (2.61), in which $C = (\partial\psi/\partial w)(0, \boldsymbol{v}^0) = 0$ since $\psi(w, \boldsymbol{v}^0) = -(A_{300}/A_{110})w^2 + \text{h.o.t.}$ by (2.86).

The stability of the solutions may be analyzed as follows. Inasmuch as

$$\frac{\partial \widehat{F}}{\partial w}(w, \widetilde{f}, 0) = 3A_{300}w^2 + A_{110}\widetilde{f} + \text{h.o.t.},$$

the fundamental path for $\widetilde{f} < 0$ is stable if

$$A_{110} < 0, \tag{2.87}$$

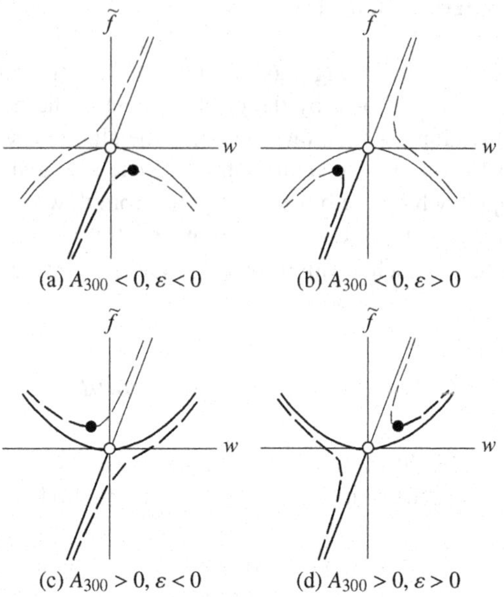

Fig. 2.7 Solution curves in the neighborhood of a pitchfork bifurcation point expressed by the leading terms of the bifurcation equation (2.85) for $A_{110} < 0$, $A_{020} > 0$, and $A_{001} > 0$. ——: path for the perfect system; — — —: path for an imperfect system; thick line: stable; thin line: unstable; ○: bifurcation point; ●: limit point.

which is assumed in the following. The fundamental path becomes unstable for $\widetilde{f} > 0$. On the bifurcated path, we have

$$\frac{\partial \widehat{F}}{\partial w} = 2A_{300}w^2 + \text{h.o.t.},$$

which is positive or negative according to whether $A_{300} > 0$ or $A_{300} < 0$. Therefore, the bifurcated path is

$$\begin{cases} \text{stable} & \text{if } A_{300} > 0, \\ \text{unstable} & \text{if } A_{300} < 0. \end{cases} \tag{2.88}$$

The stability of the bifurcation point is identical to the linear stability of the bifurcated path as long as $A_{300} \neq 0$.

For the stable bifurcation point, when f increases, the system remains stable during the shift from the fundamental path ($\widetilde{f} < 0$) to the bifurcated path ($\widetilde{f} > 0$), which is also stable. For the unstable bifurcation point, when $\widetilde{f}$ reaches the critical value $\widetilde{f} = 0$, the system becomes unstable because there is no stable path for $\widetilde{f} > 0$.

The coefficient A_{020} vanishes systematically in many physical problems due to (geometric) symmetry of the system under consideration. To be specific, suppose that $\widehat{F}(w, \widetilde{f}, 0)$ for the perfect system is an odd function in w; that is,

$$\widehat{F}(-w, \widetilde{f}, 0) = -\widehat{F}(w, \widetilde{f}, 0) \tag{2.89}$$

is satisfied. Then the bifurcation equation (2.85) with $\varepsilon = 0$ is simplified to

$$\widehat{F}(w, \widetilde{f}, 0) = w(A_{300}w^2 + A_{110}\widetilde{f} + \text{h.o.t.}) = 0, \tag{2.90}$$

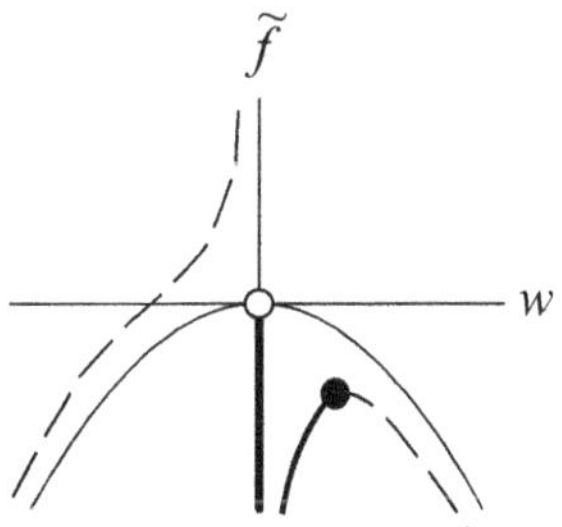

Fig. 2.8 Solution curves of a symmetric system satisfying the condition (2.89) in the neighborhood of a pitchfork bifurcation point expressed by the leading terms of the bifurcation equation (2.93) for $A_{110} < 0$, $A_{020} = 0$, and $A_{300} < 0$. ——: path for the perfect system; — — —: path for an imperfect system; thick line: stable; thin line: unstable; ○: bifurcation point; ●: limit point.

which has the solutions of

$$\begin{cases} w = 0, & \text{fundamental path,} \\ \tilde{f} = -\dfrac{A_{300}}{A_{110}} w^2 + \text{h.o.t.}, & \text{bifurcated path} \end{cases}$$

(cf., Fig. 2.8). The fundamental path for this case is often called the trivial solution, and the bifurcated path for this case has bilateral symmetry with respect to $w \mapsto -w$.

Example 2.2. Consider a system of equations

$$\begin{cases} F_1 = fx - x^3 = 0, \\ F_2 = -f + y = 0. \end{cases}$$

This system has two solution curves

$$\begin{cases} x = 0, \ f = y, & \text{trivial solution,} \\ f = x^2, \ f = y, & \text{bifurcated solution} \end{cases}$$

that intersect at the pitchfork bifurcation point $(x, y, f) = (0, 0, 0)$. The Jacobian matrix is given by

$$J(x, y, f) = \begin{pmatrix} f - 3x^2 & 0 \\ 0 & 1 \end{pmatrix}, \qquad J(0, 0, 0) = \begin{pmatrix} 0 & 0 \\ 0 & 1 \end{pmatrix}.$$

The critical eigenvector associated with the zero eigenvalue of this matrix is $\eta_1 = (1, 0)^\top$. The bifurcated solution emanates in the directions $\pm(1, 0, 0)$, which is of the form of (2.61) with $C = 0$. Thus, the bifurcated solution emanates in the direction of the critical eigenvector. □

Remark 2.10. The method of the Newton polygon is illustrated here for the solution of

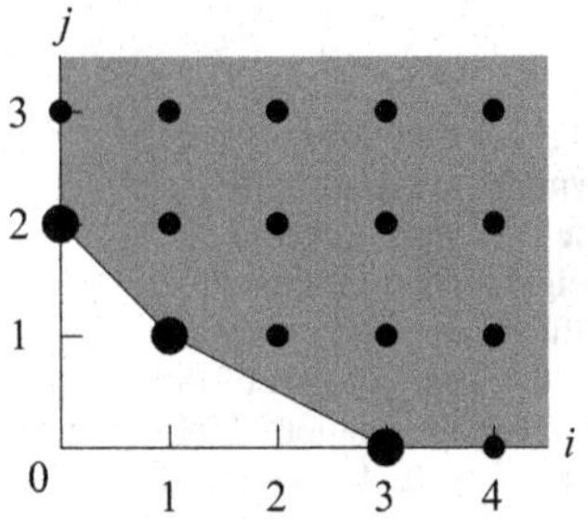

Fig. 2.9 Newton polygon for (2.91) expressed by the shaded area.

$$\sum_{i=0}\sum_{j=0} A_{ij} w^i \tilde{f}^j = 0 \qquad (A_{00} = A_{10} = A_{01} = A_{20} = 0), \tag{2.91}$$

where $A_{ij} \neq 0$ if $(i, j) \neq (0,0)$, $(1,0)$, $(0,1)$, $(2,0)$. This corresponds to (2.85) with $\varepsilon = 0$. We consider the set N of pairs (i, j) such that $w^i \tilde{f}^j$ appears in the equation; that is,

$$N = \{(i, j) \mid A_{ij} \neq 0\}.$$

The convex hull of the point set N is called the *Newton polygon* (see Fig. 2.9).

The possible dominant terms are determined from the Newton polygon. In this case, the extreme points (vertices) of the Newton polygon are given by $(3,0)$, $(1,1)$, and $(0,2)$, marked by larger symbols in this figure, which shows that the terms of w^3, $w\tilde{f}$, and $\tilde{f}^2$ are to be retained as the possible dominant terms. Accordingly, (2.91) can be replaced by

$$A_{30} w^3 + A_{11} w \tilde{f} + A_{02} \tilde{f}^2 = 0, \tag{2.92}$$

which is identical with (2.85) with $\varepsilon = 0$.

Let us assume

$$\tilde{f} \approx \alpha w^p \qquad (p > 0,\ \alpha \neq 0)$$

as the asymptotic form of the solution. Substitution of this into (2.91) or (2.92) shows that $-1/p$ must be equal to the slope of a side of the Newton polygon. In this case, $-1/p = -1$ or $-1/2$ (i.e., $p = 1$ or 2). For each value of p, the coefficient α can be determined easily. For $p = 1$, the substitution of $\tilde{f} \approx \alpha w$ into (2.92) yields

$$A_{30} w^3 + (A_{11}\alpha + A_{02}\alpha^2) w^2 = (A_{11}\alpha + A_{02}\alpha^2) w^2 + \text{h.o.t.} = 0,$$

from which $\alpha = -A_{11}/A_{02}$ results. For $p = 2$, the substitution of $\tilde{f} \approx \alpha w^2$ into (2.92) yields

$$(A_{30} + A_{11}\alpha) w^3 + A_{02}\alpha^2 w^4 = (A_{30} + A_{11}\alpha) w^3 + \text{h.o.t.} = 0,$$

from which $\alpha = -A_{30}/A_{11}$ results. To sum up, we obtain

$$\tilde{f} \approx -\frac{A_{11}}{A_{02}} w \qquad \text{or} \qquad \tilde{f} \approx -\frac{A_{30}}{A_{11}} w^2,$$

which are given in (2.86). □

Imperfect Behavior

As presented in Fig. 2.7, the imperfect system (2.85) with $\varepsilon \neq 0$ has two separate $\widetilde{f}$ versus w curves (shown by the dashed lines) in the neighborhood of the bifurcation point (shown by (∘)). A maximum point of $\widetilde{f}$ exists for $A_{300} < 0$ and a minimum point for $A_{300} > 0$ (shown by (•)).

For a symmetric system satisfying the condition (2.89) of the odd function, we have $A_{200} = 0$; accordingly, the bifurcation equation (2.85) becomes

$$\widehat{F}(w, \widetilde{f}, \varepsilon) = A_{300}w^3 + A_{110}w\widetilde{f} + A_{001}\varepsilon + \text{h.o.t.} = 0. \tag{2.93}$$

The solution curves expressed by (2.93) are shown in Fig. 2.8.

Remark 2.11. It would be in order here to mention the theory of universal unfolding described in Golubitsky and Schaeffer, 1985 [62]. This theory identifies those imperfections that produce all the qualitatively different bifurcation phenomena in a precise mathematical sense. Recall that a single equation was obtained as (2.38), which describes the local behavior around the critical point $(\boldsymbol{u}_c^0, f_c^0, \boldsymbol{v}^0)$. For concreteness, this point is assumed here as an unstable pitchfork bifurcation point described by (2.90). According to the theory, all the qualitatively different bifurcation diagrams with imperfections added to (2.90) are described by a two-parameter family of bifurcation equations, called the *universal unfolding* of (2.90). For example, the family of

$$G(w, \widetilde{f}, \beta_1, \beta_2) = w^3 + \widetilde{f}w + \beta_1 + \beta_2 w^2,$$

parametrized by (β_1, β_2), is qualified as such. Note that $(\beta_1, \beta_2) = (0, 0)$ corresponds to the perfect system. As illustrated in Fig. 2.10, the fundamental path of $G(w, \widetilde{f}, \beta_1, \beta_2) = 0$ has a kink (i.e., a pair of maximal and minimal limit points) if (β_1, β_2) belongs to the region

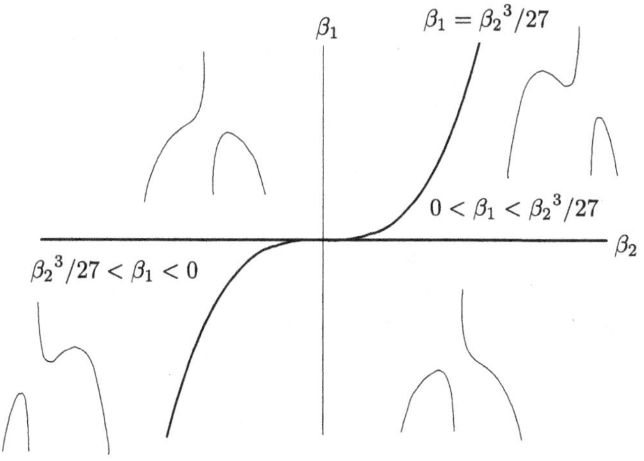

Fig. 2.10 Universal unfolding for an unstable pitchfork bifurcation point.

$$K = \{(\beta_1,\beta_2) \mid 0 < \beta_1 < {\beta_2}^3/27 \text{ or } {\beta_2}^3/27 < \beta_1 < 0\}. \tag{2.94}$$

On the other hand, for (β_1,β_2) lying outside the region K, the imperfect behavior qualitatively resembles that depicted in Fig. 2.7. □

2.6 Example of Pitchfork Bifurcation

The transcritical bifurcation point of the propped cantilever was investigated in §2.3 for the spring force

$$F_{sx} = \frac{x}{L^0} + \left(\frac{x}{L^0}\right)^2 = x + x^2$$

($L^0 = 1$), which lacks bilateral symmetry (cf., Remark 2.12 below). In this section, by investigating the propped cantilever with

$$F_{sx}(x) = \frac{x}{L^0} = x,$$

which has bilateral symmetry characterized by $|F_{sx}(-x)| = |F_{sx}(x)|$, we show that the bilateral symmetry[14] produces another type of bifurcation point, the pitchfork point of bifurcation.

To provide the cantilever with the bilateral symmetry, we set

$$\begin{aligned} \boldsymbol{v} &= \left(\frac{x_1}{L^0}, \frac{y_1}{L^0}, \frac{x_2}{L^0}, \frac{y_2}{L^0}, \beta_1, \beta_2, \beta_3, \beta_4\right)^\top \\ &= (0,1,0,0,0,1,0,2)^\top + \varepsilon(-1,0,1,0,-1,0,0,0)^\top \end{aligned} \tag{2.95}$$

and

$$\boldsymbol{v}^0 = (0,1,0,0,0,1,0,2)^\top, \qquad \boldsymbol{d} = (-1,0,1,0,-1,0,0,0)^\top. \tag{2.96}$$

Note that $\boldsymbol{v}^0$ differs from that in (2.18) used in illustrating transcritical bifurcation.

For the perfect system with $\varepsilon = 0$ (i.e., $\boldsymbol{v} = \boldsymbol{v}^0$), the spring characteristic in the x-direction becomes $F_{sx} = x$; the spring, therefore, exerts the same magnitude of force $|F_{sx}(-x)| = |F_{sx}(x)|$ for x and $-x$. The cantilever accordingly has bilateral symmetry.

Remark 2.12. The transcritical bifurcation point is characterized by the presence of the quadratic term w^2 in the bifurcation equation (2.81), that is, the term $\widetilde{x}^2$ in (2.29) for the propped cantilever. This term $\widetilde{x}^2$ arises from the absence of the bilateral symmetry (symmetry with respect to $x \mapsto -x$) of the spring force defined as (2.21): $F_{sx}(x) = x + x^2$, for which $|F_{sx}(-x)| \neq |F_{sx}(x)|$ holds and a different magnitude of force is exerted for x and $-x$. As a consequence of this asymmetry, the loci of equilibrium in Fig. 2.2 of the cantilever are not symmetric with respect to $x \mapsto -x$, as described earlier. □

[14] The relation between bifurcation and symmetry is treated systematically in Part II.

Exact Analysis

First, we refer to perfect behavior. For $\varepsilon = 0$, the equilibrium equation (2.14) reduces to

$$\begin{pmatrix} F_x \\ F_y \end{pmatrix} = EA \begin{pmatrix} \left(1 - \dfrac{1}{\sqrt{x^2+(y-1)^2}}\right)x + x \\ \left(1 - \dfrac{1}{\sqrt{x^2+(y-1)^2}}\right)(y-1) + 2y - f \end{pmatrix} = \begin{pmatrix} 0 \\ 0 \end{pmatrix}, \tag{2.97}$$

which admits an explicit solution

$$\begin{cases} x = 0, \quad f = 3y, \quad y < 1, & \text{fundamental path,} \\ y = 1 - (1/4 - x^2)^{1/2}, \quad f = 2 - (1/4 - x^2)^{1/2}, & \text{bifurcated path.} \end{cases} \tag{2.98}$$

The solution is shown by the solid lines in Fig. 2.11(a) (f versus x curves at the left and f versus y curves at the right). The solution path branches from the fundamental path at the bifurcation point, which is located at $(x_c^0, y_c^0, f_c^0) = (0, 1/2, 3/2)$. According to §2.5, such a critical point is called a pitchfork bifurcation point. It is noteworthy that the bifurcated path in (2.98) is symmetric with respect to $x \mapsto -x$.

The location of this bifurcation point on the fundamental path can be determined as a point where the Jacobian matrix on the fundamental path

$$J(0, y, f, \boldsymbol{v}^0) = \begin{pmatrix} F_{x,x} & F_{x,y} \\ F_{y,x} & F_{y,y} \end{pmatrix}\Bigg|_{(x,\boldsymbol{v})=(0,\boldsymbol{v}^0)} = EA \begin{pmatrix} 2 - \dfrac{1}{|y-1|} & 0 \\ 0 & 3 \end{pmatrix}$$

becomes singular. The critical eigenvector of $J_c^0 = J(0, 1/2, 3/2, \boldsymbol{v}^0)$ at the point is equal to $\boldsymbol{\eta}_1 = (1, 0)^\top$.

Next, we refer to imperfect bifurcation behavior. Substitution of the variables (2.95) into the equilibrium equation (2.14) yields a solution

$$y = 1 - |x + \varepsilon| \left[\left(\frac{2x}{L} - \varepsilon\right)^{-2} - 1\right]^{1/2}, \tag{2.99}$$

$$f = \frac{2}{L} + \left[\left(-\varepsilon + \frac{x - \varepsilon}{L}\right)\text{sign}(x + \varepsilon) - 2\frac{|x + \varepsilon|}{L}\right]\left[\left(\frac{2x}{L} - \varepsilon\right)^{-2} - 1\right]^{1/2}, \tag{2.100}$$

where $L = (1 + 4\varepsilon^2)^{1/2}$. Figure 2.11(a) shows a set of curves computed for $\varepsilon = 0$, ± 0.01, ± 0.05, and ± 0.1.

Asymptotic Analysis

To investigate the asymptotic properties of the solution in the neighborhood of the bifurcation point $(x_c^0, y_c^0, f_c^0) = (0, 1/2, 3/2)$, we define the increment $(\widetilde{x}, \widetilde{y}, \widetilde{f})$ by

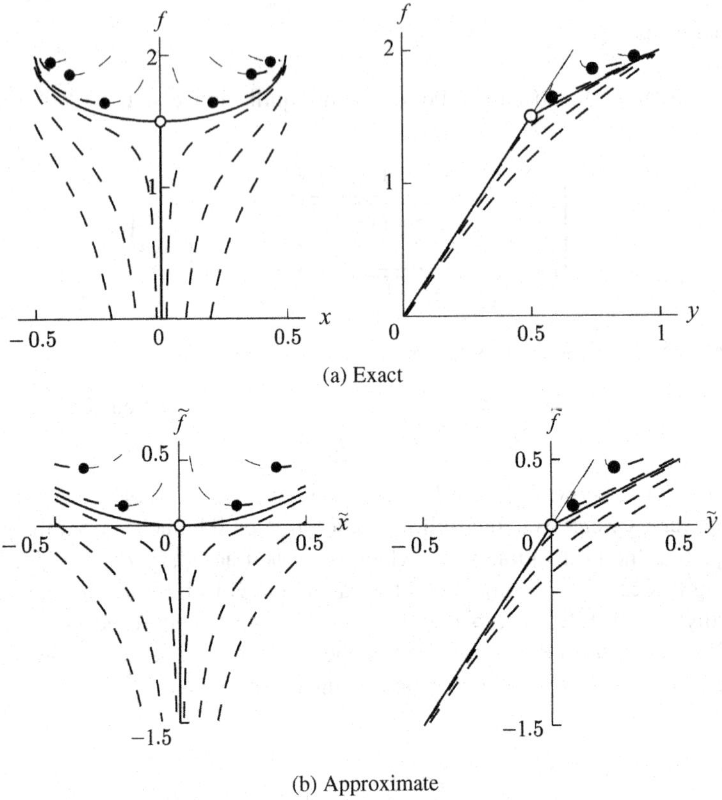

Fig. 2.11 Exact and approximate f versus x curves and f versus y curves for the propped cantilever at the stable pitchfork bifurcation point. ——: path for the perfect system; — — —: path for an imperfect system; thick line: stable; thin line: unstable; ○: bifurcation point; ●: limit point.

$$(\tilde{x},\tilde{y},\tilde{f}) = (x,y,f) - (0,1/2,3/2).$$

Expansion of the equilibrium equation (2.97) in this neighborhood produces a set of incremental equations

$$F_x(x,y,f,\varepsilon) - F_x(0,1/2,3/2,0) = EA(-4\tilde{x}\,\tilde{y} + 4\tilde{x}^3 - 3\varepsilon) + \text{h.o.t.} = 0, \quad (2.101)$$

$$F_y(x,y,f,\varepsilon) - F_y(0,1/2,3/2,0) = EA(3\tilde{y} - \tilde{f} - 2\tilde{x}^2) + \text{h.o.t.} = 0. \quad (2.102)$$

First, we obtain asymptotic f versus x curves as functions in ε. Equation (2.102) can be solved for $\tilde{y}$ as

$$\tilde{y} = \frac{1}{3}\tilde{f} + \frac{2}{3}\tilde{x}^2 + \text{h.o.t.} \quad (2.103)$$

Substitution of this equation into (2.101) gives

$$4\tilde{x}\tilde{f} - 4\tilde{x}^3 + 9\varepsilon + \text{h.o.t.} = 0. \quad (2.104)$$

The asymptotic solution of (2.104) (with the omission of higher-order terms) is shown at the left of Fig. 2.11(b), in comparison with the exact behavior in (a). The former simulates the latter quite well. Moreover, important qualitative features of the exact behavior, such as the symmetry with respect to $(x,\varepsilon) \mapsto (-x,-\varepsilon)$ is successfully inherited by the bifurcation equation.

Next, asymptotic f versus y curves are obtained as functions in ε. Equation (2.102) can be solved for $\widetilde{x}$ as

$$\widetilde{x} = \pm\sqrt{(3\widetilde{y} - \widetilde{f})/2} + \text{h.o.t.} \tag{2.105}$$

Substitution of this equation into (2.101) gives

$$+\sqrt{2(3\widetilde{y} - \widetilde{f})}(\widetilde{f} - \widetilde{y}) + 3\varepsilon + \text{h.o.t.} = 0. \tag{2.106}$$

The asymptotic solution of (2.106) (with the omission of higher-order terms) plotted at the right of Fig. 2.11(b) simulates the exact behavior in (a) well.

Remark 2.13. The f versus y curves present entirely different qualitative characteristics than the f versus x curves do, as presented in Fig. 2.11. The asymptotic behaviors of the latter curves have been described successfully by the bifurcation equation, and the description of the former curves is a topic treated in Chapter 6. □

Problems

2-1 Consider a set of equations

$$\begin{cases} x^3 + xy - 2fx + \varepsilon = 0, \\ y - f = 0. \end{cases}$$

(1) Show that $(x_c^0, y_c^0, f_c^0) = (0,0,0)$ is a singular point of the perfect system ($\varepsilon = 0$).
(2) Derive the bifurcation equation at this singular point.
Answer: $x^3 - fx + \varepsilon = 0$.

2-2 Obtain the governing equation for the system with potential function

$$U(x,y,f) = x^3 - x^2 f + yf - y^2.$$

Answer: $F_x = 3x^2 - 2xf = 0,\ F_y = f - 2y = 0$.

2-3 Show the reciprocity of the system of equations

$$\begin{cases} F_x = 4xy^2 + 2x = 0, \\ F_y = 4x^2 y + 2y = 0, \end{cases}$$

and obtain its potential function.
Answer: $U(x,y) = 2x^2y^2 + x^2 + y^2 + c$.

2-4 (1) Show that $(x,y,z) = (0,0,0)$ is an equilibrium point for the system with potential

$$U(x,y,z) = x^4 + x^2 + y^2 + z^2 + 2xy + 2yz.$$

(2) Investigate the stability at the point $(x,y,z) = (0,0,0)$.
Answer: Unstable.

2-5 Investigate the stability at the point $(x,y) = (0,0)$ for the system with potential

$$U(x,y,f) = f(x^2 + y^2 + x^4) + x^2y^2 + 3xy.$$

Answer: Stable for $f \geq 3/2$ and unstable for $f < 3/2$.

2-6 (1) Plot the curve expressed by

$$(\sin x - \sin\varepsilon)\cos x - f\sin x = 0$$

for $\varepsilon = 0.0$ and 0.03.
(2) Expand this equation into a power series in the neighborhood of the bifurcation point $(x_c^0, f_c^0) = (0,1)$ for the perfect system ($\varepsilon = 0$) to obtain its asymptotic form. Then plot the curves expressed by this asymptotic form for $\varepsilon = 0.0$ and 0.03 to compare the curves plotted in (1).
Answer: $x\widetilde{f} + 2x^3/3 + \varepsilon = 0$ with $\widetilde{f} = f - 1$.

2-7 Obtain the bifurcation equation at the bifurcation point A in Fig. 3.11(a) of the nonshallow truss arch in Fig. 3.8(b) in §3.3.2.

2-8 Derive expression (2.70) of A_{010}. Also verify, in general, that

$$\frac{\partial\widehat{F}}{\partial f}(w,\widetilde{f},\varepsilon) \neq \boldsymbol{\xi}_1^\top \frac{\partial \boldsymbol{F}}{\partial f}(\boldsymbol{u},f,\boldsymbol{v}).$$

Answer: Similar to the proof of Lemma 2.3 in §2.4.5 with $\overline{\boldsymbol{b}}$ being redefined as $\overline{b}_i = \boldsymbol{\xi}_i^\top \partial\boldsymbol{F}/\partial f$.

Summary

- The bifurcation equation has been introduced.
- Classification of simple critical points has been presented.
- Perfect and imperfect behaviors at simple critical points have been investigated.

Chapter 3
Imperfection Sensitivity Laws

3.1 Introduction

Even small imperfections can sharply reduce the strength of structures, such as shells, undergoing bifurcation. Classical linearized theory was inadequate to explain such sharp reductions for shells. This inadequacy was resolved[1] through the implementation of nonlinearity and imperfections into the governing equation. That is, the strength of a shell for given imperfections can be obtained by solving the nonlinear governing equation, which is expressed in our notation as

$$\boldsymbol{F}(\boldsymbol{u}, f, \boldsymbol{v}^0 + \epsilon \boldsymbol{d}) = \boldsymbol{0} \tag{3.1}$$

for given imperfections $\epsilon \boldsymbol{d}$. For the design of shells, we would like to know the relation of imperfection sensitivity

$$f_{\mathrm{c}} = f_{\mathrm{c}}(\varepsilon), \tag{3.2}$$

which expresses the critical load f_{c} as a function of ε. It is, however, awkward to obtain the imperfection sensitivity (3.2) in an explicit form.[2]

The mechanism of such a sharp reduction of the strength was elucidated by Koiter, 1945 [119] through the development of "imperfection sensitivity laws." Thereafter research of imperfection sensitivity was conducted extensively.[3]

---

[1] See von Kármán, Dunn, and Tsien, 1940 [204].

[2] A direct method to compute the relation (3.2) is to solve simultaneously the extended system consisting of the nonlinear governing equation (3.1) and the criticality condition $\det(\partial \boldsymbol{F}/\partial \boldsymbol{u}) = 0$. For the analysis of the extended system, see, for example, Seydel, 1979, [182, 183]; Werner and Spence, 1984 [208]; and Wriggers and Simo, 1990 [212].

[3] The imperfection sensitivity of simple structures was observed experimentally by Roorda, 1965 [172]. Thompson and Hunt, 1973 [191] formulated the imperfection sensitivity law of a system with a single imperfection parameter through the perturbation to the total potential energy function of the system. Hunt, 1977 [78] combined this approach with catastrophe theory to determine imperfection sensitivity. See also textbooks by Godoy, 2000 [61] and Ohsaki and Ikeda, 2007 [155].

K. Ikeda and K. Murota, *Imperfect Bifurcation in Structures and Materials*,
Applied Mathematical Sciences 149, DOI 10.1007/978-1-4419-7296-5_3,

This chapter introduces, in our setting, imperfection sensitivity laws for simple critical points of various kinds. We follow Koiter's approach, which is often called the Liapunov–Schmidt–Koiter approach. In a neighborhood of a simple critical point $(\boldsymbol{u}_{\rm c}^0, f_{\rm c}^0)$, we reduce the original governing equation (3.1) to the bifurcation equation (2.66):

$$\widehat{F}(w, \widetilde{f}, \varepsilon) = 0 \tag{3.3}$$

in an independent variable w and the incremental load $\widetilde{f} = f_{\rm c} - f_{\rm c}^0$. Using asymptotic approximation, we refer to a few leading terms of (3.3) and arrive at an imperfection sensitivity law that varies with the type of critical point. For example, for a pitchfork bifurcation point, we have the two-thirds power law of imperfection sensitivity, that is,

$$\widetilde{f}_{\rm c} \approx C(\boldsymbol{d})\varepsilon^{2/3},$$

which relates the asymptotic reduction $\widetilde{f}_{\rm c}$ of the critical load $f_{\rm c}$ to the imperfection magnitude ε. At the expense of the asymptotic approximation, imperfection sensitivity laws obtained in this manner are amenable to analytical derivations. Moreover, they lay a foundation of the subsequent developments in Chapters 4 to 6, and are extended to a system with group symmetry in Chapter 8.

This chapter is organized as follows.

- Imperfection sensitivity laws for simple critical points are derived from the bifurcation equation in §3.2.
- The imperfection sensitivity of examples is described in §3.3.

3.2 Imperfection Sensitivity Laws

In this section we derive the imperfection sensitivity laws from the bifurcation equation (2.66). It might be recalled from §2.4 (Lemma 2.1, in particular) that the location $(w_{\rm c}, \widetilde{f}_{\rm c})$ of a critical point for an imperfect system is determined as the simultaneous solution of the condition (2.66) for equilibrium and the condition (2.42) for criticality, that is, from

$$\widehat{F}(w_{\rm c}, \widetilde{f}_{\rm c}, \varepsilon) = 0, \qquad \frac{\partial \widehat{F}}{\partial w}(w_{\rm c}, \widetilde{f}_{\rm c}, \varepsilon) = 0. \tag{3.4}$$

Figure 3.1 depicts the local behaviors in the neighborhood of simple critical points, and $(w_{\rm c}, \widetilde{f}_{\rm c})$ corresponds to the limit point (•) on an imperfect path (dashed line). By considering a few leading terms in the expanded form (2.67),

$$\widehat{F}(w, \widetilde{f}, \varepsilon) \approx \sum_{i=0} \sum_{j=0} \sum_{k=0} A_{ijk} w^i \widetilde{f}^j \varepsilon^k, \tag{3.5}$$

where $A_{000} = A_{100} = 0$ by (2.68) and (2.69), we deduce the imperfection sensitivity laws as

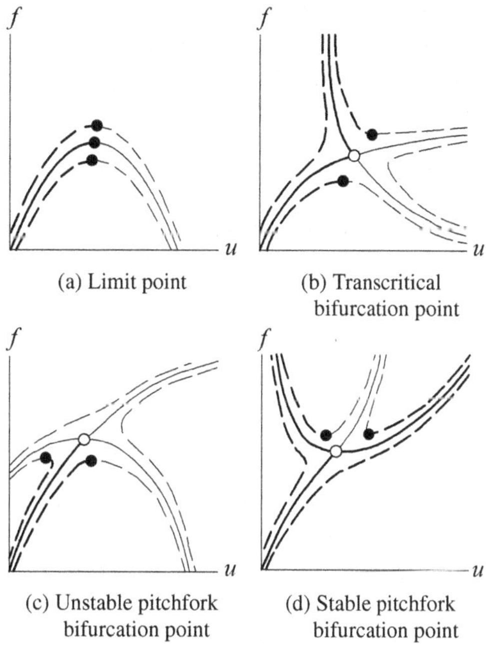

Fig. 3.1 Typical behaviors in the neighborhood of simple critical points (the origin $(u, f) = (0, 0)$ is assumed to be stable). ——: path for the perfect system; – – –: path for an imperfect system; thick line: stable; thin line: unstable; ○: bifurcation point; •: limit point.

$$\widetilde{f_c} \approx \begin{cases} C(\boldsymbol{d})\varepsilon & \text{at the limit point,} \\ C(\boldsymbol{d})|\varepsilon|^{1/2} & \text{at the transcritical bifurcation point} \\ & \text{(existent for } A_{200}A_{001}\varepsilon > 0\text{),} \\ C(\boldsymbol{d})\varepsilon^{2/3} & \text{at the pitchfork bifurcation point,} \end{cases} \tag{3.6}$$

with

$$C(\boldsymbol{d}) = \begin{cases} -\dfrac{A_{001}}{A_{010}} & \text{at the limit point,} \\ -\left(\dfrac{4|A_{200}A_{001}|}{{A_{110}}^2 - 4A_{200}A_{020}}\right)^{1/2} & \text{at the transcritical point} \\ & \text{(existent for } A_{200}A_{001}\varepsilon > 0\text{),} \\ -\dfrac{3{A_{300}}^{1/3}}{A_{110}}\left(\dfrac{A_{001}}{2}\right)^{2/3} & \text{at the pitchfork bifurcation point.} \end{cases} \tag{3.7}$$

A systematic algebraic procedure for a unified derivation is preceded below by heuristic methods for specific cases.

3.2.1 Limit Point

For a nondegenerate limit point ($A_{010} \neq 0, A_{200} \neq 0$), the bifurcation equation with (2.67) is expressed by (2.78):

$$\widehat{F}(w,\widetilde{f},\varepsilon) = A_{200}w^2 + A_{010}\widetilde{f} + A_{001}\varepsilon + A_{101}w\varepsilon + A_{110}w\widetilde{f} + \text{h.o.t.} = 0. \tag{3.8}$$

The criticality condition is

$$\frac{\partial \widehat{F}}{\partial w}(w,\widetilde{f},\varepsilon) = 2A_{200}w + A_{101}\varepsilon + A_{110}\widetilde{f} + \text{h.o.t.} = 0. \tag{3.9}$$

The simultaneous solution of (3.8) and (3.9) gives the location $(w_c, \widetilde{f}_c)$ of the limit point of the imperfect system as

$$\widetilde{f}_c = -\frac{A_{001}}{A_{010}}\varepsilon + \text{h.o.t.}, \tag{3.10}$$

and

$$w_c = -\frac{1}{2A_{200}A_{010}}(A_{101}A_{010} - A_{001}A_{110})\varepsilon + \text{h.o.t.} \tag{3.11}$$

These equations indicate that the critical load $\widetilde{f}_c$ increases or decreases in the order of ε, and so does w_c. It should be clear that the higher-order terms suppressed in (3.8) do not affect expressions (3.10) and (3.11).

3.2.2 Transcritical Bifurcation Point

For a transcritical bifurcation point, at which $A_{100} = A_{010} = 0$ and $A_{200} \neq 0$, the bifurcation equation with (2.67) is expressed by (2.81):

$$\widehat{F}(w,\widetilde{f},\varepsilon) = A_{200}w^2 + A_{110}w\widetilde{f} + A_{020}\widetilde{f}^2 + A_{001}\varepsilon + \text{h.o.t.} = 0. \tag{3.12}$$

The criticality condition is evaluated to

$$\frac{\partial \widehat{F}}{\partial w}(w,\widetilde{f},\varepsilon) = 2A_{200}w + A_{110}\widetilde{f} + \text{h.o.t.} = 0. \tag{3.13}$$

The simultaneous solution of (3.12) and (3.13) yields

$$\widetilde{f}_c^2 \approx \frac{4A_{200}A_{001}}{A_{110}^2 - 4A_{200}A_{020}}\varepsilon. \tag{3.14}$$

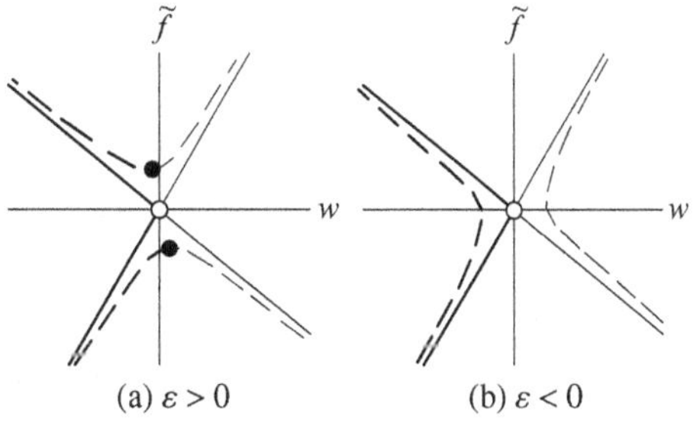

Fig. 3.2 Solution curves in the neighborhood of a transcritical bifurcation point expressed by the leading terms of the bifurcation equation (3.12) for $A_{200} > 0$, $A_{020} < 0$, and $A_{001} > 0$. ——: path for the perfect system; – – –: path for an imperfect system; thick line: stable; thin line: unstable; ○: bifurcation point; ●: limit point.

The nondegeneracy condition $A_{110}{}^2 - 4A_{200}A_{020} > 0$ in (2.83) is assumed. Then the existence of the real-valued solution of (3.14) is dependent on the sign of $A_{200}A_{001}\varepsilon$ as

$$\begin{cases} \text{existent} & \text{if } A_{200}A_{001}\varepsilon > 0, \\ \text{nonexistent} & \text{if } A_{200}A_{001}\varepsilon < 0. \end{cases}$$

For $A_{200}A_{001}\varepsilon > 0$, the solution of (3.13) and (3.14) yields

$$\widetilde{f_c} \approx \pm\left(\frac{4|A_{200}A_{001}|}{A_{110}{}^2 - 4A_{200}A_{020}}\right)^{1/2} |\varepsilon|^{1/2}, \tag{3.15}$$

$$w_c \approx \mp\frac{A_{110}}{2A_{200}}\left(\frac{4|A_{200}A_{001}|}{A_{110}{}^2 - 4A_{200}A_{020}}\right)^{1/2} |\varepsilon|^{1/2}, \tag{3.16}$$

in which $\pm$ and $\mp$ take the same order. By (3.15) and (3.16), $\widetilde{f_c}$ and w_c are of the same order of $|\varepsilon|^{1/2}$. Equation (3.15) is the *one-half power law*,[4] stating that the reduction (or increase) of the critical load is proportional to the one-half power of the imperfection magnitude $|\varepsilon|$. The imperfect system has a limit point of load f only when $A_{200}A_{001}$ and ε have the same sign. Such conditional presence of the limit point is apparent in Fig. 3.2 (repeated from Fig. 2.6 in §2.5.2). The case where $\widetilde{f_c} < 0$, with "–" on the right-hand side of (3.15), is of engineering interest in that f_c is reduced by imperfections.

3.2.3 Pitchfork Bifurcation Point

For a pitchfork bifurcation point, at which $A_{100} = A_{010} = A_{200} = 0$ and $A_{300} \neq 0$, the bifurcation equation with (2.67) is given by (2.85):

$$\widehat{F}(w, \tilde{f}, \varepsilon) = A_{300}w^3 + A_{110}w\tilde{f} + A_{020}\tilde{f}^2 + A_{001}\varepsilon + \text{h.o.t.} = 0. \tag{3.17}$$

The criticality condition is evaluated to

[4] For this law, see Koiter, 1945 [119] and Thompson and Hunt, 1973 [191].

$$\frac{\partial \widehat{F}}{\partial w}(w,\widetilde{f},\varepsilon) = 3A_{300}w^2 + A_{110}\widetilde{f} + \text{h.o.t.} = 0. \tag{3.18}$$

Elimination of $\widetilde{f}$ from (3.17) and (3.18) produces (see §3.2.4 for details)

$$w_c \approx \left(\frac{A_{001}}{2A_{300}}\right)^{1/3} \varepsilon^{1/3}. \tag{3.19}$$

Since, in (3.17), $A_{020}\widetilde{f}^2$ is of higher order than the other terms in (3.17), the coefficient A_{020} does not appear in (3.19). Substitution of (3.19) into (3.18) yields

$$\widetilde{f}_c \approx -\frac{3A_{300}{}^{1/3}}{A_{110}}\left(\frac{A_{001}}{2}\right)^{2/3} \varepsilon^{2/3}. \tag{3.20}$$

This equation is the so-called *two-thirds power law*, found by Koiter, 1945 [119], which expresses that the critical load $\widetilde{f}_c$ of the imperfect system changes in proportion with $\varepsilon^{2/3}$. Equation (3.19) indicates that the critical displacement w_c changes in proportion with $\varepsilon^{1/3}$. The variation of w_c, accordingly, is greater in order than that of $\widetilde{f}_c$.

For an unstable pitchfork bifurcation point ($A_{300} < 0$ by (2.88)), the law (3.20) offers information about the limit points of the imperfect paths with $\pm\varepsilon$ that approach the bifurcation point from below (cf., Fig. 3.1(c)). Recall our convention (2.87): $A_{110} < 0$. Because the system becomes unstable at this point, this information is of great physical importance.

In contrast, for a stable bifurcation point with $A_{300} > 0$, the law (3.20) offers information about the limit points of the imperfect paths with $\pm\varepsilon$ that approach the bifurcation point from above (cf., Fig. 3.1(d)). The law (3.20) is incapable of offering information about imperfect paths without limit points that approach the stable point from below. Such incapability is resolved in §6.3 by generalizing this law.

Remark 3.1. A subtle point of our formulation of the imperfection sensitivity law is discussed here in relation to the universal unfolding of the pitchfork bifurcation point

$$G(w,\widetilde{f},\beta_1,\beta_2) = w^3 + \widetilde{f}w + \beta_1 + \beta_2 w^2$$

explained in Remark 2.11 in §2.5.3. As Fig. 3.3 (repeated from Fig. 2.10 in §2.5.3) shows, the fundamental path of $G(w,\widetilde{f},\beta_1,\beta_2) = 0$ has a kink (i.e., a pair of maximal and minimal limit points) if (β_1,β_2) belongs to the region

$$K = \{(\beta_1,\beta_2) \mid 0 < \beta_1 < \beta_2{}^3/27 \text{ or } \beta_2{}^3/27 < \beta_1 < 0\}. \tag{3.21}$$

In our formulation of imperfection sensitivity, we have fixed pattern $\boldsymbol{d}$ of imperfection $\boldsymbol{v} = \boldsymbol{v}^0 + \varepsilon\boldsymbol{d}$ and considered the asymptotic behavior of $\widetilde{f}_c$ as $|\varepsilon|$ tends to 0. For the problem described by $G(w,\widetilde{f},\beta_1,\beta_2)$, we put $(\beta_1,\beta_2) = \varepsilon(d_1,d_2)$ and let $|\varepsilon|$ tend to 0. It is readily apparent that for any fixed (d_1,d_2), (β_1,β_2) lies outside the region K in (3.21) if $|\varepsilon|$ is sufficiently small. This explains why the fundamental path with

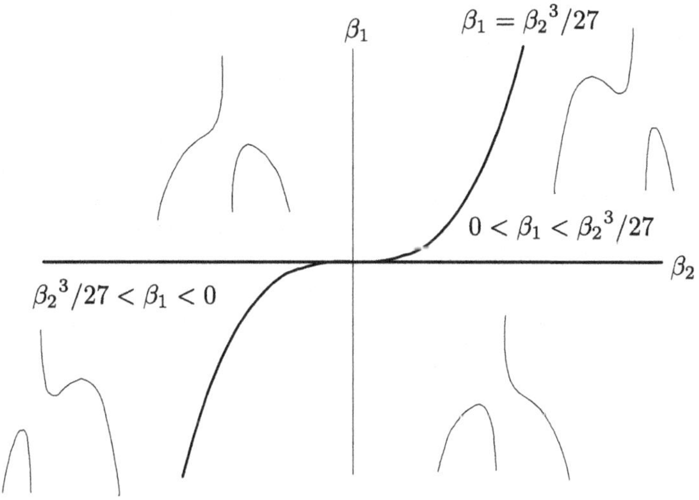

Fig. 3.3 Universal unfolding for an unstable pitchfork bifurcation point.

a kink has been ignored in our analysis. It is more likely that the *unfolding parameters* (β_1,β_2) depend on the imperfection parameter $\boldsymbol{v}$ in a complicated but smooth manner. Then we may assume expansions

$$\beta_1 = \sum_{k=1} \beta_{1k}(\boldsymbol{d})\varepsilon^k, \qquad \beta_2 = \sum_{k=1} \beta_{2k}(\boldsymbol{d})\varepsilon^k$$

for $|\varepsilon|$ sufficiently small, where $\beta_{1k}(\boldsymbol{d})$ and $\beta_{2k}(\boldsymbol{d})$ $(k = 1,2,\ldots)$ are coefficients depending on pattern $\boldsymbol{d}$. This shows that (β_1,β_2) lies outside region K in (3.21) for sufficiently small $|\varepsilon|$ provided that $\beta_{11}(\boldsymbol{d}) \neq 0$. One might generically expect that $\beta_{11}(\boldsymbol{d}) \neq 0$ for some $\boldsymbol{d}$ inasmuch as there is no reason for $\beta_{11}(\boldsymbol{d})$ to vanish for all $\boldsymbol{d}$. In other words, the quadratic term $\beta_2 w^2$ in the universal unfolding plays, generically, no role in our formulation of imperfection sensitivity. □

3.2.4 Systematic Derivation

Derivations of the imperfection sensitivity laws presented above are somewhat ad hoc; therefore, they are applicable only to lower-order critical points with nonzero lower-order terms A_{200} and/or A_{300}; see (2.74) for the classification of bifurcation points. We present here a systematic derivation for a critical point of an arbitrary order, say m. It turns out from the following derivation that it is sufficient to consider up to the mth-order terms of w in $\widehat{F}$ in (3.5) for the first-order approximation for the mth-order critical point.

The mth-order approximation to $\widehat{F}$ is expressed as

$$\widehat{F}_m = \sum_{i=0}^{m} A_i w^i,$$

where

$$A_i = \sum_{j=0}\sum_{k=0} A_{ijk}\widetilde{f}^j \varepsilon^k.$$

For $\widehat{F}_m$ expressed in terms of a polynomial of w, the solution of the condition (3.4) for a critical point can be obtained by means of a standard procedure in algebra. Elimination of w from (3.4) results in the condition that the *discriminant* of $\widehat{F}_m$ (as a polynomial in w) or, alternatively, the *resultant* of $\widehat{F}_m$ and $\partial\widehat{F}_m/\partial w$, should vanish. This condition can be written as[5]

$$D_m = \begin{vmatrix} A_0 & A_1 & A_2 & \cdots & \cdots & A_m & & & \\ & A_0 & A_1 & A_2 & \cdots & \cdots & A_m & & \\ & & \ddots & \ddots & \ddots & & \ddots & \ddots & \\ & & & A_0 & A_1 & A_2 & \cdots & \cdots & A_m \\ A_1 & 2A_2 & \cdots & \cdots & mA_m & & & & \\ & A_1 & 2A_2 & \cdots & \cdots & mA_m & & & \\ & & \ddots & \ddots & \ddots & \ddots & \ddots & & \\ & & & \ddots & \ddots & \ddots & \ddots & \ddots & \\ & & & & A_1 & 2A_2 & \cdots & \cdots & mA_m \end{vmatrix} = 0, \tag{3.22}$$

where the size of the determinant is $2m-1$. Therein, D_m is a function in ε and $\widetilde{f}$, and $\widetilde{f}$ is to be determined as a function in ε by the equation $D_m = 0$. This can be carried out with the aid of the Newton polygon described in Remark 2.10 in §2.5.3.

Limit Point

The imperfection sensitivity law (3.10) for a nondegenerate limit point ($A_{010} \neq 0, A_{200} \neq 0$) is obtained from (3.22) with $m = 2$; that is,

$$D_2 = (4A_0A_2 - A_1{}^2)A_2 = 0. \tag{3.23}$$

Since

$$\begin{aligned} A_0 &= A_{001}\varepsilon + A_{010}\widetilde{f} + \text{h.o.t.}, \\ A_1 &= A_{101}\varepsilon + A_{110}\widetilde{f} + \text{h.o.t.}, \\ A_2 &= A_{200} + \text{h.o.t.}, \end{aligned}$$

it is readily apparent that

[5] See, for example, van der Waerden, 1955 [200].

$$D_2 = 4(A_{001}\varepsilon + A_{010}\widetilde{f})A_{200}{}^2 + \text{h.o.t.} = 0,$$

from which (3.10) is derived.

Transcritical Bifurcation Point

For a transcritical bifurcation point, we have $m = 2$ and the following:

$$\begin{aligned} A_0 &= A_{001}\varepsilon + A_{020}\widetilde{f}^2 + \text{h.o.t.}, \\ A_1 &= A_{101}\varepsilon + A_{110}\widetilde{f} + \text{h.o.t.}, \\ A_2 &= A_{200} + \text{h.o.t.} \end{aligned}$$

Substitution of these into (3.23) yields

$$D_2 = [4A_{200}A_{001}\varepsilon - (A_{110}{}^2 - 4A_{200}A_{020})\widetilde{f}^2]A_{200} + \text{h.o.t.} = 0,$$

from which the imperfection sensitivity law (3.15) follows.

Pitchfork Bifurcation Point

The imperfection sensitivity law (3.20) for a pitchfork bifurcation point of order three is derived from (3.22) with $m = 3$, which is evaluated as

$$D_3 = (27A_0{}^2A_3{}^3 + 4A_1{}^3A_3{}^2) + (4A_0A_2{}^3A_3 - A_1{}^2A_2{}^2A_3 - 18A_0A_1A_2A_3{}^2) = 0.$$

Therein,

$$\begin{aligned} A_0 &= A_{001}\varepsilon + A_{020}\widetilde{f}^2 + \text{h.o.t.}, \\ A_1 &= A_{101}\varepsilon + A_{110}\widetilde{f} + \text{h.o.t.}, \\ A_2 &= A_{201}\varepsilon + A_{210}\widetilde{f} + \text{h.o.t.}, \\ A_3 &= A_{300} + \text{h.o.t.} \end{aligned}$$

The Newton polygon for D_3 is portrayed in Fig. 3.4, which demonstrates the relevant pairs (j,k) such that $\widetilde{f}^j\varepsilon^k$ is contained in D_3. For example,

$$27A_0{}^2A_3{}^3 = 27(A_{001}\varepsilon + A_{020}\widetilde{f}^2 + \text{h.o.t.})^2(A_{300} + \text{h.o.t.})^3$$

yields $(j,k) = (0,2)$, $(2,1)$, $(4,0)$, and

$$4A_1{}^3A_3{}^2 = 4(A_{101}\varepsilon + A_{110}\widetilde{f} + \text{h.o.t.})^3(A_{300} + \text{h.o.t.})^2$$

yields $(j,k) = (0,3)$, $(1,2)$, $(2,1)$, and $(3,0)$. The Newton polygon in Fig. 3.4 has a side of slope $-2/3$, which connects $(j,k) = (0,2)$ and $(3,0)$. Consequently, the first-

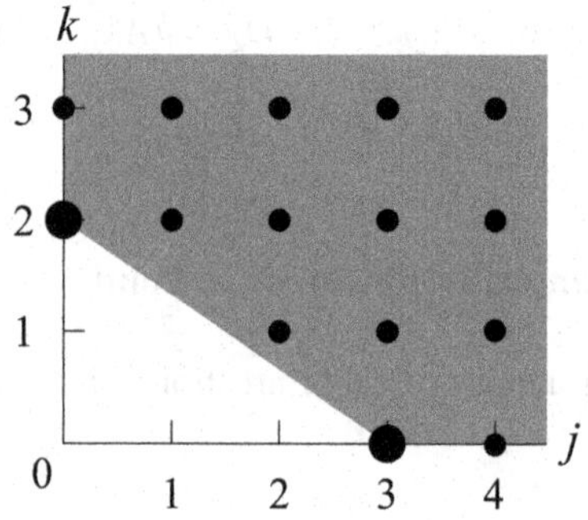

Fig. 3.4 Newton polygon for D_3 expressed by the shaded area.

order asymptotic expression of the sensitivity law in (3.20) is determined from

$$27(A_{001}\varepsilon)^2 A_{300}{}^3 + 4(A_{110}\widetilde{f})^3 A_{300}{}^2 = 0,$$

from which the two-thirds power law (3.20) follows.

Higher-Order Simple Bifurcation Point

For a simple bifurcation point of order greater than three (i.e., $m \geq 4$), it can be demonstrated that

$$D_m = A_0{}^{m-1}(mA_m)^m - (-1)^m(m-1)^{m-1}A_1{}^m A_m{}^{m-1} + \text{h.o.t.}, \qquad (3.24)$$

where

$$\begin{aligned} A_0 &= A_{001}\varepsilon + A_{020}\widetilde{f}^2 + \text{h.o.t.}, \\ A_1 &= A_{101}\varepsilon + A_{110}\widetilde{f} + \text{h.o.t.}, \\ A_m &= A_{m00} + \text{h.o.t.} \end{aligned}$$

Consequently, $D_m = 0$ gives

$$\begin{cases} \widetilde{f}_{\text{c}} \approx \pm \dfrac{m|A_{m00}|^{1/m}}{A_{110}} \left| \dfrac{A_{001}}{m-1} \right|^{1-1/m} |\varepsilon|^{1-1/m} & \text{if } A_{m00}A_{001}\varepsilon > 0, \\ \text{nonexistent} & \text{if } A_{m00}A_{001}\varepsilon < 0 \end{cases}$$

for even m (≥ 4) and

$$\widetilde{f}_{\text{c}} \approx -\frac{mA_{m00}{}^{1/m}}{A_{110}} \left| \frac{A_{001}}{m-1} \right|^{1-1/m} \varepsilon^{1-1/m}$$

for odd m (≥ 5) (A_{m00} can be positive or negative according to whether the point is stable or unstable). Recall our convention (2.87) that $A_{110} < 0$.

3.3 Examples of Imperfection Sensitivity

The imperfection sensitivity laws presented in §3.2 are explained using examples.

3.3.1 Propped Cantilever

We refer to the propped cantilever in Fig. 3.5 (repeated from §2.3 in Fig. 2.1). The transcritical bifurcation point appears on its equilibrium paths in Fig. 3.6(a) (repeated from Fig. 2.2(a)) for the perfect case with $\boldsymbol{v}^0$ in (2.18). The stable pitchfork bifurcation point appears on its equilibrium paths in Fig. 3.6(b) (repeated from Fig. 2.11(a) in §2.6) for $\boldsymbol{v}^0$ in (2.96).

Figure 3.7(a) portrays the relation between the normalized critical load f_c/f_c^0 and the magnitude ε of imperfection for the transcritical bifurcation point. Figure 3.7(b)

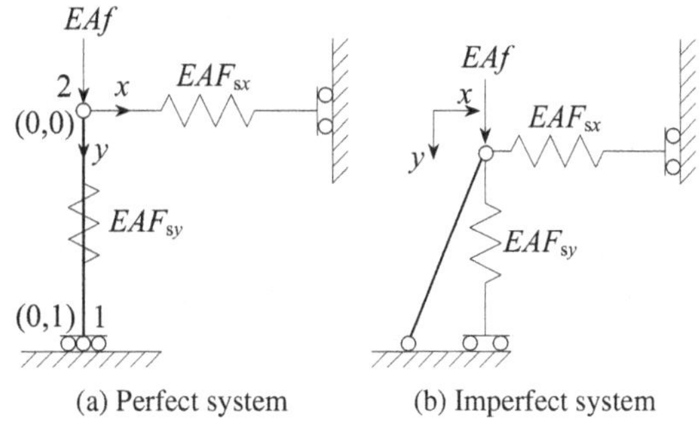

Fig. 3.5 Propped cantilever.

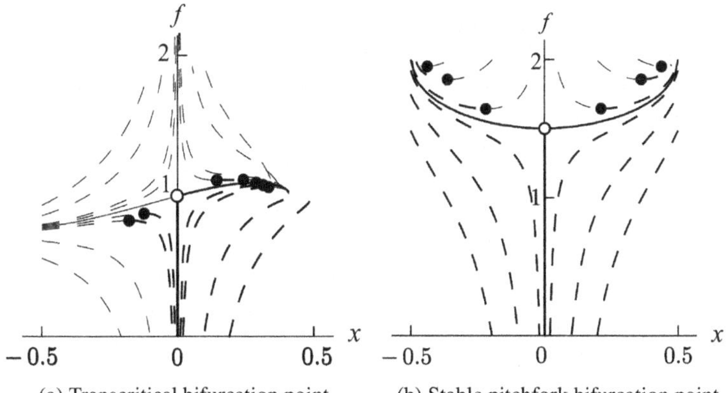

Fig. 3.6 f versus x curves for the propped cantilever. ——: path for the perfect system; — — —: path for an imperfect system; thick line: stable; thin line: unstable; ○: bifurcation point; ●: limit point.

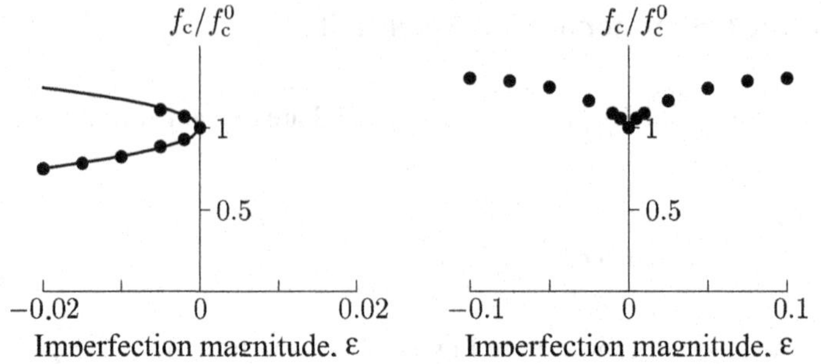

(a) Transcritical bifurcation point (b) Stable pitchfork bifurcation point

Fig. 3.7 Imperfection sensitivity of the propped cantilever expressed in terms of f_c/f_c^0 versus ε relations. ●: actual; ——: asymptotic formula.

depicts the relation for the stable pitchfork bifurcation point. Consequently, the sensitivity of f_c to ε varies according to the type of critical point, as has been summarized in (3.6).

We illustrate here the derivation (§3.2.2) of the imperfection sensitivity law for the transcritical bifurcation point. Recall the bifurcation equation (2.29) obtained for this point

$$\widehat{F} = \widetilde{x}^2 - 2\widetilde{x}\widetilde{f} - 3\varepsilon + \text{h.o.t.} = 0. \tag{3.25}$$

The critical load of an imperfect system can be characterized by the limit point on an imperfect path described by this equation. The second condition in (3.4) for the singularity of the Jacobian matrix reads as

$$\frac{\partial \widehat{F}}{\partial \widetilde{x}} = 2\widetilde{x} - 2\widetilde{f} + \text{h.o.t.} = 0. \tag{3.26}$$

The location $(\widetilde{x}_c, \widetilde{f}_c)$ of the limit point can then be determined as the simultaneous solution of the bifurcation equation (3.25) and the criticality condition (3.26). Therefore, we have the one-half power law in (3.15):

$$\widetilde{x}_c = \widetilde{f}_c = \pm \sqrt{3}(-\varepsilon)^{1/2}. \tag{3.27}$$

This law is depicted by the solid curve in Fig. 3.7(a).

3.3.2 Truss Arches

The shallow truss arch in Fig. 3.8(a) and the nonshallow truss arch in Fig. 3.8(b) are considered, respectively, as examples of a limit point and of an unstable pitchfork bifurcation point.

The equilibrium of this arch under a vertical load f is described by

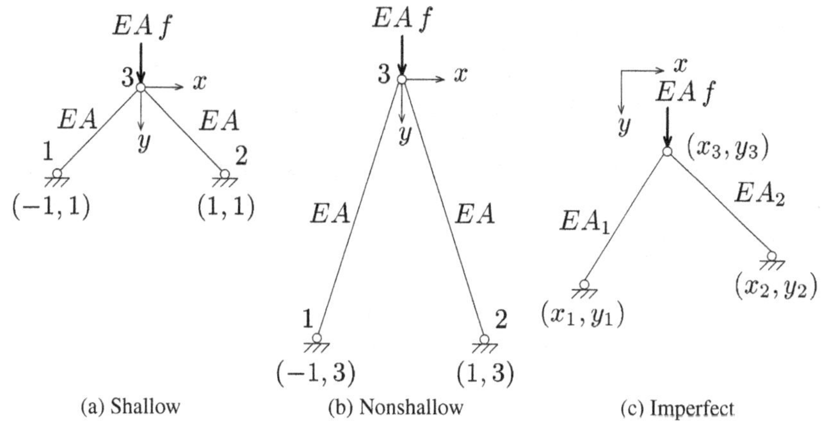

Fig. 3.8 Shallow and nonshallow truss arches.

$$\boldsymbol{F} = \begin{pmatrix} F_x \\ F_y \end{pmatrix} = EA \begin{pmatrix} \displaystyle\sum_{i=1}^{2} \frac{EA_i}{EA}\left(\frac{1}{L_i} - \frac{1}{\widehat{L}_i}\right)(x - x_i) \\ \displaystyle\sum_{i=1}^{2} \frac{EA_i}{EA}\left(\frac{1}{L_i} - \frac{1}{\widehat{L}_i}\right)(y - y_i) - f \end{pmatrix} = \begin{pmatrix} 0 \\ 0 \end{pmatrix}, \tag{3.28}$$

where EA_i expresses the product of Young's modulus E and the cross-sectional area A_i of member i $(i = 1,2)$ and EA denotes its (nominal) value for the perfect system; (x_i, y_i) is the initial location of node i $(i = 1,2,3)$; (x, y) is the location of node 3 after displacement; and

$$L_i = [(x_3 - x_i)^2 + (y_3 - y_i)^2]^{1/2}, \qquad \widehat{L}_i = [(x - x_i)^2 + (y - y_i)^2]^{1/2}, \qquad i = 1,2.$$

The Jacobian matrix is given by

$$J = \begin{pmatrix} F_{x,x} & F_{x,y} \\ F_{y,x} & F_{y,y} \end{pmatrix}, \tag{3.29}$$

where $F_{x,y} = \partial F_x / \partial y$, and so on, are evaluated to

$$F_{x,x} = \sum_{i=1}^{2} EA_i\left(\frac{1}{L_i} - \frac{(y - y_i)^2}{\widehat{L}_i^3}\right), \qquad F_{y,y} = \sum_{i=1}^{2} EA_i\left(\frac{1}{L_i} - \frac{(x - x_i)^2}{\widehat{L}_i^3}\right),$$

$$F_{x,y} = F_{y,x} = \sum_{i=1}^{2} EA_i \frac{(x - x_i)(y - y_i)}{\widehat{L}_i^3}.$$

The imperfection parameter vector is set as

$$\boldsymbol{v} = \left(\frac{x_1}{L^0}, \frac{y_1}{L^0}, \frac{x_2}{L^0}, \frac{y_2}{L^0}, \frac{x_3}{L^0}, \frac{y_3}{L^0}, \frac{EA_1}{EA}, \frac{EA_2}{EA}\right)^\top$$

($L^0 = 1$ is the unit scaling length) and

$$\boldsymbol{v}^0 = \begin{cases} (-1,1,1,1,0,0,1,1)^\top & \text{for the shallow truss arch,} \\ (-1,3,1,3,0,0,1,1)^\top & \text{for the nonshallow truss arch} \end{cases} \tag{3.30}$$

for the perfect systems shown in Fig. 3.8(a) and (b). The imperfection pattern vector $\boldsymbol{d}$ in

$$\boldsymbol{v} = \boldsymbol{v}^0 + \varepsilon \boldsymbol{d} \tag{3.31}$$

is chosen as

$$\boldsymbol{d} = \begin{cases} \left(-\dfrac{1}{\sqrt{2}}, -\dfrac{1}{\sqrt{2}}, \dfrac{1}{\sqrt{2}}, -\dfrac{1}{\sqrt{2}}, 0, 1, -1, -1\right)^\top & \text{for the shallow truss arch,} \\ (-0.73685, -0.67606, -0.73685, 0.67606, 1, 0, 1, -1)^\top & \\ & \text{for the nonshallow truss arch.} \end{cases} \tag{3.32}$$

It is readily apparent from (3.28) that the perfect system defined by (3.30) has the trivial solution, which forms the fundamental path expressed by

$$x = 0, \qquad f = 2\left(\frac{1}{(1+y_1{}^2)^{1/2}} - \frac{1}{[1+(y-y_1)^2]^{1/2}}\right)(y-y_1), \tag{3.33}$$

in which $y_1 = 1$ for the shallow arch and $y_1 = 3$ for the nonshallow arch. On this fundamental path, the Jacobian matrix in (3.29) reduces to a diagonal matrix

$$J^0 = \begin{pmatrix} (F_{x,x})^0 & 0 \\ 0 & (F_{y,y})^0 \end{pmatrix}, \tag{3.34}$$

where

$$(F_{x,x})^0 = 2EA\left\{\frac{1}{(1+y_1{}^2)^{1/2}} - \frac{(y-y_1)^2}{[1+(y-y_1)^2]^{3/2}}\right\}, \tag{3.35}$$

$$(F_{y,y})^0 = 2EA\left\{\frac{1}{(1+y_1{}^2)^{1/2}} - \frac{1}{[1+(y-y_1)^2]^{3/2}}\right\}. \tag{3.36}$$

The Jacobian matrix becomes singular when $(F_{x,x})^0$ or $(F_{y,y})^0$ vanishes; the critical eigenvector is $\boldsymbol{\xi}_1 = (1,0)^\top$ for $(F_{x,x})^0 = 0$ and is $\boldsymbol{\xi}_1 = (0,1)^\top$ for $(F_{y,y})^0 = 0$. Since the vector $(\partial \boldsymbol{F}/\partial f)^0_\mathrm{c} = (0, -EA)^\top$ satisfies

$$\boldsymbol{\xi}_1{}^\top (\partial \boldsymbol{F}/\partial f)^0_\mathrm{c} = \begin{cases} 0 & \text{for } \boldsymbol{\xi}_1 = (1,0)^\top, \\ -EA \neq 0 & \text{for } \boldsymbol{\xi}_1 = (0,1)^\top, \end{cases}$$

the classification (2.76) for simple critical points, for this case, reads as

$$\begin{cases} (F_{x,x})^0 = 0 \text{ with } \boldsymbol{\xi}_1 = (1,0)^\top, & \text{limit point,} \\ (F_{y,y})^0 = 0 \text{ with } \boldsymbol{\xi}_1 = (0,1)^\top, & \text{bifurcation point.} \end{cases}$$

Shallow Arch

For the shallow arch, $(F_{x,x})^0$ in (3.35) does not vanish for any value of y, whereas $(F_{y,y})^0$ in (3.36) vanishes for $y = 1 \pm (2^{1/3} - 1)^{1/2}$. Therefore, on the fundamental path, we see the presence of two critical points, which turn out to be the limit points of f by the observation of the curve in (3.33):

$$(x_c^0, y_c^0, f_c^0) = \begin{cases} (0, 1-(2^{1/3}-1)^{1/2}, \sqrt{2}(2^{1/3}-1)^{3/2}) = (0, 0.4902, 0.1874), \\ \qquad \text{maximum point,} \\ (0, 1+(2^{1/3}-1)^{1/2}, -\sqrt{2}(2^{1/3}-1)^{3/2}) = (0, 1.510, -0.1874), \\ \qquad \text{minimum point.} \end{cases}$$

Equilibrium paths (f versus y curves) in Fig. 3.9 are obtained by solving (3.28) for the imperfection vector of (3.31) defined by (3.30) and (3.32) with $\varepsilon = 0, \pm 0.01, \pm 0.05$, and ± 0.1. The perfect system with $\varepsilon = 0$ is stable in the original state $(x, y, f) = (0, 0, 0)$. In association with the increase in y, the perfect system becomes unstable when the load f reaches the maximum point A of f, and it becomes stable again at the minimum point of f.

We specifically examine the imperfection sensitivity of the critical load f_c at the maximum point of f, at which the system becomes unstable. Figure 3.10(a) shows the relation between the normalized critical load f_c/f_c^0 and the magnitude ε, where the straight line denotes the linear law in (3.10) with higher-order terms truncated (the values of A_{001} and A_{010} in (3.10) have been computed directly from the equilibrium equation (3.28)). This straight line correlates well with the critical loads of imperfect arches denoted by (•), although the discrepancy enlarges as ε increases due to the truncation of higher-order terms in (3.10).

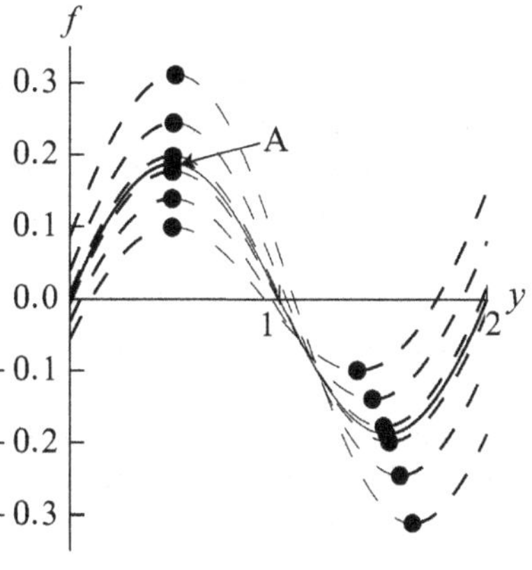

Fig. 3.9 Equilibrium paths (f versus y curves) for the shallow arch. ——: path for the perfect system; – – –: path for an imperfect system; thick line: stable; thin line: unstable; •: limit point.

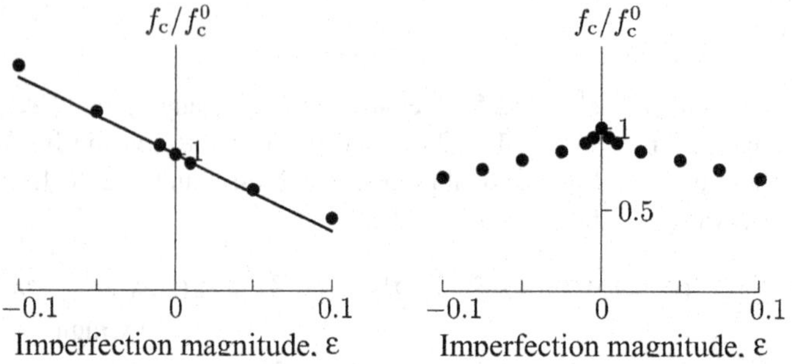

(a) Limit point (shallow truss arch)   (b) Unstable pitchfork bifurcation point (nonshallow truss arch)

Fig. 3.10 Imperfection sensitivity of the shallow truss arch and the nonshallow one expressed in terms of f_c/f_c^0 versus ε relations. ●: actual; ——: asymptotic formula.

Nonshallow Arch

For the nonshallow arch, from (3.34) with (3.35) and (3.36), the following six simple critical points are found on the fundamental path

$$
(x_c^0, y_c^0, f_c^0) = \begin{cases} (0, 0.4473, 0.2478),\ (0, 2.162, 0.7546), (0, 3.838, -0.7546), \\ (0, 5.553, -0.2478), \qquad \text{pitchfork bifurcation points}, \\ (0, 3-(10^{1/3}-1)^{1/2},\ \sqrt{10}(10^{1/3}-1)^{3/2}/5) = (0, 1.926, 0.7845), \\ \qquad \text{maximum point}, \\ (0, 3+(10^{1/3}-1)^{1/2}, -\sqrt{10}(10^{1/3}-1)^{3/2}/5) = (0, 4.074, -0.7845), \\ \qquad \text{minimum point}. \end{cases}
$$

The f versus y curves for the perfect system with $\varepsilon = 0$ in Fig. 3.11(a) give a global view. The f versus x curves in Fig. 3.11(b) give a local view for the perfect and imperfect systems around the unstable pitchfork bifurcation point A; the imperfection vector of (3.31) for the nonshallow arch defined by (3.30) and (3.32) with $\varepsilon = 0$, ± 0.01, ± 0.05, and ± 0.1 is used. The perfect system is stable in the original state $(x, y, f) = (0, 0, 0)$. In association with the increase of y, the perfect system becomes unstable at the pitchfork bifurcation point A. The imperfect systems become unstable at the limit points that are located in the neighborhood of this bifurcation point.

Figure 3.10(b) shows the f_c and ε relation for the unstable pitchfork bifurcation point A. This relation is apparently nonlinear, in association with the two-thirds power law in (3.20).

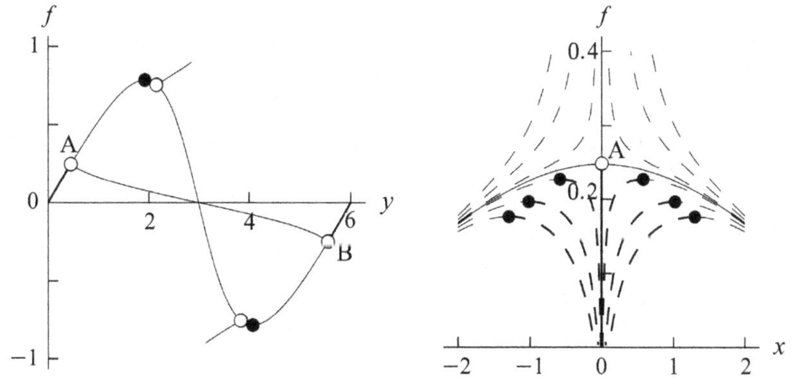

(a) f versus y curves (b) f versus x curves

Fig. 3.11 Equilibrium paths for the nonshallow arch. ——: path for the perfect system; — — —: path for an imperfect system; thick line: stable; thin line: unstable; ○: bifurcation point; ●: limit point.

Problems

3-1 Plot an f_c/f_c^0 versus $|\varepsilon|^{1/2}$ relation for the data in Fig. 3.7(a) and f_c/f_c^0 versus $\varepsilon^{2/3}$ one for the data in Fig. 3.10(b). Digitized data are shown below.

In Fig. 3.7(a)

ε	−0.02	−0.015	−0.01	−0.005	−0.002	0.00	0.002	0.005
f_c/f_c^0	0.75	0.78	0.83	0.88	0.93	1.00	1.07	1.11

In Fig. 3.10(b)

ε	±0.1	±0.075	±0.05	±0.025	±0.01	±0.005	0.0
f_c/f_c^0	0.70	0.75	0.81	0.86	0.91	0.94	1.00

3-2 Obtain imperfection sensitivity laws for a set of equations

$$\begin{cases} x^3 + xy - 2fx + \varepsilon = 0, \\ y - f = 0. \end{cases}$$

Answer: $x_c = 2^{1/3}\varepsilon^{1/3}$, $f_c = y_c = (3/2^{2/3})\varepsilon^{2/3}$.

3-3 Obtain imperfection sensitivity laws for a set of equations

$$\begin{cases} x^2 + xf + \alpha yf + \varepsilon = 0, \\ y - f = 0 \end{cases}$$

for $\varepsilon > 0$.
Answer: $x_c = \pm\sqrt{\varepsilon/(1-4\alpha)}$ and $f_c = y_c = \mp 2\sqrt{\varepsilon/(1-4\alpha)}$ for $\alpha < 1/4$, and no solutions for $\alpha \geq 1/4$, in which $\pm$ and $\mp$ take the same order.

3-4 Obtain an f_c versus ε relation for the propped cantilever in Fig. 2.1 in §2.3 for $\boldsymbol{d} = (-1,0,1,0,0,0,0,0)^\top$.

3-5 (1) Obtain an f_c versus ε relation for the cantilever supported by a linear spring in Fig. 1.2(a) in §1.2.2.
(2) Obtain an f_c versus ε relation for this cantilever for another case where member length is $L(1+\varepsilon)$ and the horizontal force $kL\varepsilon$ is absent.

3-6 Obtain an f_c versus ε relation for the nonshallow arch in Fig. 3.8(b) in §3.3.2 for $\boldsymbol{d} = (1,0,0,-1,0,1,0,0)^\top$ by actually solving (3.28).

3-7 Consider a system that follows the two-thirds power law (3.20) for imperfections. Evaluate f_c^0 for given data: $(f_c, \varepsilon) = (2.4, 0.1)$ and $(2.5, 0.01)$.
Answer: $f_c^0 \approx 2.53$.

Summary

- Imperfection sensitivity laws have been derived for simple critical points.
- Imperfection sensitivities of examples have been illustrated.

Chapter 4
Worst Imperfection (I)

4.1 Introduction

The critical load of a system undergoing bifurcation is highly sensitive to imperfections, as Chapter 3 has described. Such is particularly the case for domes and shells.[1] In their design, it is preferable to consider the "worst imperfection" that reduces the critical load most rapidly.[2]

This section presents a procedure to determine the worst pattern of imperfections.[3] For a pitchfork bifurcation point, for example, we recall the imperfection sensitivity law (3.6):

$$\widetilde{f}_{\mathrm{c}} \approx C(\boldsymbol{d})\varepsilon^{2/3}.$$

We fix ε as a scaling constant and then formulate the problem of the worst imperfection as that of finding the imperfection pattern vector $\boldsymbol{d}$ that maximizes $|C(\boldsymbol{d})|$ under the constraint that the norm of $\boldsymbol{d}$ is kept constant by

$$\boldsymbol{d}^{\top} W^{-1} \boldsymbol{d} = 1,$$

where W is a weight matrix that is arbitrary as long as it is positive-definite (symmetric). The resulting worst imperfection is substantially affected by the choice of W, which should reflect design principles and technological constraints.

This chapter is organized as follows.

- The method for obtaining the worst imperfection vector is outlined against a simple example in §4.2 to highlight the contents of this chapter.
- The formula for the worst imperfection pattern for simple critical points is derived in §4.3.

---

[1] See, for example, Hutchinson and Koiter, 1970 [80].

[2] It must be remarked that the term "critical imperfection" in the first edition is replaced by the "worst imperfection" in the second edition.

[3] This procedure is based on Ikeda and Murota, 1990, [85, 86]. A similar procedure was followed by Peek and Triantafyllidis, 1992 [164].

K. Ikeda and K. Murota, *Imperfect Bifurcation in Structures and Materials*, Applied Mathematical Sciences 149, DOI 10.1007/978-1-4419-7296-5_4,

- The formula explained in §4.3 is extended to the customary situation in practice where the imperfection parameters are divided into multiple categories in §4.4.
- The worst imperfection patterns of examples are obtained based on the formula in §4.5.

The results in this chapter are extended to a system with group symmetry in Chapter 9

4.2 Illustrative Example

The fundamental concept of the worst imperfection and the procedure to obtain it are illustrated for the nonshallow truss arch in Fig. 4.1(a); this is the same arch as that in Fig. 3.8(b) in §3.3.2 and some issues are repeated here for the reader's convenience. The general framework of the worst imperfection is presented in §4.3.

Governing Equation and Imperfection Sensitivity

The equilibrium of this arch under a vertical load is described by

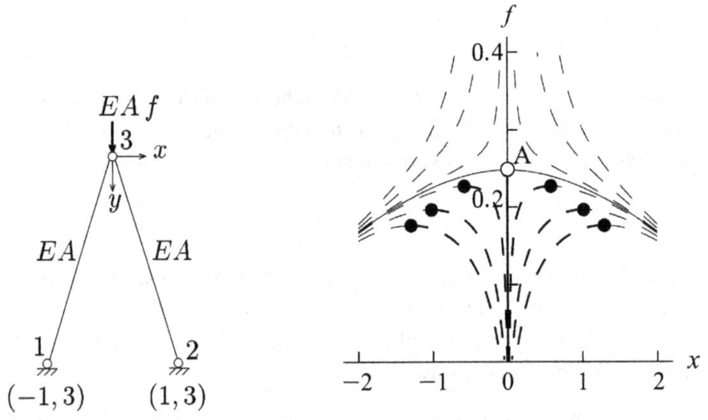

(a) Nonshallow truss arch (b) Equilibrium paths

Fig. 4.1 Nonshallow truss arch and its equilibrium paths. ——: path for the perfect system; — — —: path for an imperfect system; thick line: stable; thin line: unstable; ○: bifurcation point; ●: limit point.

$$F(u,f,v) = \begin{pmatrix} F_x \\ F_y \end{pmatrix} = EA \begin{pmatrix} \sum_{i=1}^{2} \frac{EA_i}{EA}\left(\frac{1}{L_i} - \frac{1}{\widetilde{L}_i}\right)(x - x_i) \\ \sum_{i=1}^{2} \frac{EA_i}{EA}\left(\frac{1}{L_i} - \frac{1}{\widetilde{L}_i}\right)(y - y_i) - f \end{pmatrix} = \begin{pmatrix} 0 \\ 0 \end{pmatrix} \tag{4.1}$$

(cf., §3.3.2 for notations). We employ the imperfection parameter vector

$$v = \left(\frac{x_1}{L^0}, \frac{y_1}{L^0}, \frac{x_2}{L^0}, \frac{y_2}{L^0}, \frac{x_3}{L^0}, \frac{y_3}{L^0}, \frac{EA_1}{EA}, \frac{EA_2}{EA}\right)^\top. \tag{4.2}$$

Therein, $L^0 = 1$ is the unit scaling length. As the perfect system, we choose

$$v^0 = (-1, 3, 1, 3, 0, 0, 1, 1)^\top.$$

Recall that the equilibrium paths for the perfect system shown by the solid lines in Fig. 4.1(b) have the unstable pitchfork bifurcation point A shown by (○) at $(x_c^0, y_c^0, f_c^0) = (0, 0.44735, 0.24776)$. At this point, the Jacobian matrix J is equal to

$$J_c^0 = EA \begin{pmatrix} 0 & 0 \\ 0 & 0.5354 \end{pmatrix}$$

with $\mathrm{rank}(J_c^0) = 1$, and the critical eigenvector is given by $\boldsymbol{\xi}_1 = (1, 0)^\top$. We are interested in the change $\widetilde{f_c} = f_c - f_c^0$ of the critical load f_c caused by imperfections, that is, discrepancies from the nominal values of (x_i, y_i) $(i = 1, 2, 3)$ and EA_i $(i = 1, 2)$ in (4.2).

Let us consider, for example, a pair of fixed patterns of imperfection

$$\begin{aligned} \boldsymbol{d}^{(1)} &= (-0.73685, -0.67606, -0.73685, 0.67606, 1, 0, 1, -1)^\top, \\ \boldsymbol{d}^{(2)} &= (0.73685, 0.67606, 0.73685, 0.67606, 1, 0, -1, -1)^\top \end{aligned}$$

with $\boldsymbol{d}^{(1)\top} W^{-1} \boldsymbol{d}^{(1)} = \boldsymbol{d}^{(2)\top} W^{-1} \boldsymbol{d}^{(2)} = 1$ for $W = 5\, I_8$. Recall that for pattern $\boldsymbol{d}^{(1)}$ and various values of ε, we have computed the equilibrium paths (shown by the dashed lines in Fig. 4.1(b)). Figure 4.2 shows the relation between f_c and ε for pattern $\boldsymbol{d}^{(1)}$ by (•) and that for $\boldsymbol{d}^{(2)}$ by (○); $\boldsymbol{d}^{(1)}$ is more influential than $\boldsymbol{d}^{(2)}$. The dependency of f_c on ε as well as on $\boldsymbol{d}$ for an unstable pitchfork bifurcation point is described by the two-thirds power law in (3.6):

$$\widetilde{f_c} \approx C(\boldsymbol{d})\varepsilon^{2/3} \tag{4.3}$$

with a negative coefficient $C(\boldsymbol{d})$ depending on $\boldsymbol{d}$, where this formula is valid when $|\varepsilon|$ is small. Since pattern $\boldsymbol{d}^{(1)}$ is more influential than $\boldsymbol{d}^{(2)}$, we have $|C(\boldsymbol{d}^{(1)})| > |C(\boldsymbol{d}^{(2)})|$.

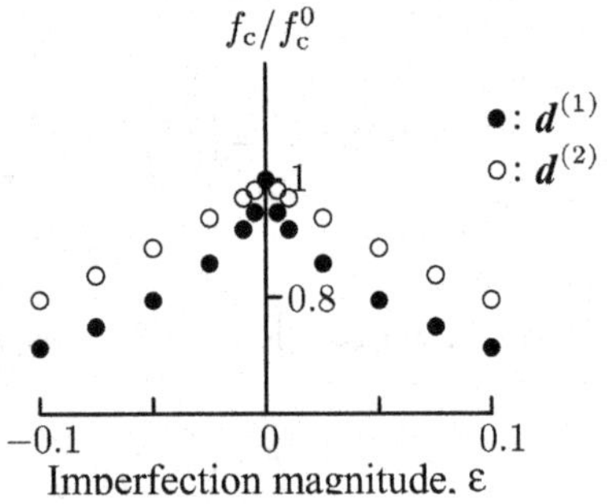

Fig. 4.2 Imperfection sensitivity for two different imperfection patterns.

Worst Imperfection

Our main task here is to determine the worst pattern of the imperfection vector that causes the maximum change (decrease) of the critical load. To put the matter more precisely, we may formulate this problem as follows. First, the imperfection pattern vector $\boldsymbol{d}$ is normalized as

$$\boldsymbol{d}^\top W^{-1}\boldsymbol{d} = 1 \tag{4.4}$$

with respect to a weight matrix W (positive-definite symmetric matrix). In the present example we choose $W = I_8$. Then our problem is to find $\boldsymbol{d}$ that maximizes $|C(\boldsymbol{d})|$ in (4.3) subject to the constraint of (4.4).

To obtain a concrete expression for the coefficients $C(\boldsymbol{d})$ in (4.3), we investigate the local properties of (4.1) around the critical point (x_c^0, y_c^0, f_c^0) using the Liapunov–Schmidt reduction explained in §2.4. Define the increment from the critical point as

$$(w, w_2, \widetilde{f}) = (x, y, f) - (x_c^0, y_c^0, f_c^0). \tag{4.5}$$

Since $\partial F/\partial y \neq 0$ at the critical point, the second equation $F_y = 0$ in (4.1) can be solved for w_2 (i.e., for y) by the implicit function theorem as

$$w_2 = \varphi(w, \widetilde{f}, \boldsymbol{v}). \tag{4.6}$$

Substituting (4.5) and (4.6) into the first equation, $F_x = 0$, in (4.1) provides a single equation

$$\widetilde{F}(w, \widetilde{f}, \boldsymbol{v}) = \sum_{i=1}^{2} EA_i\left(\frac{1}{L_i} - \frac{1}{\widehat{L_i}}\right)(x_c^0 + w - x_i) = 0, \tag{4.7}$$

where

$$\widehat{L_i} = [(x_c^0 + w - x_i)^2 + (y_c^0 + \varphi(w, \widetilde{f}, \boldsymbol{v}) - y_i)^2]^{1/2}, \qquad i = 1, 2.$$

Regarding the imperfection magnitude ε as an independent variable, we put

$$\widehat{F}(w, \widetilde{f}, \varepsilon) = \widetilde{F}(w, \widetilde{f}, \boldsymbol{v}^0 + \varepsilon\boldsymbol{d})$$

and consider its Taylor expansion around $(w, \widetilde{f}, \varepsilon) = (0, 0, 0)$:

$$\widehat{F}(w,\widetilde{f},\varepsilon) = \sum_{i=0}\sum_{j=0}\sum_{k=0} A_{ijk} w^i \widetilde{f}^j \varepsilon^k.$$

In this expansion some of the lower-order terms vanish. In fact,

$$A_{000} = \widehat{F}(0,0,0) = 0, \qquad A_{100} = \frac{\partial \widehat{F}}{\partial w}(0,0,0) = 0, \tag{4.8}$$

since $(w,\widetilde{f},\varepsilon) = (0,0,0)$ corresponds to the critical point for the perfect system (cf., (2.68) and (2.69)). We also have

$$A_{010} = 0, \qquad A_{200} = 0, \qquad A_{300} \neq 0, \qquad A_{110} < 0,$$

where the last inequality is our assumption in (2.87).

The imperfection coefficient A_{001} of the lowest-order term of ε is computed from the formula (2.71):

$$A_{001} = \boldsymbol{\xi}_1^\top B_\mathrm{c}^0 \boldsymbol{d}, \tag{4.9}$$

which gives an expression of A_{001} in terms of the imperfection sensitivity matrix B_c^0 in (2.10). We have

$$B_\mathrm{c}^0 = 10^{-2} EA \begin{pmatrix} 3.162 & 2.901 & 3.162 & -2.901 & -6.325 & 0 & -4.853 & 4.853 \\ 4.316 & -2.553 & -4.316 & -2.553 & 0 & -4.843 & 1.239 & 1.239 \end{pmatrix} \tag{4.10}$$

in the present case.

The explicit form of $C(\boldsymbol{d})$ in the two-thirds power law (4.3) is obtained as

$$C(\boldsymbol{d}) = -\frac{3A_{300}{}^{1/3}}{A_{110}} \left(\frac{A_{001}}{2}\right)^{2/3} \tag{4.11}$$

from (3.6) with (3.7). On the right-hand side of this equation, $A_{001} = \boldsymbol{\xi}_1^\top B_\mathrm{c}^0 \boldsymbol{d}$ alone is a function of $\boldsymbol{d}$; and A_{300} and A_{110} are independent of $\boldsymbol{d}$. Therefore, the maximum of $|C(\boldsymbol{d})|$ with respect to $\boldsymbol{d}$ is achieved by $\boldsymbol{d}^*$ that maximizes $|A_{001}|$ under the constraint (4.4). From expression (4.9) for A_{001}, we see that such a $\boldsymbol{d}^*$ is parallel to $W B_\mathrm{c}^{0\top} \boldsymbol{\xi}_1$ (cf., Lemma 4.1 in §4.3.2 for the proof); that is,

$$\boldsymbol{d}^* = \frac{1}{\alpha} W B_\mathrm{c}^{0\top} \boldsymbol{\xi}_1, \tag{4.12}$$

or its negative, where α is a positive scalar defined in such a way that (4.4) is satisfied. Substituting $\boldsymbol{\xi}_1 = (1,0)^\top$, $W = I_8$, and (4.10) into (4.12) yields the worst imperfection pattern

$$\boldsymbol{d}^* = (0.28404, 0.26061, 0.28404, -0.26061, -0.56812, 0, -0.43592, 0.43592)^\top. \tag{4.13}$$

In this way, the worst imperfection pattern $\boldsymbol{d}^*$ has been computed by referring only to A_{001} in (4.9). Other coefficients, such as A_{300} and A_{110} in (4.11), need not be evaluated. Figure 4.3 illustrates the influence of the worst imperfection patterns $\pm\boldsymbol{d}^*$

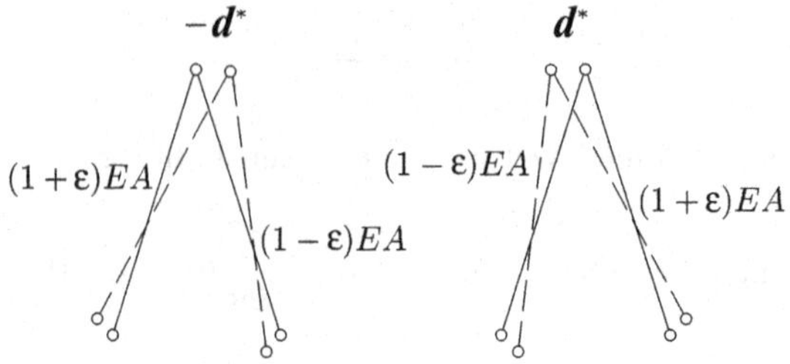

Fig. 4.3 The worst imperfection patterns of nonshallow truss arch. ——: perfect system; – – –: imperfect system.

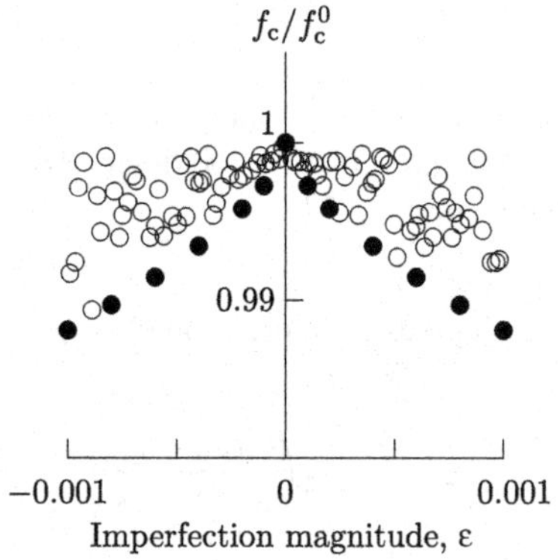

Fig. 4.4 f_c/f_c^0 versus ε relation for the nonshallow arch (unstable pitchfork bifurcation point). ○: random imperfections; ●: the worst imperfection.

on the configuration of the truss arch. These patterns apparently cause horizontal sway to trigger bifurcation.

Figure 4.4 shows an f_c/f_c^0 versus ε relation. The symbol (○) indicates the value of f_c computed from the equilibrium equation (4.1) of the arch for random imperfections, and (●) represents f_c computed for the worst imperfection in (4.13). The latter f_c value is smaller than the former one for the same value of ε, assessing the validity of the worst imperfection.

4.3 Theory of Worst Imperfection

The general framework for the worst imperfection against (structural) systems undergoing bifurcation is described in this section. The *imperfection sensitivity matrix*

$$B(\boldsymbol{u}, f, \boldsymbol{v}) = \left(\frac{\partial F_i}{\partial v_j} \,\middle|\, i = 1, \ldots, N, \, j = 1, \ldots, p \right) \tag{4.14}$$

introduced in (2.3) plays a vital role.

4.3.1 Formulation

Recall the governing equation (2.1):

$$\boldsymbol{F}(\boldsymbol{u}, f, \boldsymbol{v}) = \boldsymbol{0}. \tag{4.15}$$

The equilibrium paths and critical points are determined as functions in $\boldsymbol{v}$. We consider the critical point $(\boldsymbol{u}_{\mathrm{c}}^0, f_{\mathrm{c}}^0)$ on the fundamental path of the perfect system with $\boldsymbol{v} = \boldsymbol{v}^0$ that governs its critical load; the Jacobian matrix J of $\boldsymbol{F}$ is singular at $(\boldsymbol{u}_{\mathrm{c}}^0, f_{\mathrm{c}}^0, \boldsymbol{v}^0)$:

$$\det[J(\boldsymbol{u}_{\mathrm{c}}^0, f_{\mathrm{c}}^0, \boldsymbol{v}^0)] = 0.$$

For an imperfect system described by the imperfection parameter vector $\boldsymbol{v}$, the critical point moves to $(\boldsymbol{u}_{\mathrm{c}}, f_{\mathrm{c}})$, which is determined by (4.15) and

$$\det[J(\boldsymbol{u}_{\mathrm{c}}, f_{\mathrm{c}}, \boldsymbol{v})] = 0. \tag{4.16}$$

The imperfection is expressed in terms of the increment of $\boldsymbol{v}$ from the perfect state $\boldsymbol{v}^0$:

$$\varepsilon \boldsymbol{d} = \boldsymbol{v} - \boldsymbol{v}^0 \tag{4.17}$$

with an imperfection pattern vector $\boldsymbol{d}$ normalized as

$$\boldsymbol{d}^\top W^{-1} \boldsymbol{d} = 1, \tag{4.18}$$

where W is a positive-definite matrix to be specified in accordance with the engineering viewpoint (see Remark 4.1 below).

The asymptotic formulas for the incremental critical load $\widetilde{f_{\mathrm{c}}}$ of engineering interest are given by (3.6):

$$\widetilde{f_{\mathrm{c}}} \approx \begin{cases} C(\boldsymbol{d})\varepsilon & \text{at the limit point,} \\ C(\boldsymbol{d})|\varepsilon|^{1/2} & \text{at the transcritical bifurcation point (existent for } A_{200}A_{001}\varepsilon > 0\text{),} \\ C(\boldsymbol{d})\varepsilon^{2/3} & \text{at the pitchfork bifurcation point,} \end{cases} \tag{4.19}$$

with (3.7):

$$C(\boldsymbol{d}) = \begin{cases} -\dfrac{A_{001}}{A_{010}} & \text{at the limit point,} \\ -\left(\dfrac{4|A_{200}A_{001}|}{A_{110}{}^2 - 4A_{200}A_{020}}\right)^{1/2} & \text{at the transcritical point (existent for } A_{200}A_{001}\varepsilon > 0\text{),} \\ -\dfrac{3A_{300}{}^{1/3}}{A_{110}}\left(\dfrac{A_{001}}{2}\right)^{2/3} & \text{at the pitchfork bifurcation point.} \end{cases} \tag{4.20}$$

We formulate the problem of finding the worst imperfection as that of finding the imperfection pattern vector $\boldsymbol{d}$ that maximizes or minimizes the coefficient $C(\boldsymbol{d})$ (maximizes $|C(\boldsymbol{d})|$) under the normalization condition (4.18).

Remark 4.1. The weight W for the normalization (4.18) should be chosen to maintain (4.18) to be consistent with respect to physical dimensions. The unity on the right-hand side of (4.18) is to be understood as a nondimensional constant; the imperfection magnitude ε is also nondimensional. In the particular (but physical) case where W is diagonal, the physical dimensional consistency requires that the dimension of the ith diagonal entry W_{ii} of W be equal to that of d_i squared. □

4.3.2 Derivation

The worst imperfection is determined as follows. On the right-hand side of (4.20), A_{001} alone is a function of $\boldsymbol{d}$. By (2.71), we have

$$A_{001} = \boldsymbol{\xi}^\top B_\mathrm{c}^0 \boldsymbol{d}, \tag{4.21}$$

which is called the *imperfection coefficient*, where $\boldsymbol{\xi} = \boldsymbol{\xi}_1$ is a vector that satisfies $\boldsymbol{\xi}^\top J_\mathrm{c}^0 = \boldsymbol{0}^\top$, and

$$B_\mathrm{c}^0 = B(\boldsymbol{u}_\mathrm{c}^0, f_\mathrm{c}^0, \boldsymbol{v}^0)$$

is an $N \times p$ constant matrix, the imperfection sensitivity matrix (4.14) evaluated at $(\boldsymbol{u}_\mathrm{c}^0, f_\mathrm{c}^0, \boldsymbol{v}^0)$. The maximum of $|C(\boldsymbol{d})|$ with respect to $\boldsymbol{d}$ is, therefore, achieved by $\boldsymbol{d}$ that maximizes or minimizes $A_{001} = \boldsymbol{\xi}^\top B_\mathrm{c}^0 \boldsymbol{d}$. Throughout this chapter, it is assumed in harmony with (2.75) that $A_{001} \neq 0$ for some $\boldsymbol{d}$; that is,

$$\boldsymbol{\xi}^\top B_\mathrm{c}^0 \neq \boldsymbol{0}^\top. \tag{4.22}$$

We define

$$\alpha = (\boldsymbol{\xi}^\top B_\mathrm{c}^0 W B_\mathrm{c}^{0\top} \boldsymbol{\xi})^{1/2}, \tag{4.23}$$

and call it the *imperfection influence factor* in view of the following fact.

Lemma 4.1. *The maximum of $\boldsymbol{\xi}^\top B_\mathrm{c}^0 \boldsymbol{d}$ under the constraint $\boldsymbol{d}^\top W^{-1} \boldsymbol{d} = 1$ is equal to α in (4.23), attained by $\boldsymbol{d} = \boldsymbol{d}^*$, where*

$$\boldsymbol{d}^* = \frac{1}{\alpha} W B_\mathrm{c}^{0\top} \boldsymbol{\xi}. \tag{4.24}$$

The minimum of $\boldsymbol{\xi}^\top B_\mathrm{c}^0 \boldsymbol{d}$ under the constraint $\boldsymbol{d}^\top W^{-1} \boldsymbol{d} = 1$ is equal to $-\alpha$, attained by $\boldsymbol{d} = -\boldsymbol{d}^*$.

Proof. We decompose the weight matrix W as

$$W = V V^\top$$

and define a new imperfection vector

$$\overline{\boldsymbol{d}} = V^{-1}\boldsymbol{d}.$$

Then the constraint (4.18) becomes

$$\overline{\boldsymbol{d}}^{\top}\overline{\boldsymbol{d}} = 1, \tag{4.25}$$

and A_{001} in (4.21) becomes

$$A_{001} = \boldsymbol{\xi}^{\top} B_{\rm c}^{0}\boldsymbol{d} = \boldsymbol{\xi}^{\top} B_{\rm c}^{0} V\overline{\boldsymbol{d}}. \tag{4.26}$$

Consequently, A_{001} in (4.26) is maximized under the constraint (4.25) when $\overline{\boldsymbol{d}}$ is chosen to be a unit vector $\overline{\boldsymbol{d}}^{*}$ parallel to

$$(\boldsymbol{\xi}^{\top} B_{\rm c}^{0} V)^{\top} = V^{\top} B_{\rm c}^{0\,\top} \boldsymbol{\xi};$$

that is,

$$\overline{\boldsymbol{d}}^{*} = \frac{1}{\alpha} V^{\top} B_{\rm c}^{0\,\top} \boldsymbol{\xi}, \tag{4.27}$$

where

$$\alpha = [(V^{\top} B_{\rm c}^{0\,\top} \boldsymbol{\xi})^{\top}(V^{\top} B_{\rm c}^{0\,\top} \boldsymbol{\xi})]^{1/2} = (\boldsymbol{\xi}^{\top} B_{\rm c}^{0} W B_{\rm c}^{0\,\top} \boldsymbol{\xi})^{1/2}. \tag{4.28}$$

In the original variable, (4.27) is expressed as

$$\boldsymbol{d}^{*} = V\overline{\boldsymbol{d}}^{*} = \frac{1}{\alpha} W B_{\rm c}^{0\,\top} \boldsymbol{\xi},$$

which concludes the proof. □

In view of (4.20) and Lemma 4.1 above, we may say that $\boldsymbol{d} = \pm\boldsymbol{d}^{*}$ maximizes $|C(\boldsymbol{d})|$ under the normalization condition (4.18). More precisely, (4.20) and (4.21) show that $C(\boldsymbol{d})$ is minimized by

$$\boldsymbol{d} = \begin{cases} \mathrm{sign}(\varepsilon A_{010})\boldsymbol{d}^{*} & \text{at the limit point,} \\ \mathrm{sign}(\varepsilon A_{200})\boldsymbol{d}^{*} & \text{at the transcritical bifurcation point,} \\ \pm\boldsymbol{d}^{*} & \text{at the unstable pitchfork bifurcation point.} \end{cases} \tag{4.29}$$

Note that $A_{001} \geq 0$ for $\boldsymbol{d} = \boldsymbol{d}^{*}$ by (4.21) and (4.24).

It should be noted, however, that $\mathrm{sign}(\varepsilon A_{010})$ or $\mathrm{sign}(\varepsilon A_{200})$ is not easily obtained in practical situations. Instead of obtaining the sign, we may simply compute imperfect solution curves for both $\boldsymbol{d}^{*}$ and $-\boldsymbol{d}^{*}$ that approach the bifurcation point from downward ($\widetilde{f} < 0$) and, in turn, identify the worst imperfection as follows.

- For a limit point, $\widetilde{f}_{\rm c}$ is minimized for $\boldsymbol{d}^{*}$ and is maximized for $-\boldsymbol{d}^{*}$, or vice versa. Then that which minimizes $\widetilde{f}_{\rm c}$ serves as the worst imperfection.
- For a transcritical bifurcation point, a limit point exists on an imperfect curve for $\boldsymbol{d}^{*}$ in a neighborhood of the bifurcation point and no such point exists for

$-\boldsymbol{d}^*$, or vice versa. Consequently, that with the limit point serves as the worst imperfection.

Remark 4.2. When the governing equation $\boldsymbol{F}(\boldsymbol{u}, f, \boldsymbol{v})$ is derived from a total potential function $U(\boldsymbol{u}, f, \boldsymbol{v})$ as in (2.11), the imperfection sensitivity matrix is given by

$$B(\boldsymbol{u}, f, \boldsymbol{v}) = \frac{\partial^2 U(\boldsymbol{u}, f, \boldsymbol{v})}{\partial \boldsymbol{u}\, \partial \boldsymbol{v}}.$$

Several studies have been conducted to derive the worst imperfection using potential $U(\boldsymbol{u}, f, \boldsymbol{v})$. For example, the worst imperfection for (multiple) critical points was studied by Ho, 1974 [74] for a special case, as was reviewed by Koiter, 1976 [121]. The worst imperfection shape of structures was studied in a more general setting (e.g., Triantafyllidis and Peek, 1992 [196] and Peek and Triantafyllidis, 1992 [164]). □

4.4 Imperfection with Multiple Categories

In structural mechanics, the components of the imperfection parameter vector $\boldsymbol{v}$ (and those of $\boldsymbol{d}$) often represent physically different categories of variables, such as node location, member length, and member cross-sectional rigidity. In the design of structures, it is far more meaningful to restrict the imperfections in each category instead of controlling the imperfections of all categories altogether by a single constraint $\boldsymbol{d}^\top W^{-1} \boldsymbol{d} = 1$ in (4.4). Then it is logical to divide the components of $\boldsymbol{v}$ and, in turn, this constraint into independent multiple categories. The method explained in §4.3 can be readily adapted to such situations.

We partition the imperfection parameters into q categories as

$$\boldsymbol{v} = \begin{pmatrix} \boldsymbol{v}_1 \\ \vdots \\ \boldsymbol{v}_q \end{pmatrix}$$

and, accordingly, put

$$\boldsymbol{v}^0 = \begin{pmatrix} \boldsymbol{v}_1^0 \\ \vdots \\ \boldsymbol{v}_q^0 \end{pmatrix}, \qquad \boldsymbol{d} = \begin{pmatrix} \boldsymbol{d}_1 \\ \vdots \\ \boldsymbol{d}_q \end{pmatrix}, \qquad B_{\mathrm{c}}^0 = [(B_1)_{\mathrm{c}}^0, \ldots, (B_q)_{\mathrm{c}}^0].$$

Note that the imperfection pattern vector $\boldsymbol{d}$ and the imperfection sensitivity matrix B_{c}^0 are decomposed compatibly with the partitioning of $\boldsymbol{v}$. The imperfection patterns $\boldsymbol{d}_k$ $(k = 1, \ldots, q)$ are defined as

$$\varepsilon \boldsymbol{d}_k = \boldsymbol{v}_k - \boldsymbol{v}_k^0, \qquad k = 1, \ldots, q,$$

and normalized as

$$\boldsymbol{d}_k^\top W_k^{-1} \boldsymbol{d}_k = 1, \qquad k = 1,\dots,q, \tag{4.30}$$

with positive-definite matrices W_k $(k = 1,\dots,q)$.

For our analyses, we seek the imperfection pattern vector $\boldsymbol{d}$ that maximizes $|C(\boldsymbol{d})|$ subject to the constraint (4.30). By (4.20) this amounts to maximizing $|A_{001}|$ under the constraint (4.30). In view of the relation

$$A_{001} = \boldsymbol{\xi}^\top B_\mathrm{c}^0 \boldsymbol{d} = \sum_{k=1}^{q} \boldsymbol{\xi}^\top (B_k)_\mathrm{c}^0 \boldsymbol{d}_k$$

(cf., (4.21)), the problem of maximizing $|A_{001}|$ is decomposed into q independent problems of maximizing (or minimizing) $\boldsymbol{\xi}^\top (B_k)_\mathrm{c}^0 \boldsymbol{d}_k$ subject to $\boldsymbol{d}_k^\top W_k^{-1} \boldsymbol{d}_k = 1$ $(k = 1,\dots,q)$ in (4.30). Therefore, all the results obtained in §4.3 for one category of imperfections apply to each category, and the worst imperfection pattern for each category is given by

$$\boldsymbol{d}_k^* = \frac{1}{\alpha_k} W_k (B_k)_\mathrm{c}^{0\top} \boldsymbol{\xi}, \qquad k = 1,\dots,q, \tag{4.31}$$

where

$$\alpha_k = [\boldsymbol{\xi}^\top (B_k)_\mathrm{c}^0 W_k (B_k)_\mathrm{c}^{0\top} \boldsymbol{\xi}]^{1/2}, \qquad k = 1,\dots,q.$$

Then, for $\boldsymbol{d}^* = (\boldsymbol{d}_1^{*\top},\dots,\boldsymbol{d}_q^{*\top})^\top$, we have

$$A_{001} = \boldsymbol{\xi}^\top B_\mathrm{c}^0 \boldsymbol{d}^* = \sum_{k=1}^{q} \alpha_k. \tag{4.32}$$

The variable α_k represents the influence of the imperfection in the kth category on the critical load increment $\widetilde{f}_\mathrm{c}$; the kth category has a stronger influence on $\widetilde{f}_\mathrm{c}$ for larger α_k.

The value of $\widetilde{f}_\mathrm{c}$ is calculated from (4.20) and (4.32) as a combination of the effects from the q categories as

$$\widetilde{f}_\mathrm{c} \approx \begin{cases} C\left(\sum_{k=1}^{q} \alpha_k\right)\varepsilon & \text{at the limit point,} \\ C\left(\sum_{k=1}^{q} \alpha_k\right)^{1/2} |\varepsilon|^{1/2} & \text{at the transcritical point,} \\ C\left(\sum_{k=1}^{q} \alpha_k\right)^{2/3} \varepsilon^{2/3} & \text{at the pitchfork point,} \end{cases} \tag{4.33}$$

where C is a constant ($C = -1/[\boldsymbol{\xi}^\top (\partial \boldsymbol{F}/\partial f)_\mathrm{c}^0]$ for a limit point).

Remark 4.3. In our problem formulation the weight matrices W_k $(k = 1,\dots,q)$ are assumed to be given a priori. From the mathematical perspective, these matrices may be chosen arbitrarily as long as they are positive-definite. The choice of the weight matrices is expected to reflect design principles and technological constraints. For example, W_k may be chosen "small" if the imperfection $\boldsymbol{v}_k - \boldsymbol{v}_k^0$ in the kth category is expected to be small for some technological reason. Not surprisingly, the resulting worst imperfection pattern vectors $\boldsymbol{d}_k$ are substantially affected by the choice of W_k. It is noteworthy, however, that $(B_k)_c^0$ and $\boldsymbol{\xi}$ in (4.31) are independent of W_k. □

4.5 Examples of Worst Imperfection

As examples of the worst imperfection, we refer to truss structures.

4.5.1 *Truss Arches*

As examples of a limit point and an unstable pitchfork bifurcation point, we recall the shallow and nonshallow truss arches treated in §3.3.2 and §4.2, which are, respectively, shown in Fig. 4.5(a) and (b). Recall that the equilibrium of this arch under a vertical load f is described by (4.1):

$$\boldsymbol{F} = \begin{pmatrix} F_x \\ F_y \end{pmatrix} = EA \begin{pmatrix} \sum_{i=1}^{2} \frac{EA_i}{EA}\left(\frac{1}{L_i} - \frac{1}{\widehat{L}_i}\right)(x - x_i) \\ \sum_{i=1}^{2} \frac{EA_i}{EA}\left(\frac{1}{L_i} - \frac{1}{\widehat{L}_i}\right)(y - y_i) - f \end{pmatrix} = \begin{pmatrix} 0 \\ 0 \end{pmatrix} \tag{4.34}$$

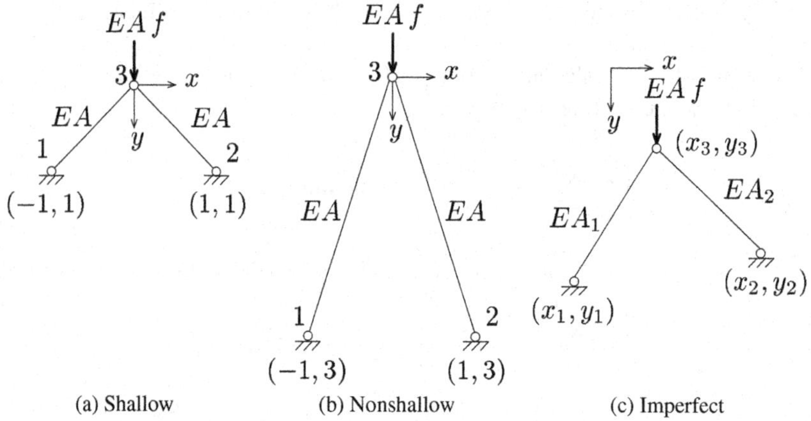

Fig. 4.5 Shallow and nonshallow truss arches.

(cf., §3.3.2 and §4.2 for notations). The imperfection parameter vector is chosen again as

$$\boldsymbol{v} = \left(\frac{x_1}{L^0}, \frac{y_1}{L^0}, \frac{x_2}{L^0}, \frac{y_2}{L^0}, \frac{x_3}{L^0}, \frac{y_3}{L^0}, \frac{EA_1}{EA}, \frac{EA_2}{EA}\right)^\top$$

($L^0 = 1$ is the unit scaling length), and

$$\boldsymbol{v}^0 = \begin{cases} (-1,1,1,1,0,0,1,1)^\top & \text{for the shallow truss arch,} \\ (-1,3,1,3,0,0,1,1)^\top & \text{for the nonshallow truss arch} \end{cases}$$

for the perfect system portrayed in Fig. 4.5(a) and (b).

For the present case, the imperfection parameter vector $\boldsymbol{v}$ might be divided into five categories. That is,

$$\boldsymbol{v} = \begin{pmatrix} \boldsymbol{v}_1 \\ \vdots \\ \boldsymbol{v}_5 \end{pmatrix},$$

where

$$\boldsymbol{v}_k = \left(\frac{x_k}{L^0}, \frac{y_k}{L^0}\right)^\top, \qquad k = 1,2,3;$$
$$\boldsymbol{v}_4 = \left(\frac{EA_1}{EA}\right), \qquad \boldsymbol{v}_5 = \left(\frac{EA_2}{EA}\right).$$

The weight matrices are set to

$$W_1 = W_2 = W_3 = \begin{pmatrix} 1 & 0 \\ 0 & 1 \end{pmatrix}, \qquad W_4 = W_5 = (1). \tag{4.35}$$

Shallow Arch (Limit Point)

The analysis of the shallow arch is recalled (cf., §3.3.2). Its curve of load versus displacement for the perfect case, shown as the solid curve in Fig. 4.6, has the limit point at $(x_c^0, y_c^0, f_c^0) = (0, 0.49018, 0.18740)$ with the relevant eigenvector of $\boldsymbol{\xi} = (0,-1)^\top$.

From (4.31) we computed the worst imperfection pattern vector

$$\boldsymbol{d}^* = \left(-\frac{1}{\sqrt{2}}, -\frac{1}{\sqrt{2}}, \frac{1}{\sqrt{2}}, -\frac{1}{\sqrt{2}}, 0, 1, -1, -1\right)^\top \tag{4.36}$$

using (4.35) and

$$B_c^0 = 10^{-1} EA \begin{pmatrix} -1.698 & 0.0695 & -1.698 & -0.0695 & -7.071 & 0 & -1.838 & 1.838 \\ 1.803 & 1.803 & -1.803 & 1.803 & 0 & -3.605 & 0.937 & 0.937 \end{pmatrix}.$$

As presented in Fig. 4.7(a), this imperfection pattern $\boldsymbol{d}^*$ makes the arch flatter. Consequently, it reduces its vertical stiffness.

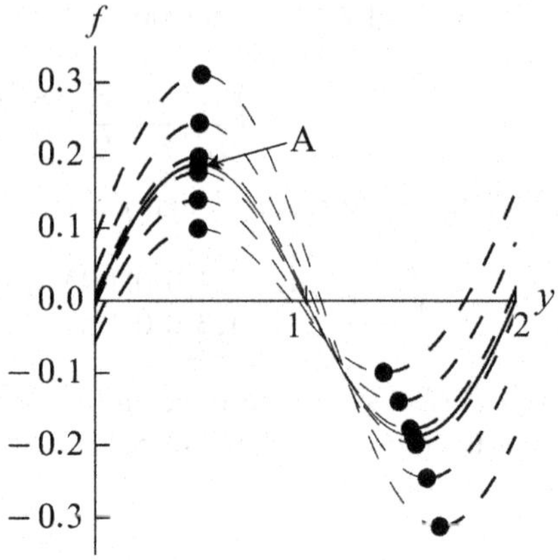

Fig. 4.6 Equilibrium paths (f versus y curves) for the shallow arch. ——: path for the perfect system; – – –: path for an imperfect system; thick line: stable; thin line: unstable; ●: limit point.

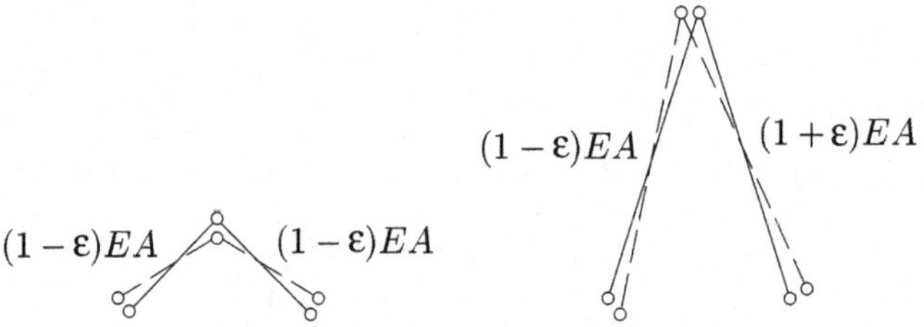

(a) Shallow arch

(b) Nonshallow arch

Fig. 4.7 The worst imperfection patterns for the truss arches. ——: perfect system; – – –: worst imperfect system.

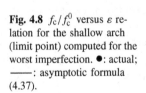

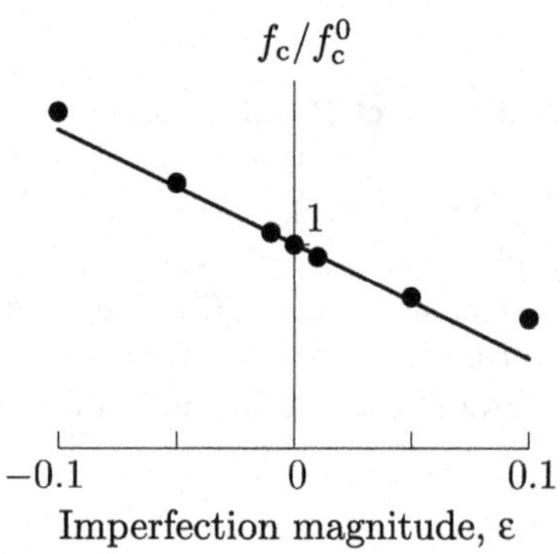

Fig. 4.8 f_c/f_c^0 versus ε relation for the shallow arch (limit point) computed for the worst imperfection. ●: actual; ——: asymptotic formula (4.37).

The increment $\widetilde{f_c}$ of f_c associated with $\boldsymbol{d}^*$ is computed from the first equation of (4.33) as

$$\widetilde{f_c} \approx -\frac{\alpha_1 + \cdots + \alpha_5}{\boldsymbol{\xi}^\top (\partial \boldsymbol{F}/\partial f)_c^0}\varepsilon = -1.0577\varepsilon, \tag{4.37}$$

where α_k $(k = 1,\ldots,5)$, denoting the influence of the imperfection in the kth category on $\widetilde{f_c}$, are

$$(\alpha_1,\ldots,\alpha_5) = EA(0.25491, 0.25491, 0.36050, 0.09370, 0.09370).$$

Since α_3 is the largest among α_k $(k = 1,\ldots,5)$, the imperfection of (x_3, y_3) reduces f_c most rapidly. The imperfections of EA_1 and EA_2 with the smallest α_ks, in contrast, are least influential.

Figure 4.8 portrays the relation between the normalized critical load f_c/f_c^0 and the magnitude of imperfection ε. The solid line shows the critical load estimated theoretically by (4.37), whereas (•) denotes that computed by the governing equation (4.34) for the worst imperfection $\boldsymbol{d}^*$ in (4.36). The theoretical estimate correlates well with the critical loads of imperfect arches, although discrepancy becomes greater as $|\varepsilon|$ increases owing to the asymptotic nature of expression (4.37), which includes only the first-order term.

The influence of x_1 and y_1 on the critical load f_c of the shallow arch is investigated by changing their values in the ranges $-1.5 \le x_1 \le -0.5$ and $0.5 \le y_1 \le 1.5$ at a fine mesh with other imperfections kept fixed. Figure 4.9(a) depicts the contour map of f_c. The numerals at the lines denote the f_c values and the arrows indicate the worst pattern vector of $(\tilde{x}_1, \tilde{y}_1)$ at each point (x_1, y_1) computed by (4.29), where $\tilde{x}_1$ and $\tilde{y}_1$, respectively, denote the increment of x_1 and y_1. The center $(x_1, y_1) = (-1, 1)$

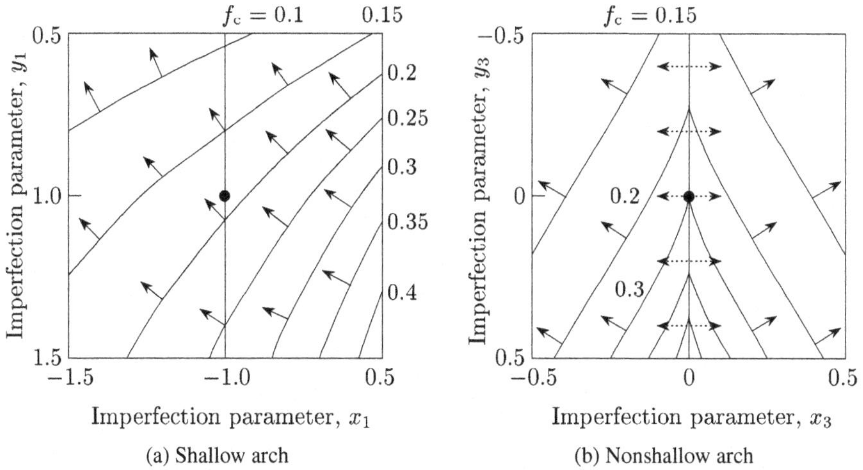

(a) Shallow arch (limit point)

(b) Nonshallow arch (unstable pitchfork bifurcation point)

Fig. 4.9 Contours of f_c. Solid arrow: the worst imperfection governed by the limit point; dotted arrow: the worst imperfection governed by the pitchfork bifurcation point; (•): perfect system.

of this figure denoted by (•) corresponds to the perfect shallow arch; the solid arrows indicate that the critical load is governed by a limit point. The theoretically computed imperfection pattern vectors are orthogonal to the contours, that is, directed toward the steepest decline of f_c. Such orthogonality verifies the validity of the theory of the worst imperfection.

Nonshallow Arch (Unstable Pitchfork Bifurcation Point)

We recall the nonshallow truss arch in Fig. 4.10(a) and its equilibrium paths in Fig. 4.10(b) (repeated from Fig. 4.1 in §4.2).

As shown, the equilibrium paths of the perfect system consist of a fundamental path and a bifurcated path that intersect at the unstable pitchfork bifurcation point. The worst imperfection pattern for this bifurcation point is computed from (4.31) as

$$\boldsymbol{d}^* = (0.73685, 0.67606, 0.73685, -0.67606, -1, 0, -1, 1)^\top \tag{4.38}$$

with the critical eigenvector $\boldsymbol{\xi} = (1, 0)^\top$, the imperfection sensitivity matrix B_c^0 in (4.10), and the weight matrices W_k in (4.35). The worst imperfection (4.38) accelerates the horizontal sway and triggers the bifurcation as shown in Fig. 4.7(b). The value of $\boldsymbol{d}^*$ of (4.38) computed for the five categories of weight matrices of (4.35) differs from $\boldsymbol{d}^*$ of (4.13) computed for the one category of weight matrix $W = I_8$.

The increment $\widetilde{f_c}$ of f_c related to $\boldsymbol{d}^*$ of (4.38) is computed as

$$\widetilde{f_c} = -C(\alpha_1 + \cdots + \alpha_5)^{2/3} \varepsilon^{2/3} \tag{4.39}$$

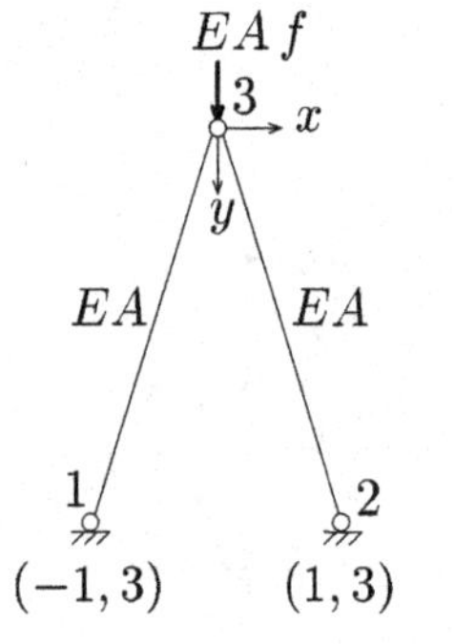

(a) Nonshallow truss arch

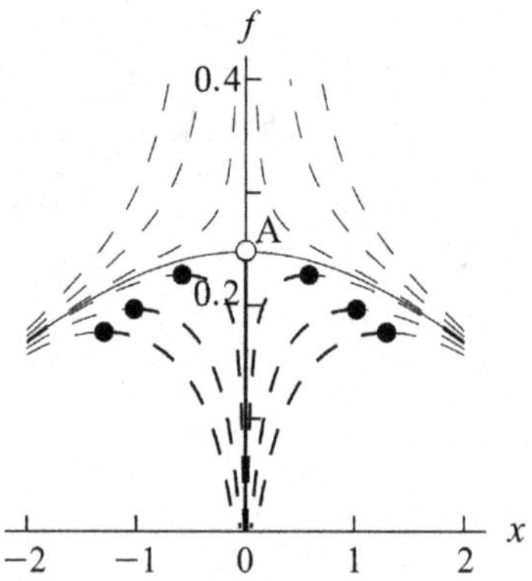

(b) Equilibrium paths

Fig. 4.10 Nonshallow truss arch and its equilibrium paths. ——: path for the perfect system; — — —: path for an imperfect system; thick line: stable; thin line: unstable; ○: bifurcation point; •: limit point.

from the last equation of (4.33), with a positive constant C independent of ε and

$$(\alpha_1, \dots, \alpha_5) = EA(0.04292, 0.04292, 0.06325, 0.04853, 0.04853).$$

Inasmuch as α_3 is the largest among α_k, the imperfection of (x_3, y_3) related to α_3 has the largest influence on f_c.

The influence of x_3 and y_3 on the critical load of the nonshallow arch is investigated by changing their values in the ranges $-0.5 \le x_3 \le 0.5$ and $-0.5 < y_3 < 0.5$. Figure 4.9(b) shows the contour map of f_c; the arrows are associated with the patterns of the worst imperfection computed by (4.29). The center $(x_3, y_3) = (0, 0)$ of this figure denoted by (•) corresponds to the perfect nonshallow arch. The dotted arrows on $x_3 = 0$ indicate that the critical loads are governed by bifurcation points. The solid arrows located elsewhere indicate that they are governed by limit points. Again, the theoretically computed imperfection pattern vectors are directed toward the steepest decline of f_c, assessing the validity of the worst imperfection.

4.5.2 Regular-Hexagonal Truss Dome

The regular-hexagonal truss dome in Fig. 4.11(a), with 24 members, is used here as a more realistic example. Its solution curves are depicted in Fig. 4.11(b). These curves are obtained for the vertical (z-directional) loadings of $0.5f$ applied at the crown node 0 and f applied at the other free nodes. The four critical points A, B, C, and D exist on the fundamental path. Point A is a limit point of the load f; B is an unstable pitchfork bifurcation point, which governs the critical load of the dome.

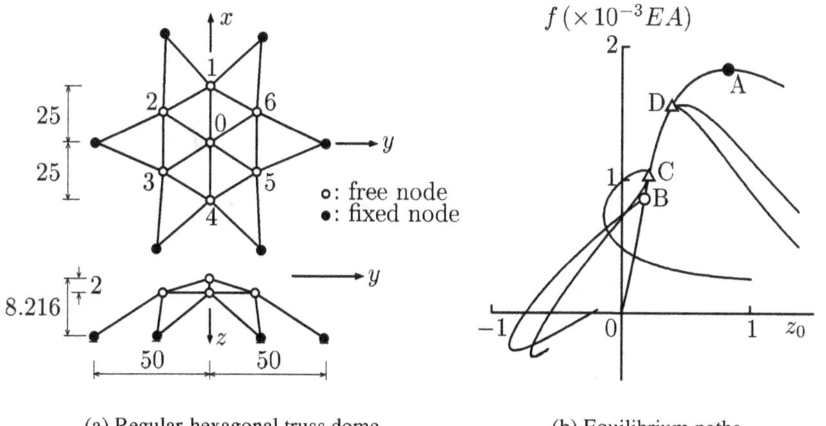

(a) Regular-hexagonal truss dome (b) Equilibrium paths

Fig. 4.11 Regular-hexagonal truss dome and its equilibrium paths. z_0: z-directional displacement of node 0; ○: pitchfork bifurcation point; △: double bifurcation point; •: limit point.

As imperfection parameters, we choose cross-sectional areas A_i $(i = 1,\ldots,24)$ of the 24 members of the dome, the perfect values of which are $A_i = A$. Weight matrices of two kinds are chosen as

$$W = W_1 = A^2 I_{24},$$
$$W = W_2 = \frac{A^2}{L^2}\mathrm{diag}({L_1}^2,\ldots,{L_{24}}^2),$$

where L_i $(i = 1,\ldots,24)$ are member lengths; and L is the representative member length.

Figure 4.12 shows the worst imperfection patterns $\boldsymbol{d}^*$ computed at the simple critical points A and B according to formula (4.24). The difference in the weight matrices has markedly altered the resulting patterns.

Limit point A has the worst imperfection pattern which is D_6-symmetric (cf., §8.2). The worst imperfection pattern of the simple bifurcation point B is less symmetric (D_3-symmetric). The worst imperfection patterns, accordingly, vary with the type of critical point. The symmetry of the worst imperfection pattern is investigated

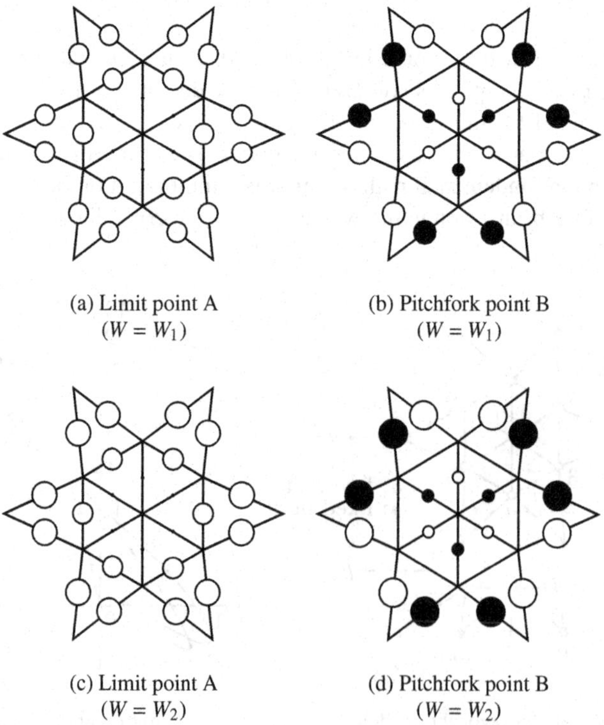

(a) Limit point A ($W = W_1$)

(b) Pitchfork point B ($W = W_1$)

(c) Limit point A ($W = W_2$)

(d) Pitchfork point B ($W = W_2$)

Fig. 4.12 Plane view of the worst imperfection patterns at simple critical points A and B of the regular-hexagonal truss dome computed for two different weight matrices $W = W_1$ and W_2. ●: positive component; ○: negative component; area of ○ or ●: magnitude of a component.

in Chapter 9. In particular, the worst imperfection for the double bifurcation points C and D is considered in §9.5.2 by extending the method to symmetric systems, for which multiple points appear generically.

Problems

4-1 Obtain the worst imperfection pattern for the truss arch shown in Fig. 4.13 for the same imperfection parameter vector used in §4.2.

4-2 Obtain the worst imperfection pattern for the propped cantilever in Fig. 2.1 in §2.3 for the imperfection parameter vector defined in (2.17) under the constraint of

$$\boldsymbol{d}_k^\top W_k{}^{-1}\boldsymbol{d}_k = 1, \qquad k = 1,2,3$$

with

$$\begin{aligned} \boldsymbol{d}_1 &= \left(\frac{x_1}{L^0}, \frac{y_1}{L^0}\right)^\top, \qquad & \boldsymbol{d}_2 = \left(\frac{x_2}{L^0}, \frac{y_2}{L^0}\right)^\top, \qquad & \boldsymbol{d}_3 = (\beta_1,\beta_2,\beta_3,\beta_4)^\top, \\ W_1 &= W_2 = I_2, & W_3 = I_4. & \end{aligned}$$

Answer: $\boldsymbol{d}^* = (-1,0,1,0,-1,0,0,0)^\top$. See Ikeda and Murota, 1990 [85] for details.

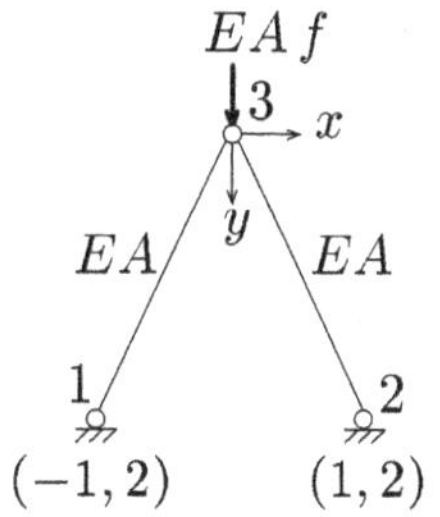

Fig. 4.13 Truss arch.

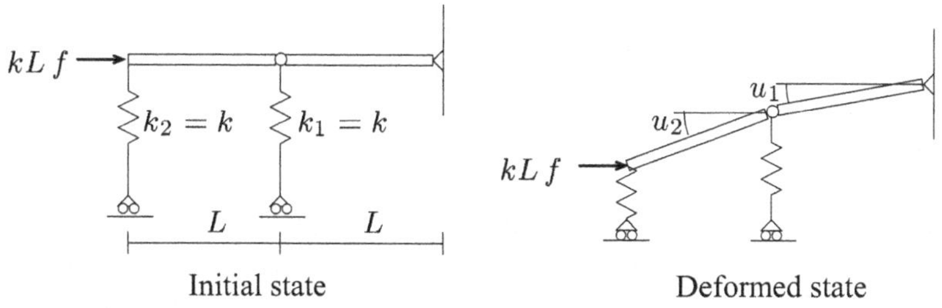

Fig. 4.14 Two-degree-of-freedom system.

4-3 Compute the fundamental and bifurcated paths of the two-degree-of-freedom system in Fig. 4.14 and obtain the worst imperfection pattern for imperfection parameters $k_i = k + \varepsilon d_i$ $(i = 1, 2)$.

4-4 Consider the equilibrium equation for a truss member that connects nodes i and j,

$$\begin{pmatrix} \boldsymbol{F}_i \\ \boldsymbol{F}_j \end{pmatrix} = EA\left(\frac{1}{L} - \frac{1}{\widehat{L}}\right)\begin{pmatrix} I_3 & -I_3 \\ -I_3 & I_3 \end{pmatrix}\begin{pmatrix} \boldsymbol{u}_i \\ \boldsymbol{u}_j \end{pmatrix},$$

where $\boldsymbol{F}_i$ and $\boldsymbol{F}_j$, respectively, represent the forces at nodes i and j; $\boldsymbol{u}_i$ and $\boldsymbol{u}_j$, respectively, represent the locations of nodes i and j;

$$L = \|\boldsymbol{u}_i^* - \boldsymbol{u}_j^*\|, \qquad \widehat{L} = \|\boldsymbol{u}_i - \boldsymbol{u}_j\|;$$

and $\boldsymbol{u}_i^*$ and $\boldsymbol{u}_j^*$, respectively, represent the initial locations of nodes i and j. Derive the imperfection sensitivity matrix for imperfections of initial locations.
Answer: See Ikeda and Murota, 1990 [86].

4-5 Compute the worst imperfection pattern for imperfections of initial locations for the regular-hexagonal truss dome in Fig. 4.11(a).
Answer: See Ikeda and Murota, 1990 [86].

Summary

- The worst imperfection pattern for simple critical points has been derived.
- The worst imperfection patterns of examples have been computed.

Chapter 5
Random Imperfection (I)

5.1 Introduction

The probabilistic variation of imperfections of structures has attracted considerable attention.[1] As first postulated by Bolotin, 1958 [15], the critical load f_c of a structure can be expressed as a function of several random imperfections d_i $(i = 1,\ldots,p)$; that is,

$$f_c = f_c(d_1,\ldots,d_p). \tag{5.1}$$

Evaluation of the probability density of f_c using (5.1) involves two major difficulties.

- The probability densities of several imperfections d_i $(i = 1,\ldots,p)$ are difficult to obtain.
- f_c is usually a complex nonlinear function, and is obtainable only implicitly using a sophisticated numerical code.

To resolve the first difficulty, the imperfections are often represented by normally distributed random variables.

Asymptotic theories of imperfections are used to tackle the second difficulty. At the expense of the asymptoticity assumption, the results obtained are quite general and simple, thus providing a deeper insight, as is true with the Koiter laws.[2]

In this chapter,[3] as a means to overcome the two difficulties associated with Bolotin's postulate in the form of (5.1), we offer a procedure to obtain the probability density function of f_c for systems with a number of imperfections with known probabilistic characteristics. The imperfection parameter vector $\boldsymbol{v}$ is expressed in the form of (2.2):

$$\boldsymbol{v} = \boldsymbol{v}^0 + \varepsilon \boldsymbol{d}.$$

---

[1] Several excellent textbooks explain this issue (e.g., Bolotin, 1969 [16], 1984 [17]; Elishakoff, 1983 [44]; Augusti, Barratta, and Casciati, 1984 [9]; Lindberg and Florence, 1987 [129]; Ben-Haim and Elishakoff, 1990 [14]; and Elishakoff, Lin, and Zhu, 1994 [47]).

[2] For example, Roorda and Hansen, 1972 [173] extended these laws to a single-mode, normally distributed imperfection.

[3] This chapter is based on Ikeda and Murota, 1991 [88], 1993 [89].

K. Ikeda and K. Murota, *Imperfect Bifurcation in Structures and Materials*, Applied Mathematical Sciences 149, DOI 10.1007/978-1-4419-7296-5_5,

Recall from (3.6) the imperfection sensitivity law, for example,

$$\widetilde{f_c} \approx C(\boldsymbol{d})\varepsilon^{2/3}$$

for a pitchfork bifurcation point, which expresses the variation of the critical load as a function of $\boldsymbol{d}$. Then the probabilistic variation of the critical load f_c is to be formulated when $\boldsymbol{d}$ varies with known probabilistic characteristics and the magnitude ε is fixed to be a small positive constant. Specifically, we emphasize the case in which $\boldsymbol{d}$ is subject to a multivariate normal distribution, as it is known to be realistic. It turns out that the distribution of f_c for a simple bifurcation point is neither a normal distribution nor the Weibull distribution,[4] but is close to the Weibull distribution.

This chapter is organized as follows.

- For an imperfection pattern vector $\boldsymbol{d}$ subject to a multivariate normal distribution, the explicit form of the probability density function of the critical load is obtained in §5.2.
- Theoretical and semiempirical evaluation procedures for determining the parameters for probability density functions are presented in §5.3.
- The distribution of the minimum value of the critical load is investigated in §5.4.
- The proposed procedure is applied to examples in §5.5.
- Scaling factors for the minimum value distribution are obtained in §5.6, the appendix of this chapter.

In this chapter, emphasis is placed on the theoretical development followed by simple and pedagogic examples; in contrast, more realistic examples are presented in Chapter 10 after further theoretical development on systems with group symmetry.

5.2 Probability Density Functions of Critical Loads

An asymptotic theory for random imperfections can be developed as a natural continuation of the results presented in the previous chapters. The objective of this section is to derive the probabilistic properties of critical loads under the assumption that the imperfections are normally distributed.

The asymptotic behaviors of the increment (increase or decrease) $\widetilde{f_c}$ of the critical load f_c of imperfect systems for simple critical points are expressed as

$$\widetilde{f_c} = f_c - f_c^0 \approx C(\boldsymbol{d})\varepsilon^{\rho} \tag{5.2}$$

when $\varepsilon > 0$ is small (cf., (3.6)). The increment $\widetilde{f_c}$ is characterized by the exponent ρ and the coefficient $C(\boldsymbol{d})$, the explicit forms of which for simple critical points[5] are given by (3.7):

---

[4] See, for example, Weibull, 1939 [206], 1951 [207].

[5] For a transcritical bifurcation point, the increment $\widetilde{f_c}$ in (5.2) and, hence, all the results in this section are defined as the conditional distribution, given that a limit point exists on the fundamental path of the imperfect system (i.e., $A_{200}A_{001}\varepsilon > 0$).

$$\begin{cases} \rho = 1, \quad C(\boldsymbol{d}) = -C_0 a & \text{at the limit point,} \\ \rho = 1/2, \ C(\boldsymbol{d}) = -C_0 |a|^{1/2} & \text{at the transcritical point,} \\ \rho = 2/3, \ C(\boldsymbol{d}) = -C_0\, a^{2/3} & \text{at the unstable pitchfork point.} \end{cases} \tag{5.3}$$

Therein, C_0 is a positive constant; and the coefficients $C(\boldsymbol{d})$ depend on $\boldsymbol{d}$ through the single variable a, which is the imperfection coefficient given as

$$a \equiv A_{001} = \boldsymbol{\xi}^\top B_{\mathrm{c}}^0 \boldsymbol{d} = \sum_{i=1}^{p} c_i d_i \tag{5.4}$$

by (2.71). Here $\boldsymbol{\xi} = \boldsymbol{\xi}_1$ is the critical (left) eigenvector and

$$(c_1, \ldots, c_p) = \boldsymbol{\xi}^\top B_{\mathrm{c}}^0. \tag{5.5}$$

5.2.1 Imperfection Coefficient

Given the joint probability density function of d_i $(i = 1, \ldots, p)$, it is possible to calculate the probability density function of the imperfection coefficient a from (5.4). Subsequently, a simple transformation from a to the critical load f_{c}, through (5.2) with (5.3), yields the probability density function of f_{c}, as shown below.

The probabilistic behavior of f_{c} is investigated when the imperfection $\boldsymbol{v} - \boldsymbol{v}^0 = \varepsilon \boldsymbol{d}$ is subject to the normal distribution $\mathrm{N}(\boldsymbol{0}, \varepsilon^2 W)$ with mean $\boldsymbol{0}$ and variance–covariance matrix $\varepsilon^2 W$, where W is a positive-definite symmetric matrix.

The following lemma gives the probabilistic variation of the imperfection coefficient a.

Lemma 5.1. *If $\boldsymbol{d} \sim \mathrm{N}(\boldsymbol{0}, W)$, then the imperfection coefficient a is subject to a normal distribution $\mathrm{N}(0, \widetilde{\sigma}^2)$ with mean 0 and variance*

$$\widetilde{\sigma}^2 = \boldsymbol{\xi}^\top B_{\mathrm{c}}^0 W B_{\mathrm{c}}^{0\top} \boldsymbol{\xi}. \tag{5.6}$$

Proof. First, note that a is subject to a normal distribution, since it is a linear combination (5.4) of normal variates $d_1, \ldots, d_p$. The mean of a is equal to 0, since

$$\overline{a} = \overline{\sum_{i=1}^{p} c_i d_i} = \sum_{i=1}^{p} c_i \overline{d_i} = 0,$$

where $\overline{\cdot}$ denotes the average (expected value) of the relevant variable. The variance $\widetilde{\sigma}^2$ of a is evaluated as

$$\widetilde{\sigma}^2 = \overline{a^2} - \overline{a}^2 = \overline{\left(\sum_{i=1}^{p} c_i d_i\right)\times\left(\sum_{j=1}^{p} c_j d_j\right)} = \overline{\sum_{i=1}^{p}\sum_{j=1}^{p} c_i d_i c_j d_j}$$

$$= \sum_{i=1}^{p}\sum_{j=1}^{p} c_i \overline{d_i d_j} c_j = \sum_{i=1}^{p}\sum_{j=1}^{p} c_i W_{ij} c_j = \boldsymbol{\xi}^{\top} B_{\mathrm{c}}^{0} W B_{\mathrm{c}}^{0\top} \boldsymbol{\xi}$$

by $W = (\overline{d_i d_j} \mid i, j = 1, \ldots, p)$ and (5.5). □

Inasmuch as $a \sim \mathrm{N}(0, \widetilde{\sigma}^2)$, a normalized variable

$$\widetilde{a} = a/\widetilde{\sigma}$$

is subject to the standard normal distribution N(0, 1), the probability density function of which is expressed as

$$\phi_{\mathrm{N}}(t) = \frac{1}{\sqrt{2\pi}} \exp\left(\frac{-t^2}{2}\right), \qquad -\infty < t < \infty. \tag{5.7}$$

5.2.2 Normalized Critical Load

With reference to (5.2) and (5.3), a *normalized critical load* (increment)

$$\zeta = \frac{\widetilde{f}_{\mathrm{c}}}{\widehat{C}} = \begin{cases} -\widetilde{a} & \text{at the limit point } (\rho = 1), \\ -|\widetilde{a}|^{1/2} & \text{at the transcritical point } (\rho = 1/2), \\ -\widetilde{a}^{2/3} & \text{at the unstable pitchfork point } (\rho = 2/3) \end{cases} \tag{5.8}$$

is introduced, where

$$\widehat{C} = C_0(\widetilde{\sigma}\varepsilon)^{\rho} \tag{5.9}$$

with $\rho = 1$, 1/2, or 2/3 as in (5.8). Then the formula

$$\phi_{\zeta}(\zeta) = \begin{cases} \phi_{\mathrm{N}}(\widetilde{a})\left|\dfrac{d\widetilde{a}}{d\zeta}\right|, & -\infty < \zeta < \infty \quad \text{at the limit point,} \\ 2\phi_{\mathrm{N}}(\widetilde{a})\left|\dfrac{d\widetilde{a}}{d\zeta}\right|, & -\infty < \zeta < 0 \quad \text{at the transcritical point or unstable pitchfork point,} \end{cases} \tag{5.10}$$

yields the probability density function of ζ:

$$\phi_\zeta(\zeta) = \begin{cases} \dfrac{1}{\sqrt{2\pi}} \exp\left(\dfrac{-\zeta^2}{2}\right), & -\infty < \zeta < \infty \quad \text{at the limit point,} \\ \dfrac{4|\zeta|}{\sqrt{2\pi}} \exp\left(\dfrac{-\zeta^4}{2}\right), & -\infty < \zeta < 0 \quad \text{at the transcritical bifurcation point,} \\ \dfrac{3|\zeta|^{1/2}}{\sqrt{2\pi}} \exp\left(\dfrac{-|\zeta|^3}{2}\right), & -\infty < \zeta < 0 \quad \text{at the unstable pitchfork point.} \end{cases} \tag{5.11}$$

The cumulative distribution function $\Phi_\zeta(\zeta)$ of ζ is obtained by

$$\Phi_\zeta(\zeta) = \int_{-\infty}^{\zeta} \phi_\zeta(\zeta)\, d\zeta. \tag{5.12}$$

Then the *reliability function* $R_\zeta(\zeta)$ of ζ is evaluated by

$$R_\zeta(\zeta) = 1 - \Phi_\zeta(\zeta), \tag{5.13}$$

which stands for the probability of the normalized critical load exceeding the designated value ζ.

The probabilistic properties of the normalized critical load ζ at simple critical points are presented in Table 5.1, including the reliability function $R_\zeta(\zeta)$, the *expected value* $\mathrm{E}[\zeta]$, the *variance* $\mathrm{Var}[\zeta]$, and so on. Therein, $\Gamma(\cdot)$ denotes the *gamma function* and

$$\Phi_\mathrm{N}(\zeta) = \int_{-\infty}^{\zeta} \frac{1}{\sqrt{2\pi}} \exp\left(\frac{-\zeta^2}{2}\right) d\zeta \tag{5.14}$$

is the cumulative distribution function of the standard normal distribution $\mathrm{N}(0,1)$.

The probability density functions $\phi_\zeta(\zeta)$ and the reliability functions $R_\zeta(\zeta)$ are depicted, respectively, in Fig. 5.1(a) and (b) for the three types of critical points. Note that $\zeta = 0$ corresponds to the critical load $f = f_\mathrm{c}^0$ for the perfect system. For

Table 5.1 Statistical properties of the normalized critical load ζ of (5.8). $\Phi_\mathrm{N}(\zeta)$: cumulative distribution function (5.14) of the standard normal distribution $\mathrm{N}(0,1)$

Type of Points	Limit	Transcritical	Unstable Pitchfork
Range of ζ	$-\infty < \zeta < \infty$	$-\infty < \zeta < 0$	$-\infty < \zeta < 0$
$\phi_\zeta(\zeta)$	$\dfrac{1}{\sqrt{2\pi}} \exp\left(\dfrac{-\zeta^2}{2}\right)$	$\dfrac{4\lvert\zeta\rvert}{\sqrt{2\pi}} \exp\left(\dfrac{-\zeta^4}{2}\right)$	$\dfrac{3\lvert\zeta\rvert^{1/2}}{\sqrt{2\pi}} \exp\left(\dfrac{-\lvert\zeta\rvert^3}{2}\right)$
$\Phi_\zeta(\zeta)$	$\Phi_\mathrm{N}(\zeta)$	$2\Phi_\mathrm{N}(-\zeta^2)$	$2\Phi_\mathrm{N}(-\lvert\zeta\rvert^{3/2})$
$R_\zeta(\zeta)$	$1-\Phi_\mathrm{N}(\zeta)$	$1-2\Phi_\mathrm{N}(-\zeta^2)$	$1-2\Phi_\mathrm{N}(-\lvert\zeta\rvert^{3/2})$
$\mathrm{E}[\zeta]$	0	$\dfrac{-2^{3/4}}{\sqrt{2\pi}}\Gamma(3/4) = -0.822$	$\dfrac{-2^{5/6}}{\sqrt{2\pi}}\Gamma(5/6) = -0.802$
$\mathrm{Var}[\zeta]$	1^2	0.349^2	0.432^2

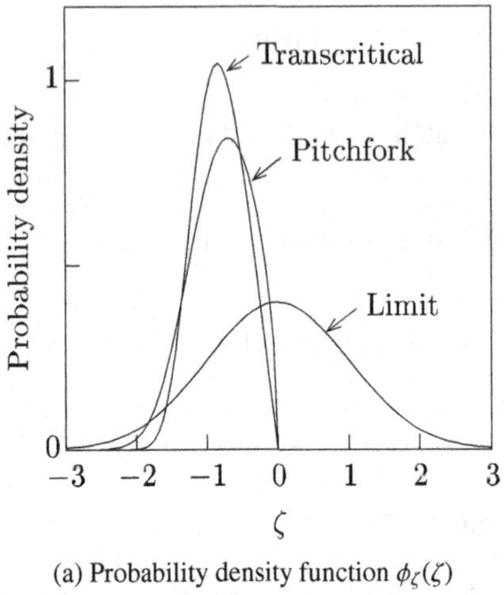

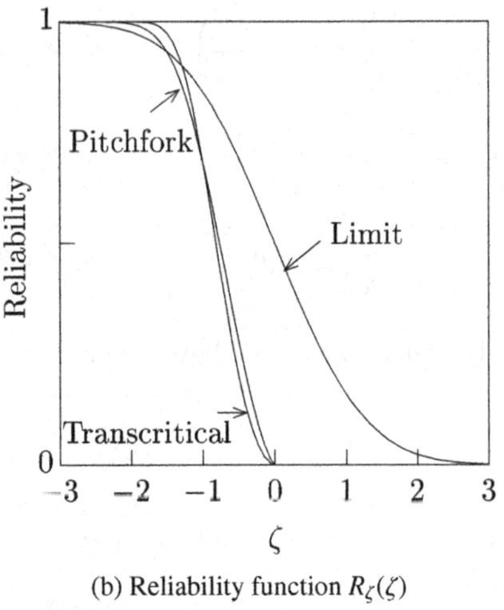

Fig. 5.1 Probability density function $\phi_\zeta(\zeta)$ and reliability function $R_\zeta(\zeta)$.

bifurcation points the reliability of the system is nullified at $\zeta = 0$ ($R_\zeta(0) = 0$). For the appropriately normalized critical load ζ, its probability density function $\phi_\zeta(\zeta)$ is independent of individual systems and is unique for the critical point of each type. Details of the systems in question do not affect the form of this function itself but only the values of f_c^0 and $\widehat{C}$, which account for linear scaling of the variable.

5.2.3 Critical Load

Transforming the formula (5.11) for the normalized critical load by (5.8) yields the probability density function of the critical load f_c:

$$\phi_{f_c}(f_c) = \begin{cases} \dfrac{1}{\sqrt{2\pi}\widehat{C}} \exp\left[\dfrac{-1}{2}\left(\dfrac{f_c - f_c^0}{\widehat{C}}\right)^2\right], & -\infty < f_c < \infty \\ & \text{at the limit point,} \\ \dfrac{4|f_c - f_c^0|}{\sqrt{2\pi}\widehat{C}^2} \exp\left[\dfrac{-1}{2}\left(\dfrac{f_c - f_c^0}{\widehat{C}}\right)^4\right], & -\infty < f_c < f_c^0 \\ & \text{at the transcritical point,} \\ \dfrac{3|f_c - f_c^0|^{1/2}}{\sqrt{2\pi}\widehat{C}^{3/2}} \exp\left[\dfrac{-1}{2}\left|\dfrac{f_c - f_c^0}{\widehat{C}}\right|^3\right], & -\infty < f_c < f_c^0 \\ & \text{at the unstable pitchfork point.} \end{cases} \tag{5.15}$$

The mean $\mathrm{E}[f_c]$ of f_c is computed as

$$\mathrm{E}[f_c] = f_c^0 + \mathrm{E}[\zeta]\widehat{C} = \begin{cases} f_c^0 & \text{at the limit point,} \\ f_c^0 - 0.822\widehat{C} & \text{at the transcritical point,} \\ f_c^0 - 0.802\widehat{C} & \text{at the unstable pitchfork point,} \end{cases} \tag{5.16}$$

and the variance $\mathrm{Var}[f_c]$ of f_c as

$$\mathrm{Var}[f_c] = \mathrm{Var}[\zeta]\widehat{C}^2 = \begin{cases} \widehat{C}^2 & \text{at the limit point,} \\ (0.349\widehat{C})^2 & \text{at the transcritical point,} \\ (0.432\widehat{C})^2 & \text{at the unstable pitchfork point.} \end{cases} \tag{5.17}$$

From Table 5.1, the reliability function of the critical load f_c becomes

$$R_{f_c}(f_c) = \begin{cases} 1 - \Phi_{\mathrm{N}}\left(\dfrac{f_c - f_c^0}{\widehat{C}}\right), & -\infty < f_c, < \infty \quad \text{at the limit point,} \\ 1 - 2\Phi_{\mathrm{N}}\left(-\left(\dfrac{f_c - f_c^0}{\widehat{C}}\right)^2\right), & -\infty < f_c < f_c^0 \\ & \qquad \text{at the transcritical point,} \\ 1 - 2\Phi_{\mathrm{N}}\left(-\left|\dfrac{f_c - f_c^0}{\widehat{C}}\right|^{3/2}\right), & -\infty < f_c < f_c^0 \\ & \qquad \text{at the unstable pitchfork point.} \end{cases} \tag{5.18}$$

Remark 5.1. The key mathematical fact in the above argument is that the imperfection coefficient a is normally distributed as a consequence of the assumed normality of $\boldsymbol{d}$. Even if $\boldsymbol{d}$ is not normally distributed, however, the imperfection coefficient a can often be regarded as being normally distributed when p is large; recall that a is a weighted sum of p imperfections and that the central limit theorem (Kendall and Stuart, 1977 [115]) says, roughly, that the distribution of the sum of many random variables can be approximated by a normal distribution under a fairly mild condition. Therefore, the obtained formulas will serve as reasonable approximations even when the assumed normality of $\boldsymbol{d}$ is not the case in the strict sense of the word. □

Remark 5.2. This section considered only the case where $\boldsymbol{d}$ is subject to a multivariate normal distribution. The distribution of $\boldsymbol{d}$, of course, is dependent on cases, and the case of uniform distribution ($\boldsymbol{d}$ is uniformly distributed on the surface defined by $\boldsymbol{d}^\top W^{-1}\boldsymbol{d} = 1$ for some weight matrix W) is studied in Ikeda and Murota, 1991 [88] and Murota and Ikeda, 1992 [144]. □

Remark 5.3. A subtle point of our analysis of random imperfection is explained here in relation to the universal unfolding

$$G(w, \widetilde{f}, \beta_1, \beta_2) = w^3 + \widetilde{f}w + \beta_1 + \beta_2 w^2$$

for an unstable pitchfork point introduced in Remark 2.11 in §2.5.3. We have $\boldsymbol{v} = (\beta_1, \beta_2)^\top = \varepsilon\boldsymbol{d}$ with $\boldsymbol{v}^0 = (0,0)^\top$ and, we may assume, for example, that (β_1, β_2) is subject to the normal distribution $\mathrm{N}(0, \varepsilon^2 I_2)$.

Our analyses are based on the asymptotic sensitivity law (5.2) for the increment $\widetilde{f}_c$ of the critical load f_c. Strictly interpreted, this asymptotic law is valid in a situation where ε tends to zero with direction $\boldsymbol{d}$ kept fixed. However, this is not the case in the present problem formulation where ε is kept fixed and (β_1,β_2), being subject to $\mathrm{N}(0,\varepsilon^2 I_2)$, is distributed axisymmetrically around $(0,0)$. Furthermore, our analysis ignores the region K of (β_1,β_2) in (2.94) where a kink appears on the fundamental path (cf., Fig. 2.10 in §2.5.3). Thus, the present analysis is admittedly lacking in mathematical rigor. It is hoped that the results are useful in engineering applications. □

5.3 Evaluation of Probability Density Functions

Theoretical and semiempirical evaluation procedures of the probability density function of the critical load are presented. These procedures are distinguished by the manner in which the parameter $\widehat{C}$ of (5.9) in the probability density function (5.15) is determined.

Theoretical Evaluation Procedure

The theoretical evaluation procedure is applicable when the imperfection sensitivity matrix B_{c}^0 in (2.10) can be obtained analytically or numerically. Then the value of $\widetilde{\sigma}$ can be computed by formula (5.6). Furthermore, we can evaluate C_0 using (5.2) and (5.3) by obtaining the critical load f_{c} from the governing equation (2.1) for a given imperfection pattern $\boldsymbol{d}$. Consequently, the theoretical evaluation of the probability density function is simple and straightforward.

Semiempirical Evaluation Procedure

The semiempirical evaluation procedure is suggested for use when B_{c}^0 cannot be computed, as is usually the case with experiments and nonlinear analyses. For a series of random imperfection patterns $\boldsymbol{d}$, taken from a known normal distribution, the critical loads f_{c} of a system are evaluated by observing or by solving the governing equation (2.1). Then the *sample mean* $\mathrm{E}_{\mathrm{sample}}[f_{\mathrm{c}}]$ and the *sample variance* $\mathrm{Var}_{\mathrm{sample}}[f_{\mathrm{c}}]$ of f_{c} are computed. Equating $\mathrm{E}_{\mathrm{sample}}[f_{\mathrm{c}}]$ with $\mathrm{E}[f_{\mathrm{c}}]$ of (5.16) and $\mathrm{Var}_{\mathrm{sample}}[f_{\mathrm{c}}]$ with $\mathrm{Var}[f_{\mathrm{c}}]$ of (5.17) produces an estimate for the (unknown) critical load f_{c}^0 for the perfect system

$$f_{\mathrm{c}}^0 = \begin{cases} \mathrm{E}_{\mathrm{sample}}[f_{\mathrm{c}}] & \text{at the limit point,} \\ \mathrm{E}_{\mathrm{sample}}[f_{\mathrm{c}}] + 2.35(\mathrm{Var}_{\mathrm{sample}}[f_{\mathrm{c}}])^{1/2} & \text{at the transcritical point,} \\ \mathrm{E}_{\mathrm{sample}}[f_{\mathrm{c}}] + 1.86(\mathrm{Var}_{\mathrm{sample}}[f_{\mathrm{c}}])^{1/2} & \text{at the unstable pitchfork point,} \end{cases} \quad (5.19)$$

and an estimate for the (unknown) variable

$$\widehat{C} = \begin{cases} (\mathrm{Var}_{\mathrm{sample}}[f_{\mathrm{c}}])^{1/2} & \text{at the limit point,} \\ (\mathrm{Var}_{\mathrm{sample}}[f_{\mathrm{c}}])^{1/2}/0.349 & \text{at the transcritical point.} \\ (\mathrm{Var}_{\mathrm{sample}}[f_{\mathrm{c}}])^{1/2}/0.432 & \text{at the unstable pitchfork point.} \end{cases} \tag{5.20}$$

Substitution of the values of $\widehat{C}$ and f_{c}^0 into (5.15) yields the probability density function $\phi_{f_{\mathrm{c}}}(f_{\mathrm{c}})$ of the critical load f_{c}. This *semiempirical procedure* is suited for practical use in experimentation and numerical analysis.

5.4 Distribution of Minimum Values

Knowledge of statistical properties of the minimum critical load achieved by a series of random imperfections $\boldsymbol{d}$ is of great assistance in making a sound engineering judgment. This section presents a study of the distribution of the minimum critical load as an application of standard results in the theory of extreme order statistics.

Let ζ_K be the minimum value of the normalized critical load ζ of (5.8) attained by K independent random imperfections, say, $\zeta_K = \min(\zeta^{(1)}, \ldots, \zeta^{(K)})$. Since $\zeta_K > \zeta$ if and only if $\zeta^{(i)} > \zeta$ for $i = 1, \ldots, K$, the cumulative distribution function Φ_K of the minimum value ζ_K satisfies

$$1 - \Phi_K(\zeta) = \Pr\{\zeta_K > \zeta\} = \prod_{i=1}^{K} \Pr\{\zeta^{(i)} > \zeta\} = (1 - \Phi_\zeta(\zeta))^K,$$

where Φ_ζ is the cumulative distribution function for ζ given by (5.12). This shows

$$\Phi_K(\zeta) = 1 - (1 - \Phi_\zeta(\zeta))^K. \tag{5.21}$$

Accordingly, the probability density function $\phi_K(\zeta)$ of ζ_K is given as

$$\phi_K(\zeta) = (1 - \Phi_\zeta(\zeta))^{K-1} \phi_\zeta(\zeta). \tag{5.22}$$

The probability density function $\phi_K(\zeta)$ is portrayed in Fig. 5.2(a) for various values of K for a transcritical bifurcation point (see Table 5.1 in §5.2.2 for the explicit form of $\phi_\zeta(\zeta)$); ζ_K tends to $-\infty$ as $K \to +\infty$. Consequently, the distribution of ζ_K must be determined along with an appropriate scaling of ζ_K. It is demonstrated that $(\zeta_K - c_K)/d_K$ indeed converges (in law) with an appropriate choice of c_K and d_K (see Fig. 5.2(b)).

The general theory of extreme order statistics provides us with a useful result to determine the asymptotic form of Φ_K as $K \to +\infty$. The following lemma[6] states that the limit distribution of the minimum value, with an appropriate scaling, is the *double exponential distribution* or the *Gumbel distribution*, which has cumulative distribution function $1 - \exp(-\mathrm{e}^x)$ in (5.27) and density function $\exp(x - \mathrm{e}^x)$.

[6] See Theorem 2.1.6 of Galambos, 1978 [54]; also Kendall and Stuart, 1977 [115].

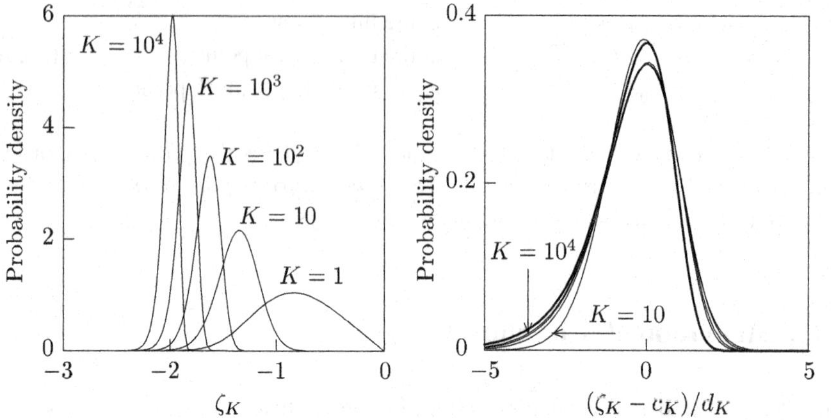

(a) Probability density function of ζ_K (b) Probability density function of $(\zeta_K - c_K)/d_K$

Fig. 5.2 Probability density functions of minimum critical loads ζ_K attained by K independent random imperfections for a transcritical bifurcation point (c_K and d_K in (5.29) are computed by (5.30) and (5.31)). The distribution of $(\zeta_K - c_K)/d_K$ in (b) is convergent (in law) to the limit probability density function $\exp(x - \mathrm{e}^x)$ expressed by the thick line; the curve for the probability density for $K = 1$ is omitted in (b).

Lemma 5.2. *Let Φ be a cumulative distribution function such that*[7]

$$\inf\{x \mid \Phi(x) > 0\} = -\infty, \tag{5.23}$$

$$\int_{-\infty}^{a} \Phi(y)\,dy < +\infty \qquad \text{for some } a \in \mathbb{R}, \tag{5.24}$$

$$\lim_{t\to-\infty} \frac{\Phi(t + xr(t))}{\Phi(t)} = \mathrm{e}^x, \qquad x \in \mathbb{R}, \tag{5.25}$$

where

$$r(t) = \frac{1}{\Phi(t)} \int_{-\infty}^{t} \Phi(y)\,dy, \qquad t \in \mathbb{R}. \tag{5.26}$$

Then the cumulative distribution function Φ_K of the minimum among K independent samples from the distribution Φ satisfies

$$\lim_{K\to+\infty} \Phi_K(c_K + d_K x) = 1 - \exp(-\mathrm{e}^x), \tag{5.27}$$

where

$$c_K = \sup\{x \mid \Phi(x) \le 1/K\}, \qquad d_K = r(c_K). \tag{5.28}$$

A straightforward application of Lemma 5.2 above to (5.12) with (5.11) yields the asymptotic distribution of the minimum of K normalized critical loads; that is,

[7] Here "inf" denotes the *infimum* of a set of numbers, which is defined to be the largest number that is not larger than any number in the set. Similarly, "sup" denotes the *supremum* of a set of numbers, which is defined to be the smallest number that is not smaller than any number in the set.

$$\lim_{K\to+\infty} \Phi_K(c_K + d_K x) = \lim_{K\to+\infty} \Pr\left\{ \frac{\zeta_K - c_K}{d_K} \le x \right\} = 1 - \exp(-e^x) \tag{5.29}$$

with the scaling factors[8] given by

$$c_K = \begin{cases} -[2\log(K/\sqrt{2\pi})]^{1/2}\left(1 - \dfrac{\log\log(K/\sqrt{2\pi}) + \log 2}{4\log(K/\sqrt{2\pi}) + 2}\right) & \\ \qquad\qquad\qquad\qquad \text{at the limit point,} & \\ -[2\log(2K/\sqrt{2\pi})]^{1/4}\left(1 - \dfrac{\log\log(2K/\sqrt{2\pi}) + \log 2}{8\log(2K/\sqrt{2\pi}) + 4}\right) & \\ \qquad\qquad\qquad\qquad \text{at the transcritical point,} & \\ -[2\log(2K/\sqrt{2\pi})]^{1/3}\left(1 - \dfrac{\log\log(2K/\sqrt{2\pi}) + \log 2}{6\log(2K/\sqrt{2\pi}) + 3}\right) & \\ \qquad\qquad\qquad\qquad \text{at the unstable pitchfork point,} & \end{cases} \tag{5.30}$$

and

$$d_K = \begin{cases} [2\log(K/\sqrt{2\pi})]^{-1/2} & \text{at the limit point,} \\ \dfrac{1}{2}[2\log(2K/\sqrt{2\pi})]^{-3/4} & \text{at the transcritical point,} \\ \dfrac{2}{3}[2\log(2K/\sqrt{2\pi})]^{-2/3} & \text{at the unstable pitchfork point.} \end{cases} \tag{5.31}$$

Consequently, the function $\phi_K((\zeta_K - c_K)/d_K)$ is convergent to the probability density function $\exp(x - e^x)$, as is demonstrated in Fig. 5.2(b).

The expression (5.29) shows that ζ_K is of the order of c_K for large K, which we designate as

$$\zeta_K \approx c_K. \tag{5.32}$$

Yet the convergence in (5.29) as $K \to +\infty$ is typically slow, so that $\Pr\{\zeta_K \le c_K + d_K x\}$ can differ significantly from $1 - \exp(-e^x)$ for a moderately large value of K.

We can rewrite (5.32) to an expression for the minimum critical load $(f_c)_K$ attained by K independent random imperfections. By

$$(f_c)_K = f_c^0 + \zeta_K \widehat{C} \approx f_c^0 + c_K \widehat{C}$$

together with (5.19) for f_c^0 and (5.20) for $\widehat{C}$, we obtain

$$(f_c)_K \approx \begin{cases} \mathrm{E}_{\mathrm{sample}}[f_c] + c_K(\mathrm{Var}_{\mathrm{sample}}[f_c])^{1/2} & \text{at the limit point,} \\ \mathrm{E}_{\mathrm{sample}}[f_c] + (2.35 + c_K/0.349)(\mathrm{Var}_{\mathrm{sample}}[f_c])^{1/2} & \text{at the transcritical point,} \\ \mathrm{E}_{\mathrm{sample}}[f_c] + (1.86 + c_K/0.432)(\mathrm{Var}_{\mathrm{sample}}[f_c])^{1/2} & \text{at the unstable pitchfork point,} \end{cases}$$

[8] The derivations of these factors are given in §5.6, the appendix of this chapter.

as $K \to +\infty$. This expression is to be employed to simulate $(f_{\rm c})_K$ using sample mean $\mathrm{E}_{\rm sample}[f_{\rm c}]$ and sample variance $\mathrm{Var}_{\rm sample}[f_{\rm c}]$. The sample mean $\mathrm{E}_{\rm sample}[f_{\rm c}]$ and the sample variance $\mathrm{Var}_{\rm sample}[f_{\rm c}]$ are computed for some sample size K_0; this size is independent of K, and therefore can be chosen to be much smaller than K. The present method is thus endowed with simplicity and efficiency.

5.5 Example of Scatter of Critical Loads

We refer again to the propped cantilever in Fig. 5.3 (repeated from Fig. 2.1 in §2.3), as an example of the scatter of critical loads. More realistic examples are given in Chapter 10 after further theoretical development.

The perfect cantilever has the governing equation (2.14) with $(\beta_1,\beta_2,\beta_3,\beta_4) = (0,1,1,1)$ in (2.16), and has a fundamental path and a bifurcated path mutually intersecting at a transcritical bifurcation point at $(x_{\rm c}^0, y_{\rm c}^0, f_{\rm c}^0) = (0, 1/2, 1)$ with the critical eigenvector $\boldsymbol{\xi} = (1,0)^\top$.

We choose (x_i, y_i) $(i = 1,2)$ as imperfection parameters, and define

$$\boldsymbol{v} = (x_1, y_1, x_2, y_2)^\top$$

as the imperfection parameter vector. In the perfect case, we have

$$\boldsymbol{v}^0 = (0,1,0,0)^\top.$$

It is assumed that $\varepsilon \boldsymbol{d} = \boldsymbol{v} - \boldsymbol{v}^0$ is subject to a multivariate normal distribution $\mathrm{N}(\boldsymbol{0}, \varepsilon^2 I_4)$; that is,

$$W = I_4. \tag{5.33}$$

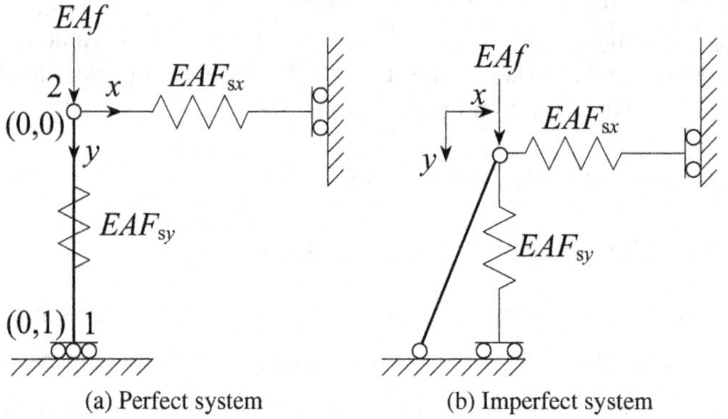

Fig. 5.3 Propped cantilever.

Theoretical Evaluation Procedure

The theoretical evaluation of the probability density function of the critical load is presented next. The imperfection sensitivity matrix

$$B_{\rm c}^{0} = EA\begin{pmatrix} 1 & 0 & -1 & 0 \\ 0 & -1 & 0 & -1 \end{pmatrix} \tag{5.34}$$

is obtained by differentiating $\boldsymbol{F}$ of the governing equation (2.14) with respect to $\boldsymbol{v}$ and evaluating it at the bifurcation point of the perfect system (cf. (2.3)). Use of (5.33) and (5.34) in (5.6) leads to

$$\widetilde{\sigma}^{2} = \boldsymbol{\xi}^{\top} B_{\rm c}^{0} W B_{\rm c}^{0\,\top} \boldsymbol{\xi} = 2(EA)^{2}. \tag{5.35}$$

Then, for an arbitrarily chosen imperfection pattern $\boldsymbol{d}$ and a sufficiently small imperfection magnitude ε, the critical load increment $\widetilde{f_{\rm c}}$ is to be obtained from the governing equation (2.14). For $\varepsilon\boldsymbol{d} = 10^{-6} \times (1,1,-1,-1)^{\top}$, which is sufficiently small

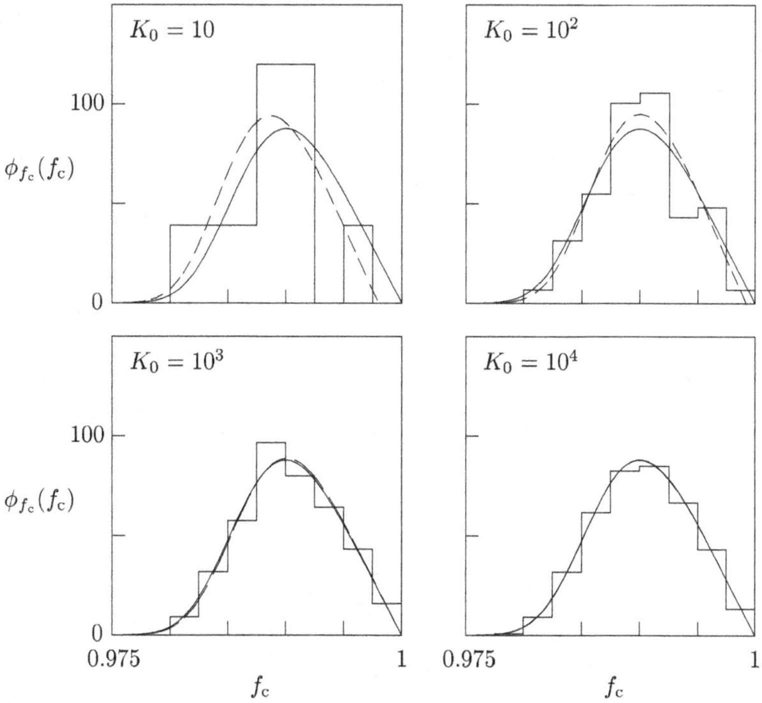

Fig. 5.4 Influence of the sample size K_0 on the semiempirical probability density function and the histogram ($\varepsilon = 10^{-4}$) for the propped cantilever (transcritical bifurcation point). ——: theoretical probability density function; – – –: semiempirical probability density function; histogram: numerical experiment.

to make the asymptotic formula (5.2) accurate, $\widetilde{f_c} = -0.00141$ is obtained, which is asymptotically equal to $-C_0(\boldsymbol{\xi}^\top B_c^0 \boldsymbol{d})^{1/2}\varepsilon^{1/2}$ by (5.2)–(5.4). Consequently, the following holds,

$$\widetilde{f_c} = -0.00141 \approx -C_0(\boldsymbol{\xi}^\top B_c^0 \boldsymbol{d})^{1/2}\varepsilon^{1/2} = -\sqrt{2}\times 10^{-3}(EA)^{1/2}C_0,$$

which yields $C_0 \approx (EA)^{-1/2}$. Therefore,

$$\widehat{C} = C_0(\widetilde{\sigma}\varepsilon)^{1/2} = 2^{1/4}\varepsilon^{1/2} \approx 1.189\varepsilon^{1/2}$$

by (5.9) and (5.35). Based on the value of $\widehat{C}$ evaluated in this manner and $f_c^0 = 1$, we compute from (5.15) the theoretical probability density function $\phi_{f_c}(f_c)$ of the critical load f_c portrayed by the solid line in Fig. 5.4.

Semiempirical Evaluation Procedure

The semiempirical evaluation of the probability density function of the critical load is illustrated. We randomly choose as many as $K_0 = 10^5$ imperfections $\varepsilon\boldsymbol{d}$ subject to the aforementioned normal distribution $\mathrm{N}(\boldsymbol{0}, \varepsilon^2 I_4)$ and compute a set of critical loads f_c for a series of imperfection magnitudes $\varepsilon = 10^{-2}$, 10^{-3}, and 10^{-4}. Table 5.2 lists the sample mean $\mathrm{E}_{\mathrm{sample}}[f_c]$ and the sample standard deviation $(\mathrm{Var}_{\mathrm{sample}}[f_c])^{1/2}$ of these critical loads. From formulas (5.19) and (5.20), we evaluated the values of f_c^0 and $\widehat{C}/\varepsilon^{1/2}$ also listed in this table. In association with the decrease in the imperfection magnitude ε, the evaluated values converge to the exact values, consistent with the asymptotic nature of (5.2). The semiempirical evaluation is apparently quite accurate. It is noteworthy that the computation of the imperfection sensitivity matrix B_c^0 is not needed in the semiempirical evaluation.

To elucidate the improvement of the semiempirical probability density function and the histogram, in association with the increase of the sample size K_0 of random imperfections, they are depicted in Fig. 5.4 based on the first $K_0 = 10$, 10^2, 10^3, and 10^4 random samples for $\varepsilon = 10^{-4}$. In contrast to the slow convergence of the histogram, the semiempirical probability density functions (shown by dashed lines)

Table 5.2 Calculated values of f_c^0 and $\widehat{C}/\varepsilon^{1/2}$ ($K_0 = 10^5$)

	$\mathrm{E}_{\mathrm{sample}}[f_c]$	$(\mathrm{Var}_{\mathrm{sample}}[f_c])^{1/2}$	f_c^0	$\widehat{C}/\varepsilon^{1/2}$
$\varepsilon = 10^{-2}$	0.905	3.87×10^{-2}	0.9964	1.11
10^{-3}	0.969	1.30×10^{-2}	0.9999	1.18
10^{-4}	0.990	4.15×10^{-3}	1.0000	1.19
Exact Values			1.0000	1.19

quickly approach the theoretical probability density function (solid line). This implies the importance of the explicit form (5.15) of the probability density functions.

Distribution of Minimum Values

Figure 5.5 shows, for $\varepsilon = 10^{-6}$, the comparison of the empirical minimum load ζ_K achieved by K random imperfections (shown by (•)) and its theoretical evaluation by (5.32) (shown by solid line). This evaluation is fairly consistent with the empirical values of ζ_K.

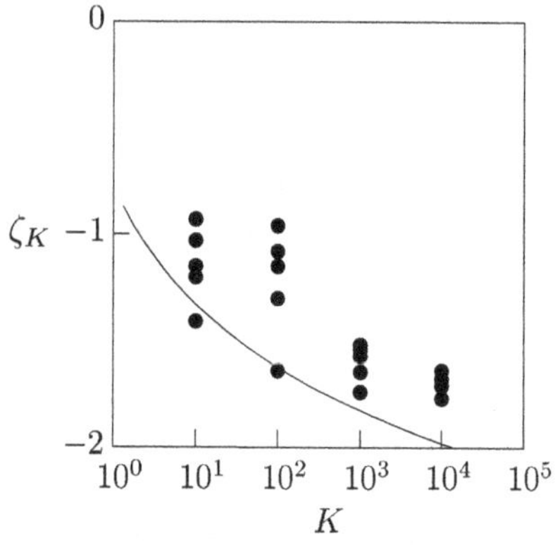

Fig. 5.5 Comparison of ζ_K and theoretical estimation $\zeta_K \approx c_K$ plotted against K in a semilogarithmic scale for the propped cantilever (transcritical bifurcation point; $\varepsilon = 10^{-6}$). •: empirical ζ_K; ——: theoretical estimation.

5.6 Appendix: Derivation of Scaling Factors

The expressions in (5.30) and (5.31) for the scaling factors c_K and d_K are derived from (5.28) for a limit point, whereas those for transcritical and pitchfork bifurcation points are addressed, respectively, in Problems 5-8 and 5-9.

In the notation of Lemma 5.2 in §5.4, we have

$$\begin{aligned}
\Phi(x) &= \int_{-\infty}^{x} \frac{1}{\sqrt{2\pi}} \exp\left(-\frac{y^2}{2}\right) dy \\
&= \int_{-x}^{\infty} \frac{1}{\sqrt{2\pi}} \frac{1}{y} \cdot y \exp\left(-\frac{y^2}{2}\right) dy \\
&= \frac{1}{\sqrt{2\pi}} \left[-\frac{1}{y} \exp\left(-\frac{y^2}{2}\right) \right]_{y=-x}^{\infty} - \frac{1}{\sqrt{2\pi}} \int_{-x}^{\infty} \frac{1}{y^2} \exp\left(-\frac{y^2}{2}\right) dy \\
&\approx \frac{1}{\sqrt{2\pi}} \exp\left(-\frac{x^2}{2}\right) \left(\frac{1}{|x|} - \frac{1}{|x|^3} + \cdots \right) \qquad (x \to -\infty)
\end{aligned} \tag{5.36}$$

through integration by parts. Substitution of (5.36) into (5.26) leads to

$$r(t) \approx \frac{1}{|t|} \qquad (t \to -\infty), \tag{5.37}$$

with which (5.25) is satisfied. The constant c_K is determined from the equation $\Phi(c_K) = 1/K$, which is equivalent, by (5.36), to

$$\log \sqrt{2\pi} + \frac{x^2}{2} + \log x - \log\left(1 - \frac{1}{x^2} + \cdots\right) = \log K \tag{5.38}$$

for $x = -c_K > 0$. Since x is large, the first-order approximation $x = x_0$ is obtained from

$$\log \sqrt{2\pi} + \frac{x^2}{2} = \log K \tag{5.39}$$

as

$$x_0 = [2\log(K/\sqrt{2\pi})]^{1/2}. \tag{5.40}$$

Substitution of $x = x_0(1-y)$ into (5.38) yields

$$\log \sqrt{2\pi} + \frac{1}{2}{x_0}^2(1-y)^2 + \log x_0 + \log(1-y) - \log\left(1 - \frac{1}{{x_0}^2(1-y)^2} + \cdots\right) = \log K;$$

that is, by (5.40),

$$\frac{1}{2}{x_0}^2(-2y+y^2) + \log x_0 + \log(1-y) - \log\left(1 - \frac{1}{{x_0}^2(1-y)^2} + \cdots\right) = 0,$$

which is approximated by

$$-{x_0}^2 y + \log x_0 - y = 0.$$

This gives

$$y = \frac{\log x_0}{{x_0}^2 + 1} = \frac{\log\log(K/\sqrt{2\pi}) + \log 2}{4\log(K/\sqrt{2\pi}) + 2}. \tag{5.41}$$

Therefore,

$$c_K = -[2\log(K/\sqrt{2\pi})]^{1/2}\left(1 - \frac{\log\log(K/\sqrt{2\pi}) + \log 2}{4\log(K/\sqrt{2\pi}) + 2}\right),$$

and, in view of (5.37), we may choose

$$d_K = r(c_K) = [2\log(K/\sqrt{2\pi})]^{-1/2}.$$

Problems

5-1 Compute the probability density functions of $\widetilde{f_c}$ for the shallow and nonshallow arches in Fig. 4.5 in §4.5.1. Choose $\boldsymbol{v} = (x_1, y_1, x_2, y_2, x_3, y_3)^\top$ and assume that $\varepsilon \boldsymbol{d} = \boldsymbol{v} - \boldsymbol{v}^0$ is subject to a normal distribution $\mathrm{N}(\boldsymbol{0}, \varepsilon^2 I_6)$ with $\varepsilon = 10^{-3}$.

5-2 Draw the histogram for the following set of critical loads

61	46	87	27	53	0	31	67	83	47	45	58	75	35	53
11	77	50	46	65	41	71	97	57	39	23	49	59	64	52

and simulate the histogram by the semiempirical procedure in §5.3.

5-3 Obtain explicit forms of probability density functions of $\widetilde{f_c}$ for simple critical points when $\boldsymbol{d}$ is distributed uniformly on the unit sphere $\|\boldsymbol{d}\| = 1$.
Answer: See Ikeda and Murota, 1991 [88].

5-4 Simulate the histograms in Fig. 5.6 by the semiempirical procedure in §5.3.

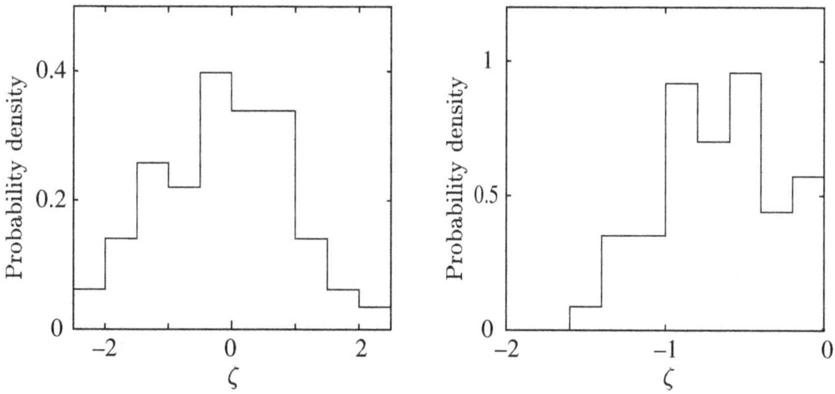

Fig. 5.6 Histograms.

5-5 Draw a histogram of $\widetilde{f_c}$ for the random data in Fig. 1.7(b) in §1.2.4 and simulate this histogram using the semiempirical procedure in §5.3.

5-6 Derive explicit forms of $\phi_\zeta(\zeta)$ in (5.11) and those of $\Phi_\zeta(\zeta)$ in Table 5.1 in §5.2.2.

5-7 Derive equations (5.15)–(5.17).

5-8 Derive the scaling factors c_K and d_K in (5.30) and (5.31) for a transcritical bifurcation point by modifying (5.36)–(5.41) in §5.6 appropriately.
Answer:

$$\Phi(x) = \int_{-\infty}^{x} \frac{4|y|}{\sqrt{2\pi}} \exp\left(-\frac{y^4}{2}\right) dy = \int_{x^2}^{\infty} \frac{2}{\sqrt{2\pi}} \exp\left(-\frac{y^2}{2}\right) dy$$

$$\approx \frac{2}{\sqrt{2\pi}} \exp\left(-\frac{x^4}{2}\right)\left(\frac{1}{x^2} - \frac{1}{x^6} + \ldots\right) \qquad (x \to -\infty),$$

$$r(t) \approx \frac{1}{2} \cdot \frac{1}{|t|^3} \qquad (t \to -\infty),$$

$$\log \frac{\sqrt{2\pi}}{2} + \frac{x^4}{2} + 2\log x - \log\left(1 - \frac{1}{x^4} + \ldots\right) = \log K,$$

$$\log \frac{\sqrt{2\pi}}{2} + \frac{x^4}{2} = \log K \quad \rightarrow \quad x_0 = [2\log(2K/\sqrt{2\pi})]^{1/4},$$

$$y = \frac{\log x_0}{x_0{}^4 + 1} = \frac{\log\log(2K/\sqrt{2\pi}) + \log 2}{8\log(2K/\sqrt{2\pi}) + 4}.$$

5-9 Derive the scaling factors c_K and d_K in (5.30) and (5.31) for a pitchfork bifurcation point.
Answer:

$$\Phi(x) = \int_{-\infty}^{x} \frac{3|y|^{1/2}}{\sqrt{2\pi}} \exp\left(-\frac{|y|^3}{2}\right) dy = \int_{(-x)^{3/2}}^{\infty} \frac{2}{\sqrt{2\pi}} \exp\left(-\frac{y^2}{2}\right) dy$$

$$\approx \frac{2}{\sqrt{2\pi}} \exp\left(-\frac{|x|^3}{2}\right)\left(\frac{1}{|x|^{3/2}} - \frac{1}{|x|^{9/2}} + \ldots\right) \qquad (x \to -\infty),$$

$$r(t) \approx \frac{2}{3} \cdot \frac{1}{|t|^2} \qquad (t \to -\infty),$$

$$\log \frac{\sqrt{2\pi}}{2} + \frac{x^3}{2} + \frac{3}{2}\log x - \log\left(1 - \frac{1}{x^3} + \ldots\right) = \log K,$$

$$\log \frac{\sqrt{2\pi}}{2} + \frac{x^3}{2} = \log K \quad \rightarrow \quad x_0 = [2\log(2K/\sqrt{2\pi})]^{1/3},$$

$$y = \frac{\log x_0}{x_0{}^3 + 1} = \frac{\log\log(2K/\sqrt{2\pi}) + \log 2}{6\log(2K/\sqrt{2\pi}) + 3}.$$

Summary

- The mechanism of the probabilistic variation of the critical load attributable to random imperfections has been investigated.
- Explicit forms of the probability density function of the critical load have been obtained.
- The distribution of minimum values of the critical load has been formulated.
- The usefulness of the procedure presented in this chapter has been demonstrated through its application to an example.

Chapter 6
Experimentally Observed Bifurcation Diagrams

6.1 Introduction

Extensive studies of structures and materials undergoing perfect and imperfect bifurcation have been conducted.[1] Nevertheless, a gap separating mathematical theory and engineering practice remains in the experimentation of materials undergoing bifurcation. Such a gap may be ascribed to the following three essential difficulties.

(1) It is hard to judge, merely from the observed curves, whether the system under consideration is undergoing bifurcation. It causes a problem in the physical interpretation of these curves.
(2) Experimentally observed displacements are influenced by unknown imperfections of various kinds, and the perfect system cannot be known, although extensive efforts have been made to reduce experimental errors.
(3) Observed diagrams of force versus displacement can differ qualitatively from bifurcation diagrams predicted by mathematics, although the influence of imperfections on the mathematical bifurcation diagram has been fully investigated in earlier chapters.

Several remarks are given below in relation to the third difficulty. Bifurcation diagrams in mathematical theory and those in engineering practice in the experimentation of materials undergoing bifurcation are different. In mathematical theory, a canonical coordinate is chosen to be the abscissa of a bifurcation diagram for mathematical convenience. In contrast, a physically meaningful variable is a natural choice of an abscissa in the bifurcation diagram obtained by an analysis or experimentation in engineering. Bifurcation diagrams observed in experiments might differ qualitatively from those in mathematics, as presented in Fig. 6.1 for a pitchfork bifurcation point (possible observed bifurcation diagrams in Fig. 6.1(a) and (b) in comparison with a mathematical diagram in Fig. 6.1(c)).

---

[1] Bifurcation of structures is highlighted, for example, by Ziegler, 1968 [214]; Thompson and Hunt, 1973 [191], 1984 [192]; Ben-Haim and Elishakoff, 1990 [14]; and Bažant and Cedolin, 1991 [12]. For bifurcation of materials, refer, for example, to Hill and Hutchinson, 1975 [73] and Vardoulakis and Sulem, 1995 [202].

K. Ikeda and K. Murota, *Imperfect Bifurcation in Structures and Materials*,
Applied Mathematical Sciences 149, DOI 10.1007/978-1-4419-7296-5_6,

Such a qualitative difference can be explained as follows. A bifurcation diagram is obtained as the projection of the solution path in a higher-dimensional space to a two-dimensional plane. The resulting picture naturally depends on the chosen projection. A canonical choice of the projection yields the mathematical bifurcation diagram (see Fig. 6.1(c)); in contrast, an arbitrary choice would result in a diagram such as that in Fig. 6.1(b), which is qualitatively similar to the mathematical diagram. If the direction of the projection is so special that it is perpendicular to the bifurcated path, then the resulting diagram resembles that in Fig. 6.1(a), which differs qualitatively from the mathematical diagram. Such an exceptional situation occurs often in engineering experiments as a result of geometrical symmetry. To fill the gap caused by this difference, the theory on imperfections is tailored to be applicable to the experimentally observed diagrams in this chapter.

The curves to be observed in experiments are under the combined influence of nonlinearity, bifurcation, and imperfection. It is highly desirable to develop a strategy to eliminate or mitigate the influence of imperfection from the experimental curves. More specific questions to be answered are the following.

- Can we construct a curve for the perfect system using a single or a number of experimental curves?

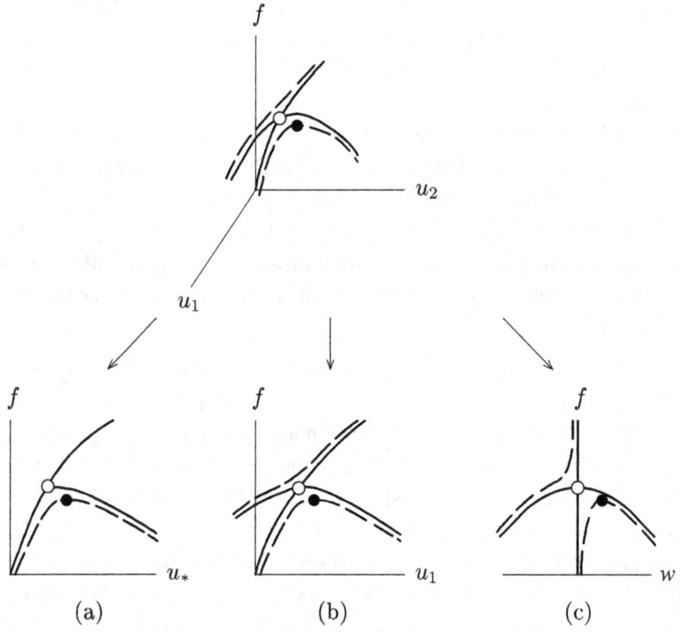

Fig. 6.1 Choice of projections in drawing bifurcation diagrams. (a) and (b) experimentally observed bifurcation diagrams and (c) a mathematical bifurcation diagram at an unstable pitchfork bifurcation point. ——: curve for the perfect system; — — —: curve for an imperfect system; ○: bifurcation point; ●: limit point.

- Can we explain the experimental curves as imperfect bifurcation phenomena in a consistent way?

A clue to answering these questions is the Koiter (imperfection sensitivity) law presented in Chapter 3. This law is extended with respect to the following two aspects (Ikeda and Murota, 1999 [92]):

- Applicability to experimentally observed bifurcation diagrams,
- The robustness against experimental errors.

This extended law is of great assistance in overcoming the three difficulties described above due to a gap separating mathematical theory and the experimentation of materials undergoing bifurcation.

This chapter is organized as follows.

- An illustrative example of bifurcation behavior is introduced in §6.2.
- State (unknown) variables of the bifurcation equation are transformed into displacements which are observable in experiments, and the explicit form of a force versus displacement curve and a pertinent power law are derived in §6.3.
- As a result of these, we present a systematic strategy to recover the curve of the perfect system from the curve of an imperfect system in §6.4. This strategy is used to detect the occurrence of bifurcation.
- The strategy is applied to a regular-hexagonal truss dome to assess its validity; furthermore, it is applied to cylindrical sand specimens to reveal that they are undergoing bifurcation in §6.5.

The results in this chapter are extended to a system with group symmetry in Chapter 8.

6.2 Illustrative Example

The propped cantilever depicted in Fig. 6.2 (cf., §2.6) is employed here as an example of an experimentally observed diagram.

Recall that its solution curves for imperfect systems are given by (2.100):

$$y = 1 - |x+\varepsilon|\left[\left(\frac{2x}{L} - \varepsilon\right)^{-2} - 1\right]^{1/2}, \tag{6.1}$$

$$f = \frac{2}{L} + \left[\left(-\varepsilon + \frac{x-\varepsilon}{L}\right)\operatorname{sign}(x+\varepsilon) - 2\frac{|x+\varepsilon|}{L}\right]\left[\left(\frac{2x}{L} - \varepsilon\right)^{-2} - 1\right]^{1/2}, \tag{6.2}$$

where $L = (1+4\varepsilon^2)^{1/2}$; and sign$(\cdot)$ denotes the sign of the variable in the parentheses. It is to be noted that (6.2) is a relation between two state variables x and y with a loading parameter f. In the two-dimensional plot of this relation, we have two choices of pairs of variables: (x,f) and (y,f). Figure 6.3 portrays a set of f versus x curves in (a) and f versus y curves in (b), both computed for $\varepsilon = 0$, ± 0.01,

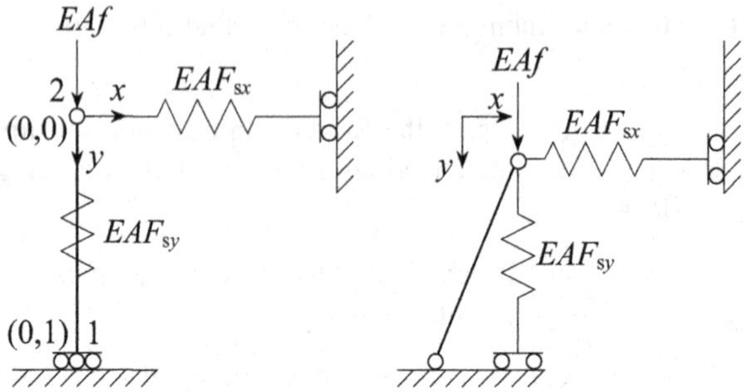

(a) Perfect system (b) Imperfect system

Fig. 6.2 Propped cantilever.

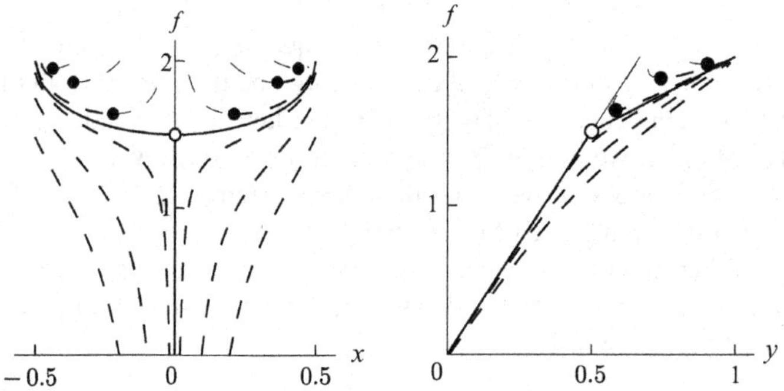

(a) f versus x curves (b) f versus y curves

Fig. 6.3 Solution curves for the propped cantilever. ——: path for the perfect system; — — —: path for an imperfect system; thick line: stable; thin line: unstable; ○: stable pitchfork bifurcation point; ●: limit point.

±0.05, and ±0.1. It is readily apparent that the former corresponds to the mathematical bifurcation diagram depicted in Fig. 6.1(c), and that the latter to a qualitatively different bifurcation diagram in Fig. 6.1(a).

Such a qualitative difference should be reflected in the reduced equation. The equation for the f versus x curves is expressed by (2.104):

$$4\widetilde{x}\widetilde{f} - 4\widetilde{x}^3 + 9\varepsilon + \text{h.o.t.} = 0. \tag{6.3}$$

Recall that imperfection sensitivity laws were developed based on this form of equation (cf., (2.85) and (2.93)).

On the other hand, the equation for the f versus y is given by (2.106):

$$\pm\sqrt{2(3\widetilde{y} - \widetilde{f})(\widetilde{y} - \widetilde{f})} + 3\varepsilon + \text{h.o.t.} = 0, \tag{6.4}$$

which differs qualitatively from (6.3). The derivation of imperfection sensitivity laws for this form of equation is the topic in the remainder of this chapter.

6.3 Imperfection Sensitivity Laws

Imperfection sensitivity laws for describing experimental (imperfect) bifurcation behaviors are introduced.

6.3.1 The Koiter Two-Thirds Power Law

The famous Koiter two-thirds power law, examined in Chapter 3, is reviewed as a prototype of an asymptotic law of imperfection sensitivity. Recall the nonlinear governing equation (2.1):

$$\boldsymbol{F}(\boldsymbol{u}, f, \boldsymbol{v}) = \boldsymbol{0}. \tag{6.5}$$

We consider an unstable pitchfork bifurcation point $(\boldsymbol{u}_{\mathrm{c}}^0, f_{\mathrm{c}}^0)$ of the perfect system. The system (6.5) of equations has been reduced to the single bifurcation equation (2.85):

$$\widetilde{F}(w, \widetilde{f}, \varepsilon) = A_{300}w^3 + A_{110}w\widetilde{f} + A_{020}\widetilde{f}^2 + A_{001}\varepsilon + \text{h.o.t.} = 0 \tag{6.6}$$

in a single variable $w \in \mathbb{R}$, which is valid in a neighborhood of $(w, \widetilde{f}, \varepsilon) = (0,0,0)$. In this equation, A_{110}, A_{020}, A_{300}, and A_{001} are constants; A_{001} depends on the imperfection pattern vector $\boldsymbol{d}$ (cf., (2.71)). It is assumed that the system is stable for $\widetilde{f} < 0$ and it becomes unstable at this point; accordingly, we have $A_{110} < 0$ and $A_{300} < 0$ (cf., (2.87) and (2.88)).

The imperfect system described by (6.6) with $\varepsilon \neq 0$ has a limit point, as depicted in Fig. 6.4, and the location $(w_{\mathrm{c}}, \widetilde{f}_{\mathrm{c}})$ of this point is calculated as

$$w_{\mathrm{c}} \approx \left(\frac{A_{001}}{2A_{300}}\right)^{1/3} \varepsilon^{1/3}, \tag{6.7}$$

$$\widetilde{f}_{\mathrm{c}} \approx -\beta\varepsilon^{2/3}, \qquad \beta = \frac{3{A_{300}}^{1/3}{A_{001}}^{2/3}}{2^{2/3}A_{110}}, \tag{6.8}$$

where β is a constant independent of ε. Recall from §3.2.4 that $A_{020}\widetilde{f}^2$ in (6.6) is negligible in order. Equation (6.8) is the famous Koiter two-thirds power law.

Eliminating ε from (6.7) and (6.8) gives an asymptotic relation

$$\widetilde{f}_{\mathrm{c}} + g_K {w_{\mathrm{c}}}^2 = 0 \tag{6.9}$$

with $g_K = 3A_{300}/A_{110} > 0$. It is important that g_K is a constant independent of the imperfection pattern vector $\boldsymbol{d}$, whereas A_{001} does depend on $\boldsymbol{d}$. The relation (6.9) is

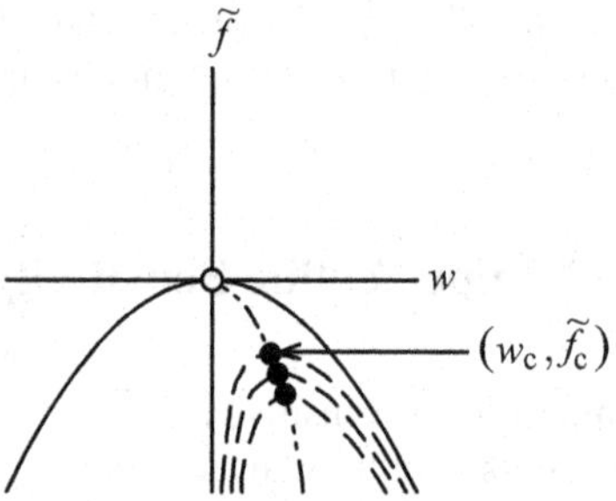

Fig. 6.4 Illustration of the Koiter two-thirds power law and the asymptotic relation (6.9). ——: curve for the perfect system; — — —: curve for an imperfect system; ----: the parabola (6.9); ○: bifurcation point; ●: limit point.

presented in Fig. 6.4 by a series of limit points (●) threaded with a dotted–dashed line. This relation (6.9) is useful for the treatment of imperfection-sensitive bifurcation behavior, but suffers in its present form from shortcomings:

(1) Because variable w is more mathematical than it is physical, the critical displacement w_c is not always observed (or observable) in experiments. The variable w is introduced in the course of mathematical reduction to obtain the bifurcation equation. It is not an f versus w curve but an f versus u_{i^*} curve (with a particular i^*) that is to be observed in customary experiments.
(2) Even if the variable w is observed, the value of w_c must be determined by identifying the limit point of an observed equilibrium path, which is necessarily blurred by various noises and errors in numerical analyses and physical experiments. In such situations the value of w_c at the limit point cannot be determined reliably, although the value of f_c at the limit point can be determined accurately.

6.3.2 Generalized Koiter Law

As a generalization of relation (6.9) for the Koiter law, we consider a parabola

$$\tilde{f} + g w^2 = 0 \tag{6.10}$$

with an arbitrary constant g portrayed by the dotted–dashed line in Fig. 6.5, and its intersection point $(w_{\cap g}, \tilde{f}_{\cap g})$ with an imperfect path, shown by (◇). Substitution of (6.10) for the parabola into the bifurcation equation (6.6) yields

$$w_{\cap g} = \left(\frac{A_{001}}{gA_{110} - A_{300}} \right)^{1/3} \varepsilon^{1/3} + \text{h.o.t.} \tag{6.11}$$

Combination of this expression with (6.8) for $\tilde{f}_c$ yields

$$\tilde{f}_c = -\frac{3A_{300}{}^{1/3}(gA_{110} - A_{300})^{2/3}}{2^{2/3}A_{110}} (w_{\cap g})^2 + \text{h.o.t.}, \tag{6.12}$$

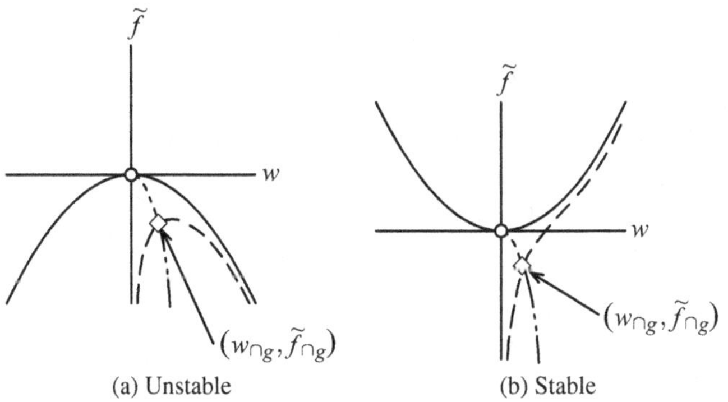

Fig. 6.5 Generalization of the Koiter law for pitchfork bifurcation points. ——: curve for the perfect system; — — —: curve for an imperfect system; — - - —: parabola (6.10); ○: bifurcation point; ◇: intersection point of the parabola and an imperfect path.

which demonstrates that $\widetilde{f}_{\rm c}$ is proportional to $(w_{\cap g})^2$ asymptotically as $\varepsilon \to 0$. The coefficient of proportionality is a constant that is independent of imperfection pattern vector $\boldsymbol{d}$.

The asymptotic law (6.12) portrays an asymptotic relation between coordinates of two distinct points: the limit point $(w_{\rm c}, \widetilde{f}_{\rm c})$ of an imperfect system and the intersection point $(w_{\cap g}, \widetilde{f}_{\cap g})$. This law (6.12) has the advantage that $w_{\cap g}$ can be determined much more reliably than $w_{\rm c}$ can be. In this sense, the introduction of the general parabola (6.10) resolves the second disadvantage of the original relation (6.9) described at the end of §6.3.1. The law (6.12) has another advantage over the original relation (6.9). As might be apparent from Fig. 6.5, the intersection point exists both for stable and unstable pitchfork bifurcation points. For that reason, the generalized asymptotic law (6.12) is applicable to points of both types, whereas the Koiter law yields physically meaningful information only for an unstable pitchfork bifurcation point.

6.3.3 Laws for Experimentally Observed Bifurcation Diagrams

To address the first problem with (6.9) concerning the observability of $w_{\rm c}$ in the experimentation presented at the end of §6.3.1, the asymptotic law (6.12) is further improved. We, accordingly, aim at an asymptotic law as well as the bifurcation equation expressed in terms of an observed variable u_{i^*} with a particular i^*.

We refer to the bifurcation equation (6.6):

$$\widehat{F}(w, \widetilde{f}, \varepsilon) = A_{300}w^3 + A_{110}w\widetilde{f} + A_{001}\varepsilon + \text{h.o.t.} = 0, \tag{6.13}$$

in which $A_{200}\widetilde{f}^2$ is suppressed, and recall the relation (2.40):

$$\boldsymbol{u} = \boldsymbol{u}_{\mathrm{c}}^0 + \sum_{j=1}^{N} w_j \boldsymbol{\eta}_j = \boldsymbol{u}_{\mathrm{c}}^0 + w\boldsymbol{\eta}_1 + \sum_{j=2}^{N} \varphi_j(w, \widetilde{f}, \boldsymbol{v})\boldsymbol{\eta}_j. \tag{6.14}$$

Equation (6.13) represents the relation between $\widetilde{f}$ and w, and the i^*th row of (6.14) represents the relation between w and u_{i^*}.

We first concentrate on (6.14) to derive an expression of the increment

$$\widetilde{u}_{i^*} \equiv u_{i^*} - (u_{i^*})_{\mathrm{c}}^0$$

of the i^*th component of $\boldsymbol{u}$ in terms of $\widetilde{f}$ and w. The term $w\boldsymbol{\eta}_1$ is the only linear term on the right-hand side of (6.14), since

$$\varphi_j(0,0,\boldsymbol{v}^0) = \frac{\partial \varphi_j}{\partial w}(0,0,\boldsymbol{v}^0) = 0, \qquad j = 2,\ldots,N$$

by (2.37) and by Lemma 2.2 in §2.4.3. For this reason, the i^*th row of (6.14) gives

$$\widetilde{u}_{i^*} = \eta_{i^*1} w + r_{i^*} \widetilde{f} + s_{i^*} w^2 + \text{h.o.t.}, \tag{6.15}$$

where η_{i^*1} is the i^*th component of the critical eigenvector $\boldsymbol{\eta}_1 = (\eta_{11},\ldots,\eta_{N1})^\top$ and r_{i^*} and s_{i^*} are constants. In (6.15), we have

$$w = \mathrm{O}(\varepsilon^{1/3}), \qquad \widetilde{f} = \mathrm{O}(\varepsilon^{2/3}),$$

in our region of interest (cf., §3.2.3), where $\mathrm{O}(\cdot)$ denotes a quantity of the same order as the term in parentheses. This implies, in particular, that the term ε can be omitted in expression (6.15). Using simplified notations

$$\widetilde{u} = \widetilde{u}_{i^*}, \qquad r = r_{i^*}, \qquad s = s_{i^*},$$

we can rewrite (6.15) as

$$\widetilde{u} = \eta_{i^*1} w + r\widetilde{f} + s w^2 + \text{h.o.t.} \tag{6.16}$$

In view of the vanishing and nonvanishing of the coefficient η_{i^*1} of w in (6.16), the asymptotic behaviors of $\widetilde{u}$ are categorized as

$$\begin{cases} \widetilde{u} = \mathrm{O}(w) & \text{if } \eta_{i^*1} \neq 0, \\ \widetilde{u} = \mathrm{O}(w^2) & \text{if } \eta_{i^*1} = 0. \end{cases} \tag{6.17}$$

The latter case ($\eta_{i^*1} = 0$) can occur not only by an accidental numerical cancellation but also, generically, as a consequence of group symmetry. Such a case, which often takes place in many physical problems, is treated in Part II.

Displacement u_{i^*} Without Symmetry

A general displacement u_{i^*} without particular symmetry ($\eta_{i^*1} \neq 0$) is treated via marginal modifications of our preliminary considerations leading to the asymptotic law (6.12). Equation (6.16) is solved for w as

$$w = \frac{1}{\eta_{i^*1}}(\widetilde{u} - \widetilde{f}/E) + \text{h.o.t.}, \tag{6.18}$$

where $E = 1/r$ is a constant. Substitution of (6.18) into (6.13) yields the asymptotic expression for the curve of an imperfect system

$$(\widetilde{u} - \widetilde{f}/E)\widetilde{f} + p^*(\widetilde{u} - \widetilde{f}/E)^3 + q^*\varepsilon + \text{o}(\varepsilon) = 0, \tag{6.19}$$

where $\text{o}(\cdot)$ denotes a quantity of a smaller order than the term in parentheses, and

$$p^* = \frac{A_{300}}{\eta_{i^*1}{}^2 A_{110}}, \qquad q^* = \frac{\eta_{i^*1} A_{001}}{A_{110}}.$$

Therein, p^* is a constant related to the curvature of the bifurcated path and q^* is a scaling factor depending on the imperfection pattern $\boldsymbol{d}$. With this notation, the Koiter law in (6.8) can be rewritten as

$$\widetilde{f}_c = -\frac{3}{2^{2/3}}(p^*)^{1/3}(q^*\varepsilon)^{2/3} + \text{h.o.t.} \tag{6.20}$$

The perfect and imperfect paths expressed by (6.19) are depicted in Fig. 6.6(a). The bifurcated path of the perfect system is a parabolic curve with a zero slope at the bifurcation point and, therefore, is almost flat in the neighborhood of this

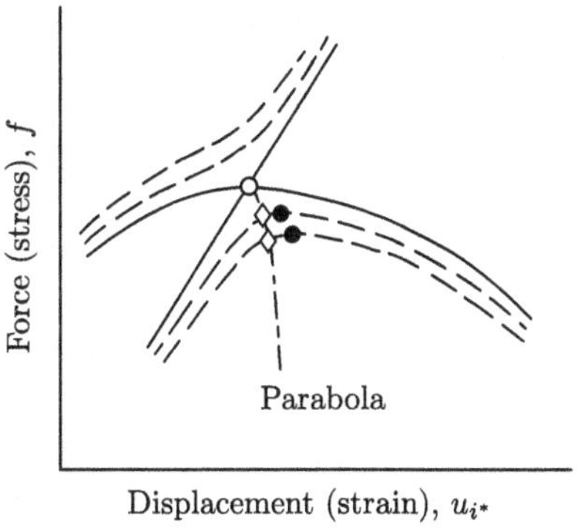

(a) Nonsymmetric displacement with $\eta_{i^*1} \neq 0$

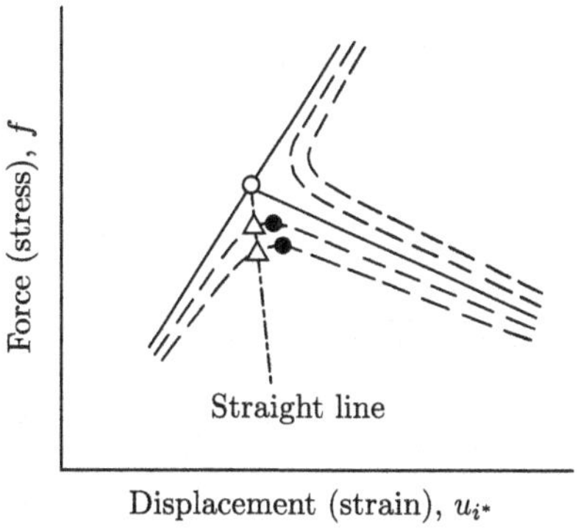

(b) Symmetric displacement with $\eta_{i^*1} = 0$

Fig. 6.6 General view of force versus displacement curves in the neighborhood of an unstable pitchfork bifurcation point. ——: curve for the perfect system; – – –: curve for an imperfect system; ○: bifurcation point; ●: limit point; ◇: intersection point of the parabola with an imperfect curve; and △: intersection point of the straight line with an imperfect curve.

point. Two *half-branches* (that form a smooth bifurcated path) bifurcate in opposite directions from the bifurcation point and two imperfect paths exist on both sides of the fundamental path for a given ε.

Instead of the parabola (6.10) in the plane of $(w, \widetilde{f})$, we consider another parabola

$$\widetilde{f} + g\widetilde{u}^2 = 0 \tag{6.21}$$

in the plane of $(\widetilde{u}, \widetilde{f})$ (where g is a constant), shown by the dotted–dashed line in Fig. 6.6(a), and its intersection point $(\widetilde{u}_{\cap g}, \widetilde{f}_{\cap g})$, portrayed as ($\diamond$) in Fig. 6.6(a), with the imperfect $\widetilde{f}$ versus $\widetilde{u}$ curve. Substitution of (6.21) into (6.19) and omission of higher-order terms yield

$$\widetilde{u}_{\cap g} = \left(\frac{q^*}{g - p^*}\right)^{1/3} \varepsilon^{1/3} + \text{h.o.t.} = \left(\frac{\eta_{i^*1}{}^3 A_{001}}{g\eta_{i^*1}{}^2 A_{110} - A_{300}}\right)^{1/3} \varepsilon^{1/3} + \text{h.o.t.} \tag{6.22}$$

Elimination of ε from (6.22) and the Koiter law (6.8) yields a power law

$$\begin{aligned} \widetilde{f}_{\mathrm{c}} &= -\frac{3(p^*)^{1/3}(g - p^*)^{2/3}}{2^{2/3}} (\widetilde{u}_{\cap g})^2 + \text{h.o.t.} \\ &= -\frac{3A_{300}{}^{1/3}(g\widetilde{A}_{110} - A_{300})^{2/3}}{2^{2/3}\widetilde{A}_{110}} (\widetilde{u}_{\cap g})^2 + \text{h.o.t.}, \end{aligned} \tag{6.23}$$

where $\widetilde{A}_{110} = \eta_{i^*1}{}^2 A_{110}$. As discussed in §6.1, the law (6.23) is qualitatively identical to (6.12).

Displacement u_{i^*} with Symmetry

For a symmetric displacement u_{i^*} with $\eta_{i^*1} = 0$, both the asymptotic relation (6.12) and the bifurcation equation (6.13) must be modified into qualitatively different forms. Equation (6.16) reduces to

$$\widetilde{u} = r\widetilde{f} + sw^2 + \text{h.o.t.} \tag{6.24}$$

From (6.13) and (6.24) we can express $(\widetilde{u}, \widetilde{f})$ as a function in w as follows.

$$\widetilde{u} = -r\left(\frac{A_{300}}{A_{110}} w^2 + \frac{A_{001}\varepsilon}{A_{110} w}\right) + sw^2 + \text{h.o.t.}, \qquad \widetilde{f} = -\frac{A_{300}}{A_{110}} w^2 - \frac{A_{001}\varepsilon}{A_{110} w} + \text{h.o.t.}$$

This serves as a parametric representation of the imperfect bifurcation diagram. Eliminating w yields

$$\pm\sqrt{\frac{1}{s}(\widetilde{u} - \widetilde{f}/E)}\left[\widetilde{f} + \frac{A_{300}}{A_{110}}\frac{1}{s}(\widetilde{u} - \widetilde{f}/E)\right] + \frac{A_{001}}{A_{110}}\varepsilon + \text{h.o.t.} = 0 \tag{6.25}$$

with an inequality condition

$$\mathrm{sign}(s)(\widetilde{u}-\widetilde{f}/E)\geq 0. \tag{6.26}$$

Here sign(s) denotes the sign of s and $E = 1/r$ denotes the slope of the $\widetilde{f}$ versus $\widetilde{u}$ curve of the fundamental path for the perfect system ($\varepsilon = 0$). Equation (6.25) can be rewritten as

$$\sqrt{\mathrm{sign}(s)(\widetilde{u}-\widetilde{f}/E)}\,[\widetilde{f}+p(\widetilde{u}-\widetilde{f}/E)]\pm q\varepsilon+\text{h.o.t.}=0, \tag{6.27}$$

where p and q are parameters, being defined, respectively, as

$$p=\frac{A_{300}}{sA_{110}},\qquad q=\frac{A_{001}|s|^{1/2}}{A_{110}}. \tag{6.28}$$

With this notation, the Koiter law (6.8) is written as

$$\widetilde{f}_{\mathrm{c}}=-\mathrm{sign}(s)\,\frac{3p^{1/3}q^{2/3}}{2^{2/3}}\varepsilon^{2/3}+\text{h.o.t.} \tag{6.29}$$

For the perfect system with $\varepsilon = 0$, (6.27) yields

$$\begin{cases}\widetilde{u}-\widetilde{f}/E+\text{h.o.t.}=0, & \text{fundamental path,}\\ \widetilde{f}+p(\widetilde{u}-\widetilde{f}/E)+\text{h.o.t.}=0\quad(\mathrm{sign}(s)(\widetilde{u}-\widetilde{f}/E)\geq 0), & \text{bifurcated path.}\end{cases} \tag{6.30}$$

Because of the inequality condition in (6.30), the solution path branches toward only one direction from the bifurcation point in the plane of $(\widetilde{u},\widetilde{f})$. This fact presents a qualitative difference from the pitchfork-type diagram.

For an imperfect system, the inequality condition (6.26) shows that all imperfect paths exist only on one side of the fundamental path; the side depends on the sign of s. For a specified value of ε, $\pm$ in (6.27) corresponds to a pair of imperfect paths: $+$ is associated with a path above the bifurcated path, and $-$ with another path below it, or vice versa (cf., Fig. 6.7).

Instead of the parabola we now consider a straight line

$$\widetilde{f}+h\widetilde{u}=0 \tag{6.31}$$

(where h is a constant) shown by the dotted–dashed line in Fig. 6.6(b), and its intersection point $(\widetilde{u}_{|h},\widetilde{f}_{|h})$, presented as ($\triangle$) in Fig. 6.6(b), with the imperfect $\widetilde{f}$ versus $\widetilde{u}$

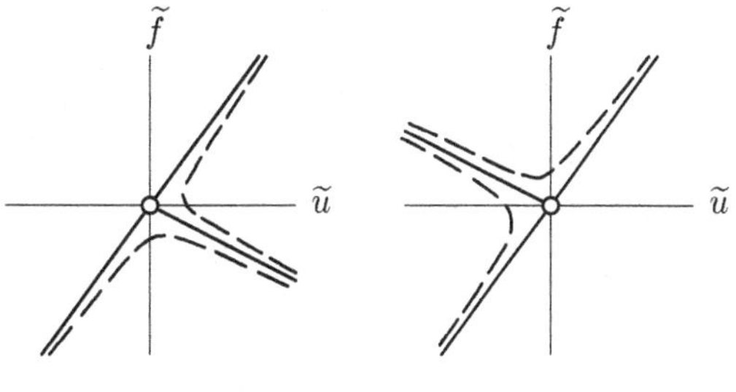

Fig. 6.7 Imperfect behaviors of two kinds in the neighborhood of an unstable pitchfork bifurcation point for a symmetric displacement $\widetilde{u}$. ——: curve for the perfect system; — — —: curve for an imperfect system; ○: bifurcation point.

curve. Substitution of (6.31) for the straight line into (6.27) yields

$$\widetilde{u}_{|h} = \gamma \varepsilon^{2/3} + \text{h.o.t.}, \tag{6.32}$$

where

$$\gamma = \frac{q^{2/3}}{\text{sign}(s)(1+h/E)^{1/3}[-h+p(1+h/E)]^{2/3}}.$$

It is noteworthy that p and E are independent of the imperfection pattern vector $\boldsymbol{d}$, whereas q varies with $\boldsymbol{d}$ (cf., (2.71) and (6.28)). Elimination of the imperfection magnitude ε from the Koiter law (6.29) and expression (6.32) leads to a *generalized asymptotic law*

$$\widetilde{f}_{\rm c} = -\gamma^* \widetilde{u}_{|h} + \text{h.o.t.} \tag{6.33}$$

with

$$\gamma^* = \frac{3}{2^{2/3}}[p(1+h/E)]^{1/3}[-h+p(1+h/E)]^{2/3}. \tag{6.34}$$

The law (6.33) denotes a linear relation between a pair of experimentally observed variables $\widetilde{u}_{|h}$ and $\widetilde{f}_{\rm c}$ that passes the origin $(\widetilde{u}_{|h}, \widetilde{f}_{\rm c}) = (0,0)$. Therefore, $(\widetilde{u}_{|h}, \widetilde{f}_{\rm c})$, for different values of ε, all lie on the line (6.33) with a common slope $-\gamma^*$. It is an important point of emphasis that the coefficient γ^* is independent of the imperfection pattern vector $\boldsymbol{d}$, although γ is not.

Remark 6.1. The procedure presented above for a pitchfork bifurcation point is applicable to a transcritical bifurcation point with minor modifications. The bifurcation equation for a transcritical bifurcation point is given by (3.12):

$$A_{020}\widetilde{f}^2 + A_{110}w\widetilde{f} + A_{200}w^2 + A_{001}\varepsilon + \text{h.o.t.} = 0, \tag{6.35}$$

and the Koiter law for this type of point is given by (3.15):

$$\widetilde{f}_{\rm c} \approx \pm\left(\frac{4A_{200}A_{001}\varepsilon}{A_{110}^2 - 4A_{200}A_{020}}\right)^{1/2}. \tag{6.36}$$

We again consider a straight line

$$\widetilde{f} + h\widetilde{u} = 0. \tag{6.37}$$

Substitution of (6.18) and (6.37) into (6.35) results in

$$\widetilde{u}_{|h} = \pm\left(\frac{-\eta_{i^*1}{}^2 A_{001}\varepsilon}{\eta_{i^*1}{}^2 A_{020}h^2 - \eta_{i^*1}A_{110}h(1+h/E) + A_{200}(1+h/E)^2}\right)^{1/2} + \text{h.o.t.} \tag{6.38}$$

Elimination of ε from the Koiter law (6.36) and expression (6.38) gives a *generalized asymptotic law*

$$\widetilde{f}_{\rm c} = \kappa \widetilde{u}_{|h} + \text{h.o.t.}$$

with

$$\kappa = \pm\left(\frac{-4A_{200}({\eta_{i^*1}}^2A_{020}h^2 - \eta_{i^*1}A_{110}h(1+h/E) + A_{200}(1+h/E)^2)}{{\eta_{i^*1}}^2({A_{110}}^2 - 4A_{200}A_{020})}\right)^{1/2}.$$

Here again κ is independent of $\boldsymbol{d}$. □

6.4 Recovering the Perfect System from Imperfect Systems

As described in the introductory part of this chapter, it is difficult to detect the occurrence of bifurcation from an imperfect curve associated with the unknown perfect system. To resolve this difficulty, a systematic procedure is presented in this section for recovering the perfect curve from experimental curves. The existence of a unique perfect system and that of its unique bifurcation point are assumed, along with the absence of the mode switching behavior and recursive bifurcation behavior; see §13.3 for a systematic procedure to sort out these behaviors.

As the first step of recovering the perfect curve, the symmetry of the displacement $u = u_{i^*}$ under consideration is investigated. We specifically examine the symmetric displacement in this section in presenting a procedure for recovering the perfect behavior; the procedure for the displacement without symmetry can be obtained simply by replacing relevant formulas.

6.4.1 Recovery from a Single Imperfect Path

For recovering the perfect curve in the plane of (u, f), the location (u_c^0, f_c^0) of the bifurcation point and the values of the parameters p and E in (6.30) are necessary. In addition, the values of $q\varepsilon$ in (6.27) are needed in the simulation of experimental (imperfect) curves. The information to be extracted from a single experimental curve is the location $\widetilde{u}_{|h}$ of the intersection point of the straight line (6.31) with the experimental curve for different values of h, where the "origin" (u_c^0, f_c^0) is unknown. This is based on the fact that the law (6.33) holds for any values of h, say $(h_i \mid i = 1, 2, \ldots)$.

Among other possibilities, the following procedure is suggested.

- Assume the location (u_c^0, f_c^0) of the bifurcation point.
- Employ four different values of h, say $(h_i \mid i = 1, 2, 3, 4)$, and estimate p and E based on (6.33) with (6.34) from the observed values of $\widetilde{u}_{|h}$. To be specific, for two different values h_i and h_j, (6.33) with (6.34) yields

$$\left(\frac{\widetilde{u}_{|h_j}}{\widetilde{u}_{|h_i}}\right)^3 = \frac{\eta_i}{\eta_j}\left(\frac{-h_i + p\eta_i}{-h_j + p\eta_j}\right)^2, \qquad (i, j) = (1, 2),\ (3, 4),$$

  where

$$\eta_i = 1 + h_i/E, \qquad i = 1, 2, 3, 4.$$

  Solve this equation to arrive at an explicit expression of p:

$$p = \frac{h_1 - \rho_{12} h_2}{\eta_1 - \rho_{12}\eta_2} = \frac{h_3 - \rho_{34} h_4}{\eta_3 - \rho_{34}\eta_4}, \tag{6.39}$$

where

$$\rho_{ij} = \pm\sqrt{\frac{\eta_j}{\eta_i}\left(\frac{\widetilde{u}_{|h_j}}{\widetilde{u}_{|h_i}}\right)^3}, \qquad (i,j) = (1,2),\ (3,4).$$

- Determine the value of E from the second equation in (6.39) using some iterative numerical method. Equation (6.39) can have more than one solution. It is suggested to observe the physical plausibility of E, which denotes the slope of the fundamental path, to select the appropriate solution. Then the value of p is determined uniquely.
- Using several sets of four different values of h_i ($i = 1,2,3,4$), obtain sufficiently many estimates of E, say ($E_k \mid k = 1,2,\ldots$), which should coincide with one another if the assumed value of (u_c^0, f_c^0) is correct. Estimate the location (u_c^0, f_c^0), which is not known, as the point where the variance among ($E_k \mid k = 1,2,\ldots$) is minimized. Then (6.32) and (6.29), respectively, yield the following pair of expressions of $q\varepsilon$:

$$q\varepsilon = \pm[\mathrm{sign}(s)(1+h/E)\{-h+p(1+h/E)\}^2(\widetilde{u}_{|h})^3]^{1/2}, \tag{6.40}$$

$$q\varepsilon = \pm\left(-\frac{4\,\mathrm{sign}(s)}{27p}(\widetilde{f}_c)^3\right)^{1/2}. \tag{6.41}$$

The values of all the parameters in the asymptotic force versus displacement curve (6.27), accordingly, can be obtained by this procedure. It must be emphasized that the procedure with (6.40) is quite robust and applicable for cases in which the peak of an experimental curve is missing and for which $\widetilde{f}_c$ cannot be observed. Such is often the case with the materials undergoing a sudden rupture or failure by cracking. In contrast, (6.41) requires the value of $\widetilde{f}_c$ and therefore is not always suitable in practical applications.

6.4.2 Recovery from a Series of Imperfect Paths

Given a series of experimental curves undergoing bifurcation at, presumably, the same bifurcation point but with different values of imperfections, one might use the following method to determine the location of the bifurcation point of the perfect system and the values of the parameters. Herein, we specifically examine the case where the displacement is symmetric, and the procedures presented below must be modified accordingly when it is not symmetric.

The location of the bifurcation point can be determined using a single value of h by repeating the procedure described below.

- Assume the location (u_c^0, f_c^0) of the bifurcation point, depicted as (∘) in Fig. 6.6(b), and obtain the incremental displacements $\widetilde{u}_{|h}$ at the intersection points, depicted

as (Δ) in Fig. 6.6(b), of the straight line $\widetilde{f} + h\widetilde{u} = 0$ with the experimental f versus u curves. Then plot $\widetilde{f}_c$ against $\widetilde{u}_{|h}$ for all the curves.
- Modify the location (u_c^0, f_c^0) so that the fitting of relation (6.33) is improved.

Then, the values of the parameters are determined as follows.

- Choose the value of parameter E, which denotes the slope of the fundamental path for the perfect system, such that the fundamental path $\widetilde{u} - \widetilde{f}/E = 0$ given by (6.30) accurately simulates experimental curves in the region sufficiently far from the bifurcation point.
- Determine the value of γ^* from the slope of the $\widetilde{f}_c$ versus $\widetilde{u}_{|h}$ relation in (6.33). Then determine the value of parameter p from formula (6.34) using the values of E and γ^*.
- Evaluate the imperfection magnitude $q\varepsilon$ for each specimen using the Koiter law (6.29) using the maximum load $\widetilde{f}_c$ of the specimen observed in the experiment.

This method is suggested for use because it is usually robust against experimental errors, as we show in §6.5.2.

6.5 Examples of Observed Bifurcation Diagrams

The validity of the generalized laws, for the experimentally observed bifurcation diagrams presented in §6.3, is confirmed based on examples, including a regular-hexagonal truss dome and a series of sand specimens.

6.5.1 Regular-Hexagonal Truss Dome

The regular-hexagonal truss dome in Fig. 6.8 is used as an example of symmetric and nonsymmetric displacements at an unstable pitchfork bifurcation point.

All members of this dome have the same Young's modulus and the same cross-section. A vertical (z-directional) load $0.5f$ is applied at the crown node 0 and a

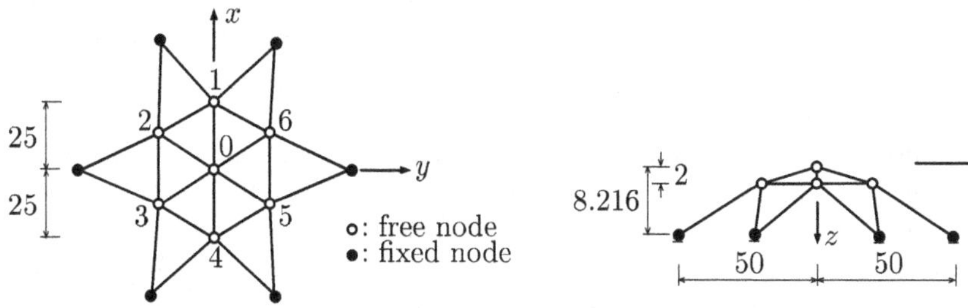

Fig. 6.8 Regular-hexagonal truss dome.

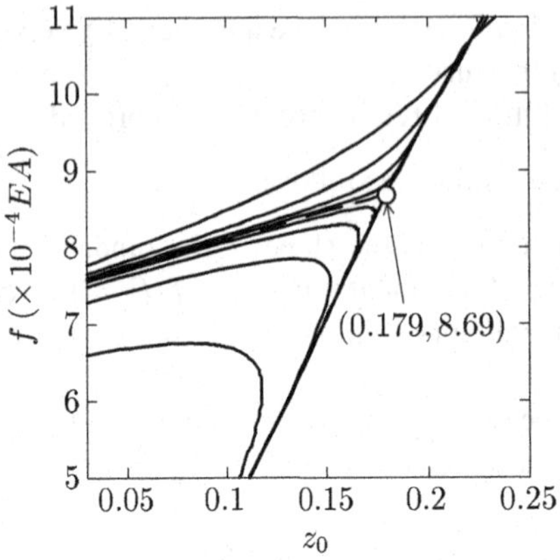

(a) f versus z_0 curves

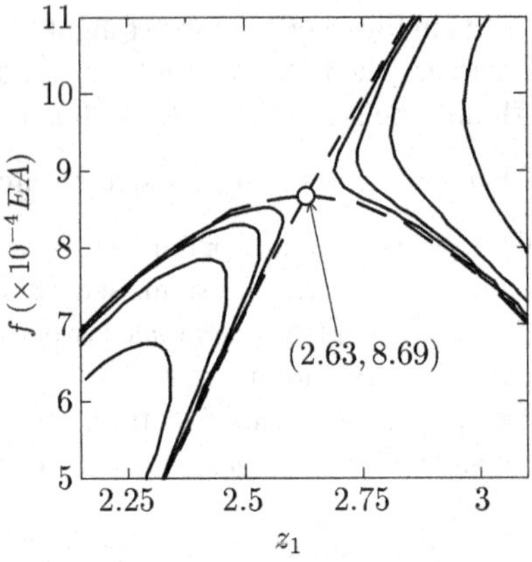

(b) f versus z_1 curves

Fig. 6.9 Equilibrium paths of the regular-hexagonal truss dome. ——: curve for an imperfect system; — — —: curve for the perfect system; ○: pitchfork bifurcation point.

uniform vertical load f at each of the other free nodes. As imperfections, the initial locations of nodes 2, 4, and 6 are lifted respectively upward in the z-direction at a length of ε. The nonlinear equilibrium equation (6.5) of the dome is solved for the imperfection magnitude $\varepsilon = 0$, 0.01, 0.03, 0.1, and 0.3 to obtain the fundamental and bifurcated paths portrayed by the dashed line and the imperfect paths portrayed by the solid lines in Fig. 6.9. An unstable pitchfork bifurcation point, shown by ($\circ$), exists on the fundamental path of the perfect system. For the critical eigenvector

$$\boldsymbol{\eta}_1 = (\eta_{x0}, \eta_{y0}, \eta_{z0}, \ldots, \eta_{x6}, \eta_{y6}, \eta_{z6})^{\top},$$

the z-directional components are given as

$$(\eta_{z0}, \eta_{z1}, \ldots, \eta_{z6}) = C(0, 1, -1, 1, -1, 1, -1)$$

for some scaling constant $C \neq 0$, as depicted in Fig. 6.10. The z-coordinate z_0 of the crown node in Fig. 6.8, which is used as the abscissa in Fig. 6.9(a), falls under the case of a symmetric displacement with $\eta_{i^*1} = 0$, since the corresponding component η_{z0} of $\boldsymbol{\eta}_1$ is equal to zero. In contrast, the z-coordinate z_1 of the first node in Fig. 6.8, which is used as the abscissa in Fig. 6.9(b), corresponds to the case of a nonsymmetric displacement with $\eta_{i^*1} \neq 0$, since the corresponding component $\eta_{z_1} = C$ of $\boldsymbol{\eta}_1$ is distinct from zero.

First, the applicability of the asymptotic laws to the symmetric displacement z_0 is investigated. Although the law (6.33) is expected to be applicable to this case, the other law (6.23) for a nonsymmetric displacement is also used for comparison. The intersection points of the imperfect paths and the straight line $\widetilde{f} + h\widetilde{z}_0 = 0$ of (6.31), and those of the paths and the parabola $\widetilde{f} + g\widetilde{z}_0^2 = 0$ of (6.21), are obtained for three values of h and g. Figure 6.11(a) portrays the $|\widetilde{f}_c|$ versus $(\widetilde{z}_0)_{|h}$ relation and

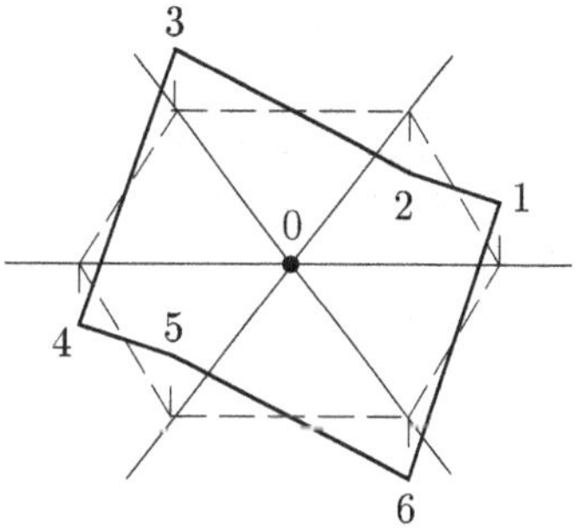

Fig. 6.10 Bird's-eye view of the z-directional components of the critical eigenvector at the bifurcation point of the regular-hexagonal truss dome.

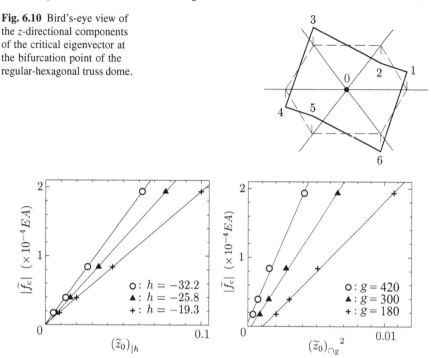

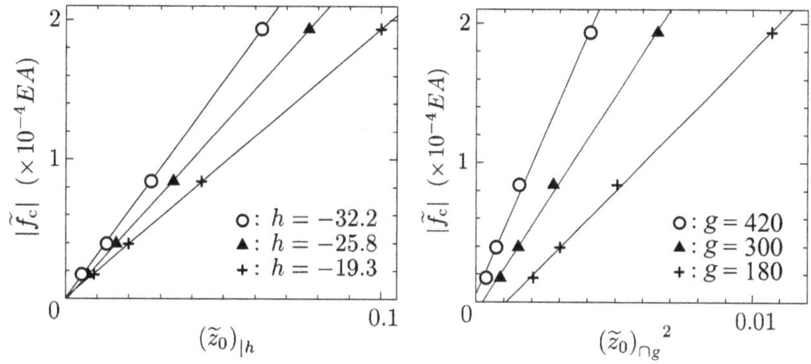

(a) $|\widetilde{f_c}|$ versus $(\widetilde{z}_0)_{|h}$ relations

(b) $|\widetilde{f_c}|$ versus $(\widetilde{z}_0)_{\cap g}{}^2$ relations

Fig. 6.11 Application of the asymptotic laws to a symmetric displacement z_0 of the regular-hexagonal truss dome.

Fig. 6.11(b) portrays the $|\widetilde{f_c}|$ versus $(\widetilde{z}_0)_{\cap g}{}^2$ relation. In this figure, the straight lines denote the least-square approximation to the data, which passes near the origin in Fig. 6.11(a) and not in Fig. 6.11(b). Consequently, the present computational results accurately follow the asymptotic law in (6.33), which represents a straight line passing the origin, and fail to satisfy the other asymptotic law in (6.23), which is not applicable to this type of displacement. The consideration of the type of displacement, accordingly, is vital in the successful application of the asymptotic laws.

Next, for the nonsymmetric displacement z_1, the $|\widetilde{f_c}|$ versus $(\widetilde{z}_1)_{\cap g}{}^2$ relation in Fig. 6.12 is obtained. The straight lines representing the least-square approximation to the presented data correlate well with these data and pass the origin. This assesses the validity and applicability of the law (6.23) that expresses the straight line passing the origin.

Finally, the equilibrium paths of the dome, portrayed by the solid lines in Fig. 6.13, are simulated by the asymptotic curves, shown by the dashed lines, which are computed, respectively, by (6.27) for Fig. 6.13(a) and (6.19) for Fig. 6.13(b). The asymptotic curves closely match the equilibrium paths for Fig. 6.13(a) and fairly match for Fig. 6.13(b). We could determine the values of the parameters E, p,

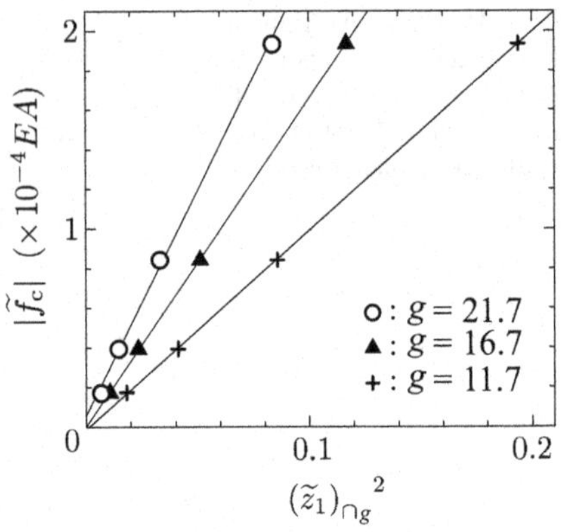

Fig. 6.12 Application of the asymptotic law (6.23) to a nonsymmetric displacement z_1 of the regular-hexagonal truss dome ($|\widetilde{f}_c|$ versus $(\widetilde{z}_1)_{\cap g}{}^2$ relation).

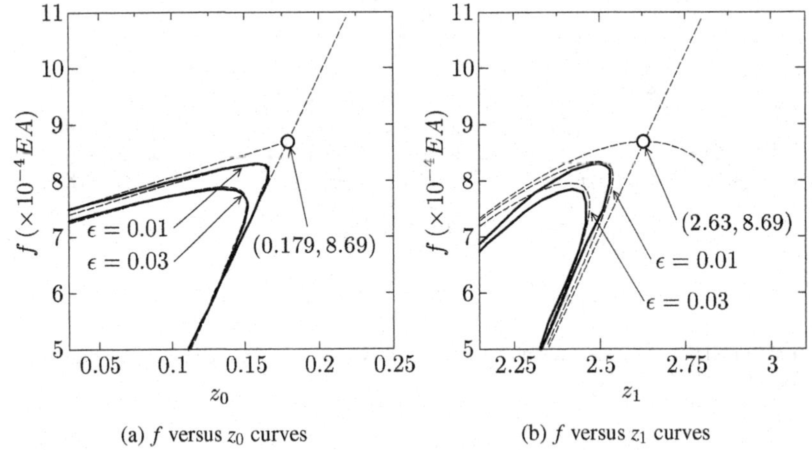

Fig. 6.13 Simulation of the equilibrium paths of the regular-hexagonal truss dome. ——: exact numerical analysis; – – –: asymptotic simulation by (6.27) for (a) and (6.19) for (b); ○: bifurcation point.

$q\varepsilon$, p^*, and $q^*\varepsilon$ in these equations using the procedure described in §6.4.2, but the following alternative procedure is adopted here to illustrate another possibility.

- The value of E, which represents the slope of the fundamental path, is chosen such that the asymptotic curve is tangential to the computational fundamental path at the bifurcation point.
- For the symmetric displacement, the value of p is chosen so that the slope of the straight line in Fig. 6.11(a) is equal to γ^* in (6.34). The value of $q\varepsilon$ is chosen based on the Koiter law (6.29).
- For the nonsymmetric displacement, the value of p^* is chosen so that the slope of the straight line in Fig. 6.12 is equal to the slope of the relation (6.23), that is, $-3(p^*)^{1/3}(g-p^*)^{2/3}/2^{2/3}$, and the value of $q^*\varepsilon$ by the Koiter law (6.20).

6.5.2 Sand Specimens

The asymptotic laws presented in §6.3 and that of the procedure for recovering the curve for the perfect system in §6.4 are applied to the results of the triaxial compression test on cylindrical sand specimens. Additional issues related to a triaxial compression test are treated in Chapter 13.

Procedure for a Single Curve

As an example of a single curve, we refer to the triaxial compression test on the cylindrical Toyoura sand specimens of Ikeda et al., 1997 [96]. These specimens have a 7 cm diameter and a 10 cm height subjected to a constant confining pressure σ_3 of 98 kPa (1 kgf/cm^2) and an increasing axial pressure (stress) of σ_1. The formulas developed in §6.3 and §6.4 are applicable for this case merely by choosing the deviatoric stress $\sigma_a = \sigma_1 - \sigma_3$ as the bifurcation parameter f and the axial strain ε_a as the displacement u.

We search for the location of the bifurcation point of specimens 4-4 and 8-1. Figure 6.14 presents the σ_a versus ε_a curves of these specimens and the rectangular areas used for the search, whereas Fig. 6.15 portrays the distribution of the inverse $1/\mathrm{Var}[E]$ of the variance $\mathrm{Var}[E]$ among E_i ($i = 1,2,\ldots$) in the rectangular areas of $(\varepsilon_a, \sigma_a)$ in Fig. 6.14. This figure clearly portrays for each specimen the presence of the local maximum of $1/\mathrm{Var}[E]$ (minimum of $\mathrm{Var}[E]$), which corresponds to the bifurcation point. In the course of the search, the values of the parameters p, E, and $q\varepsilon$, listed at the bottom of Fig. 6.15, were obtained for each specimen.

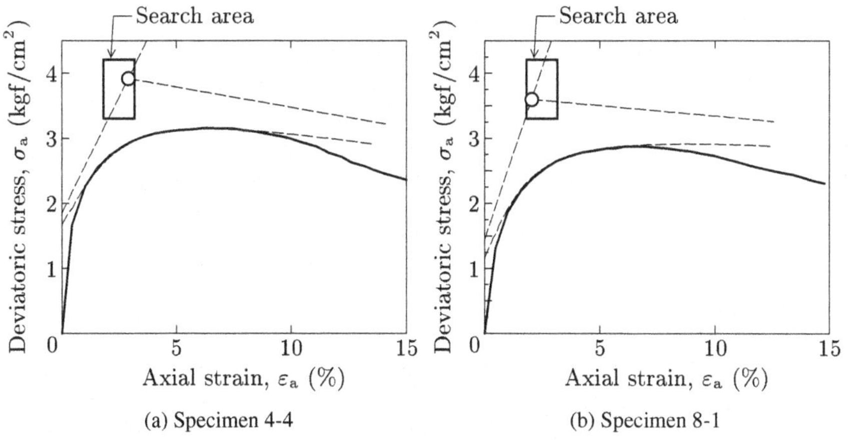

Fig. 6.14 Deviatoric stress σ_a versus axial strain ε_a curves for the sand specimens and their simulation. ——: experimental (imperfect) curves; – – –: computed curves; the solid rectangular area: area for bifurcation point search; ○: bifurcation point; 1 kgf/cm^2 = 98 kPa.

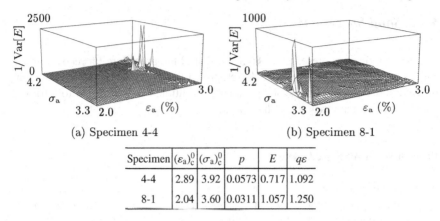

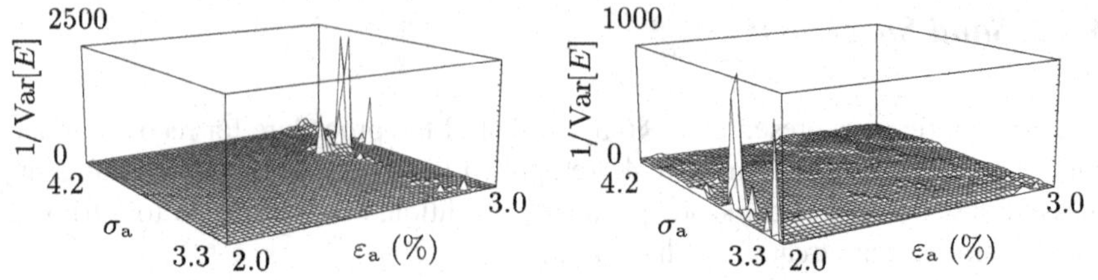

(a) Specimen 4-4 (b) Specimen 8-1

Specimen	$(\varepsilon_a)_c^0$	$(\sigma_a)_c^0$	p	E	$q\varepsilon$
4-4	2.89	3.92	0.0573	0.717	1.092
8-1	2.04	3.60	0.0311	1.057	1.250

Fig. 6.15 Distribution of the inverse of the variance Var[E] among E_i ($i = 1,2,\ldots$) in the rectangular areas of the possible bifurcation point $(\varepsilon_a, \sigma_a)$ in Fig. 6.14 (1 kgf/cm^2 = 98 kPa).

Figure 6.14 presents by dashed curves the simulation of the σ_a versus ε_a relations for the two specimens by (6.27), using the values of the parameters obtained by the search. The theoretical curves correlate fairly well with the experimental curves, especially in the neighborhood of the bifurcation point. This is consistent with the local nature of the present theory, which is more accurate in the neighborhood and less accurate away from it.

The analysis in Fig. 6.14 entails the following physical interpretation. During the first stage of the loading ($\varepsilon_a < 0.5\%$), the slope of the curve is very steep and the specimen is nearly elastic. Its slope is then greatly reduced during $0.5 < \varepsilon_a < 7.0\%$, and the specimen softens rapidly. It remains fairly constant near the peak at approximately $\varepsilon_a = 7.0\%$, and a gradual softening follows. In soil mechanics, it is customary to attribute the degradation of the slope of the curve ($0.5 < \varepsilon_a < 7.0\%$) mainly to *material softening* of soils, and the softening after the peak to the direct bifurcation.[2] This customary understanding, however, must be reconsidered in view of the analysis in Fig. 6.14; the bifurcation point is located approximately at $\varepsilon_a = 2.0\%$ (the range of rapid softening), instead of at $\varepsilon_a = 7.0\%$ (the peak). This issue is considered again in §14.5 based on the standpoint of recursive bifurcation.

Procedure for a Series of Curves

As a statistical databank for the shear behavior of sand specimens, two series of data on sand specimens have been gathered.[3] Series A consists of 50 specimens with a constant 7 cm diameter and a 15 cm height, and Series B comprises 18 specimens

[2] See, for example, Vardoulakis and Sulem, 1995 [202] for material softening.

[3] These specimens were subjected to a confining pressure σ_3 of 98 kPa (1 kgf/cm^2); see Ikeda, Chida, and Yanagisawa, 1997 [82] for details.

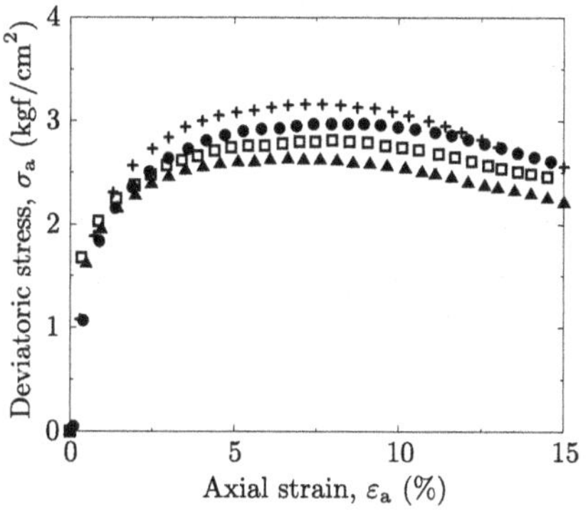

(a) σ_a versus ε_a curves

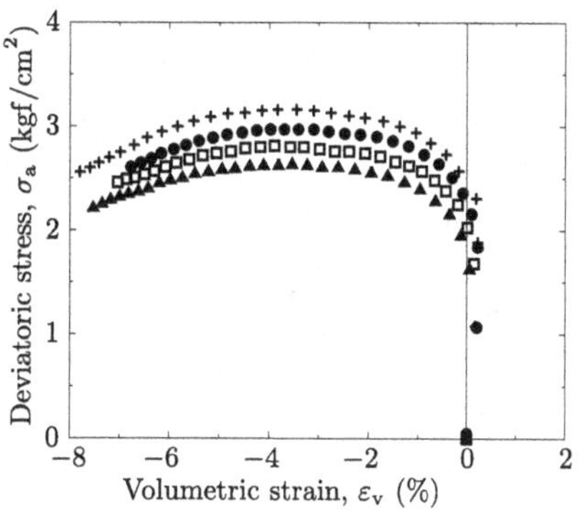

(b) σ_a versus ε_v curves

Fig. 6.16 Strength variation of sand specimens for Series B. $1\,\mathrm{kgf/cm^2} = 98\,\mathrm{kPa}$.

with a 7 cm diameter and a 10 cm height. Figure 6.16(a) portrays examples of the deviatoric stress σ_a versus axial strain ε_a curves, and Fig. 6.16(b) shows the deviatoric stress σ_a versus volumetric strain ε_v curves observed for Series B. The curves for both Series A and B apparently vary test by test.

The axial strain ε_a and the volumetric strain ε_v are symmetric displacements with $\eta_{i^*1} = 0$ for which imperfect curves exist only above or below the fundamental path, as shown in Fig. 6.6(b). Therefore, the asymptotic laws for symmetric displacements are employed in the sequel. In addition, we employ an assumption that the peaks of the curves are governed by an unstable pitchfork bifurcation point.

First, we specifically examine the variation of the σ_a versus ε_a curves of Series A. Based on the procedure in §6.4.2, the location of the bifurcation point $((\varepsilon_\mathrm{a})_\mathrm{c}^0, (\sigma_\mathrm{a})_\mathrm{c}^0) = (2.37, 5.05)$ was chosen such that the relation (6.33) holds most accurately. The parameter h for the straight line $\widetilde{\sigma}_\mathrm{a} + h\widetilde{\varepsilon}_\mathrm{a} = 0$ was chosen to be 0.18. In the course of this it was noted that this relation is not sensitive to the value of h but to the location $((\varepsilon_\mathrm{a})_\mathrm{c}^0, (\sigma_\mathrm{a})_\mathrm{c}^0)$ of the bifurcation point, consistent with the nature of the relation, which holds for any h but only for the true $((\varepsilon_\mathrm{a})_\mathrm{c}^0, (\sigma_\mathrm{a})_\mathrm{c}^0)$.

The magnitudes of the imperfection ε were computed from the two-thirds power law (6.29). The values of the incremental axial strain $(\widetilde{\varepsilon}_\mathrm{a})_{|h}$ at the intersection points of the straight line $\widetilde{\sigma}_\mathrm{a} + h\widetilde{\varepsilon}_\mathrm{a} = 0$ $(h = 0.18)$ and the experimental σ_a versus ε_a curves were obtained. For the 50 specimens, the incremental maximum stress $|(\widetilde{\sigma}_\mathrm{a})_\mathrm{c}| = |(\sigma_\mathrm{a})_\mathrm{c} - (\sigma_\mathrm{a})_\mathrm{c}^0|$ versus incremental strain $(\widetilde{\varepsilon}_\mathrm{a})_{|h}$ relation is depicted in Fig. 6.17(a). The straight line expresses the least-square approximation of this relation. This line passes near the origin; therefore, it agrees with the imperfection sensitivity law (6.33) that expresses the straight line passing the origin. That relation denoted as (•) correlates well with the straight line when $|(\widetilde{\sigma}_\mathrm{a})_\mathrm{c}|$ is small, although it is less accurate when it is large because of the asymptotic nature of the law (6.33). Such good correlation for small $|(\widetilde{\sigma}_\mathrm{a})_\mathrm{c}|$ ensures the validity of the present method to explain the variation of soil shear behavior by the variation of imperfections.

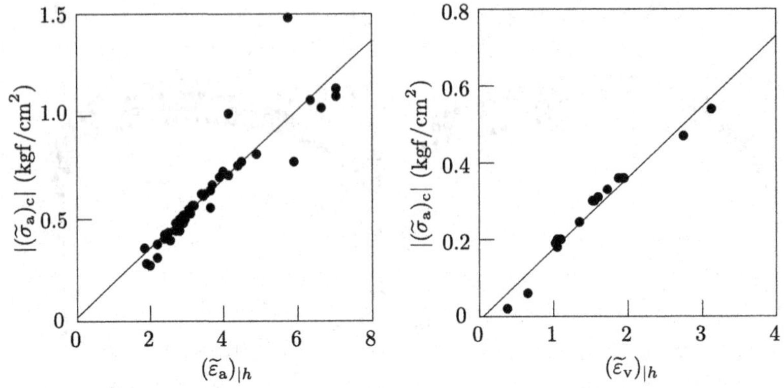

(a) $|(\widetilde{\sigma}_a)_c|$ versus $(\widetilde{\varepsilon}_a)_{|h}$ relation (Series A) (b) $|(\widetilde{\sigma}_a)_c|$ versus $(\widetilde{\varepsilon}_v)_{|h}$ relation (Series B)

Fig. 6.17 Application of asymptotic law (6.33). ——: regression line; $1\,\mathrm{kgf/cm^2} = 98\,\mathrm{kPa}$.

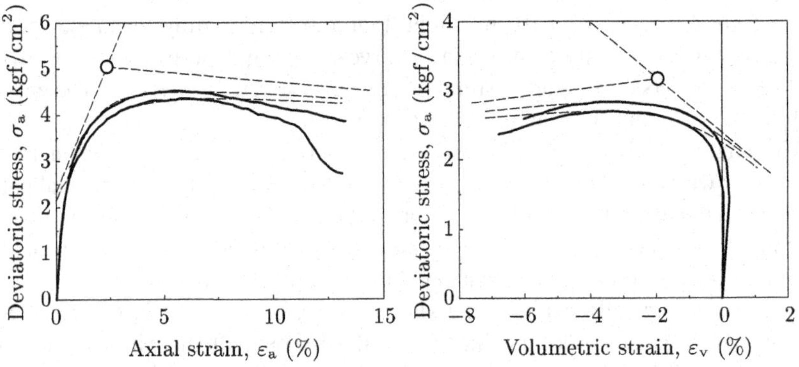

(a) σ_a versus ε_a curves for Series A (b) σ_a versus ε_v curves for Series B

	$(\varepsilon_a)_c^0$ or $(\varepsilon_v)_c^0$	$(\sigma_a)_c^0$	h	p	E
Series A	2.37	5.05	0.18	0.042	1.15
Series B	−1.95	3.17	−0.19	0.053	−0.40

Fig. 6.18 Simulation of experimental curves. ——: experimental curve; – – –: simulated curve; ○: bifurcation point; $1\,\mathrm{kgf/cm^2} = 98\,\mathrm{kPa}$.

Figure 6.18(a) presents results of the simulation of the experimental σ_a versus ε_a curves of two representative specimens by the theoretical curves (6.27). The same set of values of the parameters and different values of the imperfection ε are employed for these specimens. Here the values of the parameters p and E that are used are listed at the bottom of Fig. 6.18. The theoretical curves correlate fairly well with the experimental curves near the bifurcation point, but less accurately away from it. Such inaccuracy might be attributable to the asymptotic nature of the present method. Nevertheless, it is premature at this state of research to draw a definite con-

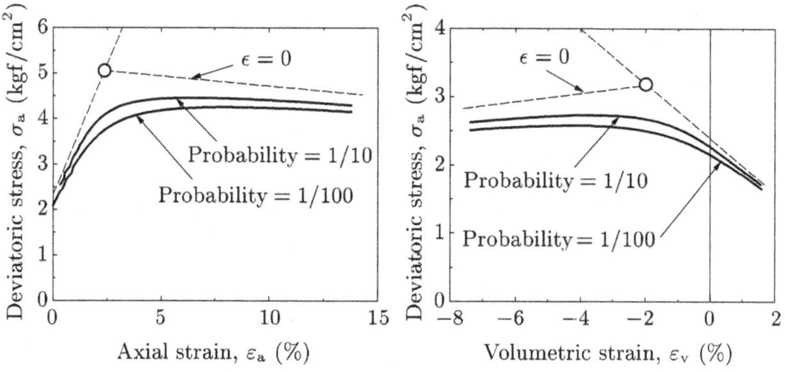

(a) σ_a versus ε_a curves for Series A (b) σ_a versus ε_v curves for Series B

Fig. 6.19 Experimental curves with the probabilities of occurrence 1/100 and 1/10. - - -: simulated curve; ○: bifurcation point of the perfect system; 1 kgf/cm^2 = 98 kPa.

clusion from these theoretical curves, as the experimental curves in reality might be obscured by the presence of mode switching and recursive bifurcation, which are widely observed for structures[4] and also for soil specimens (cf., §13.3).

Next, the variation of the volumetric strain curves of Series B consisting of 18 specimens is investigated. Figure 6.17(b) presents the $|(\widetilde{\sigma}_a)_c|$ versus $(\widetilde{\varepsilon}_v)_{|h}$ relation, which is in fair agreement with the straight line. The simulation of the two representative experimental σ_a versus ε_v curves is shown in Fig. 6.18(b). The theoretical curves correlate fairly well with the experimental curves in the neighborhood of the bifurcation point, but they are less accurate away from it again due to the asymptotic nature of the present theory.

Finally, according to formula (5.18) for the reliability function of the unstable pitchfork bifurcation point, the maximum stresses $(\sigma_a)_c$ that occur with the probabilities of 1/100 and 1/10 are computed. Then, by the Koiter law (6.29), the values of the scaled imperfection $q\varepsilon$ are evaluated for the specified values of $(\sigma_a)_c$. Figure 6.19 depicts the simulation of the σ_a versus ε_a curves and the σ_a versus ε_v curves for those possibilities. The present method, which can provide us with meaningful statistical information, even for a limited number of data, might be useful for the design of the statistical strength of soil.

Problems

6-1 Plot relation (6.12) for f versus x curves in Fig. 6.3(a) and the relation (6.33) for f versus y curves in Fig. 6.3(b).

[4] See, for example, Yamaki, 1984 [213] and Ikeda, Murota, and Fujii, 1991 [94].

6-2 Plot $\widetilde{u}_{\cap g}$ versus $\varepsilon^{1/3}$ relation for the cantilever supported by a linear spring depicted in Fig. 1.2 in §1.2.2.

6-3 Draw f versus y curves for the nonshallow truss in Fig. 3.8(b) in §3.3.2 for $\varepsilon = 0, 0.01, 0.05$, and 0.1, and obtain $\widetilde{f_{\rm c}}$ versus $\widetilde{u}_{|h}$ relation.

6-4 Consider the imperfect f versus y curves in Problem 6-3, and obtain the location of the bifurcation point from the single curve with $\varepsilon = 0.01$ using the procedure in §6.4.1, and from those curves using the procedure in §6.4.2.

Summary

- Asymptotic laws for experimentally observed bifurcation diagrams have been presented.
- The importance of identifying the symmetry of the observed displacement and the type of bifurcation point has been pointed out.
- Through the application to numerical and experimental examples, the asymptotic laws have been demonstrated as capable of describing imperfect bifurcation behaviors.

Part II
Imperfect Bifurcation of Symmetric Systems

Symmetry is found literally everywhere as has been introduced, for example, by Weyl, 1952 [209]; Stewart and Golubitsky, 1992 [186]; Rosen, 1995 [174]; and Mainzer, 2005 [132]. One might be amazed at the symmetry and orderliness of the honeycomb, which is made up of a number of hexagons arranged in order. At the expense of beauty and orderliness, symmetric systems often undergo "pattern selection" or "pattern formation" (see, e.g., Chadam et al., 1996 [22]). In fluid mechanics, nonlinear mathematics, and other fields of study, patterns are well known to be selected or formed through recursive bifurcation which "breaks" symmetry. The Couette–Taylor flow in a hollow cylinder, which is a rotating annular of fluid, displays wave patterns with various symmetries through pattern selection (e.g., Taylor, 1923 [188]). The convective motion of fluid in the Benard problem displays regularly arrayed hexagons (e.g., Koschmieder, 1974 [124]).

Symmetry is described by a group. Moreover, bifurcation structures near singular points can be investigated theoretically using group-theoretic bifurcation theory in nonlinear mathematics. We can find a group G that labels the symmetry of the system, and a hierarchy of subgroups $G \to G_1 \to G_2 \to \cdots$ that characterizes the recursive occurrence of bifurcations. Here $\to$ denotes a bifurcation, and G_i $(i = 1, 2, \ldots)$ stand for the nesting subgroups of G that label the reduced symmetry of the bifurcated solutions. Knowledge of such a hierarchy is crucial for the complete description of recursive bifurcation behavior.

In the modeling of the bifurcation phenomena of a symmetric system, we must find the group that labels the symmetry of the system because the hierarchy of subgroups presented above is dependent on the group. To avoid sophisticated mathematical concepts, in this part, Part II, we specifically address the apparent geometrical symmetry labeled by the simplest groups: the dihedral and cyclic groups. The bifurcation of systems with symmetries of various kinds labeled by other groups is studied in the next part, Part III.

Multiple critical points, at which more than one eigenvalue of the Jacobian matrix simultaneously vanish, appear in symmetric systems. The critical points of these systems accordingly consist of simple critical points and multiple critical points. This shows a sharp contrast with the case of systems without symmetries or with a reflectional symmetry, where only simple critical points appear generically.

Part II comprises six chapters. In Chapter 7 a brief account of group-theoretic bifurcation theory is presented as a basic mathematical tool to describe the bifurcation behavior of a symmetric system. In Chapter 8, using elementary calculations based on the theory in Chapter 7, the rules of the perfect and imperfect bifurcation behavior of systems with dihedral or cyclic group symmetries are obtained. These rules are put to use in the description of the perfect bifurcation behavior of regular polygonal truss domes. In Chapter 9 a procedure to obtain the worst imperfection pattern vector is presented based on the concept of the group equivariance of an imperfect system. In Chapter 10 the probabilistic variation of critical loads is formulated for imperfections subject to a multivariate normal distribution. In Chapter 11 the perfect and imperfect bifurcation behaviors of realistic symmetric systems are investigated using the tools presented above. In Chapter 12 the theory on block-diagonalization is presented.

Chapter 7
Group-Theoretic Bifurcation Theory

7.1 Introduction

Qualitative aspects of symmetry-breaking bifurcation can be described by group-theoretic bifurcation theory. In view of the symmetry of the system under consideration, possible critical points and bifurcated solutions can be classified, and the behavior of these solutions in a neighborhood of each critical point can be investigated thoroughly by the Liapunov–Schmidt reduction. An extremely important finding of this theory is that the mechanism of such bifurcation does not depend on individual material or physical properties but on the symmetry of the system under consideration.

The main ideas of group-theoretic bifurcation analysis are explained in this chapter as an informal introduction for engineers.[1] We restrict ourselves to finite-dimensional equations and finite groups to avoid the level of mathematical sophistication necessary to address differential equations and continuous groups, for which the reader is referred to standard textbooks.[2] More issues related to group symmetry in nonlinear analysis are available in the literature.[3]

At the beginning of this chapter, bifurcation behavior due to bilateral symmetry is presented in §7.2 in terms of a simple prototype example of a pitchfork bifurcation point. In the following sections mathematical tools for tackling bifurcation behavior due to symmetry are introduced.

The tools presented in this chapter include the following:

- Groups in §7.3 and linear representations of finite groups in §7.4 to formulate symmetry in mathematical terms,

---

[1] In this second edition of the book we have reinforced the theoretical background of group representation and its application, including the block-diagonalization method.

[2] See, for example, Sattinger, 1979 [176], 1980 [177]; Golubitsky and Schaeffer, 1985 [62]; and Golubitsky, Stewart, and Schaeffer, 1988 [64]).

[3] See, for example, Olver, 1986 [158], 1995 [159]; Mitropolsky and Lopatin, 1988 [139]; Allgower, Böhmer, and Golubitsky, 1992 [1]; Marsden and Ratiu, 1994 [134]; and Hoyle, 2006 [75].

K. Ikeda and K. Murota, *Imperfect Bifurcation in Structures and Materials*, Applied Mathematical Sciences 149, DOI 10.1007/978-1-4419-7296-5_7,

- Group equivariance in §7.5.1 to represent the symmetry of the governing equation (including imperfections),
- Liapunov–Schmidt reduction in §7.5.2 to derive the bifurcation equation and inheritance of symmetry by this equation as a means to exploit symmetry,
- Isotropy group in §7.6 to describe the symmetry of solutions,
- Orbit in §7.6 to represent the essential distinction of bifurcated solutions,
- Block-diagonalization in §7.7.1 and §7.7.2 to reveal the symmetry inherent in the Jacobian matrix and the imperfection sensitivity matrix.

Finally in §7.8 the tools introduced in this chapter are applied to simple examples. The tools form theoretical bases of the following chapters.

- Inheritance of symmetry is employed in the study of systems with group symmetry in Chapters 8 and 13–16.
- Block-diagonalization of the imperfect sensitivity matrix is used in the study of imperfections in Chapters 9 and 10.
- Block-diagonalization of the Jacobian matrix is utilized in Chapters 11 and 12.

7.2 Bifurcation Due to Reflection Symmetry

We explain how symmetry of a system gives rise to bifurcation behavior using a simple example with bilateral symmetry.

Consider the rigid bar model, a rigid bar of length L supported by a linear spring of spring constant k, as portrayed in Fig. 7.1. It has a single displacement variable u that denotes the tilted angle of the rigid bar; the bar stands upright for $u = 0$.

The total potential energy is expressed as

$$U(u,f) = \frac{1}{2}k(L\sin u)^2 - kLf \cdot L(1-\cos u), \tag{7.1}$$

and the equilibrium equation is given by

$$F(u,f) \equiv \frac{\partial U}{\partial u} = kL^2 \sin u(\cos u - f) = 0. \tag{7.2}$$

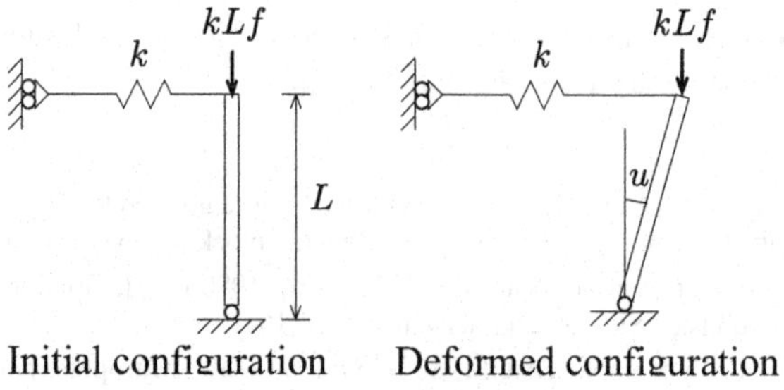

Fig. 7.1 Rigid bar model.

The system is invariant to the *reflection* σ which is defined by

$$\sigma : \; u \mapsto -u. \tag{7.3}$$

In fact, the potential (7.1) is invariant under this action, as it satisfies the condition of *invariance*

$$U(u, f) = U(-u, f). \tag{7.4}$$

As a consequence of this invariance, the equilibrium equation satisfies the condition

$$-F(u, f) = F(-u, f). \tag{7.5}$$

Equation (7.5) shows that $F(u, f)$ is an odd function in u. Hence we can write

$$F(u, f) = uF_1(u^2, f)$$

for some function F_1. Therefore, $u = 0$ is a trivial solution of (7.2) (i.e., $F(0, f) = 0$ for all f) and a bifurcated solution can arise from $F_1(u^2, f) = 0$. In this way, symmetry causes bifurcation.

The discussion presented above can be described using general notations as follows. In addition to the reflection σ, we consider the *identity transformation* $e : \; u \mapsto u$. Then we have $\sigma^2 = e$ and

$$G = \{e, \sigma\} \tag{7.6}$$

forms a *group* (see §7.3 for the definition of a group). We represent the action of σ in (7.3) by a 1×1 matrix defined as $T(\sigma) = -1$; that is,

$$T(\sigma)u = -u.$$

Setting $T(e) = 1$, we obtain a function $T : G \to \{1, -1\}$.

Then the invariance (7.4) of the potential function $U(u, f)$ can be written as

$$U(u, f) = U(T(g)u, f), \qquad g \in G, \tag{7.7}$$

and the oddness (7.5) of $F(u, f)$ with respect to u as

$$T(g)F(u, f) = F(T(g)u, f), \qquad g \in G. \tag{7.8}$$

The equation of this form, described with a general group G, turns out to be the symmetry condition of a symmetric system in general, to be called *equivariance* (cf., §7.5). The symmetry condition (7.8) for the equation F makes sense for a nonreciprocal (nonpotential) system as well. The invariance (7.7) of the potential U above plays only a supplementary role.

7.3 Group

Groups are introduced along with associated geometric transformations.

A set G is called a *group* if, for any pair of elements g and h of G, an element of G denoted as gh and called the *product* of g and h is specified, and if the following (i) through (iii) are satisfied.

(i) The *associative law* holds as

$$(g\,h)\,k = g\,(h\,k), \qquad g,h,k \in G.$$

(ii) There exists an element $e \in G$ (called the *identity element*) such that

$$e\,g = g\,e = g, \qquad g \in G.$$

(iii) For any $g \in G$ there exists $h \in G$ (called the *inverse* of g) such that

$$g\,h = h\,g = e.$$

It can be shown that the identity element e in (ii) is uniquely determined. The inverse of g is unique for each g; and denoted as g^{-1}. In this chapter we assume that G is a *finite group* (i.e., a group consisting of a finite number of elements) and denote its *order* (= the number of elements) by $|G|$. In general, a group with a larger order expresses a higher symmetry.

The structure of a group can be expressed explicitly by the *multiplication table*, which gives the product gh of elements g and h as the (g,h) entry of the table. For example, consider a group

$$G = \{g_1, g_2, g_3\}$$

consisting of three elements g_1, g_2, and g_3 with the relations

$$g_1g_1 = g_2g_3 = g_3g_2 = g_1,$$
$$g_1g_2 = g_2g_1 = g_3g_3 = g_2,$$
$$g_3g_1 = g_2g_2 = g_1g_3 = g_3.$$

This group can be represented by the following multiplication table,

	g_1	g_2	g_3
g_1	g_1	g_2	g_3
g_2	g_2	g_3	g_1
g_3	g_3	g_1	g_2

By a *subgroup* of G we means a nonempty subset H of G that also forms a group with respect to the same product operation defined in G. A subgroup H of G is called a *proper subgroup* if it is distinct from G. In Example 7.1 below, the groups G_1 and G_2 are proper subgroups of G_3. The *index* of a subgroup H of G is defined as the ratio $|G|/|H|$ of the orders of G and H. Two subgroups, say, H and K of a group G

are said to be *conjugate*, if

$$K = \{g^{-1}hg \mid h \in H\} \tag{7.9}$$

holds for some $g \in G$.

Example 7.1. Examples of groups are presented. Using the elements s and r that satisfy the defining relations,

$$s^2 = e, \qquad r^3 = e, \qquad (sr)^2 = e,$$

we can define groups

$$\begin{aligned} G_1 &= \langle s \rangle = \{e, s\}, \\ G_2 &= \langle r \rangle = \{e, r, r^2\}, \\ G_3 &= \langle r, s \rangle = \{e, r, r^2, s, sr, sr^2\}. \end{aligned}$$

Here $\langle \cdot \rangle$ denotes a *group generated* by its included element(s). For example, G_3 is generated by the two elements r and s, which, for instance, generate the element $sr^2 = s \cdot r \cdot r$, as the notation indicates. Table 7.1 lists the multiplication tables for these three groups G_1 to G_3. One can see from these tables that G_1 to G_3 satisfy the properties (i) to (iii) of a group. □

Table 7.1 Multiplication tables for the groups in Example 7.1

(a) $G_1 = \{e, s\}(= \mathrm{D}_1)$

	e	s
e	e	s
s	s	e

(b) $G_2 = \{e, r, r^2\}(= \mathrm{C}_3)$

	e	r	r^2
e	e	r	r^2
r	r	r^2	e
r^2	r^2	e	r

(c) $G_3 = \{e, r, r^2, s, sr, sr^2\}(= \mathrm{D}_3)$

	e	r	r^2	s	sr	sr^2
e	e	r	r^2	s	sr	sr^2
r	r	r^2	e	sr^2	s	sr
r^2	r^2	e	r	sr	sr^2	s
s	s	sr	sr^2	e	r	r^2
sr	sr	sr^2	s	r^2	e	r
sr^2	sr^2	s	sr	r	r^2	e

Example 7.2. The *cyclic group* of degree n, conventionally denoted as C_n, is a group of order n consisting of power products of a single element r with $r^n = e$. That is,

$$\mathrm{C}_n = \langle r \rangle = \{e, r, r^2, \ldots, r^{n-1}\}$$

with $r^i r^j = r^{i+j}$. The group G_2 in Example 7.1 above is C_3. The *dihedral group* of degree n, denoted as D_n, is a group of order $2n$ defined as

$$\mathrm{D}_n = \langle r, s \rangle = \{e, r, \ldots, r^{n-1}, s, sr, \ldots, sr^{n-1}\},$$

where r and s are assumed to satisfy the relations

$$r^i r^j = r^{i+j}, \qquad r^n = s^2 = (sr)^2 = e.$$

The groups G_1 and G_3 in Example 7.1 above are D_1 and D_3, respectively. □

Two groups G_1 and G_2 are said to be *isomorphic*, denoted as $G_1 \simeq G_2$, if there exists a one-to-one correspondence $\phi : G_1 \to G_2$ such that

$$\phi(g)\phi(h) = \phi(gh), \qquad g,h \in G_1.$$

For two groups G_1 and G_2, the set

$$G_1 \times G_2 = \{(g_1,g_2) \mid g_1 \in G_1, g_2 \in G_2\}, \tag{7.10}$$

which is the set-theoretic direct product, forms a group with the product operation defined as

$$(g_1,g_2)(h_1,h_2) = (g_1h_1,g_2h_2), \qquad g_1,h_1 \in G_1;\ g_2,h_2 \in G_2.$$

This is called the *direct product* of G_1 and G_2.

7.4 Group Representation

Group representation that expresses the action of a group is introduced, and the theory of group representation for finite groups that is sufficient for Part II is presented.[4]

7.4.1 Basic Concepts

Definition

Let V be a finite-dimensional vector space over $F = \mathbb{R}$ or $\mathbb{C}$, where $\mathbb{R}$ is the field of real numbers and $\mathbb{C}$ is the field of complex numbers. Denote by $\mathrm{GL}(V)$ the group of all nonsingular linear transformations of V onto itself, and by $\mathrm{GL}(N,F)$ the group of all nonsingular matrices over F of order N.

A *representation* of G on V means a mapping $\hat{T} : G \to \mathrm{GL}(V)$ such that[5]

$$\hat{T}(gh) = \hat{T}(g)\hat{T}(h), \qquad g,h \in G. \tag{7.11}$$

---

[4] In Part III, however, we refer to a class of infinite groups, compact groups, for which the important properties are retained. For complete accounts, the reader is referred to textbooks such as Curtis and Reiner, 1962 [34]; Hamermesh, 1962 [68]; Miller, 1972 [138]; Serre, 1977 [181]; and Jacobson, 1989 [109].

[5] In mathematical terminology the condition (7.11) says that $\hat{T}$ is a *homomorphism* from G to $\mathrm{GL}(V)$.

We call V the *representation space* and $N = \dim(V)$ the *dimension*, or the *degree*, of the representation.

The matrix counterpart of this abstract concept is a mapping $T : G \to \mathrm{GL}(N, F)$ that satisfies

$$T(gh) = T(g)T(h), \qquad g, h \in G. \tag{7.12}$$

Such a mapping T, or equivalently, a family of $N \times N$ nonsingular matrices $\{T(g) \mid g \in G\}$, is called a *matrix representation* of G.

With a representation $\hat{T}$ on V we can associate a matrix representation T in a natural way by choosing a basis $B = (\boldsymbol{u}_1, \ldots, \boldsymbol{u}_N)$ of V and by defining a family of matrices $T(g) = T_B(g) = (T_{ij}(g) \mid i, j = 1, \ldots, N)$ by

$$\hat{T}(g)\boldsymbol{u}_j = \sum_{i=1}^{N} T_{ij}(g)\boldsymbol{u}_i, \qquad g \in G. \tag{7.13}$$

A change of basis from $B = (\boldsymbol{u}_1, \ldots, \boldsymbol{u}_N)$ to $C = (\boldsymbol{v}_1, \ldots, \boldsymbol{v}_N)$ results in replacement of the matrix $T_B(g)$ by another matrix $T_C(g) = Q^{-1}T_B(g)Q$ for each $g \in G$, where $Q = (Q_{ij}) \in \mathrm{GL}(N, F)$ is a nonsingular matrix such that $\boldsymbol{v}_j = \sum_{i=1}^{N} Q_{ij}\boldsymbol{u}_i$ for $j = 1, \ldots, N$. Note that Q is a constant matrix independent of $g \in G$.

A representation $\hat{T}$ is said to be a *unitary representation* if the linear transformation $\hat{T}(g)$ is unitary (with respect to the unit metric) for each $g \in G$, that is, if

$$(\hat{T}(g)\boldsymbol{x}, \hat{T}(g)\boldsymbol{y}) = (\boldsymbol{x}, \boldsymbol{y}), \qquad \boldsymbol{x}, \boldsymbol{y} \in V, \;\; g \in G, \tag{7.14}$$

where $(\cdot, \cdot)$ denotes the inner product. This condition is equivalent to

$$T(g)^* T(g) = I_N, \qquad g \in G, \tag{7.15}$$

for the matrix representation T of $\hat{T}$ with respect to an orthonormal basis, where $(\cdot)^*$ signifies the conjugate transpose of a matrix. In the case of $F = \mathbb{R}$ we often say *orthogonal representation* instead of unitary representation.

Example 7.3. A one-dimensional representation can be obtained by defining $T(g) = 1$ for all $g \in G$. This is called the *unit representation*. □

Example 7.4. A one-dimensional representation of group $\mathrm{D}_1 = \{e, s\}$ in Table 7.1(a) is given by

$$T(e) = 1, \qquad T(s) = -1.$$

This is a matrix representation indeed, since it meets the conditions

$$\begin{aligned}
T(e)T(e) &= 1 \times 1 = 1 = T(e), \\
T(e)T(s) &= 1 \times (-1) = -1 = T(es), \\
T(s)T(e) &= (-1) \times 1 = -1 = T(se), \\
T(s)T(s) &= (-1) \times (-1) = 1 = T(ss)
\end{aligned}$$

required in (7.12). A two-dimensional representation of the group D_1 is given, for example, by

$$T(e) = \begin{pmatrix} 1 & 0 \\ 0 & 1 \end{pmatrix}, \qquad T(s) = \begin{pmatrix} 1 & 0 \\ 0 & -1 \end{pmatrix}.$$

□

Example 7.5. For the group $D_3 = \{e, r, r^2, s, sr, sr^2\}$ (cf., Table 7.1(c) in §7.3), define

$$T(e) = \begin{pmatrix} 1 & 0 \\ 0 & 1 \end{pmatrix}, \qquad T(r) = \begin{pmatrix} \alpha & -\beta \\ \beta & \alpha \end{pmatrix}, \qquad T(r^2) = \begin{pmatrix} \alpha & \beta \\ -\beta & \alpha \end{pmatrix},$$

$$T(s) = \begin{pmatrix} 1 & 0 \\ 0 & -1 \end{pmatrix}, \quad T(sr) = \begin{pmatrix} \alpha & -\beta \\ -\beta & -\alpha \end{pmatrix}, \quad T(sr^2) = \begin{pmatrix} \alpha & \beta \\ \beta & -\alpha \end{pmatrix}$$

with $\alpha = \cos(2\pi/3)$ and $\beta = \sin(2\pi/3)$. Then this T, satisfying (7.12), is a matrix representation of the group D_3. □

Example 7.6. A representation can be obtained from permutations as follows. Let P be a finite set, and assume that G acts on P through permutations, which means that a permutation $\pi(g)$ of P is given for each $g \in G$ and $\pi(gh) = \pi(g)\pi(h)$ holds for all $g, h \in G$. Consider the vector space $V = F^P$, that is, the linear space consisting of "formal" sums $\sum_{p \in P} f_p \boldsymbol{u}_p$ $(f_p \in F)$ with basis $(\boldsymbol{u}_p \mid p \in P)$, and define $\hat{T} : G \to \mathrm{GL}(V)$ by

$$\hat{T}(g)\boldsymbol{u}_p = \boldsymbol{u}_{\pi(g)p}, \qquad p \in P,$$

where $\pi(g)p$ is the element of P to which p is moved by $\pi(g)$. Then the condition (7.11) is satisfied as a consequence of $\pi(gh) = \pi(g)\pi(h)$; accordingly $\hat{T}$ is qualified as a representation of dimension $|P|$. This is called a *permutation representation*. For each $g \in G$ the matrix $T(g)$ representing $\hat{T}(g)$ is a permutation matrix. □

The *direct sum* $\hat{T}_1 \oplus \hat{T}_2$ of two representations $\hat{T}_1$ and $\hat{T}_2$ is a representation on the direct sum $V_1 \oplus V_2$ of the representation spaces V_1 and V_2, and is defined by

$$(\hat{T}_1 \oplus \hat{T}_2)(g) = \hat{T}_1(g) \oplus \hat{T}_2(g), \qquad g \in G.$$

The direct sum of two matrix representations T_1 and T_2 is given as the family of their direct sums (block-diagonal matrices)

$$T_1(g) \oplus T_2(g) = \begin{pmatrix} T_1(g) & O \\ O & T_2(g) \end{pmatrix}$$

indexed by $g \in G$. The dimension of the direct sum representation is equal to the sum of the dimensions of T_1 and T_2.

The *tensor product* $\hat{T}_1 \otimes \hat{T}_2$ of two representations $\hat{T}_1$ and $\hat{T}_2$ is a representation on the tensor product $V_1 \otimes V_2$ of the representation spaces V_1 and V_2, and is defined by

$$(\hat{T}_1 \otimes \hat{T}_2)(g) = \hat{T}_1(g) \otimes \hat{T}_2(g), \qquad g \in G.$$

The tensor product of two matrix representations T_1 and T_2 is given as the family of matrix tensor products $T_1(g) \otimes T_2(g)$ indexed by $g \in G$. The dimension of the tensor product representation is equal to the product of the dimensions of T_1 and T_2.

Remark 7.1. We may identify, mostly in this book, representations $\hat{T}$ in the abstract sense and their concrete matrix representations T with respect to particular choices of bases. Yet when we talk of computational efficiency we must distinguish T from $\hat{T}$. Choosing "good" bases for numerical computations is an interesting issue, which is addressed in Chapter 12. □

Equivalence

Let $\hat{T}_1$ and $\hat{T}_2$ be representations of G with respective representation spaces V_1 and V_2 over F. We say that $\hat{T}_1$ and $\hat{T}_2$ are *equivalent* if there exists a nonsingular (bijective) linear map $\hat{Q} : V_1 \to V_2$ such that

$$\hat{T}_1(g) = \hat{Q}^{-1}\hat{T}_2(g)\hat{Q}, \qquad g \in G.$$

We also say that two matrix representations T_1 and T_2 of G are *equivalent* if there exists a nonsingular matrix Q such that

$$T_1(g) = Q^{-1}T_2(g)Q, \qquad g \in G. \tag{7.16}$$

Any matrix representation is equivalent to a unitary representation (cf., Remark 7.2 below). Two representations are said to be *inequivalent* if they are not equivalent.

Remark 7.2. A proof for the equivalence of an arbitrary matrix representation T to a unitary representation is given here. Define a matrix

$$S = \sum_{g \in G} T(g)^* T(g)$$

and note that

$$T(g)^* S\, T(g) = S, \qquad g \in G.$$

Inasmuch as S is a positive-definite Hermitian matrix, there exists a nonsingular matrix Q such that $QQ^* = S^{-1}$, as is seen from the eigenvalue decomposition of S. Then for $T_1(g) = Q^{-1}T(g)Q$, we have

$$\begin{aligned} T_1(g)^* T_1(g) &= Q^* T(g)^* (QQ^*)^{-1} T(g) Q = Q^* T(g)^* S\, T(g) Q \\ &= Q^* S\, Q = Q^* (QQ^*)^{-1} Q = I \end{aligned}$$

for all $g \in G$. This shows that $T_1(g)$ is unitary (cf., (7.15)). □

Invariant Subspace

A subspace W of V is said to be (G-)*invariant* with respect to $\hat{T}$ if $\hat{T}(g)\mathbf{w} \in W$ for all $\mathbf{w} \in W$ and $g \in G$. For an invariant subspace W, the restriction of $\hat{T}(g)$ to W for each $g \in G$ defines a representation of G on W, called the *subrepresentation* of $\hat{T}$ on W.

It is known as the *Maschke theorem* that, for an invariant subspace W, there exists another invariant subspace W' such that

$$V = W \oplus W'. \tag{7.17}$$

Therefore, the representation matrix T can be brought into a block-diagonal form with a suitable change of basis; that is,

$$Q^{-1}T(g)Q = \begin{pmatrix} T_1(g) & O \\ O & T_2(g) \end{pmatrix}, \qquad g \in G,$$

for some nonsingular matrix Q, where T_1 and T_2, respectively, represent subrepresentations of T on W and W'. It is emphasized that the matrix Q is independent of $g \in G$. Therefore, the matrices $T(g)$, indexed by $g \in G$, are decomposed simultaneously by a single matrix Q.

If $\hat{T}$ is unitary, the *orthogonal complement* $W^{\perp}$ of an invariant subspace W is also an invariant subspace. To prove this, take any $\boldsymbol{w}' \in W^{\perp}$ and note that (cf., (7.14))

$$(\hat{T}(g)\boldsymbol{w}', \boldsymbol{w}) = (\boldsymbol{w}', \hat{T}(g^{-1})\boldsymbol{w}) = 0, \qquad \boldsymbol{w} \in W,\ g \in G,$$

since $\hat{T}(g^{-1})\boldsymbol{w} \in W$ is orthogonal to $\boldsymbol{w}'$. This shows that $\hat{T}(g)\boldsymbol{w}' \in W^{\perp}$ for all $g \in G$. Consequently, we can take $W^{\perp}$ as the complementary subspace W' in the decomposition (7.17). In other words, for an invariant subspace W, the orthogonal decomposition

$$V = W \oplus W^{\perp} \tag{7.18}$$

serves as the decomposition (7.17) into two invariant subspaces.

7.4.2 *Irreducible Representation*

Irreducible representation, which plays the central role in the description of bifurcation equation, is introduced.

Irreducibility

A representation $\hat{T}$ on V is said to be *irreducible* if there exists no nontrivial invariant subspace W, where W is nontrivial if W is neither $\{\mathbf{0}\}$ nor V. We also use the expression of F-*irreducibility* to emphasize the underlying field F, which is either $\mathbb{R}$ or $\mathbb{C}$ in this book.

There exist a finite number of mutually inequivalent irreducible representations of G (over a fixed field F). We denote by

$$\{\hat{T}^{\mu} \mid \mu \in R(G)\} \tag{7.19}$$

a family of representatives from all irreducible unitary representations of G, where $R(G) = R_F(G)$ denotes the *index set* for the irreducible representations of G over F. The dimension of representation μ is hereafter denoted by N^μ. In case of $F = \mathbb{C}$, we have an identity[6]

$$\sum_{\mu \in R_{\mathbb{C}}(G)} (N^\mu)^2 = |G|. \tag{7.20}$$

For each $\mu \in R(G)$ let T^μ be a matrix representation associated with $\hat{T}^\mu$ with respect to some orthogonal basis. Then we obtain a family

$$\{T^\mu \mid \mu \in R(G)\} \tag{7.21}$$

of representatives from all irreducible unitary matrix representations of G. For each $g \in G$, $T^\mu(g)$ is an $N^\mu \times N^\mu$ nonsingular matrix.

For $\mu \in R(G)$, we define a subgroup

$$G^\mu = \{g \in G \mid T^\mu(g) = I_{N^\mu}\}, \tag{7.22}$$

where I_{N^μ} is the identity matrix of order N^μ. This is the subgroup of elements $g \in G$ that behave in μ as if they were the identity element of G. The subgroup G^μ is sometimes called the *kernel* of μ (or of T^μ).

Remark 7.3. Determining the explicit forms of irreducible representations is a non-trivial task. For most groups that we consider in this book, such as dihedral and cyclic groups, we can use standard results available in the literature, although we need to work this out for some groups in Part III. It is worth noting that the complete set of irreducible representations of a direct product $G_1 \times G_2$ can be obtained, in the case of $F = \mathbb{C}$, as the family of tensor products of irreducible representations of G_1 and G_2. Therefore, $|R_{\mathbb{C}}(G_1 \times G_2)| = |R_{\mathbb{C}}(G_1)| \cdot |R_{\mathbb{C}}(G_2)|$. □

Decomposition into Irreducible Representations

It follows from a repeated application of the Maschke theorem (7.17) that any representation $\hat{T}$ of G on V can be expressed as a direct sum of irreducible representations. To be more precise, the representation space V is decomposed as

$$V = \bigoplus_{\mu \in R(G)} \bigoplus_{i=1}^{a^\mu} V_i^\mu, \tag{7.23}$$

where V_i^μ is an invariant subspace, the subrepresentation of $\hat{T}$ on V_i^μ is irreducible and equivalent to $\hat{T}^\mu$ in (7.19), and a^μ is a nonnegative integer, called the *multiplicity* of μ in $\hat{T}$. We designate (7.23) as the *irreducible decomposition* of the representation space V.

---

[6] The identity (7.20) implies that $|R_{\mathbb{C}}(G)| \leq |G|$. It then follows that $|R_F(G)| \leq |G|$ for $F = \mathbb{R}$ and $\mathbb{C}$, since $|R_{\mathbb{R}}(G)| \leq |R_{\mathbb{C}}(G)|$ by the definition of irreducibility and equivalence.

By defining a subspace

$$V^{\mu} = \bigoplus_{i=1}^{a^{\mu}} V_i^{\mu}, \tag{7.24}$$

which is an aggregation of the a^{μ} irreducible subspaces corresponding to the same μ, we may rewrite (7.23) to

$$V = \bigoplus_{\mu \in R(G)} V^{\mu}. \tag{7.25}$$

Decomposition (7.25) is unique and called the *isotypic decomposition*; each V^{μ} is called an *isotypic* (or *homogeneous*) component. On the other hand, decomposition (7.24) is not unique, although the multiplicity a^{μ} is uniquely determined. Consequently, decomposition (7.23) into irreducible components is not unique either.

The direct sum decomposition (7.23) means that, with a suitable nonsingular matrix Q, the matrix representation T can be put into a *block-diagonal form*

$$\overline{T}(g) \equiv Q^{-1}T(g)Q = \bigoplus_{\mu \in R(G)} \bigoplus_{i=1}^{a^{\mu}} T_i^{\mu}(g), \qquad g \in G \tag{7.26}$$

with T_i^{μ} being irreducible. It is possible and often advantageous to impose a further condition

$$T_i^{\mu}(g) = T^{\mu}(g), \qquad g \in G, \ i = 1,\dots,a^{\mu}. \tag{7.27}$$

This implies that we choose an identical matrix representation for equivalent representations (cf., (7.21)). With the choice of (7.27) in (7.26) we obtain

$$\overline{T}(g) = Q^{-1}T(g)Q = \bigoplus_{\mu \in R(G)} \bigoplus_{i=1}^{a^{\mu}} T^{\mu}(g), \qquad g \in G. \tag{7.28}$$

Here the size of the matrix $T^{\mu}(g)$ is N^{μ}, and we have $\sum_{\mu \in R(G)} a^{\mu} N^{\mu} = N$. The decomposition (7.28), as well as (7.26), is called hereafter the *irreducible decomposition* of the matrix representation T.

By defining a matrix

$$\overline{T}^{\mu}(g) = \bigoplus_{i=1}^{a^{\mu}} T^{\mu}(g), \qquad g \in G, \tag{7.29}$$

which is an aggregation of the a^{μ} irreducible representations corresponding to the same μ, we may rewrite (7.28) as

$$\overline{T}(g) = Q^{-1}T(g)Q = \bigoplus_{\mu \in R(G)} \overline{T}^{\mu}(g), \qquad g \in G. \tag{7.30}$$

We call this the *isotypic decomposition* of the representation T. Note that the size of the matrix $\overline{T}^{\mu}(g)$ is $a^{\mu}N^{\mu}$. By construction, the isotypic decomposition is coarser than the irreducible decomposition.

Suppose that T is a unitary representation. Then the transformation matrix Q in (7.26) can be chosen to be unitary (cf., (7.18)). The further condition (7.27) can also be realized by a unitary transformation because any two equivalent irreducible unitary representations are connected as (7.16) with a unitary Q (cf., Remark 7.4 in §7.4.3).

In subsequent chapters we are mostly concerned with unitary (orthogonal) representations over $\mathbb{R}$. In this case we choose an *orthogonal transformation* with an orthogonal matrix Q in the irreducible decomposition (7.28) and in the isotypic decomposition (7.30), to obtain the following forms of block-diagonalization.

$$\overline{T}(g) = Q^{\top} T(g) Q = \bigoplus_{\mu \in R(G)} \bigoplus_{i=1}^{a^{\mu}} T^{\mu}(g), \qquad g \in G, \tag{7.31}$$

$$\overline{T}(g) = Q^{\top} T(g) Q = \bigoplus_{\mu \in R(G)} \overline{T}^{\mu}(g), \qquad g \in G. \tag{7.32}$$

Example 7.7. The decomposition into irreducible representations is illustrated here for the cyclic group $C_3 = \{e, r, r^2\}$ (cf., Table 7.1(b)). We consider the case of $F = \mathbb{R}$.

The group C_3 has two inequivalent irreducible representations over $\mathbb{R}$, a one-dimensional irreducible representation, denoted as μ_1, and a two-dimensional irreducible representation, denoted as μ_2. They are defined, respectively, by

$$T^{\mu_1}(e) = 1, \qquad T^{\mu_1}(r) = 1, \qquad T^{\mu_1}(r^2) = 1$$

and

$$T^{\mu_2}(e) = \begin{pmatrix} 1 & 0 \\ 0 & 1 \end{pmatrix}, \qquad T^{\mu_2}(r) = \begin{pmatrix} -1/2 & -\sqrt{3}/2 \\ \sqrt{3}/2 & -1/2 \end{pmatrix}, \qquad T^{\mu_2}(r^2) = \begin{pmatrix} -1/2 & \sqrt{3}/2 \\ -\sqrt{3}/2 & -1/2 \end{pmatrix}.$$

We have $R(C_3) = \{\mu_1, \mu_2\}$ in our notation (7.21).

Consider an example of a 3×3 orthogonal representation of C_3 given by

$$T(e) = \begin{pmatrix} 1 & & \\ & 1 & \\ & & 1 \end{pmatrix}, \qquad T(r) = \begin{pmatrix} & & 1 \\ 1 & & \\ & 1 & \end{pmatrix}, \qquad T(r^2) = \begin{pmatrix} & 1 & \\ & & 1 \\ 1 & & \end{pmatrix}. \tag{7.33}$$

With the use of an orthogonal transformation matrix

$$Q = \begin{pmatrix} 1/\sqrt{3} & 2/\sqrt{6} & 0 \\ 1/\sqrt{3} & -1/\sqrt{6} & 1/\sqrt{2} \\ 1/\sqrt{3} & -1/\sqrt{6} & -1/\sqrt{2} \end{pmatrix},$$

the matrices in (7.33) can be transformed to

$$Q^{\top}T(e)Q = \left(\begin{array}{c|cc} 1 & 0 & 0 \\ \hline 0 & 1 & 0 \\ 0 & 0 & 1 \end{array}\right) = T^{\mu_1}(e) \oplus T^{\mu_2}(e),$$

$$Q^{\top}T(r)Q = \left(\begin{array}{c|cc} 1 & 0 & 0 \\ \hline 0 & -1/2 & -\sqrt{3}/2 \\ 0 & \sqrt{3}/2 & -1/2 \end{array}\right) = T^{\mu_1}(r) \oplus T^{\mu_2}(r),$$

$$Q^{\top}T(r^2)Q = \left(\begin{array}{c|cc} 1 & 0 & 0 \\ \hline 0 & -1/2 & \sqrt{3}/2 \\ 0 & -\sqrt{3}/2 & -1/2 \end{array}\right) = T^{\mu_1}(r^2) \oplus T^{\mu_2}(r^2),$$

all of which are of the same block-diagonal form, that is, the direct sum of a 1×1 matrix and a 2×2 matrix. Thus the matrix representation T in (7.33) splits into two irreducible representations T^{μ_1} and T^{μ_2} with multiplicities $a^{\mu_1} = a^{\mu_2} = 1$. In particular, this demonstrates that T is not irreducible. □

7.4.3 Schur's Lemma

A representation over F is said to be *absolutely irreducible* if it is irreducible as a representation over $\mathbb{C}$. In the case of $F = \mathbb{C}$, there is no distinction between irreducibility and absolute irreducibility, but they are distinguished if $F = \mathbb{R}$; see Example 7.8 below. We denote by $R_{\rm a}(G)$ the family of absolutely irreducible representations (over F); we have $R_{\rm a}(G) \subseteq R(G)$ and the difference set[7] $R(G) \setminus R_{\rm a}(G)$ consists of irreducible representations that are not absolutely irreducible. That is, we have a partition of $R(G)$ into two disjoint parts:

$$R(G) = R_{\rm a}(G) \cup [R(G) \setminus R_{\rm a}(G)]. \tag{7.34}$$

Some groups, however, have the property that every irreducible representation over $\mathbb{R}$ is absolutely irreducible. For example, the dihedral group D_n, for any n, is known to have this property: $R_{\rm a}(\mathrm{D}_n) = R(\mathrm{D}_n)$ over $\mathbb{R}$. In contrast, this is not the case with the cyclic group C_n: $R_{\rm a}(\mathrm{C}_n) \neq R(\mathrm{C}_n)$ over $\mathbb{R}$.

The following fundamental fact is known as *Schur's lemma*, which, in this book, lays the foundation of the block-diagonalization method explained in §7.7 and Chapter 12.

Lemma 7.1. *Let $F = \mathbb{R}$ or $\mathbb{C}$, and A be a (possibly rectangular) matrix over F. Assume that T_1 and T_2 are irreducible matrix representations of group G over F, and*

$$T_1(g)A = AT_2(g), \qquad g \in G. \tag{7.35}$$

(i) *$A = O$ or else A is square and nonsingular.*

[7] For two sets A and B in general, $A \setminus B$ denotes the set of elements of A that are not contained in B.

(ii) *If* T_1 *and* T_2 *are inequivalent, then* $A = O$.

(iii) *If* $T_1(g) = T_2(g)$ *for all* $g \in G$, *and* T_1 *is absolutely irreducible, then* $A = \lambda I$ *for some* $\lambda \in F$.

Proof. (i) It follows from (7.35) that, for any $\boldsymbol{x} \in \ker(A)$, we have

$$A(T_2(g)\boldsymbol{x}) = T_1(g)(A\boldsymbol{x}) = \boldsymbol{0},$$

which means $T_2(g)\boldsymbol{x} \in \ker(A)$. Hence, $\ker(A)$ is G-invariant with respect to T_2. By the irreducibility of T_2, $\ker(A)$ is either $\{\boldsymbol{0}\}$ or the entire space. On the other hand, for $\boldsymbol{y} = A\boldsymbol{x} \in \mathrm{range}(A)$, we have

$$T_1(g)\boldsymbol{y} = T_1(g)(A\boldsymbol{x}) = A(T_2(g)\boldsymbol{x}) \in \mathrm{range}(A).$$

Hence, the *range space* range(A) is G-invariant with respect to T_1. By the irreducibility of T_1, range(A) is either $\{\boldsymbol{0}\}$ or the entire space. Therefore A is square and nonsingular or $A = O$.

(ii) If A is nonsingular, then

$$T_2(g) = A^{-1}T_1(g)A, \qquad g \in G,$$

which shows the equivalence of T_2 to T_1. Hence, if T_1 and T_2 are inequivalent, then A must be singular, and therefore $A = O$ by (i).

(iii) For an eigenvalue $\lambda \in \mathbb{C}$ of A, it holds that[8]

$$T_1(g)(A - \lambda I) = (A - \lambda I)T_1(g), \qquad g \in G.$$

Since $A - \lambda I$ is singular and T_1 is irreducible over $\mathbb{C}$, we have $A - \lambda I = O$ by (i). Since $\lambda I = A$ and A is a matrix over F, the eigenvalue λ belongs to F. □

Example 7.8. This example illustrates that the condition of absolute irreducibility in Lemma 7.1(iii) above cannot be omitted. Consider the cyclic group $C_3 = \{e, r, r^2\}$ with $r^3 = e$ (cf., Table 7.1(b)). In addition to the unit representation μ_1, this group has a two-dimensional representation μ_2 given by

$$T^{\mu_2}(e) = \begin{pmatrix} 1 & 0 \\ 0 & 1 \end{pmatrix}, \qquad T^{\mu_2}(r) = \begin{pmatrix} \alpha & -\beta \\ \beta & \alpha \end{pmatrix}, \qquad T^{\mu_2}(r^2) = \begin{pmatrix} \alpha & \beta \\ -\beta & \alpha \end{pmatrix}$$

with $\alpha = \cos(2\pi/3)$ and $\beta = \sin(2\pi/3)$. This representation, also denoted as μ_2 in Example 7.7 in §7.4.2, is irreducible over $\mathbb{R}$, but it is not absolutely irreducible. Indeed, with a complex (unitary) matrix

$$Q = \frac{1}{\sqrt{2}} \begin{pmatrix} 1 & 1 \\ -\mathrm{i} & \mathrm{i} \end{pmatrix}$$

we can decompose T^{μ_2} as follows:

[8] In the case of $F = \mathbb{R}$, where A is a real matrix, we must allow λ to be a complex number at this point of the proof, although it turns out to be a real number at the end of the proof.

$$Q^{-1}T^{\mu_2}(e)Q = \begin{pmatrix} 1 & 0 \\ 0 & 1 \end{pmatrix},$$

$$Q^{-1}T^{\mu_2}(r)Q = \begin{pmatrix} \exp(2\pi i/3) & 0 \\ 0 & \exp(-2\pi i/3) \end{pmatrix},$$

$$Q^{-1}T^{\mu_2}(r^2)Q = \begin{pmatrix} \exp(4\pi i/3) & 0 \\ 0 & \exp(-4\pi i/3) \end{pmatrix}.$$

Therefore, the partition in (7.34) is given with

$$R(G) = \{\mu_1, \mu_2\}, \qquad R_a(G) = \{\mu_1\}, \qquad R(G) \setminus R_a(G) = \{\mu_2\}.$$

For a 2×2 real matrix A, the commutativity condition, $T^{\mu_2}(g)A = AT^{\mu_2}(g)$ for all $g \in C_3$, is equivalent to $T^{\mu_2}(r)A = AT^{\mu_2}(r)$. A direct calculation shows that this is the case if and only if

$$A = \begin{pmatrix} a & -b \\ b & a \end{pmatrix}, \qquad a, b \in \mathbb{R}.$$

Thus A is not restricted to be a scalar multiple of the identity matrix. It is noteworthy that

$$Q^{-1} \begin{pmatrix} a & -b \\ b & a \end{pmatrix} Q = \begin{pmatrix} a + ib & 0 \\ 0 & a - ib \end{pmatrix}.$$

□

Remark 7.4. As an application of Schur's lemma we show here that two equivalent irreducible unitary representations T_1 and T_2 over F can be transformed to each other by a unitary transformation over F. Since T_1 and T_2 are equivalent, there exists a nonsingular Q such that $T_1(g) = Q^{-1}T_2(g)Q$ for all $g \in G$. By taking the conjugate transpose of $QT_1(g) = T_2(g)Q$ we obtain $T_1(g)^*Q^* = Q^*T_2(g)^*$ for all $g \in G$. On replacing g by g^{-1} and noting that $T_i(g^{-1})^* = T_i(g)$ for $i = 1, 2$ (cf., (7.15)), we see $T_1(g)Q^* = Q^*T_2(g)$ $(g \in G)$. Therefore,

$$T_2(g)(QQ^*) = T_2(g)Q \cdot Q^* = QT_1(g) \cdot Q^* = Q \cdot T_1(g)Q^* = Q \cdot Q^*T_2(g) = (QQ^*)T_2(g)$$

for all $g \in G$. Here QQ^* is a positive-definite Hermitian matrix, and let α be a positive eigenvalue of QQ^*. Then

$$T_2(g)(QQ^* - \alpha I) = (QQ^* - \alpha I)T_2(g), \qquad g \in G.$$

Here $QQ^* - \alpha I$ is singular and T_2 is irreducible; therefore, we have $QQ^* - \alpha I = O$ by Schur's lemma (Lemma 7.1(i) above). Then $Q' = Q/\sqrt{\alpha}$ is a unitary matrix that connects T_1 and T_2. □

7.5 Symmetry of Equations

The group-theoretic method for exploiting the symmetry of the governing equation as well as that of the bifurcation equation is presented.

7.5.1 Group Equivariance of Governing Equation

Following the general framework of Part I (Chapter 2), we consider a system of nonlinear equilibrium or governing equations[9] (2.1):

$$\boldsymbol{F}(\boldsymbol{u}, f, \boldsymbol{v}) = \boldsymbol{0}, \tag{7.36}$$

where f denotes a bifurcation parameter, $\boldsymbol{u} \in \mathbb{R}^N$ indicates a state vector, and $\boldsymbol{v} \in \mathbb{R}^p$ denotes an imperfection parameter vector. We assume $\boldsymbol{F} : \mathbb{R}^N \times \mathbb{R} \times \mathbb{R}^p \to \mathbb{R}^N$ to be sufficiently smooth.

In this chapter, we are interested in the case where the system (7.36) is endowed with an additional mathematical structure of group symmetry. Following the standard setting in group-theoretic bifurcation theory, we assume that the symmetry of the perfect system (with $\boldsymbol{v} = \boldsymbol{v}^0$) is described by the *equivariance*:

$$T(g)\boldsymbol{F}(\boldsymbol{u}, f, \boldsymbol{v}^0) = \boldsymbol{F}(T(g)\boldsymbol{u}, f, \boldsymbol{v}^0), \qquad g \in G, \tag{7.37}$$

of $\boldsymbol{F}(\boldsymbol{u}, f, \boldsymbol{v}^0)$ to a group G in terms of a unitary matrix representation T of G on the N-dimensional space of the independent variable vector $\boldsymbol{u}$.

To also express the symmetry in the imperfection parameter vector $\boldsymbol{v}$, we extend the equivariance (7.37) to the following form,

$$T(g)\boldsymbol{F}(\boldsymbol{u}, f, \boldsymbol{v}) = \boldsymbol{F}(T(g)\boldsymbol{u}, f, S(g)\boldsymbol{v}), \qquad g \in G, \tag{7.38}$$

in terms of another unitary representation S of G on the p-dimensional space of the imperfection parameter vector $\boldsymbol{v}$. For the compatibility of (7.37) and (7.38) it is assumed that the imperfection vector $\boldsymbol{v}^0$ for the perfect system is *G-symmetric* in the sense that

$$S(g)\boldsymbol{v}^0 = \boldsymbol{v}^0, \qquad g \in G. \tag{7.39}$$

For a critical point $(\boldsymbol{u}_{\rm c}^0, f_{\rm c}^0)$ of the perfect system (with $\boldsymbol{v} = \boldsymbol{v}^0$), we always assume that $\boldsymbol{u}_{\rm c}^0$ is *G-symmetric* in the sense that

$$T(g)\boldsymbol{u}_{\rm c}^0 = \boldsymbol{u}_{\rm c}^0, \qquad g \in G. \tag{7.40}$$

Remark 7.5. The equivariance (7.38) is not an artificial condition for mathematical convenience, but rather a natural consequence of the objectivity of the equation, which means the observer-independence of the mathematical description (see the

---

[9] In structural mechanics, $\boldsymbol{u}$ indicates a displacement vector and f denotes a loading parameter.

examples in §7.8). It is emphasized that the equivariance does not impose symmetry on the solution $\boldsymbol{u}$ nor on the imperfection vector $\boldsymbol{v}$, but it does denote the symmetry of the system of equations $\boldsymbol{F}$ as a whole under the transformations with respect to G, which often represent geometric transformations. □

Remark 7.6. A seemingly more general formulation of equivariance (7.37) would be

$$T_1(g)\boldsymbol{F}(\boldsymbol{u}, f, \boldsymbol{v}^0) = \boldsymbol{F}(T_2(g)\boldsymbol{u}, f, \boldsymbol{v}^0), \qquad g \in G, \tag{7.41}$$

with two matrix representations T_1 and T_2. We may assume that T_1 and T_2 are equivalent in the case of our interest. To see this, note that (7.41) implies

$$T_1(g)J(\boldsymbol{u}, f, \boldsymbol{v}^0) = J(\boldsymbol{u}, f, \boldsymbol{v}^0)T_2(g), \qquad g \in G,$$

if $T_2(g)\boldsymbol{u} = \boldsymbol{u}$ for all $g \in G$, where $J = \partial \boldsymbol{F}/\partial \boldsymbol{u}$ is the Jacobian matrix (2.4) of $\boldsymbol{F}$. It is natural to assume that we do have a G-symmetric solution $\boldsymbol{u}$ at which $J(\boldsymbol{u}, f, \boldsymbol{v}^0)$ is nonsingular. This implies that T_1 and T_2 are equivalent. Therefore, we can assume that $T_1(g) = T_2(g)$ for all $g \in G$ by a suitable basis change. We can further assume that T_1 $(= T_2)$ is unitary on the basis of the fact that any representation is equivalent to a unitary representation (cf., Remark 7.2 in §7.4.1). □

Remark 7.7. When a potential function $U(\boldsymbol{u}, f, \boldsymbol{v})$ exists, the system of equilibrium equations $\boldsymbol{F}(\boldsymbol{u}, f, \boldsymbol{v})$ is derived from it as $\boldsymbol{F} = (\partial U/\partial \boldsymbol{u})^\top$. The equivariance of $\boldsymbol{F}$ to a group G is a consequence of the invariance of U to G, as explained in §7.2 in a simple case. In general, the invariance of U to G is formulated as

$$U(T(g)\boldsymbol{u}, f, S(g)\boldsymbol{v}) = U(\boldsymbol{u}, f, \boldsymbol{v}), \qquad g \in G,$$

in terms of unitary matrix representations T and S. Then $\boldsymbol{F} = (\partial U/\partial \boldsymbol{u})^\top$ satisfies

$$T(g)^{-\top}\boldsymbol{F}(\boldsymbol{u}, f, \boldsymbol{v}) = \boldsymbol{F}(T(g)\boldsymbol{u}, f, S(g)\boldsymbol{v}), \qquad g \in G,$$

where $(\cdot)^{-\top}$ means the transpose of the inverse of a matrix. This proves (7.38), since $T(g)^{-\top} = T(g)$ by the unitarity of T. □

Equivariance of Linear Parts

The equivariance (7.38) is inherited by the Jacobian matrix $J(\boldsymbol{u}, f, \boldsymbol{v})$ as follows. Recall from (2.4) that $J(\boldsymbol{u}, f, \boldsymbol{v})$ is an $N \times N$ matrix defined as

$$J(\boldsymbol{u}, f, \boldsymbol{v}) = \left(\frac{\partial F_i}{\partial u_j} \;\middle|\; i, j = 1, \ldots, N \right). \tag{7.42}$$

Differentiation of (7.38) with respect to $\boldsymbol{u}$ yields

$$T(g)J(\boldsymbol{u}, f, \boldsymbol{v}) = J(T(g)\boldsymbol{u}, f, S(g)\boldsymbol{v})T(g), \qquad g \in G, \tag{7.43}$$

which is the *equivariance* of J for general $(\boldsymbol{u}, f, \boldsymbol{v})$. For the case in which $\boldsymbol{v}$ and $\boldsymbol{u}$ are G-symmetric, that is,

$$S(g)\boldsymbol{v} = \boldsymbol{v}, \qquad T(g)\boldsymbol{u} = \boldsymbol{u}, \qquad g \in G, \tag{7.44}$$

it holds that

$$T(g)J(\boldsymbol{u}, f, \boldsymbol{v}) = J(\boldsymbol{u}, f, \boldsymbol{v})T(g), \qquad g \in G. \tag{7.45}$$

Thus the equivariance (7.43) reduces to the commutativity of J with $T(g)$ for all $g \in G$. In particular, at a critical point $(\boldsymbol{u}_{\rm c}^0, f_{\rm c}^0)$ of the perfect system (with $\boldsymbol{v} = \boldsymbol{v}^0$), we have

$$T(g)J(\boldsymbol{u}_{\rm c}^0, f_{\rm c}^0, \boldsymbol{v}^0) = J(\boldsymbol{u}_{\rm c}^0, f_{\rm c}^0, \boldsymbol{v}^0)T(g), \qquad g \in G. \tag{7.46}$$

The imperfection sensitivity matrix $B(\boldsymbol{u}, f, \boldsymbol{v})$ also inherits the equivariance (7.38). Recall from (2.3) that $B(\boldsymbol{u}, f, \boldsymbol{v})$ is an $N \times p$ matrix defined as

$$B(\boldsymbol{u}, f, \boldsymbol{v}) = \left(\frac{\partial F_i}{\partial v_j} \;\middle|\; i = 1, \ldots, N,\ j = 1, \ldots, p \right). \tag{7.47}$$

Differentiation of (7.38) with respect to $\boldsymbol{v}$ yields

$$T(g)B(\boldsymbol{u}, f, \boldsymbol{v}) = B(T(g)\boldsymbol{u}, f, S(g)\boldsymbol{v})S(g), \qquad g \in G, \tag{7.48}$$

which is the equivariance of B for general $(\boldsymbol{u}, f, \boldsymbol{v})$. Therefore, if $\boldsymbol{v}$ and $\boldsymbol{u}$ are G-symmetric in the sense of (7.44), it holds that

$$T(g)B(\boldsymbol{u}, f, \boldsymbol{v}) = B(\boldsymbol{u}, f, \boldsymbol{v})S(g), \qquad g \in G. \tag{7.49}$$

In particular, at a critical point $(\boldsymbol{u}_{\rm c}^0, f_{\rm c}^0)$ of the perfect system (with $\boldsymbol{v} = \boldsymbol{v}^0$), we have

$$T(g)B(\boldsymbol{u}_{\rm c}^0, f_{\rm c}^0, \boldsymbol{v}^0) = B(\boldsymbol{u}_{\rm c}^0, f_{\rm c}^0, \boldsymbol{v}^0)S(g), \qquad g \in G. \tag{7.50}$$

Group-Theoretic Critical Point

Let $(\boldsymbol{u}_{\rm c}^0, f_{\rm c}^0)$ be a critical point of the perfect system (with $\boldsymbol{v} = \boldsymbol{v}^0$), where the G-symmetry as in (7.39) and (7.40) is assumed. The Jacobian matrix $J_{\rm c}^0 = J(\boldsymbol{u}_{\rm c}^0, f_{\rm c}^0, \boldsymbol{v}^0)$ at this point is singular by definition. Let M be the dimension of the kernel space of $J_{\rm c}^0$; that is,

$$M = \dim[\ker(J_{\rm c}^0)] = N - \mathrm{rank}(J_{\rm c}^0).$$

The point $(\boldsymbol{u}_{\rm c}^0, f_{\rm c}^0)$ is *simple* if $M = 1$ and *multiple* if $M \geq 2$.

The kernel space of $J_{\rm c}^0$ is a G-invariant subspace, as follows.

Lemma 7.2. *The kernel space* $\ker(J_{\rm c}^0)$ *is G-invariant.*

Proof. Take any $\boldsymbol{\eta} \in \ker(J_{\rm c}^0)$ and $g \in G$. It follows from (7.46) that $J_{\rm c}^0(T(g)\boldsymbol{\eta}) = T(g)(J_{\rm c}^0\boldsymbol{\eta}) = \boldsymbol{0}$, which means $T(g)\boldsymbol{\eta} \in \ker(J_{\rm c}^0)$. Hence, $\ker(J_{\rm c}^0)$ is G-invariant. □

The critical point $(\boldsymbol{u}_{\mathrm{c}}^0, f_{\mathrm{c}}^0)$ is called *group-theoretic* if the kernel of the Jacobian matrix $J_{\mathrm{c}}^0 = J(\boldsymbol{u}_{\mathrm{c}}^0, f_{\mathrm{c}}^0, \boldsymbol{v}^0)$ is G-irreducible and *parametric* otherwise. This categorization primarily applies to multiple critical points, although any simple point is group-theoretic since a one-dimensional G-invariant subspace is necessarily G-irreducible.

In Parts II and III of this book, we devote our interest mainly to group-theoretic multiple critical points, which generically appear in a system with group symmetry. In customary terminology, "genericity" is defined in relation to a parametrized family of the systems in question.[10] Here we implicitly think of a family of physical systems with specified symmetry.

As an example of the parametric points, a hilltop bifurcation point is considered in §15.7. Parametric multiple bifurcation points were studied by Keener, 1979 [112] in relation to the secondary bifurcation; Fujii, Mimura, and Nishiura, 1982 [51] for ecological interacting and diffusing systems; Golubitsky, Stewart, and Schaeffer, 1988 [64] for systems with various symmetries; and Thompson and Hunt, 1984 [192] in connection with the hyperbolic umbilic catastrophe.

7.5.2 Liapunov–Schmidt Reduction

We explain here a standard procedure, the *Liapunov–Schmidt reduction with symmetry* (Sattinger, 1979 [176]), which is to reduce the full system of equations to a few bifurcation equations compatible with the symmetry. Before embarking upon technical arguments, we highlight two principles in this reduction process.

The first principle is concerned with symmetry and reads as follows.

Principle 7.1 (Inheritance of symmetry) *The symmetry of the original system of equations is inherited by the bifurcation equations.* □

To be specific, we consider a critical point $(\boldsymbol{u}_{\mathrm{c}}^0, f_{\mathrm{c}}^0)$ of a perfect system (with $\boldsymbol{v} = \boldsymbol{v}^0$), and let $M = N - \mathrm{rank}(J_{\mathrm{c}}^0)$. In a neighborhood of $(\boldsymbol{u}_{\mathrm{c}}^0, f_{\mathrm{c}}^0, \boldsymbol{v}^0)$, the full system of equations

$$\boldsymbol{F}(\boldsymbol{u}, f, \boldsymbol{v}) = \boldsymbol{0} \tag{7.51}$$

in $\boldsymbol{u} \in \mathbb{R}^N$ (cf., (7.36)) is reduced to a system of M bifurcation equations

$$\widetilde{\boldsymbol{F}}(\boldsymbol{w}, \widetilde{f}, \boldsymbol{v}) = \boldsymbol{0} \tag{7.52}$$

in $\boldsymbol{w} \in \mathbb{R}^M$, where $\widetilde{\boldsymbol{F}} : \mathbb{R}^M \times \mathbb{R} \times \mathbb{R}^p \to \mathbb{R}^M$ is a function and $\widetilde{f} = f - f_{\mathrm{c}}^0$ denotes the increment of f. In this reduction process the equivariance of the full system, which is formulated in (7.38) as

$$T(g)\boldsymbol{F}(\boldsymbol{u}, f, \boldsymbol{v}) = \boldsymbol{F}(T(g)\boldsymbol{u}, f, S(g)\boldsymbol{v}), \qquad g \in G, \tag{7.53}$$

is inherited by the reduced system (7.52) in the following form:

[10] Loosely speaking, the term "generically" might be replaced by "unless the parameters take special values."

$$\widetilde{T}(g)\widetilde{\boldsymbol{F}}(\boldsymbol{w},\widetilde{f},\boldsymbol{v}) = \widetilde{\boldsymbol{F}}(\widetilde{T}(g)\boldsymbol{w},\widetilde{f},S(g)\boldsymbol{v}), \qquad g \in G, \tag{7.54}$$

where $\widetilde{T}$ is the subrepresentation of T on the kernel space of $J_{\mathrm{c}}^0 = J(\boldsymbol{u}_{\mathrm{c}}^0, f_{\mathrm{c}}^0, \boldsymbol{v}^0)$. The formula (7.54) is derived in (7.77). It is this *inheritance of symmetry* that plays a key role in determining the symmetry of bifurcating solutions.

The second principle in the Liapunov–Schmidt reduction is relevant to reciprocal systems and reads as follows.

Principle 7.2 (Inheritance of reciprocity) *The reciprocity of the original system of equations is inherited by the bifurcation equations.* □

To be specific, suppose that the full system (7.51) is a reciprocal system in the sense of (2.12):

$$\frac{\partial F_i}{\partial u_j} = \frac{\partial F_j}{\partial u_i}, \qquad i,j = 1,\ldots,N. \tag{7.55}$$

Then we can choose the reduced system (7.52) to be a reciprocal system (by an appropriate choice of coordinates). That is to say, the reciprocity

$$\frac{\partial \widetilde{F}_i}{\partial w_j} = \frac{\partial \widetilde{F}_j}{\partial w_i}, \qquad i,j = 1,\ldots,M, \tag{7.56}$$

can be imposed on the bifurcation equation (7.52), as proved later in Lemma 7.5. We designate this as the *inheritance of reciprocity*.[11]

Therefore, we have two independent principles related to symmetry and reciprocity to be preserved under the Liapunov–Schmidt reduction to bifurcation equations.

Reduction Procedure

The Liapunov–Schmidt reduction procedure proceeds as follows. We consider a critical point $(\boldsymbol{u}_{\mathrm{c}}^0, f_{\mathrm{c}}^0)$ of multiplicity M for the perfect system (with $\boldsymbol{v} = \boldsymbol{v}^0$), which is either group-theoretic or parametric. We have

$$\dim[\ker(J_{\mathrm{c}}^0)] = M, \qquad \dim[\mathrm{range}(J_{\mathrm{c}}^0)] = N - M \tag{7.57}$$

for $J_{\mathrm{c}}^0 = J(\boldsymbol{u}_{\mathrm{c}}^0, f_{\mathrm{c}}^0, \boldsymbol{v}^0)$. It is assumed in (7.39) and (7.40) that both $\boldsymbol{v}^0$ and $\boldsymbol{u}_{\mathrm{c}}^0$ are G-invariant; that is,

$$S(g)\boldsymbol{v}^0 = \boldsymbol{v}^0, \qquad T(g)\boldsymbol{u}_{\mathrm{c}}^0 = \boldsymbol{u}_{\mathrm{c}}^0, \qquad g \in G. \tag{7.58}$$

It is emphasized that the resulting bifurcation equation $\widetilde{\boldsymbol{F}}(\boldsymbol{w},\widetilde{f},\boldsymbol{v}) = \boldsymbol{0}$ covers both the perfect system (with $\boldsymbol{v} = \boldsymbol{v}^0$) and an imperfect system (with $\boldsymbol{v} \neq \boldsymbol{v}^0$).

Consider a direct sum decomposition

---

[11] Reciprocity plays a significant role in the bifurcation analysis of C_n-symmetric systems (cf., Remark 8.5 in §8.6.1).

$$\mathbb{R}^N = \ker(J_\mathrm{c}^0) \oplus U \tag{7.59}$$

of the space to which $\boldsymbol{u}$ belongs, and another direct sum decomposition

$$\mathbb{R}^N = V \oplus \mathrm{range}(J_\mathrm{c}^0) \tag{7.60}$$

of the space in which $\boldsymbol{F}$ takes values. Note that $\dim(U) = N - M$ and $\dim(V) = M$. The complementary subspaces U and V here are not determined uniquely, but any choice of them is good for the derivation of the bifurcation equation. See Remark 7.8 below for a possible choice.

According to (7.59), we decompose $\boldsymbol{u} - \boldsymbol{u}_\mathrm{c}^0$ into two components as

$$\boldsymbol{u} = \boldsymbol{u}_\mathrm{c}^0 + \boldsymbol{w} + \overline{\boldsymbol{w}}, \tag{7.61}$$

where $\boldsymbol{w} \in \ker(J_\mathrm{c}^0)$ and $\overline{\boldsymbol{w}} \in U$. With reference to (7.60), let P be the *projection* to V along the subspace range(J_c^0), where $P^2 = P$. Then the full system (7.51) of equations can be decomposed into two parts[12]

$$P \cdot \boldsymbol{F}(\boldsymbol{u}_\mathrm{c}^0 + \boldsymbol{w} + \overline{\boldsymbol{w}}, f_\mathrm{c}^0 + \widetilde{f}, \boldsymbol{v}) = \boldsymbol{0}, \tag{7.62}$$
$$(I - P) \cdot \boldsymbol{F}(\boldsymbol{u}_\mathrm{c}^0 + \boldsymbol{w} + \overline{\boldsymbol{w}}, f_\mathrm{c}^0 + \widetilde{f}, \boldsymbol{v}) = \boldsymbol{0}. \tag{7.63}$$

The Jacobian matrix of (7.63) with respect to $\overline{\boldsymbol{w}}$, evaluated at $(\boldsymbol{w}, \overline{\boldsymbol{w}}, \widetilde{f}, \boldsymbol{v}) = (\boldsymbol{0}, \boldsymbol{0}, 0, \boldsymbol{v}^0)$, is invertible as a mapping from U to range(J_c^0). Consequently, by the implicit function theorem, (7.63) can be solved for $\overline{\boldsymbol{w}}$ as

$$\overline{\boldsymbol{w}} = \boldsymbol{\varphi}(\boldsymbol{w}, \widetilde{f}, \boldsymbol{v}) \tag{7.64}$$

uniquely in some neighborhood of $(\boldsymbol{w}, \overline{\boldsymbol{w}}, \widetilde{f}, \boldsymbol{v}) = (\boldsymbol{0}, \boldsymbol{0}, 0, \boldsymbol{v}^0)$. Substitution of this into (7.62) yields

$$P \cdot \boldsymbol{F}(\boldsymbol{u}_\mathrm{c}^0 + \boldsymbol{w} + \boldsymbol{\varphi}(\boldsymbol{w}, \widetilde{f}, \boldsymbol{v}), f_\mathrm{c}^0 + \widetilde{f}, \boldsymbol{v}) = \boldsymbol{0}. \tag{7.65}$$

This yields the reduced system (7.52) with

$$\widetilde{\boldsymbol{F}}(\boldsymbol{w}, \widetilde{f}, \boldsymbol{v}) = P \cdot \boldsymbol{F}(\boldsymbol{u}_\mathrm{c}^0 + \boldsymbol{w} + \boldsymbol{\varphi}(\boldsymbol{w}, \widetilde{f}, \boldsymbol{v}), f_\mathrm{c}^0 + \widetilde{f}, \boldsymbol{v}) \tag{7.66}$$

(cf., Remark 7.9).

The solutions $(\boldsymbol{w}, \widetilde{f}, \boldsymbol{v})$ to the bifurcation equation (7.52) are in one-to-one correspondence through (7.64) with the solutions $(\boldsymbol{u}, f, \boldsymbol{v})$ of the original system (7.51) in a neighborhood of $(\boldsymbol{u}_\mathrm{c}^0, f_\mathrm{c}^0, \boldsymbol{v}^0)$. For a solution $(\boldsymbol{w}, \widetilde{f}, \boldsymbol{v})$ to (7.52) the corresponding $\boldsymbol{u}$ is given by

$$\boldsymbol{u} = \boldsymbol{u}(\boldsymbol{w}, \widetilde{f}, \boldsymbol{v}) = \boldsymbol{u}_\mathrm{c}^0 + \boldsymbol{w} + \boldsymbol{\varphi}(\boldsymbol{w}, \widetilde{f}, \boldsymbol{v}). \tag{7.67}$$

Conversely, for a solution $(\boldsymbol{u}, \widetilde{f}, \boldsymbol{v})$ to (7.51), the corresponding $\boldsymbol{w}$ is given by (7.61). Hence, the qualitative picture of the solution set of the original system (7.51) is isomorphic to that of the bifurcation equation (7.52).

---

[12] Since P is the projection on an M-dimensional subspace, the equation (7.62) effectively represents M constraints and (7.63) represents $(N - M)$ constraints.

If we take a basis $\{\boldsymbol{\eta}_i \mid i = 1,\dots,M\}$ of $\ker(J_{\rm c}^0)$ and a basis $\{\boldsymbol{\eta}_j \mid j = M+1,\dots,N\}$ of its complementary subspace U, then we obtain expansions

$$\boldsymbol{w} = \sum_{i=1}^{M} w_i \boldsymbol{\eta}_i, \qquad \overline{\boldsymbol{w}} = \boldsymbol{\varphi}(\boldsymbol{w}, \widetilde{f}, \boldsymbol{v}) = \sum_{j=M+1}^{N} \varphi_j(\boldsymbol{w}, \widetilde{f}, \boldsymbol{v}) \boldsymbol{\eta}_j. \tag{7.68}$$

Accordingly, we can rewrite (7.67) as

$$\boldsymbol{u} = \boldsymbol{u}_{\rm c}^0 + \sum_{i=1}^{M} w_i \boldsymbol{\eta}_i + \sum_{j=M+1}^{N} \varphi_j(\boldsymbol{w}, \widetilde{f}, \boldsymbol{v}) \boldsymbol{\eta}_j. \tag{7.69}$$

Then, as a succinct representation of (7.62) and (7.63), we obtain

$$\boldsymbol{F}\left(\boldsymbol{u}_{\rm c}^0 + \sum_{i=1}^{M} w_i \boldsymbol{\eta}_i + \sum_{j=M+1}^{N} \varphi_j(\boldsymbol{w}, \widetilde{f}, \boldsymbol{v}) \boldsymbol{\eta}_j, f_{\rm c}^0 + \widetilde{f}, \boldsymbol{v}\right) = \boldsymbol{0}. \tag{7.70}$$

Remark 7.8. A possible (and natural) choice of U and V in (7.59) and (7.60) is

$$U = (\ker(J_{\rm c}^0))^{\perp}, \qquad V = (\mathrm{range}(J_{\rm c}^0))^{\perp}. \tag{7.71}$$

This choice is justified by the orthogonal decomposition (7.18). If we take an orthonormal basis $\{\boldsymbol{\eta}_i \mid i = 1,\dots,N\}$ of $\mathbb{R}^N$ such that $J_{\rm c}^0 \boldsymbol{\eta}_i = \boldsymbol{0}$ for $i = 1,\dots,M$, then $\{\boldsymbol{\eta}_i \mid i = 1,\dots,M\}$ forms a basis of $\ker(J_{\rm c}^0)$ and $\{\boldsymbol{\eta}_j \mid j = M+1,\dots,N\}$ a basis of U, and hence we have (7.68). For another orthonormal basis $\{\boldsymbol{\xi}_i \mid i = 1,\dots,N\}$ of $\mathbb{R}^N$ such that $\boldsymbol{\xi}_i^{\top} J_{\rm c}^0 = \boldsymbol{0}^{\top}$ for $i = 1,\dots,M$, the projection P to V is given by

$$P = \sum_{i=1}^{M} \boldsymbol{\xi}_i \boldsymbol{\xi}_i^{\top}, \tag{7.72}$$

which is an orthogonal projection. Therefore, (7.62) is equivalent to $\boldsymbol{\xi}_i^{\top} \boldsymbol{F} = 0$ for $i = 1,\dots,M$, and (7.63) is equivalent to $\boldsymbol{\xi}_j^{\top} \boldsymbol{F} = 0$ for $j = M+1,\dots,N$. □

Remark 7.9. Since P is the projection on V, which is an M-dimensional subspace, the right-hand side of the equation (7.66) effectively stands for M constraints on $(\boldsymbol{w}, \widetilde{f}, \boldsymbol{v})$. A concrete vector representation of these constraints with respect to a certain basis of V is represented by an M-dimensional vector-valued function, which is denoted here as $\widetilde{\boldsymbol{F}}$. It should be understood that the equality in (7.66) designates this correspondence, although the left-hand side is an M-dimensional vector and the right-hand side is an N-dimensional vector. □

Group Equivariance

The discussion proceeds to consideration of the group equivariance. The following facts are fundamental here.

- Both $\ker(J_c^0)$ and $\mathrm{range}(J_c^0)$ are G-invariant subspaces, and therefore
- the complementary subspaces U and V in (7.59) and (7.60) can be chosen to be G-invariant,[13] which is assumed throughout.

The G-invariance of $\ker(J_c^0)$ and $\mathrm{range}(J_c^0)$ is a consequence of the equivariance $T(g)J_c^0 = J_c^0 T(g)$ $(g \in G)$ in (7.46). In fact, this is proved for $\ker(J_c^0)$ in Lemma 7.2 in §7.5.1, whereas the claim for $\mathrm{range}(J_c^0)$ is treated in the following lemma.

Lemma 7.3. *The range space* $\mathrm{range}(J_c^0)$ *is* G*-invariant. If the complementary subspace* V *is also* G*-invariant, the projection* P *to* V *along the subspace* $\mathrm{range}(J_c^0)$ *satisfies*

$$T(g)P = PT(g), \qquad g \in G. \tag{7.73}$$

Proof. Take any $\boldsymbol{y} \in \mathrm{range}(J_c^0)$ and $g \in G$. Then $\boldsymbol{y} = J_c^0 \boldsymbol{x}$ for some $\boldsymbol{x}$. It follows from (7.46) that

$$T(g)\boldsymbol{y} = T(g)(J_c^0 \boldsymbol{x}) = J_c^0(T(g)\boldsymbol{x}) \in \mathrm{range}(J_c^0).$$

Hence, $\mathrm{range}(J_c^0)$ is G-invariant. The G-invariance of V is equivalent to

$$PT(g)P = T(g)P, \qquad g \in G,$$

whereas that of $\mathrm{range}(J_c^0)$ is equivalent to

$$(I-P)T(g)(I-P) = T(g)(I-P), \qquad g \in G.$$

Adding these two identities gives (7.73). (This proof shows that the commutativity (7.73) is, in fact, equivalent to the G-invariance of V and $\mathrm{range}(J_c^0)$.) □

As a preliminary step toward the equivariance of $\widetilde{\boldsymbol{F}}$, the equivariance of $\boldsymbol{\varphi}$ is shown first.

Lemma 7.4.

$$T(g)\boldsymbol{\varphi}(\boldsymbol{w}, \widetilde{f}, \boldsymbol{v}) = \boldsymbol{\varphi}(T(g)\boldsymbol{w}, \widetilde{f}, S(g)\boldsymbol{v}), \qquad g \in G. \tag{7.74}$$

Proof. Recall the equation (7.63) with (7.64):

$$(I-P)\cdot \boldsymbol{F}(\boldsymbol{u}_c^0 + \boldsymbol{w} + \boldsymbol{\varphi}(\boldsymbol{w}, \widetilde{f}, \boldsymbol{v}), f_c^0 + \widetilde{f}, \boldsymbol{v}) = \boldsymbol{0}, \tag{7.75}$$

which defines the function $\boldsymbol{\varphi}$ through the implicit function theorem. For the left-hand side of this equation, we have

$$\begin{aligned}
&T(g)(I-P)\cdot \boldsymbol{F}(\boldsymbol{u}_c^0 + \boldsymbol{w} + \boldsymbol{\varphi}(\boldsymbol{w}, \widetilde{f}, \boldsymbol{v}), f_c^0 + \widetilde{f}, \boldsymbol{v}) \\
&= (I-P)\cdot T(g)\boldsymbol{F}(\boldsymbol{u}_c^0 + \boldsymbol{w} + \boldsymbol{\varphi}(\boldsymbol{w}, \widetilde{f}, \boldsymbol{v}), f_c^0 + \widetilde{f}, \boldsymbol{v}) \\
&= (I-P)\cdot \boldsymbol{F}(T(g)[\boldsymbol{u}_c^0 + \boldsymbol{w} + \boldsymbol{\varphi}(\boldsymbol{w}, \widetilde{f}, \boldsymbol{v})], f_c^0 + \widetilde{f}, S(g)\boldsymbol{v}) \\
&= (I-P)\cdot \boldsymbol{F}(\boldsymbol{u}_c^0 + T(g)\boldsymbol{w} + T(g)\boldsymbol{\varphi}(\boldsymbol{w}, \widetilde{f}, \boldsymbol{v}), f_c^0 + \widetilde{f}, S(g)\boldsymbol{v})
\end{aligned}$$

[13] $U = (\ker(J_c^0))^\perp$ and $V = (\mathrm{range}(J_c^0))^\perp$ in (7.71) are valid choices, since T is assumed to be unitary (cf., Remark 7.8 above).

for all $g \in G$, where (7.73), (7.53), and (7.58) are used in this order. This means, on replacing[14] $T(g)\boldsymbol{w}$ by $\boldsymbol{w}$ and $S(g)\boldsymbol{v}$ by $\boldsymbol{v}$, that

$$(I-P)\cdot \boldsymbol{F}(\boldsymbol{u}_{\rm c}^0+\boldsymbol{w}+T(g)\boldsymbol{\varphi}(T(g^{-1})\boldsymbol{w},\widetilde{f},S(g^{-1})\boldsymbol{v}),f_{\rm c}^0+\widetilde{f},\boldsymbol{v}) = \boldsymbol{0}. \qquad (7.76)$$

Comparison of (7.75) and (7.76), together with the uniqueness[15] of the implicit function, shows that

$$T(g)\boldsymbol{\varphi}(T(g^{-1})\boldsymbol{w},\widetilde{f},S(g^{-1})\boldsymbol{v}) = \boldsymbol{\varphi}(\boldsymbol{w},\widetilde{f},\boldsymbol{v}), \qquad g \in G.$$

Since we can replace $T(g^{-1})\boldsymbol{w}$ with $\boldsymbol{w}$ and $S(g^{-1})\boldsymbol{v}$ with $\boldsymbol{v}$, we obtain (7.74). □

The equivariance of bifurcation equation $\widetilde{\boldsymbol{F}}$ is obtained from Lemmas 7.3 and 7.4 above, as follows.

$$\begin{aligned}
T(g)\widetilde{\boldsymbol{F}}(\boldsymbol{w},\widetilde{f},\boldsymbol{v}) &= T(g)P\cdot \boldsymbol{F}(\boldsymbol{u}_{\rm c}^0+\boldsymbol{w}+\boldsymbol{\varphi}(\boldsymbol{w},\widetilde{f},\boldsymbol{v}),f_{\rm c}^0+\widetilde{f},\boldsymbol{v})\\
&= P\cdot T(g)\,\boldsymbol{F}(\boldsymbol{u}_{\rm c}^0+\boldsymbol{w}+\boldsymbol{\varphi}(\boldsymbol{w},\widetilde{f},\boldsymbol{v}),f_{\rm c}^0+\widetilde{f},\boldsymbol{v})\\
&= P\cdot \boldsymbol{F}(T(g)\,[\boldsymbol{u}_{\rm c}^0+\boldsymbol{w}+\boldsymbol{\varphi}(\boldsymbol{w},\widetilde{f},\boldsymbol{v})],f_{\rm c}^0+\widetilde{f},S(g)\boldsymbol{v})\\
&= P\cdot \boldsymbol{F}(\boldsymbol{u}_{\rm c}^0+T(g)\boldsymbol{w}+\boldsymbol{\varphi}(T(g)\boldsymbol{w},\widetilde{f},S(g)\boldsymbol{v}),f_{\rm c}^0+\widetilde{f},S(g)\boldsymbol{v})\\
&= \widetilde{\boldsymbol{F}}(T(g)\boldsymbol{w},\widetilde{f},S(g)\boldsymbol{v}), \qquad (7.77)
\end{aligned}$$

in which the definition of $\widetilde{\boldsymbol{F}}$ in (7.66) is used. In this expression $T(g)\widetilde{\boldsymbol{F}}$ and $T(g)\boldsymbol{w}$ can be replaced by $\widetilde{T}(g)\widetilde{\boldsymbol{F}}$ and $\widetilde{T}(g)\boldsymbol{w}$, respectively, inasmuch as the subrepresentations of T on V and on $\ker(J_{\rm c}^0)$ are equivalent by virtue of the fact that $\ker(J_{\rm c}^0)\oplus U$ and $V \oplus {\rm range}(J_{\rm c}^0)$ are equivalent and that U and ${\rm range}(J_{\rm c}^0)$ are also equivalent. This completes the proof of the equivariance (7.54) of the bifurcation equation $\widetilde{\boldsymbol{F}}$, establishing the principle of inheritance of symmetry.

Criticality Condition

The criticality condition for the full system, $\det[J(\boldsymbol{u},f,\boldsymbol{v})] = 0$ in (2.6), is equivalent to the criticality condition for the reduced system:

$$\det[\widetilde{J}(\boldsymbol{w},\widetilde{f},\boldsymbol{v})] = 0, \qquad (7.78)$$

where

$$\widetilde{J}(\boldsymbol{w},\widetilde{f},\boldsymbol{v}) = \frac{\partial \widetilde{\boldsymbol{F}}}{\partial \boldsymbol{w}}(\boldsymbol{w},\widetilde{f},\boldsymbol{v}).$$

---

[14] We can replace $T(g)\boldsymbol{w}$ by $\boldsymbol{w}$ since $\boldsymbol{w}$ is arbitrary in $\ker(J_{\rm c}^0)$ and $\ker(J_{\rm c}^0)$ is G-invariant. We can also replace $S(g)\boldsymbol{v}$ by $\boldsymbol{v}$ since $\boldsymbol{v}$ is arbitrary.

[15] The uniqueness assertion applies inasmuch as $T(g)\boldsymbol{\varphi}(T(g^{-1})\boldsymbol{w},\widetilde{f},S(g^{-1})\boldsymbol{v}) \in U$ by the G-invariance of U and $S(g^{-1})\boldsymbol{v}$ stays in a neighborhood of $\boldsymbol{v}^0$ by $\|S(g^{-1})\boldsymbol{v}-\boldsymbol{v}^0\| = \|S(g^{-1})(\boldsymbol{v}-\boldsymbol{v}^0)\| = \|\boldsymbol{v}-\boldsymbol{v}^0\|$. Here the first equality holds by (7.58) and the second equality by the unitarity of S.

To be more precise, the following lemma states that the Jacobian matrix of the reduced system is the so-called *Schur complement* of the Jacobian matrix of the full system. Let

$$\overline{J} = \overline{J}(\boldsymbol{u}, f, \boldsymbol{v}) = (\overline{J}_{ij} \mid i, j = 1, \ldots, N)$$

be an $N \times N$ matrix defined by $\overline{J}_{ij} = \boldsymbol{\xi}_i^\top J \boldsymbol{\eta}_j$ $(i, j = 1, \ldots, N)$ and partition $\overline{J}$ as

$$\overline{J} = \begin{pmatrix} \overline{J}_{[1,1]} & \overline{J}_{[1,2]} \\ \overline{J}_{[2,1]} & \overline{J}_{[2,2]} \end{pmatrix},$$

where $\overline{J}_{[1,1]}$ is an $M \times M$ matrix and $\overline{J}_{[2,2]}$ is an $(N-M) \times (N-M)$ matrix.

Lemma 7.5. *In a neighborhood of* $(\boldsymbol{u}_{\rm c}^0, f_{\rm c}^0, \boldsymbol{v}^0)$, $\overline{J}_{[2,2]}$ *is nonsingular and*

$$\tilde{J}(\boldsymbol{w}, \widetilde{f}, \boldsymbol{v}) = \overline{J}_{[1,1]} - \overline{J}_{[1,2]}(\overline{J}_{[2,2]})^{-1}\overline{J}_{[2,1]}, \tag{7.79}$$

where $J(\boldsymbol{u}, f, \boldsymbol{v})$ *and* $\overline{J}(\boldsymbol{u}, f, \boldsymbol{v})$ *are evaluated with* (7.69) *and* $f = f_{\rm c}^0 + \widetilde{f}$. *Therefore,* $\det[\tilde{J}(\boldsymbol{w}, \widetilde{f}, \boldsymbol{v})] = 0$ *if and only if* $\det[J(\boldsymbol{u}, f, \boldsymbol{v})] = 0$.

Proof. The proof is similar to that of Lemma 2.1 in §2.4.2. □

If the original system (7.51) is reciprocal, we can take an orthonormal basis $\{\boldsymbol{\eta}_i \mid i = 1, \ldots, N\}$ and put $\boldsymbol{\xi}_i = \boldsymbol{\eta}_i$ $(i = 1, \ldots, N)$. Then $\overline{J}(\boldsymbol{u}, f, \boldsymbol{v})$ is a symmetric matrix, and so is $\tilde{J}(\boldsymbol{w}, \widetilde{f}, \boldsymbol{v})$. This shows the inheritance of reciprocity in (7.56).

Direction of Bifurcated Paths

The directions of bifurcated paths can be analyzed as follows. We assume that a solution path of the bifurcation equation (7.52) with $\boldsymbol{v} = \boldsymbol{v}^0$ is described as

$$\boldsymbol{w} = \boldsymbol{w}(s, \boldsymbol{v}^0), \qquad \widetilde{f} = \widetilde{f}(s, \boldsymbol{v}^0)$$

in terms of a scalar parameter s such that $\boldsymbol{w}(0, \boldsymbol{v}^0) = \boldsymbol{0}$, $\widetilde{f}(0, \boldsymbol{v}^0) = 0$, and $|\partial \widetilde{f}/\partial s| + \|\partial \boldsymbol{w}/\partial s\| \neq 0$. Substitution of these into (7.69) followed by differentiation with respect to s yields

$$\frac{\partial \boldsymbol{u}}{\partial s} = \sum_{i=1}^{M} \frac{\partial w_i}{\partial s}\boldsymbol{\eta}_i + \sum_{j=M+1}^{N} \left(\frac{\partial \varphi_j}{\partial w_i}\frac{\partial w_i}{\partial s} + \frac{\partial \varphi_j}{\partial \widetilde{f}}\frac{\partial \widetilde{f}}{\partial s} \right)\boldsymbol{\eta}_j, \tag{7.80}$$

which stands for the tangent (direction) vector of the solution path in the space of $\boldsymbol{u}$. To evaluate this expression at $s = 0$ we note the following fact.

Lemma 7.6.

$$\frac{\partial \varphi_j}{\partial w_i}(\boldsymbol{0}, 0, \boldsymbol{v}^0) = 0, \qquad i = 1, \ldots, M, \;\; j = M+1, \ldots, N.$$

Proof. Similar to the proof of Lemma 2.2 in §2.4.3. □

Evaluation of (7.80) at $s = 0$, using Lemma 7.6 above, shows that the solution path emanates from $\boldsymbol{u}_{\mathrm{c}}^0$ in the direction of

$$\left(\frac{\partial \boldsymbol{u}}{\partial s}\right)_{\mathrm{c}}^0 = \sum_{i=1}^{M} \frac{\partial w_i}{\partial s}(0,\boldsymbol{v}^0)\boldsymbol{\eta}_i + \frac{\partial \widetilde{f}}{\partial s}(0,\boldsymbol{v}^0) \sum_{j=M+1}^{N} \frac{\partial \varphi_j}{\partial \widetilde{f}}(\boldsymbol{0},0,\boldsymbol{v}^0)\boldsymbol{\eta}_j. \tag{7.81}$$

The direction (7.81) is not necessarily confined to the subspace spanned by the critical eigenvectors $\{\boldsymbol{\eta}_i \mid i = 1,\ldots,M\}$, but it contains an extra component in the direction of

$$\boldsymbol{\eta}_* = \sum_{j=M+1}^{N} \frac{\partial \varphi_j}{\partial \widetilde{f}}(\boldsymbol{0},0,\boldsymbol{v}^0)\boldsymbol{\eta}_j. \tag{7.82}$$

Just as (2.62) in §2.4.3 this component satisfies[16]

$$J_{\mathrm{c}}^0 \boldsymbol{\eta}_* + \left(\frac{\partial \boldsymbol{F}}{\partial f}\right)_{\mathrm{c}}^0 = \boldsymbol{0}. \tag{7.83}$$

It is worth mentioning that the special case of (7.81) with $M = 1$ and $s = w_1$ coincides with (2.58) in §2.4.3.

For a bifurcated path of a system with group symmetry it is often the case[17] that

$$\frac{\partial \widetilde{f}}{\partial s}(0,\boldsymbol{v}^0) = 0. \tag{7.84}$$

Then the direction of the solution path in the space of $\boldsymbol{u}$ is given from (7.81) as

$$\left(\frac{\partial \boldsymbol{u}}{\partial s}\right)_{\mathrm{c}}^0 = \sum_{i=1}^{M} \frac{\partial w_i}{\partial s}(0,\boldsymbol{v}^0)\boldsymbol{\eta}_i, \tag{7.85}$$

which, being free from the extra component $\boldsymbol{\eta}_*$ in (7.82), lies in the subspace spanned by the critical eigenvectors.

Remark 7.10. In this section, the Liapunov–Schmidt reduction has been explained for a finite-dimensional system of equations that is equivariant to a finite group. This reduction, however, can be carried out in a more general setting where the linearization of $\boldsymbol{F}$ in (7.51) is a "Fredholm operator of index zero" and G is a compact group. We mention that relation (7.57) is valid for a Fredholm operator of index zero in the form of

$$\dim[\ker(J_{\mathrm{c}}^0)] = \mathrm{codim}[\mathrm{range}(J_{\mathrm{c}}^0)],$$

[16] Differentiation of (7.70) with respect to $\widetilde{f}$, followed by evaluation at $(\boldsymbol{w},\widetilde{f}) = (\boldsymbol{0},0)$, yields (7.83).

[17] This is true for a certain type of double bifurcation point of a D_n-symmetric system (cf., (8.62) with $\widehat{n} \geq 4$).

and the major results on group representation are extended for a compact group. To convey the main ideas to general readers without sacrificing mathematical rigor, we have restricted ourselves to finite-dimensional equations with a finite group, although we implicitly rely on the Liapunov–Schmidt reduction in the general case in Part III. See Golubitsky and Schaeffer, 1985 [62] and Golubitsky, Stewart, and Schaeffer, 1988 [64] for the Liapunov–Schmidt reduction in full generality. □

7.6 Symmetry of Solutions

Attention is now shifted from the symmetry of equations to the symmetry of solutions. We first recall specifically that the symmetry of the governing equations is formulated in §7.5.1 as the equivariance of $\boldsymbol{F}$ to a group G.

The symmetry of a solution $\boldsymbol{u}$ is described by a subgroup of G, called the *isotropy subgroup* of $\boldsymbol{u}$, defined by

$$\Sigma(\boldsymbol{u}) = \Sigma(\boldsymbol{u};G,T) = \{g \in G \mid T(g)\boldsymbol{u} = \boldsymbol{u}\}. \tag{7.86}$$

The notation (7.86) is extended to a subset of vectors, say, W as

$$\Sigma(W) = \bigcap_{\boldsymbol{u}\in W} \Sigma(\boldsymbol{u}) = \{g \in G \mid T(g)\boldsymbol{u} = \boldsymbol{u} \text{ for all } \boldsymbol{u} \in W\}. \tag{7.87}$$

We use this extended notation primarily for $W = \ker(J_{\mathrm{c}}^0)$.

In general, $\Sigma(\boldsymbol{u})$ is equal to G for a solution $\boldsymbol{u}$ on the fundamental path and $\Sigma(\boldsymbol{u})$ is a proper subgroup of G for $\boldsymbol{u}$ on a bifurcated path. In association with the repeated occurrence of bifurcation, one obtains a hierarchy of subgroups

$$G = G_1 \rightarrow G_2 \rightarrow G_3 \rightarrow \cdots \tag{7.88}$$

that characterizes the recursive change of symmetries. Here G_{i+1} is a proper subgroup of G_i for $i = 1,2,\ldots$. Concrete forms of this hierarchy are obtained for the dihedral group in §8.3.2 and for other groups in Part III to describe physical phenomena.

Remark 7.11. The isotropy subgroup $\Sigma(\boldsymbol{u})$ of $\boldsymbol{u}$ is also referred to as the *symmetry group* of $\boldsymbol{u}$. In fact, the term "symmetry group" is used in a broader sense to mean the group formed by all operations that preserve the object of our interest. For example, if G is the largest group to which a system of equations $\boldsymbol{F}$ is equivariant as in (7.38), then it is said that G is the symmetry group of $\boldsymbol{F}$, or alternatively that the symmetry group of $\boldsymbol{F}$ is given by G. □

Near Ordinary Point

For an ordinary point the most important fact is that the symmetry of a solution remains invariant in the neighborhood of this point, at which the Jacobian matrix is nonsingular by definition. We consider the perfect system with $\boldsymbol{v} = \boldsymbol{v}^0$. Let $(\boldsymbol{u}_*, f_*)$ be an ordinary point, and let $(\boldsymbol{u}, f)$ be another solution point sufficiently close to $(\boldsymbol{u}_*, f_*)$. Then it can be shown that $\boldsymbol{u}_*$ and $\boldsymbol{u}$ share the same symmetry in the sense that

$$\Sigma(\boldsymbol{u}_*) = \Sigma(\boldsymbol{u}). \tag{7.89}$$

The proof is given in Remark 7.12 below.

Remark 7.12. Since $(\boldsymbol{u}_*, f_*)$ is an ordinary point, the Jacobian matrix $J(\boldsymbol{u}_*, f_*, \boldsymbol{v}^0)$ is nonsingular. This implies, by the implicit function theorem, that the equation (7.36) can be solved for $\boldsymbol{u}$ as

$$\boldsymbol{u} = \boldsymbol{u}_* + \boldsymbol{\varphi}(\widetilde{f}, \boldsymbol{v}^0) \tag{7.90}$$

in a neighborhood of $(\boldsymbol{u}, \widetilde{f}) = (\boldsymbol{u}_*, 0)$, where $\widetilde{f} = f - f_*$ and $\boldsymbol{\varphi}(0, \boldsymbol{v}^0) = \boldsymbol{0}$. Then we have

$$\boldsymbol{F}(\boldsymbol{u}_* + \boldsymbol{\varphi}(\widetilde{f}, \boldsymbol{v}^0), f_* + \widetilde{f}, \boldsymbol{v}^0) = \boldsymbol{0}.$$

Fix an arbitrary $g \in \Sigma(\boldsymbol{u}_*)$. Using the equivariance (7.37) we obtain

$$\begin{aligned}
\boldsymbol{0} &= T(g)\boldsymbol{F}(\boldsymbol{u}_* + \boldsymbol{\varphi}(\widetilde{f}, \boldsymbol{v}^0), f_* + \widetilde{f}, \boldsymbol{v}^0) \\
&= \boldsymbol{F}(T(g)[\boldsymbol{u}_* + \boldsymbol{\varphi}(\widetilde{f}, \boldsymbol{v}^0)], f_* + \widetilde{f}, \boldsymbol{v}^0) \\
&= \boldsymbol{F}(\boldsymbol{u}_* + T(g)\boldsymbol{\varphi}(\widetilde{f}, \boldsymbol{v}^0), f_* + \widetilde{f}, \boldsymbol{v}^0).
\end{aligned}$$

Then the uniqueness of the implicit function shows

$$T(g)\boldsymbol{\varphi}(\widetilde{f}, \boldsymbol{v}^0) = \boldsymbol{\varphi}(\widetilde{f}, \boldsymbol{v}^0),$$

and, therefore, by (7.90), we have

$$T(g)\boldsymbol{u} = T(g)[\boldsymbol{u}_* + \boldsymbol{\varphi}(\widetilde{f}, \boldsymbol{v}^0)] = \boldsymbol{u}_* + \boldsymbol{\varphi}(\widetilde{f}, \boldsymbol{v}^0) = \boldsymbol{u},$$

which shows $g \in \Sigma(\boldsymbol{u})$. Hence $\Sigma(\boldsymbol{u}_*) \subseteq \Sigma(\boldsymbol{u})$. The roles of $\boldsymbol{u}_*$ and $\boldsymbol{u}$ are interchangeable; therefore, we also have $\Sigma(\boldsymbol{u}_*) \supseteq \Sigma(\boldsymbol{u})$. The invariance of the symmetry (7.89) is thus proved. □

Near Critical Point

We go on to investigate what happens in the neighborhood of a critical point, to which (7.89) does not apply. Typically, a solution $\boldsymbol{u}$ on the bifurcated path, if any, is less symmetric than a solution on the fundamental path; that is, $\Sigma(\boldsymbol{u})$ is a proper subgroup of G. We show here that the symmetry of $\boldsymbol{u}$ is at least as large as the symmetry of the critical eigenvectors.

Let $(\boldsymbol{u}_{\mathrm{c}}^0, f_{\mathrm{c}}^0)$ be a critical point of the perfect system with $\boldsymbol{v} = \boldsymbol{v}^0$ such that $\Sigma(\boldsymbol{u}_{\mathrm{c}}^0) = G$. The Jacobian matrix $J_{\mathrm{c}}^0 = J(\boldsymbol{u}_{\mathrm{c}}^0, f_{\mathrm{c}}^0, \boldsymbol{v}^0)$ has a nontrivial kernel space; its symmetry is represented by

$$\Sigma(\ker(J_{\mathrm{c}}^0)) = \{g \in G \mid T(g)\boldsymbol{\eta} = \boldsymbol{\eta} \ \text{ for all } \ \boldsymbol{\eta} \in \ker(J_{\mathrm{c}}^0)\}. \tag{7.91}$$

For a particular critical eigenvector $\boldsymbol{\eta} \in \ker(J_{\mathrm{c}}^0)$ we have

$$\Sigma(\ker(J_{\mathrm{c}}^0)) \subseteq \Sigma(\boldsymbol{\eta}) \tag{7.92}$$

in general. At a simple point, however, the equality is guaranteed in (7.92), since $\ker(J_{\mathrm{c}}^0)$ is spanned by $\boldsymbol{\eta}$.

For a solution $(\boldsymbol{u}, f)$ of the perfect system that is sufficiently close to $(\boldsymbol{u}_{\mathrm{c}}^0, f_{\mathrm{c}}^0)$, we have

$$\Sigma(\ker(J_{\mathrm{c}}^0)) \subseteq \Sigma(\boldsymbol{w}) = \Sigma(\boldsymbol{u}), \tag{7.93}$$

as is proved in Remark 7.13 below.

In (7.93) we have a crucial technical relation $\Sigma(\boldsymbol{w}) = \Sigma(\boldsymbol{u})$, which shows that the symmetry of the solution $\boldsymbol{u}$ to the full system of equations is determined by that of the solution $\boldsymbol{w}$ to the bifurcation equation. This fact allows us to concentrate on the solution $\boldsymbol{w}$ of the bifurcation equation in discussing the symmetry of a bifurcating solution $\boldsymbol{u}$. The discrepancy between $\Sigma(\ker(J_{\mathrm{c}}^0))$ and $\Sigma(\boldsymbol{w})$ in (7.93) is attributable to the nonlinearity of the equation and, as such, the symmetry $\Sigma(\boldsymbol{w})$ of the solution $\boldsymbol{w}$ can be determined only through an analysis involving nonlinear terms.

If $(\boldsymbol{u}_{\mathrm{c}}^0, f_{\mathrm{c}}^0)$ is a group-theoretic critical point associated with an irreducible representation, say $\mu \in R(G)$, the symmetry of the kernel space of J_{c}^0 coincides with the subgroup G^{μ} introduced in (7.22). That is to say, we have

$$\Sigma(\ker(J_{\mathrm{c}}^0)) = G^{\mu}. \tag{7.94}$$

A combination of (7.93) and (7.94) yields an important relation

$$G^{\mu} = \Sigma(\ker(J_{\mathrm{c}}^0)) \subseteq \Sigma(\boldsymbol{w}) = \Sigma(\boldsymbol{u}) \subseteq G, \tag{7.95}$$

showing that the symmetry of any solution in the neighborhood of a group-theoretic critical point associated with an irreducible representation μ is at least as high as the symmetry represented by G^{μ}.

Remark 7.13. Take $g \in G$ and recall $T(g)\boldsymbol{u}_{\mathrm{c}}^0 = \boldsymbol{u}_{\mathrm{c}}^0$. By applying $T(g)$ to the identity $\boldsymbol{u} = \boldsymbol{u}_{\mathrm{c}}^0 + \boldsymbol{w} + \boldsymbol{\varphi}(\boldsymbol{w}, \widetilde{f}, \boldsymbol{v}^0)$ in (7.67), we obtain

$$\begin{aligned} T(g)\boldsymbol{u} &= T(g)[\boldsymbol{u}_{\mathrm{c}}^0 + \boldsymbol{w} + \boldsymbol{\varphi}(\boldsymbol{w}, \widetilde{f}, \boldsymbol{v}^0)] \\ &= \boldsymbol{u}_{\mathrm{c}}^0 + T(g)\boldsymbol{w} + T(g)\boldsymbol{\varphi}(\boldsymbol{w}, \widetilde{f}, \boldsymbol{v}^0) \\ &= \boldsymbol{u}_{\mathrm{c}}^0 + T(g)\boldsymbol{w} + \boldsymbol{\varphi}(T(g)\boldsymbol{w}, \widetilde{f}, \boldsymbol{v}^0), \end{aligned} \tag{7.96}$$

where the second equality follows from (7.74) for $\boldsymbol{v} = \boldsymbol{v}^0$ (Lemma 7.4 in §7.5.2).

If $T(g)\boldsymbol{w} = \boldsymbol{w}$ in (7.96), we have $T(g)\boldsymbol{u} = \boldsymbol{u}$ from (7.96). Conversely, if $T(g)\boldsymbol{u} = \boldsymbol{u}$ in (7.96), we have

$$\boldsymbol{u} = \boldsymbol{u}_{\rm c}^0 + T(g)\boldsymbol{w} + \boldsymbol{\varphi}(T(g)\boldsymbol{w}, \widetilde{f}, \boldsymbol{v}^0), \tag{7.97}$$

where $T(g)\boldsymbol{w} \in \ker(J_{\rm c}^0)$ and $\boldsymbol{\varphi}(T(g)\boldsymbol{w}, \widetilde{f}, \boldsymbol{v}^0) \in U$. The decomposition of the form of (7.61) is unique, and thus comparison of (7.97) with (7.67) shows $T(g)\boldsymbol{w} = \boldsymbol{w}$. Therefore, we have that $T(g)\boldsymbol{u} = \boldsymbol{u}$ if and only if $T(g)\boldsymbol{w} = \boldsymbol{w}$. This shows $\Sigma(\boldsymbol{w}) = \Sigma(\boldsymbol{u})$.

Finally, if $g \in \Sigma(\ker(J_{\rm c}^0))$, then we have $T(g)\boldsymbol{w} = \boldsymbol{w}$ by $\boldsymbol{w} \in \ker(J_{\rm c}^0)$ and (7.91). Hence, $\Sigma(\ker(J_{\rm c}^0)) \subseteq \Sigma(\boldsymbol{w})$. This completes the proof of (7.93). □

Orbit

The *orbit* of a vector $\boldsymbol{u}$ means the set of vectors

$$\mathrm{orb}(\boldsymbol{u}) = \{T(g)\boldsymbol{u} \mid g \in G\}. \tag{7.98}$$

Obviously, $\boldsymbol{u} \in \mathrm{orb}(\boldsymbol{u})$, since $\boldsymbol{u} = T(e)\boldsymbol{u}$ for the identity element e of G.

It follows from the G-equivariance (7.38) that if $(\boldsymbol{u}, f)$ is a solution to the system parametrized by $\boldsymbol{v}$, then $(T(g)\boldsymbol{u}, f)$ is a solution to the system parametrized by $S(g)\boldsymbol{v}$. In particular, if $(\boldsymbol{u}, f)$ is a solution to the perfect system, then $(T(g)\boldsymbol{u}, f)$ is also a solution to the perfect system. Thus solutions on the same orbit are naturally identified. It also must be emphasized that there can be two solutions, $(\boldsymbol{u}_1, f)$ and $(\boldsymbol{u}_2, f)$, not belonging to the same orbit (i.e., not related as $\boldsymbol{u}_2 = T(g)\boldsymbol{u}_1$ for any $g \in G$). Such solutions are regarded as essentially different.

For the symmetry of a solution $(T(g)\boldsymbol{u}, f)$ on the same orbit, we have

$$\Sigma(T(g)\boldsymbol{u}) = g \cdot \Sigma(\boldsymbol{u}) \cdot g^{-1}, \qquad g \in G, \tag{7.99}$$

which can be proved as follows:

$$\begin{aligned} h \in \Sigma(T(g)\boldsymbol{u}) &\iff T(h)T(g)\boldsymbol{u} = T(g)\boldsymbol{u} \iff T(g)^{-1}T(h)T(g)\boldsymbol{u} = \boldsymbol{u} \\ &\iff T(g^{-1}hg)\boldsymbol{u} = \boldsymbol{u} \iff g^{-1}hg \in \Sigma(\boldsymbol{u}) \iff h \in g \cdot \Sigma(\boldsymbol{u}) \cdot g^{-1}. \end{aligned}$$

The relation (7.99) means that the symmetry subgroups of solutions on the same orbit are mutually conjugate (cf., (7.9)).

7.7 Theory of Block-Diagonalization

According to §7.5.1, the equivariance of $\boldsymbol{F}(\boldsymbol{u}, f, \boldsymbol{v})$ implies the commutativity of the Jacobian matrix $J(\boldsymbol{u}, f, \boldsymbol{v})$ and the imperfection sensitivity matrix $B(\boldsymbol{u}, f, \boldsymbol{v})$ with the representation matrices. A standard result of group representation theory entails that $J(\boldsymbol{u}, f, \boldsymbol{v})$ and $B(\boldsymbol{u}, f, \boldsymbol{v})$ can be transformed to block-diagonal forms through suitable basis changes independent of $(\boldsymbol{u}, f, \boldsymbol{v})$. The theoretical framework of block-diagonalization is explained in this section.

7.7.1 Jacobian Matrix

It should be recalled from (7.45) that the Jacobian matrix $J(\boldsymbol{u}, f, \boldsymbol{v})$ satisfies the commutativity (or equivariance) condition

$$T(g)J(\boldsymbol{u}, f, \boldsymbol{v}) = J(\boldsymbol{u}, f, \boldsymbol{v})T(g), \qquad g \in G,$$

for all $(\boldsymbol{u}, f, \boldsymbol{v})$ possessing the symmetry (7.44).

Motivated by this, we consider a general $N \times N$ real matrix J that satisfies the condition of G-symmetry (commutativity)

$$T(g)J = JT(g), \qquad g \in G. \tag{7.100}$$

We continue to assume that T is a unitary (orthogonal) representation over $\mathbb{R}$. Recall that $R(G)$ denotes the index set for the irreducible representations of G (cf., (7.21) with $F = \mathbb{R}$), and N^{μ} is the dimension of $\mu \in R(G)$.

Construction of the block-diagonal form consists of two stages:

- Transformation to a block-diagonal form corresponding to isotypic decomposition (7.32),
- Transformation to a finer block-diagonal form with a nested block structure related to irreducible decomposition (7.31), which is especially relevant to multi-dimensional absolutely irreducible representations.

Transformation Matrix for Isotypic and Irreducible Decompositions

First, we recall the isotypic decomposition (7.32):

$$\overline{T}(g) = Q^{\top} T(g) Q = \bigoplus_{\mu \in R(G)} \overline{T}^{\mu}(g), \qquad g \in G,$$

of an orthogonal representation T, where Q is an orthogonal matrix of size N.

Compatibly with this isotypic decomposition, the matrix Q is partitioned columnwise as

$$Q = (Q^{\mu} \mid \mu \in R(G)) = \left(\ldots,\ Q^{\mu},\ \ldots\right), \tag{7.101}$$

where Q^{μ} is the submatrix, of size $N \times a^{\mu} N^{\mu}$, that corresponds to μ. For each $\mu \in R(G)$ we have

$$T(g)Q^{\mu} = Q^{\mu}\overline{T}^{\mu}(g), \qquad g \in G,$$

which implies

$$(Q^{\mu})^{\top} T(g) Q^{\mu} = \overline{T}^{\mu}(g), \qquad g \in G.$$

The column vectors of Q^μ constitute a basis of the isotypic component V^μ in (7.25) corresponding to μ.

Compatibly with the decomposition (7.29) of an isotypic component into irreducible components, the submatrix Q^μ is partitioned as

$$Q^\mu = (Q_i^\mu \mid i = 1, \ldots, a^\mu), \qquad \mu \in R(G). \tag{7.102}$$

Here Q_i^μ is an $N \times N^\mu$ matrix and the column vectors of Q_i^μ form a basis of the irreducible component V_i^μ in (7.24). We then have

$$T(g)Q_i^\mu = Q_i^\mu T^\mu(g), \qquad g \in G, \tag{7.103}$$

which implies

$$(Q_i^\mu)^\top T(g) Q_i^\mu = T^\mu(g), \qquad g \in G.$$

Substitution of (7.102) into (7.101) yields a finer partition

$$Q = (Q^\mu \mid \mu \in R(G)) = ((Q_i^\mu \mid i = 1, \ldots, a^\mu) \mid \mu \in R(G)). \tag{7.104}$$

The condition (7.103), for all i and μ, is a necessary and sufficient condition for an orthogonal matrix Q to be eligible for the irreducible decomposition (7.31) for the given T and the specified family $\{T^\mu \mid \mu \in R(G)\}$. This fact is used to determine a concrete form of Q in Chapter 12.

Example 7.9. If $R(G) = \{\mu, \nu\}$, $a^\mu = 2$, and $a^\nu = 3$, then the irreducible decomposition (7.31) is

$$\overline{T}(g) = Q^\top T(g) Q = \left(\begin{array}{cc|ccc} T^\mu(g) & & & & \\ & T^\mu(g) & & & \\ \hline & & T^\nu(g) & & \\ & & & T^\nu(g) & \\ & & & & T^\nu(g) \end{array}\right).$$

We have

$$Q = (Q_1^\mu, Q_2^\mu; Q_1^\nu, Q_2^\nu, Q_3^\nu),$$

where Q_1^μ and Q_2^μ are $N \times N^\mu$ matrices, and Q_1^ν, Q_2^ν, and Q_3^ν are $N \times N^\nu$ matrices. □

Block-Diagonalization: I

For the block-diagonal form of J, we transform J to

$$\overline{J} = Q^\top J Q \tag{7.105}$$

using the orthogonal matrix $Q = (Q^\mu \mid \mu \in R(G))$ in (7.101). The columns of Q are partitioned into blocks Q^μ; therefore, the rows and the columns of $\overline{J}$ are partitioned accordingly as

$$\overline{J} = (\overline{J}^{\mu\nu} \mid \mu, \nu \in R(G)) \tag{7.106}$$

with

$$\overline{J}^{\mu\nu} = (Q^{\mu})^{\top} J Q^{\nu}, \qquad \mu,\nu \in R(G). \tag{7.107}$$

The symmetry condition (7.100) can be rewritten as

$$Q^{\top} T(g) Q \cdot Q^{\top} J Q = Q^{\top} J Q \cdot Q^{\top} T(g) Q, \qquad g \in G, \tag{7.108}$$

in which $Q^{\top} T(g) Q$ is a block-diagonal matrix $\overline{T}(g)$ in (7.32) with diagonal blocks $\overline{T}^{\mu}(g)$; and $Q^{\top} J Q$ is represented as (7.106) with (7.107). Therefore, (7.108) is equivalent to

$$\overline{T}^{\mu}(g)\, \overline{J}^{\mu\nu} = \overline{J}^{\mu\nu}\, \overline{T}^{\nu}(g), \qquad g \in G;\ \mu,\nu \in R(G). \tag{7.109}$$

This condition for $\mu \neq \nu$ implies that

$$\overline{J}^{\mu\nu} = O \qquad \text{if} \quad \mu \neq \nu, \tag{7.110}$$

which is proved in Remark 7.14 below using Schur's lemma given in §7.4.3.

Hence $\overline{J} = Q^{\top} J Q$ is a block-diagonal matrix consisting of diagonal blocks

$$\overline{J}^{\mu} \equiv \overline{J}^{\mu\mu} = (Q^{\mu})^{\top} J Q^{\mu}, \qquad \mu \in R(G). \tag{7.111}$$

That is,

$$\overline{J} = Q^{\top} J Q = \bigoplus_{\mu \in R(G)} \overline{J}^{\mu}, \tag{7.112}$$

where $\overline{J}^{\mu}$ is of size $a^{\mu} N^{\mu}$. We have arrived at a block-diagonal form associated with the isotypic decomposition (7.32).

Remark 7.14. The key property (7.110) can be proved as follows. In accordance with the finer partition $Q^{\mu} = (Q_i^{\mu} \mid i = 1, \ldots, a^{\mu})$ in (7.102), the matrix $\overline{J}^{\mu\nu} = (Q^{\mu})^{\top} J Q^{\nu}$ with $\mu,\nu \in R(G)$ in (7.107) can be regarded as a block matrix

$$\overline{J}^{\mu\nu} = (\overline{J}_{ij}^{\mu\nu} \mid i = 1, \ldots, a^{\mu}; j = 1, \ldots, a^{\nu})$$

with

$$\overline{J}_{ij}^{\mu\nu} = (Q_i^{\mu})^{\top} J Q_j^{\nu}$$

for $i = 1, \ldots, a^{\mu}$ and $j = 1, \ldots, a^{\nu}$. Using the expression presented above and the decomposition (7.29) of $\overline{T}^{\mu}$ into irreducible components, we can rewrite the condition (7.109) to

$$T^{\mu}(g)\, \overline{J}_{ij}^{\mu\nu} = \overline{J}_{ij}^{\mu\nu}\, T^{\nu}(g), \qquad g \in G,$$

where $i = 1, \ldots, a^{\mu}$ and $j = 1, \ldots, a^{\nu}$. If $\mu \neq \nu$, then T^{μ} and T^{ν} are mutually inequivalent irreducible representations, and Schur's lemma (Lemma 7.1(ii)) in §7.4.3 shows that $\overline{J}_{ij}^{\mu\nu} = O$ for all i, j. Therefore, the claim $\overline{J}^{\mu\nu} = O$ for $\mu \neq \nu$ in (7.110) holds. □

Remark 7.15. Recall again that T is assumed to be a unitary representation. This assumption is crucial for the block-diagonalization in (7.112) using an orthogonal matrix Q. For a general (nonunitary) representation, the block-diagonalization of J

is still possible by the transformation $Q^{-1}JQ$ with a general nonsingular matrix Q. □

Block-Diagonalization: II

The diagonal block $\overline{J}^{\mu}$ in (7.112) has a further block-diagonal structure if the associated irreducible representation μ is absolutely irreducible and $N^{\mu} \geq 2$. Recall that $R_{\mathrm{a}}(G)$ designates the subset of $R(G)$ consisting of absolutely irreducible representations. The difference set $R(G) \setminus R_{\mathrm{a}}(G)$ comprises irreducible representations over $\mathbb{R}$ that are not absolutely irreducible (cf., (7.34)).

For $\mu \in R_{\mathrm{a}}(G)$ we recall $Q^{\mu} = (Q^{\mu}_{i} \mid i = 1, \ldots, a^{\mu})$ from (7.102) and define matrices

$$\overline{J}^{\mu}_{ij} = (Q^{\mu}_{i})^{\top} J Q^{\mu}_{j}, \qquad i, j = 1, \ldots, a^{\mu}.$$

Then the matrix $\overline{J}^{\mu}$ can be regarded as a block matrix consisting of these matrices. That is,

$$\overline{J}^{\mu} = (\overline{J}^{\mu}_{ij} \mid i, j = 1, \ldots, a^{\mu}). \tag{7.113}$$

It is possible to reveal the structure of these blocks $\overline{J}^{\mu}_{ij}$ as follows. Using (7.113) and the decomposition (7.29) of $\overline{T}^{\mu}$ into irreducible components, the condition (7.109) for $\mu = \nu$ can be rewritten as

$$T^{\mu}(g)\, \overline{J}^{\mu}_{ij} = \overline{J}^{\mu}_{ij}\, T^{\mu}(g), \qquad g \in G,$$

where $i, j = 1, \ldots, a^{\mu}$. This implies, by Schur's lemma (Lemma 7.1(iii) in §7.4.3),[18] that $\overline{J}^{\mu}_{ij}$ is a scalar multiple of the identity matrix. That is,

$$\overline{J}^{\mu}_{ij} = \alpha^{\mu}_{ij} I_{N^{\mu}} \tag{7.114}$$

for some $\alpha^{\mu}_{ij} \in \mathbb{R}$, where $I_{N^{\mu}}$ denotes the identity matrix of size equal to the dimension N^{μ} of μ.

Consequently, the matrix $\overline{J}^{\mu}$ is equal, up to a simultaneous permutation of rows and columns, to a block-diagonal matrix consisting of N^{μ} copies of an identical matrix

$$\tilde{J}^{\mu} = (\alpha^{\mu}_{ij} \mid i, j = 1, \ldots, a^{\mu}) \tag{7.115}$$

of size $a^{\mu} \times a^{\mu}$; that is,

$$(\Pi^{\mu})^{\top} \overline{J}^{\mu} \Pi^{\mu} = \bigoplus_{k=1}^{N^{\mu}} \tilde{J}^{\mu} \tag{7.116}$$

with a suitable permutation matrix Π^{μ} for each $\mu \in R_{\mathrm{a}}(G)$.

More specifically, Π^{μ} is the permutation matrix of order $a^{\mu} N^{\mu}$ representing the permutation

---

[18] Note that Lemma 7.1(iii) presupposes absolute irreducibility of the representation.

$$\begin{pmatrix} 1 & 2 & 3 & \cdots & a^{\mu} & ; a^{\mu}+1 & a^{\mu}+2 & a^{\mu}+3 & \cdots & 2a^{\mu} & ; \\ 1 & 1+N^{\mu} & 1+2N^{\mu} & \cdots & 1+(a^{\mu}-1)N^{\mu} & ; \quad 2 & 2+N^{\mu} & 2+2N^{\mu} & \cdots & 2+(a^{\mu}-1)N^{\mu} & ; \end{pmatrix}$$

$$\begin{pmatrix} \cdots \cdots ; \cdots \cdots ; & (N^{\mu}-1)a^{\mu}+1 & (N^{\mu}-1)a^{\mu}+2 & (N^{\mu}-1)a^{\mu}+3 & \cdots & a^{\mu}N^{\mu} \\ \cdots \cdots ; \cdots \cdots ; & N^{\mu} & 2N^{\mu} & 3N^{\mu} & \cdots & a^{\mu}N^{\mu} \end{pmatrix}.$$

Also for $\mu \in R(G) \setminus R_{\rm a}(G)$ we define Π^{μ} as such a permutation matrix. The aggregation of these permutation matrices is denoted as Π; that is,

$$\Pi = \bigoplus_{\mu \in R(G)} \Pi^{\mu}. \tag{7.117}$$

Combination of (7.112) and (7.116) yields a finer block-diagonal form

$$\begin{aligned} \Pi^{\top}\bar{J}\Pi &= \Big[\bigoplus_{\mu \in R_{\rm a}(G)} (\Pi^{\mu})^{\top}\bar{J}^{\mu}\Pi^{\mu}\Big] \bigoplus \Big[\bigoplus_{\mu \in R(G)\setminus R_{\rm a}(G)} (\Pi^{\mu})^{\top}\bar{J}^{\mu}\Pi^{\mu}\Big] \\ &= \Big[\bigoplus_{\mu \in R_{\rm a}(G)} \bigoplus_{k=1}^{N^{\mu}} \tilde{J}^{\mu}\Big] \bigoplus \Big[\bigoplus_{\mu \in R(G)\setminus R_{\rm a}(G)} \hat{J}^{\mu}\Big], \end{aligned} \tag{7.118}$$

where $\tilde{J}^{\mu}$ for $\mu \in R_{\rm a}(G)$ are given in (7.115) and, for $\mu \in R(G) \setminus R_{\rm a}(G)$, we have

$$\hat{J}^{\mu} = (\Pi^{\mu})^{\top}\bar{J}^{\mu}\Pi^{\mu}. \tag{7.119}$$

Example 7.10. Permutation Π above is illustrated. Suppose that $N = 7$, $R(G) = R_{\rm a}(G) = \{\mu, \nu\}$, $N^{\mu} = 2$, $N^{\nu} = 1$, $a^{\mu} = 2$, and $a^{\nu} = 3$ (cf., Example 7.9 above). For (7.112), we have

$$\bar{J} = \left(\begin{array}{cc|ccc} \alpha^{\mu}_{11}I_2 & \alpha^{\mu}_{12}I_2 & & & \\ \alpha^{\mu}_{21}I_2 & \alpha^{\mu}_{22}I_2 & & & \\ \hline & & \alpha^{\nu}_{11}I_1 & \alpha^{\nu}_{12}I_1 & \alpha^{\nu}_{13}I_1 \\ & & \alpha^{\nu}_{21}I_1 & \alpha^{\nu}_{22}I_1 & \alpha^{\nu}_{23}I_1 \\ & & \alpha^{\nu}_{31}I_1 & \alpha^{\nu}_{32}I_1 & \alpha^{\nu}_{33}I_1 \end{array}\right) = \left(\begin{array}{cccc|ccc} \alpha^{\mu}_{11} & & \alpha^{\mu}_{12} & & & & \\ & \alpha^{\mu}_{11} & & \alpha^{\mu}_{12} & & & \\ \alpha^{\mu}_{21} & & \alpha^{\mu}_{22} & & & & \\ & \alpha^{\mu}_{21} & & \alpha^{\mu}_{22} & & & \\ \hline & & & & \alpha^{\nu}_{11} & \alpha^{\nu}_{12} & \alpha^{\nu}_{13} \\ & & & & \alpha^{\nu}_{21} & \alpha^{\nu}_{22} & \alpha^{\nu}_{23} \\ & & & & \alpha^{\nu}_{31} & \alpha^{\nu}_{32} & \alpha^{\nu}_{33} \end{array}\right).$$

With a suitable rearrangement of rows and columns, represented by a permutation matrix Π, the matrix shown above can be put into a block-diagonal form as

$$\Pi^{\top}\bar{J}\Pi = \left(\begin{array}{cc|cc|ccc} \alpha^{\mu}_{11} & \alpha^{\mu}_{12} & & & & & \\ \alpha^{\mu}_{21} & \alpha^{\mu}_{22} & & & & & \\ \hline & & \alpha^{\mu}_{11} & \alpha^{\mu}_{12} & & & \\ & & \alpha^{\mu}_{21} & \alpha^{\mu}_{22} & & & \\ \hline & & & & \alpha^{\nu}_{11} & \alpha^{\nu}_{12} & \alpha^{\nu}_{13} \\ & & & & \alpha^{\nu}_{21} & \alpha^{\nu}_{22} & \alpha^{\nu}_{23} \\ & & & & \alpha^{\nu}_{31} & \alpha^{\nu}_{32} & \alpha^{\nu}_{33} \end{array}\right),$$

which is the block-diagonal matrix on the right-hand side of (7.118) without involving the blocks $\hat{J}^{\mu'}$ for $\mu' \in R(G) \setminus R_{\mathrm{a}}(G)$. □

With the permutation Π, the transformation matrix Q is to be modified to a new transformation matrix

$$H = Q\Pi. \tag{7.120}$$

We have

$$\Pi^{\top}\overline{J}\Pi = \Pi^{\top}(Q^{\top}JQ)\Pi = H^{\top}JH$$

by (7.105); then (7.118) reads as

$$\tilde{J} \equiv H^{\top}JH = \left[\bigoplus_{\mu\in R_{\mathrm{a}}(G)} \bigoplus_{k=1}^{N^{\mu}} \tilde{J}^{\mu}\right] \bigoplus \left[\bigoplus_{\mu\in R(G)\setminus R_{\mathrm{a}}(G)} \hat{J}^{\mu}\right], \tag{7.121}$$

where $\tilde{J}^{\mu}$ is an $a^{\mu}\times a^{\mu}$ matrix for $\mu \in R_{\mathrm{a}}(G)$ and $\hat{J}^{\mu}$ is an $(a^{\mu}N^{\mu})\times(a^{\mu}N^{\mu})$ matrix for $\mu \in R(G)\setminus R_{\mathrm{a}}(G)$.

This formula, which is of fundamental importance, reveals the intrinsic structure of a G-symmetric matrix J in (7.100). The most remarkable is the nested block-diagonal structure of the blocks for $\mu \in R_{\mathrm{a}}(G)$. On the other hand, the matrices $\hat{J}^{\mu}$ for $\mu \in R(G) \setminus R_{\mathrm{a}}(G)$ are not block-diagonalized in general, although they are endowed with a certain structure, which is demonstrated in terms of concrete instances for $G = \mathrm{C}_n$ in §7.4.3 (Example 7.8) and in §11.6.3, and not explained in detail in this book.

The block-diagonal form (7.112) looks simpler than (7.121), but in applications it is more advantageous to have finer diagonal blocks. For this reason we use (7.121) in the remainder of this book. Use of the formula (7.121) in bifurcation analysis is demonstrated in §11.6, and the construction of the transformation matrix H is discussed in Chapter 12.

Finally, we introduce some additional notations related to the formula (7.121). In (7.120) the matrix Q is partitioned as

$$Q = (Q^{\mu} \mid \mu \in R(G)) = ((Q_i^{\mu} \mid i = 1,\dots,a^{\mu}) \mid \mu \in R(G)),$$

where Q_i^{μ} is an $N\times N^{\mu}$ matrix (cf., (7.104)). For $\mu \in R(G)$ and $k = 1,\dots,N^{\mu}$, let $H^{\mu,k}$ be an $N\times a^{\mu}$ matrix consisting of the kth column vectors of Q_i^{μ} for $i = 1,\dots,a^{\mu}$, and define an $N\times(a^{\mu}N^{\mu})$ matrix

$$H^{\mu} = (H^{\mu,k} \mid k = 1,\dots,N^{\mu}) = (H^{\mu,1},\dots,H^{\mu,N^{\mu}}). \tag{7.122}$$

The matrix H in (7.120) is given as the collection of these matrices; that is,

$$\begin{aligned} H &= (H^{\mu} \mid \mu \in R(G)) = (\dots,H^{\mu},\dots) \\ &= ((H^{\mu,k} \mid k = 1,\dots,N^{\mu}) \mid \mu \in R(G)). \end{aligned} \tag{7.123}$$

With this notation, the diagonal block $\tilde{J}^{\mu}$ for $\mu \in R_{\mathrm{a}}(G)$ in (7.121) can be expressed as

$$\tilde{J}^{\mu} = (H^{\mu,k})^{\top} J H^{\mu,k}, \qquad k = 1,\ldots,N^{\mu}, \tag{7.124}$$

where the right-hand side of (7.124) gives the same matrix for any k.

For a one-dimensional irreducible representation μ, in particular, we have $\mu \in R_{\mathrm{a}}(G)$ and

$$H^{\mu,1} = H^{\mu} = Q^{\mu}, \qquad \tilde{J}^{\mu} = \overline{J}^{\mu}.$$

On the other hand, the diagonal block $\hat{J}^{\mu}$ for $\mu \in R(G) \setminus R_{\mathrm{a}}(G)$ in (7.121) is given as

$$\hat{J}^{\mu} = (H^{\mu})^{\top} J H^{\mu}. \tag{7.125}$$

7.7.2 Imperfection Sensitivity Matrix

As a consequence of the equivariance, the imperfection sensitivity matrix $B(\boldsymbol{u}, f, \boldsymbol{v})$ commutes with the group actions:

$$T(g)B(\boldsymbol{u}, f, \boldsymbol{v}) = B(\boldsymbol{u}, f, \boldsymbol{v})S(g), \qquad g \in G \tag{7.126}$$

if $\boldsymbol{u}$ and $\boldsymbol{v}$ are G-invariant (cf., §7.5.1). Then the matrix $B(\boldsymbol{u}, f, \boldsymbol{v})$ can be transformed to a block-diagonal form through suitable basis changes independent of $(\boldsymbol{u}, f, \boldsymbol{v})$ as follows. The argument below is a slight extension of the argument for $J(\boldsymbol{u}, f, \boldsymbol{v})$ in §7.7.1. Recall that T and S are assumed to be orthogonal representations, that is, unitary representations over $\mathbb{R}$.

Block-Diagonalization: I

By the isotypic decomposition (7.25), the representation spaces $\mathbb{R}^N$ and $\mathbb{R}^p$ are respectively decomposed uniquely into the direct sum of isotypic components. Let

$$Q = (Q^{\mu} \mid \mu \in R(G)), \qquad \Psi = (\Psi^{\mu} \mid \mu \in R(G))$$

be an $N \times N$ orthogonal matrix and a $p \times p$ orthogonal matrix for the bases of the isotypic decomposition of T and S (cf., (7.101)), respectively.

An isotypic component is decomposed further (not uniquely) into irreducible components (cf., (7.24)). Denote by a^{μ} and b^{μ} the multiplicities of $\mu \in R(G)$ in T and S, respectively, and let

$$Q^{\mu} = (Q_i^{\mu} \mid i = 1,\ldots,a^{\mu}), \qquad \Psi^{\mu} = (\Psi_j^{\mu} \mid j = 1,\ldots,b^{\mu}) \tag{7.127}$$

be the corresponding decompositions of the bases (cf., (7.102)).

Then (7.126) can be rewritten as

$$Q^{\top} T(g) Q \cdot Q^{\top} B \Psi = Q^{\top} B \Psi \cdot \Psi^{\top} S(g) \Psi, \qquad g \in G,$$

in which

$$Q^\top T(g)Q = \bigoplus_{\mu\in R(G)} \bigoplus_{i=1}^{a^\mu} T^\mu(g), \qquad g \in G, \tag{7.128}$$

$$\Psi^\top S(g)\Psi = \bigoplus_{\mu\in R(G)} \bigoplus_{j=1}^{b^\mu} T^\mu(g), \qquad g \in G, \tag{7.129}$$

as in (7.31). The expression above implies, by Schur's lemma (Lemma 7.1(ii) in §7.4.3), that

$$(Q^\mu)^\top B\Psi^\nu = O \qquad \text{if } \mu \neq \nu. \tag{7.130}$$

Consequently, $Q^\top B\Psi$ is a block-diagonal matrix

$$Q^\top B\Psi = \bigoplus_{\mu\in R(G)} \overline{B}^\mu \tag{7.131}$$

consisting of $(a^\mu N^\mu) \times (b^\mu N^\mu)$ diagonal blocks

$$\overline{B}^\mu = (Q^\mu)^\top B\Psi^\mu \tag{7.132}$$

indexed by $\mu \in R(G)$.

Block-Diagonalization: II

The diagonal blocks $\overline{B}^\mu$ that correspond to absolutely irreducible representations μ have a further structure by Schur's lemma (Lemma 7.1(iii) in §7.4.3). That is, for $\mu \in R_{\rm a}(G)$, we have

$$(Q_i^\mu)^\top B\Psi_j^\mu = \beta_{ij}^\mu I_{N^\mu}, \qquad i = 1,\ldots,a^\mu, \;\; j = 1,\ldots,b^\mu, \tag{7.133}$$

for some $\beta_{ij}^\mu \in \mathbb{R}$, where I_{N^μ} represents the identity matrix of size equal to the dimension N^μ of μ. In other words, the matrix $\overline{B}^\mu$ in (7.132) is equal, up to permutations of rows and columns, to a block-diagonal matrix consisting of N^μ copies of an identical matrix

$$\widetilde{B}^\mu = (\beta_{ij}^\mu \mid i = 1,\ldots,a^\mu, \; j = 1,\ldots,b^\mu)$$

of size $a^\mu \times b^\mu$. The expression above can be written as

$$(\Pi^\mu)^\top \overline{B}^\mu \Lambda^\mu = \bigoplus_{k=1}^{N^\mu} \widetilde{B}^\mu, \qquad \mu \in R_{\rm a}(G), \tag{7.134}$$

with the permutation matrices Π^μ and Λ^μ.

The argument presented above shows that, with an $N \times N$ orthogonal matrix H and a $p \times p$ orthogonal matrix Φ, which are obtained from Q and Ψ by column permutations, we can obtain the block-diagonal form

$$H^\top B\Phi = \Big[\bigoplus_{\mu\in R_{\mathrm{a}}(G)}\bigoplus_{k=1}^{N^\mu}\widetilde{B}^\mu\Big]\bigoplus\Big[\bigoplus_{\mu\in R(G)\backslash R_{\mathrm{a}}(G)}\widehat{B}^\mu\Big]. \tag{7.135}$$

Here $\widetilde{B}^\mu$ is an $a^\mu\times b^\mu$ matrix for $\mu\in R_{\mathrm{a}}(G)$, and $\widehat{B}^\mu$ for $\mu\in R(G)\backslash R_{\mathrm{a}}(G)$ is an $(a^\mu N^\mu)\times(b^\mu N^\mu)$ matrix obtained from $\overline{B}^\mu$ through permutations of rows and columns.

It must be emphasized that the transformation matrices H and Φ are constant orthogonal matrices, which are valid universally for all $(\boldsymbol{u},f,\boldsymbol{v})$ with the symmetry (7.44). The transformation matrix H in (7.135) is the same as the matrix H in (7.121) for the block-diagonalization of the Jacobian matrix J.

7.8 Examples of Symmetric Systems

Symmetry, bifurcation, and block-diagonalization of examples of symmetric systems are investigated by group-theoretic bifurcation theory.

7.8.1 Symmetry and Bifurcation of D_3*-Symmetric Truss Tent*

We consider the D_3-symmetric three-bar truss tent depicted in Fig. 7.2, which is described by a reciprocal system of equations of the form (7.36) with $\boldsymbol{u}=(x,y,z)^\top$. That is,

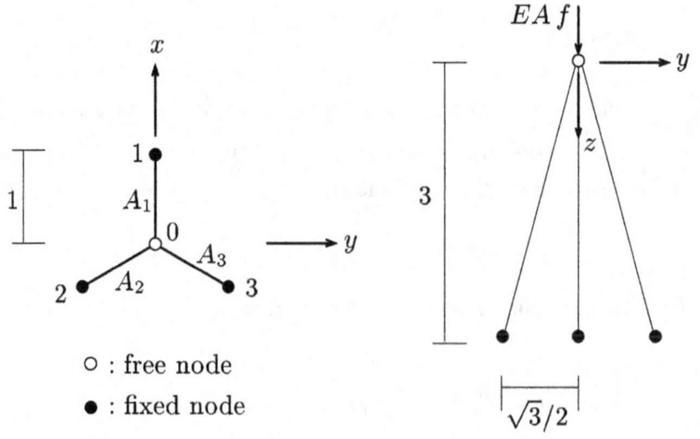

Fig. 7.2 D_3-symmetric three-bar truss tent.

$$\boldsymbol{F} = \begin{pmatrix} F_1 \\ F_2 \\ F_3 \end{pmatrix} = \sum_{i=1}^{3} EA_i\Big(\frac{1}{L_i} - \frac{1}{\hat{L}_i}\Big)\begin{pmatrix} x-x_i \\ y-y_i \\ z-z_i \end{pmatrix} - \begin{pmatrix} 0 \\ 0 \\ EAf \end{pmatrix} = \begin{pmatrix} 0 \\ 0 \\ 0 \end{pmatrix}, \tag{7.136}$$

where (x_i, y_i, z_i) is the initial location of node i $(i = 0,1,2,3)$; (x,y,z) is the independent variable denoting the location of node 0 after deformation;

$$L_i = ((x_0 - x_i)^2 + (y_0 - y_i)^2 + (z_0 - z_i)^2)^{1/2}, \qquad i = 1,2,3;$$
$$\hat{L}_i = ((x - x_i)^2 + (y - y_i)^2 + (z - z_i)^2)^{1/2}, \qquad i = 1,2,3;$$

E expresses Young's modulus; A_i denotes the cross-sectional area of the member connecting nodes 0 and i $(i = 1,2,3)$; and the nodal location (x_i, y_i, z_i) $(i = 1,2,3)$ can be read from Fig. 7.2.

We choose $\boldsymbol{v} = (A_1, A_2, A_3)^\top$ as the imperfection parameter vector, and assume $\boldsymbol{v}^0 = (A, A, A)^\top$ to make the perfect system symmetric.

Symmetry Group and Representations

This tent, which is symmetric in geometrical configuration, in stiffness distribution, and in loading, remains invariant under two geometrical transformations: the counterclockwise rotation[19] $c(2\pi/3)$ about the z-axis at an angle of $2\pi/3$ and the reflection $\sigma : y \mapsto -y$. This geometric invariance is mathematically expressed as the invariance with respect to the dihedral group of degree three

$$\mathrm{D}_3 = \langle c(2\pi/3), \sigma \rangle = \{e, c(2\pi/3), c(4\pi/3), \sigma, \sigma c(2\pi/3), \sigma c(4\pi/3)\}.$$

It is easy to show that $\boldsymbol{F}$ in the governing equation (7.136) is equivariant to D_3 in the sense of (7.38) through the unitary representation T (acting on $\boldsymbol{u}$ and $\boldsymbol{F}$) defined by

$$T(c(2\pi/3)) = \begin{pmatrix} \cos(2\pi/3) & -\sin(2\pi/3) & 0 \\ \sin(2\pi/3) & \cos(2\pi/3) & 0 \\ 0 & 0 & 1 \end{pmatrix}, \qquad T(\sigma) = \begin{pmatrix} 1 & 0 & 0 \\ 0 & -1 & 0 \\ 0 & 0 & 1 \end{pmatrix}, \tag{7.137}$$

and another unitary representation S (acting on $\boldsymbol{v}$) defined by

$$S(c(2\pi/3)) = \begin{pmatrix} 0 & 1 & 0 \\ 0 & 0 & 1 \\ 1 & 0 & 0 \end{pmatrix}, \qquad S(\sigma) = \begin{pmatrix} 1 & 0 & 0 \\ 0 & 0 & 1 \\ 0 & 1 & 0 \end{pmatrix}. \tag{7.138}$$

The group D_3 has two one-dimensional irreducible representations, say, μ_1 and μ_2, and one two-dimensional irreducible representation, say μ_3. That is,[20]

---

[19] This counterclockwise rotation appears to be clockwise in Fig. 7.2 since the z-axis is directed downward.

[20] In §8.2.2 we introduce a more systematic notation: $\mu_1 = (+,+)_{\mathrm{D}_3}$, $\mu_2 = (+,-)_{\mathrm{D}_3}$, and $\mu_3 = (1)_{\mathrm{D}_3}$. Every μ_i is absolutely irreducible, therefore, $R_{\mathrm{a}}(\mathrm{D}_3) = R(\mathrm{D}_3)$.

$$R(\mathrm{D}_3) = \{\mu_1, \mu_2, \mu_3\}$$

using the notation $R(G)$ in (7.21) for $G = \mathrm{D}_3$. The one-dimensional irreducible representations are given by

$$T^{\mu_1}(c(2\pi/3)) = 1, \qquad T^{\mu_1}(\sigma) = 1;$$
$$T^{\mu_2}(c(2\pi/3)) = 1, \qquad T^{\mu_2}(\sigma) = -1.$$

The two-dimensional representation can be chosen to be a unitary representation defined by

$$T^{\mu_3}(c(2\pi/3)) = \begin{pmatrix} \cos(2\pi/3) & -\sin(2\pi/3) \\ \sin(2\pi/3) & \cos(2\pi/3) \end{pmatrix}, \qquad T^{\mu_3}(\sigma) = \begin{pmatrix} 1 & 0 \\ 0 & -1 \end{pmatrix}. \tag{7.139}$$

It is not difficult to verify that T and S are equivalent representations, both being free from μ_2 and decomposed into the direct sum of μ_1 and μ_3:

$$a^{\mu_1} = b^{\mu_1} = 1, \qquad a^{\mu_2} = b^{\mu_2} = 0, \qquad a^{\mu_3} = b^{\mu_3} = 1$$

in (7.128) and (7.129).

Symmetry of Critical Eigenvectors

The perfect system has a double bifurcation point

$$(x_\mathrm{c}^0, y_\mathrm{c}^0, z_\mathrm{c}^0, f_\mathrm{c}^0) = (0, 0, 0.1877, 0.1586),$$

at which

$$J_\mathrm{c}^0 = EA \begin{pmatrix} 0 & 0 & 0 \\ 0 & 0 & 0 \\ 0 & 0 & 0.8359 \end{pmatrix}$$

and

$$\eta_1 = (1,0,0)^\top, \qquad \eta_2 = (0,1,0)^\top \tag{7.140}$$

are the critical eigenvectors of the (symmetric) Jacobian matrix (7.42). The subrepresentation of T on the kernel space of J_c^0, which is spanned by $\{\eta_1, \eta_2\}$, is given by the leading principal 2×2 submatrices in (7.137), and therefore equal to T^{μ_3} in (7.139). This means, in particular, that $\widetilde{T}$ in (7.54) is given by the two-dimensional irreducible representation T^{μ_3}.

A superposition (linear combination)

$$\eta(\theta) = \cos\theta \cdot \eta_1 + \sin\theta \cdot \eta_2 = \begin{pmatrix} \cos\theta \\ \sin\theta \\ 0 \end{pmatrix}, \qquad 0 \le \theta < 2\pi,$$

of the pair of critical eigenvectors $\boldsymbol{\eta}_1$ and $\boldsymbol{\eta}_2$ in (7.140) is also a critical eigenvector of the Jacobian matrix J_c^0. The superposed eigenvector $\boldsymbol{\eta}(\theta)$ is nonsymmetric; that is to say, it is symmetric with respect to the trivial group $\{e\}$, for a general angle θ $(0 \le \theta < 2\pi)$. Therefore, we have $\Sigma(\ker(J_c^0)) = \{e\}$.

However, six critical eigenvectors with nontrivial symmetry exist. For $\theta = \alpha_k$ with $\alpha_k = -\pi(k-1)/3$ $(k = 1,\dots,6)$, we have

$$\Sigma(\boldsymbol{\eta}(\alpha_k)) = \Sigma(\boldsymbol{\eta}(\alpha_{k+3})) = \mathrm{D}_1^{k,3}, \qquad k = 1,2,3,$$

with

$$\mathrm{D}_1^{1,3} = \{e,\sigma\}, \qquad \mathrm{D}_1^{2,3} = \{e,\sigma c(2\pi/3)\}, \qquad \mathrm{D}_1^{3,3} = \{e,\sigma c(4\pi/3)\}.$$

The directions related to these six critical eigenvectors are depicted in Fig. 7.3. As explained in Chapter 8, it is in these directions that the bifurcated paths do actually exist; therefore, we have $\Sigma(\boldsymbol{u}) = \mathrm{D}_1^{k,3}$ for $\boldsymbol{u}$ on bifurcated paths. Thus the bifurcated paths, accordingly, have different (higher) symmetry than the general critical eigenvectors in this case. The formula (7.95), in this case, reads as

$$G^{\mu_3} = \Sigma(\ker(J_c^0)) = \{e\} \subset \Sigma(\boldsymbol{w}) = \Sigma(\boldsymbol{u}) = \mathrm{D}_1^{k,3} \subset G = \mathrm{D}_3.$$

This difference between $\Sigma(\ker(J_c^0))$ and $\Sigma(\boldsymbol{u})$, although dependent on particular groups, often takes place.

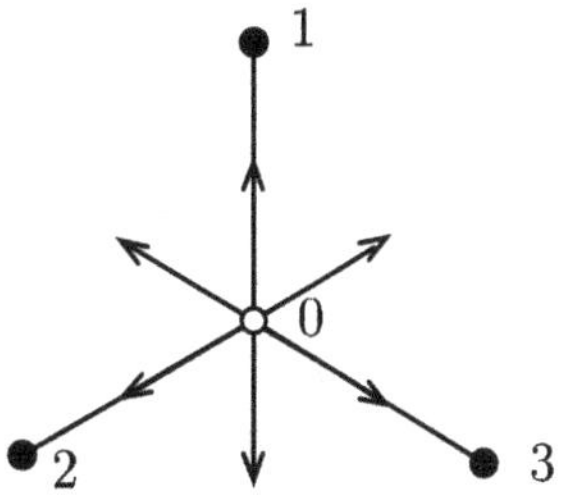

Fig. 7.3 Directions of bifurcated paths of the three-bar truss tent.

Symmetry of Imperfection Sensitivity Matrix

The imperfection sensitivity matrix B_c^0 at the double bifurcation point is evaluated as

$$B_c^0 = E\begin{pmatrix} 0.01880 & -0.00940 & -0.00940 \\ 0.00000 & 0.01628 & -0.01628 \\ 0.05288 & 0.05288 & 0.05288 \end{pmatrix},$$

which commutes with the action of D_3 as (7.50). Using orthogonal matrices

$$H = \begin{pmatrix} 1 & 0 & 0 \\ 0 & 1 & 0 \\ 0 & 0 & 1 \end{pmatrix}, \qquad \Phi = \begin{pmatrix} 2/\sqrt{6} & 0 & 1/\sqrt{3} \\ -1/\sqrt{6} & 1/\sqrt{2} & 1/\sqrt{3} \\ -1/\sqrt{6} & -1/\sqrt{2} & 1/\sqrt{3} \end{pmatrix},$$

we can diagonalize the matrix B_{c}^0 as

$$H^{\top} B_{\mathrm{c}}^0 \Phi = E \cdot \mathrm{diag}(0.0230, 0.0230, 0.0916).$$

This represents a special case of the block-diagonalization in (7.135) with $R(G) = R_{\mathrm{a}}(G) = \{\mu_1, \mu_2, \mu_3\}$ for $G = \mathrm{D}_3$. The first two identical elements (0.0230) correspond to the two-dimensional irreducible representation μ_3 and the last (0.0916) to the one-dimensional representation μ_1. We have $N^{\mu_3} = 2$, $a^{\mu_3} = b^{\mu_3} = 1$; $N^{\mu_1} = 1$, $a^{\mu_1} = b^{\mu_1} = 1$; and $N^{\mu_2} = 1$, $a^{\mu_2} = b^{\mu_2} = 0$ in (7.135).

7.8.2 Block-Diagonalization of D_4-Symmetric Plate Element

We consider a D_4-symmetric square plate element with a uniform thickness and of isotropic material property. As a discretized model of this plate, we employ a four-node finite element in the xy-plane with eight degrees of freedom, as portrayed in Fig. 7.4.

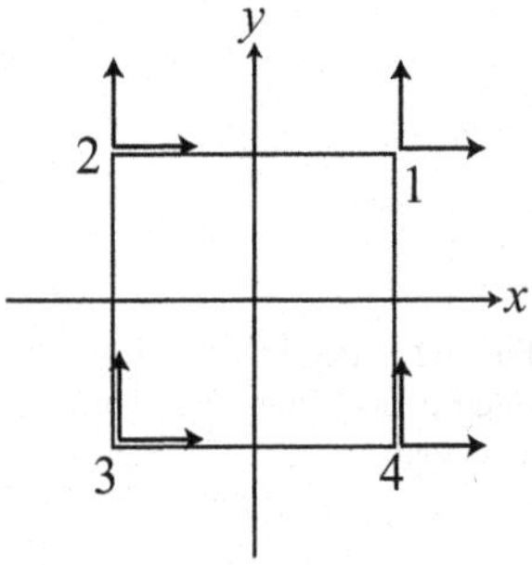

Fig. 7.4 Isotropic, four-node, square plate element.

Symmetry Group and Irreducible Representations

The symmetry of the square plate element is expressed by the dihedral group of degree four

$$\mathrm{D}_4 = \langle c(\pi/2), \sigma \rangle = \{c(\pi i/2), \sigma c(\pi i/2) \mid i = 0, 1, 2, 3\}.$$

Here $c(\pi i/2)$ denotes a counterclockwise rotation about the origin at an angle of $\pi i/2$ ($i = 0, 1, 2, 3$), and σ is the reflection $y \mapsto -y$; $c(0) = e$.

The group D_4 has four one-dimensional irreducible representations, say, $\mu_1,\dots,\mu_4$, and one two-dimensional irreducible representation, say μ_5. That is,[21]

$$R(\mathrm{D}_4) = \{\mu_1,\mu_2,\mu_3,\mu_4,\mu_5\}$$

using the notation $R(G)$ in (7.21) for $G = \mathrm{D}_4$. The one-dimensional irreducible representations are given by

$$\begin{aligned}
T^{\mu_1}(c(\pi/2)) &= 1, & T^{\mu_1}(\sigma) &= 1,\\
T^{\mu_2}(c(\pi/2)) &= 1, & T^{\mu_2}(\sigma) &= -1,\\
T^{\mu_3}(c(\pi/2)) &= -1, & T^{\mu_3}(\sigma) &= 1,\\
T^{\mu_4}(c(\pi/2)) &= -1, & T^{\mu_4}(\sigma) &= -1.
\end{aligned}$$

The two-dimensional representation can be chosen to be a unitary representation defined by

$$T^{\mu_5}(c(2\pi/n)) = \begin{pmatrix} \cos(\pi/2) & -\sin(\pi/2) \\ \sin(\pi/2) & \cos(\pi/2) \end{pmatrix} = \begin{pmatrix} 0 & -1 \\ 1 & 0 \end{pmatrix}, \qquad T^{\mu_5}(\sigma) = \begin{pmatrix} 1 & 0 \\ 0 & -1 \end{pmatrix}.$$

Symmetry of Element Stiffness Matrix

The commutativity of J and $T(g)$ reads as

$$T(g)J = JT(g), \qquad g \in \mathrm{D}_4 \tag{7.141}$$

(cf., (7.45)). That is, the element stiffness matrix J of the D_4-symmetric plate element commutes with the transformation matrices $T(c(\pi i/2))$ and $T(\sigma c(\pi i/2))$ $(i = 0,1,2,3)$. For example, $T(c(\pi/2))$ and $T(\sigma)$ are

$$T(c(\pi/2)) = \left(\begin{array}{cc|cc|cc|cc}
 & & & & & & 0 & -1 \\
 & & & & & & 1 & 0 \\ \hline
0 & -1 & & & & & & \\
1 & 0 & & & & & & \\ \hline
 & & 0 & -1 & & & & \\
 & & 1 & 0 & & & & \\ \hline
 & & & & 0 & -1 & & \\
 & & & & 1 & 0 & &
\end{array}\right) = \begin{pmatrix} & & & 1 \\ 1 & & & \\ & 1 & & \\ & & 1 & \end{pmatrix} \otimes \begin{pmatrix} & -1 \\ 1 & \end{pmatrix}, \tag{7.142}$$

[21] In §8.2.2 we introduce a more systematic notation: $\mu_1 = (+,+)_{\mathrm{D}_4}$, $\mu_2 = (+,-)_{\mathrm{D}_4}$, $\mu_3 = (-,+)_{\mathrm{D}_4}$, $\mu_4 = (-,-)_{\mathrm{D}_4}$, and $\mu_5 = (1)_{\mathrm{D}_4}$. Every μ_i is absolutely irreducible, therefore, $R_{\mathrm{a}}(\mathrm{D}_4) = R(\mathrm{D}_4)$.

$$T(\sigma) = \left(\begin{array}{cc|cc|cc|cc} & & & & & & 1 & \\ & & & & & & & -1 \\ \hline & & & & 1 & & & \\ & & & & & -1 & & \\ \hline & & 1 & & & & & \\ & & & -1 & & & & \\ \hline 1 & & & & & & & \\ & -1 & & & & & & \end{array}\right) = \begin{pmatrix} & & & 1 \\ & & 1 & \\ & 1 & & \\ 1 & & & \end{pmatrix} \otimes \begin{pmatrix} 1 & \\ & -1 \end{pmatrix}. \tag{7.143}$$

The condition (7.141) restricts the form of J as shown below.

$$J = k_1 \begin{pmatrix} 1 & & & & & & & \\ & 1 & & & & & & \\ & & 1 & & & & & \\ & & & 1 & & & & \\ & & & & 1 & & & \\ & & & & & 1 & & \\ & & & & & & 1 & \\ & & & & & & & 1 \end{pmatrix} + k_2 \begin{pmatrix} & & & & 1 & & & \\ & & & & & 1 & & \\ & & & & & & 1 & \\ & & & & & & & 1 \\ 1 & & & & & & & \\ & 1 & & & & & & \\ & & 1 & & & & & \\ & & & 1 & & & & \end{pmatrix}$$

$$+ k_3 \begin{pmatrix} & & & 1 & & & & -1 \\ & & -1 & & & & 1 & \\ & -1 & & & & 1 & & \\ 1 & & & & -1 & & & \\ & & & -1 & & & & 1 \\ & & 1 & & & & -1 & \\ & 1 & & & & -1 & & \\ -1 & & & & 1 & & & \end{pmatrix} + k_4 \begin{pmatrix} & 1 & & & & & & \\ 1 & & & & & & & \\ & & & -1 & & & & \\ & & -1 & & & & & \\ & & & & & 1 & & \\ & & & & 1 & & & \\ & & & & & & & -1 \\ & & & & & & -1 & \end{pmatrix}$$

$$+ k_5 \begin{pmatrix} & & & & & & 1 & \\ & & & 1 & & & & \\ & & & & 1 & & & \\ & 1 & & & & & & \\ & & 1 & & & & & \\ & & & & & & & 1 \\ 1 & & & & & & & \\ & & & & & 1 & & \end{pmatrix} + k_6 \begin{pmatrix} & & & & & -1 & & \\ & & & & -1 & & & \\ & & & & & & & 1 \\ & & & & & & 1 & \\ & -1 & & & & & & \\ -1 & & & & & & & \\ & & & 1 & & & & \\ & & 1 & & & & & \end{pmatrix} + k_7 \begin{pmatrix} & & 1 & & & & & \\ & & & & & & & 1 \\ 1 & & & & & & & \\ & & & & & 1 & & \\ & & & & & & 1 & \\ & & & 1 & & & & \\ & & & & 1 & & & \\ & 1 & & & & & & \end{pmatrix}, \tag{7.144}$$

where $k_1, \ldots, k_7$ are constants.

The transformation matrix H for the block-diagonalization in (7.121) reads as[22]

[22] A systematic way to construct this H is given in §12.3.

$$
H = (H^{\mu_1}, H^{\mu_2}, H^{\mu_3}, H^{\mu_4}, H^{\mu_5,1}, H^{\mu_5,2}) = \left(\begin{array}{c|c|c|c|cc|cc}
a & -a & -a & a & b & 0 & 0 & b \\
a & a & a & a & 0 & b & b & 0 \\
-a & -a & a & a & b & 0 & 0 & -b \\
a & -a & a & -a & 0 & -b & b & 0 \\
-a & a & a & -a & b & 0 & 0 & b \\
-a & -a & -a & -a & 0 & b & b & 0 \\
a & a & -a & -a & b & 0 & 0 & -b \\
-a & a & -a & a & 0 & -b & b & 0
\end{array}\right)
\tag{7.145}
$$

with $a = \sqrt{2}/4$ and $b = 1/2$, where the multiplicities of $\mu = \mu_1, \ldots, \mu_5$ in T are given by $a^{\mu_1} = a^{\mu_2} = a^{\mu_3} = a^{\mu_4} = 1$ and $a^{\mu_5} = 2$.

With the use of H in (7.145), the element stiffness matrix J in (7.144) can be put into a block-diagonal form:

$$
\tilde{J} = H^\top J H = \mathrm{diag}(\tilde{J}^{\mu_1}, \tilde{J}^{\mu_2}, \tilde{J}^{\mu_3}, \tilde{J}^{\mu_4}, \tilde{J}^{\mu_5}, \tilde{J}^{\mu_5}),
$$

where

$$
\begin{aligned}
\tilde{J}^{\mu_1} &= k_1 - k_2 + 2k_3 + k_4 + k_5 + k_6 - k_7, \\
\tilde{J}^{\mu_2} &= k_1 - k_2 + 2k_3 - k_4 - k_5 - k_6 + k_7, \\
\tilde{J}^{\mu_3} &= k_1 - k_2 - 2k_3 - k_4 + k_5 - k_6 - k_7, \\
\tilde{J}^{\mu_4} &= k_1 - k_2 - 2k_3 + k_4 - k_5 + k_6 + k_7, \\
\tilde{J}^{\mu_5} &= \begin{pmatrix} k_1 + k_2 + k_5 + k_7 & k_4 - k_6 \\ k_4 - k_6 & k_1 + k_2 - k_5 - k_7 \end{pmatrix}.
\end{aligned}
$$

Problems

7-1 For a matrix representation T of a group G, show $T(g^{-1}) = T(g)^{-1}$ for $g \in G$ and $T(e) = I$, where e is the identity element of G.

7-2 Check the equivariance (7.38) of $F(u, f, \varepsilon)$ in (1.5) to $\mathrm{D}_1 = \{e, \sigma\}$ with respect to some nonunit representations T and S.
Answer: $T(\sigma) = S(\sigma) = -1$ and $F(-u, f, -\varepsilon) = -F(u, f, \varepsilon)$.

7-3 Give $T(\sigma)$ and $T(c(2\pi/3))$ for the regular-triangular truss dome in Fig. 1.13 in §1.2.7 with 12 degrees of freedom.

7-4 Check the equivariance (7.37) of $\boldsymbol{F}$ in (7.136) to D_3.

7-5 For the representations S and T in (7.137) and (7.138), find a nonsingular matrix Q such that $S(g) = Q^{-1}T(g)Q$ for all $g \in \mathrm{D}_3$.

7-6 Complete the proof of Lemma 7.5 in §7.5.2.

7-7 Complete the proof of Lemma 7.6 in §7.5.2.

7-8 Verify (7.141) for J in (7.144) and T in (7.142) and (7.143). Obtain the matrix $Q = (Q_i^\mu)$ that corresponds to H in (7.145); then verify (7.103).

Summary

- Preliminaries related to group representation have been presented.
- Group equivariance of the governing equation of an imperfect system has been introduced.
- The Liapunov–Schmidt reduction procedure for a multiple critical point has been presented.
- The principle of inheritance of symmetry has been introduced.
- The block-diagonalization of the Jacobian matrix and that of the imperfection sensitivity matrix have been presented.

Chapter 8
Bifurcation Behavior of D_n-Equivariant Systems

8.1 Introduction

Group-theoretic bifurcation theory has been presented in the preceding chapter. In this chapter, this theory is applied to the analysis[1] of a system that is equivariant to the simplest group, the dihedral group D_n, which represents the symmetry of the regular n-sided polygon.

One can encounter dihedral group symmetries in physical, mechanical, and structural systems. The snow crystal, for example, displays hexagonal symmetry labeled by D_6. We also encounter symmetric domes and shells, which are built to be symmetric to enhance their strength, but might collapse by symmetry-breaking bifurcation. Naturally, theoretical study and numerical analysis of the bifurcation of a D_n-equivariant system are of the most importance.

In this chapter, the bifurcation of a D_n-equivariant system is investigated by the analysis of the bifurcation equation (7.52):

$$\widetilde{\boldsymbol{F}}(\boldsymbol{w}, \widetilde{f}, \boldsymbol{v}) = \boldsymbol{0}$$

at a critical point. Although simple critical points and double ones appear inherently in these systems, emphasis naturally is to be placed on the double points, because most of the results in Part I are applicable to simple critical points of these systems. By assembling the results on individual critical points, we construct the hierarchy of subgroups (7.88):

$$G \to G_2 \to G_3 \to \cdots$$

($G = D_n$) that characterizes the recursive change of symmetries. A few remarks are given about a C_n-equivariant system.

This chapter is organized as follows. The first half of this chapter is devoted to the definition of dihedral and cyclic groups and bifurcation rules of these groups.

[1] This analysis is based on Sattinger, 1979 [176], 1983 [178]; Fujii, Mimura, and Nishiura, 1982 [51]; Healey, 1985 [69], 1988 [70]; Golubitsky, Stewart, and Schaeffer, 1988 [64]; Dellnitz and Werner, 1989 [37]; and Ikeda, Murota, and Fujii, 1991 [94].

K. Ikeda and K. Murota, *Imperfect Bifurcation in Structures and Materials*, Applied Mathematical Sciences 149, DOI 10.1007/978-1-4419-7296-5_8,

- Dihedral and cyclic groups and their irreducible representations are given in §8.2.
- The bifurcation rule of a D_n-symmetric perfect system is summarized in §8.3 based on the general framework of Chapter 7.
- Recursive bifurcation behaviors of D_n- and C_n-symmetric domes are studied in §8.4.

Local behavior near a double critical point is investigated in the latter half of this chapter.

- Bifurcation equations for D_n-symmetric and/or reciprocal systems are derived in §8.5.
- Local perfect bifurcation behaviors are analyzed using bifurcation equations for perfect systems in §8.6.
- Local imperfect behaviors are investigated in §8.7.
- In §8.8 we derive imperfection sensitivity laws, which form the basis of further studies in Chapters 9 and 10.
- The experimentally observed bifurcation diagram presented in Chapter 6 is extended to a D_n-equivariant system in §8.9.
- Stability analysis near a double bifurcation point is carried out in §8.10, the appendix of this chapter.

8.2 Dihedral and Cyclic Groups

Dihedral and cyclic groups and their irreducible representations are introduced.

8.2.1 Definition of Groups

The dihedral group of degree n, which describes the symmetry of a regular n-gon in the xy-plane (a two-dimensional space $\mathbb{R}^2$), is defined by

$$D_n = \langle \sigma, c(2\pi/n) \rangle = \{c(2\pi i/n), \sigma c(2\pi i/n) \mid i = 0, 1, \ldots, n-1\}, \tag{8.1}$$

where $c(2\pi i/n)$ denotes a counterclockwise rotation about the origin at an angle of $2\pi i/n$ $(i = 0, 1, \ldots, n-1)$; σ is a reflection like $y \mapsto -y$; and $\sigma c(2\pi i/n)$ represents the combined action of the rotation $c(2\pi i/n)$ followed by the reflection σ. We have

$$c(2\pi i/n)c(2\pi j/n) = c(2\pi(i+j)/n), \qquad c(2\pi/n)^n = \sigma^2 = (\sigma c(2\pi/n))^2 = e,$$

where e is the identity element. Figure 8.1(a) depicts a D_4-symmetric figure.

We consider D_n as an abstract group and redefine it as

$$D_n = \langle r, s \rangle = \{e, r, \ldots, r^{n-1}, s, sr, \ldots, sr^{n-1}\}, \tag{8.2}$$

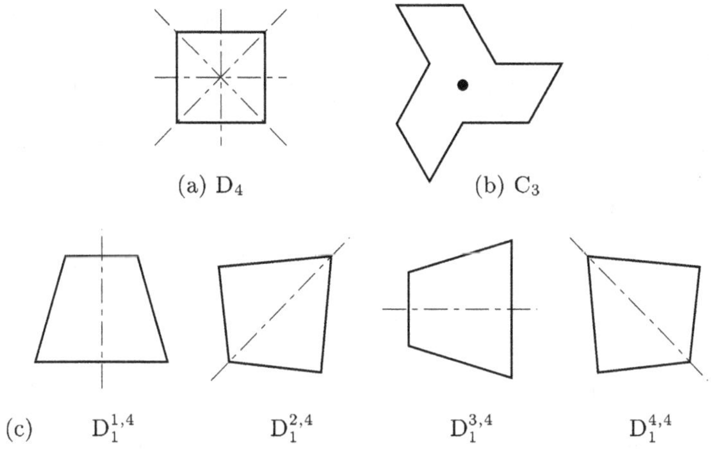

Fig. 8.1 Figures with dihedral and cyclic group symmetries. — - —: line of reflection symmetry; ●: center of rotation symmetry.

where r and s satisfy the defining relations

$$r^i r^j = r^{i+j}, \qquad r^n = s^2 = (sr)^2 = e. \tag{8.3}$$

Subgroups of D_n comprise dihedral and cyclic groups whose degree m divides n:

$$\begin{aligned} \mathrm{D}_m^{k,n} &= \{c(2\pi i/m), \sigma c(2\pi[(k-1)/n + i/m]) \mid i = 0,1,\ldots,m-1\}, \\ \mathrm{C}_m &= \{c(2\pi i/m) \mid i = 0,1,\ldots,m-1\} \end{aligned} \tag{8.4}$$

in the representation (8.1), where $1 \le m \le n$, $\mathrm{D}_m^{1,n} = \mathrm{D}_m$, and $\mathrm{C}_1 = \{e\}$. These subgroups express partial symmetries of D_n. Cyclic groups C_m denote rotation-symmetric patterns; the group C_1, in particular, signifies a completely asymmetric pattern. Dihedral groups $\mathrm{D}_m^{k,n}$ indicate reflection symmetric patterns; the superscripts are introduced here to distinguish differences in the direction of the reflection line. Figures 8.1(b) and (c), respectively, show C_3-symmetric and $\mathrm{D}_1^{k,n}$-symmetric ($k = 1,\ldots,4$) figures. Note that $\mathrm{D}_1^{k,n}$ and $\mathrm{D}_1^{k',n}$ ($k \neq k'$) are different but isomorphic groups.

Recall that the number of elements of a (sub)group is termed the order, expressing the level of symmetry (cf., §7.3). The index of a subgroup of a group is defined as the ratio of the order of the group to that of the subgroup, standing for the relative level of symmetry. For example, the order of the group D_m is $2m$, denoted as $|\mathrm{D}_m| = 2m$; the index of the subgroup D_m in the group D_n is

$$\frac{|\mathrm{D}_n|}{|\mathrm{D}_m|} = \frac{2n}{2m} = \frac{n}{m}.$$

Remark 8.1. In association with the abstract dihedral group D_n in (8.2), we can consider geometric transformations in the three-dimensional space $\mathbb{R}^3$ in several different ways. The element r, for example, can be chosen to be a rotation $c(2\pi/n)$ about the z-axis at an angle of $2\pi/n$, and s can be chosen to be either σ_x, σ_y, $\sigma_y\sigma_z$, or $\sigma_x\sigma_z$. Here $\sigma_x : x \mapsto -x$, $\sigma_y : y \mapsto -y$, and $\sigma_z : z \mapsto -z$ are reflections; therefore, $\sigma_y\sigma_z$, for example, denotes the half-rotation about the x-axis. It is easy to verify the relations in (8.3) under this geometric interpretation. In the Schoenflies notation (cf., §13.2), the group generated by a pair of elements $c(2\pi/n)$ and σ_x or σ_y is denoted as C_{nv}, and that by $c(2\pi/n)$ and $\sigma_y\sigma_z$ or $\sigma_x\sigma_z$ as D_n. These groups C_{nv} and D_n are isomorphic as abstract groups:

$$C_{nv} \simeq D_n.$$

The dihedral group D_n defined in (8.1) is identical to C_{nv} and not to D_n in the Schoenflies notation. Moreover, for odd m, the direct product of D_m and $\langle\sigma_z\rangle$, denoted as D_{mh} in the Schoenflies notation, is isomorphic to the dihedral group D_n of (8.2) for $n = 2m$. For the group

$$D_{mh} = \langle \sigma\sigma_z, c(2\pi/m)\rangle \times \langle\sigma_z\rangle = \langle \sigma, \sigma_z, c(2\pi/m)\rangle$$

with $\sigma = \sigma_x$ or σ_y, the isomorphism between D_{mh} and D_{2m} of (8.2) is given by

$$\sigma\sigma_z \leftrightarrow s, \qquad c(2\pi/m) \leftrightarrow r^2, \qquad \sigma_z \leftrightarrow r^m.$$

Figure 8.2 shows spatial symmetries expressed by the Schoenflies notation; D_{3h}, C_{6v}, and D_6 are isomorphic as abstract groups:

$$D_{3h} \simeq C_{6v} \simeq D_6.$$

A more detailed account of this notation is given in §13.2. □

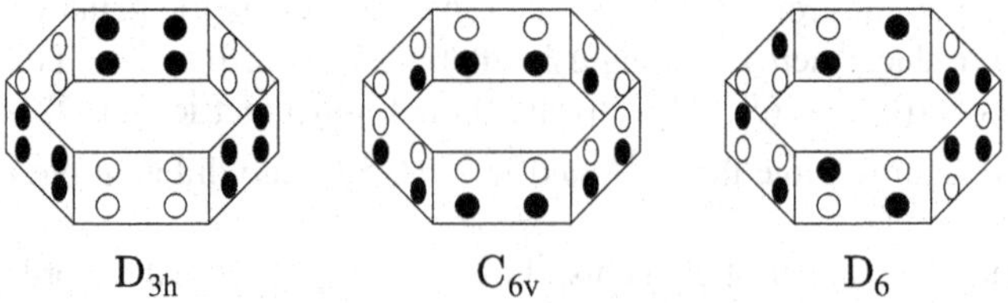

Fig. 8.2 Spatial symmetries labeled by the Schoenflies notation (cf., Dinkevich, 1991 [41]).

8.2.2 Irreducible Representations

The irreducible representations of D_n and C_n are presented.

Dihedral Group D_n

The dihedral group D_n in (8.1) has one-dimensional and two-dimensional irreducible representations over $\mathbb{R}$, which are absolutely irreducible. The number M_1 of one-dimensional irreducible representations, and the number M_2 of two-dimensional ones are given, respectively, by

$$M_1 = \begin{cases} 4 & \text{for } n \text{ even}, \\ 2 & \text{for } n \text{ odd}, \end{cases} \tag{8.5}$$

$$M_2 = \lfloor (n-1)/2 \rfloor = \begin{cases} (n-2)/2 & \text{for } n \text{ even}, \\ (n-1)/2 & \text{for } n \text{ odd}, \end{cases} \tag{8.6}$$

where, for a real number x, $\lfloor x \rfloor$ denotes the largest integer not larger than x. We have the equation

$$M_1 + 4M_2 = 2n = |\mathrm{D}_n|$$

for all n, compatible with the identity (7.20).

Let us write the one-dimensional irreducible representations as $(+,+)_{\mathrm{D}_n}$, $(+,-)_{\mathrm{D}_n}$, $(-,+)_{\mathrm{D}_n}$, and $(-,-)_{\mathrm{D}_n}$ and the two-dimensional irreducible representations as $(1)_{\mathrm{D}_n}$, $(2)_{\mathrm{D}_n}$, ... (see Remark 8.2 below for the notation). Then the index set $R(\mathrm{D}_n)$ of irreducible representations (cf., (7.21)) is given by

$$R(\mathrm{D}_n) = \begin{cases} \{(+,+)_{\mathrm{D}_n}, (+,-)_{\mathrm{D}_n}, (-,+)_{\mathrm{D}_n}, (-,-)_{\mathrm{D}_n}\} \cup \{(j)_{\mathrm{D}_n} \mid j = 1, \ldots, (n-2)/2\} \\ \qquad \text{for } n \text{ even}, \\ \{(+,+)_{\mathrm{D}_n}, (+,-)_{\mathrm{D}_n}\} \cup \{(j)_{\mathrm{D}_n} \mid j = 1, \ldots, (n-1)/2\} \qquad \text{for } n \text{ odd}, \end{cases} \tag{8.7}$$

which, for $n = 1$ and $n = 2$, reduces to

$$R(\mathrm{D}_1) = \{(+,+)_{\mathrm{D}_1}, (+,-)_{\mathrm{D}_1}\},$$
$$R(\mathrm{D}_2) = \{(+,+)_{\mathrm{D}_2}, (+,-)_{\mathrm{D}_2}, (-,+)_{\mathrm{D}_2}, (-,-)_{\mathrm{D}_2}\}.$$

The one-dimensional irreducible representations are defined by

$$\begin{aligned} T^{(+,+)_{\mathrm{D}_n}}(c(2\pi/n)) &= 1, & T^{(+,+)_{\mathrm{D}_n}}(\sigma) &= 1, \\ T^{(+,-)_{\mathrm{D}_n}}(c(2\pi/n)) &= 1, & T^{(+,-)_{\mathrm{D}_n}}(\sigma) &= -1, \\ T^{(-,+)_{\mathrm{D}_n}}(c(2\pi/n)) &= -1, & T^{(-,+)_{\mathrm{D}_n}}(\sigma) &= 1, \\ T^{(-,-)_{\mathrm{D}_n}}(c(2\pi/n)) &= -1, & T^{(-,-)_{\mathrm{D}_n}}(\sigma) &= -1, \end{aligned} \tag{8.8}$$

where $(-,+)_{\mathrm{D}_n}$ and $(-,-)_{\mathrm{D}_n}$ exist only for n even. The two-dimensional representations are chosen[2] to be the unitary representations defined by

$$T^{(j)_{\mathrm{D}_n}}(c(2\pi/n)) = \begin{pmatrix} \cos(2\pi j/n) & -\sin(2\pi j/n) \\ \sin(2\pi j/n) & \cos(2\pi j/n) \end{pmatrix}, \qquad T^{(j)_{\mathrm{D}_n}}(\sigma) = \begin{pmatrix} 1 & 0 \\ 0 & -1 \end{pmatrix},$$
$$j = 1, \ldots, M_2. \tag{8.9}$$

[2] Throughout this book, we use the matrix representation (8.9), although this is not the unique choice.

The associated subgroups G^μ of (7.22) are given as follows.

$$G^{(+,+)_{D_n}} = D_n, \quad G^{(+,-)_{D_n}} = C_n, \quad G^{(-,+)_{D_n}} = D_{n/2}, \quad G^{(-,-)_{D_n}} = D^{2,n}_{n/2},$$
$$G^{(j)_{D_n}} = C_{n/\widehat{n}}, \tag{8.10}$$

where

$$\widehat{n} = \frac{n}{\gcd(n, j)} \tag{8.11}$$

is an index associated with $\mu = (j)_{D_n}$, where $\gcd(n, j)$ denotes the greatest common divisor of n and j. We have $3 \le \widehat{n} \le n$, since $1 \le j \le \lfloor (n-1)/2 \rfloor$.

Cyclic Group C_n

The cyclic group C_n has one-dimensional and two-dimensional irreducible representations over $\mathbb{R}$. They are given by

$$R(C_n) = \begin{cases} \{(+)_{C_n}, (-)_{C_n}\} \cup \{(j)_{C_n} \mid j = 1, \ldots, (n-2)/2\} & \text{for } n \text{ even}, \\ \{(+)_{C_n}\} \cup \{(j)_{C_n} \mid j = 1, \ldots, (n-1)/2\} & \text{for } n \text{ odd}, \end{cases} \tag{8.12}$$

which, for $n = 1$ and $n = 2$, reduces to

$$R(C_1) = \{(+)_{C_1}\}, \qquad R(C_2) = \{(+)_{C_2}, (-)_{C_2}\}.$$

The one-dimensional irreducible representations are defined by

$$T^{(+)_{C_n}}(c(2\pi/n)) = 1, \qquad T^{(-)_{C_n}}(c(2\pi/n)) = -1, \tag{8.13}$$

where $(-)_{C_n}$ exists only for n even. The two-dimensional ones are defined by

$$T^{(j)_{C_n}}(c(2\pi/n)) = \begin{pmatrix} \cos(2\pi j/n) & -\sin(2\pi j/n) \\ \sin(2\pi j/n) & \cos(2\pi j/n) \end{pmatrix}, \qquad j = 1, \ldots, M_2, \tag{8.14}$$

where $M_2 = \lfloor (n-1)/2 \rfloor$. The associated subgroups G^μ of (7.22) are given by

$$G^{(+)_{C_n}} = C_n, \qquad G^{(-)_{C_n}} = C_{n/2}, \qquad G^{(j)_{C_n}} = C_{n/\widehat{n}}, \tag{8.15}$$

where $(-)_{C_n}$ exists for n even, and $\widehat{n}/n = \gcd(n, j)$ by (8.11).

The two-dimensional representation $(j)_{C_n}$ in (8.14) is not absolutely irreducible (see §7.4.3). Indeed, with the use of a unitary transformation matrix

$$U = \frac{1}{\sqrt{2}} \begin{pmatrix} 1 & 1 \\ -\mathrm{i} & \mathrm{i} \end{pmatrix},$$

the matrix $T^{(j)_{C_n}}(c(2\pi/n))$ in (8.14) can be transformed as

$$U^* T^{(j)_{C_n}}(c(2\pi/n)) U = \begin{pmatrix} \exp(\mathrm{i}2\pi j/n) & 0 \\ 0 & \exp(-\mathrm{i}2\pi j/n) \end{pmatrix},$$

where $(\cdot)^*$ means the conjugate transpose. Consequently, the two-dimensional representation $(j)_{\mathrm{C}_n}$ over $\mathbb{R}$ splits into two one-dimensional irreducible representations over $\mathbb{C}$ which are labeled by $(j+)_{\mathrm{C}_n}$ and $(j-)_{\mathrm{C}_n}$ and are defined by

$$T^{(j+)_{\mathrm{C}_n}}(c(2\pi/n)) = \exp(\mathrm{i}2\pi j/n), \qquad T^{(j-)_{\mathrm{C}_n}}(c(2\pi/n)) = \exp(-\mathrm{i}2\pi j/n). \tag{8.16}$$

Accordingly, the index set of the irreducible representations of C_n over $\mathbb{C}$ reads as

$$R_{\mathbb{C}}(\mathrm{C}_n) = \{(+)_{\mathrm{C}_n}, (-)_{\mathrm{C}_n}\} \cup \{(j+)_{\mathrm{C}_n}, (j-)_{\mathrm{C}_n} \mid j = 1, \ldots, M_2\}, \tag{8.17}$$

where $(-)_{\mathrm{C}_n}$ exists only for n even and $M_2 = \lfloor (n-1)/2 \rfloor$. We have $|R_{\mathbb{C}}(\mathrm{C}_n)| = n \times 1^2 = |\mathrm{C}_n|$, compatible with the identity (7.20).

Remark 8.2. Some conventional notations are available for the irreducible representations of D_n and C_n, which are used commonly in physics and chemistry (cf., Kettle, 1995 [116] and Kim, 1999 [117]). The correspondence of the present notation to a conventional notation for $n = 6$, for example, is given below.

Group	Present Notation	Conventional Notation
D_6	$(+,+)_{\mathrm{D}_6}$	A_1
	$(+,-)_{\mathrm{D}_6}$	A_2
	$(-,+)_{\mathrm{D}_6}$	B_1
	$(-,-)_{\mathrm{D}_6}$	B_2
	$(j)_{\mathrm{D}_6}$ $(j = 1,2)$	E_j
C_6	$(+)_{\mathrm{C}_6}$	A
	$(-)_{\mathrm{C}_6}$	B
	$(j)_{\mathrm{C}_6}$ $(j = 1,2)$	E_j

□

8.3 Perfect Bifurcation Behavior

Following the general framework of Chapter 7, we investigate the bifurcation behavior of a D_n-symmetric perfect system at a group-theoretic critical point, while deferring the details of the derivation to §8.5 and §8.6. The imperfect bifurcation behavior is treated in §8.7.

8.3.1 Symmetry of Solutions

The symmetry of the equation is formulated as the equivariance (7.37) to $G = \mathrm{D}_n$, the dihedral group defined by (8.1). That is,

$$T(g)\boldsymbol{F}(\boldsymbol{u}, f, \boldsymbol{v}^0) = \boldsymbol{F}(T(g)\boldsymbol{u}, f, \boldsymbol{v}^0), \qquad g \in \mathrm{D}_n, \tag{8.18}$$

where T is a unitary matrix representation. The symmetry of a solution $\boldsymbol{u}$ is usually lower than that of the equation and is represented by a subgroup of G, denoted by $\Sigma(\boldsymbol{u})$ in (7.86). By (8.4), the possible candidates for $\Sigma(\boldsymbol{u})$ are $D_m^{k,n}$ and C_m for some m and k.

We consider a solution $\boldsymbol{u}$ in the neighborhood of a critical point $(\boldsymbol{u}_c^0, f_c^0)$ of the perfect system. We assume that the critical point is group-theoretic. This means that $\ker(J_c^0)$, the kernel space of the Jacobian matrix at $(\boldsymbol{u}_c^0, f_c^0)$, is a G-irreducible subspace, which is either one-dimensional or two-dimensional by (8.7) in the case of $G = D_n$. In other words, a group-theoretic critical point of a D_n-symmetric system is either simple ($M = 1$) or double ($M = 2$).

The full system of equations in variable $\boldsymbol{u}$ is reduced to the bifurcation equation in another variable $\boldsymbol{w} \in \ker(J_c^0)$ by the Liapunov–Schmidt reduction (cf., §7.5.2). The type of critical point (limit, bifurcation, etc.) can be determined through the analysis of the bifurcation equation. An important relation is given by (7.95):

$$G^\mu = \Sigma(\ker(J_c^0)) \subseteq \Sigma(\boldsymbol{w}) = \Sigma(\boldsymbol{u}) \subseteq G = D_n, \tag{8.19}$$

where μ denotes the irreducible representation associated with $\ker(J_c^0)$ and G^μ is defined in (7.22).

The following points are emphasized in (8.19).

- $\Sigma(\ker(J_c^0))$ designates the symmetry shared by all critical eigenvectors, and is determined as the subgroup G^μ associated with the irreducible representation μ. The subgroups G^μ for $G = D_n$ are listed in (8.10).
- The symmetry $\Sigma(\boldsymbol{u})$ of the solution $\boldsymbol{u}$ coincides with the symmetry $\Sigma(\boldsymbol{w})$ of the corresponding solution $\boldsymbol{w}$ to the bifurcation equation.
- The symmetry of the solution $\boldsymbol{w}$ to the bifurcation equation can possibly be higher than the symmetry of the critical eigenvectors due to nonlinearity; $\Sigma(\boldsymbol{w})$ can be determined through an analysis of the bifurcation equation involving nonlinear terms. The analysis, conducted in §8.6, reveals that $\Sigma(\boldsymbol{w})$ is strictly larger than G^μ for a double bifurcation point.
- Repeated bifurcation is associated with a chain of subgroups (7.88): $G \to G_1 \to G_2 \to \cdots$, where $G_i = \Sigma(\boldsymbol{u})$ for a solution $\boldsymbol{u}$ on the bifurcated path labeled by G_i. Recall from (7.89) that the symmetry of a solution remains invariant in a neighborhood of an ordinary point.

Below we explain the relation (8.19) more specifically for simple and double critical points. Recall that a simple critical point is associated with a one-dimensional representation $\mu = (\nu_1, \nu_2)_{D_n}$ with $\nu_1, \nu_2 = +, -$, whereas a double point is associated with a two-dimensional representation $\mu = (j)_{D_n}$ for some j.

Simple Critical Point

For a simple critical point associated with the unit representation $\mu = (+,+)_{D_n}$, the equivariance (7.54) to $G = D_n$ of the bifurcation equation with $\boldsymbol{v} = \boldsymbol{v}^0$ plays no role, since $T^{(+,+)_{D_n}}(g) = 1$ $(g \in D_n)$ by (8.8) and $S(g)\boldsymbol{v}^0 = \boldsymbol{v}^0$ $(g \in D_n)$ by (7.39). Therefore, the bifurcation equation for the perfect system takes the form of (2.80):

$$\widehat{F}(w, \widetilde{f}, 0) = A_{200}w^2 + A_{010}\widetilde{f} + \text{h.o.t.} = 0,$$

in which $A_{200} \neq 0$ and $A_{010} \neq 0$ in the generic sense. Therefore, the critical point is a limit point and the results on the asymptotic behavior at a limit point presented in §2.5.1 also hold for this case.

For a simple critical point associated with $\mu = (+,-)_{D_n}$, $(-,+)_{D_n}$, or $(-,-)_{D_n}$, the equivariance (7.54) to $G = D_n$ of the bifurcation equation with $\boldsymbol{v} = \boldsymbol{v}^0$ becomes

$$\widetilde{F}(-w, \widetilde{f}, \boldsymbol{v}^0) = -\widetilde{F}(w, \widetilde{f}, \boldsymbol{v}^0) \quad \text{or} \quad \widehat{F}(-w, \widetilde{f}, 0) = -\widehat{F}(w, \widetilde{f}, 0). \tag{8.20}$$

Therefore, the bifurcation equation for the perfect system is an odd function in w, which in turn can be expressed as

$$\widehat{F}(w, \widetilde{f}, 0) = w(A_{110}\widetilde{f} + A_{300}w^2 + \text{h.o.t.}) = 0,$$

where $A_{110} \neq 0$ and $A_{300} \neq 0$ in the generic sense. Therefore, this point is a pitchfork bifurcation point (see §2.5.3). Since the left and right critical eigenvectors $\boldsymbol{\xi}_1$ and $\boldsymbol{\eta}_1$ satisfy

$$T(g)\boldsymbol{\xi}_1 = \boldsymbol{\xi}_1 T^{\mu}(g), \qquad T(g)\boldsymbol{\eta}_1 = \boldsymbol{\eta}_1 T^{\mu}(g), \qquad g \in D_n, \tag{8.21}$$

it is apparent from (8.8) that

$$\Sigma(\boldsymbol{\xi}_1) = \Sigma(\boldsymbol{\eta}_1) = G^{\mu} = \begin{cases} C_n & \text{for } \mu = (+,-)_{D_n}, \\ D_{n/2} & \text{for } \mu = (-,+)_{D_n}, \\ D_{n/2}^{2,n} & \text{for } \mu = (-,-)_{D_n}. \end{cases} \tag{8.22}$$

Note that $\Sigma(\ker(J_c^0)) = \Sigma(\boldsymbol{\eta}_1)$, since any vector in the kernel space of J_c^0 is a scalar multiple of $\boldsymbol{\eta}_1$. As for the symmetry of the solution we have

$$G^{\mu} = \Sigma(\ker(J_c^0)) = \Sigma(\boldsymbol{w}) = \Sigma(\boldsymbol{u})$$

in (8.19) in this case, inasmuch as the action of D_n on w is given by

$$\begin{aligned} g \cdot w &= w, & g &\in G^{\mu}, \\ g \cdot w &= -w, & g &\in D_n \backslash G^{\mu}, \end{aligned}$$

which shows $\Sigma(\boldsymbol{w}) = \Sigma(w) = G^{\mu}$.

Two half-branches (that form a smooth bifurcated path) bifurcate in the directions of the critical eigenvectors $\boldsymbol{\eta}_1$ and $-\boldsymbol{\eta}_1$. The relation between the two half-branches at the bifurcation point can be described by the orbit (cf., §7.6). That is, if $(\boldsymbol{u}, f)$ is a solution on a half-branch, a solution on the other half-branch can be known as $(T(g)\boldsymbol{u}, f)$ for $g = \sigma$ if the associated irreducible representation is $(+,-)_{D_n}$ or $(-,-)_{D_n}$, and for $g = c(2\pi/n)$ if it is $(-,+)_{D_n}$ or $(-,-)_{D_n}$.

Double Critical Point

A double critical (bifurcation) point on a D_n-symmetric path is associated with a two-dimensional irreducible representation $\mu = (j)_{D_n}$ for some j. The index

$$\widehat{n} = \frac{n}{\gcd(n, j)} \tag{8.23}$$

in (8.11) characterizes the critical point, where $\gcd(n, j)$ denotes the greatest common divisor of n and j. We have $3 \le \widehat{n} \le n$, inasmuch as $1 \le j \le \lfloor (n-1)/2 \rfloor$.

By our assumption (cf., §7.5.1) that T is unitary, we can choose left and right orthonormal critical eigenvectors $\{\boldsymbol{\xi}_1, \boldsymbol{\xi}_2\}$ and $\{\boldsymbol{\eta}_1, \boldsymbol{\eta}_2\}$ such that

$$T(g)[\boldsymbol{\xi}_1, \boldsymbol{\xi}_2] = [\boldsymbol{\xi}_1, \boldsymbol{\xi}_2]\, T^{(j)_{D_n}}(g), \qquad g \in D_n, \tag{8.24}$$

$$T(g)[\boldsymbol{\eta}_1, \boldsymbol{\eta}_2] = [\boldsymbol{\eta}_1, \boldsymbol{\eta}_2]\, T^{(j)_{D_n}}(g), \qquad g \in D_n, \tag{8.25}$$

for the irreducible representation $T^{(j)_{D_n}}$ in (8.9). The superposition (linear combination) of a pair of critical eigenvectors is also a critical eigenvector. We define

$$\boldsymbol{\xi}(\varphi) = \cos\varphi \cdot \boldsymbol{\xi}_1 + \sin\varphi \cdot \boldsymbol{\xi}_2, \tag{8.26}$$

$$\boldsymbol{\eta}(\theta) = \cos\theta \cdot \boldsymbol{\eta}_1 + \sin\theta \cdot \boldsymbol{\eta}_2. \tag{8.27}$$

The symmetry of $\boldsymbol{\xi}(\varphi)$ and $\boldsymbol{\eta}(\theta)$ is given by

$$\Sigma(\boldsymbol{\xi}(\varphi)) = \Sigma(\boldsymbol{\eta}(\theta)) = G^{\mu} = C_{n/\widehat{n}} \tag{8.28}$$

(cf., (8.10)) for general angles φ and θ ($0 \le \varphi < 2\pi, 0 \le \theta < 2\pi$). Some critical eigenvectors have higher symmetry than $C_{n/\widehat{n}}$. For example, $\boldsymbol{\xi}_1 = \boldsymbol{\xi}(0)$ and $\boldsymbol{\eta}_1 = \boldsymbol{\eta}(0)$ are invariant under reflection σ; that is, $T(\sigma)\boldsymbol{\xi}_1 = \boldsymbol{\xi}_1$ and $T(\sigma)\boldsymbol{\eta}_1 = \boldsymbol{\eta}_1$. There exist $2\widehat{n}$ such eigenvectors:

$$\Sigma(\boldsymbol{\xi}(\alpha_i)) = \Sigma(\boldsymbol{\xi}(\alpha_{i+\widehat{n}})) = \Sigma(\boldsymbol{\eta}(\alpha_i)) = \Sigma(\boldsymbol{\eta}(\alpha_{i+\widehat{n}})) = D^{i,n}_{n/\widehat{n}}, \qquad i = 1, \ldots, \widehat{n}, \tag{8.29}$$

where

$$\alpha_k = -\pi\, \frac{k-1}{\widehat{n}}, \qquad k = 1, \ldots, 2\widehat{n}. \tag{8.30}$$

According to the analysis of the bifurcation equations (made in §8.5 and §8.6):

- There exist $2\widehat{n}$ half-branches ($\widehat{n}$ bifurcated paths), the number of which coincides with the index $|D_n|/|D^{i,n}_{n/\widehat{n}}|$.
- They bifurcate in the directions of $\boldsymbol{\eta}(\alpha_i)$ and $\boldsymbol{\eta}(\alpha_{i+\widehat{n}})$ ($i = 1, \ldots, \widehat{n}$).
- The solutions $\boldsymbol{u}$ on the bifurcated paths for $\theta = \alpha_i$ and $\theta = \alpha_{i+\widehat{n}}$ are symmetric with respect to $D^{i,n}_{n/\widehat{n}}$, which is strictly larger than $G^{\mu} = C_{n/\widehat{n}}$.

The critical points on a D_n-symmetric path are classified in Table 8.1.

Remark 8.3. The critical points of a C_n-symmetric system (or those on a C_n-symmetric path of a D_n-symmetric system) can be treated similarly. A group-

Table 8.1 Critical points on a D_n-symmetric path

Multiplicity M	Irreducible Representation μ	Type of Points	Symmetry Groups G^μ	Bifurcated Paths
1	$(+,+)_{D_n}$	Limit	D_n	No Bifurcation
1	$(+,-)_{D_n}$	Pitchfork	C_n	C_n
	$(-,+)_{D_n}$ (n: even)		$D_{n/2}$	$D_{n/2}$
	$(-,-)_{D_n}$ (n: even)		$D^{2,n}_{n/2}$	$D^{2,n}_{n/2}$
2	$(j)_{D_n}$	Double	$C_{n/\hat{n}}$	$D^{i,n}_{n/\hat{n}}$ ($i = 1,\ldots,\hat{n}$)

Table 8.2 Critical points on a C_n-symmetric path of a reciprocal system

Multiplicity M	Irreducible Representation μ	Type of Points	Symmetry Groups G^μ	Bifurcated Paths
1	$(+)_{C_n}$	Limit	C_n	No Bifurcation
1	$(-)_{C_n}$ (n: even)	Pitchfork	$C_{n/2}$	$C_{n/2}$
2	$(j)_{C_n}$	Double	$C_{n/\hat{n}}$	$C_{n/\hat{n}}$

theoretic critical point is either simple or double by (8.12). A simple point associated with the unit irreducible representation $\mu = (+)_{C_n}$ is a limit point, and one with $\mu = (-)_{C_n}$, which exists for n even, is a pitchfork bifurcation point.

A double point is associated with $\mu = (j)_{C_n}$ for some j, and the symmetry of the critical eigenvectors is described by $C_{n/\hat{n}}$ by (8.15). It is emphasized that the reciprocity plays a primary role in a C_n-symmetric system. The analysis of the bifurcation equations, made in Remark 8.5 in §8.6.1, reveals that the occurrence of bifurcation is conditional for a nonreciprocal system. For a reciprocal system, this analysis reveals that the bifurcation occurs; the number of the half-branches is $\widehat{2n}$, and the symmetry of the bifurcated solutions is $C_{n/\hat{n}}$, coinciding with the symmetry G^μ of the critical eigenvectors. The critical points on the C_n-symmetric path of a reciprocal system are classified in Table 8.2. □

Remark 8.4. In the theoretical development, the irreducible representation μ associated with the critical point is assumed to be known. In practice, however, such μ must be identified as follows.

In an experiment of bifurcation behavior, as described in Part III, it is not the irreducible representation μ, but the symmetry $\Sigma(\boldsymbol{u})$ of a bifurcated solution that is observable. Then the multiplicity of the bifurcation point and the associated irreducible representation μ can be estimated in view of this symmetry $\Sigma(\boldsymbol{u})$.

In a numerical analysis of bifurcation behavior, as described in §8.4, the Jacobian matrix $J^0_{\rm c}$ is available and, in turn, the critical eigenvector(s) can be obtained. Then the associated irreducible representation μ can be determined by (8.21), (8.24), and (8.25). □

8.3.2 Recursive Bifurcation

The direct bifurcation of a D_n- or C_n-symmetric path has been clarified and presented in Tables 8.1 and 8.2 in §8.3.1. It is in order here to move on to the recursive bifurcation of a D_n- or C_n-equivariant system.

The symmetry of the bifurcated path is a semilocal property in that it is kept unchanged (cf., (7.89)) until the system undergoes secondary (tertiary) bifurcation. We can therefore associate each path with a group labeling its symmetry. Repeated bifurcations make up a hierarchy of symmetry groups of bifurcated paths. The hierarchy is expressed by a sequence of nested subgroups (7.88) for $G = D_n$; that is,

$$D_n \to \cdots \to C_1,$$

starting with the most symmetric $G = D_n$ and ending with the least symmetric C_1.

As listed in Table 8.1, the possible bifurcated paths from a D_n-symmetric fundamental path are invariant to either a dihedral group $D_m^{k,n}$ ($m = n/2$ or $n/\widehat{n}$) or a cyclic group C_n. Therefore, we can construct a complete rule for the recursive bifurcation of a D_n-equivariant reciprocal system through the repeated use of the bifurcation rules for D_n in Table 8.1 and for C_n in Table 8.2. (See Remark 8.5 in §8.6.1 for the role of reciprocity.) Each subgroup of D_n given in (8.4) is potentially reachable as an associated subgroup of a bifurcated path, although its actual existence depends on each problem. Figure 8.3, for example, shows the rule of recursive bifurcation for systems equivariant to D_4 and D_6. Bifurcation progresses in the direction of the arrows. A similar diagram was devised by Dellnitz and Werner, 1989 [37].

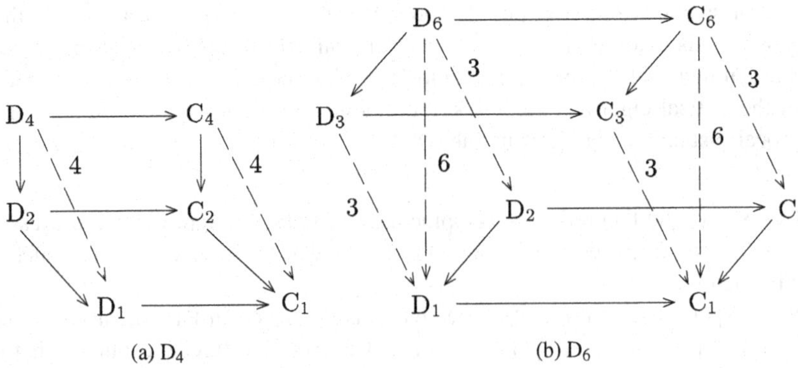

Fig. 8.3 Hierarchy of subgroups expressing the rule of bifurcation of a D_n-equivariant reciprocal system ($n = 4, 6$). Numerals denote the index $\widehat{n}$ of (8.23) for the double bifurcation points; $\longrightarrow$: simple bifurcation point; $--\to$: double bifurcation point.

8.4 Bifurcation of Domes

Bifurcations of the truss domes[3] of Fig. 8.4 are explained here based on the bifurcation rules presented in §8.3. The regular-triangular truss dome of Fig. 8.4(a) and the rotation-symmetric dome of Fig. 8.4(b) are, respectively, invariant to D_3 and C_6. Such invariance results in the group equivariance of their governing equations.

8.4.1 D_3*-Symmetric Dome*

Consider the D_3-symmetric regular-triangular truss dome in Fig. 8.4(a) subjected to the z-directional loads f applied at free nodes 1, 2, and 3.

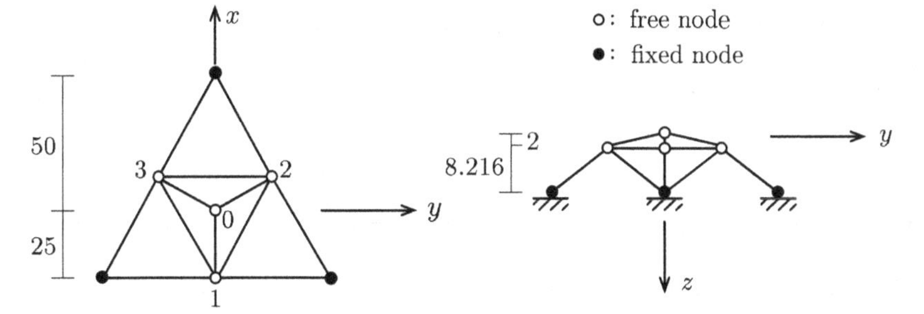

(a) Regular-triangular truss dome (D_3-symmetric)

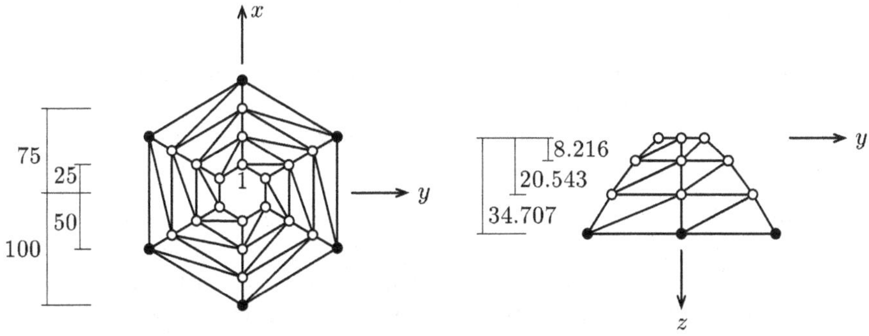

(b) Schwedler dome (C_6-symmetric)

Fig. 8.4 Symmetric truss domes.

[3] These domes are all subjected to symmetric loadings; all members of the domes have the same material and sectional properties.

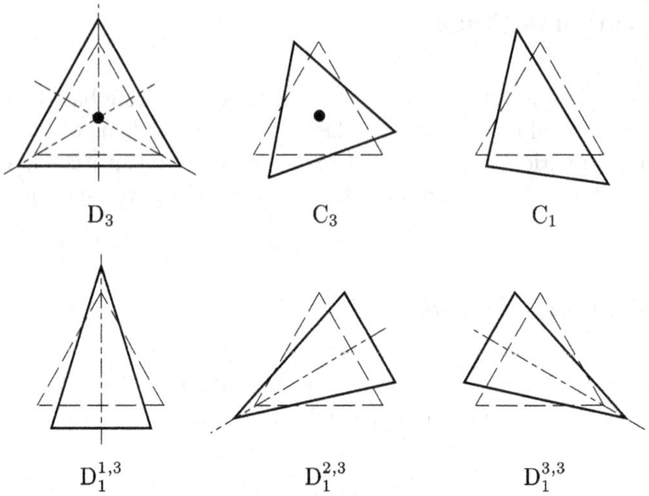

Fig. 8.5 Plane views of the deformation patterns of the regular-triangular free nodes of the regular-triangular truss dome. ——: displaced position; – – –: initial position; — - —: line of reflection symmetry; ●: center of rotation symmetry.

The deformation of this dome is described by a 12-dimensional vector $\boldsymbol{u} = (x_i, y_i, z_i \mid i = 0, 1, 2, 3)$ representing the location of the nodes 0, 1, 2, and 3. Deformation patterns of this dome are labeled by D_3 before bifurcation and by its subgroups after bifurcation; σ is chosen as the reflection: $y \mapsto -y$. Figure 8.5 presents plane views of the deformation patterns of the regular-triangular free nodes 1, 2, and 3 of this dome expressed by the subgroups:

- D_3 for a uniform expansion or shrinking of the regular triangle, accompanied by a uniform float or drop,
- C_3 for a rotated-regular-triangular pattern indicating a rotation about the z-axis, along with a uniform expansion or shrinking and a uniform float or drop,
- $D_1^{k,3}$ $(k = 1, 2, 3)$ for isosceles-triangular patterns with a reflection symmetry,
- C_1 for an asymmetric scalene-triangular pattern.

Figure 8.6 shows (a) space and (b) plane views of the equilibrium paths. The former shows the relation among f, x_0, and y_0; the latter displays the relation between f and z_0. As many as three bifurcated paths (six half-branches) branch directly at each of the two double bifurcation points with index $\widehat{n} = 3$ denoted by ($\triangle$), as portrayed in Fig. 8.6(a) for the double point A. At the simple pitchfork bifurcation points (shown as ($\circ$)) on these bifurcated paths, a bifurcated path branches further.

Such direct and secondary bifurcations correspond to a chain of subgroups

$$D_3 \to D_1^{k,3} \to C_1,$$

which is nothing other than a part of the hierarchy in Fig. 8.3(b) (note that D_1 in Fig. 8.3(b) is isomorphic to $D_1^{k,3}$).

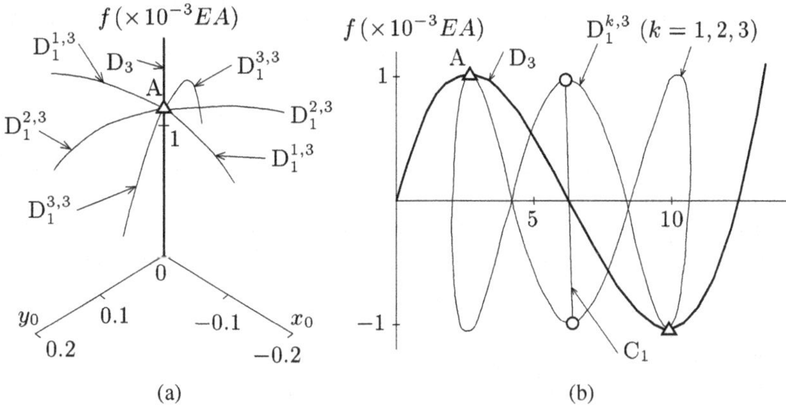

Fig. 8.6 (a) Space view and (b) plane view of the equilibrium paths for the triangular truss dome. △: double bifurcation point ($\widehat{n} = 3$); ○: simple pitchfork bifurcation point.

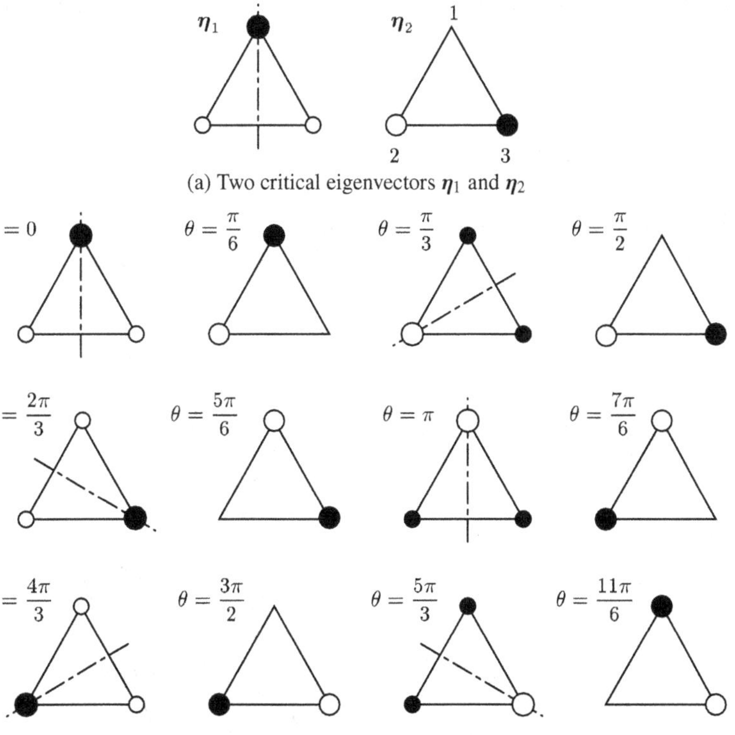

(a) Two critical eigenvectors $\boldsymbol{\eta}_1$ and $\boldsymbol{\eta}_2$

(b) Superposed eigenvector $\boldsymbol{\eta}(\theta)$ in (8.31)

Fig. 8.7 Illustration of the z-directional displacements of the triangular free nodes of the triangular truss dome associated with (a) the two critical eigenvectors and (b) their superposition. ●: positive component; ○: negative component; area of ● or ○: magnitude of a component; — - —: line of reflection symmetry.

The double bifurcation point A on the D_3-symmetric path has two orthonormal critical eigenvectors $\boldsymbol{\eta}_1$ and $\boldsymbol{\eta}_2$ satisfying (8.25). Their superposition

$$\boldsymbol{\eta}(\theta) = \cos\theta \cdot \boldsymbol{\eta}_1 + \sin\theta \cdot \boldsymbol{\eta}_2, \qquad 0 \le \theta < 2\pi, \tag{8.31}$$

serves, for any θ, as a critical eigenvector. We present, in Fig. 8.7(a), the z-directional displacements of the triangular free nodes of the dome for these two critical eigenvectors $\boldsymbol{\eta}_1$ and $\boldsymbol{\eta}_2$ and, in Fig. 8.7(b), those for their superposition $\boldsymbol{\eta}(\theta)$ for specific values of $\theta = \pi(k-1)/6$ ($k = 1,\ldots,12$). The z-directional components of $\boldsymbol{\eta}_1$ and $\boldsymbol{\eta}_2$ are found, respectively, to be equal to $(z_1,z_2,z_3) = (2/\sqrt{6}, -1/\sqrt{6}, -1/\sqrt{6})$ and $(0, -1/\sqrt{2}, 1/\sqrt{2})$. The superposed eigenvectors $\boldsymbol{\eta}(\theta)$ are generically nonsymmetric and labeled by C_1 ($= C_{n/\widehat{n}}$) but are $D_1^{k,3}$-symmetric for $\theta = (k-1)\pi/3 + j\pi$ ($k = 1,2,3$; $j = 0,1$). It is these specific $D_1^{k,3}$-symmetric eigenvectors ($k = 1,2,3$) that correspond to the directions of the bifurcated paths. Six ($= 2\widehat{n}$) half-branches exist: two of them, corresponding, for example, to $\theta = 0$ and π, are independent; and the others are known through geometric symmetry or, to be precise, through the orbit explained in (7.98). For example, the deformations on half-branches associated with $\theta = 0$, $2\pi/3$, and $4\pi/3$ in Fig. 8.7 form an orbit.

8.4.2 C_6-Symmetric Dome

Figure 8.8 shows the equilibrium paths of the C_6-symmetric Schwedler dome of Fig. 8.4(b) computed for the z-directional loads of $0.5f$ applied at the inner hexagonal nodes and f at the remaining nodes. The deformation of this dome is described by a 54-dimensional vector $\boldsymbol{u} = (x_i, y_i, z_i \mid i = 1,\ldots,18)$.

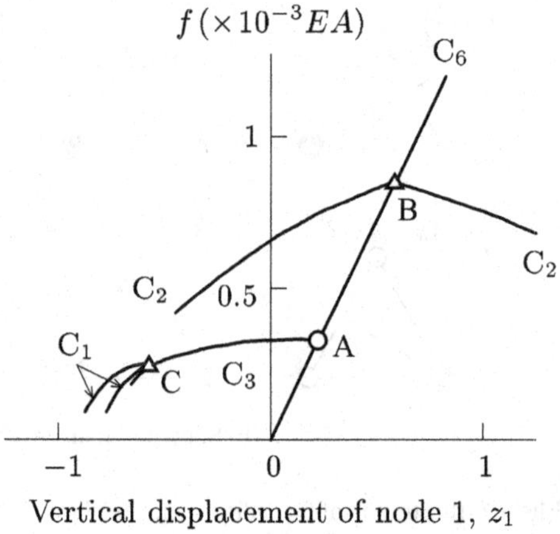

Fig. 8.8 Equilibrium paths for the Schwedler dome (C_6-symmetric). ○: simple pitchfork bifurcation point; △: double bifurcation point ($\widehat{n} = 3$).

Note that this is a reciprocal system having a potential function. In a C_n-equivariant system bifurcated paths at a double bifurcation point exist when it is a reciprocal system (cf., Remark 8.5 in §8.6.1). In fact,

- C_2-symmetric paths branch at the double bifurcation point B with the index $\widehat{n} = |C_6|/|C_2| = 3$,
- C_1-symmetric paths emanate from the C_3-symmetric path at the double point C with the index $\widehat{n} = |C_3|/|C_1| = 3$.

In addition, a C_3-symmetric path branches at the simple bifurcation point A.

8.5 Bifurcation Equations for a Double Critical Point

Following the description of major findings in §8.3 and concrete examples in §8.4, we present a detailed analysis of the bifurcation behaviors in the neighborhood of a double critical point in this and the next sections.

In this section, we obtain the generic form of the bifurcation equation in (7.52):

$$\widetilde{\boldsymbol{F}}(\boldsymbol{w}, \widetilde{f}, \boldsymbol{v}) = \boldsymbol{0} \tag{8.32}$$

at a double critical point $(\boldsymbol{u}_{\rm c}^0, f_{\rm c}^0)$ on a D_n-symmetric path of the perfect system; we have $M = 2$ in (7.52). The bifurcation equation (8.32) encompasses both the perfect system (with $\boldsymbol{v} = \boldsymbol{v}^0$) and an imperfect system (with $\boldsymbol{v} \neq \boldsymbol{v}^0$).

It should be recalled from (7.38) that the D_n-equivariance of the original governing equation is formulated as

$$T(g)\boldsymbol{F}(\boldsymbol{u}, f, \boldsymbol{v}) = \boldsymbol{F}(T(g)\boldsymbol{u}, f, S(g)\boldsymbol{v}), \qquad g \in D_n, \tag{8.33}$$

in terms of two unitary matrix representations T and S. In addition, from (7.54) it should be recalled that the D_n-equivariance of the bifurcation equation is formulated as

$$\widetilde{T}(g)\widetilde{\boldsymbol{F}}(\boldsymbol{w}, \widetilde{f}, \boldsymbol{v}) = \widetilde{\boldsymbol{F}}(\widetilde{T}(g)\boldsymbol{w}, \widetilde{f}, S(g)\boldsymbol{v}), \qquad g \in D_n, \tag{8.34}$$

where $\widetilde{T}$ is the two-dimensional irreducible matrix representation of D_n associated with the kernel of $J_{\rm c}^0 = J(\boldsymbol{u}_{\rm c}^0, f_{\rm c}^0, \boldsymbol{v}^0)$.

8.5.1 Bifurcation Equations in Complex Variables

We adopt the complex coordinates $(z, \overline{z})$ instead of (w_1, w_2); that is,

$$z = w_1 + {\rm i}w_2, \qquad \overline{z} = w_1 - {\rm i}w_2, \tag{8.35}$$

where i denotes the imaginary unit and $\overline{\cdot}$ denotes the complex conjugate. Note that the vector associated with $z = w_1 + {\rm i}w_2 = r \cdot \exp({\rm i}\theta)$, that is,

$$\frac{1}{r}(w_1\boldsymbol{\eta}_1 + w_2\boldsymbol{\eta}_2) = \cos\theta\cdot\boldsymbol{\eta}_1 + \sin\theta\cdot\boldsymbol{\eta}_2 = \boldsymbol{\eta}(\theta), \tag{8.36}$$

is exactly of the same form as the superposed critical eigenvector in (8.27).

For the components $\widetilde{F}_1$ and $\widetilde{F}_2$ of $\widetilde{\boldsymbol{F}} = (\widetilde{F}_1, \widetilde{F}_2)^\top$ we put

$$F(z,\overline{z},\widetilde{\boldsymbol{f}},\boldsymbol{v}) = \widetilde{F}_1(w_1,w_2,\widetilde{\boldsymbol{f}},\boldsymbol{v}) + \mathrm{i}\,\widetilde{F}_2(w_1,w_2,\widetilde{\boldsymbol{f}},\boldsymbol{v}). \tag{8.37}$$

Then the bifurcation equation $\widetilde{\boldsymbol{F}} = \boldsymbol{0}$ is equivalent to

$$F(z,\overline{z},\widetilde{\boldsymbol{f}},\boldsymbol{v}) = \overline{F(z,\overline{z},\widetilde{\boldsymbol{f}},\boldsymbol{v})} = 0. \tag{8.38}$$

Lemma 8.1. *For the Jacobian matrix* $\tilde{J} = (\partial\widetilde{F}_i/\partial w_j \mid i,j = 1,2)$ *of the bifurcation equation* (8.37), *we have*[4]

$$\mathrm{trace}(\tilde{J}) = 2\,\mathrm{Re}\left(\frac{\partial F}{\partial z}\right), \qquad \det(\tilde{J}) = \left|\frac{\partial F}{\partial z}\right|^2 - \left|\frac{\partial F}{\partial \overline{z}}\right|^2, \tag{8.39}$$

where trace$(\cdot)$ *means the trace of a matrix and* Re$(\cdot)$ *is the real part of a complex number. Therefore, the criticality condition* (7.78) *for the bifurcation equation is equivalent to*

$$\left|\frac{\partial F}{\partial z}\right|^2 - \left|\frac{\partial F}{\partial \overline{z}}\right|^2 = 0. \tag{8.40}$$

Proof. By (8.35), we have

$$\frac{\partial}{\partial w_1} = \frac{\partial}{\partial z} + \frac{\partial}{\partial \overline{z}}, \qquad \frac{\partial}{\partial w_2} = \mathrm{i}\left(\frac{\partial}{\partial z} - \frac{\partial}{\partial \overline{z}}\right),$$

whereas

$$\widetilde{F}_1 = \frac{1}{2}(F + \overline{F}), \qquad \widetilde{F}_2 = \frac{1}{2\mathrm{i}}(F - \overline{F}).$$

Hence we have

$$\begin{aligned} \frac{\partial \widetilde{F}_1}{\partial w_1} &= \frac{1}{2}\left(\frac{\partial}{\partial z} + \frac{\partial}{\partial \overline{z}}\right)(F + \overline{F}), & \frac{\partial \widetilde{F}_1}{\partial w_2} &= \frac{\mathrm{i}}{2}\left(\frac{\partial}{\partial z} - \frac{\partial}{\partial \overline{z}}\right)(F + \overline{F}), \\ \frac{\partial \widetilde{F}_2}{\partial w_1} &= \frac{1}{2\mathrm{i}}\left(\frac{\partial}{\partial z} + \frac{\partial}{\partial \overline{z}}\right)(F - \overline{F}), & \frac{\partial \widetilde{F}_2}{\partial w_2} &= \frac{1}{2}\left(\frac{\partial}{\partial z} - \frac{\partial}{\partial \overline{z}}\right)(F - \overline{F}). \end{aligned} \tag{8.41}$$

With the use of (8.41), we have

$$\det(\tilde{J}) = \frac{\partial \widetilde{F}_1}{\partial w_1}\frac{\partial \widetilde{F}_2}{\partial w_2} - \frac{\partial \widetilde{F}_1}{\partial w_2}\frac{\partial \widetilde{F}_2}{\partial w_1} = \frac{\partial F}{\partial z}\frac{\partial \overline{F}}{\partial \overline{z}} - \frac{\partial \overline{F}}{\partial z}\frac{\partial F}{\partial \overline{z}} = \left|\frac{\partial F}{\partial z}\right|^2 - \left|\frac{\partial F}{\partial \overline{z}}\right|^2.$$

We also have

---

[4] The formulas in (8.39) are independent of D_n-equivariance (8.34). In particular, they are applicable to both group-theoretic and parametric critical points of a D_n-symmetric system.

$$\mathrm{trace}(\tilde{J}) = \frac{\partial \widetilde{F}_1}{\partial w_1} + \frac{\partial \widetilde{F}_2}{\partial w_2} = \frac{\partial F}{\partial z} + \frac{\partial \overline{F}}{\partial \overline{z}} = 2\,\mathrm{Re}\left(\frac{\partial F}{\partial z}\right).$$

Thus we have (8.39) and and can show that the criticality condition $\det(\tilde{J}) = 0$ is equivalent to (8.40). □

8.5.2 Power Series Form

A power series form of the bifurcation equation turns out to be convenient for subsequent arguments. We expand F into a power series in $(z, \overline{z})$:

$$F(z, \overline{z}, \widetilde{f}, \boldsymbol{v}) \approx \sum_{p=0} \sum_{q=0} A_{pq}(\widetilde{f}, \boldsymbol{v}) z^p \overline{z}^q \tag{8.42}$$

involving an appropriate number of terms. Since $(z, \overline{z}, \widetilde{f}, \boldsymbol{v}) = (0, 0, 0, \boldsymbol{v}^0)$ corresponds to the double critical point $(\boldsymbol{u}_\mathrm{c}^0, f_\mathrm{c}^0, \boldsymbol{v}^0)$, we have

$$A_{00}(0, \boldsymbol{v}^0) \;=\; A_{10}(0, \boldsymbol{v}^0) \;=\; A_{01}(0, \boldsymbol{v}^0) = 0. \tag{8.43}$$

We further expand $A_{pq}(\widetilde{f}, \boldsymbol{v})$ around $(\widetilde{f}, \boldsymbol{v}) = (0, \boldsymbol{v}^0)$ as

$$A_{pq}(\widetilde{f}, \boldsymbol{v}^0 + \varepsilon \boldsymbol{d}) \approx \sum_{j=0} \sum_{k=0} A_{pqjk}(\boldsymbol{d}) \widetilde{f}^j \varepsilon^k, \qquad p, q = 0, 1, \ldots, \tag{8.44}$$

where we often write

$$A_{pqjk} = A_{pqjk}(\boldsymbol{d})$$

for brevity; note $A_{pq00} = A_{pq}(0, \boldsymbol{v}^0)$. Substitution of (8.44) into (8.42) results in

$$F(z, \overline{z}, \widetilde{f}, \boldsymbol{v}) \approx \sum_{p=0} \sum_{q=0} \sum_{j=0} \sum_{k=0} A_{pqjk} \widetilde{f}^j \varepsilon^k z^p \overline{z}^q. \tag{8.45}$$

The asymptotic influence of the imperfection on F is governed by the term $A_{0001}\varepsilon$ in (8.45). The *imperfection coefficient* A_{0001}, dependent on the imperfection pattern vector $\boldsymbol{d}$, can be expressed as follows.

Lemma 8.2.

$$A_{0001} = \boldsymbol{\xi}_1^\top B_\mathrm{c}^0 \boldsymbol{d} + \mathrm{i} \boldsymbol{\xi}_2^\top B_\mathrm{c}^0 \boldsymbol{d}, \tag{8.46}$$

where $\boldsymbol{\xi}_1$ and $\boldsymbol{\xi}_2$ are orthonormal vectors such that $\boldsymbol{\xi}_1^\top J_\mathrm{c}^0 = \boldsymbol{\xi}_2^\top J_\mathrm{c}^0 = \boldsymbol{0}^\top$.

Proof. Putting

$$\widehat{F}_i(w_1, w_2, \widetilde{f}, \varepsilon) = \widetilde{F}_i(w_1, w_2, \widetilde{f}, \boldsymbol{v}^0 + \varepsilon \boldsymbol{d}), \qquad i = 1, 2,$$

in (8.37) with $\boldsymbol{v} = \boldsymbol{v}^0 + \varepsilon \boldsymbol{d}$, we see

$$\mathrm{Re}(A_{0001}) = \frac{\partial \widehat{F}_1}{\partial \varepsilon}(0,0,0,0), \qquad \mathrm{Im}(A_{0001}) = \frac{\partial \widehat{F}_2}{\partial \varepsilon}(0,0,0,0), \tag{8.47}$$

where $\mathrm{Re}(\cdot)$ and $\mathrm{Im}(\cdot)$ mean the real and imaginary parts of a complex number, respectively. On the other hand, we have

$$\widehat{F}_i(w_1, w_2, \widetilde{f}, \varepsilon) = \boldsymbol{\xi}_i^\top \boldsymbol{F}(\boldsymbol{u}_\mathrm{c}^0 + \boldsymbol{w} + \boldsymbol{\varphi}(\boldsymbol{w}, \widetilde{f}, \boldsymbol{v}^0 + \varepsilon \boldsymbol{d}), f_\mathrm{c}^0 + \widetilde{f}, \boldsymbol{v}^0 + \varepsilon \boldsymbol{d}), \qquad i = 1,2,$$

(cf., (7.66) with (7.72)) and, therefore,

$$\frac{\partial \widehat{F}_i}{\partial \varepsilon}(0,0,0,0) = \boldsymbol{\xi}_i^\top (J_\mathrm{c}^0 \frac{\partial \boldsymbol{\varphi}}{\partial \boldsymbol{v}}(\boldsymbol{0}, 0, \boldsymbol{v}^0) + B_\mathrm{c}^0)\boldsymbol{d} = \boldsymbol{\xi}_i^\top B_\mathrm{c}^0 \boldsymbol{d}, \qquad i = 1,2 \tag{8.48}$$

with $\boldsymbol{\xi}_i^\top J_\mathrm{c}^0 = \boldsymbol{0}^\top$. Substitution of (8.48) into (8.47) completes the proof of (8.46). □

8.5.3 Equivariance

The equivariance (8.34) to $G = \mathrm{D}_n$ of the bifurcation equation at the double bifurcation point of the perfect system is expressed as follows. We assume that the double point is a group-theoretic double point and denote by $(j)_{\mathrm{D}_n}$ the associated two-dimensional irreducible representation. Recall from (8.25) with (8.9) that we have chosen an orthonormal basis $\{\boldsymbol{\eta}_1, \boldsymbol{\eta}_2\}$ of $\ker(J_\mathrm{c}^0)$ such that

$$T(\sigma)[\boldsymbol{\eta}_1, \boldsymbol{\eta}_2] = [\boldsymbol{\eta}_1, \boldsymbol{\eta}_2]\begin{pmatrix} 1 & 0 \\ 0 & -1 \end{pmatrix},$$

$$T(c(2\pi/n))[\boldsymbol{\eta}_1, \boldsymbol{\eta}_2] = [\boldsymbol{\eta}_1, \boldsymbol{\eta}_2]\begin{pmatrix} \cos(2\pi \widehat{j}/\widehat{n}) & -\sin(2\pi \widehat{j}/\widehat{n}) \\ \sin(2\pi \widehat{j}/\widehat{n}) & \cos(2\pi \widehat{j}/\widehat{n}) \end{pmatrix},$$

where $\widehat{n} = n/\gcd(n,j) \geq 3$ (cf., (8.23)) and $\widehat{j} = j/\gcd(n,j)$. Therefore, the action of D_n on (w_1, w_2), which is defined by

$$(T(g)[\boldsymbol{\eta}_1, \boldsymbol{\eta}_2])\begin{pmatrix} w_1 \\ w_2 \end{pmatrix} = [\boldsymbol{\eta}_1, \boldsymbol{\eta}_2]\left(g \cdot \begin{pmatrix} w_1 \\ w_2 \end{pmatrix}\right), \qquad g \in \mathrm{D}_n,$$

is given by

$$\sigma \cdot \begin{pmatrix} w_1 \\ w_2 \end{pmatrix} = \begin{pmatrix} w_1 \\ -w_2 \end{pmatrix},$$

$$c(2\pi/n) \cdot \begin{pmatrix} w_1 \\ w_2 \end{pmatrix} = \begin{pmatrix} \cos(2\pi \widehat{j}/\widehat{n}) & -\sin(2\pi \widehat{j}/\widehat{n}) \\ \sin(2\pi \widehat{j}/\widehat{n}) & \cos(2\pi \widehat{j}/\widehat{n}) \end{pmatrix}\begin{pmatrix} w_1 \\ w_2 \end{pmatrix}.$$

In terms of $(z, \overline{z}) = (w_1 + \mathrm{i}w_2, w_1 - \mathrm{i}w_2)$, this can be rewritten as

$$\begin{aligned} \sigma \cdot z &= \overline{z}, & \sigma \cdot \overline{z} &= z, \\ c(2\pi/n) \cdot z &= \omega z, & c(2\pi/n) \cdot \overline{z} &= \overline{\omega} \overline{z}, \end{aligned} \tag{8.49}$$

where

$$\omega = \exp(\mathrm{i} 2\pi \widehat{j}/\widehat{n}). \tag{8.50}$$

Then the equivariance (8.34) with $\boldsymbol{v} = \boldsymbol{v}^0$ is equivalent to

$$\overline{F(z, \overline{z}, \widetilde{f}, \boldsymbol{v}^0)} = F(\overline{z}, z, \widetilde{f}, \boldsymbol{v}^0), \tag{8.51}$$

$$\omega F(z, \overline{z}, \widetilde{f}, \boldsymbol{v}^0) = F(\omega z, \overline{\omega} \overline{z}, \widetilde{f}, \boldsymbol{v}^0). \tag{8.52}$$

Recall the assumption $S(g)\boldsymbol{v}^0 = \boldsymbol{v}^0$ ($g \in \mathrm{D}_n$) in (7.39).

We now express this equivariance in terms of the power series expansion (8.42). From (8.51), we see that

$$A_{pq}(\widetilde{f}, \boldsymbol{v}^0) \in \mathbb{R}, \qquad p, q = 0, 1, \ldots. \tag{8.53}$$

Substitution of (8.42) into (8.52) yields

$$\sum_{p=0} \sum_{q=0} A_{pq}(\widetilde{f}, \boldsymbol{v}^0) z^p \overline{z}^q \left[\exp\left(\mathrm{i} 2\pi \widehat{j} \frac{p-q-1}{\widehat{n}} \right) - 1 \right] \approx 0.$$

Noting that $\widehat{n}$ and $\widehat{j}$ are mutually prime, we see that

$$A_{pq}(\widetilde{f}, \boldsymbol{v}^0) = 0 \qquad \text{unless} \ \ p - q - 1 = m\widehat{n}, \ \ m \in \mathbb{Z}, \tag{8.54}$$

where $\mathbb{Z}$ is the set of integer numbers, and $p - q - 1 = m\widehat{n}$ means that $p - q - 1$ is a multiple of $\widehat{n}$. The D_n-equivariance of the perfect system, consequently, is expressed by (8.53) and (8.54).

Using (8.54) in (8.42), we obtain the bifurcation equation for the perfect system as

$$\begin{aligned} F(z, \overline{z}, \widetilde{f}, \boldsymbol{v}^0) \approx{}& \sum_{q=0} A_{q+1,q}(\widetilde{f}, \boldsymbol{v}^0) z^{q+1} \overline{z}^q \\ &+ \sum_{m=1} \sum_{q=0} [A_{q+1+m\widehat{n},q}(\widetilde{f}, \boldsymbol{v}^0) z^{q+1+m\widehat{n}} \overline{z}^q + A_{q,q-1+m\widehat{n}}(\widetilde{f}, \boldsymbol{v}^0) z^q \overline{z}^{q-1+m\widehat{n}}]. \end{aligned} \tag{8.55}$$

Therein, $A_{q+1,q}(\widetilde{f}, \boldsymbol{v}^0)$, $A_{q+1+m\widehat{n},q}(\widetilde{f}, \boldsymbol{v}^0)$, and $A_{q,q-1+m\widehat{n}}(\widetilde{f}, \boldsymbol{v}^0)$ are real by (8.53) and generically distinct from zero (as there is no reason for the disappearance of these terms).

For an imperfect system described by (8.42) with (8.44), we have $A_{10}(\widetilde{f}, \boldsymbol{v}^0) = A_{1010}\widetilde{f} + \text{h.o.t.}$ by (8.43) and therefore we have

$$F(z,\overline{z},\widetilde{f},\nu^0+\varepsilon \boldsymbol{d}) \approx A_{1010}\widetilde{f}z + \sum_{1\le q\le \widehat{n}/2-1} A_{q+1,q00}z^{q+1}\overline{z}^q + A_{0,\widehat{n}-1,00}\,\overline{z}^{\widehat{n}-1} + A_{0001}\varepsilon. \tag{8.56}$$

Therein, A_{1010}, $A_{q+1,q00}$, and $A_{0,\widehat{n}-1,00}$ in (8.56) are real by (8.53), generically distinct from zero, and independent of the imperfection pattern vector $\boldsymbol{d}$. In contrast, the imperfection coefficient A_{0001} is generically complex and distinct from zero,[5] and is dependent on $\boldsymbol{d}$ as presented in Lemma 8.2 in §8.5.2.

8.5.4 Reciprocity

In §7.5.2 reciprocity was shown to be preserved in the Liapunov–Schmidt reduction in general. For the equation (8.42) in the complex coordinates $(z,\overline{z})$, the condition of reciprocity (7.56) can be expressed as follows.

Lemma 8.3. *The bifurcation equation* (7.52) *at a double bifurcation point is reciprocal if and only if $\partial F/\partial z$ is real for F in* (8.37). *This condition is equivalent to*

$$(p+1)A_{p+1,q}(\widetilde{f},\nu) = (q+1)\overline{A_{q+1,p}(\widetilde{f},\nu)}, \qquad p,q = 0,1,2,\ldots \tag{8.57}$$

for the coefficients $A_{p,q}(\widetilde{f},\nu)$ in (8.42).

Proof. By (8.41) we have

$$\frac{\partial \widetilde{F}_2}{\partial w_1} = \frac{\partial \widetilde{F}_1}{\partial w_2} \iff \frac{\partial F}{\partial z} = \frac{\partial \overline{F}}{\partial \overline{z}}.$$

With the use of F in (8.42), the latter condition is expressed as (8.57). □

It is emphasized that the statement above is independent of group-symmetry, and in particular, whether the double point is parametric or group-theoretic. An important consequence of (8.57) is that

$$A_{p+1,p}(\widetilde{f},\nu) \in \mathbb{R}, \qquad p = 0,1,2,\ldots, \tag{8.58}$$

which fact plays a key role in the analysis of a double point on a C_n-symmetric path (see Remark 8.5 in §8.6.1). In D_n-symmetric cases, however, the reciprocity (8.57) does not add much to the D_n-equivariance represented by (8.53) and (8.54); for example, (8.58) is implied by (8.53).

[5] A group-theoretic condition for the vanishing of A_{0001} is given in Lemma 9.1 in §9.2.2.

8.6 Perfect Behavior Near a Double Critical Point

Perfect bifurcation behavior, in a neighborhood of the double critical point $(\boldsymbol{u}_{\mathrm{c}}^0, f_{\mathrm{c}}^0)$ on a D_n-symmetric path, is investigated. The stability of the solutions is also discussed.

Investigation of perfect bifurcation behavior amounts to solving the bifurcation equation (8.38) for $\boldsymbol{v} = \boldsymbol{v}^0$, which reads as

$$F(z, \overline{z}, \widetilde{f}, \boldsymbol{v}^0) = \overline{F(z, \overline{z}, \widetilde{f}, \boldsymbol{v}^0)} = 0, \tag{8.59}$$

where F is given by (8.55). In this section, we use the abbreviations

$$F(z, \overline{z}, \widetilde{f}) = F(z, \overline{z}, \widetilde{f}, \boldsymbol{v}^0), \qquad A_{pq}(\widetilde{f}) = A_{pq}(\widetilde{f}, \boldsymbol{v}^0).$$

The equation (8.59) has the trivial solution $z = 0$, corresponding to the D_n-symmetric fundamental path, since each term in (8.55) vanishes if $z = \overline{z} = 0$.

8.6.1 Bifurcated Branches

The space view of the equilibrium paths for the triangular truss dome is depicted in Fig. 8.9 (repeated from Fig. 8.6(a) in §8.4.1). The vertical axis corresponds to the trivial solution, and there is a double bifurcation point denoted by △, from which bifurcated paths branch. We would like to investigate the asymptotic behavior of such bifurcated paths through the analysis of (8.59) for a general D_n-symmetric system.

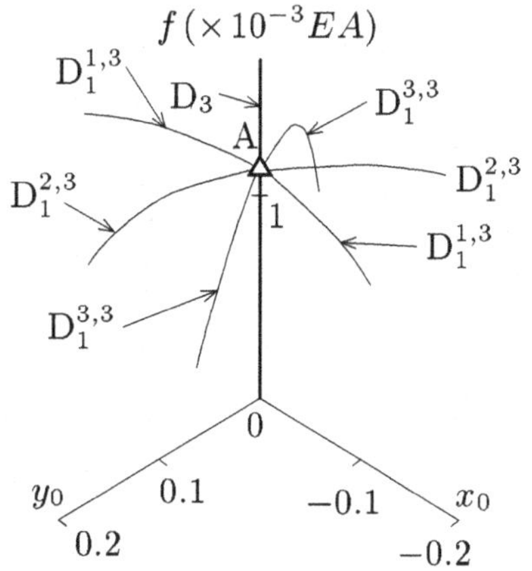

Fig. 8.9 Space view of the equilibrium paths for the triangular truss dome in §8.4.1. △: double bifurcation point ($\widehat{n} = 3$).

Asymptotic Forms of Bifurcated Branches

The nontrivial solution of (8.59) is determined from $F/z = 0$. If we put

$$\widetilde{F}(r,\theta,\widetilde{f}) = \frac{F(r\exp(\mathrm{i}\theta), r\exp(-\mathrm{i}\theta), \widetilde{f})}{r\exp(\mathrm{i}\theta)} \quad \left(= \frac{F}{z}\right)$$

using the polar coordinates $z = w_1 + \mathrm{i}w_2 = r\exp(\mathrm{i}\theta)$ $(r \geq 0)$, then we have

$$\begin{aligned}\mathrm{Re}(\widetilde{F}) \approx & \sum_{q=0} A_{q+1,q}(\widetilde{f})r^{2q} \\ & + \sum_{m=1}\sum_{q=0} [A_{q+1+m\widehat{n},q}(\widetilde{f})r^{2q+m\widehat{n}} + A_{q,q-1+m\widehat{n}}(\widetilde{f})r^{2(q-1)+m\widehat{n}}]\cos(m\widehat{n}\theta), \\ \mathrm{Im}(\widetilde{F}) \approx & \sum_{m=1}\sum_{q=0} [A_{q+1+m\widehat{n},q}(\widetilde{f})r^{2q+m\widehat{n}} - A_{q,q-1+m\widehat{n}}(\widetilde{f})r^{2(q-1)+m\widehat{n}}]\sin(m\widehat{n}\theta).\end{aligned}$$

Then the nontrivial solution of (8.59) is determined by $\mathrm{Re}(\widetilde{F}) = \mathrm{Im}(\widetilde{F}) = 0$.

Equation $\mathrm{Im}(\widetilde{F}) = 0$ is satisfied by $\theta = \alpha_k$ $(k = 1,\dots,2\widehat{n})$ with

$$\alpha_k = -\pi\,\frac{k-1}{\widehat{n}}, \qquad k = 1,\dots,2\widehat{n} \tag{8.60}$$

in (8.30), since, for $\theta = \alpha_k$, we have $\sin(m\widehat{n}\theta) = \sin(-m(k-1)\pi) = 0$.

For each k, the relation between $\widetilde{f}$ and r is determined from the other equation $\mathrm{Re}(\widetilde{F}) = 0$ as an implicit function $\widetilde{f} = f_k(r)$ in a neighborhood of $(r,\widetilde{f}) = (0,0)$. Inasmuch as $\cos(-m(k-1)\pi) = (-1)^{m(k-1)}$, equation $\mathrm{Re}(\widetilde{F}) = 0$ reads as

$$\begin{aligned}& \sum_{q=0} A_{q+1,q}(\widetilde{f})r^{2q} \\ & + \sum_{m=1}\sum_{q=0} (-1)^{m(k-1)}[A_{q+1+m\widehat{n},q}(\widetilde{f})r^{2q+m\widehat{n}} + A_{q,q-1+m\widehat{n}}(\widetilde{f})r^{2(q-1)+m\widehat{n}}] \approx 0.\end{aligned} \tag{8.61}$$

The equation for k is determined by the parity of k. Therefore, we have $f_{2k-1}(r) = f_1(r)$ and $f_{2k}(r) = f_2(r)$ for $k = 1,\dots,\widehat{n}$. That is, there exist two distinct sets of half-branches denoted by $\widetilde{f} = f_1(r)$ and $\widetilde{f} = f_2(r)$ bifurcating in the directions of $\theta = \alpha_{2k-1}$ and $\theta = \alpha_{2k}$ $(k = 1,\dots,\widehat{n})$, respectively. Altogether, there exist $2\widehat{n}$ half-branches—$\widehat{n}$ bifurcated paths.

Lemma 8.4. *The asymptotic form of the function f_k, when r is small, is given by*

$$f_k(r) = \begin{cases} (-1)^k \dfrac{A_{0200}}{A_{1010}} r + \mathrm{O}(r^2) & \text{if } \widehat{n} = 3, \\ -\dfrac{A_{2100} + (-1)^{k-1}A_{0300}}{A_{1010}} r^2 + \mathrm{O}(r^4) & \text{if } \widehat{n} = 4, \\ -\dfrac{A_{2100}}{A_{1010}} r^2 + \mathrm{O}(r^4) & \text{if } \widehat{n} \geq 5, \end{cases} \tag{8.62}$$

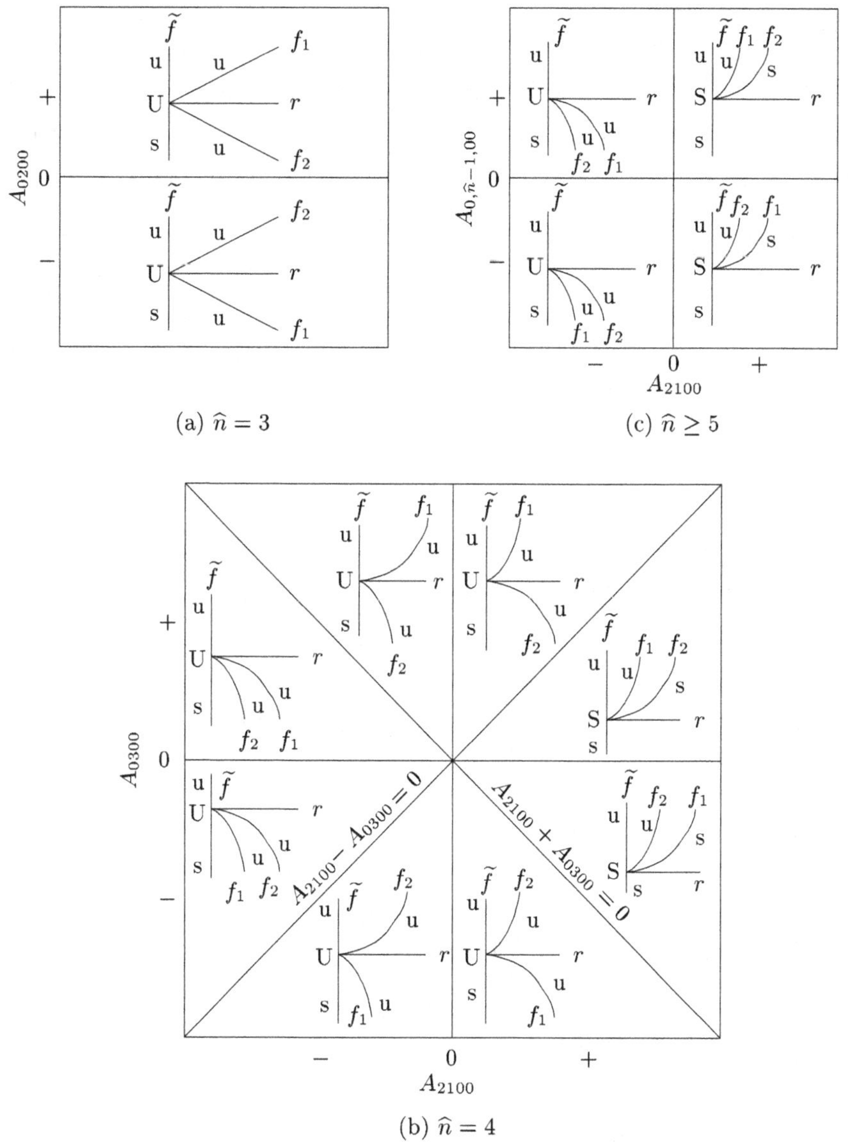

Fig. 8.10 Categorization of local bifurcation behavior at a group-theoretic double bifurcation point on a D_n-symmetric path with $A_{1010} < 0$. s: stable half-branch; u: unstable half-branch; S: stable point; U: unstable point.

where the leading terms of f_1 and f_2 are identical in the case of $\widehat{n} \geq 5$.

Proof. With the use of asymptotic relations

$$A_{10}(\widetilde{f}) \approx A_{1000} + A_{1010}\widetilde{f} = A_{1010}\widetilde{f}, \tag{8.63}$$

$$A_{pq}(\widetilde{f}) \approx A_{pq00}, \qquad p-q-1 = m\widehat{n}, \; m \in \mathbb{Z}, \; (p,q) \neq (1,0) \tag{8.64}$$

(where $A_{1000} = 0$ by (8.43)), (8.61) simplifies to

$$A_{1010}\widetilde{f} + A_{2100}r^2 + (-1)^{k-1}A_{0,\widehat{n}-1,00}r^{\widehat{n}-2} \approx 0. \tag{8.65}$$

For $\widehat{n} = 3$, we can further omit $A_{2100}r^2$, a higher-order term, in (8.65) to obtain the first of (8.62). The remaining cases $\widehat{n} = 4$ and $\widehat{n} \geq 5$ can be dealt with similarly. □

The expression (8.62) demonstrates the following.

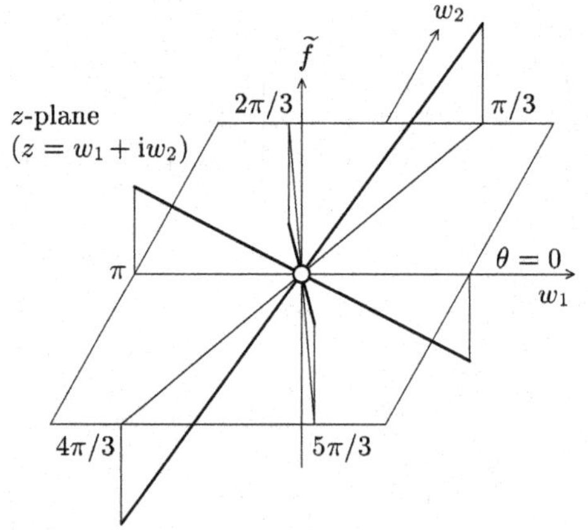

(a) Spatial view

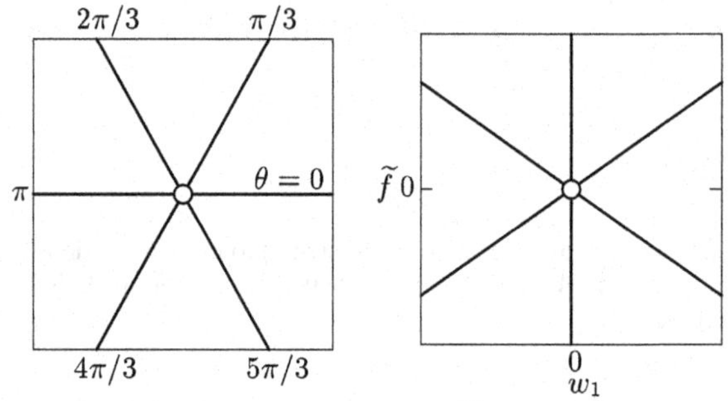

(b) z-plane view

$(\widetilde{f}, w_1)$-plane view

Fig. 8.11 Perfect bifurcation behavior in a neighborhood of the double bifurcation point (○) with $\widehat{n} = 3$ and $A_{0200}/A_{1010} > 0$.

- For $\widehat{n} = 3$, the leading terms of $f_1(r)$ and $f_2(r)$ have opposite signs, which means that f decreases toward a set of half-branches for $\widetilde{f} = f_1(r)$ and increases toward the other set for $\widetilde{f} = f_2(r)$, or vice versa.
- For $\widehat{n} = 4$, the signs of the coefficients $(A_{2100} \pm A_{0300})/A_{1010}$ determine the increase or decrease in f.
- For $\widehat{n} \geq 5$, f increases or decreases simultaneously for all half-branches according to whether A_{2100}/A_{1010} is negative or positive.

The local bifurcation behavior at the double bifurcation point investigated above is categorized in Fig. 8.10 for $A_{1010} < 0$ when it is nondegenerate in the sense that the coefficients A_{1010}, A_{0200}, A_{2100}, $A_{2100} \pm A_{0300}$, and $A_{0,\widehat{n}-1,00}$ are distinct from zero. The local bifurcation behavior at the double bifurcation point with $\widehat{n} = 3$ is depicted in Fig. 8.11, where (a) shows the spatial view, (b) portrays a plane view projected onto the z-plane, and (c) is a plane view projected onto the plane of $(w_1, \widetilde{f})$. Spatial views for $\widehat{n} = 4$ and 6 are portrayed in Fig. 8.12.

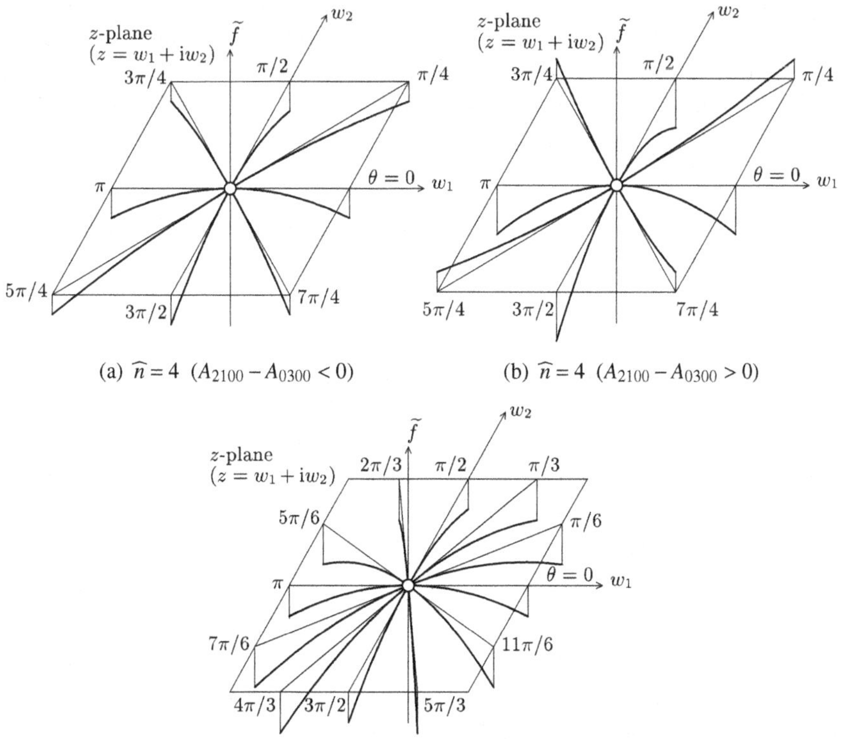

(a) $\widehat{n} = 4$ $(A_{2100} - A_{0300} < 0)$

(b) $\widehat{n} = 4$ $(A_{2100} - A_{0300} > 0)$

(c) $\widehat{n} = 6$ $(A_{1010} < 0,\ A_{2100}/A_{1010} > 0)$

Fig. 8.12 Spatial view of the perfect bifurcation behavior in a neighborhood of the double bifurcation points (○) with $\widehat{n} = 4$ and 6 $(A_{1010} < 0,\ A_{0300} < 0,\ A_{2100} + A_{0300} < 0)$.

Symmetry of Branches

The symmetry of the bifurcation solutions is considered. From (8.49) the action of D_n on z is given by

$$\sigma \cdot z = \bar{z}, \qquad c(2\pi/n)\cdot z = \omega z$$

with $\omega = \exp(\mathrm{i}2\pi \widehat{j}/\widehat{n})$. The symmetry of the solution $z = r\exp(\mathrm{i}\theta)$ with $\theta = \alpha_{i+\widehat{n}j} = -\pi(i-1+\widehat{n}j)/\widehat{n}$ is described by the isotropy subgroup

$$\Sigma(z) = \{g \in D_n \mid g\cdot z = z\} = D^{k,n}_{n/\widehat{n}} \tag{8.66}$$

for some k (for the canonical choice of $\theta = \alpha_1 (= 0)$, we have $k = 1$).

Remark 8.5. For a group-theoretic double bifurcation point on a C_n-symmetric path, the results presented above hold with minor modifications. The fundamental difference is that we have (8.54) only, and not (8.53) arising from reflection symmetry. Hence $F(z,\bar{z},\widetilde{f})$ is written as (8.55) but with the complex coefficients $A_{pq}(\widetilde{f})$.

The argument below shows that bifurcating solutions might not necessarily exist at a double bifurcation point on a C_n-symmetric path. Nonetheless, the solution does exist when the system under consideration is reciprocal, since the additional condition $A_{p+1,p}(\widetilde{f},\nu^0) \in \mathbb{R}$ $(p = 0,1,2,\ldots)$ in (8.58) implies the conditions (8.71) for $\widehat{n} = 4$ and (8.74) for $\widehat{n} \geq 5$ below. In this way, the reciprocity plays a major role for C_n-symmetric systems.[6]

For a nonreciprocal system, since $A_{q+1,q}(\widetilde{f})$ $(q = 0,1,\ldots)$ are complex in general in (8.55), the existence of the solutions for the equation $F/z = 0$ is dependent on cases. In fact, from (8.55) we have

$$F/z = A_{10}(\widetilde{f}) + A_{21}(\widetilde{f})|z|^2 + A_{0,\widehat{n}-1}(\widetilde{f})\bar{z}^{\widehat{n}}/|z|^2 + \text{h.o.t.} \tag{8.67}$$

With the use of (8.63), (8.64), and the polar coordinates $z = r\exp(\mathrm{i}\theta)$, equation $F/z = 0$ with (8.67) yields

$$\widetilde{f} = -\frac{A_{2100}}{A_{1010}} r^2 - \frac{A_{0,\widehat{n}-1,00}}{A_{1010}} \exp(-\mathrm{i}\widehat{n}\theta) r^{\widehat{n}-2} + \text{h.o.t.} \tag{8.68}$$

For $\widehat{n} = 3$, (8.68) becomes

$$\widetilde{f} \approx -\frac{A_{0200}}{A_{1010}} \exp(-3\mathrm{i}\theta) r.$$

Since $\widetilde{f}$ and r are real, there exist six half-branches in the directions of

$$\theta = -\pi\frac{k-1}{3} + \frac{1}{3}\arg\left(\frac{A_{0200}}{A_{1010}}\right), \qquad k = 1,\ldots,6, \tag{8.69}$$

where $\arg(\cdot)$ is the argument of the complex number therein.

[6] See also Krasnosel'skii, 1964 [126] and Poston and Stewart, 1978 [168].

For $\widehat{n} = 4$, (8.68) becomes

$$\widetilde{f} \approx -\left(\frac{A_{2100}}{A_{1010}} + \frac{A_{0300}}{A_{1010}} \exp(-4\mathrm{i}\theta)\right) r^2. \tag{8.70}$$

Since $\widetilde{f}$ and r are real, (8.70) has eight half-branches when

$$\left|\mathrm{Im}\left(\frac{A_{2100}}{A_{1010}}\right)\right| < \left|\frac{A_{0300}}{A_{1010}}\right| \tag{8.71}$$

is satisfied; in contrast, it has no solution if the reverse inequality holds.

For $\widehat{n} \geq 5$, (8.68) becomes

$$\widetilde{f} \approx -\frac{A_{2100}}{A_{1010}} r^2 - \frac{A_{0,\widehat{n}-1,00}}{A_{1010}} \exp(-\mathrm{i}\widehat{n}\theta)\, r^{\widehat{n}-2}. \tag{8.72}$$

Since $\widetilde{f}$ and r are real, (8.72) has $2\widehat{n}$ half-branches in the directions of

$$\theta = -\pi\frac{k-1}{\widehat{n}} + \frac{1}{\widehat{n}}\arg\left(\frac{A_{0,\widehat{n}-1,00}}{A_{1010}}\right), \qquad k = 1, \ldots, 2\widehat{n}, \tag{8.73}$$

if

$$\frac{A_{2100}}{A_{1010}} \in \mathbb{R} \tag{8.74}$$

is satisfied. In contrast, it has no solution otherwise. □

8.6.2 Stability

We investigate the stability of a double bifurcation point on a D_n-symmetric path and that of the bifurcated paths by means of an "asymptotic potential." Another method of the stability analysis free from an asymptotic potential is given in §8.10, the appendix of this chapter.

Asymptotic Potential

For $F(z, \overline{z}, \widetilde{f}, v^0)$ in (8.55) we define its leading part by

$$F_{\mathrm{L}}(z, \overline{z}, \widetilde{f}) = \sum_{0 \leq q \leq \widehat{n}/2-1} A_{q+1,q}(\widetilde{f}) z^{q+1}\overline{z}^q + A_{0,\widehat{n}-1}(\widetilde{f})\overline{z}^{\widehat{n}-1}, \tag{8.75}$$

and put

$$F_{\mathrm{L}1}(w_1, w_2, \widetilde{f}) = \mathrm{Re}(F_{\mathrm{L}}(z, \overline{z}, \widetilde{f})), \qquad F_{\mathrm{L}2}(w_1, w_2, \widetilde{f}) = \mathrm{Im}(F_{\mathrm{L}}(z, \overline{z}, \widetilde{f})).$$

A key observation here is that the leading part (8.75) satisfies the condition of reciprocity (8.57) as a result of (8.53). Therefore, we have

$$F_{\mathrm{L}i}(w_1, w_2, \widetilde{f}) = \frac{\partial U}{\partial w_i}, \qquad i = 1, 2,$$

for some function $U(w_1, w_2, \widetilde{f})$, which we name the *asymptotic potential.* Thus the D_n-equivariance implies the existence of a potential function in an asymptotic sense.

It is convenient to consider the asymptotic potential

$$\widetilde{U}(r, \theta, \widetilde{f}) = U(r\cos\theta, r\sin\theta, \widetilde{f})$$

in the polar coordinates (r, θ). It follows from

$$\begin{aligned}
\frac{\partial \widetilde{U}}{\partial r} &= \frac{\partial w_1}{\partial r}\frac{\partial U}{\partial w_1} + \frac{\partial w_2}{\partial r}\frac{\partial U}{\partial w_2} \\
&= \cos\theta \cdot F_{\mathrm{L}1} + \sin\theta \cdot F_{\mathrm{L}2} \\
&= \mathrm{Re}(\exp(-\mathrm{i}\theta) F_{\mathrm{L}}) \\
&= \sum_{0 \le q \le \widehat{n}/2-1} A_{q+1,q}(\widetilde{f}) r^{2q+1} + A_{0,\widehat{n}-1}(\widetilde{f}) r^{\widehat{n}-1} \cos(\widehat{n}\theta)
\end{aligned}$$

that

$$\begin{aligned}
\widetilde{U}(r, \theta, \widetilde{f}) &= \int_0^r \frac{\partial \widetilde{U}}{\partial r} dr \\
&= \sum_{0 \le q \le \widehat{n}/2-1} \frac{A_{q+1,q}(\widetilde{f})}{2(q+1)} r^{2(q+1)} + \frac{A_{0,\widehat{n}-1}(\widetilde{f})}{\widehat{n}} r^{\widehat{n}} \cos(\widehat{n}\theta) \\
&\approx \frac{A_{1010}}{2} \widetilde{f} r^2 + \sum_{1 \le q \le \widehat{n}/2-1} \frac{A_{q+1,q}(\widetilde{f})}{2(q+1)} r^{2(q+1)} + \frac{A_{0,\widehat{n}-1}(\widetilde{f})}{\widehat{n}} r^{\widehat{n}} \cos(\widehat{n}\theta).
\end{aligned} \tag{8.76}$$

The stability of the bifurcation point and the half-branches are considered below with reference to this asymptotic potential $\widetilde{U}$. It is assumed that the trivial solution $r = 0$ of the perfect system is stable for $\widetilde{f} < 0$, that is, that

$$A_{1010} < 0. \tag{8.77}$$

Stability of Bifurcation Point

To investigate the stability of the bifurcation point $(r, \widetilde{f}) = (0, 0)$, we set $\widetilde{f} = 0$ in the asymptotic potential $\widetilde{U}$ in (8.76) to obtain

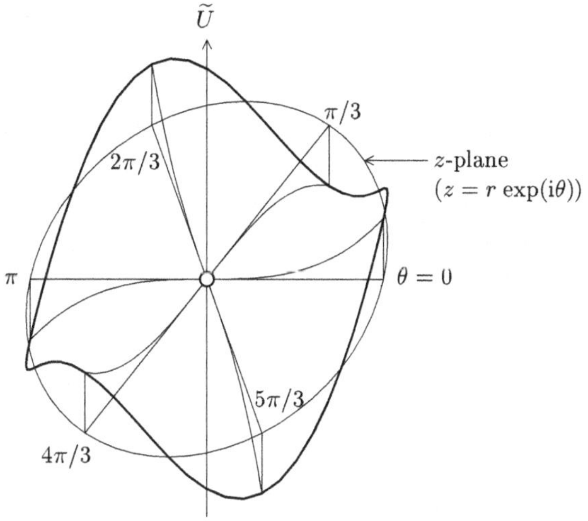

Fig. 8.13 Spatial view of an asymptotic potential function $\widetilde{U}(r,\theta,0)$ in a neighborhood of the double bifurcation point (○) with $\widehat{n} = 3$ and $A_{0200} > 0$.

$$\widetilde{U}(r,\theta,0) \approx \begin{cases} \dfrac{A_{0200}}{3}\cos(3\theta)r^3 & \text{if } \widehat{n} = 3, \\ \dfrac{1}{4}[A_{2100} + A_{0300}\cos(4\theta)]\, r^4 & \text{if } \widehat{n} = 4, \\ \dfrac{A_{2100}}{4} r^4 & \text{if } \widehat{n} \geq 5 \end{cases}$$

in a neighborhood of the bifurcation point at $r = 0$. This point is stable if $\widetilde{U}(r,\theta,0)$ is minimized at the point (cf., §2.2.4); therefore, we have the following classification.

- For $\widehat{n} = 3$, the potential $\widetilde{U}(r,\theta,0)$ has no local minimum at the point, as presented in Fig. 8.13, and is therefore unstable.
- For $\widehat{n} = 4$, it is stable if $A_{2100} \pm A_{0300} > 0$ but otherwise it is unstable.
- For $\widehat{n} \geq 5$, it is stable if A_{2100} is positive but unstable if negative.

Stability of Branches

We consider the stability of the half-branches, which are described by $\theta = \alpha_k$ in (8.60) and $\widetilde{f} = f_k(r)$ in (8.62) for $k = 1,\ldots,2\widehat{n}$. For a given $\tilde{f}$, the corresponding solution (r,θ) is determined as a stationary point of the potential $\widetilde{U}(r,\theta,\widetilde{f})$ in the space of (r,θ).

The point (r,θ) is stable if $\widetilde{U}(r,\theta,\widetilde{f})$ is minimized at the point, which is the case if the Hessian matrix of $\widetilde{U}(r,\theta,\widetilde{f})$

$$\begin{pmatrix} \dfrac{\partial^2 \widetilde{U}}{\partial r^2} & \dfrac{\partial^2 \widetilde{U}}{\partial r\,\partial\theta} \\ \dfrac{\partial^2 \widetilde{U}}{\partial\theta\,\partial r} & \dfrac{\partial^2 \widetilde{U}}{\partial\theta^2} \end{pmatrix} \tag{8.78}$$

is positive-definite at the point (cf., §2.2.4). Direct calculation from (8.76) gives its entries as

$$\frac{\partial^2 \widetilde{U}}{\partial r^2} = \sum_{0\le q\le \widehat{n}/2-1} (2q+1)A_{q+1,q}(\widetilde{f})r^{2q} + (\widehat{n}-1)A_{0,\widehat{n}-1}(\widetilde{f})r^{\widehat{n}-2}\cos(\widehat{n}\theta),$$

$$\frac{\partial^2 \widetilde{U}}{\partial \theta^2} = -\widehat{n}A_{0,\widehat{n}-1}(\widetilde{f})r^{\widehat{n}}\cos(\widehat{n}\theta),$$

$$\frac{\partial^2 \widetilde{U}}{\partial r\,\partial \theta} = \frac{\partial^2 \widetilde{U}}{\partial \theta\,\partial r} = -\widehat{n}A_{0,\widehat{n}-1}(\widetilde{f})r^{\widehat{n}-1}\sin(\widehat{n}\theta).$$

For $\theta = \alpha_k$ the Hessian matrix in (8.78) is diagonal, since $\partial^2\widetilde{U}/\partial r\,\partial\theta = \partial^2\widetilde{U}/\partial\theta\,\partial r = 0$. For the diagonal entries, we use (8.62) as well as $\theta = \alpha_k$ to obtain

$$\frac{\partial^2 \widetilde{U}}{\partial r^2}(r,\alpha_k,\widetilde{f}) \approx A_{1010}\widetilde{f} + 3A_{2100}r^2 + (-1)^{k-1}(\widehat{n}-1)A_{0,\widehat{n}-1,00}r^{\widehat{n}-2}$$
$$\approx \begin{cases} (-1)^{k-1}A_{0200}r & \text{if } \widehat{n}=3, \\ 2[A_{2100}+(-1)^{k-1}A_{0300}]r^2 & \text{if } \widehat{n}=4, \\ 2A_{2100}r^2 & \text{if } \widehat{n}\ge 5, \end{cases}$$

$$\frac{\partial^2 \widetilde{U}}{\partial \theta^2}(r,\alpha_k,\widetilde{f}) \approx (-1)^k\widehat{n}A_{0,\widehat{n}-1,00}r^{\widehat{n}}.$$

Therefore, the Hessian matrix is positive-definite (in the asymptotic sense) if and only if

$$\begin{cases} (-1)^{k-1}A_{0200} > 0 & \text{if } \widehat{n}=3, \\ A_{2100}+(-1)^{k-1}A_{0300} > 0 & \text{if } \widehat{n}=4, \\ A_{2100} > 0 & \text{if } \widehat{n}\ge 5, \end{cases} \tag{8.79}$$

and

$$(-1)^k A_{0,\widehat{n}-1,00} > 0. \tag{8.80}$$

Therefore, the half-branch for $\theta = \alpha_k$ is stable if (8.79) and (8.80) hold. Then we have the following.

- For $\widehat{n} = 3$, all half-branches are unstable.
- For $\widehat{n} = 4$, the stability of the bifurcated paths is dependent on the signs of A_{0300} and $A_{2100} \pm A_{0300}$.
- For $\widehat{n} \ge 5$, all half-branches are unstable if $A_{2100} < 0$. If $A_{2100} > 0$, the half-branches for $f_k(r)$ with k odd are unstable and those for $f_k(r)$ with k even are stable for $A_{0,\widehat{n}-1,00} > 0$, and vice versa for $A_{0,\widehat{n}-1,00} < 0$.

Discussion

Under the assumption $A_{1010} < 0$ in (8.77), the bifurcation point and the half-branches are all unstable

$$\begin{cases} \text{always} & \text{for } \widehat{n} = 3, \\ \text{if } A_{2100} - A_{0300} < 0 \ \text{ or } \ A_{2100} + A_{0300} < 0 & \text{for } \widehat{n} = 4, \\ \text{if } A_{2100} < 0 & \text{for } \widehat{n} \geq 5. \end{cases} \tag{8.81}$$

Recall that the stability of the bifurcation point and that of the half-branches are categorized in Fig. 8.10.

8.7 Imperfect Behavior Near a Double Critical Point

The local perfect bifurcation behavior near a double critical point was investigated in §8.6. Here, it is a logical sequel to investigate the imperfect behavior, which is to be obtained as the solution of the bifurcation equation

$$F(z, \overline{z}, \widetilde{f}, \boldsymbol{v}) = \overline{F(z, \overline{z}, \widetilde{f}, \boldsymbol{v})} = 0 \tag{8.82}$$

in (8.38). Emphasis is placed on the critical point on the fundamental path of an imperfect system; the imperfection sensitivity laws for this point are derived in §8.8.

It is assumed that $A_{1010} < 0$ as in (8.77), which indicates that the trivial solution $r = 0$ of the perfect system is stable for $\widetilde{f} < 0$. It is also assumed that the bifurcation point and paths are all unstable (cf., (8.81)).

8.7.1 Asymptotic Forms of Bifurcation Equations

To obtain the solution of (8.82) with (8.56), we put

$$\begin{aligned} \widehat{F}(z, \overline{z}, \widetilde{f}, \varepsilon) &\equiv F(z, \overline{z}, \widetilde{f}, \boldsymbol{v}^0 + \varepsilon \boldsymbol{d}) \\ &= A_{1010} \widetilde{f} z + \sum_{1 \leq q \leq \widehat{n}/2-1} A_{q+1,q00} z^{q+1} \overline{z}^{q} + A_{0,\widehat{n}-1,00} \overline{z}^{\widehat{n}-1} + A_{0001} \varepsilon + \text{h.o.t.} \end{aligned} \tag{8.83}$$

Rescaling the variables as

$$\begin{aligned} &z \leftarrow -A_{2100} z, && \overline{z} \leftarrow -A_{2100} \overline{z}, \\ &\widetilde{f} \leftarrow A_{1010} A_{2100} \widetilde{f}, && \varepsilon \leftarrow A_{2100}{}^2 \varepsilon, && \widehat{F} \leftarrow A_{2100}{}^2 \widehat{F}, \end{aligned} \tag{8.84}$$

and putting

$$a = A_{0001}, \qquad b = (-1)^{\widehat{n}-1}\frac{A_{0,\widehat{n}-1,00}}{{A_{2100}}^{\widehat{n}-3}}, \qquad c_q = -\frac{A_{q+1,q00}}{{A_{2100}}^{2q-1}}, \tag{8.85}$$

we simplify (8.83) to

$$\widehat{F} = -\widetilde{f}z - z^2\overline{z} + \sum_{2\le q\le \widehat{n}/2-1} c_q z^{q+1}\overline{z}^q + b\overline{z}^{\widehat{n}-1} + a\varepsilon + \text{h.o.t.} \tag{8.86}$$

We employ the polar coordinates

$$z = r\,\exp(\mathrm{i}\theta), \qquad \overline{z} = r\,\exp(-\mathrm{i}\theta), \qquad a\varepsilon = |a\varepsilon|\exp(\mathrm{i}\psi). \tag{8.87}$$

Then, with the explicit form of $\widehat{F}(z,\overline{z},\widetilde{f},\varepsilon)$ in (8.86), the bifurcation equation (8.82) for $z \neq 0$ can be rewritten as

$$\widetilde{f} \approx -r^2 + \sum_{2\le q\le \widehat{n}/2-1} c_q r^{2q} + br^{\widehat{n}-2}\cos(\widehat{n}\theta) + \frac{|a\varepsilon|}{r}\cos(\theta-\psi), \tag{8.88}$$

$$br^{\widehat{n}-1}\sin(\widehat{n}\theta) + |a\varepsilon|\sin(\theta-\psi) \approx 0. \tag{8.89}$$

In what follows, we investigate the solution paths that consist of $(r,\theta,\widetilde{f})$ satisfying these equations. For $|\varepsilon|$ small, the solution paths are close to those for the perfect system, as shown in Figs. 8.11 and 8.12.

Equation (8.89) does not contain $\widetilde{f}$ and expresses solution curves projected onto the z-plane, whereas (8.88) is solved for $\widetilde{f}$, and reduces to

$$\widetilde{f} \approx \begin{cases} br\cos(\widehat{n}\theta) + \dfrac{|a\varepsilon|}{r}\cos(\theta-\psi) & \text{for } \widehat{n}=3, \\ -r^2 + br^2\cos(\widehat{n}\theta) + \dfrac{|a\varepsilon|}{r}\cos(\theta-\psi) & \text{for } \widehat{n}=4, \\ -r^2 + \dfrac{|a\varepsilon|}{r}\cos(\theta-\psi) & \text{for } \widehat{n}\ge 5. \end{cases} \tag{8.90}$$

8.7.2 Projected Solution Curves

We concentrate first on (8.89) to investigate projected solution curves. According to the value of $\psi = \arg(a\varepsilon)$, two cases are distinguished:

- Special case: $\psi = \pi(m-1)/\widehat{n}$ for some $m = 1,\dots,2\widehat{n}$,
- General case: $\psi \neq \pi(m-1)/\widehat{n}$ for any $m = 1,\dots,2\widehat{n}$.

We assume $b<0$, since the change of the sign of b, $b \to -b$, exerts the same influence on (8.89) as $\psi \to \psi+\pi$, and the case for $b>0$ can be treated similarly.

Special Case

We consider the special case where $\psi = \pi(m-1)/\widehat{n}$ for some $m = 1,\ldots,2\widehat{n}$. Since

$$\sin(\widehat{n}\theta) = \sin(\theta - \psi) = 0 \tag{8.91}$$

has solutions $\theta = \psi,\ \psi + \pi$, equation (8.89) admits a pair of rays

$$\theta = \psi,\ \psi + \pi \qquad (r\text{: arbitrary}) \tag{8.92}$$

as solutions. Solution $\theta = \psi + \pi$ corresponds to the fundamental path since $\widetilde{f} \to -\infty$ as $r \to +0$; and the other solution $\theta = \psi$ corresponds to another path in the opposite direction since $\widetilde{f} \to +\infty$ as $r \to +0$. The $\widetilde{f}$-axis, accordingly, is an asymptote of these two paths.

Equation (8.89) has another kind of solution represented as

$$r^{\widehat{n}-1} \approx \frac{-|a\varepsilon|}{b}\,\frac{\sin(\theta-\psi)}{\sin(\widehat{n}\theta)}. \tag{8.93}$$

Inasmuch as $r > 0$, the solutions (8.93) exist for θ satisfying

$$\frac{-|a\varepsilon|}{b}\,\frac{\sin(\theta-\psi)}{\sin(\widehat{n}\theta)} > 0, \tag{8.94}$$

which is equivalent to

$$\frac{\sin(\theta-\psi)}{\sin(\widehat{n}\theta)} > 0 \tag{8.95}$$

under our assumption of $b < 0$. The set of θ satisfying (8.95) consists of $\widehat{n}-1$ intervals (modulo 2π), which correspond to $\widehat{n}-1$ sectors in the z-plane. Hence the solution of (8.93) yields $\widehat{n}-1$ paths, each of which is enclosed by a pair of asymptotes among $2(\widehat{n}-1)$ rays $\theta = \pi(k-1)/\widehat{n}$ $(k = 2,\ldots,\widehat{n},\widehat{n}+2,\ldots,2\widehat{n})$.

General Case

We consider the general case where $\psi \neq \pi(m-1)/\widehat{n}$ for any $m = 1,\ldots,2\widehat{n}$. In this case, (8.91) has no solution, and (8.89) yields the solution (8.93) only, which yields $\widehat{n}+1$ solution paths. The $\widetilde{f}$-axis and $2(\widehat{n}+1)$ rays $\theta = \psi,\ \psi+\pi$, and $\pi(k-1)/\widehat{n}$ $(k = 1,\ldots,2\widehat{n})$ are the asymptotes of these paths.

In the limit of $r \to +0$, the fundamental path is directed toward $\theta = \psi + \pi$ and $\widetilde{f} \to -\infty$ and the opposite path is directed toward $\theta = \psi$ and $\widetilde{f} \to +\infty$; the directions of these two paths vary in the θ-direction in association with the change of r, unlike the special case.

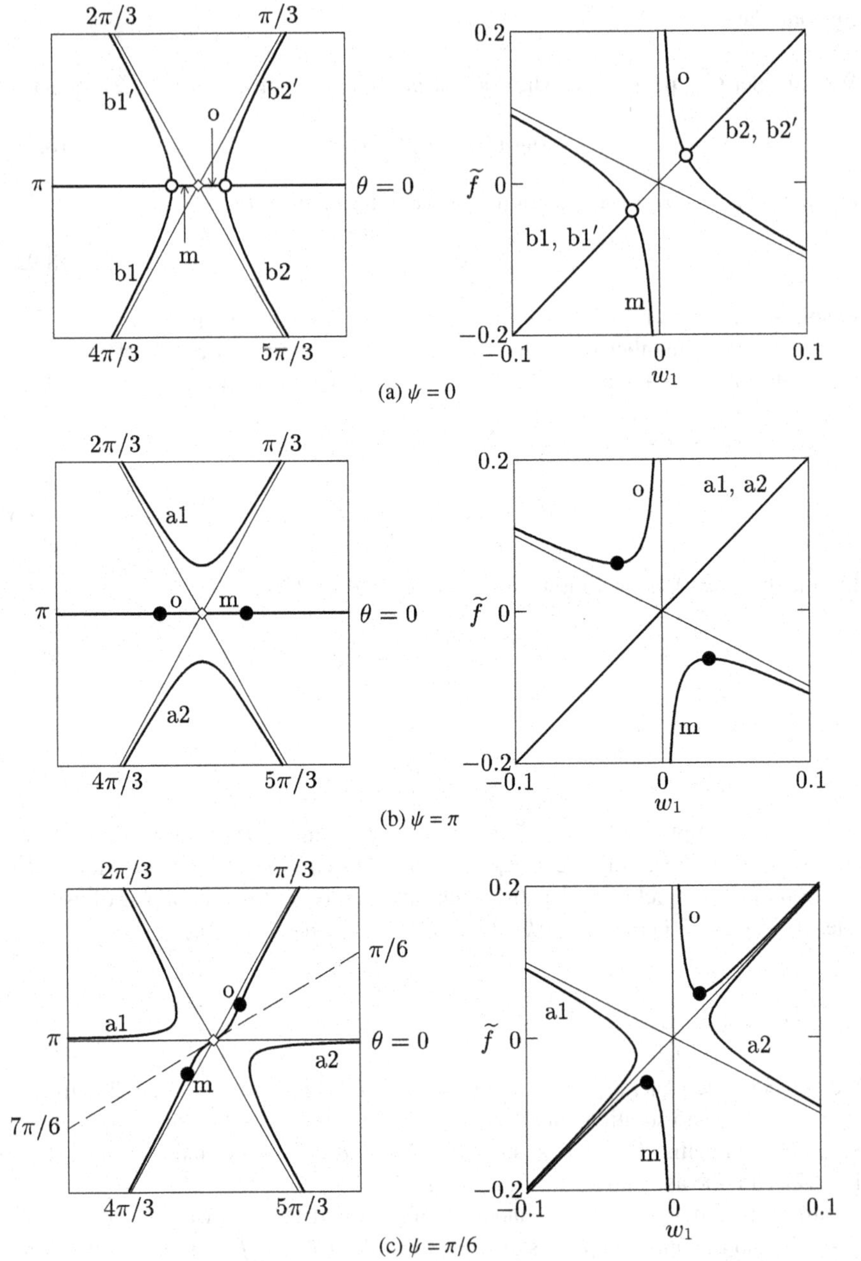

Fig. 8.14 Local imperfect behavior for $\widehat{n} = 3$ ($|a\varepsilon| = 10^{-3}$, $b = -1$). z-plane views at the left and $(w_1, \widetilde{f})$-plane views at the right. ○: simple pitchfork bifurcation point; ●: limit point; m: fundamental path; o: opposite path; ◇: asymptote on the $\widetilde{f}$-axis; thin line: solution curve for the perfect system that serves as an asymptote for a curve of an imperfect system; — — —: another asymptote.

8.7.3 Solution Curves: $\widehat{n} = 3$

For $\widehat{n} = 3$, the solution curves for the two special cases of $\psi = 0$ and $\psi = \pi$ and a general case of $\psi = \pi/6$ in Fig. 8.14 are investigated. (Recall Fig. 8.11 for the related perfect behavior.) Here the z-plane views of the solution paths are shown at the left, and the $(w_1, \widetilde{f})$-plane views are shown at the right. The z-plane views are computed by (8.92) and (8.93) for the special case and by (8.93) for the general case, and the $(w_1, \widetilde{f})$-plane views are computed by (8.90). The symbol ($\diamond$) at the origin of the z-plane views indicates the presence of an asymptote on the $\widetilde{f}$-axis.

For $\psi = 0$ in Fig. 8.14(a), the fundamental path on the ray $\theta = \pi$, denoted by "m," has an unstable pitchfork bifurcation point, at which branches a bifurcated path b1–b1′ lying in the sector of $2\pi/3 < \theta < 4\pi/3$. The opposite path on the ray $\theta = 0$ denoted by "o" also has a pitchfork bifurcation point and a bifurcated path b2–b2′, which lies in the sector of $-\pi/3 < \theta < \pi/3$.

For $\psi = \pi$ in Fig. 8.14(b), the fundamental path on the ray $\theta = 0$ has a maximum point of $\widetilde{f}$ and the opposite path on the ray $\theta = \pi$ has a minimum point. Two aloof paths a1 and a2 exist in the sectors $\pi/3 < \theta < 2\pi/3$ and $4\pi/3 < \theta < 5\pi/3$, respectively. The aloof path a1 has no limit point, since it is enclosed by an asymptote at $\theta = \pi/3$ ascending in the $\widetilde{f}$-direction ($\widetilde{f} \to +\infty$ as $r \to +\infty$) and an asymptote at $\theta = 2\pi/3$ descending in the $\widetilde{f}$-direction ($\widetilde{f} \to -\infty$ as $r \to +\infty$). Similarly, the other aloof path a2 has no limit point.

For $\psi = \pi/6$ in Fig. 8.14(c), the fundamental and opposite paths do not stay on the rays of constant θ but move in the sectors $7\pi/6 < \theta < 4\pi/3$ and $\pi/6 < \theta < \pi/3$, respectively. The fundamental path has a maximum point; the opposite path has a minimum point. Aloof paths a1 and a2 have no limit points.

8.7.4 Solution Curves: $\widehat{n} = 4$

Recall that the perfect bifurcation behavior of the double bifurcation point with $\widehat{n} = 4$ is dependent on the signs of A_{0300} and $A_{2100} \pm A_{0300}$ (cf., Fig. 8.10(b)). The value of $b = -A_{0300}/A_{2100}$ is chosen here to be $-1/2$ or $-3/2$, for which we have $A_{2100} < 0$, $A_{0300} < 0$, $A_{2100} + A_{0300} < 0$, and

$$\begin{cases} A_{2100} - A_{0300} < 0 & \text{for } b = -1/2, \\ A_{2100} - A_{0300} > 0 & \text{for } b = -3/2. \end{cases}$$

The solution curves for $\widehat{n} = 4$ for a special case of $\psi = 0$ with $b = -1/2$ and a general case of $\psi = \pi/8$ with $b = -3/2$ are portrayed in Fig. 8.15.

For $\psi = 0$ and $b = -1/2$ in Fig. 8.15(a), the fundamental path has a limit point, whereas the opposite path has a pitchfork bifurcation point and a bifurcated path moving in the sector $-\pi/4 < \theta < \pi/4$. Limit points exist on the aloof paths a1 and a2, each enclosed by a pair of asymptotes, toward which $\widetilde{f}$ decreases on the paths.

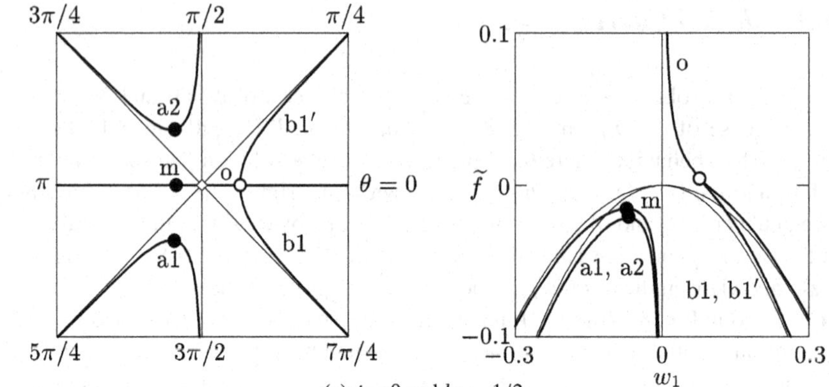

(a) $\psi = 0$ and $b = -1/2$

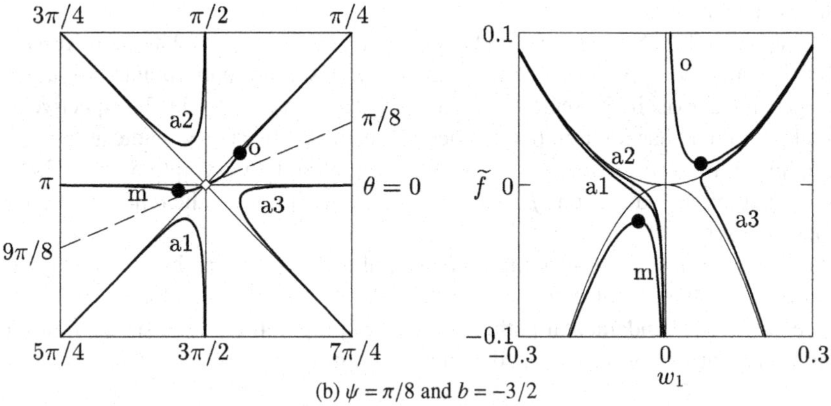

(b) $\psi = \pi/8$ and $b = -3/2$

Fig. 8.15 Local imperfect behavior for $\widehat{n} = 4$ ($|a\varepsilon| = 10^{-3}$, $b = -1/2, \ -3/2$). ○: simple pitchfork bifurcation point; ●: limit point; m: fundamental path; o: opposite path; ⋄: asymptote on the $\widetilde{f}$-axis; thin line: solution curve for the perfect system that serves as an asymptote; — — —: another asymptote.

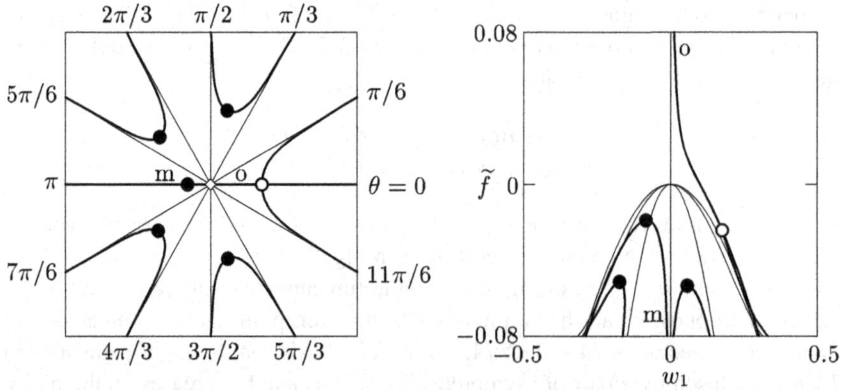

Fig. 8.16 Local imperfect behavior for $\widehat{n} = 6$ and $\psi = 0$ ($|a\varepsilon| = 10^{-3}$, $b = -1$, $c_2 = 2$). ○: simple pitchfork bifurcation point; ●: limit point; m: fundamental path; o: opposite path; ⋄: asymptote on the $\widetilde{f}$-axis; thin line: solution curve for the perfect system that serves as an asymptote.

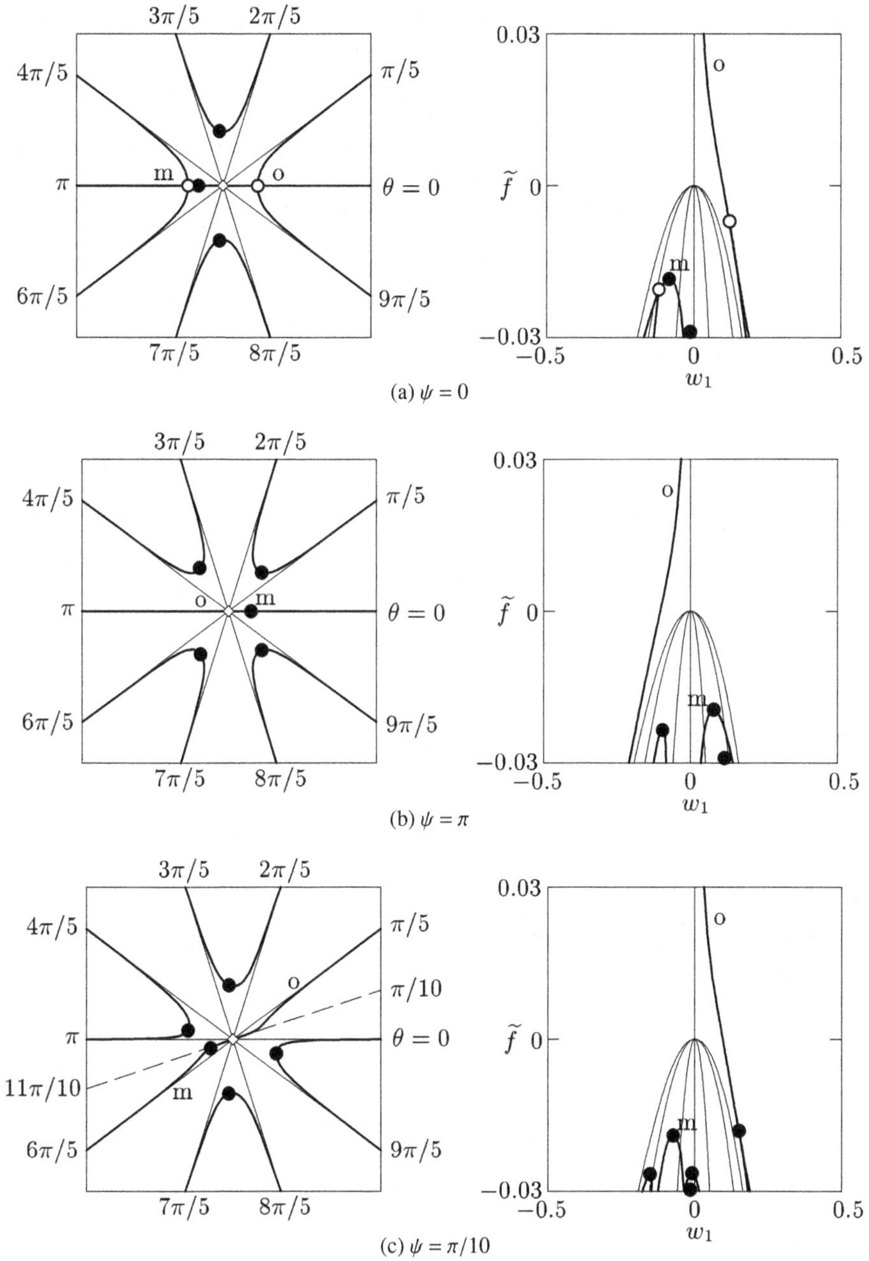

Fig. 8.17 Local imperfect behavior for $\widehat{n} = 5$ ($|a\varepsilon| = 10^{-3}$, $b = -1$, $c_2 = 2$). ○: simple pitchfork bifurcation point; •: limit point; m: fundamental path; o: opposite path; ⋄: asymptote on the $\widetilde{f}$-axis; thin line: solution curve for the perfect system that serves as an asymptote; — — —: another asymptote.

This occurs in association with the related perfect behavior shown in Fig. 8.12(a), for which all eight half-branches have negative slopes.

For $\psi = \pi/8$ and $b = -3/2$ in Fig. 8.15(b), the fundamental path has a limit point, and the opposite path also has a limit point. Limit points do not exist on the aloof paths a1, a2, and a3. For each aloof path $\widetilde{f}$ increases toward an asymptote and decreases toward another. See Fig. 8.12(b) for the related perfect behavior, for which four half-branches have negative slopes and the other four half-branches have positive ones.

8.7.5 Solution Curves: $\widehat{n} \geq 5$

For $\widehat{n} \geq 5$, recall that we consider the case in which the bifurcation point is unstable; that is, $A_{2100} < 0$ by (8.81). Imperfect behaviors for $\widehat{n} = 6$ and $\psi = 0$ are depicted in Fig. 8.16; those for $\widehat{n} = 5$ and $\psi = 0$, π, and $\pi/10$ are depicted in Fig. 8.17. These imperfect behaviors vary depending on the values of $\widehat{n}$ and ψ. Nonetheless, these behaviors are mutually similar in that the fundamental path has a limit point for all cases, and limit points exist on all aloof paths inasmuch as $\widetilde{f}$ decreases toward all asymptotes.

8.8 Imperfection Sensitivity Laws

For the double bifurcation point on a D_n-symmetric path, we present imperfection sensitivity laws, expressing the asymptotic sensitivity of $\widetilde{f}_{\rm c}$ to the imperfection magnitude ε. These laws are employed in the formulation of the worst imperfection in Chapter 9 and in the formulation of random imperfections in Chapter 10.

We assume that $A_{1010} < 0$ in (8.77) and that the double bifurcation point is unstable. In the remainder of this subsection, we derive imperfection sensitivity laws

$$\widetilde{f}_{\rm c} \approx C(\boldsymbol{d})|\varepsilon|^{\rho}, \tag{8.96}$$

where ρ and $C(\boldsymbol{d})$ are given, depending on the value of index $\widehat{n}$ in (8.23), as follows:

$$\begin{cases} \rho = 1/2, & C(\boldsymbol{d}) = -\tau(\psi)C_0 \cdot |a|^{1/2} & \text{if } \widehat{n} = 3, \\ \rho = 2/3, & C(\boldsymbol{d}) = -\hat{\tau}(\psi)C_0 \cdot |a|^{2/3} & \text{if } \widehat{n} = 4, \\ \rho = 2/3, & C(\boldsymbol{d}) = -C_0 \cdot |a|^{2/3} & \text{if } \widehat{n} \geq 5. \end{cases} \tag{8.97}$$

Here

- $a = A_{0001}(\boldsymbol{d})$ is a complex number, called the imperfection coefficient, depending on $\boldsymbol{d}$ as in (8.46) in Lemma 8.2 in §8.5.2.
- $\psi = \arg(a\varepsilon)$.
- $\tau(\psi)$ is a nonlinear function in ψ and is positive.

- $\hat{\tau}(\psi) = \hat{\tau}(\psi; A_{0200})$ is a nonlinear function in ψ and A_{0200} and is dependent on individual systems.

Formula (8.96) with (8.97) is derived below from the bifurcation equation (8.38) and the criticality condition (8.40) with the expression of the bifurcation equations such as (8.86).

8.8.1 Case $\widehat{n} \geq 5$

We start with the case $\widehat{n} \geq 5$, which admits the simplest formulation. As presented in Figs. 8.16 and 8.17, the fundamental path for an imperfect system has a maximum point for $\widehat{n} \geq 5$.

Special Case

Let us consider the special case with $\psi = 0$ as a preliminary stage to obtain the imperfection sensitivity laws. From (8.88) with (8.91) the fundamental path is described by $\theta = \pi$ with

$$\widetilde{f} \approx -r^2 - \frac{|a\varepsilon|}{r},$$

which has a maximum point of $\widetilde{f}$ at

$$\widetilde{f}_{\mathrm{c}} \approx -\frac{3}{4^{1/3}}|a\varepsilon|^{2/3}, \qquad r_{\mathrm{c}} \approx \frac{1}{2^{1/3}}|a\varepsilon|^{1/3}. \tag{8.98}$$

General Case

To deal with the general case, it is more convenient to work with the bifurcation equation (8.38) and the criticality condition (8.40) using (8.86). It turns out that we may omit some higher-order terms of (8.86) to get a simplified equation

$$\widehat{F} = -\widetilde{f}z - z^2\overline{z} + a\varepsilon.$$

Then the bifurcation equation (8.38) and the criticality condition (8.40) become

$$-\widetilde{f}z - z^2\overline{z} + a\varepsilon = 0, \tag{8.99}$$

$$-\widetilde{f}\overline{z} - z\overline{z}^2 + \overline{a}\varepsilon = 0, \tag{8.100}$$

$$\widetilde{f}^2 + 4\widetilde{f}z\overline{z} + 3z^2\overline{z}^2 = 0. \tag{8.101}$$

The desired relation between $\widetilde{f}$ and ε is obtained by eliminating z and $\overline{z}$ using these three equations. With some calculations (cf., Remark 8.6 below) we obtain

$$\widetilde{f}_{\mathrm{c}} \approx -\frac{3}{4^{1/3}}|a\varepsilon|^{2/3}, \tag{8.102}$$

or

$$\widetilde{f}_{\mathrm{c}} \approx -\frac{3A_{2100}{}^{1/3}}{4^{1/3}A_{1010}}|A_{0001}|^{2/3}\varepsilon^{2/3} \tag{8.103}$$

in the original variables (cf., (8.85)). We see also that

$$z_{\mathrm{c}} \approx -\frac{1}{2^{1/3}}\frac{a\varepsilon}{|a\varepsilon|^{2/3}} = \frac{A_{0001}}{2^{1/3}A_{2100}{}^{1/3}|A_{0001}|^{2/3}}\varepsilon^{1/3}. \tag{8.104}$$

The law (8.102) and (8.104) contains the law (8.98) for the special case.

Remark 8.6. The expressions (8.102) and (8.104) are derived here from (8.99)–(8.101). First observe that $[(8.99)\times\overline{z}-(8.100)\times z]$ shows that $a\overline{z}=\overline{a}z$, that is, $a\overline{z}$ is real.

Put

$$\alpha = (-a\varepsilon\overline{z}/|z|)^{1/3}, \tag{8.105}$$

which is also real. It then follows from (8.99) that

$$\widetilde{f} = -|z|^2 - \frac{\alpha^3}{|z|}. \tag{8.106}$$

Since (8.101) factors as

$$(\widetilde{f}+|z|^2)(\widetilde{f}+3|z|^2)=0,$$

we obtain

$$\widetilde{f}=-|z|^2 \qquad \text{or} \qquad \widetilde{f}=-3|z|^2,$$

of which the first is incompatible with (8.106). The combination of the second with (8.106) yields

$$\widetilde{f}=-3|z|^2=-|z|^2-\frac{\alpha^3}{|z|}.$$

This implies

$$\widetilde{f}=-\frac{3}{4^{1/3}}\alpha^2=-\frac{3}{4^{1/3}}|a\varepsilon|^{2/3}$$

as well as

$$|z|=\frac{1}{2^{1/3}}\alpha,$$

which means $\alpha>0$; that is, $\arg(z)=\arg(a\varepsilon)+\pi$ in (8.105). Hence

$$z=-\frac{1}{2^{1/3}}\frac{a\varepsilon}{|a\varepsilon|^{2/3}}.$$

□

Remark 8.7. As Figs. 8.15 to 8.17 portray, the opposite path and the aloof paths also have critical points. These critical points are at a distance of $O(|\varepsilon|^{1/(\hat{n}-1)})$ from the double bifurcation point of the perfect system, as might be readily apparent from (8.93). □

8.8.2 Case $\widehat{n} = 3$

For $\widehat{n} = 3$, we may omit some higher-order terms in (8.86) to obtain

$$\widehat{F} = -\widetilde{f}z + b\overline{z}^2 + a\varepsilon. \tag{8.107}$$

Note that $a = A_{0001}$ is complex and $b = A_{0200}$ is real (cf., (8.85)). Using the polar coordinates (8.87), we obtain (8.88) and (8.89) with $\widehat{n} = 3$.

Special Case

Consider the special case where $\psi = \arg(a\varepsilon) = 0$ or π, that is, where a is real. Then the solution $\theta = \pi + \psi$ corresponds to the fundamental path, which is given by

$$\widetilde{f} \approx -\mathrm{sign}(a\varepsilon) \cdot br - \frac{|a\varepsilon|}{r}.$$

In the case of $ab\varepsilon > 0$, as portrayed in Fig. 8.14(b) ($\psi = \pi$ for $b = -1$), the fundamental path has a limit (maximum) point of $\widetilde{f}$ at

$$\widetilde{f}_c \approx -2|ab\varepsilon|^{1/2}, \qquad r_c \approx \left(\frac{a\varepsilon}{b}\right)^{1/2}. \tag{8.108}$$

In the other case of $ab\varepsilon < 0$, as depicted in Fig. 8.14(a) ($\psi = 0$ for $b = -1$), the fundamental path has an unstable pitchfork bifurcation point at

$$\widetilde{f}_c \approx -\frac{2}{3^{1/2}}|ab\varepsilon|^{1/2}, \qquad r_c \approx \left(\frac{-a\varepsilon}{3b}\right)^{1/2}, \tag{8.109}$$

where r_c is determined by the limit of (8.93) as $\theta \to \psi + \pi$.

General Case

We consider general ψ. Just as in Case $\widehat{n} \geq 5$, the asymptotic expression of the critical load can be derived as in Lemma 8.5 below.

Lemma 8.5.

$$\widetilde{f}_c \approx -\tau(\psi)\frac{|A_{0200}|^{1/2}|A_{0001}|^{1/2}}{|A_{1010}|}|\varepsilon|^{1/2}, \tag{8.110}$$

where $\tau(\psi)$ is a function lying in a bounded range $2/\sqrt{3} \leq \tau(\psi) \leq 2$.

Proof. For $\widehat{F}$ of (8.107), equations (8.38) and (8.40) become

$$\begin{aligned} -\widetilde{f}z + b\overline{z}^2 + a\varepsilon &= 0, \\ -\widetilde{f}\overline{z} + bz^2 + \overline{a}\varepsilon &= 0, \\ \widetilde{f}^2 - 4b^2 z\overline{z} &= 0. \end{aligned}$$

On eliminating z and $\overline{z}$ from these three equations we obtain

$$\frac{27}{256}\widetilde{f}^8 - \frac{9}{8}|a|^2 b^2 \varepsilon^2 \widetilde{f}^4 - (a^3 + \overline{a}^3) b^3 \varepsilon^3 \widetilde{f}^2 - |a|^4 b^4 \varepsilon^4 = 0. \tag{8.111}$$

If we put

$$t = \frac{\widetilde{f}^2}{|a\varepsilon|b}$$

and use $\psi = \arg(a\varepsilon)$, this equation is written as

$$\frac{27}{256}t^3 - \frac{9}{8}t - \frac{1}{t} = 2\cos(3\psi). \tag{8.112}$$

For $t > 0$, the left-hand side of this equation is a monotone increasing continuous function ranging from $-\infty$ to $+\infty$; the same is true for $t < 0$. Hence the equation (8.112), for each ψ, has exactly one positive solution and one negative solution. If we designate the positive solution as $t = \tau_3(\psi)^2$ with $\tau_3(\psi) > 0$, then the negative solution is given by $t = -\tau_3(\psi + \pi)^2$, since the left-hand side of (8.112) is an odd function in t and $\cos(3(\psi + \pi)) = -\cos(3\psi)$. Hence,

$$\frac{\widetilde{f}^2}{|a\varepsilon|b} = \begin{cases} \tau_3(\psi)^2 & \text{if } b > 0, \\ -\tau_3(\psi + \pi)^2 & \text{if } b < 0. \end{cases} \tag{8.113}$$

Therefore, equation (8.111) is solved for $\widetilde{f}$ as

$$\widetilde{f}_c \approx \pm\tau(\psi)|ab\varepsilon|^{1/2}, \tag{8.114}$$

where

$$\tau(\psi) = \tau_3(\psi + \arg(b)) \tag{8.115}$$

and $\arg(b) = 0$ or π inasmuch as b is real. In the original variables, this gives the critical load increment in (8.110). It is not difficult to see from (8.112) that the function $\tau_3(\psi)$ satisfies

$$\tau_{\min} \le \tau_3(\psi) \le \tau_{\max} \tag{8.116}$$

with $\tau_{\min} = \tau_3(\pi) = 2/\sqrt{3}$ and $\tau_{\max} = \tau_3(0) = 2$. □

In (8.116), the maximum $\tau_{\max}$ of $\tau_3(\psi)$ is achieved by $\psi = 0$ and $\pm 2\pi/3$ (i.e., $\cos(3\psi) = 1$), whereas the minimum $\tau_{\min}$ is achieved by $\psi = \pi$ and $\pm\pi/3$ (i.e., $\cos(3\psi) = -1$). Accordingly, the maximum $\tau_{\max}$ of $\tau(\psi)$ is achieved by

$$\psi = \arg(b) + 2\pi j/3, \qquad j = 0, \pm 1, \tag{8.117}$$

whereas the minimum $\tau_{\min}$ is achieved by

$$\psi = \arg(b) + \pi + 2\pi j/3, \qquad j = 0, \pm 1.$$

The maximization of $\tau(\psi)$ is considered again in §9.4.2 in determining the worst imperfection.

Remark 8.8. It is noteworthy that for $\widehat{n} = 3$ the general case contains the special case. First, we consider the case of $a\varepsilon > 0$ (with $\psi = 0$), then (8.115) gives

$$\tau(0) = \tau_3(\arg(b)) = \begin{cases} \tau_3(0) = 2 & \text{if } b > 0, \\ \tau_3(\pi) = 2/\sqrt{3} & \text{if } b < 0. \end{cases}$$

Then the formula (8.113) for the general case becomes

$$\frac{\widetilde{f}^2}{|a\varepsilon| b} = \begin{cases} 4 & \text{if } b > 0, \\ -4/3 & \text{if } b < 0, \end{cases}$$

and (8.114) becomes

$$\widetilde{f}_{\mathrm{c}} \approx \begin{cases} \pm 2|ab\varepsilon|^{1/2}, & \text{if } b > 0, \\ \pm \frac{2}{3^{1/2}}|ab\varepsilon|^{1/2}, & \text{if } b < 0. \end{cases}$$

The law for $b > 0$ in this equation is identical to the law (8.108) for the special case; that for $b < 0$ is identical to the law (8.109). Thus the law (8.113) for the general case contains the special case. A similar argument holds for the other case $a\varepsilon < 0$ with $\psi = \pi$. □

8.8.3 Case $\widehat{n} = 4$

For $\widehat{n} = 4$ we consider the general case only. To get the critical load, we can simplify (8.86) to

$$\widehat{F} = -\widetilde{f}z - z^2\overline{z} + b\overline{z}^3 + a\varepsilon.$$

Note that $b = -A_{0300}/A_{2100}$ is real and $a = A_{0001}$ is complex.

Then equations (8.38) and (8.40) become

$$\begin{aligned} -\widetilde{f}z - z^2\overline{z} + b\overline{z}^3 + a\varepsilon &= 0, \\ -\widetilde{f}\overline{z} - z\overline{z}^2 + bz^3 + \overline{a}\varepsilon &= 0, \\ \widetilde{f}^2 + 4\widetilde{f}z\overline{z} + (3 - 9b^2)z^2\overline{z}^2 + 3b(z^4 + \overline{z}^4) &= 0. \end{aligned}$$

The relation between $\widetilde{f}$ and ε is obtained by eliminating z and $\overline{z}$ from these three equations.

By a scaling argument similar to the one in Case $\widehat{n} = 3$, we see that

$$\widetilde{f}_c \approx -\hat{\tau}(\psi)\frac{A_{2100}{}^{1/3}|A_{0001}|^{2/3}}{A_{1010}}|\varepsilon|^{2/3}, \tag{8.118}$$

where $\hat{\tau}(\psi) = \hat{\tau}(\psi; b)$ depends on $\psi = \arg(a\varepsilon)$ and b, having a period of $\psi = \pi/2$; that is, $\hat{\tau}(\psi + \pi/2) = \hat{\tau}(\psi)$. Unlike Case $\widehat{n} = 3$, no nontrivial lower bound exists for $|\hat{\tau}(\psi)|$. For example, if $b = -1/3$ and $\psi = 0$, we have $\hat{\tau}(\psi) = 0$ (and $z = \bar{z} = -(3|a\varepsilon|/4)^{1/3}$).

8.9 Experimentally Observed Bifurcation Diagrams

A gap between the mathematical theory and engineering practice in the experiment of materials undergoing bifurcation has been pointed out in Chapter 6 with reference to the difference between observed diagrams of force versus displacement and bifurcation diagrams predicted by mathematics. The following three points are addressed.

- It is difficult to judge merely from the observed curves whether the system under consideration is undergoing bifurcation.
- Experimentally observed displacements are under the influence of unknown imperfections of various kinds, and the perfect system cannot be known.
- Observed diagrams of force versus displacement can differ qualitatively from the bifurcation diagrams predicted by mathematics.

Regarding the last point, in particular, the reason for the qualitative discrepancy has been explained in terms of the symmetry of the displacement used in the experimentally observed bifurcation diagram. On the basis of these considerations, a systematic strategy for recovering the curve of the perfect system from that of an imperfect system has been presented

This section extends the strategy for the observed displacements in Chapter 6 to a D_n-equivariant system. For a D_n-symmetric displacement, it is shown that the formulas in Chapter 6 are also applicable to:

- A simple (pitchfork) bifurcation point,
- A double bifurcation point with the index $\widehat{n} \geq 5$.

8.9.1 Simple Bifurcation Point

A simple bifurcation point of a D_n-equivariant system is associated with a one-dimensional irreducible representation $(+,-)_{D_n}$, $(-,+)_{D_n}$, or $(-,-)_{D_n}$. In each case, the bifurcation equation is expressed in the form of (2.93):

$$\widehat{F}(w,\widetilde{f},\varepsilon) = A_{110}w\widetilde{f} + A_{300}w^3 + A_{001}\varepsilon + \text{h.o.t.} = 0. \tag{8.119}$$

For an experimentally observed displacement component, say u_{i^*} (with a particular i^*) with D_n-symmetry, we have the following lemma[7] that connects the variables w with u_{i^*}.

Lemma 8.6. *If u_{i^*} has D_n-symmetry in the sense that*

$$c(2\pi/n),\ \sigma\ :\ u_{i^*} \mapsto u_{i^*}, \tag{8.120}$$

the incremental displacement $\widetilde{u}_{i^*} = u_{i^*} - (u_{i^*})_{\mathrm{c}}^0$ is expressed as

$$\widetilde{u}_{i^*} = r_{i^*}\widetilde{f} + s_{i^*}w^2 + \text{h.o.t.}, \tag{8.121}$$

where r_{i^*}, r_{i^*}, and s_{i^*} are constants.

Proof. The incremental displacement $\widetilde{u}_{i^*}$ is given by (6.15):

$$\widetilde{u}_{i^*} = \eta_{i^*1}w + r_{i^*}\widetilde{f} + s_{i^*}w^2 + \text{h.o.t.}, \tag{8.122}$$

where η_{i^*1} is the i^*th component of the critical eigenvector $\boldsymbol{\eta}_1$. We deal only with the case of $(-,+)_{\mathrm{D}_n}$; however, the other cases $(+,-)_{\mathrm{D}_n}$ and $(-,-)_{\mathrm{D}_n}$ can be treated similarly.

The associated displacement w satisfies

$$c(2\pi/n)\ :\ w \mapsto -w, \tag{8.123}$$

because $(-,+)_{\mathrm{D}_n}$ is the associated irreducible representation. The use of (8.120) and (8.123) in (8.122) leads to $\eta_{i^*1} = 0$; and then we have (8.121). □

The relation between $\widetilde{u}_{i^*}$ and $\widetilde{f}$, which constitutes the experimentally observed bifurcation diagram, can be obtained from (8.119) and (8.121) by eliminating w. The resulting equation is

$$\sqrt{\mathrm{sign}(s)(\widetilde{u}_{i^*} - \widetilde{f}/E)}\,[\widetilde{f} + p(\widetilde{u}_{i^*} - \widetilde{f}/E)] \pm q\varepsilon + \text{h.o.t.} = 0, \tag{8.124}$$

which is nothing but the equation (6.27) presented in §6.3.3. Consequently, all the formulas (6.27)–(6.34) presented for the displacement with symmetry are applicable to the present case. In particular, we refer to (6.33):

$$\widetilde{f}_{\mathrm{c}} = -\gamma^*\widetilde{u}_{|h} + \text{h.o.t.}, \tag{8.125}$$

which relates the critical load increment $\widetilde{f}_{\mathrm{c}}$ with the incremental displacement $\widetilde{u}_{|h}$. This displacement is defined by the intersection point $(\widetilde{u}_{|h}, \widetilde{f}_{|h})$ between the imperfect $\widetilde{f}$ versus $\widetilde{u}$ curve and the straight line (6.31):

$$\widetilde{f} + h\widetilde{u} = 0, \tag{8.126}$$

[7] This lemma is the group-theoretic explanation for the second case (symmetric displacement) of the categorization in (6.17).

which passes the bifurcation point $(\widetilde{u}, \widetilde{f}) = (0,0)$.

8.9.2 Double Bifurcation Point

We consider a group-theoretic double bifurcation point on a D_n-symmetric path associated with the two-dimensional irreducible representation $(j)_{D_n}$ for some j ($< n/2$) defined in (8.9); recall the index $\widehat{n} = n/\gcd(n, j)$.

For D_n-symmetric experimentally observed displacement u_{i^*} (with a particular i^*) for $\widehat{n} \geq 5$, we demonstrate below that all the results presented in Chapter 6 for a symmetric displacement for a pitchfork bifurcation point are applicable.

As stated in Remark 8.9 below, the bifurcation equation can be approximated by

$$A_{1010}\, r\widetilde{f} + A_{2100}\, r^3 + |A_{0001}\varepsilon| + \text{h.o.t.} = 0. \tag{8.127}$$

If u_{i^*} is D_n-symmetric in the sense of (8.120), then $\widetilde{u}_{i^*}$ is expressed (cf., Remark 8.10 below) as

$$\widetilde{u}_{i^*} = R_{i^*}\widetilde{f} + S_{i^*} r^2 + \text{h.o.t.}, \tag{8.128}$$

where $r = |z| = ({w_1}^2 + {w_2}^2)^{1/2}$ and R_{i^*} and S_{i^*} are constants.

It is interesting that (8.128) is of exactly the same form as (8.121) for a D_n-symmetric displacement of a pitchfork bifurcation point. In addition, the bifurcation equation (8.127) for the double bifurcation point is essentially of the same form as (8.119) for a pitchfork bifurcation point. Hence the results for the pitchfork bifurcation point in Chapter 6 are applicable to the present case. However, the results are not applicable when $\widehat{n} = 3$ or 4, or when u_{i^*} is not D_n-symmetric (cf., Remark 8.11 below).

Remark 8.9. To derive (8.127), recall the set of bifurcation equations (8.88) and (8.89):

$$r\widetilde{f} \approx -r^3 + \sum_{2 \leq q \leq \widehat{n}/2-1} c_q r^{2q+1} + b r^{\widehat{n}-1}\cos(\widehat{n}\theta) + |a\varepsilon|\cos(\theta - \psi), \tag{8.129}$$

$$b r^{\widehat{n}-1}\sin(\widehat{n}\theta) + |a\varepsilon|\sin(\theta - \psi) \approx 0. \tag{8.130}$$

When $\widehat{n} \geq 5$, the critical point on the fundamental path of an imperfect system is given by (8.102) and (8.104):

$$\widetilde{f}_c \approx -\frac{3}{4^{1/3}}|a\varepsilon|^{2/3}, \qquad z_c \approx -\frac{1}{2^{1/3}}\frac{a\varepsilon}{|a\varepsilon|^{2/3}}.$$

Combining these expressions with (8.130), we can assume

$$r = \mathrm{O}(\varepsilon^{1/3}), \qquad \theta - \psi = \pi + \mathrm{O}(\varepsilon^{(\widehat{n}-4)/3})$$

for the fundamental path. Therefore, we can approximate (8.129) by

$$r\widetilde{f} + r^3 + |a\varepsilon| \approx 0, \tag{8.131}$$

which is nothing but (8.127) when expressed in the original variables (cf., (8.84)). □

Remark 8.10. The derivation of (8.128) refers to the isotypic decomposition in (7.25) with $G = \mathrm{D}_n$. Let $\mu = (+,+)_{\mathrm{D}_n}$ be the unit representation, and $\boldsymbol{\eta}_k^{\mu}$ ($k = 1,\dots,a^{\mu}$) be the basis vectors of the isotypic component for the unit representation μ, where a^{μ} is its multiplicity in T in (8.33). Since the observed variable u_{i^*} is D_n-symmetric, it can be expressed as a linear combination of the i^*th components $\eta_{i^*k}^{\mu}$ of those vectors $\boldsymbol{\eta}_k^{\mu}$ ($k = 1,\dots,a^{\mu}$):

$$\widetilde{u}_{i^*} \equiv u_{i^*} - (u_{i^*})_{\mathrm{c}}^0 = \sum_{k=1}^{a^{\mu}} \eta_{i^*k}^{\mu} w_k^{\mu}. \tag{8.132}$$

It must be clear that w_1 and w_2 are disjoint from the coefficients w_k^{μ} ($k = 1,\dots,a^{\mu}$) above, since w_1 and w_2 are the coefficients associated with the critical eigenvectors $\boldsymbol{\eta}_1$ and $\boldsymbol{\eta}_2$ that belong to the isotypic component labeled by $(j)_{\mathrm{D}_n}$. We now consider the part of equations (7.63) that corresponds to the unit representation μ; denote those equations by $\widehat{F}_k(\boldsymbol{w},\widetilde{f},\varepsilon)$ ($k = 1,\dots,a^{\mu}$). The block-diagonal structure (7.121) of the Jacobian matrix, which is a consequence of the group-equivariance, means that the possible linear terms appearing in $\widehat{F}_k(\boldsymbol{w},\widetilde{f},\varepsilon)$ are restricted to those in w_l^{μ} ($l = 1,\dots,a^{\mu}$) as well as in $\widetilde{f}$ and ε. Hence, with a suitable choice of the basis vectors $\boldsymbol{\eta}_k^{\mu}$ ($k = 1,\dots,a^{\mu}$), we may assume that

$$\widehat{F}_k(\boldsymbol{w},\widetilde{f},\varepsilon) = e_k w_k^{\mu} + a_k\widetilde{f} + b_k\varepsilon + c_k^{11}{w_1}^2 + c_k^{12}w_1w_2 + c_k^{22}{w_2}^2 + \text{h.o.t.} = 0, \quad k = 1,\dots,a^{\mu}, \tag{8.133}$$

where a_k, b_k, e_k, c_k^{11}, c_k^{12}, and c_k^{22} ($k = 1,\dots,a^{\mu}$) are some constants (b_k depends on the imperfection pattern $\boldsymbol{d}$). Here, $e_k \neq 0$ for $k = 1,\dots,a^{\mu}$. The D_n-equivariance of (8.133) has a further implication for the coefficients c_k^{11}, c_k^{12}, and c_k^{22}. By (8.8) and (8.9), the actions of the elements $c(2\pi/n)$ and σ of the dihedral group D_n on the equations and variables of (8.133) are expressed as

$$\begin{aligned}
c(2\pi/n) &: \begin{pmatrix} w_1 \\ w_2 \end{pmatrix} \mapsto \begin{pmatrix} \cos(2\pi j/n) & -\sin(2\pi j/n) \\ \sin(2\pi j/n) & \cos(2\pi j/n) \end{pmatrix} \begin{pmatrix} w_1 \\ w_2 \end{pmatrix}, \\
\sigma &: \begin{pmatrix} w_1 \\ w_2 \end{pmatrix} \mapsto \begin{pmatrix} w_1 \\ -w_2 \end{pmatrix}, \\
c(2\pi/n),\ \sigma &: \widehat{F}_k \mapsto \widehat{F}_k, \\
c(2\pi/n),\ \sigma &: w_k^{\mu} \mapsto w_k^{\mu}.
\end{aligned}$$

By virtue of these, the coefficients c_k^{11}, c_k^{22}, and c_k^{12} must satisfy

$$c_k^{11} = c_k^{22} (\equiv C_k), \qquad c_k^{12} = 0.$$

Equation (8.133), accordingly, takes the form of

$$\widehat{F}_k(\boldsymbol{w},\widetilde{f},\varepsilon) = e_k w_k^{\mu} + a_k\widetilde{f} + b_k\varepsilon + C_k r^2 + \text{h.o.t.} = 0, \qquad k = 1,\ldots,a^{\mu}.$$

Hence, the variable w_k^{μ} can be evaluated to

$$w_k^{\mu} = -\frac{1}{e_k}(a_k\widetilde{f} + C_k r^2) + \text{h.o.t.}, \tag{8.134}$$

where $b_k\varepsilon$ is suppressed in this equation as it turns out to be a higher-order term. Substitution of (8.134) into (8.132) results in

$$\widetilde{u}_{i^*} = R_{i^*}\widetilde{f} + S_{i^*}r^2 + \text{h.o.t.}$$

with

$$R_{i^*} = -\sum_{k=1}^{a^{\mu}} \frac{\eta^{\mu}_{i^*k}a_k}{e_k}, \qquad S_{i^*} = -\sum_{k=1}^{a^{\mu}} \frac{\eta^{\mu}_{i^*k}C_k}{e_k}.$$

This shows (8.128). □

Remark 8.11. We describe the cases of $\widehat{n} = 3$ and $\widehat{n} = 4$, for which (8.129) becomes

$$\begin{cases} r\widetilde{f} = br^2\cos(3\theta) + |a\varepsilon|\cos(\theta - \psi) & \text{if } \widehat{n} = 3, \\ r\widetilde{f} = -r^3 + br^3\cos(4\theta) + |a\varepsilon|\cos(\theta - \psi) & \text{if } \widehat{n} = 4. \end{cases} \tag{8.135}$$

The orders of the terms in the bifurcation equation (8.135) are evaluated to

$$\begin{cases} r = \mathrm{O}(\varepsilon^{1/2}), \quad \widetilde{f} = \mathrm{O}(\varepsilon^{1/2}), \quad \sin(\theta - \psi) = \mathrm{O}(1) & \text{if } \widehat{n} = 3, \\ r = \mathrm{O}(\varepsilon^{1/3}), \quad \widetilde{f} = \mathrm{O}(\varepsilon^{2/3}), \quad \sin(\theta - \psi) = \mathrm{O}(1) & \text{if } \widehat{n} = 4. \end{cases}$$

Thus the asymptotic curve expressed by (8.135) for $\widehat{n} = 3$ or 4 is dependent on ψ and on the imperfection pattern vector $\boldsymbol{d}$ even when ε is infinitesimal. For $\widehat{n} = 3$ or 4, the bifurcation equation is more complex than that for $\widehat{n} \geq 5$ in (8.127), and hence the results presented in Chapter 6 are not applicable. □

8.10 Appendix: Alternative Stability Analysis

To supplement §8.6.2, we derive here conditions (8.79) and (8.80) for the stability of half-branches without referring to the "asymptotic potential" $\widetilde{U}$, while working with the full expansion (8.55). Let $\tilde{J} = (\partial\widetilde{F}_i/\partial w_j \mid i,j = 1,2)$ be the Jacobian matrix of the reduced equations and note that the system is stable if the real part of each eigenvalue of $\tilde{J}$ is positive, and is unstable if at least one eigenvalue has a negative real part. Since $\tilde{J}$ is two-dimensional, both eigenvalues have positive real parts if and only if

$$\text{trace}(\tilde{J}) > 0 \qquad \text{and} \qquad \det(\tilde{J}) > 0. \tag{8.136}$$

On the half-branch with $\theta = \alpha_k = -\pi(k-1)/\widehat{n}$, we obtain (see below) the following expressions when r is small:

$$\frac{1}{2}\text{trace}(\tilde{J}) \approx \begin{cases} (-1)^k A_{0200}\, r & \text{if } \widehat{n} = 3, \\ [A_{2100} + (-1)^k A_{0300}]\, r^2 & \text{if } \widehat{n} = 4, \\ A_{2100}\, r^2 & \text{if } \widehat{n} \geq 5, \end{cases} \tag{8.137}$$

$$\det(\tilde{J}) \approx \begin{cases} -3 A_{0200}{}^2\, r^2 & \text{if } \widehat{n} = 3, \\ (-1)^k 8 A_{0300}[A_{2100} - (-1)^k A_{0300}]\, r^4 & \text{if } \widehat{n} = 4, \\ (-1)^k 2\widehat{n} A_{0,\widehat{n}-1,00} A_{2100}\, r^{\widehat{n}} & \text{if } \widehat{n} \geq 5. \end{cases} \tag{8.138}$$

Substitution of (8.137) and (8.138) into (8.136) reveals the stability of the half-branches, which is studied in §8.6.2 and is categorized in Fig. 8.10 in §8.6.1.

The expressions (8.137) and (8.138) of the trace and determinant of $\tilde{J}$ on the half-branch with $\theta = \alpha_k = -\pi(k-1)/\widehat{n}$ are derived below. First, it must be noted that

$$z^{m\widehat{n}} = \bar{z}^{m\widehat{n}} = (-1)^{m(k-1)} r^{m\widehat{n}}$$

for $z = r\exp(\mathrm{i}\theta)$ with $\theta = \alpha_k$. With the abbreviations

$$A_{q+1,q} = A_{q+1,q}(\widetilde{f}, \boldsymbol{v}^0), \qquad B_{mq} = A_{q+1+m\widehat{n},q}(\widetilde{f}, \boldsymbol{v}^0), \qquad C_{qm} = A_{q,q-1+m\widehat{n}}(\widetilde{f}, \boldsymbol{v}^0),$$

it follows from (8.55) that

$$\begin{aligned} \frac{\partial F}{\partial z} &= \sum_{q=0} (q+1) A_{q+1,q} z^q \bar{z}^q + \sum_{m=1} \sum_{q=0} [(q+1+m\widehat{n}) B_{mq} z^{q+m\widehat{n}} \bar{z}^q + q C_{qm} z^{q-1} \bar{z}^{q-1+m\widehat{n}}] \\ &= \sum_{q=0} (q+1) A_{q+1,q} r^{2q} \\ &\quad + \sum_{m=1} \sum_{q=0} (-1)^{m(k-1)} [(q+1+m\widehat{n}) B_{mq} r^{2q+m\widehat{n}} + q C_{qm} r^{2(q-1)+m\widehat{n}}], \end{aligned} \tag{8.139}$$

$$\begin{aligned} \frac{\partial F}{\partial \bar{z}} &= \sum_{q=0} q A_{q+1,q} z^{q+1} \bar{z}^{q-1} + \sum_{m=1} \sum_{q=0} [q B_{mq} z^{q+1+m\widehat{n}} \bar{z}^{q-1} + (q-1+m\widehat{n}) C_{qm} z^q \bar{z}^{q-2+m\widehat{n}}] \\ &= \mathrm{e}^{2\mathrm{i}\theta} \{ \sum_{q=0} q A_{q+1,q} r^{2q} \\ &\quad + \sum_{m=1} \sum_{q=0} (-1)^{m(k-1)} [q B_{mq} r^{2q+m\widehat{n}} + (q-1+m\widehat{n}) C_{qm} r^{2(q-1)+m\widehat{n}}] \}. \end{aligned} \tag{8.140}$$

Expressions of trace($\tilde{J}$) and det($\tilde{J}$) in terms of r and $\widetilde{f}$ are obtained by the substitution of (8.139) and (8.140) into (8.39):

$$\text{trace}(\tilde{J}) = 2\,\text{Re}\left(\frac{\partial F}{\partial z}\right), \qquad \det(\tilde{J}) = \left|\frac{\partial F}{\partial z}\right|^2 - \left|\frac{\partial F}{\partial \bar{z}}\right|^2.$$

To obtain expressions in r, not containing $\widetilde{f}$, we are to eliminate the variable $\widetilde{f}$ using (8.61):

$$\sum_{q=0} A_{q+1,q} r^{2q} + \sum_{m=1}\sum_{q=0} (-1)^{m(k-1)} [A_{q+1+m\widehat{n},q} r^{2q+m\widehat{n}} + A_{q,q-1+m\widehat{n}} r^{2(q-1)+m\widehat{n}}] \approx 0. \tag{8.141}$$

For the trace we subtract (8.141) from (8.139) to eliminate A_{10}, which is asymptotically equal to $A_{1010}\widetilde{f}$ as shown in (8.63). Then we obtain

$$\begin{aligned}
\frac{1}{2}\mathrm{trace}(\tilde{J}) &= \mathrm{Re}\left(\frac{\partial F}{\partial z}\right) \\
&= \frac{\partial F}{\partial z} \\
&= \sum_{q=1} q A_{q+1,q} r^{2q} \\
&\quad + \sum_{m=1}\sum_{q=0} (-1)^{m(k-1)} [(q+m\widehat{n}) B_{mq} r^{2q+m\widehat{n}} + (q-1) C_{qm} r^{2(q-1)+m\widehat{n}}] \\
&\approx A_{2100} r^2 + (-1)^k A_{0,\widehat{n}-1,00} r^{\widehat{n}-2}.
\end{aligned} \tag{8.142}$$

This shows (8.137).

For the determinant we define

$$\begin{aligned}
G_1 &= \frac{\partial F}{\partial z} \\
&= \sum_{q=0} (q+1) A_{q+1,q} r^{2q} \\
&\quad + \sum_{m=1}\sum_{q=0} (-1)^{m(k-1)} [(q+1+m\widehat{n}) B_{mq} r^{2q+m\widehat{n}} + q C_{qm} r^{2(q-1)+m\widehat{n}}], \\
G_2 &= \frac{\partial F}{\partial \bar{z}} / \mathrm{e}^{2\mathrm{i}\theta} \\
&= \sum_{q=0} q A_{q+1,q} r^{2q} \\
&\quad + \sum_{m=1}\sum_{q=0} (-1)^{m(k-1)} [q B_{mq} r^{2q+m\widehat{n}} + (q-1+m\widehat{n}) C_{qm} r^{2(q-1)+m\widehat{n}}]
\end{aligned}$$

referring to (8.139) and (8.140). Since G_1 and G_2 are real, we have

$$\det(\tilde{J}) = |G_1|^2 - |G_2|^2 = {G_1}^2 - {G_2}^2 = (G_1 - G_2)(G_1 + G_2).$$

Using (8.141) we can evaluate the two factors; that is,

$$
\begin{aligned}
G_1 - G_2 &= \sum_{q=0} A_{q+1,q} r^{2q} \\
&\quad + \sum_{m=1} \sum_{q=0} (-1)^{m(k-1)} [(1 + m\widehat{n}) B_{mq} r^{2q+m\widehat{n}} + (1 - m\widehat{n}) C_{qm} r^{2(q-1)+m\widehat{n}}] \\
&= \sum_{m=1} \sum_{q=0} (-1)^{m(k-1)} (m\widehat{n} B_{mq} r^{2q+m\widehat{n}} - m\widehat{n} C_{qm} r^{2(q-1)+m\widehat{n}}) \\
&\approx (-1)^k \widehat{n} A_{0,\widehat{n}-1,00} r^{\widehat{n}-2} = \begin{cases} (-1)^k 3A_{0200} r & \text{if } \widehat{n} = 3, \\ (-1)^k 4A_{0300} r^2 & \text{if } \widehat{n} = 4, \\ (-1)^k \widehat{n} A_{0,\widehat{n}-1,00} r^{\widehat{n}-2} & \text{if } \widehat{n} \geq 5, \end{cases}
\end{aligned} \tag{8.143}
$$

and

$$
\begin{aligned}
&G_1 + G_2 \\
&= \sum_{q=0} (2q+1) A_{q+1,q} r^{2q} \\
&\quad + \sum_{m=1} \sum_{q=0} (-1)^{m(k-1)} [(2q + 1 + m\widehat{n}) B_{mq} r^{2q+m\widehat{n}} + (2q - 1 + m\widehat{n}) C_{qm} r^{2(q-1)+m\widehat{n}}] \\
&= \sum_{q=1} 2q A_{q+1,q} r^{2q} \\
&\quad + \sum_{m=1} \sum_{q=0} (-1)^{m(k-1)} [(2q + m\widehat{n}) B_{mq} r^{2q+m\widehat{n}} + (2q - 2 + m\widehat{n}) C_{qm} r^{2(q-1)+m\widehat{n}}] \\
&\approx 2A_{2100} r^2 + (-1)^{k-1} (\widehat{n} - 2) A_{0,\widehat{n}-1,00} r^{\widehat{n}-2} \\
&\approx \begin{cases} (-1)^{k-1} A_{0200} r & \text{if } \widehat{n} = 3, \\ 2[A_{2100} + (-1)^{k-1} A_{0300}] r^2 & \text{if } \widehat{n} = 4, \\ 2A_{2100} r^2 & \text{if } \widehat{n} \geq 5. \end{cases}
\end{aligned} \tag{8.144}
$$

Then the claim (8.138) follows from (8.143) and (8.144).

Problems

8-1 Draw hierarchies of subgroups for D_3, D_8, and D_{12} (cf., Fig. 8.3 in §8.3.2).

8-2 Draw spatial views of asymptotic potential function $\widetilde{U}(r,\theta,0)$ in the neighborhood of the double bifurcation point with $\widehat{n} = 4$ and 5 for $A_{0200} > 0$ (cf., Fig. 8.13 in §8.6.2).

8-3 Obtain the total potential function of the n-bar truss tent (cf., Fig. 9.1 in §9.5.1) and a double bifurcation point on the fundamental path. Expand this function in the neighborhood of this point to obtain an explicit form similar to (8.76).
Answer: See Ikeda et al., 1988 [98].

8-4 Draw the local imperfect behavior similar to the one in Fig. 8.14 in §8.7.3 for $\widehat{n} = 8$ and $\psi = 0, \pi/10$, and π.

8-5 Consider the regular-hexagonal truss dome depicted in Fig. 6.8 in §6.5.1. Illustrate z-directional displacements of free nodes of this dome for the two critical eigenvectors at bifurcation point B and their superposition (cf., Fig. 8.7 in §8.4.1).

8-6 Derive equation (8.111) and show equation (8.116).

8-7 Derive equation (8.118).

Summary

- A complete bifurcation diagram of a D_n-equivariant system has been presented.
- Asymptotic behaviors of D_n and C_n-equivariant systems in the neighborhood of double bifurcation points have been investigated.
- Imperfection sensitivity laws for double bifurcation points of a D_n-equivariant system have been obtained.
- The experimentally observed bifurcation diagram presented in Chapter 6 has been extended to a D_n-equivariant system.

Chapter 9
Worst Imperfection (II)

9.1 Introduction

The critical load of a structural system undergoing bifurcation is sensitive to imperfections (cf., Chapter 3). In view of the engineering demand to consider the imperfection that reduces the critical load most rapidly, the worst imperfection[1] has been formulated in Chapter 4 for simple critical points.

In this chapter, we present the theory[2] on the worst imperfection for D_n-symmetric systems, such as domes and shells, for which simple and double critical points appear generically. Naturally, emphasis is to be placed on double critical points.

In the formulation of the worst imperfection for simple critical points in Chapter 4, we have referred to the imperfection sensitivity law, when $|\varepsilon|$ is small,

$$\widetilde{f_{\mathrm{c}}} \approx C(\boldsymbol{d})|\varepsilon|^{\rho}, \tag{9.1}$$

which gives the asymptotic expression of the critical load increment $\widetilde{f_{\mathrm{c}}}$, and formulated the problem of the worst imperfection as that of maximizing $|C(\boldsymbol{d})|$ under the constraint that the norm of the imperfection pattern vector $\boldsymbol{d}$ is kept constant by

$$\boldsymbol{d}^{\top} W^{-1} \boldsymbol{d} = 1,$$

where W is a weight matrix (positive-definite symmetric matrix). Then the worst imperfection is given by $\boldsymbol{d}^*$ and/or $-\boldsymbol{d}^*$ with

$$\boldsymbol{d}^* = \frac{1}{\alpha} W {B_{\mathrm{c}}^{0}}^{\top} \xi_1, \tag{9.2}$$

---

[1] It should be remarked that the term "critical imperfection" in the first edition is replaced by the term "worst imperfection" in the second edition.

[2] This theory is based on Murota and Ikeda, 1991 [143].

K. Ikeda and K. Murota, *Imperfect Bifurcation in Structures and Materials*, Applied Mathematical Sciences 149, DOI 10.1007/978-1-4419-7296-5_9,

which is the product of the weight matrix W, the critical (left) eigenvector ξ_1, and the transpose of the imperfection sensitivity matrix $B_{\rm c}^0$ at the critical point of the perfect system (see (2.3) and (2.10) for the notation of $B_{\rm c}^0$).

A double point is associated with two independent critical (left) eigenvectors ξ_1 and ξ_2; however, it may be natural, in view of (9.2), to consider the form of

$$\boldsymbol{d}^*(\varphi) = \frac{1}{\alpha} W B_{\rm c}^{0\top} (\cos\varphi \cdot \xi_1 + \sin\varphi \cdot \xi_2) \tag{9.3}$$

for the worst imperfections. As described in this chapter, the imperfection in (9.3), in fact, serves as the worst imperfection for all φ for $\widehat{n} \geq 5$ and for some φ for $\widehat{n} = 3$ and 4. It should be recalled that the index $\widehat{n} = n/\gcd(n, j)$ is a parameter characterizing the properties of a double critical point.

The group-theoretic bifurcation theory is used to exploit symmetry. In particular, the block-diagonal form of $B_{\rm c}^0$ (cf., §7.7.2), which is a consequence of the group-equivariance of an imperfect system, is utilized in deriving the worst imperfection. Results show that the symmetry of the critical eigenvectors of the Jacobian matrix is inherited by the worst imperfection pattern vectors. This is another major finding of this chapter and is called the "resonance of symmetry."

This chapter is organized as follows.

- Formulation of the worst imperfection in Chapter 4 is extended in §9.2 to a system with group-equivariance using the framework of Chapter 7.
- Procedure to obtain the worst imperfection pattern vector is presented by considering the D_n-equivariance for an imperfect system and resonance of symmetry is explained for simple critical points in §9.3 and for double critical points in §9.4.
- Regular-polygonal truss tents and domes are used as examples in §9.5.

9.2 Formulation of Worst Imperfection

The formulation of the worst imperfection in Chapter 4 is extended to a system with group-equivariance using the theoretical framework of Chapter 7.

9.2.1 Group Equivariance

We review the theoretical framework related to the worst imperfection and introduce the symmetry condition on the weight matrix W.

Recall the governing equation in (2.1):

$$\boldsymbol{F}(\boldsymbol{u}, f, \boldsymbol{v}) = \boldsymbol{0}, \tag{9.4}$$

where $\boldsymbol{u} \in \mathbb{R}^N$, $f \in \mathbb{R}$, and $\boldsymbol{v} \in \mathbb{R}^p$. This equation is assumed to be equivariant to a group G, which is represented by (7.38):

$$T(g)\boldsymbol{F}(\boldsymbol{u}, f, \boldsymbol{v}) = \boldsymbol{F}(T(g)\boldsymbol{u}, f, S(g)\boldsymbol{v}), \qquad g \in G, \tag{9.5}$$

in terms of a unitary matrix representation T of G on the N-dimensional space of the independent variable vector $\boldsymbol{u}$, and another unitary representation S of G on the p-dimensional space of the imperfection parameter vector $\boldsymbol{v}$. For the G-equivariance of the perfect system, the imperfection parameter vector for the perfect system is assumed to be G-symmetric as in (7.39); that is,

$$\Sigma(\boldsymbol{v}^0; G, S) = G. \tag{9.6}$$

Let $(\boldsymbol{u}_c^0, f_c^0)$ be a critical point of the perfect system. The Jacobian matrix J of $\boldsymbol{F}$ is singular at $(\boldsymbol{u}_c^0, f_c^0, \boldsymbol{v}^0)$ (i.e., $\det[J(\boldsymbol{u}_c^0, f_c^0, \boldsymbol{v}^0)] = 0$). For an imperfect system described by the imperfection parameter vector $\boldsymbol{v}$, the critical point moves to $(\boldsymbol{u}_c, f_c)$ which is determined similarly by

$$\det[J(\boldsymbol{u}_c, f_c, \boldsymbol{v})] = 0 \tag{9.7}$$

and (9.4). Recall that the imperfection is expressed in terms of the increment of $\boldsymbol{v}$ from the perfect state $\boldsymbol{v}^0$:

$$\varepsilon \boldsymbol{d} = \boldsymbol{v} - \boldsymbol{v}^0$$

with imperfection pattern vector $\boldsymbol{d}$ and imperfection magnitude ε.

To formulate the problem of the worst imperfection we introduce a weight matrix W to normalize the imperfection pattern vector $\boldsymbol{d}$ as

$$\boldsymbol{d}^\top W^{-1} \boldsymbol{d} = 1. \tag{9.8}$$

The matrix W is a positive-definite matrix to be specified in accordance with an engineering viewpoint (see Remark 4.1 in §4.3.1). An additional condition is imposed on the weight matrix W for the compatibility with the group symmetry:

$$S(g) W S(g)^\top = W, \qquad g \in G, \tag{9.9}$$

which is apparently satisfied if W is equal to the unit matrix I_p.

We hereinafter consider a D_n-equivariant system that satisfies the equivariance (9.5) as well as the symmetry conditions (9.6) and (9.9) for $G = \mathrm{D}_n$.

9.2.2 Imperfection Sensitivity Law

As in Chapter 4, we make use of the imperfection sensitivity law to formulate the problem of the worst imperfection. The imperfection sensitivity laws for a D_n-equivariant system (cf., §8.8) are reviewed here.

We consider a simple or a double critical point $(\boldsymbol{u}_c^0, f_c^0)$ of the perfect D_n-equivariant system that is assumed to be group-theoretic. We denote by μ the irreducible representation of D_n associated with the critical point.

As shown in §3.2 and §8.8, the change $\widetilde{f}_{\rm c}$ of the critical load, when $\varepsilon > 0$ is small, is expressed by the imperfection sensitivity law

$$\widetilde{f}_{\rm c} \approx C(\boldsymbol{d})\varepsilon^{\rho}, \tag{9.10}$$

where ρ is an exponent determined by the type of critical point $(\boldsymbol{u}_{\rm c}^0, f_{\rm c}^0)$ of the perfect system; and $C(\boldsymbol{d})$ is a coefficient depending on the imperfection pattern $\boldsymbol{d}$. When the critical point is a bifurcation point, we restrict ourselves to the unstable bifurcation point, for which $C(\boldsymbol{d}) < 0$.

For simple critical points, the explicit forms of ρ and $C(\boldsymbol{d})$ in (9.10) are given as (cf., §3.2)

$$\begin{cases} \rho = 1, \quad C(\boldsymbol{d}) = -C_0 \cdot a & \text{at the limit point,} \\ \rho = 2/3, \ C(\boldsymbol{d}) = -C_0 \cdot a^{2/3} & \text{at the pitchfork bifurcation point,} \end{cases} \tag{9.11}$$

where C_0 is a constant and a is the imperfection coefficient dependent on $\boldsymbol{d}$ through the formula

$$a = \boldsymbol{\xi}_1^{\top} B_{\rm c}^0 \boldsymbol{d}, \tag{9.12}$$

where $\boldsymbol{\xi}_1$ is the critical (left) eigenvector of $J_{\rm c}^0$ with $\boldsymbol{\xi}_1^{\top}\boldsymbol{\xi}_1 = 1$.

Recall that group-theoretic double bifurcation points are classified on the basis of the associated two-dimensional irreducible representation $\mu = (j)_{{\rm D}_n}$. In the sensitivity law (9.10), ρ and $C(\boldsymbol{d})$ vary with the index $\widehat{n} = n/\gcd(n, j)$ as given by (8.97) in §8.8:

$$\begin{cases} \rho = 2/3, \ \ C(\boldsymbol{d}) = -C_0 \cdot |a|^{2/3} & \text{if } \widehat{n} \geq 5, \\ \rho = 1/2, \ \ C(\boldsymbol{d}) = -\tau(\psi) C_0 \cdot |a|^{1/2} & \text{if } \widehat{n} = 3, \\ \rho = 2/3, \ \ C(\boldsymbol{d}) = -\hat{\tau}(\psi) C_0 \cdot |a|^{2/3} & \text{if } \widehat{n} = 4, \end{cases} \tag{9.13}$$

where a is the imperfection coefficient that is expressed by (8.46):

$$a = A_{0001} = \boldsymbol{\xi}_1^{\top} B_{\rm c}^0 \boldsymbol{d} + {\rm i}\boldsymbol{\xi}_2^{\top} B_{\rm c}^0 \boldsymbol{d} \tag{9.14}$$

(cf., Lemma 8.2 in §8.5.2) with the orthonormal critical (left) eigenvectors $\boldsymbol{\xi}_1$ and $\boldsymbol{\xi}_2$ of $J_{\rm c}^0$; $\psi = \arg(a\varepsilon)$; $\tau(\psi) > 0$ is a nonlinear function in ψ; and $\hat{\tau}(\psi) = \hat{\tau}(\psi; A_{0200})$ is a nonlinear function in ψ and A_{0200} and is dependent on individual systems.

We can make a unified statement for simple and double points by introducing the orthogonal projection matrix

$$P = \sum_{i=1}^{M} \boldsymbol{\xi}_i \boldsymbol{\xi}_i^{\top} = \begin{cases} \boldsymbol{\xi}_1 \boldsymbol{\xi}_1^{\top} & \text{at simple points } (M = 1), \\ \boldsymbol{\xi}_1 \boldsymbol{\xi}_1^{\top} + \boldsymbol{\xi}_2 \boldsymbol{\xi}_2^{\top} & \text{at double points } (M = 2) \end{cases} \tag{9.15}$$

on the kernel space of $J_{\rm c}^{0\top}$. The coefficient $C(\boldsymbol{d})$ in (9.10) is governed primarily by the vector

$$PB_{\rm c}^0 \boldsymbol{d} = \sum_{i=1}^{M} a_i \boldsymbol{\xi}_i, \tag{9.16}$$

where

$$a_i = \boldsymbol{\xi}_i^\top B_{\rm c}^0 \boldsymbol{d}, \qquad i = 1,\ldots,M,$$

although we must consider the factor of $\tau(\psi)$ or $\hat{\tau}(\psi)$ for double points with $\widehat{n} = 3$ or 4. It is noteworthy that

$$|a|^2 = ||PB_{\rm c}^0 \boldsymbol{d}||^2 = \boldsymbol{d}^\top B_{\rm c}^{0\top} PB_{\rm c}^0 \boldsymbol{d} \tag{9.17}$$

for the variable a in (9.12) and (9.14).

As might be expected, $PB_{\rm c}^0 \boldsymbol{d} \neq \boldsymbol{0}$ holds for some $\boldsymbol{d}$ (actually, for almost all $\boldsymbol{d}$) if and only if

$$PB_{\rm c}^0 \neq O. \tag{9.18}$$

We assume (9.18) in our subsequent arguments. The assumption (9.18) enables us to carry out the asymptotic analysis on the basis of the linear term of the imperfection in the bifurcation equation, and holds in most applications by Lemma 9.1 below. Recalling that μ denotes the irreducible representation associated with the critical point, we say that μ is contained in S if $b^\mu \neq 0$, where b^μ denotes the multiplicity of μ in S (cf., §7.7.2).

Lemma 9.1. *We have $PB_{\rm c}^0 \neq O$ only if the associated irreducible representation μ is contained in S. The converse is also true in the generic sense.*

Proof. See Remark 9.1 in §9.2.4. □

9.2.3 Maximization Problem for Worst Imperfection

We formulate the problem of finding the worst imperfection as that of finding the imperfection pattern vector $\boldsymbol{d}$ that maximizes $|C(\boldsymbol{d})|$ in the imperfection sensitivity law (9.10), as is conducted in Chapter 4.

By (9.13), the maximization of $|C(\boldsymbol{d})|$ is replaced by the maximization of $|a|^2$ in (9.17) for $\widehat{n} \geq 5$; on the other hand, it is replaced by the simultaneous maximization of $|a|^2$ and $\tau(\psi)$ or $\hat{\tau}(\psi)$ for $\widehat{n} = 3$ or 4. It turns out that the variation of $|a|^2$ is predominant over the variation of $\tau(\psi)$ or $\hat{\tau}(\psi)$. Hence, our problem is reduced primarily to

$$\text{Maximize} \quad \boldsymbol{d}^\top B_{\rm c}^{0\top} PB_{\rm c}^0 \boldsymbol{d} \quad \text{subject to} \quad \boldsymbol{d}^\top W^{-1} \boldsymbol{d} = 1. \tag{9.19}$$

We have assumed $PB_{\rm c}^0 \neq O$ in (9.18). It is usually the case in practical examples that an irreducible representation contained in T is also contained in S. It then follows, by Lemma 9.1 above, that the assumption (9.18) is satisfied in the generic case.

9.2.4 Block-Diagonal Form of Imperfection Sensitivity Matrix

The equivariance (7.50) of the imperfection sensitivity matrix $B_{\rm c}^0$, which reads as

$$T(g)B_{\rm c}^0 = B_{\rm c}^0 S(g), \qquad g \in {\rm D}_n \tag{9.20}$$

is crucial in our theoretical development. It is emphasized that (9.20) is a consequence of the equivariance (9.5) of the governing equation.

This equivariance (9.20) entails block-diagonalization (7.135) with $R_{\rm a}(G) = R(G) = R({\rm D}_n)$, that is,

$$H^\top B_{\rm c}^0 \varPhi = \bigoplus_{\mu' \in R({\rm D}_n)} \bigoplus_{k=1}^{N^{\mu'}} \tilde{B}^{\mu'} \tag{9.21}$$

for a pair of $N \times N$ and $p \times p$ orthogonal matrices H and $\varPhi$ (i.e., $H^\top H = I_N$, $\varPhi^\top \varPhi = I_p$). Here μ' is used as a running index that is distinguished from the irreducible representation μ associated with the critical point $(\boldsymbol{u}_{\rm c}^0, f_{\rm c}^0)$; $N^{\mu'}$ (= 1 or 2) denotes the dimension of an irreducible representation μ', and $\tilde{B}^{\mu'}$ is an $a^{\mu'} \times b^{\mu'}$ matrix, where $a^{\mu'}$ and $b^{\mu'}$ are the multiplicities of μ' in T and S, respectively, defined in (7.127). The entries of $\tilde{B}^{\mu'}$ are generically distinct from zero for $\mu' \in R({\rm D}_n)$.

We may assume that the matrix H contains the critical eigenvector(s) $\boldsymbol{\xi}_1$ (and $\boldsymbol{\xi}_2$) as its column vector(s) (i.e., $H = [\boldsymbol{\xi}_1, \boldsymbol{\xi}_2, \ldots]$). Then, we have

$$H^\top B_{\rm c}^0 \varPhi = \begin{cases} \begin{pmatrix} \boldsymbol{\beta}^\top & 0 \\ * & (B_1)_{\rm c}^0 \end{pmatrix} & \text{at the simple point } (M=1), \\ \begin{pmatrix} \boldsymbol{\beta}^\top & 0 & 0 \\ 0 & \boldsymbol{\beta}^\top & 0 \\ * & * & (B_1)_{\rm c}^0 \end{pmatrix} & \text{at the double point } (M=2), \end{cases} \tag{9.22}$$

where $\boldsymbol{\beta}$ is a b^μ-dimensional vector which is generically distinct from zero, $(B_1)_{\rm c}^0$ is an $(N-M) \times (p - Mb^\mu)$ matrix, and the off-diagonal blocks marked $(*)$ are possibly nonzero.

Remark 9.1. Lemma 9.1 in §9.2.2 is an immediate consequence of (9.22). From the expression of P in (9.15) and the first M rows in (9.22), we see that

$$PB_{\rm c}^0 \varPhi = \begin{cases} \boldsymbol{\xi}_1 \begin{pmatrix} \boldsymbol{\beta}^\top & 0 \end{pmatrix} & \text{at the simple point } (M=1), \\ [\boldsymbol{\xi}_1, \boldsymbol{\xi}_2] \begin{pmatrix} \boldsymbol{\beta}^\top & 0 & 0 \\ 0 & \boldsymbol{\beta}^\top & 0 \end{pmatrix} & \text{at the double point } (M=2). \end{cases}$$

Since $\boldsymbol{\beta}$ is a b^μ-dimensional vector, which is generically nonzero provided $b^\mu > 0$, we have $PB_{\rm c}^0 \varPhi = O$ if $b^\mu = 0$; and $PB_{\rm c}^0 \varPhi \neq O$ (generically) if $b^\mu > 0$. □

9.3 Simple Critical Points

The worst imperfection and the resonance of symmetry at simple critical points are investigated.

9.3.1 Worst Imperfection

The worst imperfection for simple critical points can be determined as in §4.3. The answer to the maximization problem (9.19) is given by

$$\boldsymbol{d}^* = \frac{1}{\alpha} W B_{\mathrm{c}}^{0\top} \boldsymbol{\xi}_1 \tag{9.23}$$

with the imperfection influence factor

$$\alpha = (\boldsymbol{\xi}_1^\top B_{\mathrm{c}}^0 W B_{\mathrm{c}}^{0\top} \boldsymbol{\xi}_1)^{1/2}. \tag{9.24}$$

Then we see from (4.29) that $\widetilde{f}_{\mathrm{c}}$ is minimized (i.e., $|C(\boldsymbol{d})|$ is maximized) by

$$\boldsymbol{d}_{\min} = \begin{cases} \mathrm{sign}(\varepsilon A_{010})\boldsymbol{d}^* & \text{at the limit point,} \\ \pm \boldsymbol{d}^* & \text{at the pitchfork bifurcation point,} \end{cases} \tag{9.25}$$

where $\mathrm{sign}(\varepsilon A_{010})$ depends on individual systems.

The present assumption (9.18) coincides with assumption (4.22) in §4.3.2 and is justified by Lemma 9.1 in §9.2.2.

9.3.2 Resonance of Symmetry

The worst imperfection pattern displays a special geometrical characteristic by inheriting the symmetry of the critical eigenvectors. This inheritance, which is called the *resonance of symmetry*, is explained here.

We investigate the symmetry of the worst imperfection $\boldsymbol{d}_{\min}$ (cf., (9.25)), which is given by $\boldsymbol{d}^*$ and/or $-\boldsymbol{d}^*$ with $\boldsymbol{d}^*$ in (9.23). Noting that $S(g)W = WS(g)$ by (9.9) and $S(g)B_{\mathrm{c}}^{0\top} = B_{\mathrm{c}}^{0\top} T(g)$ by (9.20), we see

$$S(g)\boldsymbol{d}^* = S(g) W B_{\mathrm{c}}^{0\top} \boldsymbol{\xi}_1/\alpha = WS(g) B_{\mathrm{c}}^{0\top} \boldsymbol{\xi}_1/\alpha = W B_{\mathrm{c}}^{0\top} T(g) \boldsymbol{\xi}_1/\alpha \tag{9.26}$$

for $g \in \mathrm{D}_n$. It then follows that $S(g)\boldsymbol{d}^* = \boldsymbol{d}^*$ if $T(g)\boldsymbol{\xi}_1 = \boldsymbol{\xi}_1$. The converse is also true, since, if $T(g)\boldsymbol{\xi}_1 \neq \boldsymbol{\xi}_1$, then $T(g)\boldsymbol{\xi}_1 = -\boldsymbol{\xi}_1$ and therefore $S(g)\boldsymbol{d}^* = -\boldsymbol{d}^*$. Consequently, we obtain

$$\Sigma(\boldsymbol{d}^*; \mathrm{D}_n, S) = \Sigma(\boldsymbol{\xi}_1; \mathrm{D}_n, T),$$

which implies

$$\Sigma(\boldsymbol{d}_{\min}; \mathrm{D}_n, S) = \Sigma(\ker(J_{\mathrm{c}}^0); \mathrm{D}_n, T) \tag{9.27}$$

for the worst imperfection $\boldsymbol{d}_{\min}$, where the notation Σ is defined in (7.86). Thus the symmetry of the critical eigenvector $\boldsymbol{\xi}_1$ is reflected in the symmetry of the worst imperfection pattern $\boldsymbol{d}_{\min}$.

The relation (9.27) remains valid for a general group G with some modifications.

9.4 Double Critical Points

The worst imperfection and the resonance of symmetry at group-theoretic double critical points of a D_n-equivariant system are investigated.

9.4.1 Technical Preliminary

The block-diagonalizations of B_{c}^0 and W are the key technical ingredient in the analysis for double critical points. We recall the block-diagonalization of B_{c}^0 in (9.22); that is,

$$H^\top B_{\mathrm{c}}^0 \Phi = \begin{pmatrix} \boldsymbol{\beta}^\top & 0 & 0 \\ 0 & \boldsymbol{\beta}^\top & 0 \\ * & * & (B_1)_{\mathrm{c}}^0 \end{pmatrix}, \tag{9.28}$$

where $\boldsymbol{\beta}$ is a b^μ-dimensional vector which is generically distinct from zero; $(B_1)_{\mathrm{c}}^0$ is an $(N-2)\times(p-2b^\mu)$ matrix; and the off-diagonal blocks marked $(*)$ are possibly nonzero.

The weight matrix W is also block-diagonalized by virtue of the assumed symmetry (9.9). For the double point, we have

$$\Phi^\top W \Phi = \begin{pmatrix} W^\mu & 0 & 0 \\ 0 & W^\mu & 0 \\ 0 & 0 & W_1 \end{pmatrix}, \tag{9.29}$$

where Φ is the orthogonal matrix in (9.21), and W^μ and W_1, respectively, represent $b^\mu \times b^\mu$ and $(p-2b^\mu)\times(p-2b^\mu)$ positive-definite matrices.

The block-diagonalizations of B_{c}^0 and W play a pivotal role in the following two lemmas.

Lemma 9.2. *The two vectors* ${B_{\mathrm{c}}^0}^\top \boldsymbol{\xi}_1$ *and* ${B_{\mathrm{c}}^0}^\top \boldsymbol{\xi}_2$ *appearing in* (9.14) *are orthogonal to each other with respect to the weight matrix* W*; that is, we have*

$$\boldsymbol{\xi}_i^\top B_{\mathrm{c}}^0 W {B_{\mathrm{c}}^0}^\top \boldsymbol{\xi}_j = \alpha^2 \delta_{ij}, \qquad i,j = 1,2, \tag{9.30}$$

where

$$\alpha^2 = \boldsymbol{\beta}^\top W^\mu \boldsymbol{\beta}, \tag{9.31}$$

δ_{ij} *denotes Kronecker's delta, and* $\boldsymbol{\beta}$ *is the vector in* (9.28).

Proof. We employ (9.28) and (9.29) to obtain

$$\xi_1^\top B_c^0 W B_c^{0\top} \xi_1 = \xi_1^\top B_c^0 \Phi \cdot \Phi^\top W \Phi \cdot (\xi_1^\top B_c^0 \Phi)^\top$$
$$= \begin{pmatrix} \beta^\top & 0 & 0 \end{pmatrix} \begin{pmatrix} W^\mu & 0 & 0 \\ 0 & W^\mu & 0 \\ 0 & 0 & W_1 \end{pmatrix} \begin{pmatrix} \beta \\ 0 \\ 0 \end{pmatrix} = \beta^\top W^\mu \beta,$$

$$\xi_2^\top B_c^0 W B_c^{0\top} \xi_2 = \begin{pmatrix} 0 & \beta^\top & 0 \end{pmatrix} \begin{pmatrix} W^\mu & 0 & 0 \\ 0 & W^\mu & 0 \\ 0 & 0 & W_1 \end{pmatrix} \begin{pmatrix} 0 \\ \beta \\ 0 \end{pmatrix} = \beta^\top W^\mu \beta,$$

$$\xi_1^\top B_c^0 W B_c^{0\top} \xi_2 = \begin{pmatrix} \beta^\top & 0 & 0 \end{pmatrix} \begin{pmatrix} W^\mu & 0 & 0 \\ 0 & W^\mu & 0 \\ 0 & 0 & W_1 \end{pmatrix} \begin{pmatrix} 0 \\ \beta \\ 0 \end{pmatrix} = 0.$$

□

The maximization problem (9.19) for a double point admits an explicit solution as follows.

Lemma 9.3. *A vector $\boldsymbol{d}$ maximizes $\boldsymbol{d}^\top B_c^{0\top} P B_c^0 \boldsymbol{d}$ subject to $\boldsymbol{d}^\top W^{-1} \boldsymbol{d} = 1$ if and only if $\boldsymbol{d}$ is equal to*

$$\boldsymbol{d}^*(\varphi) = \frac{1}{\alpha} W B_c^{0\top} (\cos\varphi \cdot \xi_1 + \sin\varphi \cdot \xi_2) \tag{9.32}$$

for some φ.

Proof. First note that $\boldsymbol{d}^*(\varphi)^\top W^{-1} \boldsymbol{d}^*(\varphi) = 1$ by Lemma 9.2 above and $\boldsymbol{d}^*(\varphi)$ satisfies the constraint (9.8) for $\boldsymbol{d}$. In view of the block-diagonal forms in (9.28) and (9.29), we change the variable $\boldsymbol{d}$ to

$$\overline{\boldsymbol{d}} = \Phi^\top \boldsymbol{d} = \begin{pmatrix} \overline{\boldsymbol{d}}_1 \\ \overline{\boldsymbol{d}}_2 \\ \overline{\boldsymbol{d}}_3 \end{pmatrix}, \tag{9.33}$$

where $\overline{\boldsymbol{d}}_1$ and $\overline{\boldsymbol{d}}_2$ are of dimension b^μ, and $\overline{\boldsymbol{d}}_3$ is of dimension $p - 2b^\mu$. Using (9.28) and (9.29), we obtain

$$\boldsymbol{d}^\top B_c^{0\top} P B_c^0 \boldsymbol{d} = \overline{\boldsymbol{d}}^\top \begin{pmatrix} \beta\beta^\top & 0 & 0 \\ 0 & \beta\beta^\top & 0 \\ 0 & 0 & 0 \end{pmatrix} \overline{\boldsymbol{d}} = \sum_{i=1}^{2} \overline{\boldsymbol{d}}_i^\top \beta\beta^\top \overline{\boldsymbol{d}}_i, \tag{9.34}$$

$$\boldsymbol{d}^\top W^{-1} \boldsymbol{d} = \overline{\boldsymbol{d}}^\top \begin{pmatrix} W^\mu & 0 & 0 \\ 0 & W^\mu & 0 \\ 0 & 0 & W_1 \end{pmatrix}^{-1} \overline{\boldsymbol{d}} = \sum_{i=1}^{2} \overline{\boldsymbol{d}}_i^\top (W^\mu)^{-1} \overline{\boldsymbol{d}}_i + \overline{\boldsymbol{d}}_3^\top W_1{}^{-1} \overline{\boldsymbol{d}}_3. \tag{9.35}$$

For a maximizer $\boldsymbol{d}$ of (9.34), we must have $\overline{\boldsymbol{d}}_3 = \boldsymbol{0}$, since $\overline{\boldsymbol{d}}_3$ appears in (9.35) and not in (9.34). Then the constraint $\boldsymbol{d}^\top W^{-1} \boldsymbol{d} = 1$ is expressed as

$$\overline{\boldsymbol{d}}_1^\top (W^\mu)^{-1} \overline{\boldsymbol{d}}_1 = \cos^2\varphi, \qquad \overline{\boldsymbol{d}}_2^\top (W^\mu)^{-1} \overline{\boldsymbol{d}}_2 = \sin^2\varphi$$

for some φ. For a fixed φ, the maxima of $\overline{\boldsymbol{d}}_1^\top \beta\beta^\top \overline{\boldsymbol{d}}_1$ and $\overline{\boldsymbol{d}}_2^\top \beta\beta^\top \overline{\boldsymbol{d}}_2$ subject to these constraints are attained (cf., Lemma 4.1 in §4.3.2) by

$$\overline{\boldsymbol{d}}_1 = \frac{\cos\varphi}{\alpha} W^{\mu}\boldsymbol{\beta}, \qquad \overline{\boldsymbol{d}}_2 = \frac{\sin\varphi}{\alpha} W^{\mu}\boldsymbol{\beta} \tag{9.36}$$

with α in (9.31). For $\overline{\boldsymbol{d}}_1$ and $\overline{\boldsymbol{d}}_2$ in (9.36), the expression in (9.34) is equal to α^2, independent of φ. Noting

$$\Phi\begin{pmatrix}\overline{\boldsymbol{d}}_1\\0\\0\end{pmatrix} = \Phi\cdot\frac{\cos\varphi}{\alpha}\begin{pmatrix}W^{\mu} & 0 & 0\\0 & W^{\mu} & 0\\0 & 0 & W_1\end{pmatrix}\begin{pmatrix}\boldsymbol{\beta}\\0\\0\end{pmatrix} = \frac{\cos\varphi}{\alpha} W B_{\mathrm{c}}^{0\top}\boldsymbol{\xi}_1,$$

$$\Phi\begin{pmatrix}0\\\overline{\boldsymbol{d}}_2\\0\end{pmatrix} = \Phi\cdot\frac{\sin\varphi}{\alpha}\begin{pmatrix}W^{\mu} & 0 & 0\\0 & W^{\mu} & 0\\0 & 0 & W_1\end{pmatrix}\begin{pmatrix}0\\\boldsymbol{\beta}\\0\end{pmatrix} = \frac{\sin\varphi}{\alpha} W B_{\mathrm{c}}^{0\top}\boldsymbol{\xi}_2,$$

which follow from (9.28) and (9.29), we see that the solution $\boldsymbol{d}$ of the present maximization problem can be expressed as

$$\boldsymbol{d} = \Phi\begin{pmatrix}\overline{\boldsymbol{d}}_1\\\overline{\boldsymbol{d}}_2\\0\end{pmatrix} = \frac{\cos\varphi}{\alpha} W B_{\mathrm{c}}^{0\top}\boldsymbol{\xi}_1 + \frac{\sin\varphi}{\alpha} W B_{\mathrm{c}}^{0\top}\boldsymbol{\xi}_2 = \boldsymbol{d}^*(\varphi).$$

This expression gives (9.32). □

9.4.2 Worst Imperfection

For double critical points, the expression for the coefficient $C(\boldsymbol{d})$ in (9.10) is dependent on the value of $\widehat{n}$, as (9.13) shows; therefore, the worst imperfection is also dependent on this value. In what follows we present the results for three cases: $\widehat{n} \geq 5$, $\widehat{n} = 3$, and $\widehat{n} = 4$.

Case $\widehat{n} \geq 5$

The imperfection sensitivity law for $\widehat{n} \geq 5$ is given by the first equation of (9.13), and the minimum of the critical load $\widetilde{f}_{\mathrm{c}}$ (i.e., the maximum of $|C(\boldsymbol{d})|$) is attained by $\boldsymbol{d}$ that maximizes $|a|^2 = \boldsymbol{d}^\top B_{\mathrm{c}}^{0\top} P B_{\mathrm{c}}^0 \boldsymbol{d}$ in (9.17). By Lemma 9.3 in §9.4.1, the worst imperfection is given by $\boldsymbol{d}_{\min} = \boldsymbol{d}^*(\varphi)$ in (9.32) for any φ.

Case $\widehat{n} = 3$

The imperfection sensitivity law for $\widehat{n} = 3$ is given by the second equation of (9.13). Accordingly, the problem of worst imperfection is reduced to maximization of

$\tau(\psi)|a|^{1/2}$ with respect to $\boldsymbol{d}$ under constraint (9.8), where $\psi = \arg(a\varepsilon) = \arg(a)$ for $\varepsilon > 0$.

By (9.14) and (9.33) we have

$$a = \boldsymbol{\xi}_1^\top B_c^0 \boldsymbol{d} + \mathrm{i}\boldsymbol{\xi}_2^\top B_c^0 \boldsymbol{d} = \boldsymbol{\beta}^\top \overline{\boldsymbol{d}}_1 + \mathrm{i}\boldsymbol{\beta}^\top \overline{\boldsymbol{d}}_2.$$

This shows, in particular, that we can assume $\overline{\boldsymbol{d}}_3 = \boldsymbol{0}$, just as in the proof of Lemma 9.3 in §9.4.1. By Lemma 9.3, the maximum value of $|a|$ is attained by $\boldsymbol{d} = \boldsymbol{d}^*(\varphi)$ in (9.32) with any φ. For $\boldsymbol{d} = \boldsymbol{d}^*(\varphi)$ we have, by Lemma 9.2 in §9.4.1, that

$$a = \boldsymbol{\xi}_1^\top B_c^0 \boldsymbol{d}^*(\varphi) + \mathrm{i}\boldsymbol{\xi}_2^\top B_c^0 \boldsymbol{d}^*(\varphi) = \alpha(\cos\varphi + \mathrm{i}\sin\varphi),$$

which shows $\psi = \varphi \pmod{2\pi}$. On the other hand, $\tau(\psi)$ is maximized by $\psi = \varphi_0 + 2\pi j/3$ $(j = 0, \pm 1)$, as shown in (8.117), where $\varphi_0 = \arg(b) = \arg(A_{0200}) = 0$ or π, Consequently, the worst imperfection $\boldsymbol{d}_{\min}$ is given by

$$\boldsymbol{d}_{\min} = \boldsymbol{d}^*(\varphi_0 + 2\pi j/3), \qquad j = 0, \pm 1. \tag{9.37}$$

Remark 9.2. Even when φ_0 is unknown, the imperfection $\boldsymbol{d}^* = \boldsymbol{d}^*(\varphi)$ in (9.32) affords useful information about the critical load. Let $C(\boldsymbol{d}_{\min})$ and $C(\boldsymbol{d}^*)$ denote the coefficients in (9.10) for the imperfections $\boldsymbol{d}_{\min}$ and $\boldsymbol{d}^*$, respectively. Based on the relations (cf., (8.116))

$$|A_{0001}(\boldsymbol{d}_{\min})| \le |A_{0001}(\boldsymbol{d}^*)|, \qquad \tau(\psi_{\min}) \le \frac{\tau_{\max}}{\tau_{\min}}\tau(\psi^*) = \sqrt{3}\tau(\psi^*),$$

where $\psi_{\min} = \arg(A_{0001}(\boldsymbol{d}_{\min}))$ and $\psi^* = \arg(A_{0001}(\boldsymbol{d}^*))$, we can obtain an estimate of $C(\boldsymbol{d}_{\min})$ up to a factor of $\tau_{\max}/\tau_{\min} = \sqrt{3}$ as

$$|C(\boldsymbol{d}_{\min})| \le \frac{\tau_{\max}}{\tau_{\min}}|C(\boldsymbol{d}^*)| = \sqrt{3}\,|C(\boldsymbol{d}^*)|. \tag{9.38}$$

□

Case $\widehat{n} = 4$

The imperfection sensitivity law for $\widehat{n} = 4$ is given by the third equation of (9.13). Since $C(\boldsymbol{d})$ is dependent on A_{0200}, which varies with individual systems, no explicit means exists to determine the worst imperfection for this case.

We suggest a practical procedure for numerical analysis. Since $C(\boldsymbol{d})$ depends on $\boldsymbol{d}$ only through $A_{0001}(\boldsymbol{d})$, the worst imperfection $\boldsymbol{d}_{\min}$ is given by $\boldsymbol{d}^*(\varphi)$ for some φ in (9.32). We might compute $\widetilde{f}_c$ for the imperfection pattern $\boldsymbol{d} = \boldsymbol{d}^*(\varphi)$ for sufficiently many values of φ $(0 \le \varphi < \pi/2)$ to identify a nearly worst imperfection pattern. Such a procedure is applicable to all cases, irrespective of the value of $\widehat{n}$.

9.4.3 Resonance of Symmetry

For the symmetry of the worst imperfection at a double point, we have

$$\Sigma(\boldsymbol{d}_{\min};\mathrm{D}_n,S) \supseteq \Sigma(\ker(J_{\mathrm{c}}^0);\mathrm{D}_n,T). \tag{9.39}$$

This statement is similar to, but weaker than, the statement (9.27) for a simple critical point.

The inclusion relation (9.39) can be shown as follows. Just as (9.26), we can verify

$$S(g)\boldsymbol{d}^*(\varphi) = \frac{1}{\alpha} W B_{\mathrm{c}}^{0\top} T(g)[\boldsymbol{\xi}_1,\boldsymbol{\xi}_2]\begin{pmatrix}\cos\varphi\\ \sin\varphi\end{pmatrix}, \qquad g \in \mathrm{D}_n,$$

for $\boldsymbol{d}^*(\varphi)$ in (9.32). For $g \in \Sigma(\ker(J_{\mathrm{c}}^0);\mathrm{D}_n,T)$, we have

$$T(g)[\boldsymbol{\xi}_1,\boldsymbol{\xi}_2] = [\boldsymbol{\xi}_1,\boldsymbol{\xi}_2].$$

Therefore,

$$S(g)\boldsymbol{d}^*(\varphi) = \boldsymbol{d}^*(\varphi),$$

which means $g \in \Sigma(\boldsymbol{d}^*(\varphi);\mathrm{D}_n,S)$. This implies (9.39), since $\boldsymbol{d}_{\min} = \boldsymbol{d}^*(\varphi)$ for some φ, as shown in §9.4.2.

For $\widehat{n} = 3$, the worst imperfection enjoys a higher symmetry. It should be recalled that $\boldsymbol{\xi}_1$ is chosen to be invariant with respect to $\mathrm{D}_{n/3}$ by (8.29). For that reason, $\boldsymbol{d}_{\min}$ is also reflection invariant; that is,

$$\Sigma(\boldsymbol{d}_{\min};\mathrm{D}_n,S) = \mathrm{D}_{n/3}^{i,n} \qquad \text{for some} \quad i = 1,2,3. \tag{9.40}$$

9.5 Examples of Worst Imperfection

The D_n-equivariant truss tents and domes with n = 3, 5, and 6 are used here to illustrate the worst imperfection and the resonance of symmetry.

9.5.1 Truss Tents

We refer to the D_n-symmetric n-bar truss tents in Fig. 9.1 ($n = 3, 5$), which consist of n elastic truss members connecting a free crown node 0 to fixed nodes 1 to n. Recall that the three-bar truss tent (n = 3) in Fig. 9.1(a) was used as a simple example in §7.8.

As imperfection parameters, we choose cross-sectional areas A_i ($i = 1,\ldots,n$), the perfect values of which are $A_i = A$. We introduce the imperfection parameter vector

$$\boldsymbol{v} = (A_1,\ldots,A_n)^\top,$$

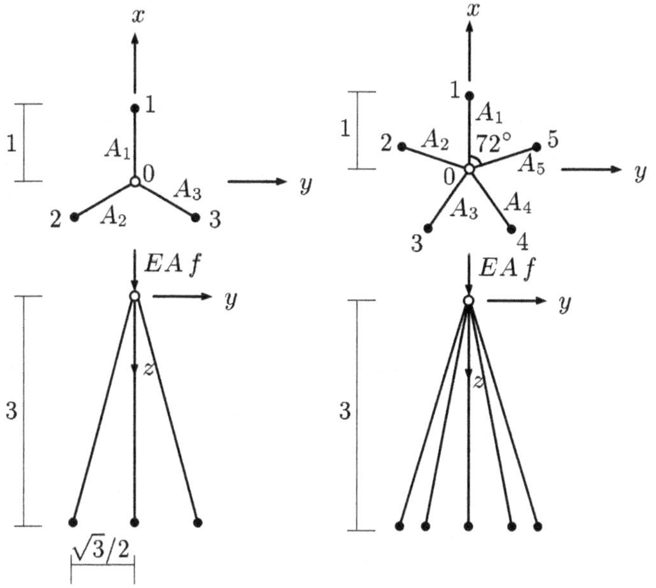

(a) $n = 3$ (b) $n = 5$

Fig. 9.1 n-bar truss tents ($n = 3, 5$).

which is equal to

$$\boldsymbol{v}^0 = (A, \ldots, A)^\top$$

for the perfect system. Recall $\boldsymbol{v} = \boldsymbol{v}^0 + \varepsilon \boldsymbol{d}$. For the weight matrix we take

$$W = A^2 \cdot I_n = A^2 \,\mathrm{diag}(1, \ldots, 1),$$

which meets requirement (9.9). Then the constraint on $\boldsymbol{d}$ in (9.8) reduces to

$$\boldsymbol{d}^\top \boldsymbol{d} = A^2. \tag{9.41}$$

For the perfect system with $\boldsymbol{v} = \boldsymbol{v}^0$, we computed the equilibrium paths (see Fig. 9.2(a) for the paths for $n = 5$) and found a pair of (group-theoretic) double bifurcation points

$$(x_{\mathrm{c}}^0, y_{\mathrm{c}}^0, z_{\mathrm{c}}^0, f_{\mathrm{c}}^0) = \begin{cases} (0, 0, 0.1877, 0.1586) & \text{for } n = 3, \\ (0, 0, 0.1877, 0.2644) & \text{for } n = 5. \end{cases}$$

The two critical eigenvectors of the Jacobian matrix at these double points are

$$\boldsymbol{\xi}_1 = \begin{pmatrix} 1 \\ 0 \\ \boldsymbol{0} \end{pmatrix}, \qquad \boldsymbol{\xi}_2 = \begin{pmatrix} 0 \\ 1 \\ \boldsymbol{0} \end{pmatrix}.$$

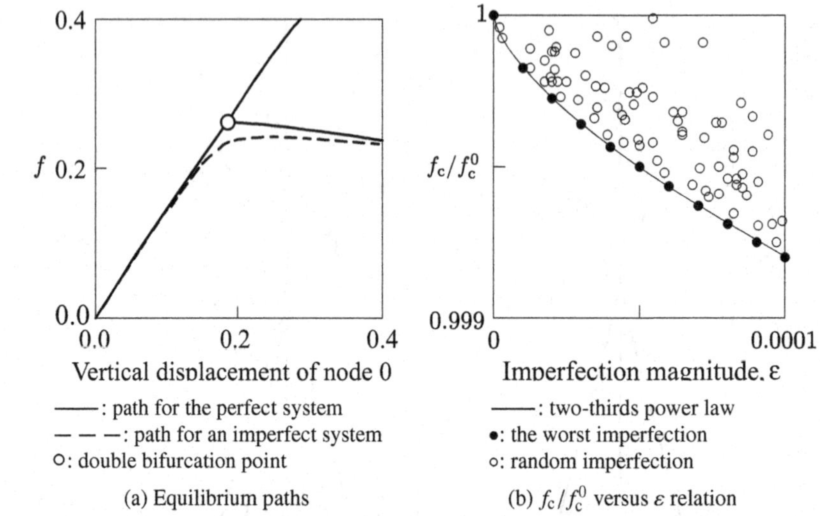

Fig. 9.2 Equilibrium paths of the five-bar truss tent and its normalized critical load f_c/f_c^0 versus imperfection magnitude ε relation (unstable double bifurcation point with $\widehat{n} = 5$).

These double points are associated with the two-dimensional irreducible representation $\mu = (1)_{D_n}$ ($n = 3, 5$) and, accordingly, the index $\widehat{n}$ is equal to n (= 3 or 5).

Five-Bar Truss Tent ($\widehat{n} = n = 5$)

We next consider the D_5-symmetric five-bar truss tent shown in Fig. 9.1(b), which corresponds to Case $\widehat{n} \geq 5$ of §9.4.2.

For a fixed ε, $\widetilde{f}_c$ is asymptotically minimized under the constraint (9.41) by $\boldsymbol{d} = \boldsymbol{d}^*(\varphi)$ in (9.32) for any φ $(0 \leq \varphi < 2\pi)$. The space of $\boldsymbol{d}^*(\varphi)$ is a two-dimensional subspace of $\boldsymbol{d}$. Note that $\widetilde{f}_c$ asymptotically stays in the range

$$C_{\min}\varepsilon^{2/3} \leq \widetilde{f}_c \leq 0,$$

where $C_{\min}$ is the minimum of $C(\boldsymbol{d})$ achieved by the worst imperfection $\boldsymbol{d} = \boldsymbol{d}^*(\varphi)$ for any φ $(0 \leq \varphi < 2\pi)$. The equilibrium path for $\boldsymbol{d} = \boldsymbol{d}^*(\varphi)$ for $\varphi = 0$ with $\varepsilon = 0.1$ is represented in Fig. 9.2(a) by the dashed line.

We numerically computed the values of $\widetilde{f}_c$ for various imperfections $(\boldsymbol{d}, \varepsilon)$. Figure 9.2(b) portrays the relation between the critical load f_c/f_c^0 and the imperfection magnitude ε. The solid curve depicts the two-thirds power law ((9.10) with (9.13)) for the worst imperfection. The values of f_c computed for the worst imperfection $\boldsymbol{d}^*(0)$ represented by (•) are smaller than those for random imperfections shown by (○) for the same imperfection magnitude ε, as it should be by the definition of the worst imperfection.

Three-Bar Truss Tent ($\widehat{n} = n = 3$)

We consider the D_3-symmetric, three-bar truss tent depicted in Fig. 9.1(a). This corresponds to Case $\widehat{n} = 3$ of §9.4.2.

The critical eigenvectors $\boldsymbol{\xi}_1$ and $\boldsymbol{\xi}_2$ and α in (9.30) are

$$\boldsymbol{\xi}_1 = (1,0,0)^\top, \qquad \boldsymbol{\xi}_2 = (0,1,0)^\top, \qquad \alpha^2 = 0.0005303 \cdot (EA)^2.$$

Consequently, $\boldsymbol{d}^*(\varphi)$ of (9.32) is given by

$$\boldsymbol{d}^*(\varphi) = \cos\varphi \cdot \frac{A}{\sqrt{6}} \begin{pmatrix} 2 \\ -1 \\ -1 \end{pmatrix} + \sin\varphi \cdot \frac{A}{\sqrt{2}} \begin{pmatrix} 0 \\ 1 \\ -1 \end{pmatrix}, \tag{9.42}$$

and the worst imperfection $\boldsymbol{d}_{\min}$ is given by (9.37) as

$$\boldsymbol{d}_{\min} = \frac{A}{\sqrt{6}} \begin{pmatrix} 2 \\ -1 \\ -1 \end{pmatrix}, \quad \frac{A}{\sqrt{6}} \begin{pmatrix} -1 \\ 2 \\ -1 \end{pmatrix}, \quad \frac{A}{\sqrt{6}} \begin{pmatrix} -1 \\ -1 \\ 2 \end{pmatrix}$$

with $\varphi_0 = 0$ for this particular case. The symmetry of $\boldsymbol{d}_{\min}$ is given by $D_1^{i,3}$ $(i = 1,2,3)$ compatibly with (9.40).

Figure 9.3 shows the relation between the critical load $\widetilde{f}_c$ and the square root of the imperfection magnitude ε. The symbols (○), (□), and (●) denote the values of $\widetilde{f}_c$ computed for the imperfection patterns $\boldsymbol{d}^*(\varphi)$ with angle φ in (9.32) of π, $3\pi/2$, and 0, where $\boldsymbol{d}^*(0) = \boldsymbol{d}_{\min}$. As might be readily apparent, $\widetilde{f}_c$ is linearly proportional to $\varepsilon^{1/2}$ for each imperfection pattern, following the one-half power law in (9.10) with (9.13). In addition, the ratio of the slopes of the $\widetilde{f}_c$ versus $\varepsilon^{1/2}$ relations for $\varphi = 0$ and π converges to $\sqrt{3}$ when ε becomes smaller, as is expected from (9.38). All these features follow the theory presented in §9.4.2 for $\widehat{n} = 3$.

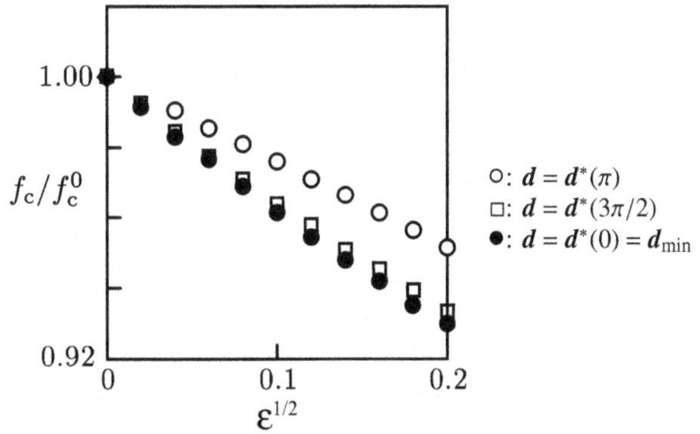

Fig. 9.3 Critical load $\widetilde{f}_c/f_c^0$ versus imperfection magnitude $\varepsilon^{1/2}$ relation for the three-bar truss tent (unstable double bifurcation point with $\widehat{n} = 3$).

9.5.2 Regular-Hexagonal Truss Dome

The D_6-symmetric regular-hexagonal truss dome portrayed in Fig. 9.4(a) is used here to illustrate the resonance of symmetry between the critical eigenvectors and the worst imperfection patterns. Recall that the worst imperfection of the simple critical points of this dome was studied in §4.5.2.

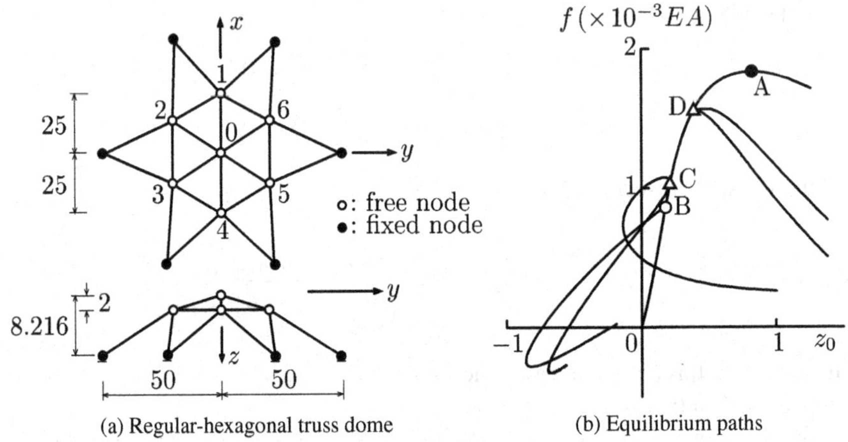

Fig. 9.4 Regular-hexagonal truss dome and its equilibrium paths. z_0: z-directional displacement of node 0; ∘: simple bifurcation point; △: double bifurcation point; •: limit point.

Figure 9.4(b) depicts equilibrium paths obtained for a set of D_6-symmetric z-directional loads of $0.5f$ applied at the crown node 0 and f applied at other free nodes. On the D_6-symmetric fundamental path, four critical points exist:

- Point A is a limit point of f with a D_6-symmetric critical eigenvector.
- Point B is a simple (pitchfork) bifurcation point with a D_3-symmetric critical eigenvector.
- Point C is a double bifurcation point with a two-dimensional C_2-symmetric kernel space ($\widehat{n} = 3$) from which D_2-symmetric branches emanate (cf., §8.6.1).
- Point D is a double point with a two-dimensional C_1-symmetric kernel space ($\widehat{n} = 6$) and with D_1-symmetric branches.

As imperfection parameters, we choose A_i $(i = 1,\dots,24)$ of the 24 members of the dome, the perfect values of which are $A_i = A$. Figure 9.5 presents the worst imperfection patterns computed at these four critical points for $W = A^2 \cdot I_{24}$.

The limit point A has the D_6-symmetric worst imperfection pattern as depicted in Fig. 9.5(a), whereas the simple bifurcation point B has the D_3-symmetric worst imperfection pattern (cf., Fig. 9.5(b)). For each of the two simple points, the symmetry of the critical eigenvector is identical to that of the worst imperfection pattern, in agreement with (9.27) expressing the resonance of symmetry.

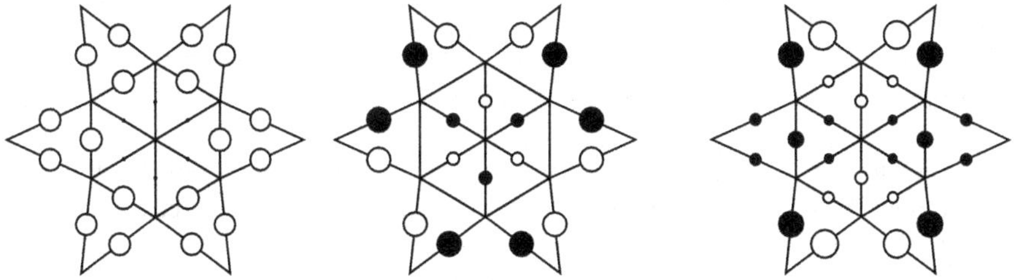

(a) Limit point A (b) Simple bifurcation point B (c) Double bifurcation point C ($\widehat{n} = 3$)

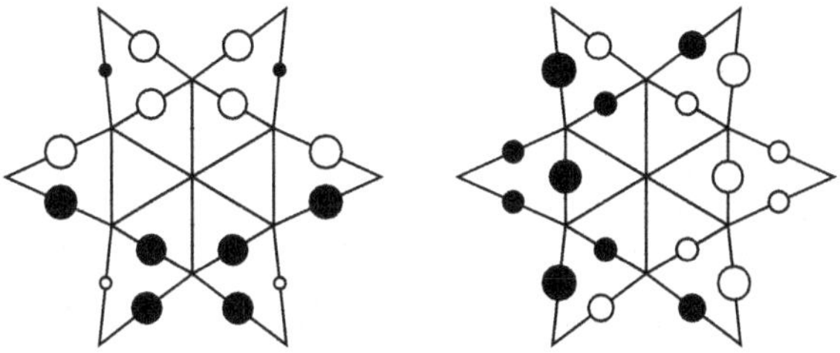

(d) Double bifurcation point D ($\widehat{n} = 6$)

Fig. 9.5 Plane view of the worst imperfection patterns at the critical points A, B, C, and D of the regular-hexagonal truss dome. ●: positive component; ○: negative component; area of ○ or ●: magnitude of a component.

The double point C, with a C_2-symmetric kernel with $\widehat{n} = 3$, has the (essentially) unique worst imperfection pattern of (9.37) in spite of the fact that the critical eigenvectors span a two-dimensional subspace. Figure 9.5(c) shows the D_2-symmetric worst imperfection pattern $\boldsymbol{d}_{\min}$, being more symmetric than the critical eigenvectors, which have C_2-symmetry. Hence the resonance of symmetry (9.40) holds.

The double point D, with a C_1-symmetric kernel with $\widehat{n} = 6$, has infinitely many worst imperfection patterns, which are linear combinations (cf., (9.32)) of the two patterns in Fig. 9.5(d). The resonance of symmetry (9.39) is observed again.

Problems

9-1 Compute the worst imperfection for the four-bar truss tent in Fig. 10.4(b) in §10.4.1 by choosing the cross-sectional areas, which are equal to A for the perfect system, as imperfection variables.

9-2 For the three-bar truss tent shown in Fig. 9.1(a) in §9.5.1, compute $f_{\rm c}$ for $\boldsymbol{d}^*(\varphi)$ $(0 \leq \varphi < 2\pi)$ in (9.42) and plot the $f_{\rm c}/f_{\rm c}^0$ versus φ relation for $\varepsilon = 0.1$.

9-3 (1) For the five-bar truss tent presented in Fig. 9.1(b), plot the $f_{\rm c}/f_{\rm c}^0$ versus φ relation for $\varepsilon = 0.1$.
(2) Compare this relation with that obtained in Problem 9-2.

9-4 Compute the worst imperfection for the regular-triangular truss dome in Fig. 8.4(a) in §8.4.1 by choosing the cross-sectional areas, which are equal to A for the perfect system, as imperfection variables.

9-5 (1) Obtain the equilibrium paths for the regular-octagonal truss dome in Fig. 9.6 and compute the worst imperfection by choosing the cross-sectional areas, which are equal to A for the perfect system, as imperfection variables.
(2) Check that the resonance of symmetry holds for this case.

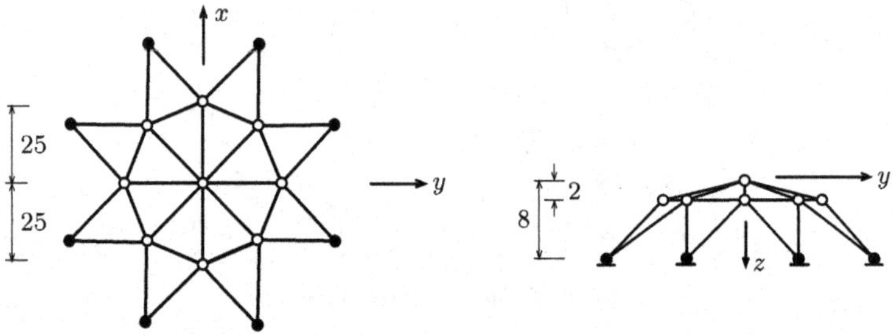

Fig. 9.6 Regular-octagonal truss dome.

Summary

- Formulas for the worst imperfection pattern of D_n-equivariant systems have been derived.
- Resonance of symmetry has been explained.

Chapter 10
Random Imperfection (II)

10.1 Introduction

It was clarified in Chapter 5, for simple critical points, that the probabilistic properties of critical loads can be formulated in an asymptotic sense (when imperfections are small). In this chapter, this formulation is extended to a D_n-equivariant system that potentially has simple and double bifurcation points.[1] For a simple critical point of a D_n-equivariant system, which is either a limit point or a pitchfork bifurcation point (cf., §8.3.1), the relevant results presented in Chapter 5 are applicable. Therefore, specific examination of a group-theoretic double bifurcation point is performed. The formulation for double points is much more complex than that for simple critical points, but the analysis can be done by exploiting the symmetry of an imperfect system.

As §8.8 shows, the change $\widetilde{f_c}$ of the critical load, when $\varepsilon > 0$ is small, is expressed by the imperfection sensitivity law (8.96):

$$\widetilde{f_c} \approx C(\boldsymbol{d})\varepsilon^{\rho}, \tag{10.1}$$

where ρ and $C(\boldsymbol{d})$ vary with the values of index[2] $\widehat{n}$, and are given by (8.97):

$$\begin{cases} \rho = 2/3, & C(\boldsymbol{d}) = -C_0 \cdot |a|^{2/3} & \text{if } \widehat{n} \geq 5, \\ \rho = 1/2, & C(\boldsymbol{d}) = -\tau(\psi) C_0 \cdot |a|^{1/2} & \text{if } \widehat{n} = 3, \\ \rho = 2/3, & C(\boldsymbol{d}) = -\hat{\tau}(\psi) C_0 \cdot |a|^{2/3} & \text{if } \widehat{n} = 4. \end{cases}$$

Here C_0 is a positive constant, a is the imperfection coefficient A_{0001} in (8.46), $\psi = \arg(a\varepsilon) = \arg(a)$, $\tau(\psi) > 0$ is a nonlinear function in ψ, and $\hat{\tau}(\psi) = \hat{\tau}(\psi; A_{0200})$ is a nonlinear function in ψ and A_{0200}.

---

[1] This extended formulation was presented in Murota and Ikeda, 1992 [144] and Ikeda and Murota, 1993 [89].

[2] The index $\widehat{n} = n/\gcd(n, j)$ (cf., (8.11)) characterizes the critical point through the associated irreducible representation $\mu = (j)_{D_n}$.

K. Ikeda and K. Murota, *Imperfect Bifurcation in Structures and Materials*, Applied Mathematical Sciences 149, DOI 10.1007/978-1-4419-7296-5_10,

The change $\widetilde{f_c}$ of the critical load in (10.1) is primarily governed by the probabilistic variation of $a = a(\boldsymbol{d})$. Under the assumption that $\boldsymbol{d}$ is normally distributed, we can show that a^2 is subject to an exponential distribution. Then, for example, for $\widehat{n} \geq 5$ the probability density function of $\widetilde{f_c}$ is shown to follow a Weibull-like distribution.

This chapter is organized as follows.

- A procedure to obtain the probability density function of the critical load for an imperfection pattern vector $\boldsymbol{d}$ subject to a multivariate normal distribution is presented in §10.2.
- Distribution of the minimum value of critical loads is investigated in §10.3.
- The procedure presented in §10.2 is applied to structural models and to the experimental results on sands and concretes in §10.4.

10.2 Probability Density Function of Critical Loads

We derive the probability density function of critical loads at a double bifurcation point of the D_n-equivariant system subjected to random imperfections.

10.2.1 Formulation

The formulation in Chapter 5 is extended to a system with group equivariance using the framework of Chapter 7.

We recall the governing equation (2.1):

$$\boldsymbol{F}(\boldsymbol{u}, f, \boldsymbol{v}) = \boldsymbol{0}, \tag{10.2}$$

where $\boldsymbol{u} \in \mathbb{R}^N$, $f \in \mathbb{R}$, and $\boldsymbol{v} \in \mathbb{R}^p$. This equation is assumed to be equivariant to a group G, which is represented by (7.53):

$$T(g)\boldsymbol{F}(\boldsymbol{u}, f, \boldsymbol{v}) = \boldsymbol{F}(T(g)\boldsymbol{u}, f, S(g)\boldsymbol{v}), \qquad g \in G \tag{10.3}$$

in terms of a unitary matrix representation T of G on the N-dimensional space of the independent variable vector $\boldsymbol{u}$, and another unitary representation S of G on the p-dimensional space of the imperfection parameter vector $\boldsymbol{v}$. For the G-equivariance of the perfect system, the imperfection vector for the perfect system is assumed to be G-symmetric. That is,

$$\Sigma(\boldsymbol{v}^0; G, S) = G$$

as in (7.39). Let $(\boldsymbol{u}_c^0, f_c^0)$ be a critical point of the perfect system.

Following Chapter 5, we consider the random variation of the critical load $\widetilde{f_c}$ when the imperfection pattern vector $\boldsymbol{d}$ is a random variable subject to a normal distribution $\mathrm{N}(\boldsymbol{0}, W)$ with a variance–covariance matrix W. Recall that the imperfection

$\boldsymbol{v}$ is expressed as (2.2)

$$\boldsymbol{v} = \boldsymbol{v}^0 + \varepsilon \boldsymbol{d}$$

in terms of the perfect state $\boldsymbol{v}^0$, the imperfection pattern vector $\boldsymbol{d}$, and the imperfection magnitude ε. We assume in this chapter that ε is fixed to a small positive constant. For consistency with group symmetry, we naturally adopt an additional assumption

$$S(g)WS(g)^\top = W, \qquad g \in G. \tag{10.4}$$

This condition is apparently satisfied if W is equal to the unit matrix I_p, that is, if the components of $\boldsymbol{d}$ are independent, each being subject to the standard normal distribution.

The equivariance (10.4) of the variance–covariance matrix W, as well as the equivariance (10.3) of the governing equation, plays a pivotal role in the theoretical development. Note that (10.4) is exactly of the same form as (9.9) for the weight matrix for the problem of critical imperfection and, accordingly, entails a block-diagonal form of W in (9.29).

We hereinafter restrict ourselves to a D_n-equivariant system that satisfies the equivariance (10.3) for $G = \mathrm{D}_n$, and consider only an unstable group-theoretic double critical point $(\boldsymbol{u}_\mathrm{c}^0, f_\mathrm{c}^0)$ of the perfect system because the results for a simple critical point (limit or pitchfork bifurcation point) in Chapter 5 are applicable to a simple point of this system.

10.2.2 Derivation of Probability Density Functions

The probability density functions of the critical load at the double critical point $(\boldsymbol{u}_\mathrm{c}^0, f_\mathrm{c}^0)$ are derived. An explicit formula that is related to the Weibull distribution can be obtained for $\widehat{n} \geq 5$. Here $\widehat{n} = n/\gcd(n, j)$ by (8.11).

As shown in §8.8, the change $\widetilde{f}_\mathrm{c}$ of the critical load, when $\varepsilon > 0$ is small, is expressed by the imperfection sensitivity law in (8.96):

$$\widetilde{f}_\mathrm{c} \approx C(\boldsymbol{d})\varepsilon^\rho, \tag{10.5}$$

where ρ and $C(\boldsymbol{d})$ vary with the values of index $\widehat{n}$ as given by (8.97):

$$\begin{cases} \rho = 2/3, & C(\boldsymbol{d}) = -C_0 \cdot |a|^{2/3} & \text{if } \widehat{n} \geq 5, \\ \rho = 1/2, & C(\boldsymbol{d}) = -\tau(\psi) C_0 \cdot |a|^{1/2} & \text{if } \widehat{n} = 3, \\ \rho = 2/3, & C(\boldsymbol{d}) = -\hat{\tau}(\psi) C_0 \cdot |a|^{2/3} & \text{if } \widehat{n} = 4. \end{cases} \tag{10.6}$$

Here C_0 is a positive constant, and a is given by (8.46):

$$a = A_{0001} = \boldsymbol{\xi}_1^\top B_\mathrm{c}^0 \boldsymbol{d} + \mathrm{i} \boldsymbol{\xi}_2^\top B_\mathrm{c}^0 \boldsymbol{d}$$

with the orthonormal critical (left) eigenvectors $\boldsymbol{\xi}_1$ and $\boldsymbol{\xi}_2$ of J_c^0; $\psi = \arg(a\varepsilon) = \arg(a)$ since $\varepsilon > 0$; $\tau(\psi) > 0$ is a nonlinear function in ψ; and $\hat{\tau}(\psi) = \hat{\tau}(\psi; A_{0200})$ is a nonlinear function in ψ and A_{0200}. An important implication is that the change $\widetilde{f}_c$ of the critical load is primarily governed by (cf., (9.17))

$$|a|^2 = \|PB_c^0 \boldsymbol{d}\|^2 = |\boldsymbol{\xi}_1^\top B_c^0 \boldsymbol{d}|^2 + |\boldsymbol{\xi}_2^\top B_c^0 \boldsymbol{d}|^2, \tag{10.7}$$

where $P = \xi_1 \xi_1^\top + \xi_2 \xi_2^\top$. Recall that $a = A_{0001}$ is named the imperfection coefficient.

By the assumption, $\boldsymbol{d}$ is normally distributed with mean $\boldsymbol{0}$ and variance–covariance W. By virtue of the group equivariance, the two row vectors $\boldsymbol{\xi}_1^\top B_c^0$ and $\boldsymbol{\xi}_2^\top B_c^0$ in (10.7) are orthogonal and have the same norm with respect to W as presented in Lemma 9.2 in §9.4.1. This makes possible the derivation of an explicit formula for the probability density function of $|a|^2 = \|PB_c^0 \boldsymbol{d}\|^2$, which turns out, in Lemma 10.2 below, to be the exponential distribution up to scaling. This distribution is then transformed to that of the critical load on the basis of the asymptotic relation given by (10.5). The distribution of the critical load that is so derived varies with the index $\widehat{n}$ as a consequence of the dependence of expression (10.5) on the index $\widehat{n}$.

Lemma 10.1.

$$\|PB_c^0 \boldsymbol{d}\|^2 = |\boldsymbol{\xi}_1^\top B_c^0 \boldsymbol{d}|^2 + |\boldsymbol{\xi}_2^\top B_c^0 \boldsymbol{d}|^2 = \widetilde{\sigma}^2(|\boldsymbol{\eta}_1^\top \widetilde{\boldsymbol{d}}|^2 + |\boldsymbol{\eta}_2^\top \widetilde{\boldsymbol{d}}|^2). \tag{10.8}$$

Therein, $\widetilde{\boldsymbol{d}}$ is a random variable subject to the standard normal distribution (i.e., $\widetilde{\boldsymbol{d}} \sim \mathrm{N}(\boldsymbol{0}, I_p)$), $\boldsymbol{\eta}_1$ and $\boldsymbol{\eta}_2$ are mutually orthogonal p-dimensional unit vectors independent of $\widetilde{\boldsymbol{d}}$, and

$$\widetilde{\sigma}^2 = \boldsymbol{\xi}_1^\top B_c^0 W B_c^{0\top} \boldsymbol{\xi}_1 = \boldsymbol{\xi}_2^\top B_c^0 W B_c^{0\top} \boldsymbol{\xi}_2.$$

Proof. If we decompose W^{-1} as

$$W^{-1} = V^\top V$$

and define a transformation

$$\widetilde{\boldsymbol{d}} = V\boldsymbol{d},$$

then we have $\widetilde{\boldsymbol{d}} \sim \mathrm{N}(\boldsymbol{0}, I_p)$, since the variance–covariance matrix for $\widetilde{\boldsymbol{d}}$ reads as

$$\mathrm{Var}[\widetilde{\boldsymbol{d}}] = \mathrm{E}[\widetilde{\boldsymbol{d}}\widetilde{\boldsymbol{d}}^\top] = \mathrm{E}[V\boldsymbol{d}\boldsymbol{d}^\top V] = V \cdot \mathrm{E}[\boldsymbol{d}\boldsymbol{d}^\top] \cdot V = V \cdot \mathrm{Var}[\boldsymbol{d}] \cdot V^\top = VWV^\top = I_p.$$

In view of (9.28), we define

$$\boldsymbol{\eta}_1 = \frac{1}{\widetilde{\sigma}}(V^\top)^{-1}\Phi \begin{pmatrix} \boldsymbol{\beta} \\ 0 \\ 0 \end{pmatrix}, \qquad \boldsymbol{\eta}_2 = \frac{1}{\widetilde{\sigma}}(V^\top)^{-1}\Phi \begin{pmatrix} 0 \\ \boldsymbol{\beta} \\ 0 \end{pmatrix}.$$

Then it follows from (9.29) that

$$\|\boldsymbol{\eta}_1\| = \|\boldsymbol{\eta}_2\| = 1, \qquad \boldsymbol{\eta}_1^\top \boldsymbol{\eta}_2 = 0.$$

Furthermore, relation (10.8) can be verified using (9.22), (9.29), and (9.30) in a way similar to the proof of Lemma 9.2 in §9.4.1. □

Lemma 10.2. *The variable*

$$x = \left(\frac{|a|}{\widetilde{\sigma}}\right)^2 \tag{10.9}$$

is subject to the exponential distribution, or the χ^2 *distribution of two degrees of freedom. The probability density function of x is therefore given as*

$$\phi_x(x) = \frac{1}{2}\exp\left(\frac{-x}{2}\right), \qquad x > 0. \tag{10.10}$$

Proof. It follows from Lemma 10.1 above that $(\boldsymbol{\eta}_1^\top \widetilde{\boldsymbol{d}}, \boldsymbol{\eta}_2^\top \widetilde{\boldsymbol{d}}) \sim \mathrm{N}(\mathbf{0}, I_2)$. The sum of squares of two mutually independent standard normal variables is well known[3] to be subject to the χ^2 distribution of two degrees of freedom. Hence $|a|^2$ in (10.7) is subject to this distribution. □

Using x of (10.9) in (10.5) and (10.6), we introduce the *normalized critical load increment* as

$$\zeta = \frac{\widetilde{f}_\mathrm{c}}{\widehat{C}} = \begin{cases} -x^{1/3} & \text{if } \widehat{n} \geq 5, \\ -\tau(\psi)x^{1/4} & \text{if } \widehat{n} = 3, \\ -\hat{\tau}(\psi)x^{1/3} & \text{if } \widehat{n} = 4, \end{cases} \tag{10.11}$$

where $\widehat{C} = C_0(\widetilde{\sigma}\varepsilon)^\rho$ with $\rho = 1/2$ for $\widehat{n} = 3$ and $\rho = 2/3$ for $\widehat{n} \geq 4$. As might be apparent from this equation, the explicit form of the probability density function $\phi_\zeta(\zeta)$ of ζ is determined entirely from the value of index $\widehat{n}$, although the scaling factor $\widehat{C}$ depends on individual systems.

Case $\widehat{n} \geq 5$

For an unstable double bifurcation point with $\widehat{n} \geq 5$, the combination of (10.10) with (10.11) gives the probability density function $\phi_\zeta(\zeta)$, the cumulative distribution function $\Phi_\zeta(\zeta)$, and the reliability function $R_\zeta(\zeta)$:

$$\phi_\zeta(\zeta) = \frac{3\zeta^2}{2}\exp\left(\frac{-|\zeta|^3}{2}\right), \qquad -\infty < \zeta < 0, \tag{10.12}$$

$$\Phi_\zeta(\zeta) = \exp\left(\frac{-|\zeta|^3}{2}\right), \qquad -\infty < \zeta < 0, \tag{10.13}$$

$$R_\zeta(\zeta) = 1 - \exp\left(\frac{-|\zeta|^3}{2}\right), \qquad -\infty < \zeta < 0. \tag{10.14}$$

See (5.13) for the definition of the reliability function.

The expected value $\mathrm{E}[\zeta]$ and the variance $\mathrm{Var}[\zeta]$ of ζ are

[3] See, for example, Kendall and Stuart, 1977 [115].

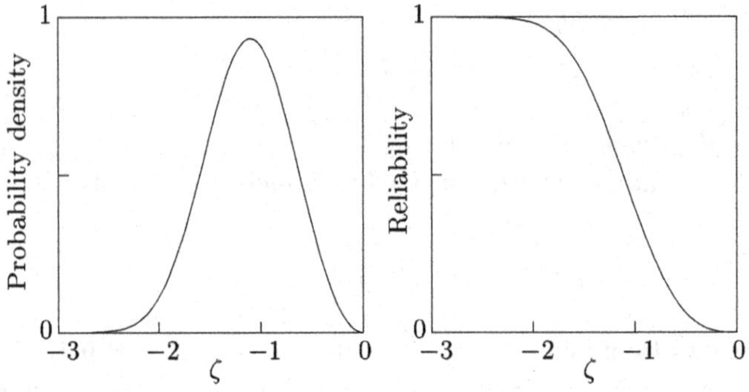

(a) Probability density function $\phi_\zeta(\zeta)$ (b) Reliability function $R_\zeta(\zeta)$

Fig. 10.1 Probability density function $\phi_\zeta(\zeta)$ and reliability function $R_\zeta(\zeta)$ of the normalized critical load increment ζ of (10.11) for an unstable double bifurcation point with $\widehat{n} \geq 5$.

$$\mathrm{E}[\zeta] = -2^{1/3}\Gamma(4/3) = -1.13, \tag{10.15}$$
$$\mathrm{E}[\zeta^2] = 2^{2/3}\Gamma(5/3) = 1.43, \tag{10.16}$$
$$\mathrm{Var}[\zeta] = 2^{2/3}\{\Gamma(5/3) - [\Gamma(4/3)]^2\} = 0.409^2. \tag{10.17}$$

Note that $|\zeta|$ is subject to a Weibull distribution. Figure 10.1(a) and (b), respectively, show the shape of the probability density function $\phi_\zeta(\zeta)$ and the reliability function $R_\zeta(\zeta)$.

Simple calculations yield various statistical properties of the critical load

$$f_c = \widehat{C}\zeta + f_c^0$$

as presented below:

$$\phi_{f_c}(f_c) = \frac{3(f_c - f_c^0)^2}{2\widehat{C}^3} \exp\left(\frac{-|f_c - f_c^0|^3}{2\widehat{C}^3}\right), \quad -\infty < f_c < f_c^0, \tag{10.18}$$

$$\Phi_{f_c}(f_c) = \exp\left(\frac{-|f_c - f_c^0|^3}{2\widehat{C}^3}\right), \quad -\infty < f_c < f_c^0, \tag{10.19}$$

$$R_{f_c}(f_c) = 1 - \exp\left(\frac{-|f_c - f_c^0|^3}{2\widehat{C}^3}\right), \quad -\infty < f_c < f_c^0, \tag{10.20}$$

$$\mathrm{E}[f_c] = f_c^0 - 2^{1/3}\Gamma(4/3)\widehat{C} = f_c^0 - 1.13\widehat{C}, \tag{10.21}$$

$$\mathrm{Var}[f_c] = 2^{2/3}\{\Gamma(5/3) - [\Gamma(4/3)]^2\}\widehat{C}^2 = (0.409\widehat{C})^2. \tag{10.22}$$

Case $\widehat{n} = 3$

As presented in (10.11), the normalized critical load increment ζ for $\widehat{n} = 3$ is given by

$$\zeta = -\tau(\psi)y, \tag{10.23}$$

where $y = x^{1/4}$, and $\tau(\psi) > 0$ is a solution to the equation[4]

$$g(\tau) \equiv \frac{27}{256}\tau^6 - \frac{9}{8}\tau^2 - \frac{1}{\tau^2} = 2\cos(3\psi).$$

Inasmuch as the variables y and $\tau(\psi)$ are statistically independent, we can derive the distributions of y and $\tau(\psi)$ separately, and then combine these distributions to obtain the distribution of ζ.

The probability density function of $y = x^{1/4}$ is computed using (10.10) as

$$\phi_y(y) = 2y^3 \exp\left(\frac{-y^4}{2}\right), \qquad 0 < y < \infty. \tag{10.24}$$

On the other hand, ψ is uniformly distributed in the range of $0 < \psi < 2\pi$, and the probability density function of $\tau = \tau(\psi)$ is therefore given as

$$\phi_\tau(\tau) = 6 \times \frac{1}{2\pi}\left|\frac{d\psi}{d\tau}\right| = \frac{g'(\tau)}{2\pi\sqrt{1 - g(\tau)^2/4}}, \qquad \tau_{\min} < \tau < \tau_{\max}, \tag{10.25}$$

where

$$\tau_{\min} = 2/\sqrt{3}, \qquad \tau_{\max} = 2.$$

The probability density functions $\phi_y(y)$ in (10.24) and $\phi_\tau(\tau)$ in (10.25) are depicted, respectively, in Fig. 10.2(a) and (b).

The statistical independence of y and τ enables us to combine (10.24) and (10.25) to obtain the probability density function $\phi_\zeta(\zeta)$ of $\zeta = -\tau y$:

$$\phi_\zeta(\zeta) = \int_{\tau_{\min}}^{\tau_{\max}} \phi_y\left(\frac{|\zeta|}{\tau}\right)\frac{\phi_\tau(\tau)}{\tau}\,d\tau, \qquad -\infty < \zeta < 0, \tag{10.26}$$

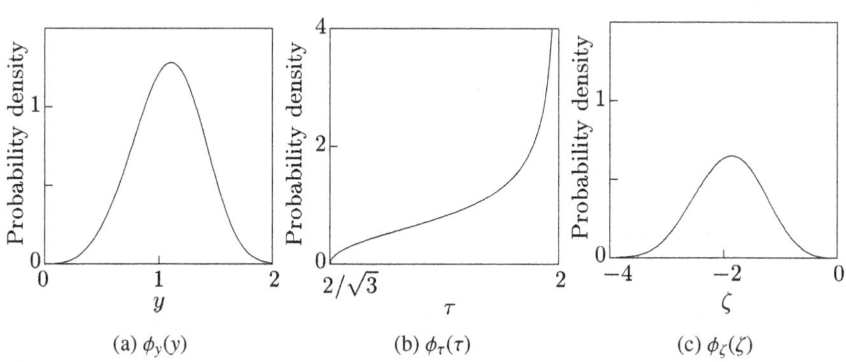

Fig. 10.2 Probability density functions for a double bifurcation point with $\widehat{n} = 3$.

[4] See §8.8.2. We assume that $b > 0$. The other case $b < 0$ can be treated similarly.

showing that the probability density function of ζ is independent of individual systems, just as in Case $\widehat{n} \geq 5$. The function $\phi_\zeta(\zeta)$ is portrayed in Fig. 10.2(c). In fact, τ lies in a bounded positive interval away from zero and, therefore, plays only a secondary role in (10.26) compared with y.

The mean and the variance of the relevant variables are evaluated as follows. From the probability density function (10.24) of y, we have

$$\mathrm{E}[y] = 2^{1/4}\Gamma(5/4) = 1.08, \qquad \mathrm{Var}[y] = \sqrt{\pi/2} - \sqrt{2}\,[\Gamma(5/4)]^2 = 0.302^2.$$

For τ, we resort to numerical integration for the density function in (10.25) to obtain

$$\mathrm{E}[\tau] = 1.77, \qquad \mathrm{Var}[\tau] = 0.221^2.$$

Numerical evaluations using (10.26) yield

$$\mathrm{E}[\zeta] = 1.91, \qquad \mathrm{Var}[\zeta] = 0.590^2,$$

which give the mean and the variance of the critical load f_c:

$$\mathrm{E}[f_\mathrm{c}] = f_\mathrm{c}^0 - 1.91\widehat{C}, \qquad \mathrm{Var}[f_\mathrm{c}] = (0.590\widehat{C})^2. \tag{10.27}$$

Case $\widehat{n} = 4$

As shown in (10.11), the normalized critical load increment ζ for $\widehat{n} = 4$ is given by

$$\zeta = -\hat{\tau}(\psi)y,$$

where $y = x^{1/3}$ and $\hat{\tau}(\psi) = \hat{\tau}(\psi; A_{0200})$ varies with individual systems (cf., §8.8.3 for details).

The variables $y = x^{1/3}$ and $\hat{\tau} = \hat{\tau}(\psi)$ are statistically independent, like the case of $\widehat{n} = 3$. The density function of $y = x^{1/3}$ is obtained from (10.10) as

$$\phi_y(y) = \frac{3y^2}{2}\exp\left(\frac{-y^3}{2}\right), \qquad 0 < y < \infty.$$

Since $\hat{\tau}$ is a function depending also on parameter A_{0200}, the probability distribution of $\hat{\tau}$ and therefore that of ζ vary with individual systems.

Remark 10.1. In numerical analyses, the probability distribution of ζ for $\widehat{n} = 4$, can be evaluated as follows. Compute $\widetilde{f}_\mathrm{c}$ for the imperfection patterns

$$\boldsymbol{d}^*(\varphi) = \cos\varphi \frac{{B_\mathrm{c}^0}^\top \xi_1}{\|{B_\mathrm{c}^0}^\top \xi_1\|} + \sin\varphi \frac{{B_\mathrm{c}^0}^\top \xi_2}{\|{B_\mathrm{c}^0}^\top \xi_2\|} \tag{10.28}$$

for sufficiently many values of φ ($0 \leq \varphi < \pi/2$). Since $\|PB_\mathrm{c}^0\boldsymbol{d}^*(\varphi)\|$ is independent of φ, this gives the distribution of $\hat{\tau}$. Then the distribution of ζ is computed from formula (10.26). The values of mean E[ζ] and variance Var[ζ] must be obtained through numerical integration for each case. □

10.2.3 Semiempirical Evaluation

A *semiempirical* evaluation similar to that explained in §5.3 is applicable also to double bifurcation points with $\widehat{n} = 3$ and $\widehat{n} \geq 5$, for which the form of the probability density function $\phi_\zeta(\zeta)$ is independent of individual systems. From (10.21), (10.22), and (10.27), the critical load f_c^0 for the perfect system is computed as

$$f_c^0 = \begin{cases} \mathrm{E_{sample}}[f_c] + 2.75(\mathrm{Var_{sample}}[f_c])^{1/2} & \text{if } \widehat{n} \geq 5, \\ \mathrm{E_{sample}}[f_c] + 3.24(\mathrm{Var_{sample}}[f_c])^{1/2} & \text{if } \widehat{n} = 3, \end{cases} \tag{10.29}$$

and the scale factor $\widehat{C}$ is computed as

$$\widehat{C} = \begin{cases} (\mathrm{Var_{sample}}[f_c])^{1/2}/0.409 & \text{if } \widehat{n} \geq 5, \\ (\mathrm{Var_{sample}}[f_c])^{1/2}/0.590 & \text{if } \widehat{n} = 3. \end{cases} \tag{10.30}$$

Substitution of the values of f_c^0 and $\widehat{C}$ computed in this manner into (10.18) yields the semiempirical probability density function of the critical load f_c.

This procedure is not applicable to the case of $\widehat{n} = 4$, in which the form of the probability density function $\phi_\zeta(\zeta)$ is dependent on individual systems.

10.3 Distribution of Minimum Values

As shown in §5.4, the minimum critical load attained by a series of random imperfections $\boldsymbol{d}$ for simple critical points can be estimated asymptotically. In this section, we extend this to an unstable double bifurcation point with $\widehat{n} \geq 5$.

Let ζ_K denote the minimum of the normalized critical load ζ achieved by K independent random imperfections. In the case of $\widehat{n} \geq 5$, the cumulative distribution function of ζ_K is expressed from (10.12) as

$$\Phi_K(\zeta_K) = 1 - \left[1 - \exp\left(\frac{-|\zeta_K|^3}{2}\right)\right]^K, \qquad -\infty < \zeta_K < 0 \tag{10.31}$$

and the differentiation of (10.31) yields the probability density function

$$\phi_K(\zeta_K) = \frac{3K{\zeta_K}^2}{2} \exp\left(\frac{-|\zeta_K|^3}{2}\right)\left[1 - \exp\left(\frac{-|\zeta_K|^3}{2}\right)\right]^{K-1}, \tag{10.32}$$

which is plotted in Fig. 10.3 for various values of K. The peak of this function shifts toward $-\infty$ and sharpens as K increases.

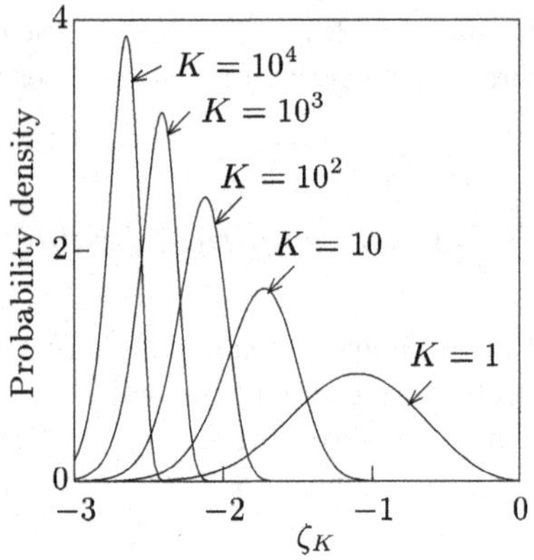

Fig. 10.3 Probability density functions of minimum normalized critical loads ζ_K attained by K independent random imperfections for a double bifurcation point with $\widehat{n} \geq 5$.

For the asymptotic form of (10.31), formula (5.29),

$$\lim_{K\to\infty} \Phi_K(c_K + d_K x) = \lim_{K\to\infty} \Pr\left\{\frac{\zeta_K - c_K}{d_K} \leq x\right\} = 1 - \exp(-\mathrm{e}^x)$$

also holds (see Remark 10.2 below) with

$$c_K = -(2\log K)^{1/3}, \qquad d_K = \frac{2}{3}(2\log K)^{-2/3}.$$

Remark 10.2. The derivation of the scaling factors c_K and d_K above is sketched here. Refer to §5.6 for a similar argument. By (10.13) we see that

$$r(t) \approx \frac{2}{3} \cdot \frac{1}{|t|^2} \qquad (t \to -\infty)$$

in the notation of Lemma 5.2 in §5.4, and that (5.25) is satisfied. The constant c_K is determined to be $c_K = -(2\log K)^{1/3}$ from the equation $\Phi(c_K) = 1/K$ with $\Phi = \Phi_\zeta$ of (10.13), whereas

$$d_K = r(c_K) \approx \frac{2}{3}(2\log K)^{-2/3}.$$

□

10.4 Examples of Scatter of Critical Loads

We offer in this section examples of the probabilistic variation of critical loads of D_n-symmetric systems undergoing bifurcation, and describe this variation using the theory presented in this chapter. Numerical examples are presented in §10.4.1; experimental examples are presented in §10.4.2.

10.4.1 Regular-Polygonal Truss Tents and Domes

As examples for double bifurcation points we consider the n-bar truss tents (n = 3,4,5) presented in Fig. 10.4, and the regular polygonal truss domes (n = 3,4,5) in Fig. 10.5. The former is subjected to a vertical load f applied at the top node; the latter is subject to uniform vertical loads f applied to the other free nodes.

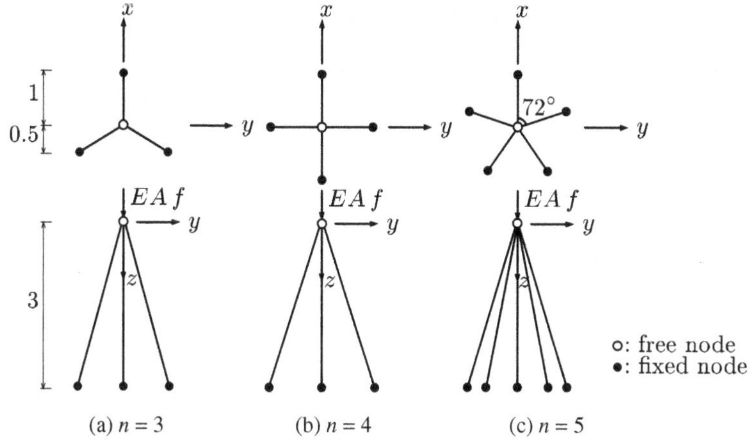

Fig. 10.4 n-bar truss tents (n = 3,4,5).

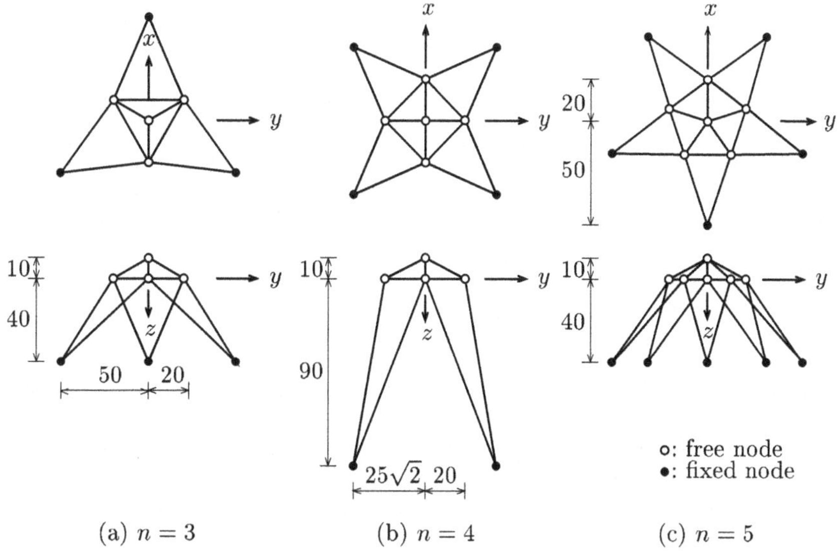

Fig. 10.5 Regular n-sided polygonal truss domes (n = 3,4,5).

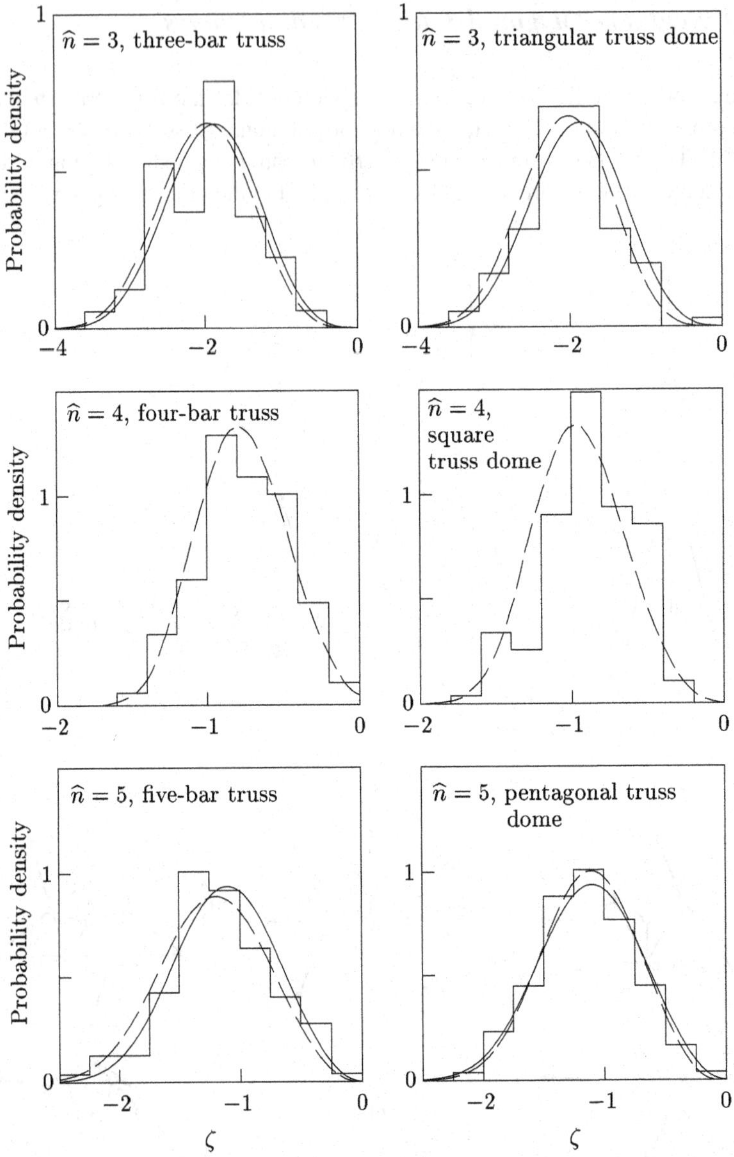

Fig. 10.6 Comparison of the empirical histograms (100 samples) and probability density functions $\phi_\zeta(\zeta)$ at unstable double bifurcation points. ——: theoretical probability density function; — —: semiempirical probability density function for $\widehat{n}=3$ and 5 and numerically estimated probability density function for $\widehat{n}=4$; histogram: numerical experiment.

We define an imperfection parameter vector as

$$\boldsymbol{v} = (A_1,\dots,A_p)^\top,$$

where A_i is the cross-sectional area of member i for $i = 1,\dots,p$. For the perfect system, we have

$$\boldsymbol{v}^0 = (A,\dots,A)^\top.$$

We assume that $\varepsilon \boldsymbol{d} = \boldsymbol{v} - \boldsymbol{v}^0$ is subject to a multivariate normal distribution $\mathrm{N}(\boldsymbol{0}, \varepsilon^2 I_p)$. Then the group equivariance (10.4) is satisfied by $W = I_p$.

The critical loads of these tents and domes for the perfect cases ($\varepsilon = 0$) are governed by unstable (group-theoretic) double bifurcation points, with $\Sigma(\ker(J_\mathrm{c}^0)) = \mathrm{C}_1$. These double points for $n = 3$, 4, and 5 respectively correspond to the three cases $\widehat{n} = 3$, $\widehat{n} = 4$, and $\widehat{n} \geq 5$ in §10.2.2.

For each case, we have randomly chosen $K = 100$ imperfection patterns of $\varepsilon \boldsymbol{d} = \boldsymbol{v} - \boldsymbol{v}^0$ subject to a multivariate normal distribution $\mathrm{N}(\boldsymbol{0}, \varepsilon^2 I_p)$ and computed a set of normalized critical loads ζ for a constant imperfection magnitude $\varepsilon = 10^{-4}$. In Fig. 10.6 the empirical histograms obtained from these 100 imperfections are compared with the theoretical and semiempirical probability density functions $\phi_\zeta(\zeta)$ for $\widehat{n} = 3$ and 5, and the numerically estimated probability density function $\phi_\zeta(\zeta)$ for $\widehat{n} = 4$ (cf., Remark 10.1 in §10.2.2). The theoretical and semiempirical functions show fair agreement with the histogram in each case of $\widehat{n} = 3$ and 5.

10.4.2 Cylindrical Material Specimens

The strength variation of cylindrical sand and concrete specimens under compression is explained using the present method. These specimens are cylindrical; therefore, we assume that they are D_n-symmetric[5] for large n. They, for example, can undergo a bifurcation process

$$\mathrm{D}_n \to \mathrm{D}_{n/2} \quad \text{or} \quad \mathrm{C}_n$$

associated with a pitchfork bifurcation point, and

$$\mathrm{D}_n \to \mathrm{D}_{n/\widehat{n}}$$

related to a double bifurcation point with $\widehat{n} \geq 5$ corresponding to Case $\widehat{n} \geq 5$ in §10.2.2. In the remainder of this subsection, we consider a pitchfork bifurcation point and a double bifurcation point with $\widehat{n} \geq 5$ referring to other types of critical points for comparison.

[5] We further exploit the upside-down symmetry of the cylindrical specimens in §13.3.

Sand Specimens

We refer to the test results on sand specimens presented in §6.5.2. The histograms of the maximum deviatoric stress $(\sigma_a)_c$ for Series A and B are compared with the probability density functions for critical points of various kinds in Fig. 10.7. The values of the sample mean $E_{sample}[(\sigma_a)_c]$ and the sample variance $Var_{sample}[(\sigma_a)_c]$ of the maximum deviatoric stress for these series are listed in Table 10.1(a). The use of these values in (5.19), (5.20), (10.29), and (10.30) results in the values of $(\sigma_a)_c^0$ and $\widehat{C}$ also listed in this table. Substitution of the values of $(\sigma_a)_c^0$ and $\widehat{C}$ into (5.15) and (10.18) gives the curves of the semiempirical probability density function of $(\sigma_a)_c$ depicted in Fig. 10.7. These curves are fairly consistent with the empirical histogram.

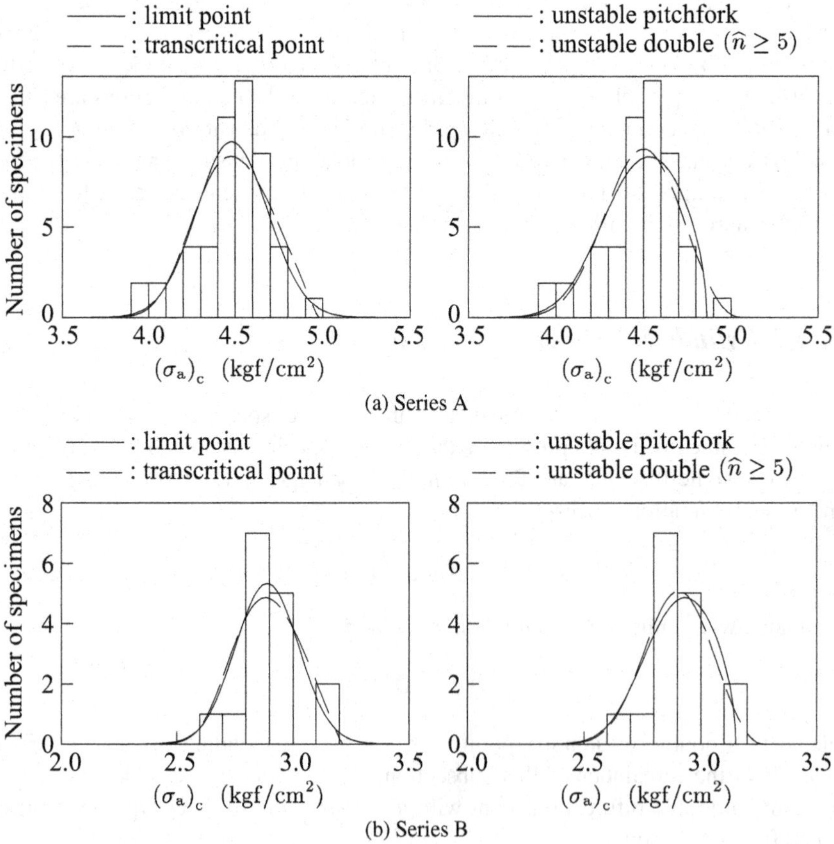

Fig. 10.7 Comparison of the empirical histograms and semiempirical probability density functions of the maximum deviatoric stress $(\sigma_a)_c$ for sand specimens ($1\,\text{kgf/cm}^2 = 98\,\text{kPa}$).

Table 10.1 Statistical parameters (unit in kgf/cm^2; 1 kgf/cm^2 = 98 kPa)

(a) Sand specimens

Series	Type of Point	$E_{sample}[(\sigma_a)_c]$	$Var_{sample}[(\sigma_a)_c]$	$(\sigma_a)_c^0$	$\widehat{C}$
A	Limit	4.48	0.205^2	4.48	0.205
	Transcritical			4.97	0.586
	Pitchfork			4.86	0.474
	Double ($\widehat{n} \geq 5$)			5.05	0.500
B	Limit	2.89	0.135^2	2.89	0.135
	Transcritical			3.21	0.388
	Pitchfork			3.15	0.314
	Double ($\widehat{n} \geq 5$)			3.27	0.331

(b) Concrete specimens

Series	Type of Point	$E_{sample}[\sigma_c]$	$Var_{sample}[\sigma_c]$	σ_c^0	$\widehat{C}$
A	Pitchfork	262.4	22.5^2	304.1	52.0
	Double ($\widehat{n} \geq 5$)			324.5	54.9
B	Pitchfork	222.1	25.8^2	270.0	59.7
	Double ($\widehat{n} \geq 5$)			293.4	63.1

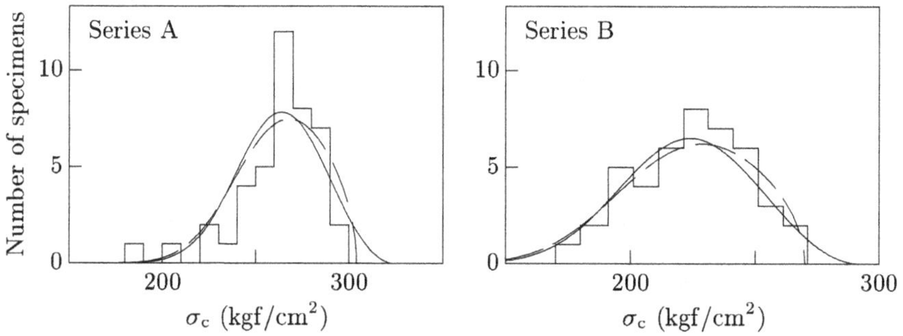

Fig. 10.8 Comparison of the histograms and semiempirical probability density functions of the compressive strength σ_c of concrete specimens (1 kgf/cm^2 = 98 kPa). ——: unstable pitchfork bifurcation point; — — —: unstable double bifurcation point ($\widehat{n} \geq 5$).

Concrete Specimens

A series of uniaxial compression tests on concrete specimens is conducted to estimate the random variation of the compressive strength, which varies considerably test by test. As an example of this, we refer here to two series of data,[6] called A and B, which, respectively, comprise 46 and 44 specimens. The histograms of the compressive strength $\sigma_{\rm c}$ for those specimens are compared with the semiempirical probability density functions in Fig. 10.8. In the calculation of these functions, the sample mean $\mathrm{E_{sample}}[\sigma_{\rm c}]$ and the sample variance $\mathrm{Var_{sample}}[\sigma_{\rm c}]$ of the maximum stress $\sigma_{\rm c}$ listed in Table 10.1(b) are used. The curves of semiempirical probability density functions again are fairly consistent with the histograms.

Problems

10-1 Draw the histogram for the set of critical loads given in Problem 5-2, and simulate the histogram using the semiempirical procedure in §10.2.3.

10-2 Simulate the histograms in Fig. 5.6 in Problem 5-4 by the semiempirical procedure in §10.2.3.

10-3 Derive equations (10.12)–(10.17).

10-4 Derive equation (10.25).

10-5 Obtain explicit forms of probability density functions of $\widetilde{f_{\rm c}}$ for double critical points when $\boldsymbol{d}$ is distributed uniformly on the unit sphere $\|\boldsymbol{d}\| = 1$.
Answer: See Murota and Ikeda, 1991 [143].

Summary

- Explicit forms of the probability density functions of the critical loads for D_n-symmetric systems have been obtained.
- Formulas to evaluate the distribution of minimum values have been presented.
- The procedure to obtain the probability density function of the critical load presented in this chapter has been applied to the numerical and experimental examples.

[6] Details of these data are given in Ikeda et al., 1997 [83].

Chapter 11
Description and Computation of Bifurcation Behaviors

11.1 Introduction

In the preceding chapters in Part II, several procedures to tackle the perfect and imperfect behaviors of a D_n-equivariant system were introduced and applied to simple examples. Each procedure, however, can throw light on only a fraction of all the bifurcation behaviors, especially in experiments. It might be in order, in this chapter, to investigate the perfect and imperfect behaviors of realistic systems, by a synthetic application of those procedures, to demonstrate their capability.

The procedures employed include the following:

- Block-diagonalization of the Jacobian matrix J in §7.7.1,
- Rules for recursive bifurcation and perfect bifurcation at double bifurcation points summarized in §8.3,
- The procedure for the experimentally observed bifurcation diagram presented in Chapter 6 and extended to D_n-equivariant systems in §8.9,
- The theory of random imperfections in Chapter 10.

The rules for recursive bifurcation and perfect bifurcation at double bifurcation points are of great importance in the numerical analysis of the perfect system in tracing primary branches, secondary branches, and so on in a recursive manner. The procedure for the experimentally observed bifurcation diagram enables us to construct the curve for the perfect system using a single or a number of experimental curves and to explain the experimental curves as imperfect bifurcation phenomena. The theory of random imperfections is used in the description of the experimental scatter of critical loads and in the estimation of the critical load for the perfect system. The block-diagonalization of the Jacobian matrix serves as a systematic and powerful tool in bifurcation analysis. In addition, a numerical procedure of bifurcation analysis is presented in this chapter.

This chapter is organized as follows.

- A numerical procedure of bifurcation analysis is presented in §11.2.

K. Ikeda and K. Murota, *Imperfect Bifurcation in Structures and Materials*,
Applied Mathematical Sciences 149, DOI 10.1007/978-1-4419-7296-5_11,

- A revised scaled-corrector method is presented as an application of block-diagonalization to bifurcation analysis in §11.3.
- Recursive bifurcation of a truss dome is investigated in §11.4.
- Imperfect bifurcation behaviors of a truss dome and sands are studied in §11.5.
- Use of block-diagonalization in bifurcation analysis of symmetric systems is demonstrated in §11.6.

11.2 Numerical Bifurcation Analysis

A procedure for numerical bifurcation analysis is presented.

11.2.1 Path Tracing

In the numerical procedure called *path tracing*, a sequence of equilibrium points as presented in Fig. 11.1 is obtained; these points are interpolated to approximate a smooth equilibrium path.

In this regard, we recall that the governing equation (2.1):

$$\boldsymbol{F}(\boldsymbol{u}, f, \boldsymbol{v}) = \boldsymbol{0} \tag{11.1}$$

for a given imperfection $\boldsymbol{v} \in \mathbb{R}^p$, has N equations, but has $N+1$ independent variables, consisting of $\boldsymbol{u} = (u_1, \ldots, u_N)^\top \in \mathbb{R}^N$ and $f \in \mathbb{R}$. Hence, to obtain a solution $(\boldsymbol{u}, f)$, an additional interrelationship between independent variables must be specified. For example, we can specify the value of f, u_i, or the arc length. The path-tracing method, accordingly, is classified as

$$\begin{cases} \text{load control method:} & \text{value of } f \text{ is specified,} \\ \text{displacement control method:} & \text{value of } u_i \text{ is specified,} \\ \text{arc-length control method:} & \text{arc length is specified.} \end{cases}$$

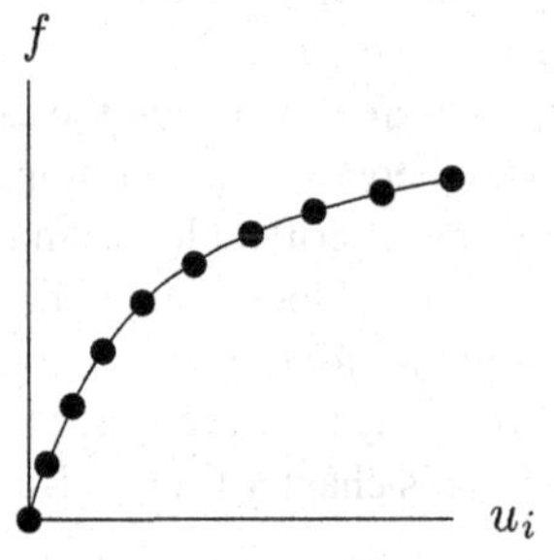

Fig. 11.1 A sequence of equilibrium points and a smooth equilibrium path obtained by interpolating these points.

Then an equilibrium point $(\boldsymbol{u}, f)$ is to be obtained for the value specified. Repeated application of this procedure will produce the sequence of equilibrium points and an interpolated equilibrium path as depicted in Fig. 11.1. If the path obtained in this manner is not smooth, then the interval of equilibrium points is reduced, or another path-tracing method is used.

A procedure to obtain an equilibrium point by the Newton method is illustrated below, for example, for the load control method, in which f is specified to be f_*. With $f = f_*$, the governing equation (11.1) becomes a set of N equations

$$\boldsymbol{F}(\boldsymbol{u}, f_*, \boldsymbol{v}) = \boldsymbol{0} \tag{11.2}$$

with N unknowns. We employ $(\boldsymbol{u}^{(0)}, f_*)$ as the initial value for the Newton iteration and obtain an increment $\widetilde{\boldsymbol{u}}$ such that $(\boldsymbol{u}^{(0)} + \widetilde{\boldsymbol{u}}, f_*)$ satisfies the equilibrium condition (11.2). The condition for $\widetilde{\boldsymbol{u}}$ reads as

$$\boldsymbol{F}(\boldsymbol{u}^{(0)} + \widetilde{\boldsymbol{u}}, f_*, \boldsymbol{v}) = \boldsymbol{F}(\boldsymbol{u}^{(0)}, f_*, \boldsymbol{v}) + J(\boldsymbol{u}^{(0)}, f_*, \boldsymbol{v})\widetilde{\boldsymbol{u}} + \text{h.o.t.} = \boldsymbol{0}.$$

At an ordinary point with nonsingular J, this equation yields

$$\widetilde{\boldsymbol{u}} \approx -J(\boldsymbol{u}^{(0)}, f_*, \boldsymbol{v})^{-1}\boldsymbol{F}(\boldsymbol{u}^{(0)}, f_*, \boldsymbol{v}),$$

and an improved approximation $\boldsymbol{u}^{(1)}$ of $\boldsymbol{u}$ is given by

$$\boldsymbol{u}^{(1)} = \boldsymbol{u}^{(0)} + \widetilde{\boldsymbol{u}} = \boldsymbol{u}^{(0)} - J(\boldsymbol{u}^{(0)}, f_*, \boldsymbol{v})^{-1}\boldsymbol{F}(\boldsymbol{u}^{(0)}, f_*, \boldsymbol{v}).$$

Therefore, we arrive at the Newton iteration

$$\boldsymbol{u}^{(\nu+1)} = \boldsymbol{u}^{(\nu)} - J(\boldsymbol{u}^{(\nu)}, f_*, \boldsymbol{v})^{-1}\boldsymbol{F}(\boldsymbol{u}^{(\nu)}, f_*, \boldsymbol{v}), \qquad \nu = 0, 1, \ldots, \tag{11.3}$$

which is repeated until convergence.

For the displacement control method and the arc-length control method, we have schemes of the Newton iteration other than (11.3), which produces $\{(\boldsymbol{u}^{(\nu)}, f^{(\nu)}) \mid \nu = 0, 1, \ldots\}$ starting from some initial value $(\boldsymbol{u}^{(0)}, f^{(0)})$. We can express the iteration in a generic form as

$$(\boldsymbol{u}^{(\nu+1)}, f^{(\nu+1)}) = \Phi(\boldsymbol{u}^{(\nu)}, f^{(\nu)}), \qquad \nu = 0, 1, \ldots, \tag{11.4}$$

where the parameter $\boldsymbol{v}$ is suppressed for notational simplicity.

11.2.2 Singularity Detection

At the equilibrium points, the singularity of the Jacobian matrix J is detected to find the location of a critical point, and to test the stability of the equilibrium points.

The numerical determination of the location of a critical point can be conducted by eigenanalysis as in the following. Find two equilibrium points, say A and B, at

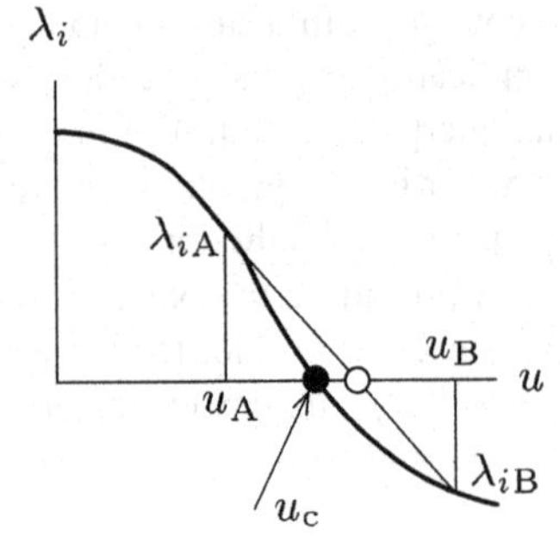

Fig. 11.2 Interpolation of zero eigenvalue $\lambda_i = 0$ in the neighborhood of a critical point. •: location of bifurcation point; ∘: approximated location.

which

$$\lambda_{iA}\lambda_{iB} < 0$$

is satisfied; that is, the sign of an eigenvalue λ_i for some i changes between points A and B, where the numbering i of λ_i is chosen such that its value changes continuously between A and B. Then, at least one critical point exists between the points. As depicted in Fig. 11.2, the location of a critical point shown by (•) is approximated by the location (∘):

$$\boldsymbol{u}_{\mathrm{c}} \approx \frac{\lambda_{i\mathrm{B}}\boldsymbol{u}_{\mathrm{A}} - \lambda_{i\mathrm{A}}\boldsymbol{u}_{\mathrm{B}}}{\lambda_{i\mathrm{B}} - \lambda_{i\mathrm{A}}}, \qquad f_{\mathrm{c}} \approx \frac{\lambda_{i\mathrm{B}} f_{\mathrm{A}} - \lambda_{i\mathrm{A}} f_{\mathrm{B}}}{\lambda_{i\mathrm{B}} - \lambda_{i\mathrm{A}}}.$$

The accuracy of this approximation is improved by reducing the distance separating points A and B.

11.2.3 Branch-Switching Analysis

An equilibrium point $(\boldsymbol{u}, f)$ on a bifurcated path is obtained by conducting a *branch-switching* analysis at a bifurcation point $(\boldsymbol{u}_{\mathrm{c}}, f_{\mathrm{c}})$. Once such an equilibrium point is obtained, equilibrium points on the bifurcated path are to be obtained by the path-tracing analysis procedure in §11.2.1.

Finding a point on a bifurcation path necessitates setting an appropriate initial point $(\boldsymbol{u}^{(0)}, f^{(0)})$ of the iteration (11.4). It is rational to set this initial point in the direction of the bifurcated path. That is, we set

$$(\boldsymbol{u}^{(0)}, f^{(0)}) = (\boldsymbol{u}_{\mathrm{c}}, f_{\mathrm{c}}) + (\widetilde{\boldsymbol{u}}, \widetilde{f}) \tag{11.5}$$

with $(\widetilde{\boldsymbol{u}}, \widetilde{f})$ headed toward the bifurcated path.

The direction $(\widetilde{\boldsymbol{u}}, \widetilde{f})$ varies with the type of the bifurcation point $(\boldsymbol{u}_{\mathrm{c}}, f_{\mathrm{c}})$ as follows, which is a summary of our analyses for a simple critical point in §2.4.3 and for a group-theoretic double bifurcation point in §7.5.2.

Transcritical Bifurcation Point

For a transcritical bifurcation point $(\boldsymbol{u}_c, f_c)$, the direction $(\widetilde{\boldsymbol{u}}, \widetilde{f})$ in (11.5) is given by

$$(\widetilde{\boldsymbol{u}}, \widetilde{f}) = (C\boldsymbol{\eta}_1 + \boldsymbol{\eta}_*, 1)\,\widetilde{f}, \tag{11.6}$$

where $\boldsymbol{\eta}_1$ is the critical eigenvector of $J(\boldsymbol{u}_c, f_c, \boldsymbol{v})$; that is, the solution to $J(\boldsymbol{u}_c, f_c, \boldsymbol{v})\boldsymbol{\eta}_1 = \boldsymbol{0}$; $\boldsymbol{\eta}_*$ is the solution[1] to

$$J(\boldsymbol{u}_c, f_c, \boldsymbol{v})\boldsymbol{\eta}_* = -\frac{\partial \boldsymbol{F}}{\partial f}(\boldsymbol{u}_c, f_c, \boldsymbol{v}); \tag{11.7}$$

and the values of C and $\widetilde{f}$ are to be given appropriately.

Remark 11.1. The expressions (11.6) and (11.7) are essentially equivalent to (2.61) and (2.62) in §2.4.3, as follows. Although (2.61) and (2.62) are obtained for the perfect system with $\boldsymbol{v} = \boldsymbol{v}^0$, their derivation is valid also for general $\boldsymbol{v}$. Consequently, (2.61) and (2.62) remain valid when J_c^0 and $(\partial \boldsymbol{F}/\partial f)_c^0$ are replaced, respectively, by $J_c = J(\boldsymbol{u}_c, f_c, \boldsymbol{v})$ and $(\partial \boldsymbol{F}/\partial f)_c = (\partial \boldsymbol{F}/\partial f)(\boldsymbol{u}_c, f_c, \boldsymbol{v})$. Observe also that the coefficient C in (2.61), being nonzero for a transcritical point, can be replaced by its reciprocal $1/C$, which yields (11.6). □

Pitchfork Bifurcation Point

For a pitchfork bifurcation point $(\boldsymbol{u}_c, f_c)$, we have

$$(\widetilde{\boldsymbol{u}}, \widetilde{f}) = (C\boldsymbol{\eta}_1, 0) \tag{11.8}$$

in (11.5), where the value of C is to be chosen appropriately. In fact, this is a consequence of (2.61), in which we have $C = 0$ by (2.86).

Double Bifurcation Point

For a double bifurcation point $(\boldsymbol{u}_c, f_c)$, with two critical eigenvectors $\boldsymbol{\eta}_1$ and $\boldsymbol{\eta}_2$, we have

$$(\widetilde{\boldsymbol{u}}, \widetilde{f}) = (C\,[\cos\theta\cdot\boldsymbol{\eta}_1 + \sin\theta\cdot\boldsymbol{\eta}_2] + \boldsymbol{\eta}_*, 1)\,\widetilde{f}$$

in (11.5), where C and θ are constants. In the search for a bifurcated path, the values of C and θ must be chosen pertinently. In addition, such a search must be conducted repeatedly to exhaust all bifurcated paths. A more systematic choice of the value of θ for a D_n-symmetric system was explained in §8.6.1.

[1] The equation (11.7) has a unique solution, since the rank deficiency of $J_c = J(\boldsymbol{u}_c, f_c, \boldsymbol{v})$ is one and $\boldsymbol{\xi}_1^\top(\partial \boldsymbol{F}/\partial f)_c = 0$ for the left critical eigenvector $\boldsymbol{\xi}_1$ of J_c (cf., Remark 2.8 in §2.5); the latter implies that $(\partial \boldsymbol{F}/\partial f)_c$ lies in the range space of J_c.

11.3 Revised Scaled-Corrector Method

In the bifurcation analysis for large-scale systems, the standard eigenanalysis of the Jacobian matrix yields important information. But it demands a large amount of computation, despite extensive study of the solution methods of finite element eigenanalysis. A scaled-corrector, which is a normalized correction vector in the Newton iteration, has been found to simulate the critical eigenvector quite well in the vicinity of a critical point. Yet it involves the following difficulties.

- Nearly coincidental bifurcation points cannot be separated.
- Simple and double bifurcation points cannot be distinguished.
- For a double bifurcation point, which has two critical eigenvectors, only a single critical eigenvector can be obtained.

In this section, block-diagonalization is used to overcome these difficulties. To be precise, the critical eigenvector, which is used as an initial value of the Newton iteration, is approximated accurately by decomposing a scaled-corrector vector into a number of vectors and, in turn, by choosing the predominant one.

We assume hereinafter that the Jacobian matrix J is symmetric. Then we can consider the *eigenpairs* $(\lambda_i, \boldsymbol{\eta}_i)$ $(i = 1, \ldots, N)$ of J that satisfy

$$J\boldsymbol{\eta}_i = \lambda_i \boldsymbol{\eta}_i, \qquad i = 1, \ldots, N,$$

$\boldsymbol{\eta}_i^{\top}\boldsymbol{\eta}_i = 1$, and $\boldsymbol{\eta}_i^{\top}\boldsymbol{\eta}_j = 0$ if $i \neq j$. It then follows that

$$\lambda_i = \boldsymbol{\eta}_i^{\top} J \boldsymbol{\eta}_i, \qquad i = 1, \ldots, N. \tag{11.9}$$

When J is nonsingular, we have

$$J^{-1} = \sum_{i=1}^{N} \frac{1}{\lambda_i} \boldsymbol{\eta}_i \boldsymbol{\eta}_i^{\top}. \tag{11.10}$$

11.3.1 Original Scaled-Corrector Method

We briefly explain the original *scaled-corrector method*,[2] by considering, for simplicity, the path-tracing by the load control method.[3] In the Newton iteration in (11.3), the correction vector for the νth step is given by

$$\widetilde{\boldsymbol{u}} = \boldsymbol{u}^{(\nu+1)} - \boldsymbol{u}^{(\nu)} = -J(\boldsymbol{u}^{(\nu)}, f_*, \boldsymbol{v})^{-1} \boldsymbol{F}(\boldsymbol{u}^{(\nu)}, f_*, \boldsymbol{v}). \tag{11.11}$$

---

[2] The scaled-corrector method was presented in Noguchi and Hisada, 1993 [151] and Fujii and Noguchi, 2002 [50].

[3] The adaptation to other control methods is straightforward.

The value of $\widetilde{\boldsymbol{u}}$ is available during the iteration; therefore, we seek to extract the information about the eigenvectors of J from (11.11) with (11.10).

Simple Critical Point

Suppose that we are to identify a simple critical point and denote by λ_1 the eigenvalue vanishing at this point. In the vicinity of this point, the eigenvalue λ_1 approaches zero; in turn, $|1/\lambda_1|$ becomes very large. Then (11.10) can be approximated as

$$J^{-1} \approx \frac{1}{\lambda_1} \boldsymbol{\eta}_1 \boldsymbol{\eta}_1^\top. \tag{11.12}$$

With the use of (11.12) in (11.11), we obtain

$$\widetilde{\boldsymbol{u}} \approx \frac{-\boldsymbol{\eta}_1^\top \boldsymbol{F}(\boldsymbol{u}^{(\nu)}, f_*, \boldsymbol{v})}{\lambda_1} \boldsymbol{\eta}_1, \tag{11.13}$$

or, conversely,

$$\boldsymbol{\eta}_1 \approx \frac{-\lambda_1}{\boldsymbol{\eta}_1^\top \boldsymbol{F}(\boldsymbol{u}^{(\nu)}, f_*, \boldsymbol{v})} \widetilde{\boldsymbol{u}},$$

which indicates that the eigenvector $\boldsymbol{\eta}_1$ for the eigenvalue λ_1 in question can be approximated by the corrector $\widetilde{\boldsymbol{u}}$ at hand. Then, by defining the *scaled-corrector*

$$\boldsymbol{\eta}^{\text{SC}} = \frac{\widetilde{\boldsymbol{u}}}{\|\widetilde{\boldsymbol{u}}\|} \tag{11.14}$$

from the corrector $\widetilde{\boldsymbol{u}}$ in (11.11), we see

$$\boldsymbol{\eta}_1 \approx \pm \boldsymbol{\eta}^{\text{SC}}. \tag{11.15}$$

Therefore, we can use $\boldsymbol{\eta}^{\text{SC}}$ as an approximation of $\boldsymbol{\eta}_1$.

It is convenient to introduce the *Rayleigh quotient* as an approximation of λ_1; that is,

$$\hat{\lambda} = (\boldsymbol{\eta}^{\text{SC}})^\top J \boldsymbol{\eta}^{\text{SC}}, \tag{11.16}$$

which is called the *pseudo-eigenvalue* in Ikeda et al., 2007 [97]. We have

$$\lambda_1 \approx \hat{\lambda}.$$

In numerical analysis, the location of a bifurcation point can be monitored according to the vanishing of the value of pseudo-eigenvalue $\hat{\lambda}$.

By a well-known fact in linear algebra (cf., Gantmacher, 1959 [55]), we have

$$\hat{\lambda} \geq \min_i \lambda_i. \tag{11.17}$$

Therefore, $\hat{\lambda}$ is an approximation of λ_1 from the above in the most customary case where λ_1 is the smallest eigenvalue.

Difficulty for Coincidental Critical Point

The scaled-corrector method is not so successful for a (nearly) coincidental critical point. Assume that M (≥ 2) eigenvalues, say, $\lambda_1, \ldots, \lambda_M$, are close to 0 in the region of our interest. In this case, (11.12) should be replaced by

$$J^{-1} \approx \sum_{i=1}^{M} \frac{1}{\lambda_i} \boldsymbol{\eta}_i \boldsymbol{\eta}_i^{\top};$$

accordingly, (11.13) becomes

$$\widetilde{\boldsymbol{u}} \approx -\sum_{i=1}^{M} \frac{\boldsymbol{\eta}_i^{\top} \boldsymbol{F}(\boldsymbol{u}^{(\nu)}, f_*, \boldsymbol{v})}{\lambda_i} \boldsymbol{\eta}_i.$$

Then the scaled-corrector (11.14) and the pseudo-eigenvalue (11.16) are expressed, respectively, as

$$\boldsymbol{\eta}^{\mathrm{sc}} \approx \sum_{i=1}^{M} C_i \boldsymbol{\eta}_i, \qquad \hat{\lambda} \approx \sum_{i=1}^{M} C_i^{\,2} \lambda_i \tag{11.18}$$

with some constants C_i ($i = 1, \ldots, M$) satisfying $\sum_{i=1}^{M} C_i^{\,2} = 1$.

Consider, as a particular case, a double bifurcation point at which two eigenvalues, say, λ_1 and λ_2, vanish. For this point, (11.18) becomes

$$\boldsymbol{\eta}^{\mathrm{sc}} \approx C_1 \boldsymbol{\eta}_1 + C_2 \boldsymbol{\eta}_2, \qquad \hat{\lambda} \approx C_1^{\,2} \lambda_1 + C_2^{\,2} \lambda_2.$$

The single vector $\boldsymbol{\eta}^{\mathrm{sc}}$ approximates the mixture of $\boldsymbol{\eta}_1$ and $\boldsymbol{\eta}_2$, but the two vectors $\boldsymbol{\eta}_1$ and $\boldsymbol{\eta}_2$ cannot be separated.

11.3.2 Revised Scaled-Corrector Method

As described in §11.3.1, the scaled-corrector method has difficulty at coincidental or nearly coincidental bifurcation points. However, for a symmetric system, these problems can be addressed by the *revised scaled-corrector method* that implements the block-diagonalization of the Jacobian matrix J discussed in §7.7.1. It is assumed that the Jacobian matrix is symmetric and that it satisfies the group equivariance (7.45) for a general group G. For simplicity of presentation we assume that all the irreducible representations of G are absolutely irreducible, which is true of $G = \mathrm{D}_n$.

We continue to use the scaled-corrector $\boldsymbol{\eta}^{\mathrm{sc}}$ in (11.14), defined as the normalized correction vector $\widetilde{\boldsymbol{u}}$ in the Newton iteration (11.11). Using the transformation matrix

$$H = ((H^{\mu,k} \mid k = 1, \ldots, N^{\mu}) \mid \mu \in R(G))$$

introduced in (7.123) in §7.7.1, we define a family of vectors as

$$\boldsymbol{\eta}^{\mu,k} = H^{\mu,k}(H^{\mu,k})^\top \boldsymbol{\eta}^{\mathrm{sc}}, \qquad \mu \in R(G), \quad k = 1,\dots,N^\mu. \tag{11.19}$$

These vectors give a decomposition of the scaled-corrector $\boldsymbol{\eta}^{\mathrm{sc}}$ as

$$\boldsymbol{\eta}^{\mathrm{sc}} = (HH^\top)\boldsymbol{\eta}^{\mathrm{sc}} = \sum_{\mu\in R(G)} \sum_{k=1}^{N^\mu} \boldsymbol{\eta}^{\mu,k}.$$

Accordingly we define the *pseudo-eigenvalues* of J by

$$\hat{\lambda}^{\mu,k} = \frac{(\boldsymbol{\eta}^{\mu,k})^\top J \boldsymbol{\eta}^{\mu,k}}{(\boldsymbol{\eta}^{\mu,k})^\top \boldsymbol{\eta}^{\mu,k}}, \qquad \mu \in R(G), \quad k = 1,\dots,N^\mu, \tag{11.20}$$

which are the Rayleigh quotients for the matrix J by the vectors $\boldsymbol{\eta}^{\mu,k}$. For a one-dimensional irreducible representation μ, $\hat{\lambda}^{\mu,1}$ is sometimes abbreviated to $\hat{\lambda}^{\mu}$.

The pseudo-eigenvalues introduced above are motivated by the block-diagonalization (7.121) of the Jacobian matrix J, which reads as

$$H^\top J H = \bigoplus_{\mu\in R(G)} \bigoplus_{k=1}^{N^\mu} \tilde{J}^\mu$$

by virtue of our assumption that all the irreducible representations of G are absolutely irreducible. The pseudo-eigenvalue $\hat{\lambda}^{\mu,k}$ is equal to the Rayleigh quotient for the diagonal block $\tilde{J}^\mu$ by the vector $(H^{\mu,k})^\top \boldsymbol{\eta}^{\mathrm{sc}}$, inasmuch as $\tilde{J}^\mu = (H^{\mu,k})^\top J H^{\mu,k}$ by (7.124) and

$$(\boldsymbol{\eta}^{\mu,k})^\top J \boldsymbol{\eta}^{\mu,k} = (\boldsymbol{\eta}^{\mathrm{sc}})^\top H^{\mu,k} \cdot (H^{\mu,k})^\top J H^{\mu,k} \cdot (H^{\mu,k})^\top \boldsymbol{\eta}^{\mathrm{sc}},$$
$$(\boldsymbol{\eta}^{\mu,k})^\top \boldsymbol{\eta}^{\mu,k} = (\boldsymbol{\eta}^{\mathrm{sc}})^\top H^{\mu,k} \cdot (H^{\mu,k})^\top \boldsymbol{\eta}^{\mathrm{sc}}.$$

Then the relation

$$\hat{\lambda}^{\mu,k} \geq \lambda^\mu_{\mathrm{min}}, \qquad k = 1,\dots,N^\mu, \tag{11.21}$$

is satisfied as a ramification of (11.17), where $\lambda^\mu_{\mathrm{min}}$ is the smallest eigenvalue of $\tilde{J}^\mu$.

Suppose that we are in a neighborhood of a group-theoretic critical point and denote by μ^* the irreducible representation associated with this critical point. Then the critical eigenvalue, say λ_1, is an eigenvalue of $\tilde{J}^{\mu^*}$ and we have

$$\min_{1\leq k\leq N^\mu} \hat{\lambda}^{\mu^*,k} \approx \lambda_1 \approx 0.$$

In practice, a particular μ^* must be identified. A procedure suggested herein is the following.

- Compute $\hat{\lambda}^{\mu,k}$ for all (μ,k) by (11.20) using (11.14) and (11.19).
- For each (μ,k), plot $\hat{\lambda}^{\mu,k}$ against a pertinent displacement component for sufficiently many equilibrium points near the critical point.
- Find a particular $\mu = \mu^*$ for which zero-crossing of the values of $\hat{\lambda}^{\mu,k}$ is encountered for any k.

- For $\mu = \mu^*$, we can obtain using (11.19) the approximation $\eta^{\mu^*,k}$ $(k = 1,\dots,N^\mu)$ to the eigenvector(s) for the zero eigenvalue(s).

11.3.3 Regular-Hexagonal Truss Dome

We explain below how equilibrium paths are obtained by the revised scaled-corrector method. The D_6-symmetric *regular-hexagonal truss dome* presented in Fig. 11.3(a) is considered. All members of the dome have the same cross-sectional area A and the same Young's modulus E; we set $EA = 1$ for normalization. The vertical load of $0.2f$ is applied to the center node and the vertical loads of $0.4f$ to the regular-hexagonal nodes surrounding the center node.

Figure 11.3(b) shows the equilibrium paths for this set of loads. On the fundamental path with D_6-symmetric deformation of this truss dome, four critical points exist:

- Bifurcation point A with D_3-symmetric bifurcated path,
- Bifurcation point B with $D_2^{k,6}$-symmetric $(k = 1,2,3)$ bifurcated paths,
- Bifurcation point C with $D_1^{k,6}$-symmetric $(k = 1,\dots,6)$ bifurcated paths,
- Limit point D of f.

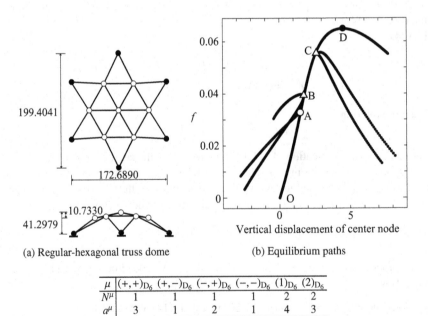

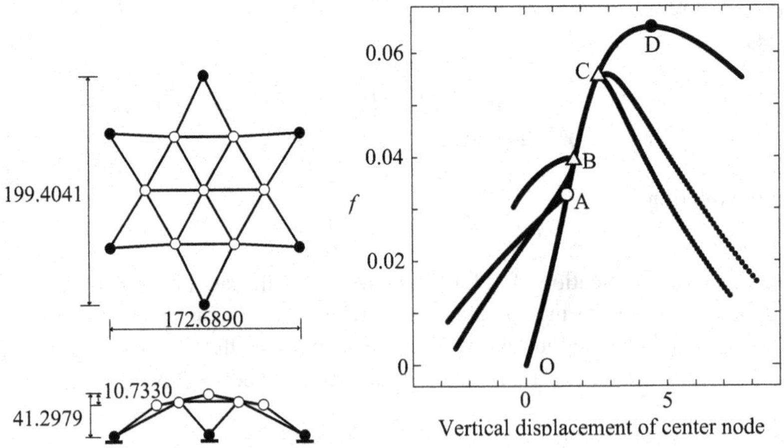

(a) Regular-hexagonal truss dome (b) Equilibrium paths

μ	$(+,+)_{D_6}$	$(+,-)_{D_6}$	$(-,+)_{D_6}$	$(-,-)_{D_6}$	$(1)_{D_6}$	$(2)_{D_6}$
N^μ	1	1	1	1	2	2
a^μ	3	1	2	1	4	3

(c) Degree N^μ and multiplicity a^μ for irreducible representation μ of D_6

Fig. 11.3 Regular-hexagonal truss dome and its equilibrium paths. ○: simple bifurcation point; △: double bifurcation point; ●: limit point of f.

Singularity Detection

The original scaled-corrector method in §11.3.1 and the revised one in §11.3.2 are used for the singularity detection, without resort to the eigenanalysis, of the bifurcation points A, B, and C of the regular-hexagonal truss dome in Fig. 11.3.

First, the original scaled-corrector method is used. The pseudo-eigenvalues $\hat{\lambda}$ for the scaled-corrector $\boldsymbol{\eta}^{\mathrm{sc}}$ in (11.14) are computed using (11.16) and shown by (•) in Fig. 11.4 against the vertical displacement of the center node. The values of $\boldsymbol{\eta}^{\mathrm{sc}}$ oscillate in association with an increase of the displacement, and (•)s are scattered.

For comparison, the eigenvalues λ_i $(i = 1, 2, \ldots)$ computed by the eigenanalysis are also shown in this figure by the dashed curve for the simple eigenvalue and by the solid curves for double eigenvalues; both curves are monotone decreasing, and the zero crossing points of these curves correspond to the locations of those bifurcation points. The pseudo-eigenvalues presented in this manner display a large scatter, especially away from those bifurcation points. The zero crossing of the pseudo-eigenvalues $\hat{\lambda}$ indicated by (•) is not very clear; it is difficult to determine the locations of bifurcation points.

Next, the revised scaled-corrector method is employed;[4] the number of column vectors of $H^{\mu,k}$, which is equal to the multiplicity a^{μ}, is listed in Fig. 11.3(c). The pseudo-eigenvalues $\hat{\lambda}^{\mu,k}$ in (11.20) for $\mu \in R(\mathrm{D}_6)$ are computed as depicted in Fig. 11.5, in which the pseudo-eigenvalues for $\mu = (+,-)_{\mathrm{D}_6}$ and $\mu = (-,-)_{\mathrm{D}_6}$ are omitted as they do not cross zero. The pseudo-eigenvalues for μ other than $(+,-)_{\mathrm{D}_6}$ and $(-,-)_{\mathrm{D}_6}$ cross zero with scatters.

The zero crossing of the pseudo-eigenvalues $\hat{\lambda}^{\mu,k}$s is recognized for

- $\mu = (-,+)_{\mathrm{D}_6}$ corresponding to the bifurcation point A,
- $\mu = (2)_{\mathrm{D}_6}$ to the bifurcation point B,
- $\mu = (1)_{\mathrm{D}_6}$ to the bifurcation point C.

From Table 8.1 in §8.3.1, point A is classified as a simple bifurcation point; point B, at which pseudo-eigenvalues for $\mu = (2)_{\mathrm{D}_6}$ cross zero, is a double bifurcation point; similarly, point C is a double bifurcation point. The multiplicity of bifurcation points has thus been identified by the revised scaled-corrector method.

In addition, Fig. 11.5 shows that the computed $\hat{\lambda}^{\mu}$ all serve as an upper bound on the true eigenvalue; it assesses the validity of (11.21).

Determination of Critical Eigenvectors

For the simple bifurcation point A, the accuracy of $\boldsymbol{\eta}^{\mathrm{sc}}$ computed by the scaled-corrector method and that of $\boldsymbol{\eta}^{(-,+)_{\mathrm{D}_6}}$ by the revised method are compared with the exact eigenvector $\boldsymbol{\eta}_1$ for the smallest eigenvalue λ_1 computed by the eigenanalysis.

As an index of the accuracy of $\boldsymbol{\eta}^{\mathrm{sc}}$ and $\boldsymbol{\eta}^{(-,+)_{\mathrm{D}_6}}$ as approximations to $\boldsymbol{\eta}_1$, we introduce

[4] The transformation matrix H for this dome can be constructed based on the procedure given in Chapter 12.

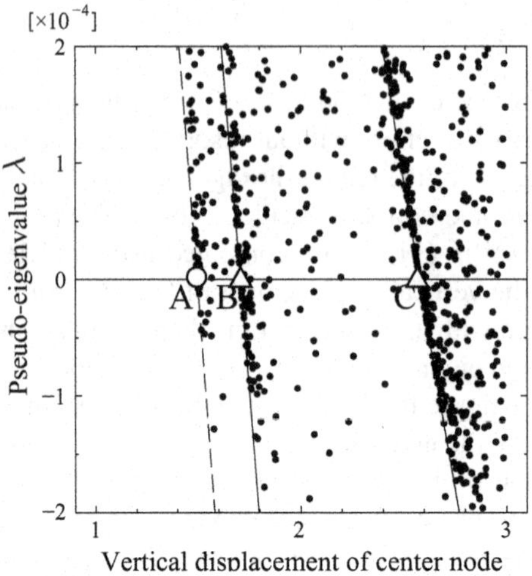

Fig. 11.4 Pseudo-eigenvalue $\hat{\lambda}$ computed by the original scaled-corrector method for the regular-hexagonal truss dome. •: $\hat{\lambda}$; dashed and solid curves: simple and double eigenvalues λ_i $(i = 1, 2, \ldots)$, respectively, computed by the eigenanalysis.

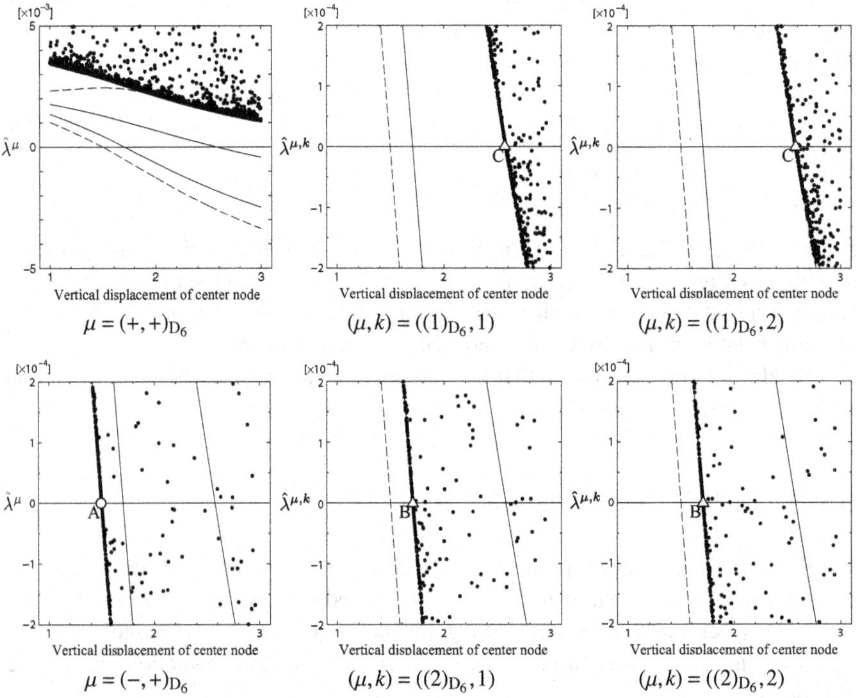

Fig. 11.5 Pseudo-eigenvalue $\hat{\lambda}^{\mu,k}$ for each irreducible representation computed by the revised scaled-corrector method for the regular-hexagonal truss dome. •: $\hat{\lambda}^{\mu,k}$; dashed and solid curves: simple and double eigenvalues λ_i, respectively, computed by the eigenanalysis.

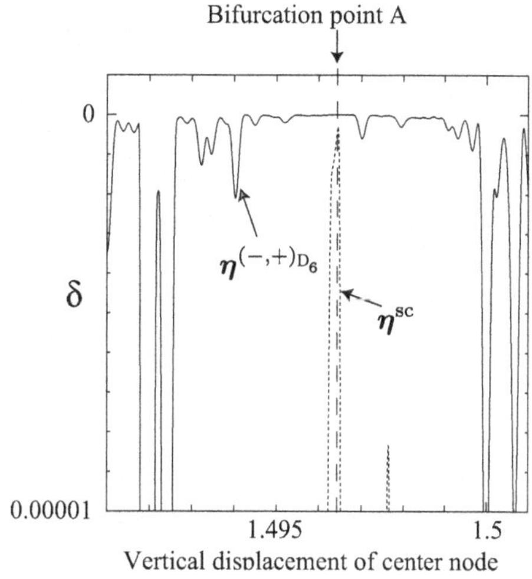

Fig. 11.6 Comparison of the accuracy of $\eta^{(-,+)D_6}$ and η^{sc}.

$$\delta = \begin{cases} 1 - |(\eta^{sc})^\top \eta_1| & \text{for original scaled-corrector method,} \\ 1 - |(\eta^{(-,+)D_6})^\top \eta_1| / \|\eta^{(-,+)D_6}\| & \text{for revised scaled-corrector method,} \end{cases}$$

which should coincide with zero when the computed approximate critical eigenvectors are exact.

In Fig. 11.6 the values of this index δ are plotted against the vertical displacement of the crown node in the neighborhood of the bifurcation point A. The original scaled-corrector method, shown by the dotted line, achieves accuracy of up to 5 digits only in a very close neighborhood of the bifurcation point A. The revised method, shown by the solid line, attains such accuracy in a wide range of the center node displacement; this result demonstrates the superiority of the revised method.

For the double bifurcation point B, two critical eigenvectors are to be obtained. The critical eigenvectors obtained by the scaled-corrector method and the revised method are compared in Fig. 11.7. For the point B, the original scaled-corrector method gives only a single critical eigenvector η^{sc}, which does not necessarily coincide with the direction of a bifurcated path. In contrast, the revised method affords two critical eigenvectors $\eta^{(2)D_6,1}$ and $\eta^{(2)D_6,2}$.

Also, for the double bifurcation point C, only a single critical eigenvector η^{sc} can be obtained again by the original scaled-corrector method. In contrast, a pair of critical eigenvectors $\eta^{(1)D_6,1}$ and $\eta^{(1)D_6,2}$ can be obtained by the revised method.

As described, the revised method is consistent with the bifurcation analysis at a double bifurcation point and is therefore superior to the original scaled-corrector method.

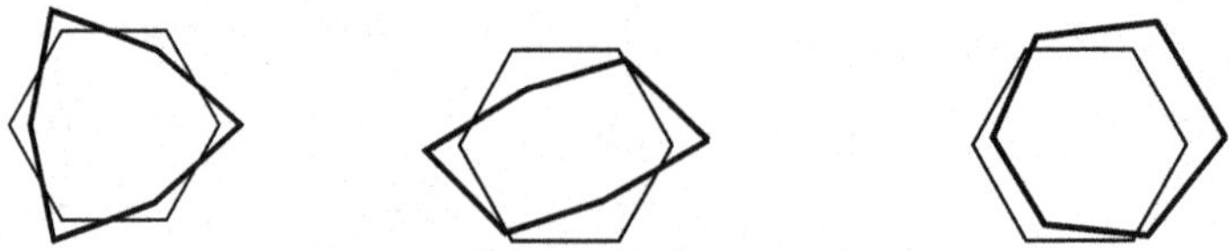

Bifurcation point A Bifurcation point B Bifurcation point C

(a) Original scaled-corrector method

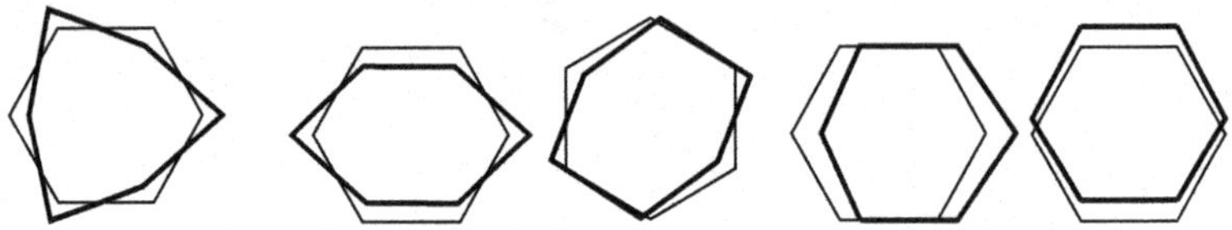

Bifurcation point A Bifurcation point B Bifurcation point C

(b) Revised scaled-corrector method

Fig. 11.7 Critical eigenvectors for the regular-hexagonal truss dome expressed in terms of the plane view of the hexagonal nodes.

11.3.4 Large-Scale Truss Dome

We refer to the D_6-symmetric large-scale truss dome with $N = 885$ that is depicted in Fig. 11.8(a). All members of the dome have the same cross-sectional area A and the same Young's modulus E; we set $EA = 1$ for normalization. The vertical load of $f \times 6.28 \times 10^{-16}$ is applied to the center node, the vertical loads of $f \times 2 \times 6.28 \times 10^{-16}$, $f \times 2^2 \times 6.28 \times 10^{-16}, \ldots$ are applied to all layers of the regular-hexagonal nodes from the inside toward the outside. Figure 11.8(b) shows the equilibrium paths of this dome.

On the fundamental path with D_6-symmetric deformation, four critical points exist:

- Limit point A,
- Bifurcation point B with $D_1^{k,6}$-symmetric ($k = 1,\ldots,6$) bifurcated paths,
- Bifurcation point C with $D_2^{k,6}$-symmetric ($k = 1,2,3$) bifurcated paths,
- Bifurcation point D with D_3-symmetric bifurcated path.

Critical points A and B are nearly coincidental.

The original scaled-corrector method and the revised one are used for the singularity detection of the critical points A to D of the dome without resort to the eigenanalysis.

First, the original scaled-corrector method is employed to obtain the pseudo-eigenvalues $\hat{\lambda}$ plotted by (•) in Fig. 11.9. The zero crossing of the pseudo-eigenvalues is apparent at the nearly coincidental critical points A and B. However, it is not possible to distinguish these points. It is also difficult to locate the bifurcation points C and D.

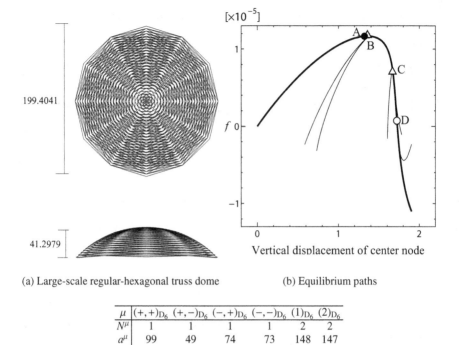

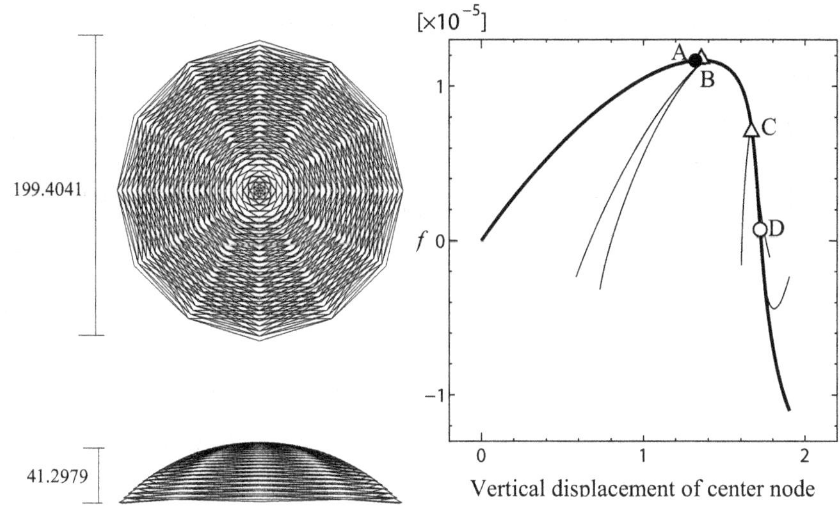

(a) Large-scale regular-hexagonal truss dome (b) Equilibrium paths

μ	$(+,+)_{\mathrm{D}_6}$	$(+,-)_{\mathrm{D}_6}$	$(-,+)_{\mathrm{D}_6}$	$(-,-)_{\mathrm{D}_6}$	$(1)_{\mathrm{D}_6}$	$(2)_{\mathrm{D}_6}$
N^μ	1	1	1	1	2	2
a^μ	99	49	74	73	148	147

(c) Degree N^μ and multiplicity a^μ for irreducible representation μ of D_6

Fig. 11.8 Large-scale truss dome and its equilibrium paths. ∘: simple bifurcation point; △: double bifurcation point; •: limit point of f.

Next, the revised scaled-corrector method is used[5] to compute the pseudo-eigenvalues $\hat{\lambda}^\mu$ for $\boldsymbol{\eta}^\mu$ in Fig. 11.10, in which the pseudo-eigenvalues for $\mu = (+,-)_{\mathrm{D}_6}$ and $\mu = (-,-)_{\mathrm{D}_6}$ are omitted as they do not cross zero. Each of the four critical points A to D can be recognized by the zero crossing of the pseudo-eigenvalues $\hat{\lambda}^\mu$s for

- $\mu = (+,+)_{\mathrm{D}_6}$ corresponding to the limit point A,
- $\mu = (1)_{\mathrm{D}_6}$ to the double bifurcation point B,
- $\mu = (2)_{\mathrm{D}_6}$ to the double bifurcation point C,
- $\mu = (-,+)_{\mathrm{D}_6}$ corresponding to the simple bifurcation point D.

The multiplicity of bifurcation points has therefore been identified by the revised method. The bifurcation analysis can be conducted in a systematic manner using the critical eigenvectors obtained by the revised scaled-corrector method.

[5] The transformation matrix H for this dome can be constructed based on the procedure that is explained in Chapter 12. The number of column vectors of $H^{\mu,k}$, which is equal to the multiplicity a^μ, is listed in Fig. 11.8(c).

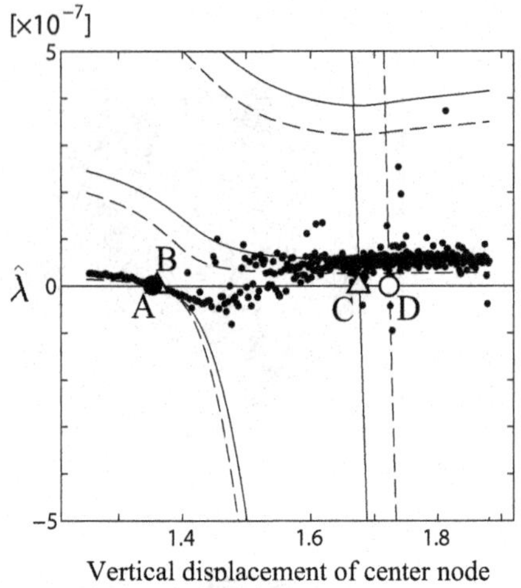

Fig. 11.9 Pseudo-eigenvalue $\hat{\lambda}$ computed by the original scaled-corrector method for the large-scale truss dome. •: $\hat{\lambda}$; dashed and solid curves: simple and double eigenvalues λ_i, respectively, computed by the eigenanalysis.

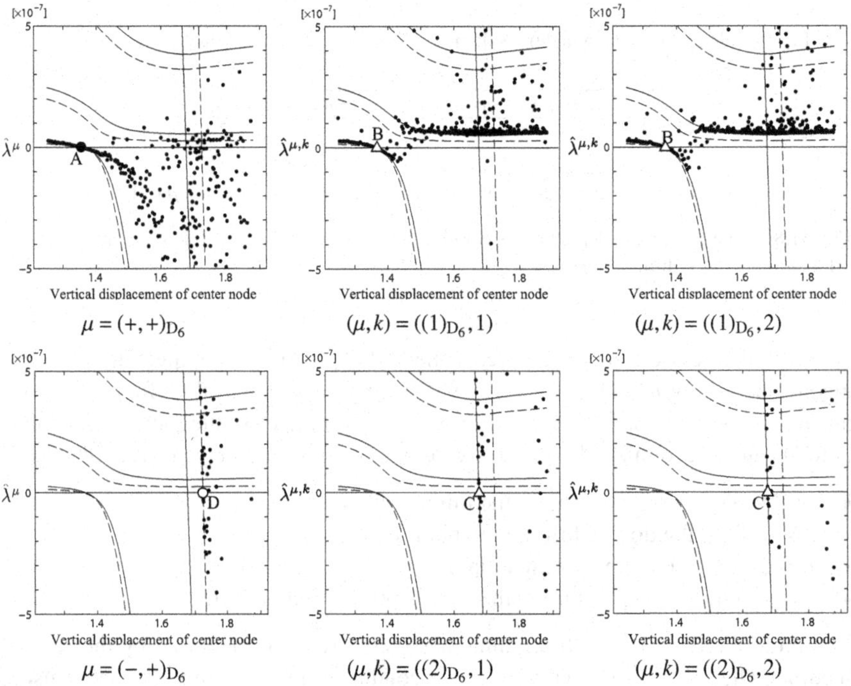

Fig. 11.10 Pseudo-eigenvalue $\hat{\lambda}^{\mu,k}$ for each irreducible representation computed by the revised scaled-corrector method for the large-scale truss dome with 50 layers. •: $\hat{\lambda}^{\mu,k}$; dashed and solid curves: simple and double eigenvalues λ_i, respectively, computed by the eigenanalysis.

11.4 Example of Recursive Bifurcation

As an example of recursive bifurcation (cf., §8.3.2), we consider here the regular-hexagonal truss dome (D_6-symmetric) shown in Fig. 11.11(a). The deformation of this dome is described by a 57-dimensional vector $\boldsymbol{u} = (x_i, y_i, z_i \mid i = 0, 1, \ldots, 18)$. We computed two sets of equilibrium paths for this dome for two D_6-symmetric z-directional loadings:

Loading (a): $0.5f$ is applied at the center node and f at the other free nodes.
Loading (b): f is applied at the six nodes of the second layer at $z = 20.543$.

By the bifurcation analysis procedure presented in §11.2, the equilibrium paths shown in Fig. 11.12 for loading (a) and in Fig. 11.13 for loading (b) are computed. Results show that the use of different loadings has engendered completely different equilibrium paths. Nonetheless, both paths do follow the same rule of recursive bifurcation for D_6-symmetric systems in Fig. 11.11(b), which is presented in Fig. 8.3(b). Such a rule is important for the analysis of recursive bifurcation behavior of symmetric systems. In particular, the information about double bifurcation points summarized in §8.3.2 is of great assistance in tracing the bifurcated paths, because one can determine whether all bifurcated paths emanating from the double bifurcation points have been found.

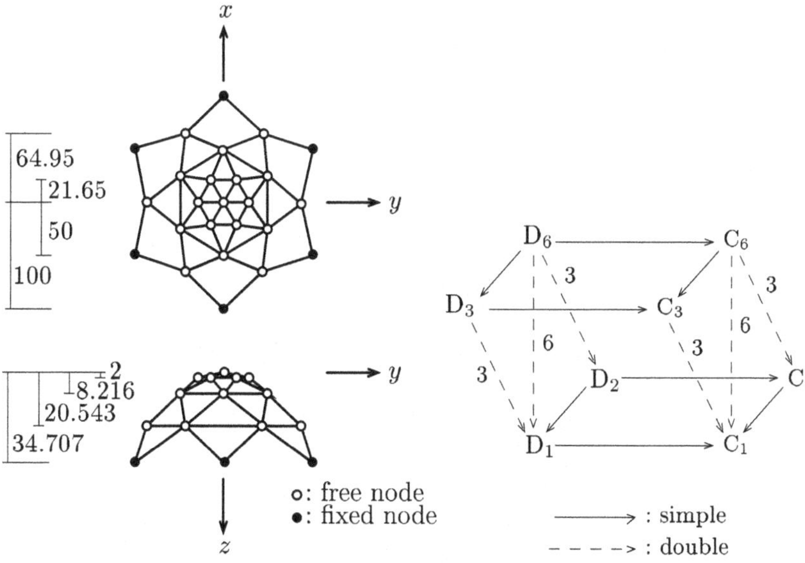

(a) Regular-hexagonal truss dome (b) Bifurcation rule

Fig. 11.11 Regular-hexagonal truss dome (D_6-symmetric) and its bifurcation rule.

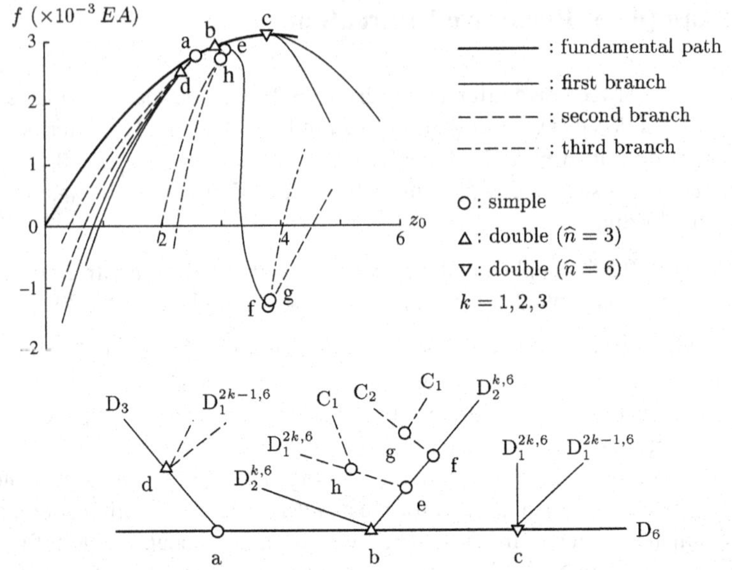

Fig. 11.12 Equilibrium paths of the regular-hexagonal truss dome (D_6-symmetric) and associated schematic diagram for the loading (a). Here, $0.5f$ is applied at the center node and f at the other free nodes.

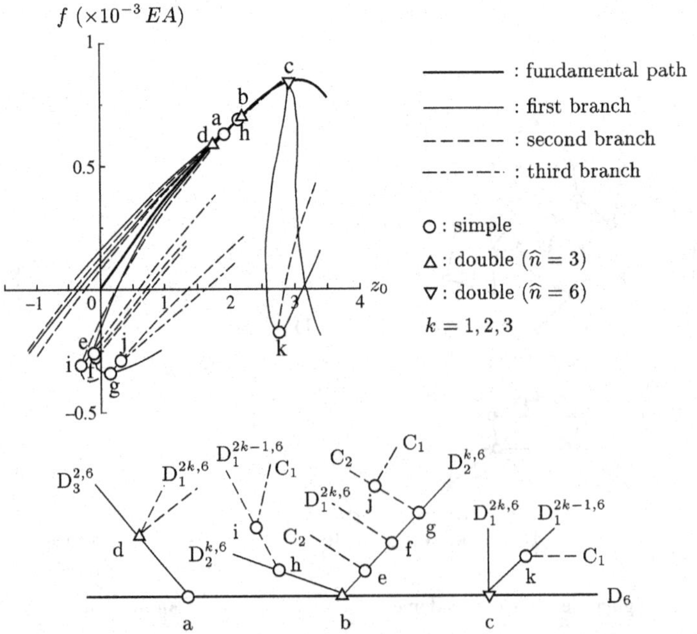

Fig. 11.13 Equilibrium paths of the regular-hexagonal truss dome (D_6-symmetric) and associated schematic diagram for the loading (b). Here, f is applied at the nodes of the second layer at $z = 20.543$.

11.5 Examples of Imperfect Behavior

A combined procedure of the experimentally observed bifurcation diagram presented in §8.9 and the statistical approach presented in Chapter 10 is illustrated for a few examples of imperfect bifurcation behavior.

11.5.1 Regular-Pentagonal Truss Dome

The D_5-symmetric regular-pentagonal dome in Fig. 11.14 is considered for an example of a double bifurcation point with index $\widehat{n} \geq 5$ (cf., $\widehat{n} = n/\gcd(n, j)$ in (8.23)). The center node is designated as node 0 and the inner pentagonal nodes surrounding the center node are designated as nodes 1 to 5.

The equilibrium paths for the perfect system are shown by the dotted curves in Fig. 11.15. The vertical displacement z_0 of the center node used as the abscissa in Fig. 11.15(a) is D_5-symmetric, whereas z_1 used as the abscissa in (b) is not. There is a double bifurcation point shown by (∘) on these paths. The z-directional components of the inner pentagonal nodes of a pair of critical eigenvectors $\boldsymbol{\eta}_1$ and $\boldsymbol{\eta}_2$ at this point are depicted in Fig. 11.16. The mode of $\boldsymbol{\eta}_1$, which has reflection symmetry with respect to a vertical plane, is D_1-symmetric, whereas that of $\boldsymbol{\eta}_2$, lacking reflection symmetry, is C_1-symmetric. This bifurcation point is associated with the two-dimensional irreducible representation $(2)_{D_5}$ defined by (8.9) and has the index $\widehat{n} = 5$. Therefore, it is related to a symmetry-breaking process: $D_5 \to D_1$.

As imperfections, the initial locations of nodes 1 to 5 are displaced randomly in the z-direction to simulate experiments in which the pattern and magnitude of the imperfections cannot be known a priori. The solid lines in Fig. 11.15 show equilibrium paths computed for some of those imperfections.

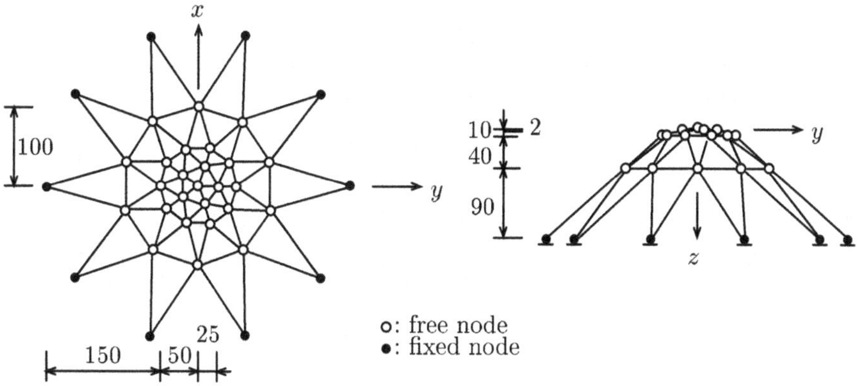

Fig. 11.14 Regular-pentagonal truss dome (D_5-symmetric).

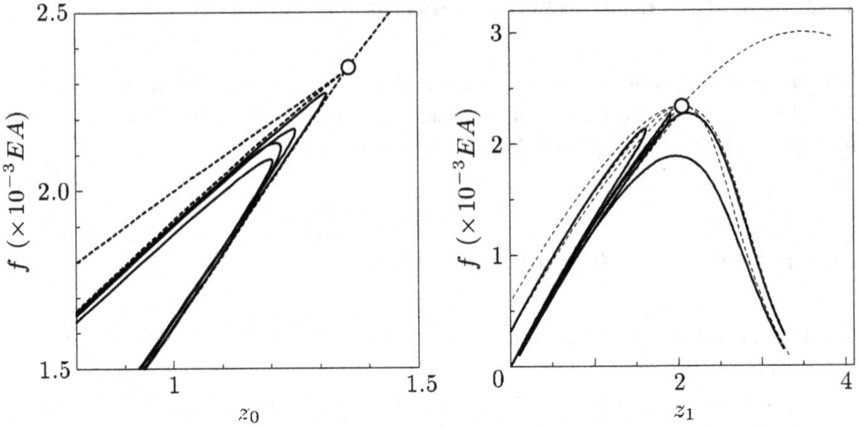

(a) f versus z_0 curves (b) f versus z_1 curves

Fig. 11.15 Equilibrium paths of the regular-pentagonal truss dome. ——: curve for an imperfect system; - - - -: curve for the perfect system; ○: double bifurcation point.

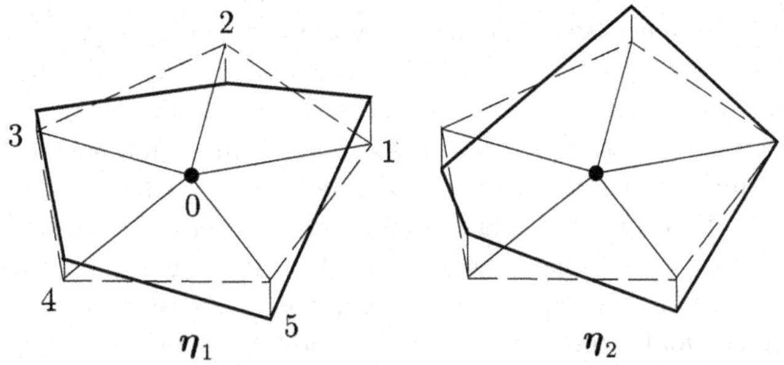

Fig. 11.16 Bird's-eye view of the z-directional components of the inner pentagonal nodes for a pair of bifurcation modes at the double bifurcation point ($\widehat{n} = 5$) of the pentagonal dome.

The asymptotic laws in §8.9.1 are applicable[6] to the D_5-symmetric vertical displacement z_0 of node 0. In particular, we recall the asymptotic law in (8.125):

$$\widetilde{f_c} = -\gamma^* \widetilde{u}_{|h} + \text{h.o.t.} \tag{11.22}$$

which relates the critical load increment $\widetilde{f_c}$ with the incremental displacement $\widetilde{u}_{|h}$. This displacement is defined by the intersection point $(\widetilde{u}_{|h}, \widetilde{f}_{|h})$ between the imperfect $\widetilde{f}$ versus $\widetilde{u}$ curve and the straight line (8.126): $\widetilde{f} + h\widetilde{u} = 0$. With the correspondences: $\widetilde{f_c} \leftrightarrow -|\widetilde{f_c}|$ and $\widetilde{u} \leftrightarrow \widetilde{z}_0$, (11.22) becomes

$$|\widetilde{f_c}| = \gamma^* (\widetilde{z}_0)_{|h} + \text{h.o.t.}, \tag{11.23}$$

[6] These laws are not applicable to the non-D_5-symmetric displacement z_1 of node 1.

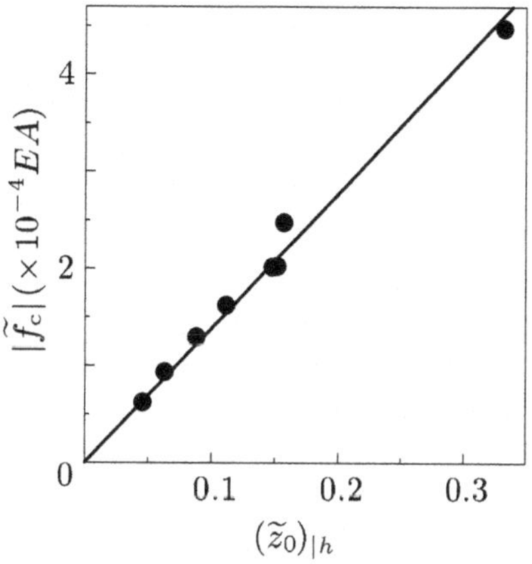

Fig. 11.17 Application of the asymptotic law (11.23) for a D_5-symmetric displacement to the pentagonal truss dome (the $|\widetilde{f}_c|$ versus $(\widetilde{z}_0)_{|h}$ relation for $h = -0.1388$).

which represents a straight line passing the origin. The $|\widetilde{f}_c|$ versus $(\widetilde{z}_0)_{|h}$ relation for $h = -0.1388$ portrayed in Fig. 11.17 shows good agreement with (11.23).

A set of 50 imperfect equilibrium paths are obtained by imposing normally distributed imperfections on the z-directional components of the free nodes of the pentagonal dome. The theory of random imperfections presented in §10.2 is applied to this case. In Fig. 11.18, the histogram of the critical loads f_c is compared with the probability density function (5.15) for an unstable (pitchfork) bifurcation point and the function (10.18) for an unstable double bifurcation point with $\widehat{n} \geq 5$, where the abscissa is a normalized critical load $\zeta = (f_c - f_c^0)/\widehat{C}$. The sample mean $\mathrm{E}[f_c]$ and sample variance $\mathrm{Var}[f_c]$ of the critical load f_c are

$$\mathrm{E}[f_c] = 2.216 \times 10^{-3} EA, \qquad \mathrm{Var}[f_c] = 12.62 \times (10^{-3} EA)^2.$$

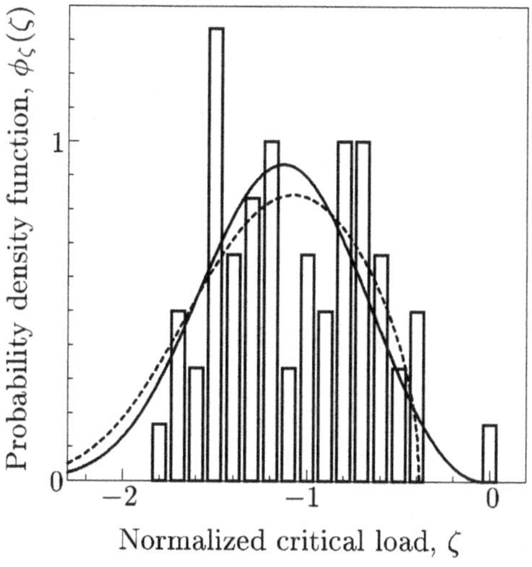

Fig. 11.18 Comparison of a histogram and the probability density functions of the normalized critical load $\zeta = (f_c - f_c^0)/\widehat{C}$ for 50 imperfect regular-pentagonal domes. ----: semiempirical curve for an unstable pitchfork bifurcation point; ——: semiempirical curve for an unstable double bifurcation point.

The use of these values in (5.16), (5.17), (10.29), and (10.30) yields the following:

$$
\begin{cases}
f_{\rm c}^0 = 2.282\times 10^{-3}EA, \quad \widehat{C} = 8.214\times 10^{-5}EA, \\
\qquad\qquad \text{assuming the simple (pitchfork) bifurcation point,} \\
f_{\rm c}^0 = 2.314\times 10^{-3}EA, \quad \widehat{C} = 8.686\times 10^{-5}EA, \\
\qquad\qquad \text{assuming the double bifurcation point.}
\end{cases}
\tag{11.24}
$$

Substitution of these values of $f_{\rm c}^0$ and $\widehat{C}$ into (5.15) and (10.18) gives the semiempirical probability density function of $f_{\rm c}$ shown by the dashed curve for an unstable pitchfork bifurcation point and the solid curve for an unstable double one in Fig. 11.18. The statistical estimate $f_{\rm c}^0 = 2.314\times 10^{-3}EA$ for the double point in (11.24) is much closer to the exact value $f_{\rm c}^0 = 2.332\times 10^{-3}EA$, in comparison with $f_{\rm c}^0 = 2.282\times 10^{-3}EA$ estimated for the simple (pitchfork) bifurcation point. Therefore, the bifurcation point is more likely to be a double point. The theory of random variation of critical loads, in this manner, enables us to determine the multiplicity of a bifurcation point if the value of $f_{\rm c}^0$ (or its estimate somehow obtained) is available.

11.5.2 Sand Specimens

We now move on to physical examples: a set of 32 experimental curves of deviatoric stress $\sigma_{\rm a}$ versus axial strain $\varepsilon_{\rm a}$ of cylindrical sand specimens. See Fig. 11.19 for some of these curves.

The initial states for the specimens are assumed to be D_n-symmetric with n large, although a more complete account using a larger group is given in §13.3.1. Among

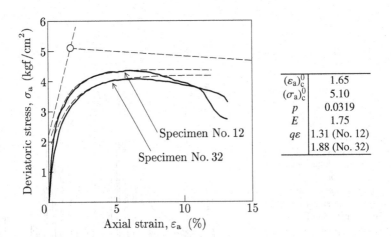

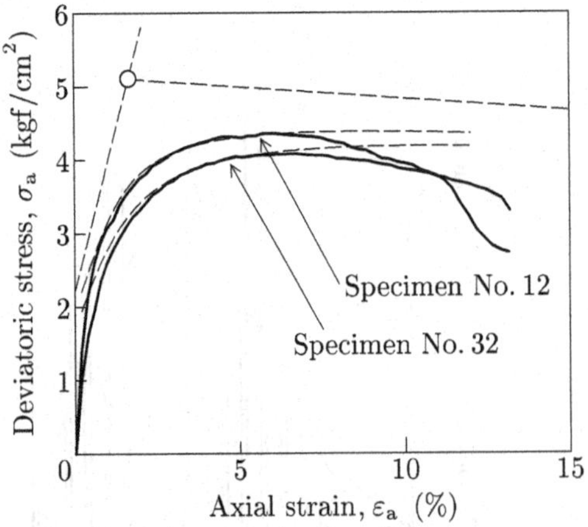

$(\varepsilon_{\rm a})_{\rm c}^0$	1.65
$(\sigma_{\rm a})_{\rm c}^0$	5.10
p	0.0319
E	1.75
$q\varepsilon$	1.31 (No. 12)
	1.88 (No. 32)

Fig. 11.19 Curves of deviatoric stress $\sigma_{\rm a}$ versus axial strain $\varepsilon_{\rm a}$ for the sand specimens and their simulation by the present method. ——: experimental (imperfect) curve; - - -: simulated curve; ○: bifurcation point; 1 kgf/cm^2 = 98 kPa.

the possible bifurcation points of this system, we restrict ourselves to the possibility of a simple (pitchfork) bifurcation point and a double bifurcation point with $\widehat{n} \geq 5$, which are associated, respectively, with the symmetry-breaking processes:

$$\begin{cases} \mathrm{D}_n \to \mathrm{C}_n & \text{at the pitchfork bifurcation point,} \\ \mathrm{D}_n \to \mathrm{D}_{n/\widehat{n}} & \text{at the double bifurcation point with } \widehat{n} \geq 5. \end{cases}$$

To determine the type of displacement and bifurcation point, we refer to the asymptotic laws and the theory of random imperfections. Since the observed variable ε_{a} is D_n-symmetric, the asymptotic law (11.22) for a D_n-symmetric displacement of a simple (pitchfork) bifurcation point or a double bifurcation point with $\widehat{n} \geq 5$ is expected to be applicable in the present case with the correspondences: $\widetilde{f}_{\mathrm{c}} \leftrightarrow -|(\widetilde{\sigma}_{\mathrm{a}})_{\mathrm{c}}|$ and $\widetilde{u} \leftrightarrow \widetilde{\varepsilon}_{\mathrm{a}}$. That is, we have

$$|(\widetilde{\sigma}_{\mathrm{a}})_{\mathrm{c}}| = \gamma^*(\widetilde{\varepsilon}_{\mathrm{a}})_{|h} + \text{h.o.t.} \tag{11.25}$$

The paths for the perfect system and the location of the bifurcation point are not known a priori; therefore, a random search in the two-dimensional space of $(\varepsilon_{\mathrm{a}}, \sigma_{\mathrm{a}})$ with a sufficiently fine mesh is conducted and a bifurcation point $((\varepsilon_{\mathrm{a}})_{\mathrm{c}}^0, (\sigma_{\mathrm{a}})_{\mathrm{c}}^0)$ is chosen to be the location where the relation (11.25) holds most accurately. The $|(\widetilde{\sigma}_{\mathrm{a}})_{\mathrm{c}}|$ versus $(\widetilde{\varepsilon}_{\mathrm{a}})_{|h}$ relation with $h = 0.17$ for the 32 sets of data shown in Fig. 11.20 is in good agreement with the asymptotic law in (11.25) expressing the straight line passing the origin.

Figure 11.19 presents the simulation by (6.27) of the σ_{a} versus ε_{a} curves for the two specimens, using the values of the parameters listed at the right of the figure that are chosen based on the procedure presented in §6.4.2. The theoretical curves correlate fairly well with the experimental curves.

The multiplicity of the bifurcation point cannot be determined merely from the asymptotic law (11.25). To determine the multiplicity, we employ the theory of random imperfections. The histogram of the maximum deviatoric stress for those spec-

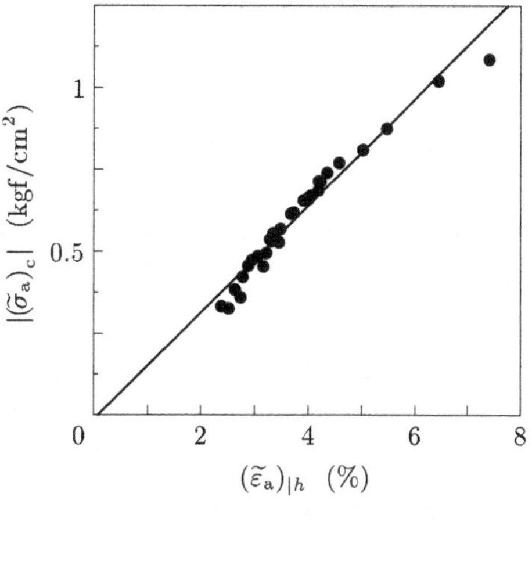

Fig. 11.20 $|(\widetilde{\sigma}_{\mathrm{a}})_{\mathrm{c}}|$ versus $(\widetilde{\varepsilon}_{\mathrm{a}})_{|h}$ relation (h = 0.17) of 32 sand specimens. $((\varepsilon_{\mathrm{a}})_{\mathrm{c}}^0, (\sigma_{\mathrm{a}})_{\mathrm{c}}^0) = (1.65, 5.10)$.

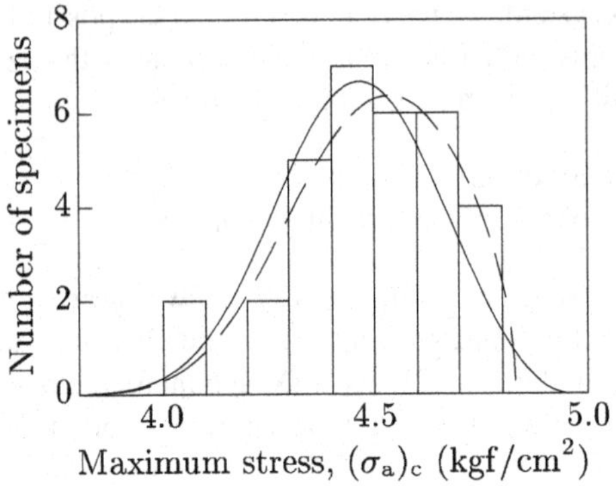

Type	Simple	Double
$E[(\sigma_a)_c]$ kgf/cm^2	4.49	
$Var[(\sigma_a)_c]$ $(kgf/cm^2)^2$	0.183^2	
$(\sigma_a)_c^0$ kgf/cm^2	4.83	4.96
$\widehat{C}$	0.424	0.448

Fig. 11.21 Comparison of a histogram and the probability density functions of the maximum deviatoric stress $(\sigma_a)_c$ for 32 sand specimens. — — —: semiempirical curve for a simple, unstable (pitchfork) bifurcation point; ——: semiempirical curve for an unstable double bifurcation point.

imens is compared in Fig. 11.21 with the probability density functions for a simple (pitchfork) and for a double bifurcation point that are computed based on the same procedure taken in §11.5.1. The two statistically estimated values of $(\sigma_a)_c^0$ listed at the right of Fig. 11.21 are compared with the target value $(\sigma_a)_c^0 = 5.10$ obtained from the $|(\widetilde{\sigma}_a)_c|$ versus $(\widetilde{\varepsilon}_a)_{|h}$ relation in Fig. 11.20. Evidently, the statistically estimated value $(\sigma_a)_c^0 = 4.96$ for the double point is closer to the target value $(\sigma_a)_c^0 = 5.10$ than $(\sigma_a)_c^0 = 4.83$ for the simple (pitchfork) bifurcation point. The bifurcation point, accordingly, is more likely to be a double point.

11.6 Use of Block-Diagonalization in Bifurcation Analysis

To exploit symmetry, several methodologies, which range from very primitive ones to sophisticated ones, have been developed. Among others, block-diagonalization analysis has emerged as a systematic and rigorous procedure for symmetry exploitation for the following two purposes:

- To get insight into bifurcation behaviors via blockwise singularity detection,
- To enhance the computational efficiency and accuracy of the numerical analysis.

Algorithms for block-diagonalization are available.[7] Computational use of symmetry, or block-diagonalization, in bifurcation analysis is discussed in references.[8]

---

[7] See, for example, Murota et al., 2010 [146] and Maehara and Murota, 2010 [131].

[8] See, for example, Bossavit, 1986 [18]; Healey, 1988 [70]; Zloković, 1989 [215]; Chen and Sameh, 1989 [24]; Healey and Treacy, 1991 [71]; Murota and Ikeda, 1991 [142]; Ikeda and Murota, 1991 [87]; Dinkevich, 1991 [41]; Gatermann and Hohmann, 1991 [57]; Gatermann and Werner, 1994 [59]; Govaerts, 2000 [65]; Gatermann, 2000 [56]; and Gatermann and Hosten, 2005 [58].

In this section, a procedure of block-diagonalization in the bifurcation analysis of symmetric systems is presented on the basis of the theoretical foundation given in §7.7.1. We consider the perfect system only.

11.6.1 Eigenanalysis Versus Block-Diagonalization

Use of block-diagonalization of the Jacobian matrix is illustrated for a simple symmetric system in comparison with the diagonalization based on eigenanalysis.

As a simple example, the regular-triangular truss dome shown in Fig. 11.22(a) is used. All members of this dome have the samc Young's modulus E and the same cross-sectional area A; we set $EA = 1$ for simplicity. The displacement vector

$$\boldsymbol{u} = (x_1, y_1, z_1, x_2, y_2, z_2, x_3, y_3, z_3)^\top$$

has nine degrees of freedom. The z-directional load f is applied to nodes 1 to 3.

Figure 11.22(b) portrays the equilibrium paths, consisting of the fundamental path with the two limit points C and F and bifurcated paths branching from the four bifurcation points B, D, E, and G. A critical point on the fundamental path is to be identified as a point where the Jacobian matrix $J = J(\boldsymbol{u}, f)$ becomes singular; that is, an eigenvalue(s) of J vanishes ($\lambda_i = 0$ for some i). The Jacobian matrix $J(\boldsymbol{u}, f)$ is a function of the solution $(\boldsymbol{u}, f)$; therefore, the eigenanalysis of $J(\boldsymbol{u}, f)$ is to be conducted at each solution point.

The Jacobian matrix of this truss dome, for example, at the initial undeformed state $(\boldsymbol{u}, f) = (\boldsymbol{0}, 0)$ reads as

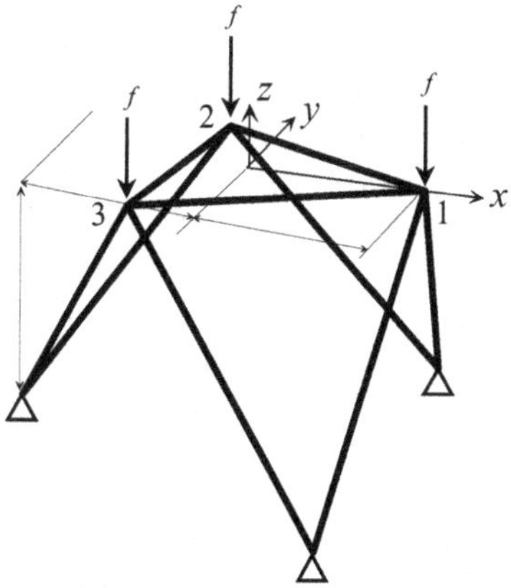

(a) Regular-triangular truss dome

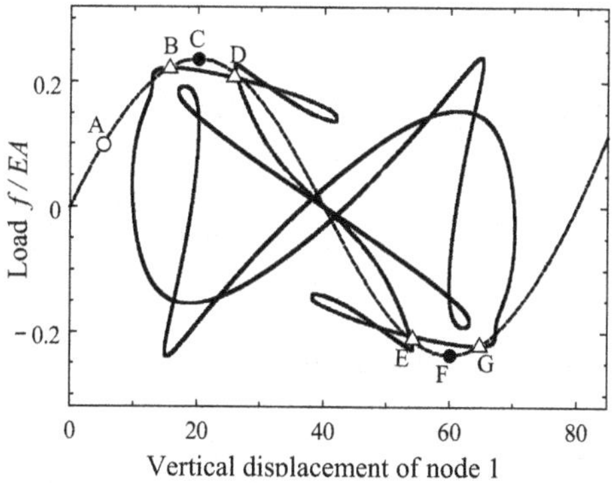

(b) Equilibrium paths

Fig. 11.22 Regular-triangular truss dome and its equilibrium paths. ○: reference point; △: double bifurcation point; •: limit point of f.

$$J(\mathbf{0},0) \approx 10^{-2} \times \begin{pmatrix} 3.50 & 0 & 0.27 & -1.73 & 1.00 & 0 & -1.73 & -1.00 & 0 \\ 0 & 2.75 & 0 & 1.00 & -0.58 & 0 & -1.00 & -0.58 & 0 \\ 0.27 & 0 & 2.13 & 0 & 0 & 0 & 0 & 0 & 0 \\ -1.73 & 1.00 & 0 & 2.94 & -0.32 & -0.13 & 0 & 0 & 0 \\ 1.00 & -0.58 & 0 & -0.32 & 3.31 & 0.23 & 0 & -2.31 & 0 \\ 0 & 0 & 0 & -0.13 & 0.23 & 2.13 & 0 & 0 & 0 \\ -1.73 & -1.00 & 0 & 0 & 0 & 0 & 2.94 & 0.32 & -0.13 \\ -1.00 & -0.58 & 0 & 0 & -2.31 & 0 & 0.32 & 3.31 & -0.23 \\ 0 & 0 & 0 & 0 & 0 & 0 & -0.13 & -0.23 & 2.13 \end{pmatrix}. \tag{11.26}$$

We conduct the diagonalization and block-diagonalization of this Jacobian matrix J below by a pertinent orthogonal transformation $H^\top JH$.

Eigenanalysis and Diagonalization

Diagonalization analysis of J is based on the standard eigenanalysis of J. Let $\boldsymbol{\eta}_i = \boldsymbol{\eta}_i(\boldsymbol{u}, f)$ and $\lambda_i = \lambda_i(\boldsymbol{u}, f)$, respectively, be orthonormal eigenvectors and eigenvalues of $J(\boldsymbol{u}, f)$. Then the transformation matrix is given by

$$H(\boldsymbol{u}, f) = (\boldsymbol{\eta}_1, \dots, \boldsymbol{\eta}_9).$$

This matrix $H(\boldsymbol{u}, f)$ varies with $(\boldsymbol{u}, f)$. It is therefore to be reconstructed for each solution $(\boldsymbol{u}, f)$.

The transformation $H^\top JH$ with this $H(\boldsymbol{u}, f)$ produces a diagonal form of $J(\boldsymbol{u}, f)$. For example, at the initial undeformed state $(\boldsymbol{u}, f) = (\mathbf{0}, 0)$ the transformation matrix reads as

$$H(\mathbf{0},0) \approx 10^{-1} \times \begin{pmatrix} 0.33 & 6.81 & 0 & -0.32 & -0.02 & 0.73 & 0.01 & -4.45 & 5.77 \\ 4.34 & 0.15 & 5.77 & 0 & -1.08 & 0.04 & 6.83 & 0 & 0 \\ -0.06 & -1.20 & 0 & 5.77 & -0.19 & 8.05 & 0 & -0.51 & 0.32 \\ -0.83 & 4.92 & -5.00 & 0.16 & -0.17 & 0.99 & 4.87 & 4.01 & -2.88 \\ 6.14 & -0.85 & -2.89 & -0.28 & -0.82 & 0.18 & -1.64 & 4.88 & 4.99 \\ -1.01 & 0.56 & 0 & 5.77 & -6.88 & -3.75 & -0.44 & 0.26 & 0.32 \\ 1.31 & 4.99 & 5.00 & 0.16 & 0.13 & 0.98 & -4.89 & 4.01 & -2.88 \\ 6.24 & 1.29 & -2.89 & 0.28 & -0.81 & -0.12 & -1.61 & -4.88 & -4.99 \\ 1.06 & 0.63 & 0 & 5.77 & 7.08 & -4.31 & 0.44 & 0.25 & 0.32 \end{pmatrix},$$

and the diagonalized form of $J(\mathbf{0},0)$ reads as

$$\begin{aligned} \tilde{J}(\mathbf{0},0) &= H(\mathbf{0},0)^\top J(\mathbf{0},0) H(\mathbf{0},0) \\ &\approx 10^{-2} \times \mathrm{diag}(0.614,\ 0.614,\ 1.60,\ 2.12,\ 2.16,\ 2.16,\ 4.46,\ 4.46,\ 6.98). \end{aligned}$$

Among the nine eigenvalues $\lambda_1, \dots, \lambda_9$, we can see six distinct values. Three of these are simple

$$\lambda_3 = 0.0160, \qquad \lambda_4 = 0.0212, \qquad \lambda_9 = 0.0698. \tag{11.27}$$

The remaining three are repeated twice

$$\lambda_1 = \lambda_2 = 0.00614, \qquad \lambda_5 = \lambda_6 = 0.0216, \qquad \lambda_7 = \lambda_8 = 0.0446. \tag{11.28}$$

These double eigenvalues are an outcome of structural degeneracy due to symmetry; its mechanism is explained below using block-diagonalization.

It is emphasized that the transformation matrix $H(\mathbf{0},0)$ constructed for a particular point $(\boldsymbol{u},f) = (\mathbf{0},0)$ is not applicable, in general, to another point $(\boldsymbol{u},f)$. In fact, for the solution $(\boldsymbol{u}_{\mathrm{A}},f_{\mathrm{A}})$ at point A in Fig. 11.22(b), the transformation by $H(\mathbf{0},0)$ cannot diagonalize $J(\boldsymbol{u}_{\mathrm{A}},f_{\mathrm{A}})$ as shown below:

$$H(\mathbf{0},0)^{\top} J(\boldsymbol{u}_{\mathrm{A}},f_{\mathrm{A}}) H(\mathbf{0},0)$$
$$\approx 10^{-2} \times \begin{pmatrix} 0.44 & -0.04 & 0 & 0 & -0.07 & 0 & 0.16 & -0.01 & 0 \\ -0.04 & 0.44 & 0 & 0 & 0 & 0.07 & -0.01 & 0.16 & 0 \\ 0 & 0 & 1.77 & 0 & 0 & 0 & 0 & 0 & 0 \\ 0 & 0 & 0 & 1.71 & 0 & 0 & 0 & 0 & 0.03 \\ -0.07 & 0 & 0 & 0 & 1.82 & -0.12 & -0.03 & 0 & 0 \\ 0 & 0.07 & 0 & 0 & -0.12 & 1.82 & 0 & 0.03 & 0 \\ 0.16 & -0.01 & 0 & 0 & -0.03 & 0 & 4.48 & -0.01 & 0 \\ -0.01 & 0.16 & 0 & 0 & 0 & 0.03 & -0.01 & 4.48 & 0 \\ 0 & 0 & 0 & 0.03 & 0 & 0 & 0 & 0 & 6.70 \end{pmatrix},$$

in which nonzero off-diagonal entries are present. In this way, the transformation matrix for the diagonalization is nonuniversal in the sense that the matrix $H(\boldsymbol{u},f)$ depends on the solution $(\boldsymbol{u},f)$.

Block-Diagonalization

We explain the block-diagonalization of the Jacobian matrix $J(\boldsymbol{u},f)$ according to the theoretical recipe given in §7.7.1.

The geometry of the regular-triangular truss dome and the loading are invariant to the dihedral group of degree three $\mathrm{D}_3 = \langle c(2\pi/3), \sigma \rangle$, where $c(2\pi/3)$ stands for the counterclockwise rotation around the z-axis at an angle of $2\pi/3$ and σ for the reflection with respect to the xz-plane.

The irreducible representations of group D_3 consist of two one-dimensional irreducible representations $\mu_1 = (+,+)_{\mathrm{D}_3}$ and $\mu_2 = (+,-)_{\mathrm{D}_3}$ and one two-dimensional irreducible representation $\mu_3 = (1)_{\mathrm{D}_3}$ (cf., §8.2.2). Therefore, we have (cf., (8.7))

$$R(\mathrm{D}_3) = \{(+,+)_{\mathrm{D}_3}, (+,-)_{\mathrm{D}_3}, (1)_{\mathrm{D}_3}\} = \{\mu_1, \mu_2, \mu_3\}. \tag{11.29}$$

Recall the block-diagonal form (7.121) for the Jacobian matrix J equivariant to a general group G. This form reduces for $G = \mathrm{D}_3$ to

$$H^\top JH = \bigoplus_{\mu \in R(\mathrm{D}_3)} \bigoplus_{k=1}^{N^\mu} \tilde{J}^\mu = \mathrm{diag}(\tilde{J}^{\mu_1}, \tilde{J}^{\mu_2}, \tilde{J}^{\mu_3}, \tilde{J}^{\mu_3}) \tag{11.30}$$

with a suitable orthogonal matrix H. Therein $N^{\mu_1} = N^{\mu_2} = 1$ and $N^{\mu_3} = 2$. Compatibly with the block-diagonal form (11.30), the transformation matrix H is decomposed into four blocks as

$$H = (H^{\mu_1}, H^{\mu_2}, H^{\mu_3,1}, H^{\mu_3,2}). \tag{11.31}$$

On the basis of a systematic procedure for constructing the transformation matrix H, to be given in §12.2, the entries of H of (11.31) can be chosen as

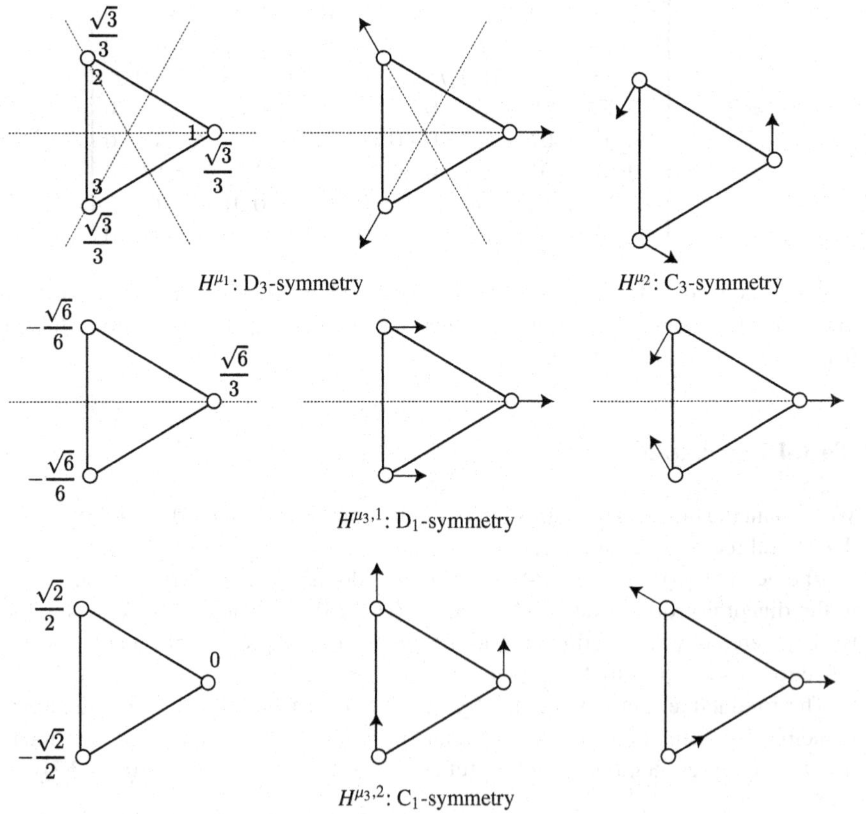

Fig. 11.23 Deformation patterns for the column vectors of the transformation matrix H. Numerals: z-directional components.

$$H = (H^{\mu_1}, H^{\mu_2}, H^{\mu_3,1}, H^{\mu_3,2})$$
$$= \left(\begin{array}{cc|c|ccc|ccc} 0 & \frac{\sqrt{3}}{3} & 0 & 0 & \frac{\sqrt{3}}{3} & \frac{\sqrt{3}}{3} & 0 & 0 & 0 \\ 0 & 0 & \frac{\sqrt{3}}{3} & 0 & 0 & 0 & 0 & \frac{\sqrt{3}}{3} & -\frac{\sqrt{3}}{3} \\ \frac{\sqrt{3}}{3} & 0 & 0 & \frac{\sqrt{6}}{3} & 0 & 0 & 0 & 0 & 0 \\ 0 & -\frac{\sqrt{3}}{6} & -\frac{1}{2} & 0 & \frac{\sqrt{3}}{3} & -\frac{\sqrt{3}}{6} & 0 & 0 & -\frac{1}{2} \\ 0 & \frac{1}{2} & -\frac{\sqrt{3}}{6} & 0 & 0 & -\frac{1}{2} & 0 & \frac{\sqrt{3}}{3} & \frac{\sqrt{3}}{6} \\ \frac{\sqrt{3}}{3} & 0 & 0 & -\frac{\sqrt{6}}{6} & 0 & 0 & \frac{\sqrt{2}}{2} & 0 & 0 \\ 0 & -\frac{\sqrt{3}}{6} & \frac{1}{2} & 0 & \frac{\sqrt{3}}{3} & -\frac{\sqrt{3}}{6} & 0 & 0 & \frac{1}{2} \\ 0 & -\frac{1}{2} & -\frac{\sqrt{3}}{6} & 0 & 0 & \frac{1}{2} & 0 & \frac{\sqrt{3}}{3} & \frac{\sqrt{3}}{6} \\ \frac{\sqrt{3}}{3} & 0 & 0 & -\frac{\sqrt{6}}{6} & 0 & 0 & -\frac{\sqrt{2}}{2} & 0 & 0 \end{array}\right), \quad (11.32)$$

which is independent of solutions $(\boldsymbol{u}, f)$. The submatrices H^{μ_1}, H^{μ_2}, $H^{\mu_3,1}$, and $H^{\mu_3,2}$, respectively, have 2, 1, 3, and 3 columns and are D_3-, C_3-, D_1-, and C_1-symmetric, as presented in Fig. 11.23.

With the use of the matrix H in (11.32), the Jacobian matrix at the initial state $(\boldsymbol{u}, f) = (\boldsymbol{0}, 0)$ is transformed to a block-diagonal form:

$$H^\top J(\boldsymbol{0},0) H$$
$$= \mathrm{diag}(\tilde{J}^{\mu_1}(\boldsymbol{0},0), \tilde{J}^{\mu_2}(\boldsymbol{0},0), \tilde{J}^{\mu_3}(\boldsymbol{0},0), \tilde{J}^{\mu_3}(\boldsymbol{0},0))$$
$$\approx 10^{-2} \times \left(\begin{array}{ccccccccc} 2.13 & 0.27 & & & & & & & \\ 0.27 & 6.96 & & & & & & O & \\ & & 1.60 & & & & & & \\ & & & 2.13 & 0.19 & 0.19 & & & \\ & & & 0.19 & 0.82 & -0.78 & & & \\ & & & 0.19 & -0.78 & 4.28 & & & \\ & & & & & & 2.13 & 0.19 & 0.19 \\ & O & & & & & 0.19 & 0.82 & -0.78 \\ & & & & & & 0.19 & -0.78 & 4.28 \end{array}\right). \quad (11.33)$$

The eigenanalysis of the four diagonal blocks yields the same set of eigenvalues as those in (11.27) and (11.28) of the original Jacobian matrix $J(\boldsymbol{0},0)$. The last two 3×3 diagonal blocks are coincidental and yield three double eigenvalues. Such coincidence is not accidental but occurs systematically due to symmetry.

It is the characteristic of the block-diagonalization that the transformation matrix H is universal in the sense that it is valid for every point $(\boldsymbol{u}, f)$. This matrix H, which works for the initial state $(\boldsymbol{u}, f) = (\boldsymbol{0}, 0)$, also works for any solution $(\boldsymbol{u}, f)$ on the fundamental path.

Indeed, for the solution $(\boldsymbol{u}_{\mathrm{A}}, f_{\mathrm{A}})$ at the point A in Fig. 11.22(b), the same H brings the Jacobian matrix to a block-diagonal form as

$$H^\top J(\boldsymbol{u}_{\mathrm{A}}, f_{\mathrm{A}})H \approx 10^{-2} \times \begin{pmatrix} \begin{array}{|cc|}\hline 1.72 & 0.30 \\ 0.30 & 6.69 \\ \hline \end{array} & & & O \\ & \begin{array}{|c|}\hline 1.77 \\ \hline \end{array} & & \\ & & \begin{array}{|ccc|}\hline 1.78 & 0.21 & 0.21 \\ 0.21 & 0.73 & -0.97 \\ 0.21 & -0.97 & 4.23 \\ \hline \end{array} & \\ O & & & \begin{array}{|ccc|}\hline 1.78 & 0.21 & 0.21 \\ 0.21 & 0.73 & -0.97 \\ 0.21 & -0.97 & 4.23 \\ \hline \end{array} \end{pmatrix},$$

where numerical values of the nonzero entries are different from those in (11.33) at the initial state $(\boldsymbol{u}, f) = (\boldsymbol{0}, 0)$.

Similarly, the Jacobian matrix $J(\boldsymbol{u}, f)$ can be put into a block-diagonal form for any $(\boldsymbol{u}, f)$:

$$H^\top J(\boldsymbol{u}, f)H = \mathrm{diag}(\tilde{J}^{\mu_1}(\boldsymbol{u}, f), \tilde{J}^{\mu_2}(\boldsymbol{u}, f), \tilde{J}^{\mu_3}(\boldsymbol{u}, f), \tilde{J}^{\mu_3}(\boldsymbol{u}, f)). \tag{11.34}$$

These diagonal blocks consist of:

- Two distinct matrices $\tilde{J}^{\mu_1}(\boldsymbol{u}, f)$ and $\tilde{J}^{\mu_2}(\boldsymbol{u}, f)$ of sizes 2 and 1, respectively,
- A twice-repeated matrix $\tilde{J}^{\mu_3}(\boldsymbol{u}, f)$ of size 3, which yields three pairs of double eigenvalues.

In this way, the singularity detection of 9×9 matrix $J(\boldsymbol{u}, f)$ can be replaced by mutually independent tests for the three smaller matrices. The singularity detection must be conducted at each point $(\boldsymbol{u}, f)$ as the values of the entries of the diagonal blocks change along the fundamental path.

11.6.2 Block-Diagonal Form for D_n*-Symmetric System*

The block-diagonal form (7.121) of the Jacobian matrix is given explicitly for a D_n-symmetric system.

Recall the dihedral group $\mathrm{D}_n = \langle c(2\pi/n), \sigma \rangle$ in (8.1) and the index set of irreducible representations in (8.7):

$$R(\mathrm{D}_n) = \begin{cases} \{(+,+)_{\mathrm{D}_n}, (+,-)_{\mathrm{D}_n}, (-,+)_{\mathrm{D}_n}, (-,-)_{\mathrm{D}_n}\} \cup \{(j)_{\mathrm{D}_n} \mid j = 1, \ldots, (n-2)/2\} & \\ \qquad\qquad\qquad\qquad\qquad\qquad\qquad\qquad\qquad\qquad \text{for } n \text{ even}, \\ \{(+,+)_{\mathrm{D}_n}, (+,-)_{\mathrm{D}_n}\} \cup \{(j)_{\mathrm{D}_n} \mid j = 1, \ldots, (n-1)/2\} \qquad \text{for } n \text{ odd}. \end{cases}$$

The number M_1 of the one-dimensional irreducible representations and the number M_2 of the two-dimensional ones are given respectively by (8.5) and (8.6):

$$M_1 = \begin{cases} 4 & \text{for } n \text{ even}, \\ 2 & \text{for } n \text{ odd}, \end{cases} \qquad M_2 = \begin{cases} (n-2)/2 & \text{for } n \text{ even}, \\ (n-1)/2 & \text{for } n \text{ odd}. \end{cases}$$

In addition, all the irreducible representations of D_n over $\mathbb{R}$ are absolutely irreducible; that is, $R(\mathrm{D}_n) = R_{\mathrm{a}}(\mathrm{D}_n)$.

Hence the block-diagonal form (7.121) of J with D_n-symmetry in general reads as

$$\begin{aligned} & H^\top J H \\ &= \bigoplus_{\mu \in R(\mathrm{D}_n)} \bigoplus_{k=1}^{N^\mu} \tilde{J}^\mu \\ &= \mathrm{diag}(\tilde{J}^{(+,+)\mathrm{D}_n}, \tilde{J}^{(+,-)\mathrm{D}_n}, \tilde{J}^{(-,+)\mathrm{D}_n}, \tilde{J}^{(-,-)\mathrm{D}_n}; \tilde{J}^{(1)\mathrm{D}_n}, \tilde{J}^{(1)\mathrm{D}_n}; \ldots; \tilde{J}^{(M_2)\mathrm{D}_n}, \tilde{J}^{(M_2)\mathrm{D}_n}), \end{aligned} \tag{11.35}$$

where $\tilde{J}^{(-,+)\mathrm{D}_n}$ and $\tilde{J}^{(-,-)\mathrm{D}_n}$ exist only for n even.

The number of possible blocks for a D_n-symmetric system is equal to

$$M_1 + 2M_2 = \begin{cases} 4 + (n-2) = n+2, & n \text{ even}, \\ 2 + (n-1) = n+1, & n \text{ odd}, \end{cases}$$

and increases in proportion to n as $n \to \infty$. The sizes of the diagonal blocks vary with individual systems, being equal to the multiplicity of the corresponding irreducible representation denoted a^μ in §7.7.1.

Example 11.1. We consider the regular n-gonal truss dome as depicted in Fig. 11.24. With a pertinent transformation matrix H, the Jacobian matrix J of this dome can be put into a block-diagonal form in (11.35). The sizes of the diagonal blocks are given as below.

Diagonal block	$\tilde{J}^{(+,+)\mathrm{D}_n}$	$\tilde{J}^{(+,-)\mathrm{D}_n}$	$\tilde{J}^{(-,+)\mathrm{D}_n}$	$\tilde{J}^{(-,-)\mathrm{D}_n}$	$\tilde{J}^{(1)\mathrm{D}_n}$	$\tilde{J}^{(2)\mathrm{D}_n}$	$\cdots$	$\tilde{J}^{(M_2)\mathrm{D}_n}$
Size a^μ	3	1	2	1	4	3	$\cdots$	3

Since the number of total degrees of freedom is $3n$, the average size of a block is equal asymptotically to $3n/n = 3$ for large n. In association with an increase of the

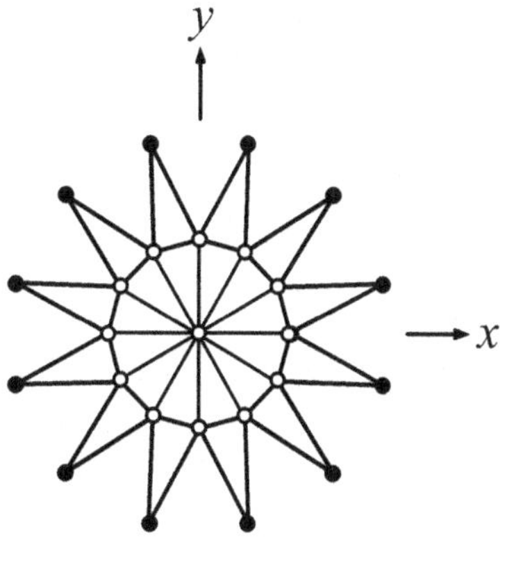

Fig. 11.24 Plane view of regular n-gonal truss dome ($n = 12$). ∘: free nodes; •: fixed nodes.

number n, the number of blocks increases almost linearly, but the average size of the blocks remains almost invariant. □

11.6.3 Block-Diagonal Form for C_n-Symmetric System

The block-diagonalization for C_n-symmetric systems is markedly different from that of D_n-symmetric systems in §11.6.2. This difference stems from the fact that $R(\mathrm{C}_n) \neq R_{\mathrm{a}}(\mathrm{C}_n)$ over $\mathbb{R}$ (cf., $R(\mathrm{D}_n) = R_{\mathrm{a}}(\mathrm{D}_n)$).

The block-diagonalization is conducted for the C_3-symmetric truss dome portrayed in Fig. 11.25. The members shown by thick lines have cross-sectional rigidity of $EA = 1$ and those shown by thin lines that of $EA = 0.5$. This dome has the same geometry as, but different cross-sectional rigidity from, the regular-triangular truss dome used in §11.6.1.

The Jacobian matrix of this truss dome in the initial undeformed state at $(\boldsymbol{u}, f) = (\boldsymbol{0}, 0)$ reads as

$$J(\boldsymbol{0},0) \approx 10^{-2} \times \begin{pmatrix} 3.49 & -0.06 & 0.20 & -1.73 & 1.00 & 0 & -1.73 & -1.00 & 0 \\ -0.06 & 2.35 & -0.46 & 1.00 & -0.58 & 0 & -1.00 & -0.58 & 0 \\ 0.20 & -0.46 & 1.60 & 0 & 0 & 0 & 0 & 0 & 0 \\ -1.73 & 1.00 & 0 & 2.59 & -0.46 & 0.30 & 0 & 0 & 0 \\ 1.00 & -0.58 & 0 & -0.46 & 3.26 & 0.40 & 0 & -2.31 & 0 \\ 0 & 0 & 0 & 0.30 & 0.40 & 1.60 & 0 & 0 & 0 \\ -1.73 & -1.00 & 0 & 0 & 0 & 0 & 2.69 & 0.52 & -0.50 \\ -1.00 & -0.58 & 0 & 0 & -2.31 & 0 & 0.52 & 3.16 & 0.06 \\ 0 & 0 & 0 & 0 & 0 & 0 & -0.50 & 0.06 & 1.60 \end{pmatrix}.$$

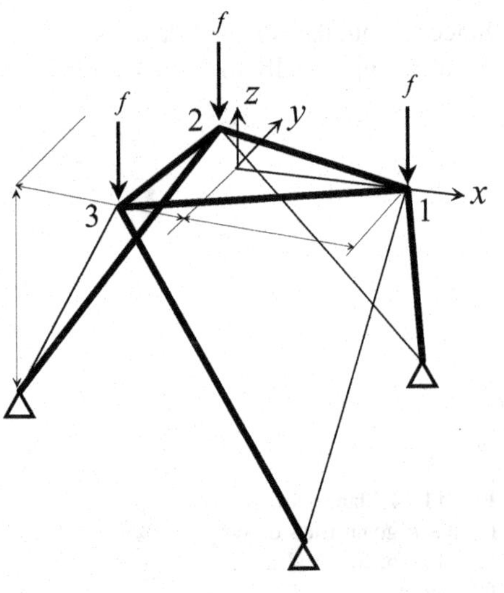

Fig. 11.25 Cyclically symmetric truss dome.

The nine eigenvalues $\lambda_1,\dots,\lambda_9$ of $J(\mathbf{0},0)$ are computed. We can see six distinct eigenvalues. Three of these are simple

$$\lambda_3 = 0.00895, \qquad \lambda_6 = 0.0189, \qquad \lambda_9 = 0.0696. \tag{11.36}$$

The remaining three are repeated twice

$$\lambda_1 = \lambda_2 = 0.00429, \qquad \lambda_4 = \lambda_5 = 0.0162, \qquad \lambda_7 = \lambda_8 = 0.0423. \tag{11.37}$$

Actually, this is an outcome of *structural degeneracy* due to C_3-symmetry, as illustrated below via block-diagonalization.

Block-Diagonalization over Real Numbers

We conduct block-diagonalization over real numbers. Recall that C_3 has two irreducible representations over $\mathbb{R}$, $\mu_1 = (+)_{C_3}$ and $\mu_2 = (1)_{C_3}$ (cf., (8.12)). Note that $\mu_1 \in R_a(C_3)$ and $\mu_2 \in R(C_3) \setminus R_a(C_3)$ in (7.121) for $G = C_3$.

The transformation matrix[9] H in (7.121) takes the form of

$$H = (H^{\mu_1}, H^{\mu_2}) = \left(\begin{array}{ccc|cccccc}
0 & \frac{\sqrt{3}}{3} & 0 & 0 & \frac{\sqrt{3}}{3} & \frac{\sqrt{3}}{3} & 0 & 0 & 0 \\
0 & 0 & \frac{\sqrt{3}}{3} & 0 & 0 & 0 & 0 & \frac{\sqrt{3}}{3} & -\frac{\sqrt{3}}{3} \\
\frac{\sqrt{3}}{3} & 0 & 0 & \frac{\sqrt{6}}{3} & 0 & 0 & 0 & 0 & 0 \\
0 & -\frac{\sqrt{3}}{6} & -\frac{1}{2} & 0 & \frac{\sqrt{3}}{3} & -\frac{\sqrt{3}}{6} & 0 & 0 & -\frac{1}{2} \\
0 & \frac{1}{2} & -\frac{\sqrt{3}}{6} & 0 & 0 & -\frac{1}{2} & 0 & \frac{\sqrt{3}}{3} & \frac{\sqrt{3}}{6} \\
\frac{\sqrt{3}}{3} & 0 & 0 & -\frac{\sqrt{6}}{6} & 0 & 0 & \frac{\sqrt{2}}{2} & 0 & 0 \\
0 & -\frac{\sqrt{3}}{6} & \frac{1}{2} & 0 & \frac{\sqrt{3}}{3} & -\frac{\sqrt{3}}{6} & 0 & 0 & \frac{1}{2} \\
0 & -\frac{1}{2} & -\frac{\sqrt{3}}{6} & 0 & 0 & \frac{1}{2} & 0 & \frac{\sqrt{3}}{3} & \frac{\sqrt{3}}{6} \\
\frac{\sqrt{3}}{3} & 0 & 0 & -\frac{\sqrt{6}}{6} & 0 & 0 & -\frac{\sqrt{2}}{2} & 0 & 0
\end{array}\right). \tag{11.38}$$

With this H, we can block-diagonalize the Jacobian matrix $J(\mathbf{0},0)$ as

9 It is possible to obtain this transformation matrix H by the reassemblage of the transformation matrix (11.32).

$$
\begin{aligned}
&H^\top J(\mathbf{0},0)H \\
&= \mathrm{diag}(\tilde{J}^{\mu_1}(\mathbf{0},0), \hat{J}^{\mu_2}(\mathbf{0},0)) \\
&\approx 10^{-2} \times \left(\begin{array}{ccc|ccc|ccc}
1.60 & 0.20 & -0.46 & & & & & & \\
0.20 & 6.95 & -0.06 & & & & & O & \\
-0.46 & -0.06 & 1.20 & & & & & & \\
\hline
 & & & 1.60 & 0.14 & 0.14 & 0.00 & -0.33 & 0.33 \\
 & & & 0.14 & 0.61 & -0.59 & 0.33 & 0.00 & 0.06 \\
 & & & 0.14 & -0.59 & 4.08 & -0.33 & -0.06 & 0.00 \\
\hline
 & & & 0.00 & 0.33 & -0.33 & 1.60 & 0.14 & 0.14 \\
 & O & & -0.33 & 0.00 & -0.06 & 0.14 & 0.61 & -0.59 \\
 & & & 0.33 & 0.06 & 0.00 & 0.14 & -0.59 & 4.08
\end{array}\right).
\end{aligned}
\tag{11.39}
$$

We see the structure

$$
\hat{J}^{\mu_2}(\mathbf{0},0) = \begin{pmatrix} A & -B \\ B & A \end{pmatrix}, \tag{11.40}
$$

where

$$
A = \begin{pmatrix} 1.60 & 0.14 & 0.14 \\ 0.14 & 0.61 & -0.59 \\ 0.14 & -0.59 & 4.08 \end{pmatrix}, \qquad B = \begin{pmatrix} 0.00 & 0.33 & -0.33 \\ -0.33 & 0.00 & -0.06 \\ 0.33 & 0.06 & 0.00 \end{pmatrix}.
$$

This is the mechanism to yield the coincident eigenvalues in (11.37).

Block-Diagonalization over Complex Numbers

By working with block-diagonalization over complex numbers $\mathbb{C}$ we can further exploit the symmetry of the system. Recall that C_3 has three irreducible representations over $\mathbb{C}$, $\mu_1 = (+)_{C_3}$, $\mu_2 = (1+)_{C_3}$, and $\mu_3 = (1-)_{C_3}$ (cf., (8.17)).

The transformation matrix takes the form of

$$
H = (H^{\mu_1}, H^{\mu_2}, H^{\mu_3}).
$$

Submatrix H^{μ_1} is given by the first three column vectors of (11.38). Submatrix H^{μ_2} is a 6×3 matrix and can be obtained as

$$
H^{\mu_2} = \frac{1}{\sqrt{6}} \begin{pmatrix}
0 & 1 & 1 \\
0 & -\mathrm{i} & \mathrm{i} \\
\sqrt{2} & 0 & 0 \\
0 & 1 & \omega \\
0 & -\mathrm{i} & \mathrm{i}\omega \\
\sqrt{2}\omega & 0 & 0 \\
0 & 1 & \omega^2 \\
0 & -\mathrm{i} & \mathrm{i}\omega^2 \\
\sqrt{2}\omega^2 & 0 & 0
\end{pmatrix}
$$

with $\omega = \exp(\mathrm{i}2\pi/3)$, and $H^{\mu_3} = \overline{H^{\mu_2}}$. With the use of this unitary (complex) transformation matrix H, the Jacobian matrix can be block-diagonalized as

$$H^* J(\mathbf{0},0)H = \mathrm{diag}(\tilde{J}^{\mu_1}, \tilde{J}^{\mu_2}, \tilde{J}^{\mu_3}) = \mathrm{diag}(\tilde{J}^{\mu_1}, A + \mathrm{i}B, A - \mathrm{i}B)$$

with $\tilde{J}^{\mu_2} = A + \mathrm{i}B$, $\tilde{J}^{\mu_3} = A - \mathrm{i}B$, and $\tilde{J}^{\mu_1}$ is identical with the first diagonal block of (11.39). These three diagonal blocks have the same set of eigenvalues as (11.36) and (11.37). More specifically, $\tilde{J}^{\mu_2}$ and $\tilde{J}^{\mu_3}$ have the same set of eigenvalues, yielding double eigenvalues for J. Therefore, we only have to carry out the eigenanalysis for matrix $\tilde{J}^{\mu_2}$. Consequently, the block-diagonalization over $\mathbb{C}$ has an advantage at the expense of involving complex numbers. A systematic procedure to construct the transformation matrix H for C_n-symmetric trusses is given in Remark 12.2 in §12.4.

Problems

11-1 (1) Show that the governing equation

$$F(u, f) = (\sin 2u - 1 + \cos u)(\cos 2u + \frac{3}{4}\sin 2u - f) = 0$$

has a transcritical bifurcation point.
(2) Carry out the path-tracing and branch-switching analysis of this governing equation.

11-2 For $(1)_{\mathrm{D}_5}$ of the regular-pentagonal truss dome in Fig. 11.14 in §11.5.1, draw a bird's-eye view similar to the one in Fig. 11.16.

11-3 For the f versus z_0 curves of the regular-pentagonal truss dome in Fig. 11.15(a) in §11.5.1, plot $|\widetilde{f_{\mathrm{c}}}|$ versus $(\widetilde{z_0})_{|h}$ relations like Fig. 11.17 for a few values of h.

11-4 Simulate the histograms in Fig. 11.18 in §11.5.1 and Fig. 11.21 in §11.5.2 by the probability density functions for an asymmetric simple bifurcation point and a limit point, respectively.

Summary

- A procedure for numerical bifurcation analysis has been presented.
- A revised scaled-corrector method has been presented and put to use in computing the locations of double bifurcation points and nearly coincidental bifurcation points of symmetric systems.
- Perfect and imperfect behaviors of realistic systems have been investigated by means of a synthetic application of the procedures presented in the previous chapters in Part II.

- Use of block-diagonalization in bifurcation analysis of a symmetric system has been demonstrated.

Chapter 12
Efficient Transformation for Block-Diagonalization

12.1 Introduction

Group representation theory guarantees that the Jacobian matrix of symmetric systems can be transformed to a block-diagonal form (cf., §7.7.1). This chapter presents an efficient computational method for this block-diagonalization.[1]

As our main target of application we presume truss structures that are symmetric with respect to the dihedral group D_n. By taking advantage of the underlying geometrical structure, we give a systematic procedure to determine the transformation matrix H and an efficient method to compute the block-diagonal form $H^\top JH$ of the Jacobian matrix J. We describe the procedure in a slightly general form so that the main idea can be extended, mutatis mutandis, to systems other than truss structures that possibly have symmetries other than D_n.

Reviewing the theoretical procedure described in §7.7.1, we assume that J is an $N \times N$ real matrix that satisfies the G-symmetry in (7.100):

$$T(g)J = JT(g), \qquad g \in G, \tag{12.1}$$

where T is a unitary (orthogonal) representation over $\mathbb{R}$. For a general group G, the block-diagonalization reads as (7.121)

$$H^\top JH = \Big[\bigoplus_{\mu \in R_{\mathrm{a}}(G)} \bigoplus_{k=1}^{N^\mu} \tilde{J}^\mu \Big] \bigoplus \Big[\bigoplus_{\mu \in R(G) \setminus R_{\mathrm{a}}(G)} \hat{J}^\mu \Big], \tag{12.2}$$

where $R(G)$ denotes the index set for the irreducible representations of G, $R_{\mathrm{a}}(G)$ is that for absolutely irreducible representations, and N^μ denotes the dimension of $\mu \in R(G)$. If all the $\mathbb{R}$-irreducible representations are absolutely irreducible, which is the case with $G = \mathrm{D}_n$, the expression (12.2) is simplified to

[1] This chapter is based on Murota and Ikeda, 1991 [142].

$$H^{\top}JH = \bigoplus_{\mu\in R(G)} \bigoplus_{k=1}^{N^{\mu}} \tilde{J}^{\mu}. \tag{12.3}$$

In the construction of the orthogonal matrix H for the block-diagonalization (12.3), we first construct an orthogonal matrix Q for the irreducible decomposition of T in (7.31):

$$Q^{\top}T(g)Q = \bigoplus_{\mu\in R(G)} \bigoplus_{i=1}^{a^{\mu}} T^{\mu}(g), \qquad g \in G. \tag{12.4}$$

This matrix Q takes the form (7.104):

$$Q = (Q^{\mu} \mid \mu \in R(G)) = ((Q_i^{\mu} \mid i = 1,\ldots,a^{\mu}) \mid \mu \in R(G)). \tag{12.5}$$

Therein each submatrix Q_i^{μ} is an $N \times N^{\mu}$ matrix associated with an irreducible representation μ of G, and is determined[2] by the relation (7.103):

$$T(g)Q_i^{\mu} = Q_i^{\mu}T^{\mu}(g), \qquad g \in G. \tag{12.6}$$

Next we fabricate the matrix H based on the following three procedures.

- For $\mu \in R(G)$ and $k = 1,\ldots,N^{\mu}$, construct an $N \times a^{\mu}$ matrix $H^{\mu,k}$ by gathering the kth column vectors of Q_i^{μ} for $i = 1,\ldots,a^{\mu}$.
- Define an $N \times (a^{\mu}N^{\mu})$ matrix

$$H^{\mu} = (H^{\mu,k} \mid k = 1,\ldots,N^{\mu}) = (H^{\mu,1},\ldots,H^{\mu,N^{\mu}}). \tag{12.7}$$

  By construction, H^{μ} has the same set of column vectors as Q^{μ}; a rearrangement of the columns of Q^{μ} gives H^{μ}.
- Collect these matrices H^{μ} by (7.123) to arrive at the matrix H in (12.3); that is,

$$H = (H^{\mu} \mid \mu \in R(G)) = ((H^{\mu,k} \mid k = 1,\ldots,N^{\mu}) \mid \mu \in R(G)). \tag{12.8}$$

The transformation matrix H is not uniquely determined in the theory of group representation. The amount of computation for the block-diagonalization (12.3) and the task to construct H heavily depend on the choice of H; therefore, we aim at a good choice of H from a computational perspective. It turns out that a geometrically natural choice is pertinent in a systematic fabrication of H that enjoys sparsity.

This chapter is organized as follows.

- The basic idea of the construction of the transformation matrix H is illustrated for a simple pedagogic example in §12.2.
- A general procedure to obtain the transformation matrix is described in §12.3.
- Formulas of the local transformation matrix for D_n are presented in §12.4.
- Derivation of these formulas is given in §12.5.

[2] Matrix Q_i^{μ} satisfying (12.6) can be determined by the method of projection (see Miller, 1972 [138]). The method presented in this chapter is geometrically natural.

12.2 Construction of Transformation Matrix: Illustration

The basic idea of a systematic construction of the transformation matrix H for block-diagonalization is illustrated here for a simple example. A more complete description is given in §12.3 and §12.5.

12.2.1 Regular-Triangular Truss

As an illustration of the main idea we consider the D_3-symmetric, regular-triangular truss depicted in Fig. 12.1. The members of this truss have the same Young's modulus E and the same cross-sectional area A. The truss has four nodes i $(= 0,1,2,3)$, each with 3 degrees of freedom ($N = 4 \times 3 = 12$), and is described by a 12-dimensional vector

$$\boldsymbol{u} = (x_0, y_0, z_0, x_1, y_1, z_1, x_2, y_2, z_2, x_3, y_3, z_3)^\top.$$

The Jacobian matrix J in (7.42) for this truss is a 12×12 matrix.

The symmetry of the truss is described by the group

$$D_3 = \langle c(2\pi/3), \sigma \rangle,$$

where $c(2\pi/3)$ denotes the counterclockwise rotation about the z-axis at an angle of $2\pi/3$, and σ is the reflection $y \mapsto -y$. The index set of irreducible representations of group D_3 is given (cf., (8.7)) by

$$R(D_3) = \{(+,+)_{D_3}, (+,-)_{D_3}, (1)_{D_3}\} = \{\mu_1, \mu_2, \mu_3\}. \tag{12.9}$$

Accordingly, the transformation matrix H in (12.3) consists of four submatrices (cf., (7.123)) as

$$H = (H^{\mu_1}, H^{\mu_2}, H^{\mu_3,1}, H^{\mu_3,2}),$$

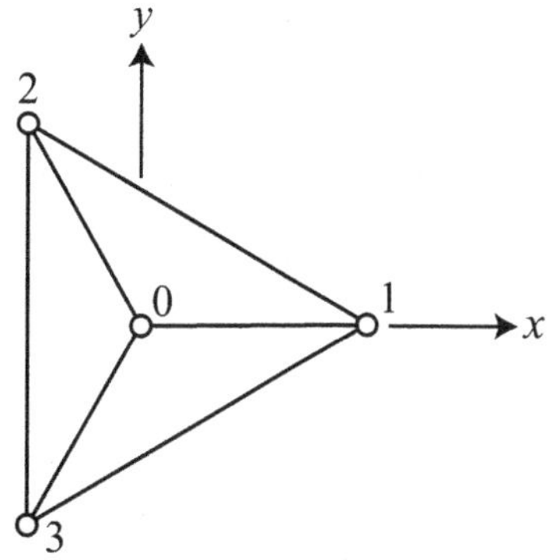

Fig. 12.1 Regular-triangular truss.

the concrete form of which reads as

$$
\begin{aligned}
H &= (H^{\mu_1}, H^{\mu_2}, H^{\mu_3,1}, H^{\mu_3,2}) \\
&= \left(\begin{array}{ccc|c|cccc|cccc}
0 & 0 & 0 & 0 & 1 & 0 & 0 & 0 & 0 & 0 & 0 & 0 \\
0 & 0 & 0 & 0 & 0 & 0 & 0 & 0 & 1 & 0 & 0 & 0 \\
1 & 0 & 0 & 0 & 0 & 0 & 0 & 0 & 0 & 0 & 0 & 0 \\
0 & 0 & \frac{\sqrt{3}}{3} & 0 & 0 & 0 & \frac{\sqrt{3}}{3} & \frac{\sqrt{3}}{3} & 0 & 0 & 0 & 0 \\
0 & 0 & 0 & \frac{\sqrt{3}}{3} & 0 & 0 & 0 & 0 & 0 & 0 & \frac{\sqrt{3}}{3} & -\frac{\sqrt{3}}{3} \\
0 & \frac{\sqrt{3}}{3} & 0 & 0 & 0 & \frac{\sqrt{6}}{3} & 0 & 0 & 0 & 0 & 0 & 0 \\
0 & 0 & -\frac{\sqrt{3}}{6} & -\frac{1}{2} & 0 & 0 & \frac{\sqrt{3}}{3} & -\frac{\sqrt{3}}{6} & 0 & 0 & 0 & -\frac{1}{2} \\
0 & 0 & \frac{1}{2} & -\frac{\sqrt{3}}{6} & 0 & 0 & 0 & -\frac{1}{2} & 0 & 0 & \frac{\sqrt{3}}{3} & \frac{\sqrt{3}}{6} \\
0 & \frac{\sqrt{3}}{3} & 0 & 0 & 0 & -\frac{\sqrt{6}}{6} & 0 & 0 & 0 & \frac{\sqrt{2}}{2} & 0 & 0 \\
0 & 0 & -\frac{\sqrt{3}}{6} & \frac{1}{2} & 0 & 0 & \frac{\sqrt{3}}{3} & -\frac{\sqrt{3}}{6} & 0 & 0 & 0 & \frac{1}{2} \\
0 & 0 & -\frac{1}{2} & -\frac{\sqrt{3}}{6} & 0 & 0 & 0 & \frac{1}{2} & 0 & 0 & \frac{\sqrt{3}}{3} & \frac{\sqrt{3}}{6} \\
0 & \frac{\sqrt{3}}{3} & 0 & 0 & 0 & -\frac{\sqrt{6}}{6} & 0 & 0 & 0 & -\frac{\sqrt{2}}{2} & 0 & 0
\end{array}\right).
\end{aligned}
\tag{12.10}
$$

This section is intended to explain the derivation of this transformation matrix H.

12.2.2 Representation Matrix

We start with the representation matrix $T(g)$, which we need in (12.6) to determine the transformation matrix.

Concrete Form

The concrete form of the representation matrix $T(g)$ for this truss, expressing the action of $g \in \mathrm{D}_3$ on $\boldsymbol{u}$, is given by

$$
T(c(2\pi/3)) = \left(\begin{array}{cc|cc|cc|cc}
R_3 & & & & & & & \\
 & 1 & & & & & & \\
\hline
 & & & & & & R_3 & \\
 & & & & & & & 1 \\
\hline
 & & R_3 & & & & & \\
 & & & 1 & & & & \\
\hline
 & & & & R_3 & & & \\
 & & & & & 1 & &
\end{array}\right), \qquad
T(\sigma) = \left(\begin{array}{cc|cc|cc|cc}
S & & & & & & & \\
 & 1 & & & & & & \\
\hline
 & & S & & & & & \\
 & & & 1 & & & & \\
\hline
 & & & & & & S & \\
 & & & & & & & 1 \\
\hline
 & & & & S & & & \\
 & & & & & 1 & &
\end{array}\right)
\tag{12.11}
$$

with

$$
R_3 = \begin{pmatrix} \cos(2\pi/3) & -\sin(2\pi/3) \\ \sin(2\pi/3) & \cos(2\pi/3) \end{pmatrix}, \qquad S = \begin{pmatrix} 1 & 0 \\ 0 & -1 \end{pmatrix}.
$$

It suffices to consider $T(g)$ for $g = c(2\pi/3)$ and σ, inasmuch as D_3 is generated by these two elements.

Decomposition by Rearrangement

Inspection of the representation matrices in (12.11) motivates us to partition and rearrange the components of the displacement vector $\boldsymbol{u}$ as

$$\hat{\boldsymbol{u}} = \begin{pmatrix} \hat{\boldsymbol{u}}_{0z} \\ \hat{\boldsymbol{u}}_{0xy} \\ \hat{\boldsymbol{u}}_{1z} \\ \hat{\boldsymbol{u}}_{1xy} \end{pmatrix} = \left(\begin{array}{c} z_0 \\ \hline x_0 \\ y_0 \\ \hline z_1 \\ z_2 \\ z_3 \\ \hline x_1 \\ y_1 \\ x_2 \\ y_2 \\ x_3 \\ y_3 \end{array}\right). \tag{12.12}$$

The rearrangement above can be expressed with a *permutation matrix* Ξ as

$$\boldsymbol{u} = \Xi \hat{\boldsymbol{u}}.$$

The representation matrices $\hat{T}(g)$ for the rearranged vector $\hat{\boldsymbol{u}}$ can be obtained from the matrices in (12.11) by permuting the rows and columns as

$$\hat{T}(g) = \Xi^{\top} T(g) \Xi, \qquad g \in \mathrm{D}_3.$$

By construction, the matrices $\hat{T}(g)$ are endowed with a block-diagonal structure:

$$\hat{T}(c(2\pi/3)) = \left(\begin{array}{c|c|c|c} 1 & & & \\ \hline & R_3 & & \\ \hline & & \begin{matrix} 0 & 0 & 1 \\ 1 & 0 & 0 \\ 0 & 1 & 0 \end{matrix} & \\ \hline & & & \begin{matrix} O & O & R_3 \\ R_3 & O & O \\ O & R_3 & O \end{matrix} \end{array}\right), \quad \hat{T}(\sigma) = \left(\begin{array}{c|c|c|c} 1 & & & \\ \hline & S & & \\ \hline & & \begin{matrix} 1 & 0 & 0 \\ 0 & 0 & 1 \\ 0 & 1 & 0 \end{matrix} & \\ \hline & & & \begin{matrix} S & O & O \\ O & O & S \\ O & S & O \end{matrix} \end{array}\right) \tag{12.13}$$

with four diagonal blocks of sizes 1, 2, 3 and 6. We denote this block-diagonal form as

$$\hat{T}(g) = \mathrm{diag}(\hat{T}_{0z}(g), \hat{T}_{0xy}(g), \hat{T}_{1z}(g), \hat{T}_{1xy}(g)). \tag{12.14}$$

It is noted, however, that each diagonal block $\hat{T}_\kappa(g)$, where $\kappa = 0z, 0xy, 1z, 1xy$, need not be an irreducible representation, but it should be decomposed further into irreducible components through a suitable orthogonal transformation matrix, say, $\hat{Q}_\kappa$, which we call the *local transformation matrix*. Then the transformation matrix Q can be obtained as an aggregation of such local transformations $\hat{Q}_\kappa$, as is explained below.

12.2.3 Local Transformation Matrix

We decompose each diagonal block $\hat{T}_\kappa(g)$ in (12.14) into irreducible components. Let $\hat{Q}_\kappa$ be an orthogonal matrix for this decomposition, that is, such an orthogonal matrix that

$$(\hat{Q}_\kappa)^\top \hat{T}_\kappa(g)\hat{Q}_\kappa = \bigoplus_{\mu\in R(G)} \bigoplus_{i=1}^{d_\kappa^\mu} T^\mu(g), \qquad g \in \mathrm{D}_3,$$

where d_κ^μ denotes the multiplicity of μ in $\hat{T}_\kappa$. We denote by N_κ the size of the matrix $\hat{T}_\kappa(g)$, which is equal to the dimension of the vector $\hat{\boldsymbol{u}}_\kappa$.

We specifically examine the last block $\hat{T}_{1xy}(g)$ in (12.14) to explain the decomposition procedure. For notational convenience we put $\kappa = 1xy$. Then we have[3]

$$\hat{T}_\kappa(c(2\pi/3)) = \begin{pmatrix} O & O & R_3 \\ R_3 & O & O \\ O & R_3 & O \end{pmatrix}, \qquad \hat{T}_\kappa(\sigma) = \begin{pmatrix} S & O & O \\ O & O & S \\ O & S & O \end{pmatrix}. \tag{12.15}$$

We consider the condition (12.6) for $T(g) = \hat{T}_\kappa(g)$ to obtain the decomposition of $\hat{T}_\kappa(g)$ into irreducible representations $T^\mu(g)$. The irreducible representations of D_3 in (12.9) are given by (8.8) and (8.9) with $n = 3$.

- For $\mu = \mu_1 = (+,+)_{\mathrm{D}_3}$ we have $T^{\mu_1}(c(2\pi/3)) = 1$, $T^{\mu_1}(\sigma) = 1$ by (8.8). Therefore, (12.6) reduces to

$$\hat{T}_\kappa(c(2\pi/3))\boldsymbol{q} = \boldsymbol{q}, \qquad \hat{T}_\kappa(\sigma)\boldsymbol{q} = \boldsymbol{q}$$

  for a vector $\boldsymbol{q}$. The solution $\boldsymbol{q}$ to the above equations is essentially unique, and is given, for example, by

$$\boldsymbol{q}_\kappa^{\mu_1} = (\sqrt{3}/3, 0, -\sqrt{3}/6, 1/2, -\sqrt{3}/6, -1/2)^\top.$$

- For $\mu = \mu_2 = (+,-)_{\mathrm{D}_3}$ we have $T^{\mu_2}(c(2\pi/3)) = 1$, $T^{\mu_2}(\sigma) = -1$ by (8.8). Therefore, (12.6) reduces to

$$\hat{T}_\kappa(c(2\pi/3))\boldsymbol{q} = \boldsymbol{q}, \qquad \hat{T}_\kappa(\sigma)\boldsymbol{q} = -\boldsymbol{q}.$$

  The solution $\boldsymbol{q}$ is essentially unique again, and is given, e.g., by

[3] In the notation of §12.3.2 we have $\xi = 1Vxy$ for the type associated with $\kappa = 1xy$.

$$q_\kappa^{\mu_2} = (0, \sqrt{3}/3, -1/2, -\sqrt{3}/6, 1/2, -\sqrt{3}/6)^\top.$$

- For two-dimensional irreducible representation $\mu = \mu_3 = (1)_{D_3}$ the equations take the forms of

$$\hat{T}_\kappa(c(2\pi/3))\left(q^1, q^2\right) = \left(q^1, q^2\right)R_3, \qquad \hat{T}_\kappa(\sigma)\left(q^1, q^2\right) = \left(q^1, q^2\right)S,$$

where q^1 and q^2 are unknown vectors. There exist two (essentially different or linearly independent) pairs of solutions, for example,

$$\left(q_{\kappa,1}^{\mu_3,1}, q_{\kappa,1}^{\mu_3,2}\right) = \begin{pmatrix} \frac{\sqrt{3}}{3} & 0 \\ 0 & \frac{\sqrt{3}}{3} \\ \frac{\sqrt{3}}{3} & 0 \\ 0 & \frac{\sqrt{3}}{3} \\ \frac{\sqrt{3}}{3} & 0 \\ 0 & \frac{\sqrt{3}}{3} \end{pmatrix}, \qquad \left(q_{\kappa,2}^{\mu_3,1}, q_{\kappa,2}^{\mu_3,2}\right) = \begin{pmatrix} \frac{\sqrt{3}}{3} & 0 \\ 0 & -\frac{\sqrt{3}}{3} \\ -\frac{\sqrt{3}}{6} & -\frac{1}{2} \\ -\frac{1}{2} & \frac{\sqrt{3}}{6} \\ -\frac{\sqrt{3}}{6} & \frac{1}{2} \\ \frac{1}{2} & \frac{\sqrt{3}}{6} \end{pmatrix}.$$

The multiplicity a_κ^μ of μ in $\hat{T}_\kappa$ is equal to the number of linearly independent (pairs of) solutions to (12.6) for μ. We have seen that $a_\kappa^{\mu_1} = a_\kappa^{\mu_2} = 1$ and $a_\kappa^{\mu_3} = 2$.

By aggregating the vectors obtained above, we arrive at the transformation matrix

$$\hat{Q}_{1xy} = \left(q_\kappa^{\mu_1}, q_\kappa^{\mu_2}, q_{\kappa,1}^{\mu_3,1}, q_{\kappa,1}^{\mu_3,2}, q_{\kappa,2}^{\mu_3,1}, q_{\kappa,2}^{\mu_3,2}\right)$$
$$= \begin{pmatrix} \frac{\sqrt{3}}{3} & 0 & \frac{\sqrt{3}}{3} & 0 & \frac{\sqrt{3}}{3} & 0 \\ 0 & \frac{\sqrt{3}}{3} & 0 & \frac{\sqrt{3}}{3} & 0 & -\frac{\sqrt{3}}{3} \\ -\frac{\sqrt{3}}{6} & -\frac{1}{2} & \frac{\sqrt{3}}{3} & 0 & -\frac{\sqrt{3}}{6} & -\frac{1}{2} \\ \frac{1}{2} & -\frac{\sqrt{3}}{6} & 0 & \frac{\sqrt{3}}{3} & -\frac{1}{2} & \frac{\sqrt{3}}{6} \\ -\frac{\sqrt{3}}{6} & \frac{1}{2} & \frac{\sqrt{3}}{3} & 0 & -\frac{\sqrt{3}}{6} & \frac{1}{2} \\ -\frac{1}{2} & -\frac{\sqrt{3}}{6} & 0 & \frac{\sqrt{3}}{3} & \frac{1}{2} & \frac{\sqrt{3}}{6} \end{pmatrix} \tag{12.16}$$

for the decomposition of $\hat{T}_{1xy}$ into irreducible representations.

Similarly we can obtain the local transformation matrices for the decomposition of $\hat{T}_{0z}$, $\hat{T}_{0xy}$, and $\hat{T}_{1z}$, respectively, as

$$\hat{Q}_{0z} = \left(q_{0z}^{\mu_1}\right) = \left(1\right), \qquad \hat{Q}_{0xy} = \left(q_{0xy}^{\mu_3,1}, q_{0xy}^{\mu_3,2}\right) = \begin{pmatrix} 1 & 0 \\ 0 & 1 \end{pmatrix},$$

$$\hat{Q}_{1z} = \left(q_{1z}^{\mu_1}, q_{1z}^{\mu_3,1}, q_{1z}^{\mu_3,2}\right) = \begin{pmatrix} \frac{\sqrt{3}}{3} & \frac{\sqrt{6}}{3} & 0 \\ \frac{\sqrt{3}}{3} & -\frac{\sqrt{6}}{6} & \frac{\sqrt{2}}{2} \\ \frac{\sqrt{3}}{3} & -\frac{\sqrt{6}}{6} & -\frac{\sqrt{2}}{2} \end{pmatrix}.$$

12.2.4 Assemblage of Local Transformations

Local transformation matrices $\hat{Q}_\kappa$ are assembled to form Q and H.

Transformation Matrix Q

The transformation matrix Q can be obtained through an assemblage of local transformations $\hat{Q}_\kappa$ as follows. Make a block-diagonal matrix

$$\begin{aligned}
&\mathrm{diag}\left(\hat{Q}_{0z}, \hat{Q}_{0xy}, \hat{Q}_{1z}, \hat{Q}_{1xy}\right) \\
&= \begin{pmatrix}
\boldsymbol{q}_{0z}^{\mu_1} & 0 & 0 & 0 & 0 & 0 & 0 & 0 & 0 & 0 & 0 & 0 \\
\mathbf{0} & \boldsymbol{q}_{0xy}^{\mu_3,1} & \boldsymbol{q}_{0xy}^{\mu_3,2} & \mathbf{0} & \mathbf{0} & \mathbf{0} & \mathbf{0} & \mathbf{0} & \mathbf{0} & \mathbf{0} & \mathbf{0} & \mathbf{0} \\
\mathbf{0} & \mathbf{0} & \mathbf{0} & \boldsymbol{q}_{1z}^{\mu_1} & \boldsymbol{q}_{1z}^{\mu_3,1} & \boldsymbol{q}_{1z}^{\mu_3,2} & \mathbf{0} & \mathbf{0} & \mathbf{0} & \mathbf{0} & \mathbf{0} & \mathbf{0} \\
\mathbf{0} & \mathbf{0} & \mathbf{0} & \mathbf{0} & \mathbf{0} & \mathbf{0} & \boldsymbol{q}_{1xy}^{\mu_1} & \boldsymbol{q}_{1xy}^{\mu_2} & \boldsymbol{q}_{1xy,1}^{\mu_3,1} & \boldsymbol{q}_{1xy,1}^{\mu_3,2} & \boldsymbol{q}_{1xy,2}^{\mu_3,1} & \boldsymbol{q}_{1xy,2}^{\mu_3,2}
\end{pmatrix}
\end{aligned}$$

and permute its columns compatibly with the indices $\{\mu_1, \mu_2, \mu_3\}$ of irreducible representations to arrive at

$$\hat{Q} = \left(\begin{array}{ccc|c|cccccccc}
\boldsymbol{q}_{0z}^{\mu_1} & 0 & 0 & 0 & 0 & 0 & 0 & 0 & 0 & 0 & 0 & 0 \\
\mathbf{0} & \mathbf{0} & \mathbf{0} & \mathbf{0} & \boldsymbol{q}_{0xy}^{\mu_3,1} & \boldsymbol{q}_{0xy}^{\mu_3,2} & \mathbf{0} & \mathbf{0} & \mathbf{0} & \mathbf{0} & \mathbf{0} & \mathbf{0} \\
\mathbf{0} & \boldsymbol{q}_{1z}^{\mu_1} & \mathbf{0} & \mathbf{0} & \mathbf{0} & \mathbf{0} & \boldsymbol{q}_{1z}^{\mu_3,1} & \boldsymbol{q}_{1z}^{\mu_3,2} & \mathbf{0} & \mathbf{0} & \mathbf{0} & \mathbf{0} \\
\mathbf{0} & \mathbf{0} & \boldsymbol{q}_{1xy}^{\mu_1} & \boldsymbol{q}_{1xy}^{\mu_2} & \mathbf{0} & \mathbf{0} & \mathbf{0} & \mathbf{0} & \boldsymbol{q}_{1xy,1}^{\mu_3,1} & \boldsymbol{q}_{1xy,1}^{\mu_3,2} & \boldsymbol{q}_{1xy,2}^{\mu_3,1} & \boldsymbol{q}_{1xy,2}^{\mu_3,2}
\end{array}\right). \tag{12.17}$$

Permute the rows of the matrix $\hat{Q}$ compatibly with the rearrangement into the original coordinate $\boldsymbol{u}$, to arrive at

$$Q = \Xi \hat{Q} = (Q^{\mu_1}, Q^{\mu_2}, Q^{\mu_3}).$$

We have

$$Q^\top T(g) Q = \hat{Q}^\top \hat{T}(g) \hat{Q} = \bigoplus_{\mu \in R(\mathrm{D}_3)} \bigoplus_{i=1}^{a^\mu} T^\mu(g),$$

where the multiplicities a^μ are given as $a^\mu = \sum_\kappa a_\kappa^\mu$. For example,

$$a^{\mu_1} = a_{0z}^{\mu_1} + a_{0xy}^{\mu_1} + a_{1z}^{\mu_1} + a_{1xy}^{\mu_1} = 1 + 0 + 1 + 1 = 3.$$

We have $a^{\mu_1} = 3$, $a^{\mu_2} = 1$, and $a^{\mu_3} = 4$.

Transformation Matrix H

The transformation matrix H can be obtained from Q via a permutation of columns of Q (i.e., $H = Q\Pi$ with a permutation matrix Π). More specifically, the submatrix of H^{μ_3} corresponding to the two-dimensional irreducible representation μ_3 is divided into two parts as $H^{\mu_3} = (H^{\mu_3,1}, H^{\mu_3,2})$; the first part $H^{\mu_3,1}$ comprises odd-numbered columns of the submatrix Q^{μ_3} and the second part $H^{\mu_3,2}$ comprises even-numbered columns of Q^{μ_3}.

Alternatively, the matrix H can be constructed directly from $\hat{Q}$, without going through Q. First, we permute the columns of $\hat{Q}$ in (12.17) via permutation Π as

$$\begin{aligned}\hat{H} &= \hat{Q}\Pi \\ &= \left(\hat{H}^{\mu_1}, \hat{H}^{\mu_2}, \hat{H}^{\mu_3,1}, \hat{H}^{\mu_3,2}\right) \\ &= \left(\begin{array}{ccc|c|cccc|cccc} \boldsymbol{q}_{0z}^{\mu_1} & 0 & 0 & 0 & 0 & 0 & 0 & 0 & 0 & 0 & 0 & 0 \\ \boldsymbol{0} & \boldsymbol{0} & \boldsymbol{0} & \boldsymbol{0} & \boldsymbol{q}_{0xy}^{\mu_3,1} & \boldsymbol{0} & \boldsymbol{0} & \boldsymbol{0} & \boldsymbol{q}_{0xy}^{\mu_3,2} & \boldsymbol{0} & \boldsymbol{0} & \boldsymbol{0} \\ \boldsymbol{0} & \boldsymbol{q}_{1z}^{\mu_1} & \boldsymbol{0} & \boldsymbol{0} & \boldsymbol{0} & \boldsymbol{q}_{1z}^{\mu_3,1} & \boldsymbol{0} & \boldsymbol{0} & \boldsymbol{0} & \boldsymbol{q}_{1z}^{\mu_3,2} & \boldsymbol{0} & \boldsymbol{0} \\ \boldsymbol{0} & \boldsymbol{0} & \boldsymbol{q}_{1xy}^{\mu_1} & \boldsymbol{q}_{1xy}^{\mu_2} & \boldsymbol{0} & \boldsymbol{0} & \boldsymbol{q}_{1xy,1}^{\mu_3,1} & \boldsymbol{q}_{1xy,2}^{\mu_3,1} & \boldsymbol{0} & \boldsymbol{0} & \boldsymbol{q}_{1xy,1}^{\mu_3,2} & \boldsymbol{q}_{1xy,2}^{\mu_3,2} \end{array}\right).\end{aligned}$$

Next, we permute the rows to arrive at

$$H = \Xi \hat{H} = \Xi \hat{Q}\Pi,$$

which is equal to $Q\Pi$ since $Q = \Xi\hat{Q}$. This expression gives the explicit form of the transformation matrix H in (12.10).

12.3 Construction of Transformation Matrix: General Procedure

A systematic construction of the transformation matrix H for block-diagonalization is presented in this section as a generalization of the procedure presented in §12.2 for the simple pedagogic example, the triangular truss. Although we presume D_n-symmetric trusses as our main target of application, we describe the procedure in such a form that is applicable, mutatis mutandis, to other types of systems with symmetries other than D_n.

12.3.1 Representation Matrix

We provide a general setting that contains the example in §12.2 as a special case.

Tensor Form

Consider, as our canonical example for motivation, a D_n-symmetric truss that is rotationally symmetric with respect to the z-axis, and denote the set of nodes by P. Each node $i \in P$ has three displacement components and is described by a three-dimensional vector

$$\boldsymbol{u}_i = (x_i, y_i, z_i)^\top. \tag{12.18}$$

The vector $\boldsymbol{u}_i$ is a member of the three-dimensional Euclidean space $\mathbb{R}^E$ with $E = \{x, y, z\}$. The displacement vector $\boldsymbol{u}$ of the whole structure is $|P| \cdot |E|$-dimensional, and is given as a collection of such nodal components; that is,

$$\boldsymbol{u} = (\boldsymbol{u}_i \mid i \in P). \tag{12.19}$$

For a general treatment it is convenient to regard $\boldsymbol{u}$ as a member of the space

$$V = \mathbb{R}^P \otimes \mathbb{R}^E, \tag{12.20}$$

the tensor product of $\mathbb{R}^P$ and $\mathbb{R}^E$. Mathematically, P and E can be any finite sets, which renders our subsequent discussion fairly general. For trusses we have $E = \{x, y, z\}$, but for other systems such as frames and plates, we can take a larger set for E to represent other degrees of freedom such as rotations.

We assume that the structure in question has the symmetry represented by a group G. In accordance with the tensor structure (12.20) the representation matrix $T(g)$ is assumed to be a tensor product,

$$T(g) = T_P(g) \otimes T_E(g), \qquad g \in G, \tag{12.21}$$

where T_P is a permutation representation on P (cf., Example 7.6 in §7.4.1) and T_E is a representation on $\mathbb{R}^E$. The matrix $T(g)$ is of size $N = |P| \cdot |E|$ and its row (or column) set is indexed by $P \times E = \{(i, e) \mid i \in P,\ e \in E\}$.

Example 12.1. The assumed tensor form of (12.21) is verified here for regular n-gonal nodes of a truss. Denote the nodal displacements at the node i by (x_i, y_i, z_i) $(i = 1, \ldots, n)$, and put $\boldsymbol{u} = (x_1, y_1, z_1, \ldots, x_n, y_n, z_n)^\top$, where the nodes are numbered 1 to n counterclockwise, as shown in Fig. 12.2(a) for $n = 5$. It turns out to be convenient to partition these displacements into two parts as $\boldsymbol{u}_z = (z_1, \ldots, z_n)^\top$ and $\boldsymbol{u}_{xy} = (x_1, y_1, \ldots, x_n, y_n)^\top$.

The set of regular n-gonal nodes is invariant to the dihedral group $\mathrm{D}_n = \langle c(2\pi/n), \sigma \rangle$ of degree n, generated by two elements:

- Counterclockwise rotation around the z-axis at an angle of $2\pi/n$, denoted by $c(2\pi/n)$,
- Reflection $y \mapsto -y$ with respect to the xz-plane, denoted by σ.

See Fig. 12.2 for an example for $n = 5$.

The z-directional displacement vector $\boldsymbol{u}_z = (z_1, \ldots, z_n)^\top$ is transformed by $c(2\pi/n)$ and σ, respectively, as

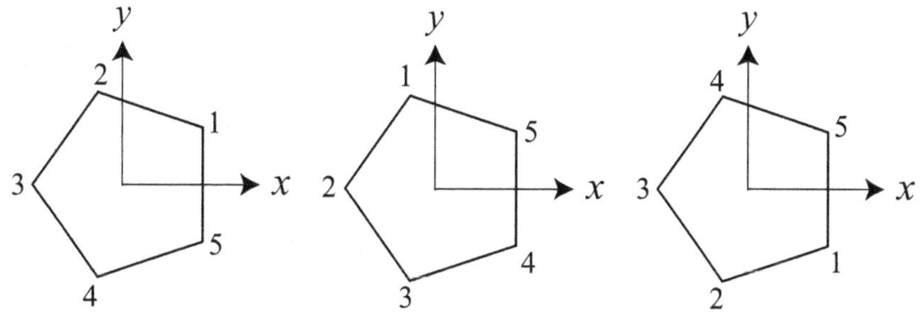

(a) Original configuration (b) Rotated by $c(2\pi/5)$ (c) Reflected by σ

Fig. 12.2 Set of regular $n(=5)$-gonal nodes that is rotated by $c(2\pi/5)$ and reflected by σ. The z-axis is directed upward from the xy-plane.

$$c(2\pi/n)\cdot \boldsymbol{u}_z = c(2\pi/n)\cdot \begin{pmatrix} z_1 \\ z_2 \\ \vdots \\ z_n \end{pmatrix} = \begin{pmatrix} z_n \\ z_1 \\ \vdots \\ z_{n-1} \end{pmatrix} = \begin{pmatrix} & & & 1 \\ 1 & & & \\ & \ddots & & \\ & & 1 & \end{pmatrix} \begin{pmatrix} z_1 \\ z_2 \\ \vdots \\ z_n \end{pmatrix},$$

$$\sigma \cdot \boldsymbol{u}_z = \sigma \cdot \begin{pmatrix} z_1 \\ z_2 \\ \vdots \\ z_n \end{pmatrix} = \begin{pmatrix} z_n \\ z_{n-1} \\ \vdots \\ z_1 \end{pmatrix} = \begin{pmatrix} & & & 1 \\ & & 1 & \\ & \cdot^{\cdot^{\cdot}} & & \\ 1 & & & \end{pmatrix} \begin{pmatrix} z_1 \\ z_2 \\ \vdots \\ z_n \end{pmatrix}.$$

In other words, the actions of these two elements on $\boldsymbol{u}_z$ are represented by the permutation matrices below.

$$T_P(c(2\pi/n)) = \begin{pmatrix} & & & 1 \\ 1 & & & \\ & \ddots & & \\ & & 1 & \end{pmatrix}, \qquad T_P(\sigma) = \begin{pmatrix} & & & 1 \\ & & 1 & \\ & \cdot^{\cdot^{\cdot}} & & \\ 1 & & & \end{pmatrix}. \tag{12.22}$$

Similarly, the actions of $c(2\pi/n)$ and σ on the xy-directional displacement vector $\boldsymbol{u}_{xy} = (x_1, y_1, \ldots, x_n, y_n)^\top$ are expressed, respectively, as

$$c(2\pi/n)\cdot \boldsymbol{u}_{xy} = \begin{pmatrix} & & & R_n \\ R_n & & & \\ & \ddots & & \\ & & R_n & \end{pmatrix} \boldsymbol{u}_{xy} = (T_P(c(2\pi/n)) \otimes R_n)\boldsymbol{u}_{xy}, \tag{12.23}$$

$$\sigma \cdot \boldsymbol{u}_{xy} = \begin{pmatrix} & & & S \\ & & S & \\ & \cdot^{\cdot^{\cdot}} & & \\ S & & & \end{pmatrix} \boldsymbol{u}_{xy} = (T_P(\sigma) \otimes S)\boldsymbol{u}_{xy}, \tag{12.24}$$

where

$$R_n = \begin{pmatrix} \cos(2\pi/n) & -\sin(2\pi/n) \\ \sin(2\pi/n) & \cos(2\pi/n) \end{pmatrix}, \qquad S = \begin{pmatrix} 1 & 0 \\ 0 & -1 \end{pmatrix}. \tag{12.25}$$

We summarize the relations shown above as follows. Define a three-dimensional matrix representation T_E of D_n by

$$T_E(c(2\pi/n)) = \begin{pmatrix} R_n & \\ & 1 \end{pmatrix}, \qquad T_E(\sigma) = \begin{pmatrix} S & \\ & 1 \end{pmatrix}. \tag{12.26}$$

This represents the action of D_n on the three-dimensional space of the displacements $\mathbb{R}^E$ attached to each node, where $E = \{x,y,z\}$. The action of each $g \in \mathrm{D}_n$ on $\boldsymbol{u}$ can be represented as $T(g)\boldsymbol{u}$ with a $3n$-dimensional representation matrix $T(g)$. The discussion above can be restated that $T(g)$ is decomposed as a tensor product of $T_P(g)$ in (12.22) and $T_E(g)$ in (12.26). That is, we have $T(g) = T_P(g) \otimes T_E(g)$ in (12.21). Note also that we have $V = \mathbb{R}^P \otimes \mathbb{R}^E$ in (12.20) for $P = \{1,\dots,n\}$, and that $E = \{x,y,z\}$ is partitioned as $E = \{z\} \cup \{x,y\}$. □

Decomposition by Rearrangement

Appropriate partitions of P and E yield a decomposition of representation matrix $T(g)$ into a block-diagonal form $\hat{T}(g)$ as in (12.13). The block-diagonal form $\hat{T}(g)$ is obtained from $T(g)$ by a permutation of rows and columns, corresponding to a rearrangement of the components of the vector $\boldsymbol{u}$, although the resulting diagonal blocks of $\hat{T}(g)$ are not necessarily irreducible representations.

Let

$$P = \bigcup_{l \in L} P_l \tag{12.27}$$

be the decomposition of P into disjoint *orbits* with respect to the permutation representation in (12.21). That is, two nodes i and j belong to the same P_l if and only if i is moved to j by $T_P(g)$ for some $g \in G$. We refer to this as the *orbit decomposition*. Obviously, $\mathbb{R}^{P_l}$ is a G-invariant subspace for each l.

An orbit decomposition is naturally induced from the geometric symmetry. For example, the regular-triangular truss in §12.2.1 has node set $P = \{0,1,2,3\}$, which is partitioned into two orbits, $P_1 = \{0\}$ and $P_2 = \{1,2,3\}$. As another example, Fig 12.3(a) shows a D_6-symmetric truss dome, spherical diamond shell, and Fig 12.3(b) shows the decomposition into five orbits P_1 to P_5, where "types" associated with orbits are explained later in §12.3.2.

Also let

$$E = \bigcup_{m \in M} E_m \tag{12.28}$$

be a decomposition of E into disjoint subsets such that $\mathbb{R}^{E_m}$ is a G-invariant subspace for each m. For example, in a D_n-symmetric truss that is rotationally symmetric with respect to the z-axis, $E = \{x,y,z\}$ can be decomposed into two parts:

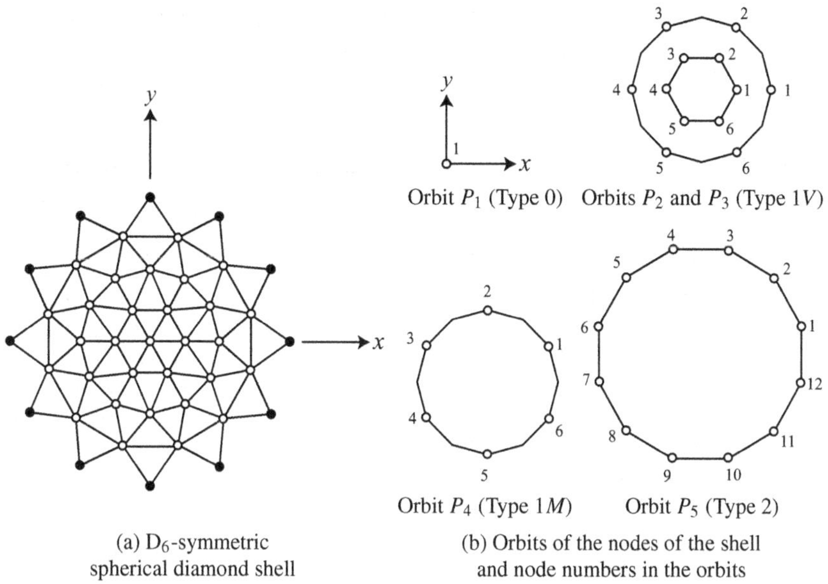

Fig. 12.3 Decomposition of nodes of D_6-symmetric truss dome into orbits. ∘: free nodes; •: fixed nodes.

$$E_1 = \{z\}, \qquad E_2 = \{x, y\}. \tag{12.29}$$

Under any geometrical transformation represented by an element of D_n, the z-directional displacement at a node is permuted to the z-directional displacement at another node. Such is also the case for the displacements in $E_2 = \{x, y\}$.

The two kinds of partitioning, (12.27) and (12.28), are combined to a refined partitioning of displacements indexed by

$$\kappa = (l, m), \qquad l \in L,\ m \in M. \tag{12.30}$$

Compatibly with this partitioning, we permute the components of the displacement vector $\boldsymbol{u}$ to

$$\hat{\boldsymbol{u}} = (\hat{\boldsymbol{u}}_\kappa \mid \kappa = (l, m)) = \begin{pmatrix} \vdots \\ \hat{\boldsymbol{u}}_{(l,m)} \\ \vdots \end{pmatrix}. \tag{12.31}$$

Here $\kappa = (l, m)$ runs over all $l \in L$ and $m \in M$, and the dimension of the subvector $\hat{\boldsymbol{u}}_\kappa$, denoted by N_κ, is given by $N_\kappa = |P_l| \cdot |E_m|$. A concrete example of (12.31) was presented in (12.12).

The rearrangement of the components of $\boldsymbol{u}$ to $\hat{\boldsymbol{u}}$ can be expressed as

$$\boldsymbol{u} = \Xi\, \hat{\boldsymbol{u}}, \qquad \text{or} \qquad \hat{\boldsymbol{u}} = \Xi^\top \boldsymbol{u} \tag{12.32}$$

with a *permutation matrix* Ξ. The representation matrix $\hat{T}(g)$ for the rearranged vector $\hat{\boldsymbol{u}}$ can be obtained as

$$\hat{T}(g) = \Xi^\top T(g) \Xi, \qquad g \in G. \tag{12.33}$$

Then, by construction, the representation matrix $\hat{T}(g)$ takes a block-diagonal form

$$\hat{T}(g) = \bigoplus_{\kappa} \hat{T}_\kappa(g), \qquad g \in G, \tag{12.34}$$

where $\hat{T}_\kappa(g)$ is the representation matrix of size N_κ associated with κ. Each diagonal block $\hat{T}_\kappa(g)$ is decomposed further into irreducible components with an orthogonal matrix $\hat{Q}_\kappa$ of size N_κ. We refer to such $\hat{Q}_\kappa$ as the *local transformation matrix.*

12.3.2 Local Transformation Matrix

Each diagonal block $\hat{T}_\kappa(g)$ in (12.34) is decomposed further into irreducible components through the local transformation matrix $\hat{Q}_\kappa$ as

$$(\hat{Q}_\kappa)^\top \hat{T}_\kappa(g) \hat{Q}_\kappa = \bigoplus_{\mu \in R(G)} \bigoplus_{i=1}^{a_\kappa^\mu} T^\mu(g), \qquad g \in G, \tag{12.35}$$

where a_κ^μ denotes the multiplicity of $\mu \in R(G)$ in $\hat{T}_\kappa$.

Classification of Orbits

Local transformation matrix $\hat{Q}_\kappa$ can be constructed systematically by classifying $\kappa = (l, m)$ into several types in relation to the associated orbit P_l and the coordinate subset E_m.

To begin with, recall that we have in mind a truss structure that is symmetric with respect to the dihedral group $\mathrm{D}_n = \langle c(2\pi/n), \sigma \rangle$, where $c(2\pi/n)$ denotes a counter-clockwise rotation about the z-axis at an angle of $2\pi/n$ and σ is a reflection $y \mapsto -y$.

An orbit P_l of such a D_n-symmetric truss consists of either 1, n, or $2n$ nodes. This is indeed true of the examples in Fig. 12.3 ($n = 6$) and in Fig. 12.4 ($n = 1, 3, 4$). Accordingly, we define three types of orbits:

- Type 0 orbit, consisting of a single node at the center,
- Type 1 orbit, consisting of n nodes,
- Type 2 orbit, consisting of $2n$ nodes.

By convention, Type 0 is absent for $n = 1$. For a Type 1 orbit, we can distinguish two further types, Type $1V$ and Type $1M$, as follows.

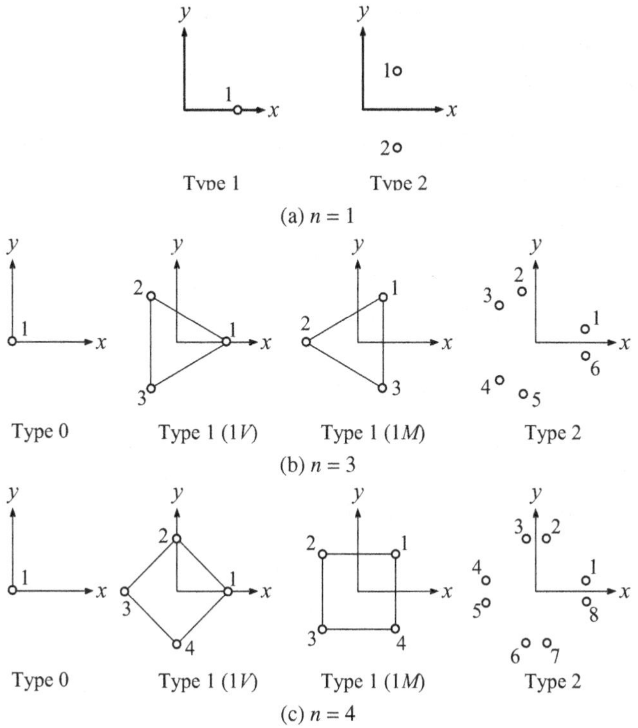

Fig. 12.4 Classification of orbits.

- For n even, an orbit of Type $1V$ has two nodes on the xz-plane and one of Type $1M$ has no nodes on the xz-plane (cf., Fig 12.4(c)).
- For n odd, an orbit of Type $1V$ has the node 1 on the xz-plane and one of Type $1M$ has the node $(n+1)/2$ on the xz-plane (cf., Fig 12.4(b)).

Consequently, we have the classification of orbits P_l into the four types of

$$0,\ 1V,\ 1M,\ 2. \tag{12.36}$$

Combination of two classifications, (12.36) for P_l and (12.29) for E_m, yields eight types for $\kappa = (l,m)$, which we name as

$$\xi = 0z,\ 0xy,\ 1Vz,\ 1Vxy,\ 1Mz,\ 1Mxy,\ 2z,\ 2xy. \tag{12.37}$$

The type ξ associated with κ is denoted as $\xi(\kappa)$. For each ξ we define N_ξ by $N_\xi = N_\kappa$ using (any) κ such that $\xi = \xi(\kappa)$:

$$N_\xi = N_\kappa = |P_l| \cdot |E_m| \tag{12.38}$$

if $\xi = \xi(\kappa)$ and $\kappa = (l,m)$. Recall our convention.

- l is the index for an orbit P_l of nodes (cf., (12.27)).
- $m = z$ or xy in E_m (cf., (12.28), (12.29)).
- $\kappa = (l,m)$ (cf., (12.30)).
- Each $\kappa = (l,m)$ is associated with a type $\xi = \xi(\kappa)$ in (12.37).
- The components of the vector $\hat{\boldsymbol{u}}_\kappa$ are ordered consistently with the node numbers presented in Fig. 12.4.

Determination of Local Transformations

The representation matrix $\hat{T}_\kappa(g)$ for κ is determined by the type $\xi(\kappa)$ of κ; that is,

$$\hat{T}_\kappa(g) = \hat{T}_\xi(g), \qquad g \in G \tag{12.39}$$

for $\xi = \xi(\kappa)$ in (12.37), where $\hat{T}_\xi(g)$ denotes the representation matrix[4] for the type ξ. Consequently, finding the local transformation matrix $\hat{Q}_\kappa$ in (12.35) is reduced to finding an orthogonal matrix $\hat{Q}_\xi$ such that

$$(\hat{Q}_\xi)^\top \hat{T}_\xi(g) \hat{Q}_\xi = \bigoplus_{\mu \in R(G)} \bigoplus_{i=1}^{a_\xi^\mu} T^\mu(g), \qquad g \in G, \tag{12.40}$$

where a_ξ^μ denotes the multiplicity of μ in $\hat{T}_\xi$. In accordance with this block-diagonal structure the matrix $\hat{Q}_\xi$ is partitioned columnwise as

$$\hat{Q}_\xi = (\hat{Q}_\xi^\mu \mid \mu \in R(G)) = ((\hat{Q}_{\xi,i}^\mu \mid i = 1,\ldots,a_\xi^\mu) \mid \mu \in R(G)), \tag{12.41}$$

where $\hat{Q}_\xi$, $\hat{Q}_\xi^\mu$, and $\hat{Q}_{\xi,i}^\mu$ are $N_\xi \times N_\xi$, $N_\xi \times (a_\xi^\mu N^\mu)$, and $N_\xi \times N^\mu$ matrices, respectively; see (12.38) for the notation N_ξ.

It follows from (12.40) and (12.41) that

$$\hat{T}_\xi(g) \hat{Q}_{\xi,i}^\mu = \hat{Q}_{\xi,i}^\mu T^\mu(g), \qquad g \in G, \tag{12.42}$$

for $i = 1,\ldots,a_\xi^\mu$. Since $\hat{Q}_\xi$ is an orthogonal matrix, its submatrices $\hat{Q}_{\xi,i}^\mu$ $(i = 1,\ldots,a_\xi^\mu)$ should satisfy a normalization condition

$$(\hat{Q}_{\xi,i}^\mu)^\top \hat{Q}_{\xi,j}^\mu = \begin{cases} I_{N^\mu} & \text{for } i = j, \\ O & \text{for } i \neq j. \end{cases} \tag{12.43}$$

For each $\mu \in R(G)$, the submatrix $\hat{Q}_\xi^\mu = (\hat{Q}_{\xi,i}^\mu \mid i = 1,\ldots,a_\xi^\mu)$ can be determined from (12.42) and (12.43). In particular, the multiplicity a_ξ^μ can be determined as the maximum number of $N_\xi \times N^\mu$ matrices $\hat{Q}_{\xi,i}^\mu$ that satisfy (12.42) and (12.43). Then the

[4] Concrete forms of $\hat{T}_\xi(g)$ are given in §12.5.

local transformation matrix $\hat{Q}_\xi = (\hat{Q}^\mu_\xi \mid \mu \in R(G))$ is given as a collection of these submatrices $\hat{Q}^\mu_\xi$.

In §12.2.3 we have, in effect,[5] conducted the procedure described above for $\xi = 1Vxy$ and $n = 3$. The representation matrix $\hat{T}_\xi(g)$ is given by (12.15), and the local transformation matrix $\hat{Q}_\xi$ is obtained in (12.16).

It is noteworthy that $\hat{T}_\xi(g)$ is not dependent on individual truss structures, but is determined exclusively by ξ and n. Moreover, only eight possibilities exist for ξ. This fact motivates us to make a complete list of orthogonal matrices $\hat{Q}_\xi$ for all ξ and n. The formulas of submatrices $\hat{Q}^\mu_\xi$ are listed in §12.4, and are derived in §12.5. Once the matrices $\hat{Q}_\xi$ are tabulated, $\hat{Q}_\kappa$ in (12.35) can be easily obtained as

$$\hat{Q}_\kappa = \hat{Q}_{\xi(\kappa)}. \tag{12.44}$$

For $k = 1, \dots, N^\mu$, we designate by $\hat{Q}^{\mu,k}_{\xi,i}$ the kth column of the $N_\xi \times N^\mu$ matrix $\hat{Q}^\mu_{\xi,i}$, and define $N_\xi \times a^\mu$ matrices

$$\hat{H}^{\mu,k}_\xi = (\hat{Q}^{\mu,k}_{\xi,i} \mid i = 1, \dots, a^\mu_\xi), \qquad k = 1, \dots, N^\mu. \tag{12.45}$$

These matrices are used in constructing the matrix H in §12.3.3.

12.3.3 Assemblage of Local Transformations

Local transformation matrices $\hat{Q}_\kappa$ treated in §12.3.2 are assembled here to the transformation matrices Q and H.

Transformation Matrix Q

We assume that the local transformation matrices

$$\hat{Q}_\kappa = (\hat{Q}^\mu_\kappa \mid \mu \in R(G)) = ((\hat{Q}^\mu_{\kappa,i} \mid i = 1, \dots, a^\mu_\kappa) \mid \mu \in R(G)) \tag{12.46}$$

for all κ are available by (12.44). Recall that the sizes of the matrices $\hat{Q}_\kappa$, $\hat{Q}^\mu_\kappa$, and $\hat{Q}^\mu_{\kappa,i}$ are $N_\kappa \times N_\kappa$, $N_\kappa \times (a^\mu_\kappa N^\mu)$, and $N_\kappa \times N^\mu$, respectively.

We construct block-diagonal matrices

$$\hat{Q}^\mu = \bigoplus_\kappa \hat{Q}^\mu_\kappa, \qquad \mu \in R(G), \tag{12.47}$$

and arrange them compatibly with the index set $R(G)$ to arrive at

[5] For $\kappa = 1xy$ in §12.2.3, we have $\xi(\kappa) = 1Vxy$. Note also that $\kappa = 1xy$ should be written as $\kappa = (l,m) = (1,xy)$ in our general notation if we take $P_0 = \{0\}$ and $P_1 = \{1,2,3\}$.

$$\hat{Q}^{\mu} = (\hat{Q}^{\mu}_i \mid i = 1,\dots,a^{\mu}), \qquad \mu \in R(G), \tag{12.48}$$

$$\hat{Q} = (\hat{Q}^{\mu} \mid \mu \in R(G)) = ((\hat{Q}^{\mu}_i \mid i = 1,\dots,a^{\mu}) \mid \mu \in R(G)). \tag{12.49}$$

Here

$$a^{\mu} = \sum_{\kappa} a^{\mu}_{\kappa} \tag{12.50}$$

means the multiplicity of μ in $\hat{T}$. The sizes of the matrices $\hat{Q}$, $\hat{Q}^{\mu}$, and $\hat{Q}^{\mu}_i$ are $N \times N$, $N \times (a^{\mu}N^{\mu})$, and $N \times N^{\mu}$, respectively. See Fig.12.5 for illustration for the case of $\kappa = \kappa_1, \kappa_2, \kappa_3$ and $\mu = \mu_1, \mu_2$.

Then we permute the rows of the matrix $\hat{Q}$ compatibly with the original vector $\boldsymbol{u}$ to arrive at

$$Q = (Q^{\mu} \mid \mu \in R(G)) = ((Q^{\mu}_i \mid i = 1,\dots,a^{\mu}) \mid \mu \in R(G)). \tag{12.51}$$

With the permutation matrix Ξ in (12.32) we have the relations

$$Q = \Xi \hat{Q}, \qquad Q^{\mu} = \Xi \hat{Q}^{\mu}, \qquad Q^{\mu}_i = \Xi \hat{Q}^{\mu}_i.$$

Then we obtain the irreducible decomposition in (12.4); that is,

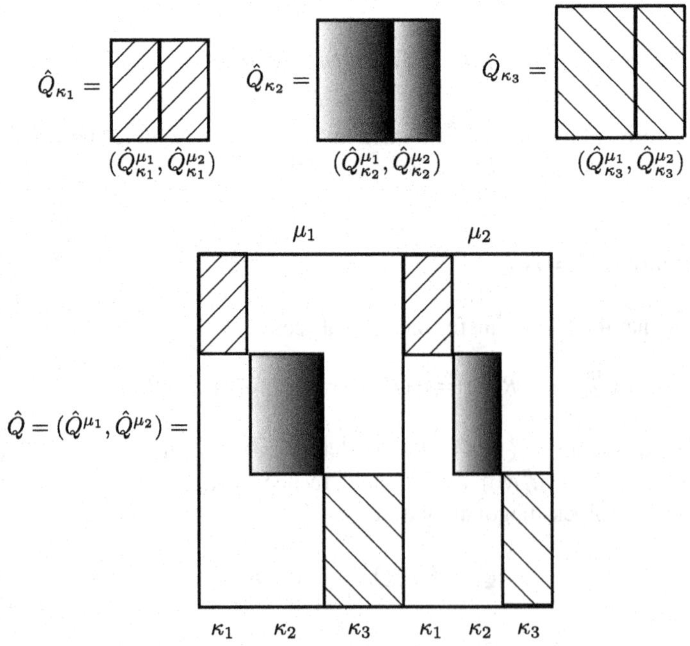

Fig. 12.5 Assemblage of transformation matrix $\hat{Q}$.

$$Q^\top T(g)Q = \hat{Q}^\top \hat{T}(g)\hat{Q} = \bigoplus_{\mu\in R(G)} \bigoplus_{i=1}^{a^\mu} T^\mu(g). \tag{12.52}$$

Transformation Matrix H

The transformation matrix H for the block-diagonalization (12.3) can be assembled from the submatrices of $\hat{Q}$ obtained above.

Recall that $\hat{Q}_i^\mu$ in (12.48) is an $N\times N^\mu$ matrix for $i = 1,\dots,a^\mu$. For $k = 1,\dots,N^\mu$, denote the kth column of $\hat{Q}_i^\mu$ by $\hat{Q}_i^{\mu,k}$; that is, put

$$\hat{Q}_i^\mu = (\hat{Q}_i^{\mu,k} \mid k = 1,\dots,N^\mu), \qquad i = 1,\dots,a^\mu.$$

Using these column vectors, we define $N\times a^\mu$ matrices

$$\hat{H}^{\mu,k} = (\hat{Q}_i^{\mu,k} \mid i = 1,\dots,a^\mu), \qquad k = 1,\dots,N^\mu, \tag{12.53}$$

to construct

$$\hat{H}^\mu = (\hat{H}^{\mu,k} \mid k = 1,\dots,N^\mu), \qquad \mu\in R(G), \tag{12.54}$$

$$\hat{H} = (\hat{H}^\mu \mid \mu\in R(G)) = ((\hat{H}^{\mu,k} \mid k = 1,\dots,N^\mu) \mid \mu\in R(G)), \tag{12.55}$$

where $\hat{H}^\mu$ and $\hat{H}$, respectively, represent $N\times(a^\mu N^\mu)$ and $N\times N$ matrices. The matrix $\hat{H}$ is obtained from $\hat{Q}$ via a permutation of columns as

$$\hat{H} = \hat{Q}\Pi$$

for some permutation matrix Π.

A row-permutation of $\hat{H}$, represented by the permutation matrix Ξ in (12.32), yields the transformation matrix

$$H = \Xi\hat{H} = \Xi\hat{Q}\Pi = Q\Pi \tag{12.56}$$

for the block-diagonalization (12.3). We then have

$$H = (H^\mu \mid \mu\in R(G)) = ((H^{\mu,k} \mid k = 1,\dots,N^\mu) \mid \mu\in R(G)) \tag{12.57}$$

with

$$H^\mu = \Xi\hat{H}^\mu, \qquad H^{\mu,k} = \Xi\hat{H}^{\mu,k}.$$

It is noteworthy that the matrix $\hat{H}^{\mu,k}$ inherits the block-diagonal structure of $\hat{Q}^\mu$ in (12.47). That is,

$$\hat{H}^{\mu,k} = \bigoplus_{\kappa} \hat{H}_\kappa^{\mu,k}, \tag{12.58}$$

where $\hat{H}^{\mu,k}_{\kappa}$ is given as $\hat{H}^{\mu,k}_{\kappa} = \hat{H}^{\mu,k}_{\xi(\kappa)}$ using $\hat{H}^{\mu,k}_{\xi}$ in (12.45) with $\xi = \xi(\kappa)$. We take advantage of this expression for efficient computation of the block-diagonal form of the Jacobian matrix, as explained later in (12.62).

Diagonal Block of J

The diagonal blocks $\tilde{J}^{\mu}$ of $H^{\top}JH = \bigoplus_{\mu}\bigoplus_{k}\tilde{J}^{\mu}$ in (12.3) can be computed efficiently by exploiting the sparsity of H and J.

It follows from (12.55) that $\tilde{J}^{\mu}$ can be expressed as

$$\tilde{J}^{\mu} = (H^{\mu,k})^{\top} J H^{\mu,k} = (\Xi^{\top} H^{\mu,k})^{\top} (\Xi^{\top} J \Xi)(\Xi^{\top} H^{\mu,k}) = (\hat{H}^{\mu,k})^{\top} \hat{J} \hat{H}^{\mu,k}, \tag{12.59}$$

where $\hat{J} = \Xi^{\top} J \Xi$ is the Jacobian matrix for the rearranged vector $\hat{\boldsymbol{u}}$, which can be obtained from J without any multiplication operations.

The matrix $\hat{H}^{\mu,k}$ has a block-diagonal form

$$\hat{H}^{\mu,k} = \bigoplus_{\kappa} \hat{H}^{\mu,k}_{\kappa},$$

as noted in (12.58). The matrices $\hat{J}$ and $\tilde{J}^{\mu}$ are partitioned in accordance with (12.59) as

$$\hat{J} = (\hat{J}_{\kappa\lambda} \mid \kappa, \lambda = (l,m)), \tag{12.60}$$

$$\tilde{J}^{\mu} = (\tilde{J}^{\mu}_{\kappa\lambda} \mid \kappa, \lambda = (l,m)). \tag{12.61}$$

Instead of computing $\tilde{J}^{\mu}$ by (12.59) directly, we can compute the submatrices $\tilde{J}^{\mu}_{\kappa\lambda}$ in (12.61) from the expression

$$\tilde{J}^{\mu}_{\kappa\lambda} = (\hat{H}^{\mu,k}_{\kappa})^{\top} \hat{J}_{\kappa\lambda} \hat{H}^{\mu,k}_{\lambda}. \tag{12.62}$$

The matrix multiplication of (12.62) is local in that it is conducted within the submatrices associated with the pair of indices (κ, λ). This *local multiplication* is far more efficient numerically than the direct multiplication $H^{\top}JH$ in (12.3), and is recommended for practical use.

Remark 12.1. It should be understood that the partition in (12.61) does not put $\tilde{J}^{\mu}$ into a block-diagonal form. That is, the submatrix $\tilde{J}^{\mu}_{\kappa\lambda}$ of $\tilde{J}^{\mu}$ in (12.62) does not vanish in general even if $\kappa \neq \lambda$. This is because the representation matrices $\hat{T}_{\kappa}$ and $\hat{T}_{\lambda}$ for κ and λ may have some irreducible components in common. Nevertheless, $\tilde{J}^{\mu}_{\kappa\lambda}$ is likely to vanish for many pairs of (κ, λ) in practical applications. This is not attributable to the group symmetry but to the sparsity of J and the locality of our transformation H. In fact, if $\hat{J}_{\kappa\lambda} = O$, then $\tilde{J}^{\mu}_{\kappa\lambda} = O$ by (12.62). Therefore, each $\tilde{J}^{\mu}$ inherits the sparsity of J in the sense that $\tilde{J}^{\mu}_{\kappa\lambda} = O$ if $\hat{J}_{\kappa\lambda} = O$. □

12.4 Local Transformation Matrix: Formulas

We give concrete formulas of the local transformation matrices in (12.45):

$$\hat{H}^{\mu,k}_{\xi} = (\hat{Q}^{\mu,k}_{\xi,i} \mid i = 1,\dots,a^{\mu}_{\xi}), \qquad k = 1,\dots,N^{\mu}$$

for all irreducible representations $\mu \in R(\mathrm{D}_n)$ in (8.7) and all types ξ of orbits in (12.37). The matrices $\hat{H}^{\mu,k}_{\xi}$ are to be employed for the assemblage of the transformation matrix H, as described in §12.3.3. The multiplicity a^{μ}_{ξ} of the irreducible representation μ in $\hat{T}_{\xi}$ of (12.39) is presented in Table 12.1. The formulas are targeted specifically at D_n-symmetric trusses.

Table 12.1 Multiplicity a^{μ}_{ξ} for D_n-symmetric truss structures. An entry with /- is not applicable for n odd; $j = 2,\dots,M_2(=\lfloor (n-1)/2 \rfloor)$

	Type ξ				$1z$		$1xy$	$2z$	$2xy$
$n=1$	$\mu = (+,+)_{\mathrm{D}_1}$				1		1	1	2
	$\mu = (+,-)_{\mathrm{D}_1}$				0		1	1	2
	Type ξ	$0z$	$0xy$	$1Vz$	$1Mz$	$1Vxy$	$1Mxy$	$2z$	$2xy$
	$\mu = (+,+)_{\mathrm{D}_2}$	1	0	1	1	1	1	1	2
$n=2$	$\mu = (+,-)_{\mathrm{D}_2}$	0	0	0	0	1	1	1	2
	$\mu = (-,+)_{\mathrm{D}_2}$	0	1	1	0	1	1	1	2
	$\mu = (-,-)_{\mathrm{D}_2}$	0	1	0	1	1	1	1	2
	Type ξ	$0z$	$0xy$	$1Vz$	$1Mz$	$1Vxy$	$1Mxy$	$2z$	$2xy$
	$\mu = (+,+)_{\mathrm{D}_n}$	1	0	1	1	1	1	1	2
	$\mu = (+,-)_{\mathrm{D}_n}$	0	0	0	0	1	1	1	2
$n \geq 3$	$\mu = (-,+)_{\mathrm{D}_n}$	0/-	0/-	1/-	0/-	1/-	1/-	1/-	2/-
	$\mu = (-,-)_{\mathrm{D}_n}$	0/-	0/-	0/-	1/-	1/-	1/-	1/-	2/-
	$\mu = (1)_{\mathrm{D}_n}$	0	1	1	1	2	2	2	4
	$\mu = (j)_{\mathrm{D}_n}$	0	0	1	1	2	2	2	4

For complex numbers a and b we define an n-dimensional complex vector $\gamma_n(a,b)$ and a $2n$-dimensional real vector $\Gamma_n(a,b)$ by[6]

$$\gamma_n(a,b) = \sqrt{\frac{2}{n}} \begin{pmatrix} a \\ ab \\ \vdots \\ ab^{n-1} \end{pmatrix}, \qquad \Gamma_n(a,b) = \frac{1}{\sqrt{n}} \begin{pmatrix} \mathrm{Re}(a) \\ \mathrm{Im}(a) \\ \mathrm{Re}(ab) \\ \mathrm{Im}(ab) \\ \vdots \\ \mathrm{Re}(ab^{n-1}) \\ \mathrm{Im}(ab^{n-1}) \end{pmatrix}. \tag{12.63}$$

[6] The vectors $\gamma_n(a,b)$ and $\Gamma_n(a,b)$ are normalized as follows. We have $\|\gamma_n(a,b)\| = \sqrt{2}|a|$ if $|b| = 1$. If $|a| = 1$, $b^2 \neq 1$, and $b^{2n} = 1$, we have $\mathrm{Re}\,\gamma_n(a,b)^{\top}\mathrm{Im}\,\gamma_n(a,b) = 0$ and $\|\mathrm{Re}\,\gamma_n(a,b)\| = \|\mathrm{Im}\,\gamma_n(a,b)\| = 1$. We have $\|\Gamma(a,b)\| = 1$ if $|a| = |b| = 1$.

For $n = 1$, we have

$$\hat{H}_{1z}^{(+,+)\mathrm{D}_1} = 1;$$

$$\hat{H}_{1xy}^{(+,+)\mathrm{D}_1} = \begin{pmatrix} 1 \\ 0 \end{pmatrix}, \qquad \hat{H}_{1xy}^{(+,-)\mathrm{D}_1} = \begin{pmatrix} 0 \\ 1 \end{pmatrix};$$

$$\hat{H}_{2z}^{(+,+)\mathrm{D}_1} = \gamma_2(1/\sqrt{2}, 1), \qquad \hat{H}_{2z}^{(+,-)\mathrm{D}_1} = \gamma_2(1/\sqrt{2}, -1);$$

$$\hat{H}_{2xy}^{(+,+)\mathrm{D}_1} = (\Gamma_2(1,1), \Gamma_2(\mathrm{i},-1)), \qquad \hat{H}_{2xy}^{(+,-)\mathrm{D}_1} = (\Gamma_2(1,-1), \Gamma_2(\mathrm{i},1)).$$

For $n = 2$, we have

$$\hat{H}_{0z}^{(+,+)\mathrm{D}_2} = 1;$$

$$\hat{H}_{0xy}^{(-,+)\mathrm{D}_2} = \begin{pmatrix} 1 \\ 0 \end{pmatrix}, \qquad \hat{H}_{0xy}^{(-,-)\mathrm{D}_2} = \begin{pmatrix} 0 \\ 1 \end{pmatrix};$$

$$\hat{H}_{1Vz}^{(+,+)\mathrm{D}_2} = \gamma_2(1/\sqrt{2}, 1), \qquad \hat{H}_{1Vz}^{(-,+)\mathrm{D}_2} = \gamma_2(1/\sqrt{2}, -1);$$

$$\hat{H}_{1Vxy}^{(+,+)\mathrm{D}_2} = \Gamma_2(1,-1), \qquad \hat{H}_{1Vxy}^{(+,-)\mathrm{D}_2} = \Gamma_2(\mathrm{i},-1),$$

$$\hat{H}_{1Vxy}^{(-,+)\mathrm{D}_2} = \Gamma_2(1,1), \qquad \hat{H}_{1Vxy}^{(-,-)\mathrm{D}_2} = \Gamma_2(\mathrm{i},1);$$

$$\hat{H}_{1Mz}^{(+,+)\mathrm{D}_2} = \gamma_2(1/\sqrt{2}, 1), \qquad \hat{H}_{1Mz}^{(-,-)\mathrm{D}_2} = \gamma_2(1/\sqrt{2}, -1);$$

$$\hat{H}_{1Mxy}^{(+,+)\mathrm{D}_2} = \Gamma_2(\mathrm{i},-1), \qquad \hat{H}_{1Mxy}^{(+,-)\mathrm{D}_2} = \Gamma_2(-1,-1),$$

$$\hat{H}_{1Mxy}^{(-,+)\mathrm{D}_2} = \Gamma_2(-1,1), \qquad \hat{H}_{1Mxy}^{(-,-)\mathrm{D}_2} = \Gamma_2(\mathrm{i},1);$$

$$\hat{H}_{2z}^{(+,+)\mathrm{D}_2} = \gamma_4(1/\sqrt{2}, 1), \qquad \hat{H}_{2z}^{(+,-)\mathrm{D}_2} = \gamma_4(1/\sqrt{2}, -1),$$

$$\hat{H}_{2z}^{(-,+)\mathrm{D}_2} = \mathrm{Re}\,\gamma_{2n}(\tfrac{1+\mathrm{i}}{\sqrt{2}}, \mathrm{i}), \qquad \hat{H}_{2z}^{(-,-)\mathrm{D}_2} = \mathrm{Im}\,\gamma_{2n}(\tfrac{1+\mathrm{i}}{\sqrt{2}}, \mathrm{i});$$

$$\hat{H}_{2xy}^{(+,+)\mathrm{D}_2} = (\Gamma_4(\tfrac{1+\mathrm{i}}{\sqrt{2}}, \mathrm{i}), \Gamma_4(\tfrac{-1+\mathrm{i}}{\sqrt{2}}, -\mathrm{i})), \qquad \hat{H}_{2xy}^{(+,-)\mathrm{D}_2} = (\Gamma_4(\tfrac{-1+\mathrm{i}}{\sqrt{2}}, \mathrm{i}), \Gamma_4(\tfrac{1+\mathrm{i}}{\sqrt{2}}, -\mathrm{i})),$$

$$\hat{H}_{2xy}^{(-,+)\mathrm{D}_2} = (\Gamma_4(1,1), \Gamma_4(\mathrm{i},-1)), \qquad \hat{H}_{2xy}^{(-,-)\mathrm{D}_2} = (\Gamma_4(\mathrm{i},1), \Gamma_4(1,-1)).$$

For $n \geq 3$, with the notations $j^* = n - j$ and

$$\omega = \exp(\mathrm{i}2\pi/n), \tag{12.64}$$

we have the following:

$$\hat{H}_{0z}^{(+,+)\mathrm{D}_n} = 1;$$

$$\hat{H}_{0xy}^{(1)\mathrm{D}_n,1} = \begin{pmatrix}1\\0\end{pmatrix}, \qquad \hat{H}_{0xy}^{(1)\mathrm{D}_n,2} = \begin{pmatrix}0\\1\end{pmatrix};$$

$$\hat{H}_{1Vz}^{(+,+)\mathrm{D}_n} = \gamma_n(1/\sqrt{2},1), \qquad \hat{H}_{1Vz}^{(-,+)\mathrm{D}_n} = \gamma_n(1/\sqrt{2},-1),$$

$$\hat{H}_{1Vz}^{(j)\mathrm{D}_n,1} = \mathrm{Re}\ \gamma_n(1,\omega^j), \qquad \hat{H}_{1Vz}^{(j)\mathrm{D}_n,2} = \mathrm{Im}\ \gamma_n(1,\omega^j);$$

$$\hat{H}_{1Vxy}^{(+,+)\mathrm{D}_n} = \Gamma_n(1,\omega), \qquad \hat{H}_{1Vxy}^{(+,-)\mathrm{D}_n} = \Gamma_n(\mathrm{i},\omega),$$

$$\hat{H}_{1Vxy}^{(-,+)\mathrm{D}_n} = \Gamma_n(1,-\omega), \qquad \hat{H}_{1Vxy}^{(-,-)\mathrm{D}_n} = \Gamma_n(\mathrm{i},-\omega),$$

$$\hat{H}_{1Vxy}^{(j)\mathrm{D}_n,1} = (\Gamma_n(1,\omega^{-j+1}),\Gamma_n(1,\omega^{j+1})), \qquad \hat{H}_{1Vxy}^{(j)\mathrm{D}_n,2} = (\Gamma_n(\mathrm{i},\omega^{-j+1}),\Gamma_n(-\mathrm{i},\omega^{j+1}));$$

$$\hat{H}_{1Mz}^{(+,+)\mathrm{D}_n} = \gamma_n(1/\sqrt{2},1), \qquad \hat{H}_{1Mz}^{(-,-)\mathrm{D}_n} = \gamma_n(1/\sqrt{2},-1),$$

$$\hat{H}_{1Mz}^{(j)\mathrm{D}_n,1} = \mathrm{Re}\ \gamma_n(\omega^{j/2},\omega^j), \qquad \hat{H}_{1Mz}^{(j)\mathrm{D}_n,2} = \mathrm{Im}\ \gamma_n(\omega^{j/2},\omega^j);$$

$$\hat{H}_{1Mxy}^{(+,+)\mathrm{D}_n} = \Gamma_n(\omega^{1/2},\omega), \qquad \hat{H}_{1Mxy}^{(+,-)\mathrm{D}_n} = \Gamma_n(\mathrm{i}\omega^{1/2},\omega),$$

$$\hat{H}_{1Mxy}^{(-,+)\mathrm{D}_n} = \Gamma_n(\mathrm{i}\omega^{1/2},-\omega), \qquad \hat{H}_{1Mxy}^{(-,-)\mathrm{D}_n} = \Gamma_n(\omega^{1/2},-\omega),$$

$$\hat{H}_{1Mxy}^{(j)\mathrm{D}_n,1} = (\Gamma_n(\omega^{(-j+1)/2},\omega^{-j+1}),\Gamma_n(\omega^{(j+1)/2},\omega^{j+1})),$$

$$\hat{H}_{1Mxy}^{(j)\mathrm{D}_n,2} = (\Gamma_n(\mathrm{i}\omega^{(-j+1)/2},\omega^{-j+1}),\Gamma_n(-\mathrm{i}\omega^{(j+1)/2},\omega^{j+1}));$$

$$\hat{H}_{2z}^{(+,+)\mathrm{D}_n} = \gamma_{2n}(1/\sqrt{2},1), \qquad \hat{H}_{2z}^{(+,-)\mathrm{D}_n} = \gamma_{2n}(1/\sqrt{2},-1),$$

$$\hat{H}_{2z}^{(-,+)\mathrm{D}_n} = \mathrm{Re}\ \gamma_{2n}(\tfrac{1+\mathrm{i}}{\sqrt{2}},\mathrm{i}), \qquad \hat{H}_{2z}^{(-,-)\mathrm{D}_n} = \mathrm{Im}\ \gamma_{2n}(\tfrac{1+\mathrm{i}}{\sqrt{2}},\mathrm{i}),$$

$$\hat{H}_{2z}^{(j)\mathrm{D}_n,1} = (\mathrm{Re}\ \gamma_{2n}(\omega^{j/4},\omega^{j/2}),\mathrm{Re}\ \gamma_{2n}(\omega^{j^*/4},\omega^{j^*/2})),$$

$$\hat{H}_{2z}^{(j)\mathrm{D}_n,2} = (\mathrm{Im}\ \gamma_{2n}(\omega^{j/4},\omega^{j/2}),\mathrm{Im}\ \gamma_{2n}(\omega^{j^*/4},\omega^{j^*/2}));$$

$$\hat{H}_{2xy}^{(+,+)\mathrm{D}_n} = (\Gamma_{2n}(\omega^{1/4},\omega^{1/2}),\Gamma_{2n}(\mathrm{i}\omega^{1/4},-\omega^{1/2})),$$

$$\hat{H}_{2xy}^{(+,-)\mathrm{D}_n} = (\Gamma_{2n}(\mathrm{i}\omega^{1/4},\omega^{1/2}),\Gamma_{2n}(\omega^{1/4},-\omega^{1/2})),$$

$$\hat{H}_{2xy}^{(-,+)\mathrm{D}_n} = (\Gamma_{2n}(\tfrac{1-\mathrm{i}}{\sqrt{2}}\omega^{1/4},-\mathrm{i}\omega^{1/2}),\Gamma_{2n}(\tfrac{1+\mathrm{i}}{\sqrt{2}}\omega^{1/4},\mathrm{i}\omega^{1/2})),$$

$$\hat{H}_{2xy}^{(-,-)\mathrm{D}_n} = (\Gamma_{2n}(\tfrac{1+\mathrm{i}}{\sqrt{2}}\omega^{1/4},-\mathrm{i}\omega^{1/2}),\Gamma_{2n}(\tfrac{1-\mathrm{i}}{\sqrt{2}}\omega^{1/4},\mathrm{i}\omega^{1/2})),$$

$$\hat{H}_{2xy}^{(j)\mathrm{D}_n,1} = (\Gamma_{2n}(\omega^{(-j+1)/4},\omega^{(-j+1)/2}),\Gamma_{2n}(\omega^{(j+1)/4},\omega^{(j+1)/2}),$$
$$\Gamma_{2n}(\omega^{(-j^*+1)/4},\omega^{(-j^*+1)/2}),\Gamma_{2n}(\omega^{(j^*+1)/4},\omega^{(j^*+1)/2})),$$

$$\hat{H}_{2xy}^{(j)\mathrm{D}_n,2} = (\Gamma_{2n}(\mathrm{i}\omega^{(-j+1)/4},\omega^{(-j+1)/2}),\Gamma_{2n}(-\mathrm{i}\omega^{(j+1)/4},\omega^{(j+1)/2}),$$
$$\Gamma_{2n}(-\mathrm{i}\omega^{(-j^*+1)/4},\omega^{(-j^*+1)/2}),\Gamma_{2n}(\mathrm{i}\omega^{(j^*+1)/4},\omega^{(j^*+1)/2})).$$

Example 12.2. The formulas presented above give the following matrices $\hat{H}^{\mu,k}_{1Vz}$ and $\hat{H}^{\mu,k}_{1Vxy}$ for $n = 6$, where $\omega = \exp(\mathrm{i}\pi/3)$:

$$
\begin{aligned}
&(\hat{H}^{(+,+)\mathrm{D}_6}_{1Vz}, \hat{H}^{(-,+)\mathrm{D}_6}_{1Vz}, \hat{H}^{(1)\mathrm{D}_6,1}_{1Vz}, \hat{H}^{(1)\mathrm{D}_6,2}_{1Vz}, \hat{H}^{(2)\mathrm{D}_6,1}_{1Vz}, \hat{H}^{(2)\mathrm{D}_6,2}_{1Vz}) \\
&= (\gamma_6(1/\sqrt{2}, 1), \gamma_6(1/\sqrt{2}, -1); \\
&\qquad \mathrm{Re}\,\gamma_6(1,\omega), \mathrm{Im}\,\gamma_6(1,\omega); \mathrm{Re}\,\gamma_6(1,\omega^2), \mathrm{Im}\,\gamma_6(1,\omega^2)) \\
&= \frac{1}{\sqrt{3}}\left(\begin{array}{c|c|cc|cc}
\frac{1}{\sqrt{2}} & \frac{1}{\sqrt{2}} & 1 & 0 & 1 & 0 \\
\frac{1}{\sqrt{2}} & -\frac{1}{\sqrt{2}} & \frac{1}{2} & \frac{\sqrt{3}}{2} & -\frac{1}{2} & \frac{\sqrt{3}}{2} \\
\frac{1}{\sqrt{2}} & \frac{1}{\sqrt{2}} & -\frac{1}{2} & \frac{\sqrt{3}}{2} & -\frac{1}{2} & -\frac{\sqrt{3}}{2} \\
\frac{1}{\sqrt{2}} & -\frac{1}{\sqrt{2}} & -1 & 0 & 1 & 0 \\
\frac{1}{\sqrt{2}} & \frac{1}{\sqrt{2}} & -\frac{1}{2} & -\frac{\sqrt{3}}{2} & -\frac{1}{2} & \frac{\sqrt{3}}{2} \\
\frac{1}{\sqrt{2}} & -\frac{1}{\sqrt{2}} & \frac{1}{2} & -\frac{\sqrt{3}}{2} & -\frac{1}{2} & -\frac{\sqrt{3}}{2}
\end{array}\right),
\end{aligned}
$$

$$
\begin{aligned}
&(\hat{H}^{(+,+)\mathrm{D}_6}_{1Vxy}, \hat{H}^{(+,-)\mathrm{D}_6}_{1Vxy}, \hat{H}^{(-,+)\mathrm{D}_6}_{1Vxy}, \hat{H}^{(-,-)\mathrm{D}_6}_{1Vxy}, \hat{H}^{(1)\mathrm{D}_6,1}_{1Vxy}, H^{(1)\mathrm{D}_6,2}_{1Vxy}, \hat{H}^{(2)\mathrm{D}_6,1}_{1Vxy}, H^{(2)\mathrm{D}_6,2}_{1Vxy}) \\
&= (\varGamma_6(1,\omega), \varGamma_6(\mathrm{i},\omega), \varGamma_6(1,-\omega), \varGamma_6(\mathrm{i},-\omega); \varGamma_6(1,1), \varGamma_6(1,\omega^2); \\
&\qquad \varGamma_6(\mathrm{i},1), \varGamma_6(-\mathrm{i},\omega^2); \varGamma_6(1,\omega^{-1}), \varGamma_6(1,-1); \varGamma_6(\mathrm{i},\omega^{-1}), \varGamma_6(-\mathrm{i},-1)) \\
&= \frac{1}{\sqrt{6}}\left(\begin{array}{c|c|c|c|cc|cc|cc|cc}
1 & 0 & 1 & 0 & 1 & 1 & 0 & 0 & 1 & 1 & 0 & 0 \\
0 & 1 & 0 & 1 & 0 & 0 & 1 & -1 & 0 & 0 & 1 & -1 \\
\frac{1}{2} & -\frac{\sqrt{3}}{2} & -\frac{1}{2} & \frac{\sqrt{3}}{2} & 1 & -\frac{1}{2} & 0 & \frac{\sqrt{3}}{2} & \frac{1}{2} & -1 & \frac{\sqrt{3}}{2} & 0 \\
\frac{\sqrt{3}}{2} & \frac{1}{2} & -\frac{\sqrt{3}}{2} & -\frac{1}{2} & 0 & \frac{\sqrt{3}}{2} & 1 & \frac{1}{2} & -\frac{\sqrt{3}}{2} & 0 & \frac{1}{2} & 1 \\
-\frac{1}{2} & -\frac{\sqrt{3}}{2} & -\frac{1}{2} & -\frac{\sqrt{3}}{2} & 1 & -\frac{1}{2} & 0 & -\frac{\sqrt{3}}{2} & -\frac{1}{2} & 1 & \frac{\sqrt{3}}{2} & 0 \\
\frac{\sqrt{3}}{2} & -\frac{1}{2} & \frac{\sqrt{3}}{2} & -\frac{1}{2} & 0 & -\frac{\sqrt{3}}{2} & 1 & \frac{1}{2} & -\frac{\sqrt{3}}{2} & 0 & -\frac{1}{2} & -1 \\
-1 & 0 & 1 & 0 & 1 & 1 & 0 & 0 & -1 & -1 & 0 & 0 \\
0 & -1 & 0 & 1 & 0 & 0 & 1 & -1 & 0 & 0 & -1 & 1 \\
-\frac{1}{2} & \frac{\sqrt{3}}{2} & -\frac{1}{2} & \frac{\sqrt{3}}{2} & 1 & -\frac{1}{2} & 0 & \frac{\sqrt{3}}{2} & -\frac{1}{2} & 1 & -\frac{\sqrt{3}}{2} & 0 \\
-\frac{\sqrt{3}}{2} & -\frac{1}{2} & -\frac{\sqrt{3}}{2} & -\frac{1}{2} & 0 & \frac{\sqrt{3}}{2} & 1 & \frac{1}{2} & \frac{\sqrt{3}}{2} & 0 & -\frac{1}{2} & -1 \\
\frac{1}{2} & \frac{\sqrt{3}}{2} & -\frac{1}{2} & -\frac{\sqrt{3}}{2} & 1 & -\frac{1}{2} & 0 & -\frac{\sqrt{3}}{2} & \frac{1}{2} & -1 & -\frac{\sqrt{3}}{2} & 0 \\
-\frac{\sqrt{3}}{2} & \frac{1}{2} & \frac{\sqrt{3}}{2} & -\frac{1}{2} & 0 & -\frac{\sqrt{3}}{2} & 1 & \frac{1}{2} & \frac{\sqrt{3}}{2} & 0 & \frac{1}{2} & 1
\end{array}\right).
\end{aligned}
$$

The column vectors of the above matrices correspond to deformation modes depicted in Fig. 12.6. □

Remark 12.2. We give here the local transformation matrices for C_n-symmetric truss structures that are rotationally symmetric with respect to the z-axis, where $n \geq 2$. In §11.6.3 we have described that the block-diagonalization for such a system depends on whether we work with $F = \mathbb{R}$ or $\mathbb{C}$.

The procedure presented in §12.3 for D_n-symmetric trusses remains valid with minor modifications. Orbits are classified into two types:

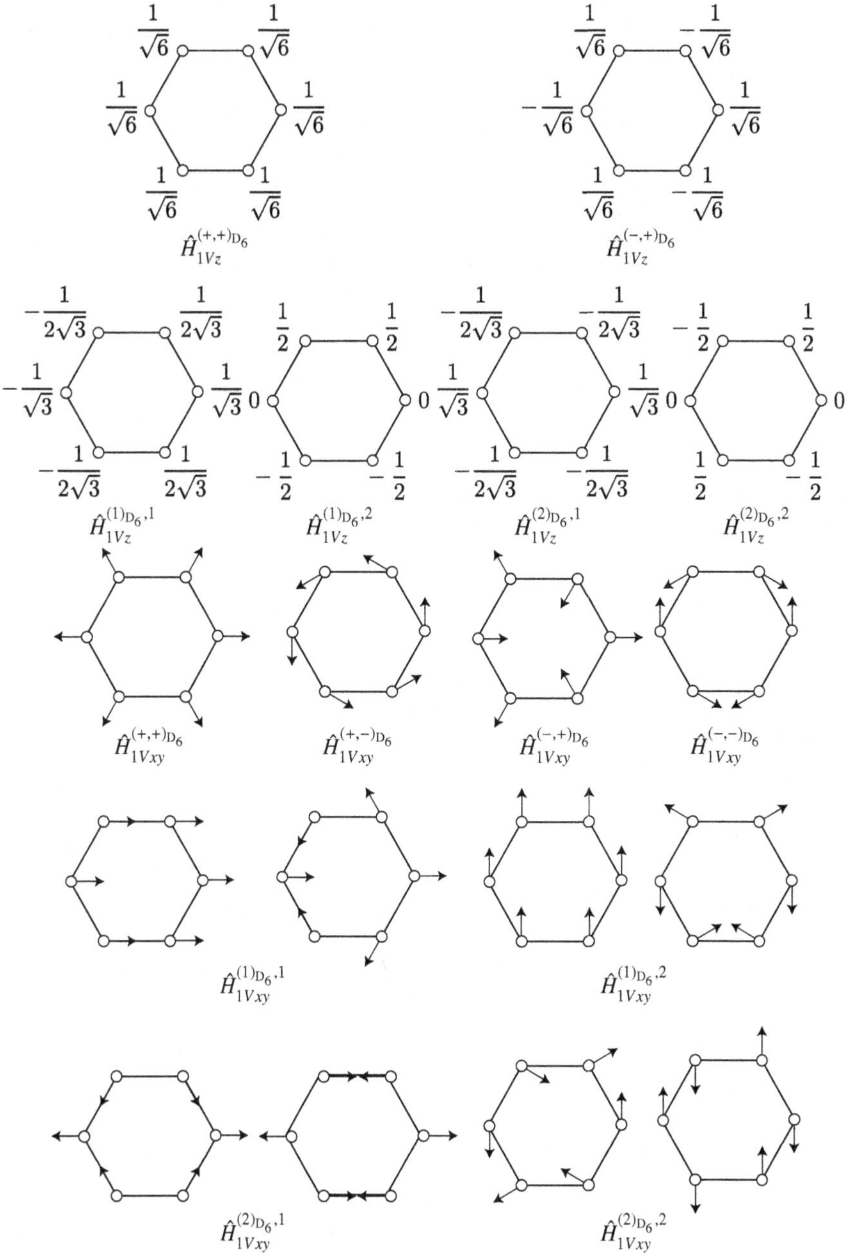

Fig. 12.6 Deformation modes of the column vectors of $\hat{H}_{\xi}^{\mu,k}$ of an orbit of type $1V$ ($n = 6$). Numerals: the values of the z-directional components.

Table 12.2 Multiplicity a_ξ^μ for C_n-symmetric truss structures. An entry with /- is not applicable for n odd; $j = 2, \ldots, M_2 (= \lfloor (n-1)/2 \rfloor)$

(a) Over $\mathbb{R}$

	Type ξ	$0z$	$0xy$	$1z$	$1xy$
$n=2$	$\mu=(+)_{C_2}$	1	0	1	2
	$\mu=(-)_{C_2}$	0	2	1	2
	Type ξ	$0z$	$0xy$	$1z$	$1xy$
$n \geq 3$	$\mu=(+)_{C_n}$	1	0	1	2
	$\mu=(-)_{C_n}$	0/-	0/-	1/-	2/-
	$\mu=(1)_{C_n}$	0	1	1	2
	$\mu=(j)_{C_n}$	0	0	1	2

(b) Over $\mathbb{C}$

	Type ξ	$0z$	$0xy$	$1z$	$1xy$
$n=2$	$\mu=(+)_{C_2}$	1	0	1	2
	$\mu=(-)_{C_2}$	0	2	1	2
	Type ξ	$0z$	$0xy$	$1z$	$1xy$
$n \geq 3$	$\mu=(+)_{C_n}$	1	0	1	2
	$\mu=(-)_{C_n}$	0/-	0/-	1/-	2/-
	$\mu=(1+)_{C_n}$	0	1	1	2
	$\mu=(1-)_{C_n}$	0	1	1	2
	$\mu=(j+)_{C_n}$	0	0	1	2
	$\mu=(j-)_{C_n}$	0	0	1	2

- Type 0 orbit consisting of a single node at the center,
- Type 1 orbit consisting of n nodes.

Accordingly, (12.37) is replaced by $\xi = 0z,\ 0xy,\ 1z,\ 1xy$.

In the case of $\mathbb{R}$, the two-dimensional irreducible representations (8.14) are not absolutely irreducible. Therefore, we have the general block-diagonal form (12.2) with $R(C_n) \setminus R_a(C_n) = \{(j)_{C_n} \mid j = 1, \ldots, M_2\}$, where $M_2 = \lfloor (n-1)/2 \rfloor$. The multiplicity a_ξ^μ of $\mu \in R(C_n)$ in $\hat{T}_\xi$ of (12.39) is given in Table 12.2(a), and the transformation matrices read as follows.

For $n = 2$, we have

$$\hat{H}_{0z}^{(+)C_2} = 1; \qquad \hat{H}_{0xy}^{(-)C_2} = \begin{pmatrix} 1 & 0 \\ 0 & 1 \end{pmatrix};$$

$$\hat{H}_{1z}^{(+)C_2} = \gamma_2(1/\sqrt{2}, 1), \qquad \hat{H}_{1z}^{(-)C_2} = \gamma_2(1/\sqrt{2}, -1);$$
$$\hat{H}_{1xy}^{(+)C_2} = (\Gamma_2(1,-1), \Gamma_2(\mathrm{i},-1)), \qquad \hat{H}_{1xy}^{(-)C_2} = (\Gamma_2(1,1), \Gamma_2(\mathrm{i},1)).$$

For $n \geq 3$, we have

$$\hat{H}_{0z}^{(+)C_n} = 1;$$

$$\hat{H}_{0xy}^{(1)C_n} = \left(\begin{array}{c|c} 1 & 0 \\ 0 & 1 \end{array}\right);$$

$$\hat{H}_{1z}^{(+)C_n} = \gamma_n(1/\sqrt{2}, 1), \qquad \hat{H}_{1z}^{(-)C_n} = \gamma_n(1/\sqrt{2}, -1),$$
$$\hat{H}_{1z}^{(j)C_n} = (\mathrm{Re}\ \gamma_n(1, \omega^j); \mathrm{Im}\ \gamma_n(1, \omega^j));$$
$$\hat{H}_{1xy}^{(+)C_n} = (\Gamma_n(1,\omega), \Gamma_n(\mathrm{i},\omega)), \qquad \hat{H}_{1xy}^{(-)C_n} = (\Gamma_n(1,-\omega), \Gamma_n(\mathrm{i},-\omega)),$$
$$\hat{H}_{1xy}^{(j)C_n} = (\Gamma_n(1,\omega^{-j+1}), \Gamma_n(1,\omega^{j+1}); \Gamma_n(\mathrm{i},\omega^{-j+1}), \Gamma_n(-\mathrm{i},\omega^{j+1})).$$

In the case of $\mathbb{C}$, the block-diagonalization takes the form of

$$H^* J H = \bigoplus_{\mu \in R_{\mathbb{C}}(\mathrm{C}_n)} \tilde{J}^\mu.$$

Here $R_{\mathbb{C}}(\mathrm{C}_n)$ denotes the index set (8.17) of the irreducible representations over $\mathbb{C}$, which are all one-dimensional.[7] The multiplicity a_ξ^μ of $\mu \in R_{\mathbb{C}}(\mathrm{C}_n)$ in $\hat{T}_\xi$ of (12.39) is given in Table 12.2(b). The transformation matrices $\hat{H}_\xi^{(+)_{\mathrm{C}_n}}$ and $\hat{H}_\xi^{(-)_{\mathrm{C}_n}}$ associated with the irreducible representations $(+)_{\mathrm{C}_n}$ and $(-)_{\mathrm{C}_n}$ for $\mathbb{C}$ are the same for all ξ as those over $\mathbb{R}$ given above. The transformation matrices for the irreducible representations $(j+)_{\mathrm{C}_n}$ and $(j-)_{\mathrm{C}_n}$ are complex matrices and they are given by

$$\hat{H}_{0xy}^{(1+)_{\mathrm{C}_n}} = \frac{1}{\sqrt{2}}\begin{pmatrix} 1 \\ \mathrm{i} \end{pmatrix}, \qquad \hat{H}_{1z}^{(j+)_{\mathrm{C}_n}} = \frac{1}{\sqrt{2}}\gamma_n(1, \omega^j),$$

$$\hat{H}_{1xy}^{(j+)_{\mathrm{C}_n}} = \frac{1}{\sqrt{2}}(\Gamma_n(1, \omega^{-j+1}) - \mathrm{i}\Gamma_n(\mathrm{i}, \omega^{-j+1}),\ \Gamma_n(1, \omega^{j+1}) - \mathrm{i}\Gamma_n(-\mathrm{i}, \omega^{j+1})),$$

$$\hat{H}_{0xy}^{(1-)_{\mathrm{C}_n}} = \overline{\hat{H}_{0xy}^{(1+)_{\mathrm{C}_n}}}, \qquad \hat{H}_{1z}^{(j-)_{\mathrm{C}_n}} = \overline{\hat{H}_{1z}^{(j+)_{\mathrm{C}_n}}}, \qquad \hat{H}_{1xy}^{(j-)_{\mathrm{C}_n}} = \overline{\hat{H}_{1xy}^{(j+)_{\mathrm{C}_n}}},$$

where $j = 1, \ldots, M_2$.

In a reciprocal system (cf., §2.2.3) with real and symmetric $J (= J^\top = J^*)$, each diagonal block $\tilde{J}^\mu$ is Hermitian and $\tilde{J}^{(j-)_{\mathrm{C}_n}} = (\tilde{J}^{(j+)_{\mathrm{C}_n}})^\top$ for each $j = 1, \ldots, M_2$. In this way, the block-diagonalization over $\mathbb{C}$ reveals the mechanism for the emergence of real double eigenvalues in a C_n-symmetric reciprocal system. □

12.5 Local Transformation Matrix: Derivation

We derive local transformation matrices for D_n-symmetric trusses by elementary calculations. The matrices $\hat{Q}_\xi^\mu$ in (12.41) and $\hat{H}_\xi^{\mu,k}$ in (12.45) are constructed for irreducible representations $\mu \in R(\mathrm{D}_n)$ in (8.7) and types ξ in (12.37). The cases $\xi =$ 0z, 0xy, 1Mz, 1Mxy, 2z and 2xy with $n \geq 3$ are treated, whereas the cases $\xi =$ 1Vz and 1Vxy are omitted because they can be respectively treated similarly to the cases $\xi =$ 1Mz and 1Mxy.

For each ξ and $\mu \in R(\mathrm{D}_n)$ we are to determine

$$\hat{Q}_\xi^\mu = (\hat{Q}_{\xi,i}^\mu \mid i = 1, \ldots, a_\xi^\mu)$$

in (12.41), where a_ξ^μ is the multiplicity of μ, $\hat{Q}_\xi^\mu$ and $\hat{Q}_{\xi,i}^\mu$ are $N_\xi \times (a_\xi^\mu N^\mu)$ and $N_\xi \times N^\mu$ matrices, respectively; N^μ is the degree of irreducible representation μ, and N_ξ is defined in (12.38).

Recall from (12.42) and (12.43) that matrices $\hat{Q}_{\xi,i}^\mu$ are determined by

[7] The two-dimensional representation $(j)_{\mathrm{C}_n}$ over $\mathbb{R}$ splits into two one-dimensional irreducible representations over $\mathbb{C}$, labeled $(j+)_{\mathrm{C}_n}$ and $(j-)_{\mathrm{C}_n}$; see (8.16).

$$\hat{T}_\xi(g)\hat{Q}^\mu_{\xi,i} = \hat{Q}^\mu_{\xi,i}\, T^\mu(g), \qquad g \in \mathrm{D}_n,\ i = 1,\dots,a^\mu_\xi \tag{12.65}$$

with the orthonormality condition

$$(\hat{Q}^\mu_{\xi,i})^\top \hat{Q}^\mu_{\xi,j} = \begin{cases} I_{N^\mu} & \text{for } i = j, \\ O & \text{for } i \neq j. \end{cases} \tag{12.66}$$

The multiplicity a^μ_ξ can be determined as the maximum number of $N_\xi \times N^\mu$ matrices $\hat{Q}^\mu_{\xi,i}$ that satisfy these equations.

We make frequent use of following notations:

$$R_n = \begin{pmatrix} \cos(2\pi/n) & -\sin(2\pi/n) \\ \sin(2\pi/n) & \cos(2\pi/n) \end{pmatrix}, \qquad S = \begin{pmatrix} 1 & 0 \\ 0 & -1 \end{pmatrix}, \tag{12.67}$$

$$\omega = \exp(\mathrm{i}2\pi/n), \tag{12.68}$$

$$U_n = \begin{pmatrix} & & & 1 \\ 1 & & & \\ & \ddots & & \\ & & 1 & \end{pmatrix}, \qquad L_n = \begin{pmatrix} & & & 1 \\ & & 1 & \\ & \iddots & & \\ 1 & & & \end{pmatrix}, \tag{12.69}$$

where U_n and L_n are $n \times n$ matrices. Also recall $\gamma_n(a,b)$ and $\Gamma_n(a,b)$ introduced in (12.63).

12.5.1 *Case* 0z

We construct $\hat{Q}^\mu_{0z}$ and $\hat{H}^\mu_{0z}$ for $\mu \in R(\mathrm{D}_n)$, where $N_{0z} = 1$. The representation matrices for this case are given by

$$\hat{T}_{0z}(c(2\pi/n)) = \hat{T}_{0z}(\sigma) = 1.$$

For $\mu = (+,+)_{\mathrm{D}_n}$, we have $T^\mu(c(2\pi/n)) = T^\mu(\sigma) = 1$ from (8.8). Then (12.65) becomes a trivial condition, $1 \times q = q \times 1$ (repeated twice), for an unknown scalar q, which is determined uniquely up to scaling ($a^{(+,+)_{\mathrm{D}_n}}_{0z} = 1$). We can choose $q = 1$ as a solution, and can take

$$\hat{Q}^{(+,+)_{\mathrm{D}_n}}_{0z} = \hat{H}^{(+,+)_{\mathrm{D}_n}}_{0z} = 1.$$

Since $\hat{T}_{0z}(g)$ is one-dimensional and $a^{(+,+)_{\mathrm{D}_n}}_{0z} = 1$, we have $a^\mu_{0z} = 0$ for $\mu \neq (+,+)_{\mathrm{D}_n}$.

12.5.2 *Case* 0xy

We construct $\hat{Q}^\mu_{0xy}$ and $\hat{H}^{\mu,k}_{0xy}$ for $\mu \in R(\mathrm{D}_n)$, where $N_{0xy} = 2$. The representation matrices for this case are given (cf., (12.67)) by

$$\hat{T}_{0xy}(c(2\pi/n)) = R_n = \begin{pmatrix} \cos(2\pi/n) & -\sin(2\pi/n) \\ \sin(2\pi/n) & \cos(2\pi/n) \end{pmatrix}, \qquad \hat{T}_{0xy}(\sigma) = S = \begin{pmatrix} 1 & 0 \\ 0 & -1 \end{pmatrix}.$$

For $\mu = (1)_{\mathrm{D}_n}$, we have (8.9) for $T^\mu(g)$ and the condition (12.65) becomes

$$R_n(\boldsymbol{q}^1, \boldsymbol{q}^2) = (\boldsymbol{q}^1, \boldsymbol{q}^2)R_n, \qquad S(\boldsymbol{q}^1, \boldsymbol{q}^2) = (\boldsymbol{q}^1, \boldsymbol{q}^2)S$$

for a set of unknown vectors $(\boldsymbol{q}^1, \boldsymbol{q}^2)$. As a nontrivial solution we can choose

$$\boldsymbol{q}^1 = \begin{pmatrix} 1 \\ 0 \end{pmatrix}, \qquad \boldsymbol{q}^2 = \begin{pmatrix} 0 \\ 1 \end{pmatrix}.$$

Therefore,

$$\hat{Q}_{0xy}^{(1)_{\mathrm{D}_n}} = \begin{pmatrix} 1 & 0 \\ 0 & 1 \end{pmatrix},$$

and

$$\hat{H}_{0xy}^{(1)_{\mathrm{D}_n},1} = \begin{pmatrix} 1 \\ 0 \end{pmatrix}, \qquad \hat{H}_{0xy}^{(1)_{\mathrm{D}_n},2} = \begin{pmatrix} 0 \\ 1 \end{pmatrix}.$$

Since $\hat{T}_{0xy}(g)$ is two-dimensional and $a_{0xy}^{(1)_{\mathrm{D}_n}} = 1$, we have $a_{0xy}^{\mu} = 0$ for $\mu \neq (1)_{\mathrm{D}_n}$.

12.5.3 *Case* 1Mz

The representation matrices for Case $1Mz$ are given by

$$\hat{T}_{1Mz}(c(2\pi/n)) = U_n, \qquad \hat{T}_{1Mz}(\sigma) = L_n \tag{12.70}$$

with U_n and L_n in (12.69). We have $N_{1Mz} = n$.

One-Dimensional Irreducible Representations

We construct $\hat{Q}_{1Mz}^{\mu} = \hat{H}_{1Mz}^{\mu}$ for one-dimensional irreducible representations $\mu = (+,+)_{\mathrm{D}_n}$, $(+,-)_{\mathrm{D}_n}$, $(-,+)_{\mathrm{D}_n}$, and $(-,-)_{\mathrm{D}_n}$.

For $\mu = (+,+)_{\mathrm{D}_n}$, we have $T^\mu(c(2\pi/n)) = T^\mu(\sigma) = 1$ from (8.8). Then the condition (12.65) reduces to

$$U_n\boldsymbol{q} = \boldsymbol{q}, \qquad L_n\boldsymbol{q} = \boldsymbol{q}$$

for an unknown vector $\boldsymbol{q}$, which is determined uniquely up to scaling ($a_{1Mz}^{(+,+)_{\mathrm{D}_n}} = 1$). We can choose

$$\boldsymbol{q} = \frac{1}{\sqrt{n}}(1, \ldots, 1)^\top = \gamma_n(1/\sqrt{2}, 1),$$

and therefore

$$\hat{Q}_{1Mz}^{(+,+)_{\mathrm{D}_n}} = \hat{H}_{1Mz}^{(+,+)_{\mathrm{D}_n}} = \gamma_n(1/\sqrt{2}, 1).$$

For $\mu = (+,-)_{\mathrm{D}_n}$, we have $T^{\mu}(c(2\pi/n)) = 1$ and $T^{\mu}(\sigma) = -1$ from (8.8). Then the condition (12.65) reduces to

$$U_n \boldsymbol{q} = \boldsymbol{q}, \qquad L_n \boldsymbol{q} = -\boldsymbol{q}.$$

No solution $\boldsymbol{q}$ exists for this case; that is, $a_{1Mz}^{(+,-)_{\mathrm{D}_n}} = 0$.

For $\mu = (-,+)_{\mathrm{D}_n}$, which is existent only for n even, no solution $\boldsymbol{q}$ exists; that is, $a_{1Mz}^{(-,+)_{\mathrm{D}_n}} = 0$.

For $\mu = (-,-)_{\mathrm{D}_n}$, which is existent only for n even, we have $T^{\mu}(c(2\pi/n)) = T^{\mu}(\sigma) = -1$ from (8.8). Then the condition (12.65) reduces to

$$U_n \boldsymbol{q} = -\boldsymbol{q}, \qquad L_n \boldsymbol{q} = -\boldsymbol{q},$$

which has essentially a unique solution

$$\boldsymbol{q} = \frac{1}{\sqrt{n}}(1,-1,\ldots,1,-1)^{\top} = \gamma_n(1/\sqrt{2},-1),$$

showing $a_{1Mz}^{(-,-)_{\mathrm{D}_n}} = 1$. Therefore, we can take

$$\hat{Q}_{1Mz}^{(-,-)_{\mathrm{D}_n}} = \hat{H}_{1Mz}^{(-,-)_{\mathrm{D}_n}} = \gamma_n(1/\sqrt{2},-1).$$

Two-Dimensional Irreducible Representations

We construct $\hat{Q}_{1Mz}^{\mu}$ and $\hat{H}_{1Mz}^{\mu,k}$ for a two-dimensional irreducible representation $\mu = (j)_{\mathrm{D}_n}$ for a fixed j, where $j = 1,\ldots,M_2 (= \lfloor (n-1)/2 \rfloor)$.

We have

$$T^{(j)_{\mathrm{D}_n}}(c(2\pi/n)) = R_n{}^j = \begin{pmatrix} \cos(2\pi j/n) & -\sin(2\pi j/n) \\ \sin(2\pi j/n) & \cos(2\pi j/n) \end{pmatrix},$$

$$T^{(j)_{\mathrm{D}_n}}(\sigma) = S = \begin{pmatrix} 1 & 0 \\ 0 & -1 \end{pmatrix}$$

from (8.9). Then (12.65) takes the form of

$$U_n(\boldsymbol{q}^1,\boldsymbol{q}^2) = (\boldsymbol{q}^1,\boldsymbol{q}^2)R_n{}^j, \qquad L_n(\boldsymbol{q}^1,\boldsymbol{q}^2) = (\boldsymbol{q}^1,\boldsymbol{q}^2)S$$

for a pair of vectors $(\boldsymbol{q}^1,\boldsymbol{q}^2)$. Using a complex vector $\boldsymbol{z} = \boldsymbol{q}^1 + \mathrm{i}\boldsymbol{q}^2$, we can rewrite the above equations as

$$U_n \boldsymbol{z} = \omega^{-j}\boldsymbol{z}, \qquad L_n \boldsymbol{z} = \overline{\boldsymbol{z}}, \tag{12.71}$$

where $\omega = \exp(\mathrm{i}2\pi/n)$ and $\overline{\boldsymbol{z}}$ is the (componentwise) complex conjugate of $\boldsymbol{z}$.

The solution of (12.71) is essentially unique (i.e., $a_{1Mz}^{(j)_{\mathrm{D}_n}} = 1$), and can be chosen as

$$z = \sqrt{\frac{2}{n}}\,\omega^{j/2}\begin{pmatrix} 1 \\ \omega^j \\ \vdots \\ \omega^{(n-1)j} \end{pmatrix} = \gamma_n(\omega^{j/2}, \omega^j).$$

We have

$$\boldsymbol{q}^1 = \mathrm{Re}\, z = \mathrm{Re}\, \gamma_n(\omega^{j/2}, \omega^j), \qquad \boldsymbol{q}^2 = \mathrm{Im}\, z = \mathrm{Im}\, \gamma_n(\omega^{j/2}, \omega^j),$$

which satisfy the orthonormality in (12.66). Finally, we obtain

$$\hat{Q}_{1Mz}^{(j)\mathrm{D}_n} = (\mathrm{Re}\, \gamma_n(\omega^{j/2}, \omega^j), \mathrm{Im}\, \gamma_n(\omega^{j/2}, \omega^j)),$$

and

$$\hat{H}_{1Mz}^{(j)\mathrm{D}_n,1} = \mathrm{Re}\, \gamma_n(\omega^{j/2}, \omega^j), \qquad \hat{H}_{1Mz}^{(j)\mathrm{D}_n,2} = \mathrm{Im}\, \gamma_n(\omega^{j/2}, \omega^j).$$

12.5.4 Case $1Mxy$

We have $N_{1Mxy} = 2n$ and can set (cf., Example 12.1 in §12.3.1)

$$\hat{T}_{1Mxy}(c(2\pi/n)) = \begin{pmatrix} & & & R_n \\ R_n & & & \\ & \ddots & & \\ & & R_n & \end{pmatrix} = U_n \otimes R_n, \tag{12.72}$$

$$\hat{T}_{1Mxy}(\sigma) = \begin{pmatrix} & & S \\ & S & \\ & \cdot^{\cdot^{\cdot}} & \\ S & & \end{pmatrix} = L_n \otimes S. \tag{12.73}$$

One-Dimensional Irreducible Representations

We construct $\hat{Q}_{1Mxy}^{\mu} = \hat{H}_{1Mxy}^{\mu}$ for one-dimensional irreducible representations $\mu = (+,+)_{\mathrm{D}_n}$ and $(+,-)_{\mathrm{D}_n}$. The cases for $\mu = (-,+)_{\mathrm{D}_n}$ and $(-,-)_{\mathrm{D}_n}$, which can be treated similarly, are omitted.

For $\mu = (+,+)_{\mathrm{D}_n}$, we have $T^{\mu}(c(2\pi/n)) = 1$ and $T^{\mu}(\sigma) = 1$; and we can rewrite the condition (12.65) as

$$(U_n \otimes R_n)\boldsymbol{q} = \boldsymbol{q}, \qquad (L_n \otimes S)\boldsymbol{q} = \boldsymbol{q}, \tag{12.74}$$

where

$$\boldsymbol{q} = (q_{1x}, q_{1y}, \ldots, q_{nx}, q_{ny})^{\top}.$$

If we set

$$\zeta = (q_{1x} + \mathrm{i}q_{1y}, \ldots, q_{nx} + \mathrm{i}q_{ny})^\top,$$

the condition (12.74) can be rewritten as

$$\omega U_n \zeta = \zeta, \qquad L_n \overline{\zeta} = \zeta, \tag{12.75}$$

where $\omega = \exp(\mathrm{i}2\pi/n)$. The solution ζ of (12.75) is essentially unique (i.e., $a_{1Mxy}^{(+,+)_{\mathrm{D}_n}} = 1$), and can be chosen to be a complex vector

$$\zeta = \frac{1}{\sqrt{2}} \gamma_n(\omega^{1/2}, \omega).$$

Consequently, we have

$$\hat{Q}_{1Mxy}^{(+,+)_{\mathrm{D}_n}} = \hat{H}_{1Mxy}^{(+,+)_{\mathrm{D}_n}} = \Gamma_n(\omega^{1/2}, \omega).$$

The case of $\mu = (+,-)_{\mathrm{D}_n}$ can be treated similarly. First, (12.74) is replaced by

$$(U_n \otimes R_n)\boldsymbol{q} = \boldsymbol{q}, \qquad (L_n \otimes S)\boldsymbol{q} = -\boldsymbol{q}.$$

Accordingly, (12.75) is replaced by

$$\omega U_n \zeta = \zeta, \qquad L_n \overline{\zeta} = -\zeta,$$

the solution of which is given by

$$\zeta = \frac{1}{\sqrt{2}} \gamma_n(\mathrm{i}\omega^{1/2}, \omega).$$

Therefore,

$$\hat{Q}_{1Mxy}^{(+,-)_{\mathrm{D}_n}} = \hat{H}_{1Mxy}^{(+,-)_{\mathrm{D}_n}} = \Gamma_n(\mathrm{i}\omega^{1/2}, \omega).$$

Two-Dimensional Irreducible Representations

We construct $\hat{Q}_{1Mxy}^{\mu}$ and $\hat{H}_{1Mxy}^{\mu,k}$ for a two-dimensional irreducible representation $\mu = (j)_{\mathrm{D}_n}$ for a fixed j, where $j = 1, \ldots, M_2$.

The relation (12.65) becomes

$$(U_n \otimes R_n)(\boldsymbol{q}^1, \boldsymbol{q}^2) = (\boldsymbol{q}^1, \boldsymbol{q}^2) R_n{}^j, \qquad (L_n \otimes S)(\boldsymbol{q}^1, \boldsymbol{q}^2) = (\boldsymbol{q}^1, \boldsymbol{q}^2) S \tag{12.76}$$

for a pair of unknown vectors $(\boldsymbol{q}^1, \boldsymbol{q}^2)$. Here

$$\boldsymbol{q}^k = (q_{1x}^k, q_{1y}^k, \ldots, q_{nx}^k, q_{ny}^k)^\top, \qquad k = 1, 2,$$

are $2n$-dimensional real vectors. By introducing n-dimensional complex vectors

$$\zeta^k = (q_{1x}^k + \mathrm{i}q_{1y}^k, \ldots, q_{nx}^k + \mathrm{i}q_{ny}^k)^\top, \qquad k = 1, 2,$$

we can rewrite the conditions of (12.76) as

$$\omega U_n(\zeta^1,\zeta^2) = (\zeta^1,\zeta^2)R_n{}^j, \qquad L_n(\overline{\zeta^1},\overline{\zeta^2}) = (\zeta^1,\zeta^2)S. \tag{12.77}$$

With the change of variables from (ζ^1,ζ^2) to

$$(z^1,z^2) = (\zeta^1 + \mathrm{i}\zeta^2, \zeta^1 - \mathrm{i}\zeta^2),$$

the equations in (12.77) can be rewritten as

$$\omega U_n z^1 = \omega^{-j} z^1, \qquad L_n \overline{z^1} = z^1, \tag{12.78}$$

$$\omega U_n z^2 = \omega^{j} z^2, \qquad L_n \overline{z^2} = z^2. \tag{12.79}$$

The solutions to these equations are given by

$$z^1 = \alpha^1 \gamma_n(\omega^{(j+1)/2}, \omega^{j+1}), \qquad z^2 = \alpha^2 \gamma_n(\omega^{(-j+1)/2}, \omega^{-j+1}),$$

where α^1 and α^2 are real numbers. Noting

$$\zeta^1 = \frac{z^1 + z^2}{2}, \qquad \zeta^2 = \frac{z^1 - z^2}{2\mathrm{i}},$$

we see that there exist two independent solutions (i.e., $a_{1Mxy}^{(j)\mathrm{D}_n} = 2$) satisfying the additional condition of orthonormality in (12.66). They are given, for example, by choosing $(\alpha^1,\alpha^2) = (0, 1/\sqrt{2}), (1/\sqrt{2}, 0)$.

For $(\alpha^1,\alpha^2) = (0, 1/\sqrt{2})$ we obtain

$$\zeta^1 = \frac{1}{\sqrt{2}} \gamma_n(\omega^{(-j+1)/2}, \omega^{-j+1}), \qquad \zeta^2 = \mathrm{i}\zeta^1$$

whereas, for $(\alpha^1,\alpha^2) = (1/\sqrt{2}, 0)$, we have

$$\zeta^1 = \frac{1}{\sqrt{2}} \gamma_n(\omega^{(j+1)/2}, \omega^{j+1}), \qquad \zeta^2 = -\mathrm{i}\zeta^1.$$

Then we obtain

$$\begin{aligned}\hat{Q}_{1Mxy}^{(j)\mathrm{D}_n} = (&\Gamma_n(\omega^{(-j+1)/2}, \omega^{-j+1}), \Gamma_n(\mathrm{i}\omega^{(-j+1)/2}, \omega^{-j+1}),\\ &\Gamma_n(\omega^{(j+1)/2}, \omega^{j+1}), \Gamma_n(-\mathrm{i}\omega^{(j+1)/2}, \omega^{j+1})),\end{aligned}$$

and

$$\hat{H}_{1Mxy}^{(j)\mathrm{D}_n,1} = (\Gamma_n(\omega^{(-j+1)/2}, \omega^{-j+1}), \Gamma_n(\omega^{(j+1)/2}, \omega^{j+1})),$$

$$\hat{H}_{1Mxy}^{(j)\mathrm{D}_n,2} = (\Gamma_n(\mathrm{i}\omega^{(-j+1)/2}, \omega^{-j+1}), \Gamma_n(-\mathrm{i}\omega^{(j+1)/2}, \omega^{j+1})).$$

12.5.5 Case $2z$

We have $N_{2z} = 2n$ and assume the node numbering as presented in Fig. 12.7 for $n = 6$.

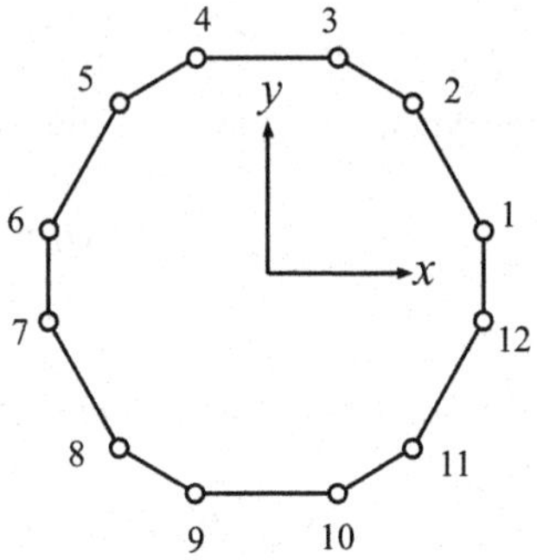

Fig. 12.7 Node numbers for Cases $2z$ and $2xy$ ($n = 6$).

One-Dimensional Irreducible Representations

We construct matrices $\hat{Q}^{\mu}_{2z} = \hat{H}^{\mu}_{2z}$ for $\mu = (+,+)_{D_n}$ and $(+,-)_{D_n}$. The cases for $\mu = (-,+)_{D_n}$ and $(-,-)_{D_n}$ are omitted as they can be treated similarly.

We consider a $2n$-dimensional vector

$$\boldsymbol{q} = (q_1, q_2, \ldots, q_{2n-1}, q_{2n})^{\top} \tag{12.80}$$

to represent $\hat{Q}^{\mu}_{2z,i}$. It turns out to be convenient to employ two n-dimensional vectors

$$\boldsymbol{q}_1 = (q_1, q_3, \ldots, q_{2n-1})^{\top}, \qquad \boldsymbol{q}_2 = (q_2, q_4, \ldots, q_{2n})^{\top}. \tag{12.81}$$

For $\mu = (+,+)_{D_n}$, the relation (12.65) can be expressed as

$$\begin{pmatrix} U_n & O \\ O & U_n \end{pmatrix} \begin{pmatrix} \boldsymbol{q}_1 \\ \boldsymbol{q}_2 \end{pmatrix} = \begin{pmatrix} \boldsymbol{q}_1 \\ \boldsymbol{q}_2 \end{pmatrix}, \qquad \begin{pmatrix} O & L_n \\ L_n & O \end{pmatrix} \begin{pmatrix} \boldsymbol{q}_1 \\ \boldsymbol{q}_2 \end{pmatrix} = \begin{pmatrix} \boldsymbol{q}_1 \\ \boldsymbol{q}_2 \end{pmatrix}.$$

The solution to the equations presented above can be determined uniquely up to scaling ($a_{2z}^{(+,+)_{D_n}} = 1$). We can choose

$$\boldsymbol{q}_1 = \boldsymbol{q}_2 = \frac{1}{\sqrt{2n}}(1, \ldots, 1)^{\top}$$

as a solution, and take

$$\hat{Q}_{2z}^{(+,+)_{D_n}} = \hat{H}_{2z}^{(+,+)_{D_n}} = \frac{1}{\sqrt{2n}}(1, 1, \ldots, 1, 1)^{\top} = \gamma_{2n}(1/\sqrt{2}, 1).$$

For $\mu = (+,-)_{\mathrm{D}_n}$, the relation (12.65) becomes

$$\begin{pmatrix} U_n & O \\ O & U_n \end{pmatrix}\begin{pmatrix} q_1 \\ q_2 \end{pmatrix} = \begin{pmatrix} q_1 \\ q_2 \end{pmatrix}, \qquad \begin{pmatrix} O & L_n \\ L_n & O \end{pmatrix}\begin{pmatrix} q_1 \\ q_2 \end{pmatrix} = -\begin{pmatrix} q_1 \\ q_2 \end{pmatrix}.$$

The solution to the equations presented above can be determined uniquely up to scaling ($a_{2z}^{(+,-)_{\mathrm{D}_n}} = 1$). We can choose

$$q_1 = -q_2 = \frac{1}{\sqrt{2n}}(1,\dots,1)^\top.$$

We interleave the components of q_1 and q_2 compatibly with q in (12.80) to obtain

$$\hat{Q}_{2z}^{(+,-)_{\mathrm{D}_n}} = \hat{H}_{2z}^{(+,-)_{\mathrm{D}_n}} = \frac{1}{\sqrt{2n}}(1,-1,\dots,1,-1)^\top = \gamma_{2n}(1/\sqrt{2},-1).$$

Two-Dimensional Irreducible Representations

We construct $\hat{Q}_{2z}^{\mu}$ and $\hat{H}_{2z}^{\mu,k}$ for $\mu = (j)_{\mathrm{D}_n}$ with a fixed j, where $j = 1,\dots,M_2$.

Similarly to (12.80) and (12.81) we define $2n$-dimensional vectors

$$q^k = (q_1^k, q_2^k, \dots, q_{2n-1}^k, q_{2n}^k)^\top, \qquad k = 1,2 \tag{12.82}$$

for $\hat{Q}_{2z,i}^{\mu} = (q^1, q^2)$ as well as n-dimensional vectors

$$q_1^k = (q_1^k, q_3^k, \dots, q_{2n-1}^k)^\top, \qquad q_2^k = (q_2^k, q_4^k, \dots, q_{2n}^k)^\top, \qquad k = 1,2.$$

Then the relation (12.65) becomes

$$\begin{pmatrix} U_n & O \\ O & U_n \end{pmatrix}\begin{pmatrix} q_1^1 & q_1^2 \\ q_2^1 & q_2^2 \end{pmatrix} = \begin{pmatrix} q_1^1 & q_1^2 \\ q_2^1 & q_2^2 \end{pmatrix} R_n^{\ j}, \qquad \begin{pmatrix} O & L_n \\ L_n & O \end{pmatrix}\begin{pmatrix} q_1^1 & q_1^2 \\ q_2^1 & q_2^2 \end{pmatrix} = \begin{pmatrix} q_1^1 & q_1^2 \\ q_2^1 & q_2^2 \end{pmatrix} S.$$

With n-dimensional complex vectors $z_i = q_i^1 + \mathrm{i}\, q_i^2$ for $i = 1,2$, the equations presented above can be rewritten as

$$U_n z_i = \omega^{-j} z_i, \qquad i = 1,2; \qquad L_n z_1 = \overline{z_2}.$$

As the solution to this set of equations, we obtain

$$z_1 = \alpha \gamma_n(1,\omega^j), \qquad z_2 = \overline{\alpha}\omega^j \gamma_n(1,\omega^j),$$

where α is a complex number.

There exist two independent solutions (i.e., $a_{2z}^{(j)_{\mathrm{D}_n}} = 2$) satisfying the additional condition of orthonormality (12.66). They are given, for example, by the choice of $\alpha = \omega^{j/4}/\sqrt{2}$ and $\alpha = \omega^{j^*/4}/\sqrt{2}$, where $j^* = n - j$.

For $\alpha = \omega^{j/4}/\sqrt{2}$ we interleave the components of z_1 and z_2 compatibly with $\boldsymbol{q}^k$ in (12.82) to obtain

$$\boldsymbol{q}^1 + \mathrm{i}\boldsymbol{q}^2 = \gamma_{2n}(\omega^{j/4}, \omega^{j/2}),$$

which represents the basis

$$(\boldsymbol{q}^1, \boldsymbol{q}^2) = (\mathrm{Re}\ \gamma_{2n}(\omega^{j/4}, \omega^{j/2}), \mathrm{Im}\ \gamma_{2n}(\omega^{j/4}, \omega^{j/2}))$$

of the two-dimensional irreducible representation. With $\alpha = \omega^{j^*/4}/\sqrt{2}$, the other basis is obtained as

$$\boldsymbol{q}^1 + \mathrm{i}\boldsymbol{q}^2 = \gamma_{2n}(\omega^{j^*/4}, \omega^{j^*/2}),$$

which represents

$$(\boldsymbol{q}^1, \boldsymbol{q}^2) = (\mathrm{Re}\ \gamma_{2n}(\omega^{j^*/4}, \omega^{j^*/2}), \mathrm{Im}\ \gamma_{2n}(\omega^{j^*/4}, \omega^{j^*/2})).$$

Therefore, we obtain

$$\begin{aligned}\hat{Q}_{2z}^{(j)\mathrm{D}_n} &= (\mathrm{Re}\ \gamma_{2n}(\omega^{j/4}, \omega^{j/2}), \mathrm{Im}\ \gamma_{2n}(\omega^{j/4}, \omega^{j/2}),\\ &\qquad \mathrm{Re}\ \gamma_{2n}(\omega^{j^*/4}, \omega^{j^*/2}), \mathrm{Im}\ \gamma_{2n}(\omega^{j^*/4}, \omega^{j^*/2})),\end{aligned}$$

and

$$\begin{aligned}\hat{H}_{2z}^{(j)\mathrm{D}_n,1} &= (\mathrm{Re}\ \gamma_{2n}(\omega^{j/4}, \omega^{j/2}), \mathrm{Re}\ \gamma_{2n}(\omega^{j^*/4}, \omega^{j^*/2})),\\ \hat{H}_{2z}^{(j)\mathrm{D}_n,2} &= (\mathrm{Im}\ \gamma_{2n}(\omega^{j/4}, \omega^{j/2}), \mathrm{Im}\ \gamma_{2n}(\omega^{j^*/4}, \omega^{j^*/2})).\end{aligned}$$

12.5.6 Case $2xy$

We have $N_{2xy} = 4n$ and assume the node numbering as portrayed in Fig. 12.7 for $n = 6$.

One-Dimensional Irreducible Representations

We construct $\hat{Q}_{2xy}^{\mu} = \hat{H}_{2xy}^{\mu}$ for $\mu = (+,+)_{\mathrm{D}_n}$ and $\mu = (+,-)_{\mathrm{D}_n}$. The cases of $\mu = (-,+)_{\mathrm{D}_n}$ and $(-,-)_{\mathrm{D}_n}$ are omitted because they can be treated similarly.

We consider a $4n$-dimensional vector

$$\boldsymbol{q} = (q_{1x}, q_{1y}, q_{2x}, q_{2y}, \ldots, q_{2n,x}, q_{2n,y})^{\top}$$

to represent $\hat{Q}_{2xy,i}^{\mu}$. It is convenient to employ two $2n$-dimensional vectors

$$\begin{aligned}\boldsymbol{q}_1 &= (q_{1x}, q_{1y}, q_{3x}, q_{3y}, \ldots, q_{2n-1,x}, q_{2n-1,y})^{\top},\\ \boldsymbol{q}_2 &= (q_{2x}, q_{2y}, q_{4x}, q_{4y}, \ldots, q_{2n,x}, q_{2n,y})^{\top}\end{aligned}$$

as well as two n-dimensional complex vectors

$$\zeta_1 = (q_{1x} + \mathrm{i}q_{1y}, q_{3x} + \mathrm{i}q_{3y}, \ldots, q_{2n-1,x} + \mathrm{i}q_{2n-1,y})^\top,$$
$$\zeta_2 = (q_{2x} + \mathrm{i}q_{2y}, q_{4x} + \mathrm{i}q_{4y}, \ldots, q_{2n,x} + \mathrm{i}q_{2n,y})^\top.$$

For $\mu = (+,+)_{\mathrm{D}_n}$, the relation (12.65) can be expressed as

$$\begin{pmatrix} U_n \otimes R_n & O \\ O & U_n \otimes R_n \end{pmatrix} \begin{pmatrix} \boldsymbol{q}_1 \\ \boldsymbol{q}_2 \end{pmatrix} = \begin{pmatrix} \boldsymbol{q}_1 \\ \boldsymbol{q}_2 \end{pmatrix}, \tag{12.83}$$

$$\begin{pmatrix} O & L_n \otimes S \\ L_n \otimes S & O \end{pmatrix} \begin{pmatrix} \boldsymbol{q}_1 \\ \boldsymbol{q}_2 \end{pmatrix} = \begin{pmatrix} \boldsymbol{q}_1 \\ \boldsymbol{q}_2 \end{pmatrix} \tag{12.84}$$

(cf., (12.72) and (12.73)). Using the complex vectors ζ_1 and ζ_2 we can rewrite the expressions above as

$$\omega U_n \zeta_i = \zeta_i, \qquad i = 1,2; \qquad L_n \overline{\zeta_1} = \zeta_2. \tag{12.85}$$

As the solution to this set of equations, we obtain

$$\zeta_1 = \alpha \gamma_n(1,\omega), \qquad \zeta_2 = \overline{\alpha} \omega \gamma_n(1,\omega),$$

where α is a complex number.

There exist two independent solutions (i.e., $a_{2xy}^{(+,+)_{\mathrm{D}_n}} = 2$) satisfying the additional condition of orthonormality (12.66). They are given, for example, by the choice of $\alpha = \omega^{1/4}/\sqrt{2}$ and $\alpha = \mathrm{i}\omega^{1/4}/\sqrt{2}$. By interleaving the components of ζ_1 and ζ_2 compatibly with $\boldsymbol{q}$ we arrive at two orthogonal basis vectors in the complex form:

$$\frac{1}{\sqrt{2}} \gamma_{2n}(\omega^{1/4}, \omega^{1/2}), \qquad \frac{1}{\sqrt{2}} \gamma_{2n}(\mathrm{i}\omega^{1/4}, -\omega^{1/2}),$$

which yield

$$\hat{Q}_{2xy}^{(+,+)_{\mathrm{D}_n}} = \hat{H}_{2xy}^{(+,+)_{\mathrm{D}_n}} = (\Gamma_{2n}(\omega^{1/4}, \omega^{1/2}), \Gamma_{2n}(\mathrm{i}\omega^{1/4}, -\omega^{1/2})).$$

The case of $\mu = (+,-)_{\mathrm{D}_n}$ can be treated similarly. First, (12.84) is replaced by

$$\begin{pmatrix} O & L_n \otimes S \\ L_n \otimes S & O \end{pmatrix} \begin{pmatrix} \boldsymbol{q}_1 \\ \boldsymbol{q}_2 \end{pmatrix} = -\begin{pmatrix} \boldsymbol{q}_1 \\ \boldsymbol{q}_2 \end{pmatrix},$$

whereas (12.83) remains unchanged. Accordingly, (12.85) is replaced by

$$\omega U_n \zeta_i = \zeta_i, \qquad i = 1,2; \qquad L_n \overline{\zeta_1} = -\zeta_2,$$

the solution of which is given by

$$\zeta_1 = \alpha \gamma_n(1,\omega), \qquad \zeta_2 = -\overline{\alpha} \omega \gamma_n(1,\omega)$$

with a complex number α. There exist two independent solutions (i.e., $a_{2xy}^{(+,-)\mathrm{D}_n} = 2$) satisfying the additional condition of orthonormality (12.66). They are given, for example, by the choice of $\alpha = \mathrm{i}\omega^{1/4}/\sqrt{2}$ and $\alpha = \omega^{1/4}/\sqrt{2}$. We therefore obtain

$$\hat{Q}_{2xy}^{(+,-)\mathrm{D}_n} = \hat{H}_{2xy}^{(+,-)\mathrm{D}_n} = (\Gamma_{2n}(\mathrm{i}\omega^{1/4}, \omega^{1/2}), \Gamma_{2n}(\omega^{1/4}, -\omega^{1/2})).$$

Two-Dimensional Irreducible Representations

We construct $\hat{Q}_{2xy}^{\mu}$ and $\hat{H}_{2xy}^{\mu,k}$ for $\mu = (j)\mathrm{D}_n$ with a fixed j, where $j = 1, \ldots, M_2$.

With the node numbering as shown in Fig. 12.7, we consider a pair of $4n$-dimensional vectors

$$\boldsymbol{q}^k = (q_{1x}^k, q_{1y}^k, q_{2x}^k, q_{2y}^k, \ldots, q_{2n,x}^k, q_{2n,y}^k)^\top, \qquad k = 1, 2,$$

to represent $\hat{Q}_{2xy,i}^{\mu} = (\boldsymbol{q}^1, \boldsymbol{q}^2)$. For $k = 1, 2$ we introduce $2n$-dimensional vectors

$$\boldsymbol{q}_1^k = (q_{1x}^k, q_{1y}^k, q_{3x}^k, q_{3y}^k, \ldots, q_{2n-1,x}^k, q_{2n-1,y}^k)^\top,$$
$$\boldsymbol{q}_2^k = (q_{2x}^k, q_{2y}^k, q_{4x}^k, q_{4y}^k, \ldots, q_{2n,x}^k, q_{2n,y}^k)^\top,$$

as well as n-dimensional complex vectors

$$\zeta_1^k = (q_{1x}^k + \mathrm{i}q_{1y}^k, q_{3x}^k + \mathrm{i}q_{3y}^k, \ldots, q_{2n-1,x}^k + \mathrm{i}q_{2n-1,y}^k)^\top,$$
$$\zeta_2^k = (q_{2x}^k + \mathrm{i}q_{2y}^k, q_{4x}^k + \mathrm{i}q_{4y}^k, \ldots, q_{2n,x}^k + \mathrm{i}q_{2n,y}^k)^\top.$$

For this case, the relation (12.65) reads as

$$\begin{pmatrix} U_n \otimes R_n & O \\ O & U_n \otimes R_n \end{pmatrix} \begin{pmatrix} \boldsymbol{q}_1^1 & \boldsymbol{q}_1^2 \\ \boldsymbol{q}_2^1 & \boldsymbol{q}_2^2 \end{pmatrix} = \begin{pmatrix} \boldsymbol{q}_1^1 & \boldsymbol{q}_1^2 \\ \boldsymbol{q}_2^1 & \boldsymbol{q}_2^2 \end{pmatrix} R_n{}^j,$$

$$\begin{pmatrix} O & L_n \otimes S \\ L_n \otimes S & O \end{pmatrix} \begin{pmatrix} \boldsymbol{q}_1^1 & \boldsymbol{q}_1^2 \\ \boldsymbol{q}_2^1 & \boldsymbol{q}_2^2 \end{pmatrix} = \begin{pmatrix} \boldsymbol{q}_1^1 & \boldsymbol{q}_1^2 \\ \boldsymbol{q}_2^1 & \boldsymbol{q}_2^2 \end{pmatrix} S,$$

which can be rewritten as

$$\omega U_n(\zeta_i^1, \zeta_i^2) = (\zeta_i^1, \zeta_i^2) R_n{}^j, \qquad i = 1, 2,$$
$$L_n(\overline{\zeta_1^1}, \overline{\zeta_1^2}) = (\zeta_2^1, \zeta_2^2) S.$$

With the change of variables from ζ_i^k $(i, k = 1, 2)$ to

$$(z_i^1, z_i^2) = (\zeta_i^1 + \mathrm{i}\zeta_i^2, \zeta_i^1 - \mathrm{i}\zeta_i^2), \qquad i = 1, 2,$$

the equations presented above can be rewritten further as

$$\omega U_n z_i^1 = \omega^{-j} z_i^1, \qquad \omega U_n z_i^2 = \omega^{j} z_i^2, \qquad i = 1,2,$$
$$L_n \overline{z_1^k} = z_2^k, \qquad k = 1,2.$$

The solutions to these equations are given by

$$z_1^1 = \alpha^1 \gamma_n(1, \omega^{j+1}), \qquad z_2^1 = \overline{\alpha^1}\, \omega^{j+1} \gamma_n(1, \omega^{j+1}), \tag{12.86}$$
$$z_1^2 = \alpha^2 \gamma_n(1, \omega^{-j+1}), \qquad z_2^2 = \overline{\alpha^2}\, \omega^{-j+1} \gamma_n(1, \omega^{-j+1}) \tag{12.87}$$

for some complex numbers α^1 and α^2. There exist four independent solutions (i.e., $a_{2xy}^{(j)\mathrm{D}_n} = 4$) satisfying the additional condition of orthonormality (12.66). They are given, for example, by the choice of

$$(\alpha^1, \alpha^2) = (0, \sqrt{2}\omega^{(-j+1)/4}),\ (\sqrt{2}\omega^{(j+1)/4}, 0),\ (-\mathrm{i}\sqrt{2}\omega^{(j+1)/4}, 0),\ (0, \mathrm{i}\sqrt{2}\omega^{(-j+1)/4}).$$

Note that

$$\zeta_i^1 = \frac{z_i^1 + z_i^2}{2}, \qquad \zeta_i^2 = \frac{z_i^1 - z_i^2}{2\mathrm{i}}, \qquad i = 1,2.$$

For $(\alpha^1, \alpha^2) = (0, \sqrt{2}\omega^{(-j+1)/4})$, (12.86) and (12.87) yield

$$\zeta_1^1 = \frac{1}{2} z_1^2 = \frac{1}{\sqrt{2}} \gamma_n(\omega^{(-j+1)/4}, \omega^{-j+1}), \qquad \zeta_1^2 = \mathrm{i}\zeta_1^1,$$
$$\zeta_2^1 = \frac{1}{2} z_2^2 = \frac{1}{\sqrt{2}} \gamma_n(\omega^{3(-j+1)/4}, \omega^{-j+1}), \qquad \zeta_2^2 = \mathrm{i}\zeta_2^1.$$

By interleaving the components of ζ_1^k and ζ_2^k compatibly with $\boldsymbol{q}$, we arrive at two orthogonal basis vectors in the complex form:

$$\frac{1}{\sqrt{2}} \gamma_{2n}(\omega^{(-j+1)/4}, \omega^{(-j+1)/2}), \qquad \frac{1}{\sqrt{2}} \gamma_{2n}(\mathrm{i}\omega^{(-j+1)/4}, \omega^{(-j+1)/2}),$$

which give a pair of basis vectors

$$\varGamma_{2n}(\omega^{(-j+1)/4}, \omega^{(-j+1)/2}), \qquad \varGamma_{2n}(\mathrm{i}\omega^{(-j+1)/4}, \omega^{(-j+1)/2}).$$

We can similarly compute three other pairs of basis vectors for $(\alpha^1, \alpha^2) = (\sqrt{2}\omega^{(j+1)/4}, 0)$, $(-\mathrm{i}\sqrt{2}\omega^{(j+1)/4}, 0)$, and $(0, \mathrm{i}\sqrt{2}\omega^{(-j+1)/4})$.

To sum up these results, we arrive at

$$\begin{aligned}\hat{Q}_{2xy}^{(j)\mathrm{D}_n} = (&\varGamma_{2n}(\omega^{(-j+1)/4}, \omega^{(-j+1)/2}), \varGamma_{2n}(\mathrm{i}\omega^{(-j+1)/4}, \omega^{(-j+1)/2}),\\ &\varGamma_{2n}(\omega^{(j+1)/4}, \omega^{(j+1)/2}), \varGamma_{2n}(-\mathrm{i}\omega^{(j+1)/4}, \omega^{(j+1)/2}),\\ &\varGamma_{2n}(\omega^{(-j^*+1)/4}, \omega^{(-j^*+1)/2}), \varGamma_{2n}(-\mathrm{i}\omega^{(-j^*+1)/4}, \omega^{(-j^*+1)/2}),\\ &\varGamma_{2n}(\omega^{(j^*+1)/4}, \omega^{(j^*+1)/2})), \varGamma_{2n}(\mathrm{i}\omega^{(j^*+1)/4}, \omega^{(j^*+1)/2})),\end{aligned}$$

and

$$\hat{H}_{2xy}^{(j)\mathrm{D}_n,1} = (\Gamma_{2n}(\omega^{(-j+1)/4},\omega^{(-j+1)/2}),\Gamma_{2n}(\omega^{(j+1)/4},\omega^{(j+1)/2}),$$
$$\Gamma_{2n}(\omega^{(-j^*+1)/4},\omega^{(-j^*+1)/2}),\Gamma_{2n}(\omega^{(j^*+1)/4},\omega^{(j^*+1)/2})),$$
$$\hat{H}_{2xy}^{(j)\mathrm{D}_n,2} = (\Gamma_{2n}(\mathrm{i}\omega^{(-j+1)/4},\omega^{(-j+1)/2}),\Gamma_{2n}(-\mathrm{i}\omega^{(j+1)/4},\omega^{(j+1)/2}),$$
$$\Gamma_{2n}(-\mathrm{i}\omega^{(-j^*+1)/4},\omega^{(-j^*+1)/2}),\Gamma_{2n}(\mathrm{i}\omega^{(j^*+1)/4},\omega^{(j^*+1)/2})).$$

Problems

12-1 Sketch the deformation modes for the column vectors of $\hat{H}_{2xy}^{\mu,k}$ for $\mu \in R(\mathrm{D}_4)$ in §12.4.
Answer: See Fig. 12.8.

12-2 Decompose 25 nodes of the square plate shown in Fig. 12.9 into orbits.
Answer: The nodes are decomposed into six orbits P_1 to P_6, as presented in Fig. 12.10:

- P_1 of Type 0 consisting of node 13,
- P_2 of Type $1V$ consisting of nodes 3, 11, 15, and 23,
- P_3 of Type $1V$ consisting of nodes 8, 12, 14, and 18,
- P_4 of Type $1M$ consisting of nodes 1, 5, 21, and 25,
- P_5 of Type $1M$ consisting of nodes 7, 9, 17, and 19,
- P_6 of Type 2 consisting of nodes 2, 4, 6, 10, 16, 20, 22, and 24.

12-3 Determine the sizes a^μs of diagonal blocks for the square plate presented in Fig. 12.9 for a four-node finite element in the xy-plane with eight degrees of freedom, as portrayed in Fig. 7.4 in §7.8.2.
Answer:

$$a^\mu = \begin{cases} 6 & \text{for } \mu = (+,+)_{\mathrm{D}_4},(+,-)_{\mathrm{D}_4},(-,+)_{\mathrm{D}_4},(-,-)_{\mathrm{D}_4}, \\ 13 & \text{for } \mu = (1)_{\mathrm{D}_4}. \end{cases}$$

12-4 Prove $\tilde{J}^{(j-)\mathrm{C}_n} = (\tilde{J}^{(j+)\mathrm{C}_n})^\top$ for a C_n-symmetric reciprocal system (cf., Remark 12.2 in §12.4).
Answer: Since $H^{(j-)\mathrm{C}_n} = \overline{H^{(j+)\mathrm{C}_n}}$ and $J^\top = J$, we have

$$\tilde{J}^{(j-)\mathrm{C}_n} = (H^{(j-)\mathrm{C}_n})^* J H^{(j-)\mathrm{C}_n} = (H^{(j+)\mathrm{C}_n})^\top J\, \overline{H^{(j+)\mathrm{C}_n}}$$
$$= ((H^{(j+)\mathrm{C}_n})^* J H^{(j+)\mathrm{C}_n})^\top = (\tilde{J}^{(j+)\mathrm{C}_n})^\top.$$

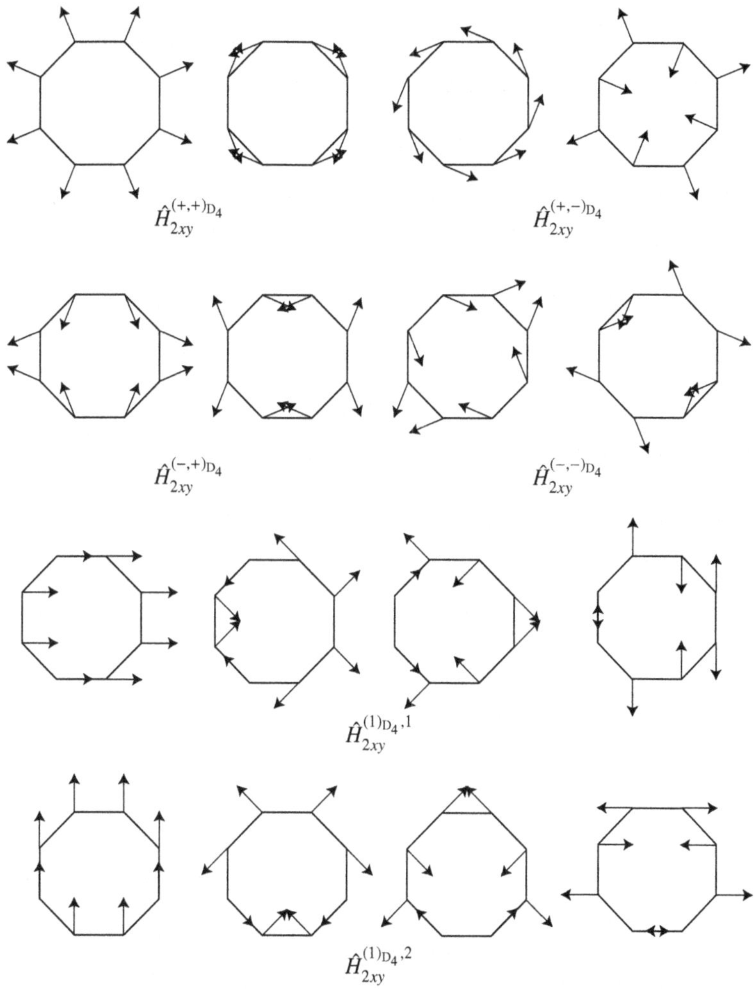

Fig. 12.8 Deformation modes for the column vectors of $\hat{H}_{2xy}^{\mu,k}$ $(n = 4)$.

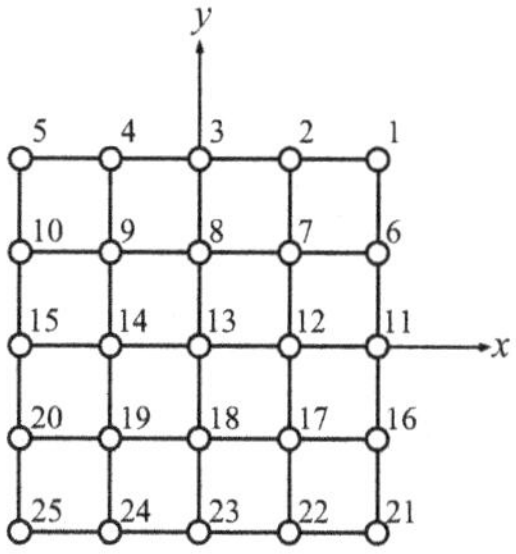

Fig. 12.9 Square plate divided into 4×4 elements.

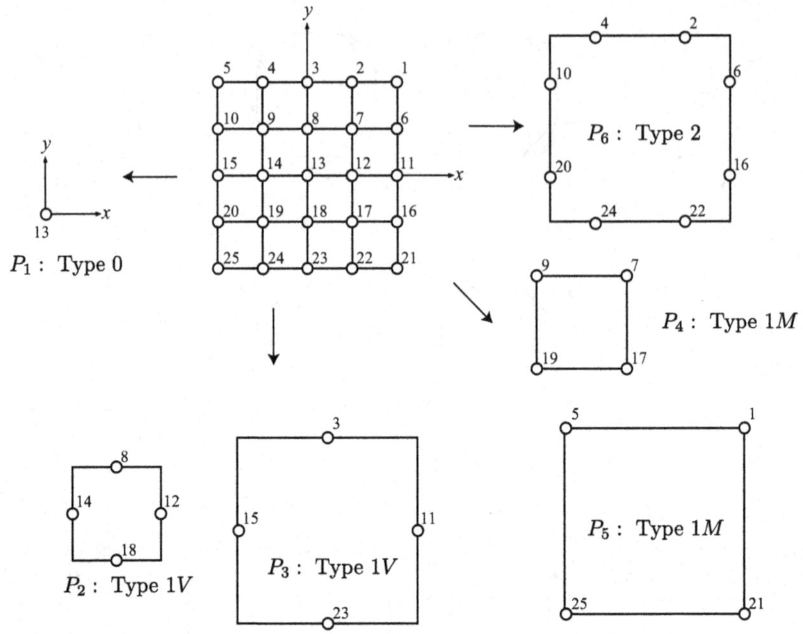

Fig. 12.10 Decomposition of nodes of the square plate into orbits.

Summary

- The theoretical foundation of block-diagonalization analysis of discrete systems has been presented on the basis of group-representation theory.
- A systematic procedure to construct transformation matrix H has been presented.
- Formulas for local transformation matrices have been presented.

Part III
Modeling of Bifurcation Phenomena

The asymptotic, statistical, and group-theoretic approaches were presented in Parts I and II as basic tools to investigate imperfect bifurcation phenomena from an engineering standpoint. In Part III, using these approaches, the bifurcation phenomena of physical and structural systems (soil, sand, kaolin, plate, steel, etc.) are treated. They involve essential uncertainty and complexity.

Imperfections are the major source of uncertainty. To overcome this uncertainty, the occurrence of bifurcation is detected by the imperfection sensitivity laws for experimentally observed bifurcation diagrams and the bifurcation point search technique, which were presented in Chapter 6 for simple bifurcation points and extended to double bifurcation points of a D_n-equivariant system in Chapter 8.

The recursive bifurcation attributable to the symmetry of a system is a major source of complexity. Although it is customary to observe mainly the force versus displacement (stress versus strain) curves in experiments, deformation pattern change is observed and described using group-theoretic bifurcation theory to identify recursive bifurcation. Groups labeling symmetries vary with particular systems; therefore, the group-theoretic approach, developed in Part II for the dihedral and cyclic groups, should be adapted to the relevant groups. The symmetry is classified into two types: apparent *geometrical symmetry* and hidden *periodic symmetry*. It is easy to determine the group that labels the geometrical symmetry, as is typically the case for the truss domes presented in Part II. The bifurcation of systems with symmetries of various kinds is studied in this part, Part III, with emphasis on the hidden periodic symmetry.

As Chapters 13 and 14 describe, the bifurcation of cylindrical soil specimens, for example, is classified into two stages:

(1) The earlier stage, in which patterns with high spatial frequencies are formed,
(2) The later stage, in which the deformation of the cylindrical shape progresses.

The geometrical symmetry alone is insufficient, however, for the explanation of the mechanism of patterns with high spatial frequencies at the earlier stage. A basic strategy to overcome such insufficiency is to enlarge this symmetry appropriately by implementing the local uniformity of the system. This strategy is demonstrated for symmetry groups that frequently appear in the physical and structural systems. The successful modeling of symmetries turns out to be crucial for the proper understanding of bifurcation behaviors.

Part III is organized as follows. In Chapter 13 the recursive change of shapes of cylindrical sand specimens undergoing bifurcation at the later stage is investigated. In Chapter 14, as an example of the bifurcation at the earlier stage that produces patterns with high spatial frequencies, the mechanism of echelon-mode formation is explained as the bifurcation of an $O(2)\times O(2)$-equivariant system. Such a formation is studied for the experiment on cylindrical soil specimens, computational analysis of a rectangular plate, and image simulation analysis of kaolin and steel specimens. In Chapter 15 the recursive bifurcation of rectangular parallelepiped steel specimens is investigated through theoretical, experimental, and computational standpoints. In Chapter 16 flower patterns appearing on honeycomb structures are investigated.

Chapter 13
Bifurcation of Cylindrical Sand Specimens

13.1 Introduction

Triaxial compression tests of soils and sands are conducted to ascertain their mechanical properties in soil mechanics, as shown in Fig. 13.1. A cylindrical soil or

Fig. 13.1 Photograph of a cylindrical soil specimen during a triaxial compression test.

K. Ikeda and K. Murota, *Imperfect Bifurcation in Structures and Materials*,
Applied Mathematical Sciences 149, DOI 10.1007/978-1-4419-7296-5_13,

sand specimen under compression exhibits the concentration of strain into a narrow damaged zone, which is called a *shear band*. Under constant uniform circumferential pressure (stress) σ_3, the axial strain ε_a or the deviatoric stress $\sigma_a = \sigma_1 - \sigma_3$ (σ_1 is the axial pressure) is increased to shear the specimen.[1]

The formation of shear bands is highlighted in experiments as a key phenomenon that leads to the final failure. Yet it has been a long-standing paradox that "The patterns of shear bands are so diverse that every test appears to be unique."[2] For example, Fig. 13.2 depicts sketches of diverse shear bands of soil specimens.[3] The frustrating and surprising aspect is that experimental efforts to make granular material specimens as homogeneous as possible invariably yield unpredictable and diverse responses.

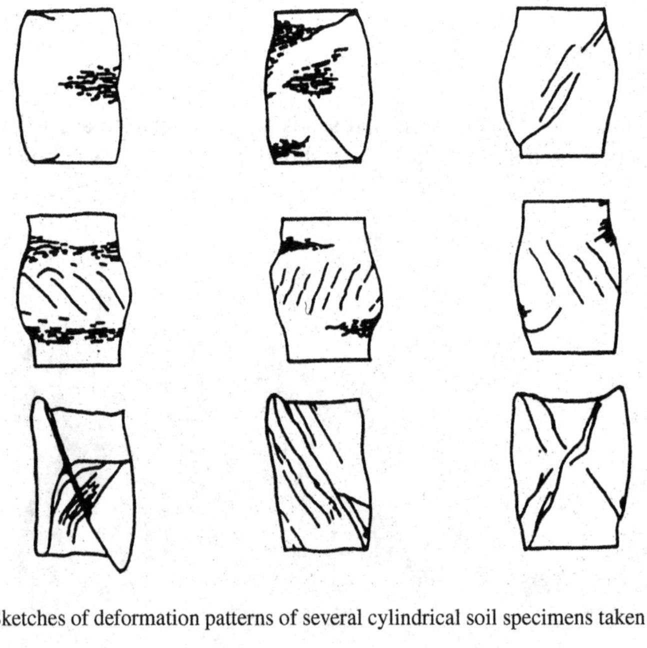

Fig. 13.2 Sketches of deformation patterns of several cylindrical soil specimens taken by Nakano, 1993 [147].

[1] The axial strain ε_a, axial stress σ_1, and volumetric strain ε_v are measured to plot experimental curves, which are used for engineering-related decisions. The axial strain is obtained as the average of the shortening of the whole specimen, and the axial stress is obtained as the average on the top surface force of the specimen. For more issues on this test, refer, for example, to Terzaghi and Peck, 1967 [189].

[2] This statement is based on Desrues and Viggiani, 2004 [39].

[3] Diverse shear bands and deformation patterns of soils and granular materials have been observed in the literature describing experimental studies; see, for example, Desrues, Lanier, and Stutz, 1985 [38]; Nakano, 1993 [147]; Melo, Umbanhowar, and Swinney, 1995 [137]; Venkataramani and Ott, 1998 [203]; Andersen et al., 2002 [2]; Wolf, König, and Triantafyllidis, 2003 [211]; and Ikeda, Sasaki, and Ichimura, 2006 [103].

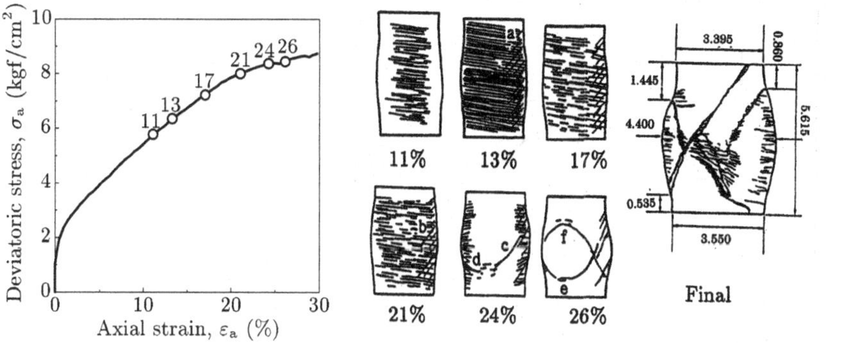

(a) Deviatoric stress σ_a versus axial strain ε_a relation (b) Sketches of deformations (c) Sketch of deformation at the final stage

Fig. 13.3 Progress of deformation of a cylindrical Kawasaki soil specimen (Asaoka and Nakano, 1996 [7]). $1\,\mathrm{kgf/cm^2} = 98\,\mathrm{kPa}$.

Plastic bifurcation theory[4] has elucidated the mechanism for formation of a single shear band or a series of parallel shear bands on uniform material by direct bifurcation from the uniform state. However, this theory alone cannot explain the mechanism engendering such diverse shear band patterns as those depicted in Fig. 13.2.

We must accordingly consider broader symmetry-breaking bifurcations than the direct bifurcation considered in the plastic bifurcation theory. In this regard, we refer to the progress of deformation patterns of a cylindrical soil specimen presented in Fig. 13.3, which displays patterns with high spatial frequencies for $\varepsilon_a = 11 \sim 21\%$ and the barreling of the whole specimen for $\varepsilon_a = 24 \sim 26\%$. These patterns and barreling are possibly the result of recursive symmetry-breaking bifurcations, which are classified into:

1. The earlier stage, in which patterns with high spatial frequencies are formed,
2. The later stage, in which the deformation of the shape progresses.

In this chapter,[5] the mechanism of the bifurcation at the later stage is revealed with emphasis on two characteristic behaviors, recursive bifurcation and mode switching portrayed in Fig. 13.4, whereas the bifurcation mechanism at the earlier stage is dealt with in Chapter 14.

Here, recursive bifurcation means repeated occurrence of symmetry-breaking bifurcations, as has been introduced in §8.3.2. On the other hand, *mode switching* means the change of the bifurcated path that the actual behavior follows because

[4] Plastic bifurcation theory (Hill and Hutchinson, 1975 [73]) was derived on the basis of Hill's theory of the uniqueness and stability of solutions of elastic–plastic solids (Hill, 1958 [72]). This theory lays a foundation of the numerical bifurcation analyses of soils developed thereafter; see, for example, Kolymbas, 1981 [122]; Vardoulakis and Sulem, 1995 [202]; and Asaoka and Noda, 1995 [8].

[5] The study in this chapter is based on Ikeda et al., 1997 [96].

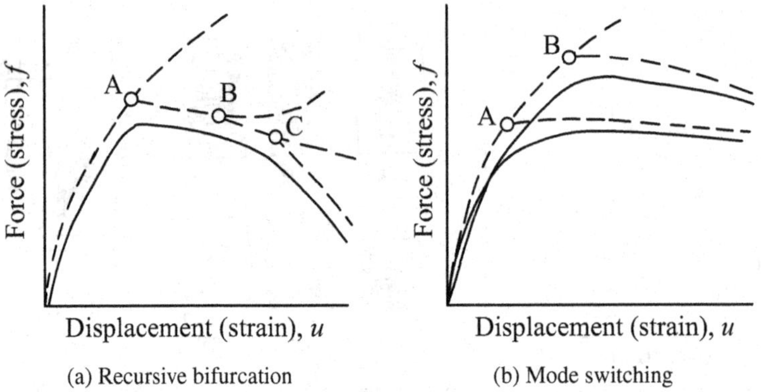

Fig. 13.4 General views of recursive bifurcation and mode switching. ——: experimental (imperfect) curve; — — —: perfect curve; ○: bifurcation point.

of the difference in imperfections. The change of the bifurcated path results in the change of deformation mode to be observed.

Mathematical tools employed in this chapter are readily available as presented below:[6]

1. Asymptotic approach for the experimentally observed bifurcation diagram presented in Chapter 6 for a simple (pitchfork) bifurcation point and extended to a double bifurcation point of a D_n-equivariant system in §8.9,
2. Group-theoretic bifurcation theory presented in Chapter 7 and its application to the dihedral group in Chapter 8.

The cylindrical domain under consideration has the symmetry labeled by the group $D_{\infty h}$ in the Schoenflies notation, which denotes the combination of upside-down symmetry and axisymmetry. The group-theoretic approach is applied to the group $D_{\infty h}$ to arrive at an exhaustive list of possible bifurcating modes of the cylindrical domain.

This chapter is organized as presented below.

- Group $D_{\infty h}$ and its subgroups are introduced to label the symmetry of the initial and deformed configurations of cylindrical specimens in §13.2.
- Deformation modes of cylindrical soil specimens under shear are observed, and recursive bifurcation and mode switching behaviors are identified by the asymptotic and group-theoretic approaches in §13.3.
- The bifurcation rule for a $D_{\infty h}$-equivariant system is derived in §13.4, the appendix of this chapter.

[6] In applying these mathematical tools, the concrete form of the governing equations for the soil specimen need not be identified. It might, however, be mentioned that the *Cam clay model* is popular in soil mechanics (e.g., Schofield and Wroth, 1968 [180]),

13.2 Groups for Spatial Symmetry

The symmetry of crystals and molecules was studied in chemical crystallography, and led to the recognition of the point groups, which describe the symmetry around a point.[7] Among possible alternatives, we refer to the Schoenflies notation in describing the spatial symmetry because it is common and pertinent.

13.2.1 Symmetry of Cylindrical Domain

Of primary concern are the deformation patterns of some uniform (homogeneous and isotropic) material in a cylindrical domain (see Fig. 13.5):

$$\Omega = \{(x,y,z) \in \mathbb{R}^3 \mid x^2 + y^2 \leq R^2, -L/2 \leq z \leq L/2\}, \tag{13.1}$$

where $R = R(f)$ is the radius and $L = L(f)$ is the length of the cylindrical domain, each of which might vary with a parameter f.

The cylindrical domain has natural geometrical symmetry, as presented in Fig. 13.5, characterized by geometrical transformations of three kinds:

- Reflection σ_{v} with respect to a vertical plane (containing the z-axis), which, for example, is

$$\sigma_{\mathrm{v}} : x \mapsto -x, \text{ or } y \mapsto -y,$$

- Reflection with respect to the horizontal plane

$$\sigma_{\mathrm{h}} : z \mapsto -z,$$

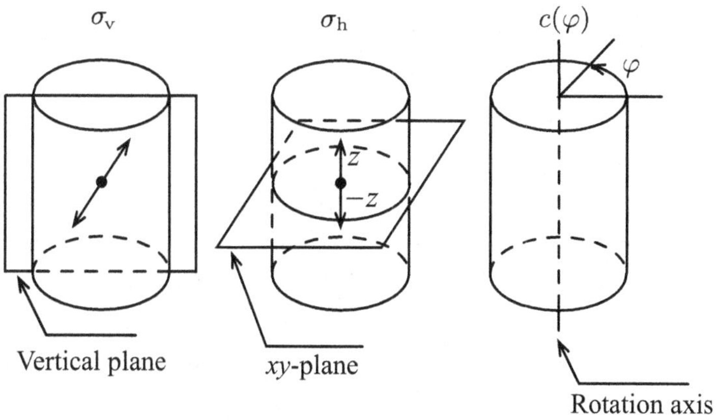

Fig. 13.5 Geometrical transformations for spatial symmetries of the cylindrical domain.

[7] See, for example, Prince, 1994 [169]; Kettle, 1995 [116]; and Ludwig and Falter, 1996 [130].

- Counterclockwise rotation about the z-axis at an angle of φ $(0 \le \varphi < 2\pi)$

$$c(\varphi) : \theta \mapsto \theta + \varphi, \qquad \text{where} \quad \theta = \tan^{-1}(y/x). \tag{13.2}$$

Accordingly, the symmetry of the cylindrical domain is expressed by the invariance under the group

$$\mathrm{D}_{\infty\mathrm{h}} = \langle \sigma_{\mathrm{v}}, \sigma_{\mathrm{h}}, c(\varphi) \rangle, \tag{13.3}$$

where the right-hand side is a shorthand notation for the group generated by σ_{v}, σ_{h}, and $c(\varphi)$ with $0 \le \varphi < 2\pi$. We have the relations

$$\begin{array}{lll} {\sigma_{\mathrm{v}}}^2 = (\sigma_{\mathrm{v}} c(\varphi))^2 = {\sigma_{\mathrm{h}}}^2 = e, & \sigma_{\mathrm{v}}\sigma_{\mathrm{h}} = \sigma_{\mathrm{h}}\sigma_{\mathrm{v}}, & c(\varphi)\sigma_{\mathrm{h}} = \sigma_{\mathrm{h}} c(\varphi), \\ c(\varphi_1)c(\varphi_2) = c(\varphi_1 + \varphi_2), & c(\varphi + 2\pi) = c(\varphi), & \end{array} \tag{13.4}$$

where e is the identity element.

The group $\mathrm{D}_{\infty\mathrm{h}}$ is the direct product of the group $\langle \sigma_{\mathrm{v}}, c(\varphi) \rangle$ and $\langle \sigma_{\mathrm{h}} \rangle$, which are given, respectively, by[8]

$$\langle \sigma_{\mathrm{v}}, c(\varphi) \rangle = \{ c(\varphi), \sigma_{\mathrm{v}} c(\varphi) \mid 0 \le \varphi < 2\pi \} \simeq \mathrm{O}(2), \tag{13.5}$$

$$\langle \sigma_{\mathrm{h}} \rangle = \{ e, \sigma_{\mathrm{h}} \} \simeq \mathbb{Z}_2. \tag{13.6}$$

Here $\simeq$ denotes that the groups are isomorphic. Using these notations, the symmetry of the cylindrical domain is represented by

$$\mathrm{D}_{\infty\mathrm{h}} = \langle \sigma_{\mathrm{v}}, c(\varphi) \rangle \times \langle \sigma_{\mathrm{h}} \rangle \simeq \mathrm{O}(2) \times \mathbb{Z}_2. \tag{13.7}$$

The direct product structure in (13.7) is used in the theoretical analysis of bifurcation behavior under the symmetry of $\mathrm{D}_{\infty\mathrm{h}}$.

13.2.2 Subgroups of $\mathrm{D}_{\infty\mathrm{h}}$

The subgroups of $\mathrm{D}_{\infty\mathrm{h}}$ are

$$\begin{array}{llll} \mathrm{D}_{\infty\mathrm{h}} = \langle \sigma_{\mathrm{v}}, \sigma_{\mathrm{h}}, c(\varphi) \rangle, & & \mathrm{D}_{n\mathrm{h}} = \langle \sigma_{\mathrm{v}}, \sigma_{\mathrm{h}}, c(2\pi/n) \rangle, \\ & & \mathrm{D}_{n\mathrm{d}} = \langle \sigma_{\mathrm{v}}\sigma_{\mathrm{h}}, \sigma_{\mathrm{h}} c(\pi/n) \rangle, \\ \mathrm{C}_{\infty\mathrm{v}} = \langle \sigma_{\mathrm{v}}, c(\varphi) \rangle, & & \mathrm{C}_{n\mathrm{v}} = \langle \sigma_{\mathrm{v}}, c(2\pi/n) \rangle, \\ \mathrm{C}_{\infty\mathrm{h}} = \langle \sigma_{\mathrm{h}}, c(\varphi) \rangle, & & \mathrm{C}_{n\mathrm{h}} = \langle \sigma_{\mathrm{h}}, c(2\pi/n) \rangle, \\ & & \mathrm{S}_{n} = \langle \sigma_{\mathrm{h}} c(2\pi/n) \rangle, \\ \mathrm{D}_{\infty} = \langle \sigma_{\mathrm{v}}\sigma_{\mathrm{h}}, c(\varphi) \rangle, & & \mathrm{D}_{n} = \langle \sigma_{\mathrm{v}}\sigma_{\mathrm{h}}, c(2\pi/n) \rangle, \\ \mathrm{C}_{\infty} = \langle c(\varphi) \rangle, & & \mathrm{C}_{n} = \langle c(2\pi/n) \rangle, \end{array} \tag{13.8}$$

[8] The notations O(2) and $\mathbb{Z}_2$ are commonly used in the mathematical literature to signify groups isomorphic ($\simeq$) to $\mathrm{C}_{\infty\mathrm{v}} = \langle \sigma_{\mathrm{v}}, c(\varphi) \rangle$ and $\langle \sigma_{\mathrm{h}} \rangle$, respectively. To be more specific, O(2) is the two-dimensional *orthogonal group* consisting of 2×2 orthogonal matrices and $\mathbb{Z}_2 = \{1, -1\}$ is the *two-element group*.

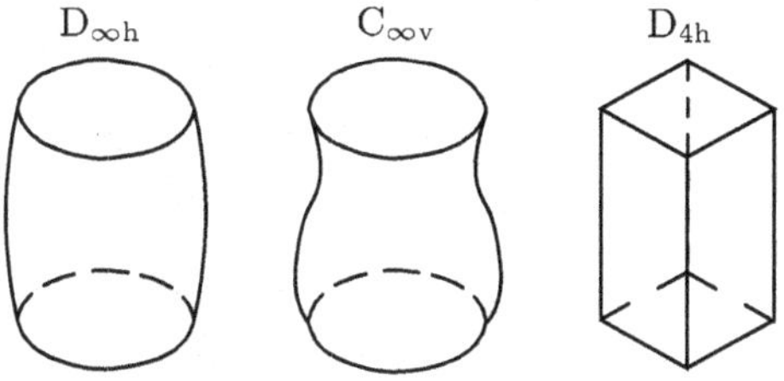

Fig. 13.6 Spatial symmetries labeled by groups in the Schoenflies notation.

where the subscript ∞ denotes the axisymmetry.[9] The notation S_n is used only for n even because $S_n = C_{nh}$ for n odd, and S_2 is usually designated as

$$C_i = \{e, \sigma_h c(\pi)\}, \tag{13.9}$$

which is often designated as the *group of inversion*. The orders of these groups are given by

$$|D_{nh}| = |D_{nd}| = 4n, \qquad |C_{nv}| = |C_{nh}| = |D_n| = 2n, \qquad |S_n| = |C_n| = n.$$

Figure 13.6 portrays spatial symmetries labeled by such groups; see also Fig. 8.2 in §8.2.1.

When loading is compatible with the natural symmetry of the cylindrical domain, the governing equations for the deformation are equivariant to $D_{\infty h}$. Such is the case, for example, with a cylindrical soil specimen in a triaxial compression test.

The bifurcation equation for $D_{\infty h}$ is solved to find the subgroups that can appear via direct bifurcation. Then bifurcation equations for these subgroups are solved in a similar manner to find subgroups that can appear via further bifurcations. In this manner, we can exhaust all possible bifurcation processes $G_i \to G_j$ indicating that a G_j-symmetric state branches from a G_i-symmetric state. By assembling these processes, we arrive at the recursive bifurcation rule. For reciprocal systems, Fig. 13.7 depicts the rule expressed in terms of a hierarchy of subgroups that exhausts all possible bifurcation processes. See §13.4 for mathematical derivation and the rule for nonreciprocal systems.

Remark 13.1. The dihedral group that is denoted as D_n in (8.1) is denoted as C_{nv} in (13.8), whereas D_n in the Schoenflies notation means another (isomorphic) group defined in (13.8). It is noteworthy that, in the Schoenflies notation, the definition of σ_v is not unique, and that σ_v and $\sigma_v c(\varphi)$ are sometimes identified. For example, the subgroups defined as $D_m^{k,n}$ in (8.4) with different values of k are not distinguished but are identified with C_{mv}. □

Remark 13.2. The subgroups of $D_{\infty h}$ listed in (13.8) are isomorphic to dihedral groups, cyclic groups, or the direct product of these groups with the group of inversion C_i of (13.9). To be specific (cf., Kim, 1999 [117]), we have

[9] The notation SO(2) is used in the mathematical literature to indicate a group isomorphic to $C_\infty = \langle c(\varphi) \rangle$. To be more specific, SO(2) is the two-dimensional *special orthogonal group* consisting of 2×2 orthogonal matrices with determinant equal to one.

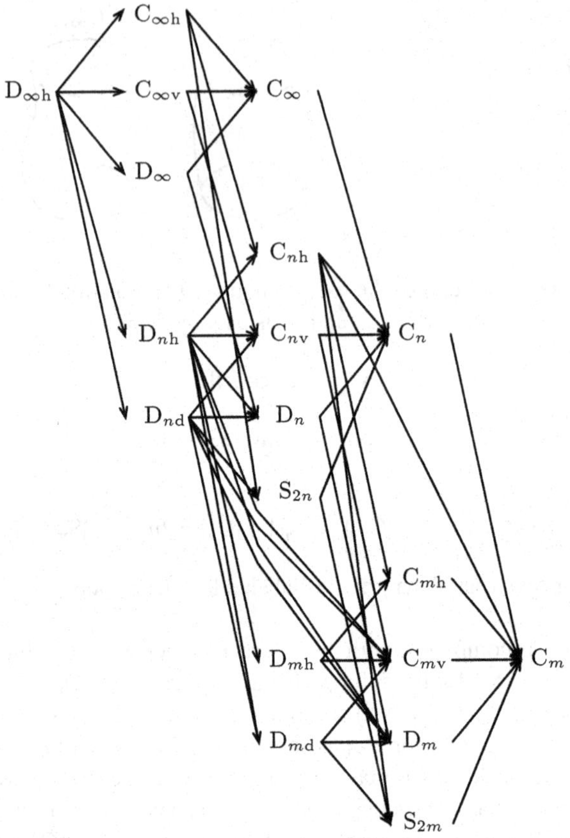

Fig. 13.7 Bifurcation rule of $D_{\infty h}$-invariant reciprocal systems that are expressed in terms of a hierarchy of subgroups (n/m is odd for $D_{nh} \rightarrow C_{mv}$, $D_{nh} \rightarrow D_m$, $D_{nd} \rightarrow D_{md}$, $C_{nh} \rightarrow C_m$, and $S_{2n} \rightarrow S_{2m}$; n/m is even for $D_{nh} \rightarrow D_{md}$ and $C_{nh} \rightarrow S_{2m}$).

$$D_{\infty h} \simeq D_{\infty} \times C_i, \tag{13.10}$$

$$C_{\infty v} \simeq D_{\infty}, \tag{13.11}$$

$$C_{\infty h} \simeq C_{\infty} \times C_i, \tag{13.12}$$

$$D_{nh} \simeq \begin{cases} D_n \times C_i \simeq D_{2n} & (n\text{: odd}), \\ D_n \times C_i & (n\text{: even}), \end{cases} \tag{13.13}$$

$$D_{nd} \simeq \begin{cases} D_n \times C_i \simeq D_{2n} & (n\text{: odd}), \\ D_{2n} & (n\text{: even}), \end{cases} \tag{13.14}$$

$$C_{nv} \simeq D_n, \tag{13.15}$$

$$C_{nh} \simeq \begin{cases} C_n \times C_i \simeq C_{2n} & (n\text{: odd}), \\ C_n \times C_i & (n\text{: even}), \end{cases} \tag{13.16}$$

$$S_{2n} \simeq \begin{cases} C_n \times C_i & (n\text{: odd}), \\ C_{2n} & (n\text{: even}). \end{cases} \tag{13.17}$$

The isomorphism of $\mathrm{D}_{n\mathrm{d}}$ to $\mathrm{D}_n \times \mathrm{C_i}$ or D_{2n}, for example, enables us to reduce the analysis of $\mathrm{D}_{n\mathrm{d}}$-equivariant bifurcation equations (essentially) to the case of the dihedral group, $\mathrm{C}_{n\mathrm{v}} \simeq \mathrm{D}_n$, which is done in Chapter 8. See §13.4 for details. □

13.2.3 Example of Description of Cylindrical Sand Deformation

Using the Schoenflies notation presented in §13.2.2, we classify deformation modes of sand specimens with three different sizes:

(a) $H = 5$ cm and $D = 5$ cm,
(b) $H = 10$ cm and $D = 5$ cm,
(c) $H = 10$ cm and $D = 3.5$ cm,

where H is the height and D is the diameter of the specimen. The deformation modes in the earlier stage are shown at the left of Fig. 13.8 and the later stage are shown at the right. All these modes are axisymmetric, and hence are invariant under the rotation $c(\varphi)$ around the z-axis and the reflection σ_{v} with respect to a vertical plane. The modes for (a) and (c) have upside-down symmetry labeled by σ_{h}, whereas the mode for (b) does not. Consequently, these modes are labeled by the groups

$$\begin{cases} \mathrm{D}_{\infty\mathrm{h}} = \langle \sigma_{\mathrm{v}}, \sigma_{\mathrm{h}}, c(\varphi) \rangle & \text{for (a) and (c),} \\ \mathrm{C}_{\infty\mathrm{v}} = \langle \sigma_{\mathrm{v}}, c(\varphi) \rangle & \text{for (b).} \end{cases}$$

More diverse deformation modes are introduced in the next section.

13.3 Experiments on Cylindrical Sand Specimens

We observe the deformation behaviors of Toyoura sand specimens in triaxial compression tests, and describe these behaviors from dual viewpoints of mode switching and recursive bifurcation presented in Fig. 13.4.

The sand specimens are made up of a few horizontal layers as portrayed in Fig. 13.9; these layers might be interpreted as artificial imperfections that trigger diverse bifurcation modes. Specimens of five different kinds, with the number of layers of 1, 3, 4, 6, and 8, each with five cases are used. The jth specimen with m layers is designated as m-j.

Figure 13.10 shows curves of deviatoric stress σ_{a} versus axial strain ε_{a} categorized based on the number of layers. These curves vary test by test; the variation is greater for the specimens with a single layer than for the specimens with multiple layers.

(a) $D_{\infty h}$-symmetric deformation ($H = 5$ cm and $D = 5$ cm)

(b) $C_{\infty v}$-symmetric deformation ($H = 10$ cm and $D = 5$ cm)

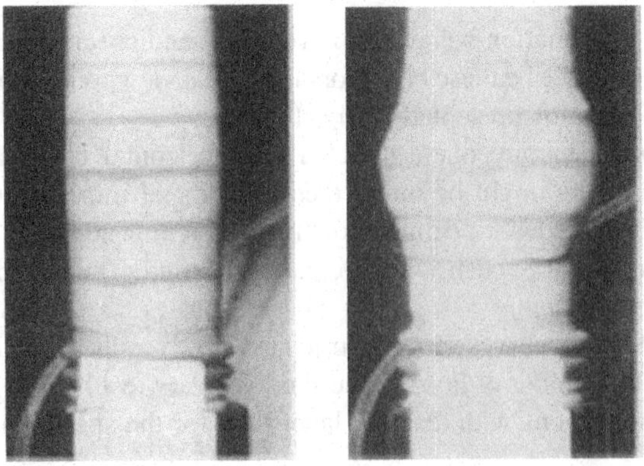

(c) $D_{\infty h}$-symmetric deformation ($H = 10$ cm and $D = 3.5$ cm)

Fig. 13.8 Deformation modes of cylindrical sand specimens with various sizes. Earlier stages at the left and later stages at the right.

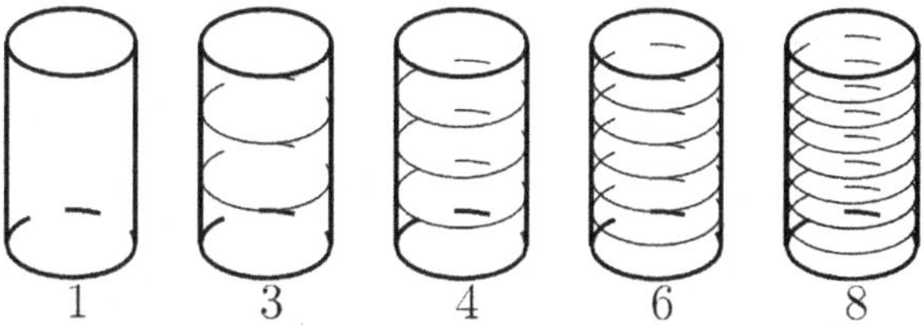

Fig. 13.9 Cylindrical sand specimens with different number of layers.

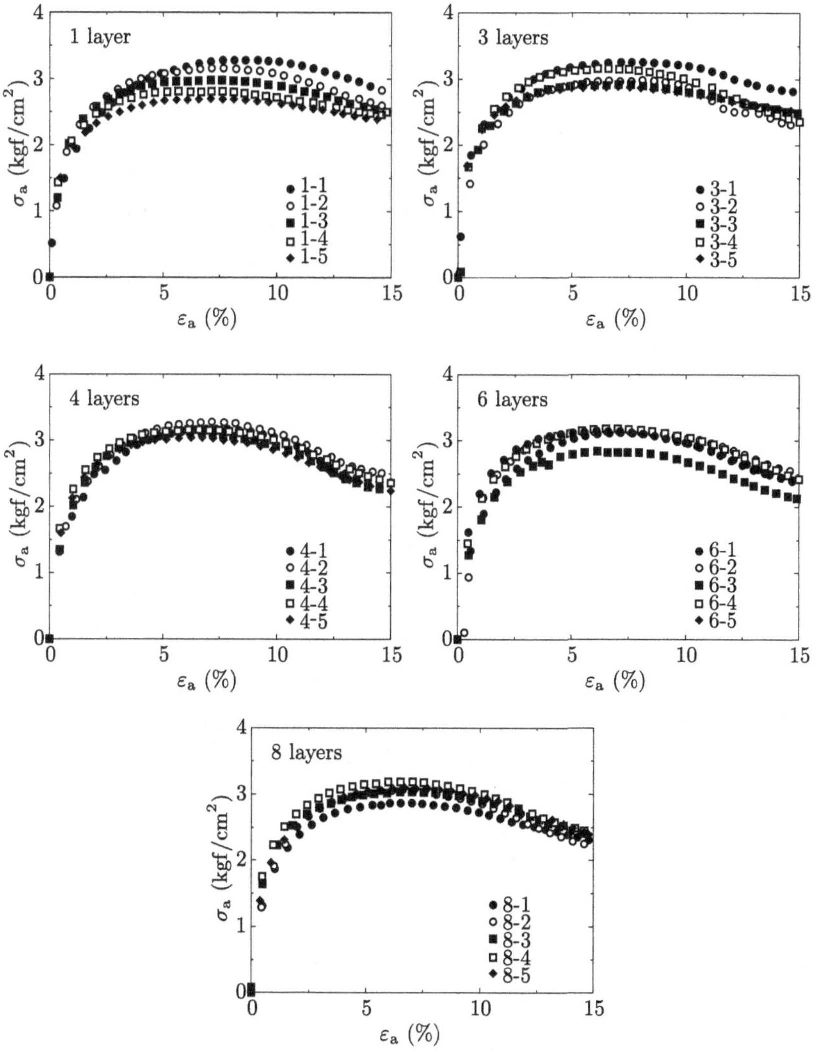

Fig. 13.10 Deviatoric stress σ_a versus axial strain ε_a curves categorized based on the number of layers. $1\,\mathrm{kgf/cm^2} = 98\,\mathrm{kPa}$.

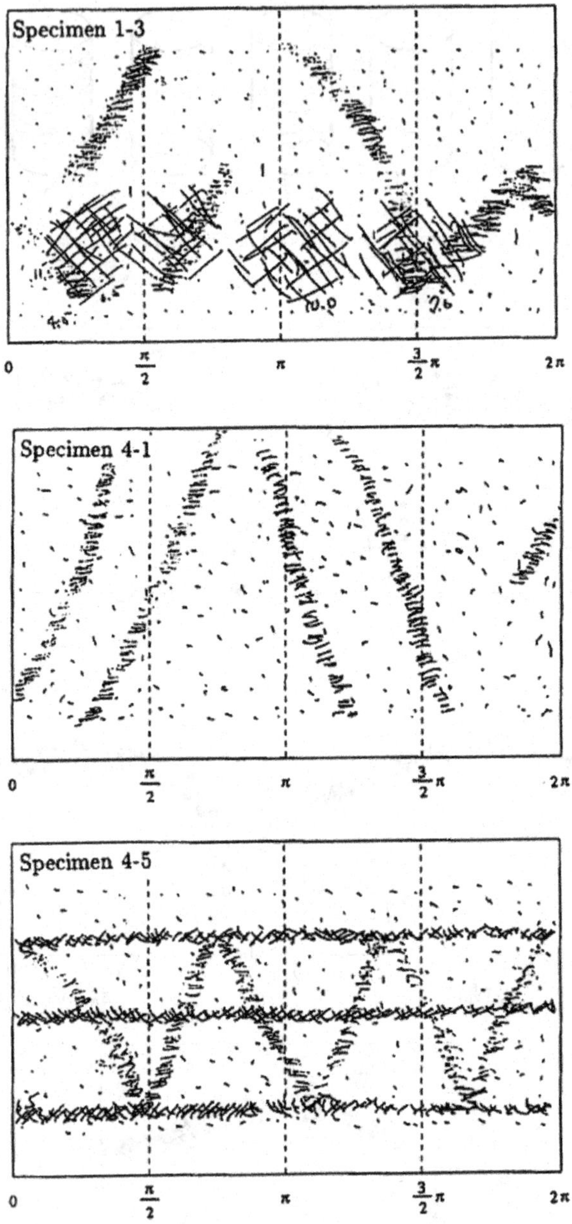

Fig. 13.11 Sketches of expansion plans of wrinkles that appeared on specimens.

13.3.1 Recursive Bifurcation Behavior

The mechanism of the larger variation in the experimental curves for the specimens with a single layer depicted in Fig. 13.9 is elucidated from a standpoint of recursive bifurcation behavior using the group-theoretic approach presented in § 13.2.

Deformation patterns of the specimens are observed and categorized using the subgroups in (13.8) and the hierarchical bifurcation rule in Fig. 13.7. Figure 13.11 portrays sketches of expansion plans of the wrinkles that appeared on the circumferential surfaces of Specimens 1-3, 4-1, and 4-5. The upside-down symmetry is lost for Specimen 1-3; Specimen 4-1 has a C_{1v}-symmetric pattern; and a D_{3d}-symmetric pattern is observed for Specimen 4-5.

Figure 13.12 depicts typical deformation patterns of specimens labeled by the subgroups of $D_{\infty h}$ below D_{2h}. For example, the group D_{2h} denotes the symmetry of the barreling of a specimen; the group C_{2v} expresses the symmetry of the swelling at the bottom and shrink at the top; the groups D_{1d}, C_{1v}, and D_1 represent the symmetry of a shear band formation.

Through the categorization of patterns of wrinkles and deformations, we have identified the emergence of five sets of bifurcation processes listed in Table 13.1 among the whole sets of processes in Fig. 13.7. It is noteworthy that various bifurcation processes exist even for specimens with the same number of layers. In particular, the single-layered specimens have as many as four sets of processes, but those with eight layers have only two. The introduction of the layers therefore has limited possible bifurcation processes by imposing imperfections.

A typical example of a recursive pattern change associated with the bifurcation of Process B is illustrated in Fig. 13.13.

- The upside-down symmetry of the specimen is lost at the onset of the bifurcation $D_{\infty h} \to C_{\infty v}$.
- Its cross-section becomes elliptic at $C_{\infty v} \to C_{2v}$.
- The elliptic cross-section further deforms and an oblique shear band is formed at $C_{2v} \to C_{1v}$.

There appear numerous parallel wrinkles with fine intervals that are labeled by the group D_n with the half-rotation symmetry $\sigma_v \sigma_h$ for Process E, and by the group C_n without this symmetry for Process D.

Table 13.1 Bifurcation processes for the cylindrical sand specimens

Process	Recursive Bifurcation	Name of Specimens
A	$D_{\infty h} \to D_{2h} \to D_{1h} \to C_{1v}$	1-2, 3-4, 4-1, 4-2, 4-3, 4-4, 6-2, 6-4, 6-5, 8-2, 8-3, 8-4, 8-5
B	$D_{\infty h} \to C_{\infty v} \to C_{2v} \to C_{1v}$	1-3, 1-5, 3-2, 3-3, 3-5, 6-3, 8-1
C	$D_{\infty h} \to D_{3d} \to S_6$	4-5, 6-1
D	$D_{\infty h} \to C_{\infty v} \to C_n \to C_4$ (n large)	1-4
E	$D_{\infty h} \to D_n \to D_2$ (n large)	1-1, 3-1

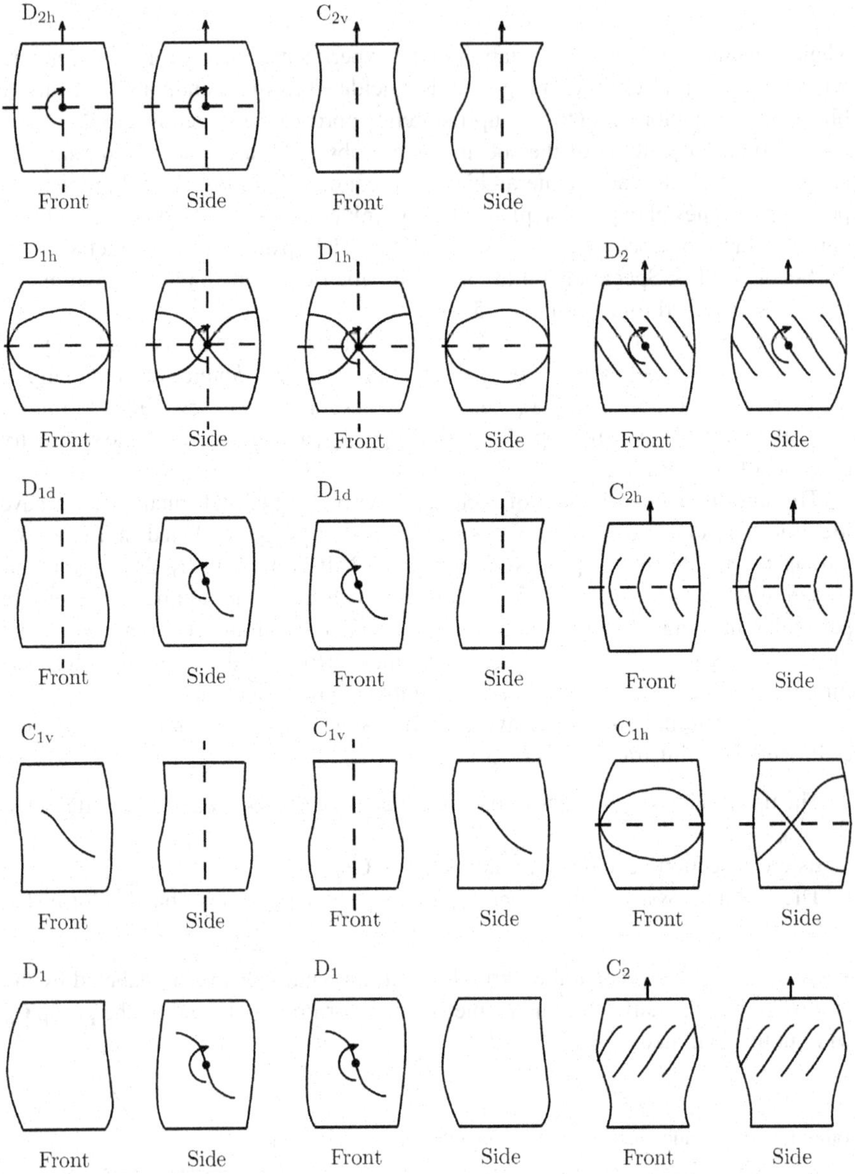

Fig. 13.12 Deformation patterns of cylindrical sand specimens labeled by subgroups of $D_{\infty h}$. ↑: axis of half-rotation symmetry; dashed line: plane of reflection symmetry.

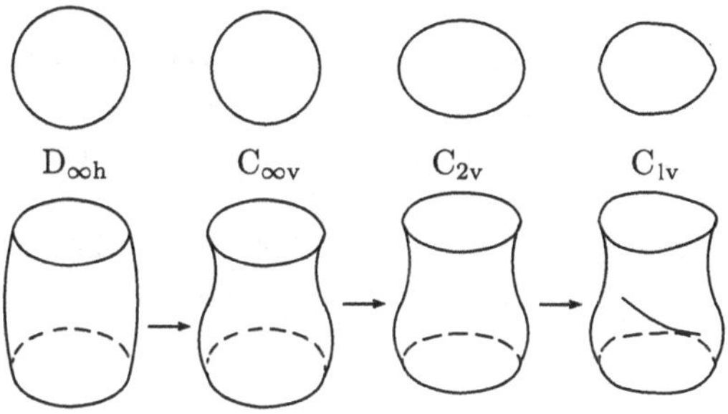

Fig. 13.13 Shape change associated with the bifurcation of Process B.

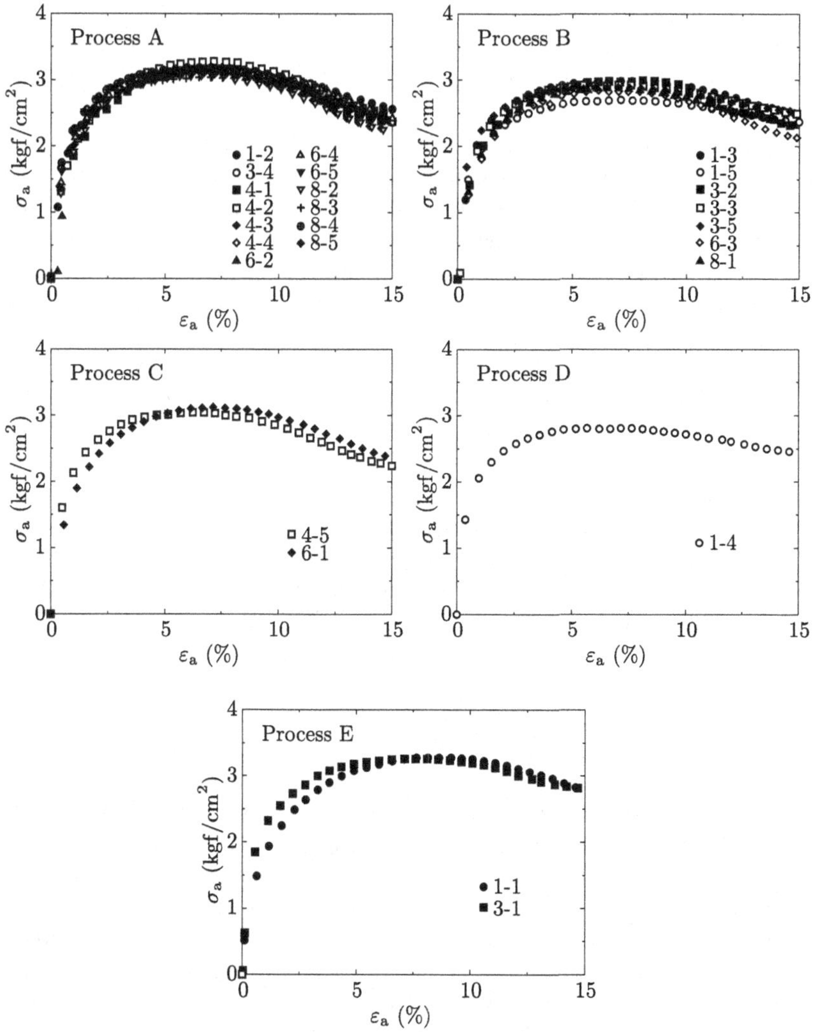

Fig. 13.14 Deviatoric stress σ_a versus axial strain ε_a curves categorized based on the bifurcation processes. $1\,\text{kgf/cm}^2 = 98\,\text{kPa}$.

The curves of deviatoric stress versus axial strain categorized in this manner are portrayed in Fig. 13.14. The variation among the curves for Process A is discernibly smaller than that in Fig. 13.10 categorized based on the number of layers. Such is also the case for Process B. The curves for Process A have markedly higher peaks (strength) than those of Process B. This might suffice to demonstrate the importance of identifying the bifurcation process, which has greater influence on the strength variation than the number of layers does, in studying the bifurcation behavior of sand specimens.

13.3.2 Mode Switching Behavior

Mode switching means that a specimen might take different courses of bifurcation dependent on imperfections. Such courses are called Processes A to E in Table 13.1 in §13.3.1.

Mode switching behavior is simulated by the bifurcation equation (6.27). Figure 13.15(a) shows the simulation of the deviatoric stress versus axial strain curves of two specimens for Process A; Figure 13.15(b) shows the simulation for Process B. Here the values of the parameters p and E that are used for the simulation are listed at the bottom-right of this figure. The location $((\varepsilon_{\mathrm{a}})_{\mathrm{c}}^{0}, (\sigma_{\mathrm{a}})_{\mathrm{c}}^{0})$ of the bifurcation point has been obtained by the bifurcation point search procedure conducted in §13.3.3. Note that the same set of values of the parameters but different values of imperfection ε are used for the specimens belonging to each process.

Figure 13.15(c) shows a superposition of the curves for Processes A and B (one curve is chosen from each process). The theoretical curves correlate fairly well with the experimental curves. Furthermore, the two bifurcation points are located on the fundamental path to trigger mode switching. Of course, actual shear behavior of sands might be more complex because it is expected to involve several bifurcation modes and undergo recursive bifurcation, as Table 13.1 shows. In the study of deformation properties of sands, it is vital to understand the mechanism of mode switching and recursive bifurcation.

13.3.3 Bifurcation Point Search

The location of a bifurcation point, for Specimen 6-3 belonging to Process B, is searched for by the asymptotic procedure presented in §6.4.1. Figure 13.16(a) shows the deviatoric stress versus axial strain curve of this specimen and the rectangular area used for the search, and Fig. 13.16(b) shows the contour map of the sample variance among parameter values p_i $(i = 1,\dots,4)$ in a two-dimensional space of ε_{a} and σ_{a} of the rectangular area in Fig. 13.16(a). This figure clearly portrays the presence of three local minima, where three bifurcation points a, b, and c are located. By combining this result with the bifurcation process B:

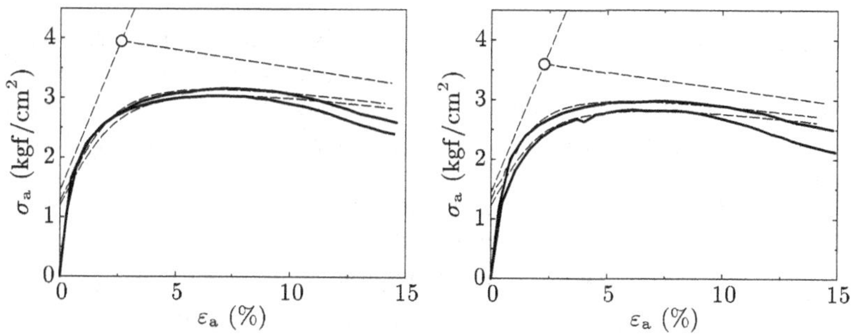

(a) Process A

(b) Process B

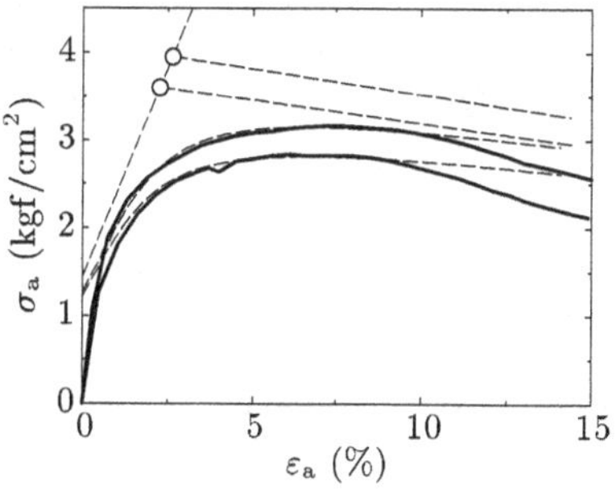

(c) Processes A and B

Process	A	B
$(\varepsilon_a)_c^0$	2.64	2.26
$(\sigma_a)_c^0$	3.95	3.60
h	0.19	0.15
p	0.055	0.050
E	0.95	0.95

Fig. 13.15 Simulation of mode switching of deviatoric stress σ_a versus axial strain ε_a curves. ——: experimental curve; - - -: simulated curve; ○: bifurcation point; 1 kgf/cm^2 = 98 kPa.

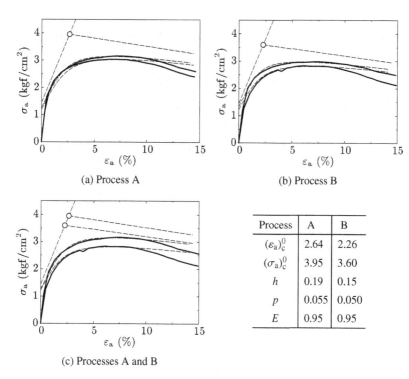

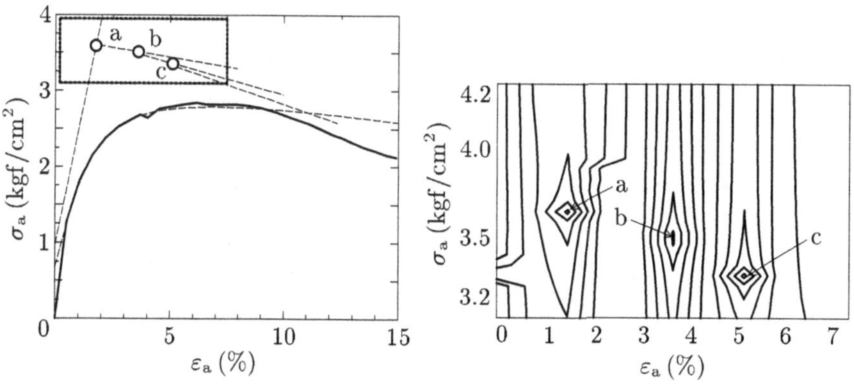

(a) Deviatoric stress σ_a versus axial strain ε_a curve and the rectangular search area

(b) Contour map of the variance of $(p_i \mid i = 1, \ldots, 4)$ in the search area

Fig. 13.16 Bifurcation point search for Specimen 6-3 for Process B. ○: bifurcation point; 1 kgf/cm^2 = 98 kPa.

$$\mathrm{D}_{\infty\mathrm{h}} \to \mathrm{C}_{\infty\mathrm{v}} \to \mathrm{C}_{2\mathrm{v}} \to \mathrm{C}_{1\mathrm{v}}$$

in Table 13.1, we can interpret that these three bifurcation points are related to $\mathrm{D}_{\infty\mathrm{h}} \to \mathrm{C}_{\infty\mathrm{v}}$, $\mathrm{C}_{\infty\mathrm{v}} \to \mathrm{C}_{2\mathrm{v}}$, and $\mathrm{C}_{2\mathrm{v}} \to \mathrm{C}_{1\mathrm{v}}$. These symmetry changes reveal (cf., Tables 13.2(a) and (b), and 13.3(c) in §13.4) that the bifurcation points a and c are simple and b is double. Therefore, we can presume the occurrence of a recursive bifurcation:

$$\begin{cases} \mathrm{D}_{\infty\mathrm{h}} \to \mathrm{C}_{\infty\mathrm{v}} & \text{at simple bifurcation point a,} \\ \mathrm{C}_{\infty\mathrm{v}} \to \mathrm{C}_{2\mathrm{v}} & \text{at double bifurcation point b,} \\ \mathrm{C}_{2\mathrm{v}} \to \mathrm{C}_{1\mathrm{v}} & \text{at simple bifurcation point c.} \end{cases}$$

Specimen 6-3 is therefore undergoing recursive bifurcation. Note that the search procedure presented in §6.4.1 is applicable to the simple bifurcation points a and c and also to the double bifurcation point b. The results for a double bifurcation point with $\widehat{n} \geq 5$ of the dihedral group in §8.9 are applicable to the double point b, because the process $\mathrm{C}_{\infty\mathrm{v}} \to \mathrm{C}_{2\mathrm{v}}$ can be treated as the limiting case of $\mathrm{C}_{n\mathrm{v}} \to \mathrm{C}_{2\mathrm{v}}$ with $n \to \infty$.

13.3.4 Application of Asymptotic Law

The occurrence of bifurcation is further ensured by the asymptotic method developed in Chapter 6. This method, which was explained for a D_n-symmetric system in §8.9, is adapted to a $\mathrm{D}_{\infty\mathrm{h}}$-symmetric system. It turns out, in particular, that the method of recovering perfect system behavior from the imperfect system behavior presented in §6.4.1 is applicable to the simple bifurcation point and also to the double bifurcation point treated below.

We specifically examine the first bifurcation point of Process A associated with $\mathrm{D}_{\infty\mathrm{h}} \to \mathrm{D}_{2\mathrm{h}}$ and that of Process B with $\mathrm{D}_{\infty\mathrm{h}} \to \mathrm{C}_{\infty\mathrm{v}}$. In view of the change of symmetry (cf., Table 13.2(a) in §13.4), we note that the former is a double bifurcation point and the latter is a simple one. That is, we particularly examine

$$\begin{cases} \text{Process A:} & \mathrm{D}_{\infty\mathrm{h}} \to \mathrm{D}_{2\mathrm{h}} & \text{at the double bifurcation point,} \\ \text{Process B:} & \mathrm{D}_{\infty\mathrm{h}} \to \mathrm{C}_{\infty\mathrm{v}} & \text{at the simple bifurcation point.} \end{cases}$$

The location of the bifurcation point for each process is chosen according to the procedure described in §11.5.2 as follows. We recall the asymptotic law (11.25):

$$|(\widetilde{\sigma}_{\mathrm{a}})_{\mathrm{c}}| = \gamma^{*}(\widetilde{\varepsilon}_{\mathrm{a}})_{|h} + \text{h.o.t.} \tag{13.18}$$

The location is chosen such that (13.18) holds most accurately. The incremental strain $(\widetilde{\varepsilon}_{\mathrm{a}})_{|h}$ versus incremental maximum stress $|(\widetilde{\sigma}_{\mathrm{a}})_{\mathrm{c}}| = |(\sigma_{\mathrm{a}})_{\mathrm{c}} - (\sigma_{\mathrm{a}})_{\mathrm{c}}^{0}|$ relation for Process B is shown in Fig. 13.17 for three values of $h = 0.15$, 0.18, and 0.21. The straight lines in Fig. 13.17, which are the least-square approximations of these relations, fairly accurately follow the law (13.18), which denotes a straight line passing the origin. This ensures that the variation of sand shear behaviors among specimens results from the variation of imperfections.

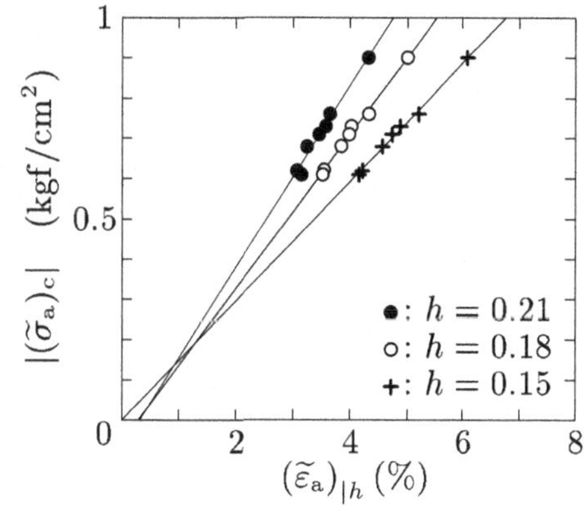

Fig. 13.17 $(\widetilde{\varepsilon}_a)_{|h}$ versus $|(\widetilde{\sigma}_a)_c|$ relations for Process B. $(\varepsilon_a)_c^0 = 2.26$ and $(\sigma_a)_c^0 = 3.60$; ——: relation (6.33); $1\,\text{kgf/cm}^2 = 98\,\text{kPa}$.

13.4 Appendix: Derivation of Bifurcation Rules

To supplement the discussion in this chapter, the rule of recursive bifurcation of a $D_{\infty h}$-equivariant system is presented in this appendix by determining the subgroups, such as D_{nh} and D_{nd}, for symmetries of the direct bifurcated paths of this system. We obtain the symmetries of secondary and further bifurcated paths, and arrive at the hierarchy of subgroups in Fig. 13.7. In this connection, we make use of the following facts (with slight modifications).

- The subgroups of our interest, listed in (13.8), are isomorphic to C_{nv}, C_n, or to their direct product with $C_i \simeq \mathbb{Z}_2$, as mentioned in Remark 13.2 in §13.2.2. The symmetries of the direct bifurcated paths for C_{nv}- and C_n-equivariant systems were determined already in §8.3.1.
- $C_{\infty v}$ and D_∞ are isomorphic to O(2), and a thorough study of O(2)-equivariant systems is available in the literature.[10]

The symmetries of the recursive bifurcated paths of a $D_{\infty h}$-equivariant system are presented in Tables 13.2–13.4, in which some associated bifurcated paths are nonexistent for nonreciprocal systems;

$$Q = \begin{pmatrix} \cos(\pi j/n) & -\sin(\pi j/n) \\ \sin(\pi j/n) & \cos(\pi j/n) \end{pmatrix}, \qquad R = \begin{pmatrix} \cos(n\varphi) & -\sin(n\varphi) \\ \sin(n\varphi) & \cos(n\varphi) \end{pmatrix},$$

$$S = \begin{pmatrix} 1 & 0 \\ 0 & -1 \end{pmatrix},$$

$$\widehat{j} = \frac{j}{\gcd(n,j)}, \qquad \widehat{n} = \frac{n}{\gcd(n,j)}, \qquad n = 1, 2, \ldots,$$

and $\gcd(n,j)$ is the greatest common divisor of n and j.

[10] See Sattinger, 1983 [178] and Golubitsky, Stewart, and Schaeffer, 1988 [64].

Table 13.2 Classification of critical points of systems equivariant to $D_{\infty h}$, $C_{\infty v}$, $C_{\infty h}$, D_{∞}, or C_{∞}. (∗) denotes that the associated bifurcated path is nonexistent for a nonreciprocal system

(a) $D_{\infty h} = \langle c(\varphi), \sigma_v, \sigma_h \rangle$

Multiplicity	Irreducible	$T^{\mu}(g)$			Symmetry Groups	
M	Representation, μ	$c(\varphi)$	σ_v	σ_h	G^{μ}	Bifurcated Paths
1	$(+,+)_{D_{\infty h}}$	1	1	1	$D_{\infty h}$	No Bifurcation
	$(+,-)_{D_{\infty h}}$	1	1	−1	$C_{\infty v}$	$C_{\infty v}$
	$(-,+)_{D_{\infty h}}$	1	−1	1	$C_{\infty h}$	$C_{\infty h}$
	$(-,-)_{D_{\infty h}}$	1	−1	−1	D_{∞}	D_{∞}
2	$(n,+)_{D_{\infty h}}$	R	S	I_2	C_{nh}	D_{nh}
	$(n,-)_{D_{\infty h}}$	R	S	$-I_2$	S_{2n}	D_{nd}
	$n = 1, 2, \ldots$					

(b) $C_{\infty v} = \langle c(\varphi), \sigma_v \rangle$

Multiplicity	Irreducible	$T^{\mu}(g)$		Symmetry Groups	
M	Representation, μ	$c(\varphi)$	σ_v	G^{μ}	Bifurcated Paths
1	$(+)_{C_{\infty v}}$	1	1	$C_{\infty v}$	No Bifurcation
	$(-)_{C_{\infty v}}$	1	−1	C_{∞}	C_{∞}
2	$(n)_{C_{\infty v}}$	R	S	C_n	C_{nv} (∗)
	$n = 1, 2, \ldots$				

(c) $C_{\infty h} = \langle c(\varphi), \sigma_h \rangle$

Multiplicity	Irreducible	$T^{\mu}(g)$		Symmetry Groups	
M	Representation, μ	$c(\varphi)$	σ_h	G^{μ}	Bifurcated Paths
1	$(+,+)_{C_{\infty h}}$	1	1	$C_{\infty h}$	No Bifurcation
	$(+,-)_{C_{\infty h}}$	1	−1	C_{∞}	C_{∞}
2	$(n,+)_{C_{\infty h}}$	R	I_2	C_{nh}	C_{nh} (∗)
	$(n,-)_{C_{\infty h}}$	R	$-I_2$	S_{2n}	S_{2n} (∗)
	$n = 1, 2, \ldots$				

(d) $D_{\infty} = \langle c(\varphi), \sigma_h\sigma_v \rangle$

Multiplicity	Irreducible	$T^{\mu}(g)$		Symmetry Groups	
M	Representation, μ	$c(\varphi)$	$\sigma_h\sigma_v$	G^{μ}	Bifurcated Paths
1	$(+)_{D_{\infty}}$	1	1	D_{∞}	No Bifurcation
	$(-)_{D_{\infty}}$	1	−1	C_{∞}	C_{∞}
2	$(n)_{D_{\infty}}$	R	S	C_n	D_n
	$n = 1, 2, \ldots$				

(e) $C_{\infty} = \langle c(\varphi) \rangle$

Multiplicity	Irreducible	$T^{\mu}(g)$	Symmetry Groups	
M	Representation, μ	$c(\varphi)$	G^{μ}	Bifurcated Paths
1	$(+)_{C_{\infty}}$	1	C_{∞}	No Bifurcation
2	$(n)_{C_{\infty}}$	R	C_n	C_n (∗)
	$n = 1, 2, \ldots$			

Table 13.3 Classification of critical points of systems equivariant to $\mathrm{D}_{n\mathrm{h}}$, $\mathrm{D}_{n\mathrm{d}}$, or $\mathrm{C}_{n\mathrm{v}}$. $\widehat{n} = n/\mathrm{gcd}(n,j)$; $\widehat{j} = j/\mathrm{gcd}(n,j)$

(a) $\mathrm{D}_{n\mathrm{h}} = \langle c(2\pi/n), \sigma_{\mathrm{v}}, \sigma_{\mathrm{h}} \rangle$

Multiplicity	Irreducible	$T^{\mu}(g)$			Symmetry Groups	
M	Representation, μ	$c(2\pi/n)$	σ_{v}	σ_{h}	G^{μ}	Bifurcated Paths
1	$(+,+,+)_{\mathrm{D}_{n\mathrm{h}}}$	1	1	1	$\mathrm{D}_{n\mathrm{h}}$	No Bifurcation
	$(+,+,-)_{\mathrm{D}_{n\mathrm{h}}}$	1	1	−1	$\mathrm{C}_{n\mathrm{v}}$	$\mathrm{C}_{n\mathrm{v}}$
	$(+,-,+)_{\mathrm{D}_{n\mathrm{h}}}$	1	−1	1	$\mathrm{C}_{n\mathrm{h}}$	$\mathrm{C}_{n\mathrm{h}}$
	$(+,-,-)_{\mathrm{D}_{n\mathrm{h}}}$	1	−1	−1	D_{n}	D_{n}
1	$(-,+,+)_{\mathrm{D}_{n\mathrm{h}}}$	−1	1	1	$\mathrm{D}_{(n/2)\mathrm{h}}$	$\mathrm{D}_{(n/2)\mathrm{h}}$
(n: even)	$(-,+,-)_{\mathrm{D}_{n\mathrm{h}}}$	−1	1	−1	$\mathrm{D}_{(n/2)\mathrm{d}}$	$\mathrm{D}_{(n/2)\mathrm{d}}$
	$(-,-,+)_{\mathrm{D}_{n\mathrm{h}}}$	−1	−1	1	$\mathrm{D}_{(n/2)\mathrm{h}}$	$\mathrm{D}_{(n/2)\mathrm{h}}$
	$(-,-,-)_{\mathrm{D}_{n\mathrm{h}}}$	−1	−1	−1	$\mathrm{D}_{(n/2)\mathrm{d}}$	$\mathrm{D}_{(n/2)\mathrm{d}}$
2	$(j,+)_{\mathrm{D}_{n\mathrm{h}}}$	Q^2	S	I_2	$\mathrm{C}_{(n/\widehat{n})\mathrm{h}}$	$\mathrm{D}_{(n/\widehat{n})\mathrm{h}}$
	$(j,-)_{\mathrm{D}_{n\mathrm{h}}}$	Q^2	S	$-I_2$	$\mathrm{S}_{2n/\widehat{n}}$ ($\widehat{n}$: even)	$\mathrm{D}_{(n/\widehat{n})\mathrm{d}}$
	$1 \leq j \leq \lfloor (n-1)/2 \rfloor$				$\mathrm{C}_{n/\widehat{n}}$ ($\widehat{n}$: odd)	$\mathrm{C}_{(n/\widehat{n})\mathrm{v}}$ or $\mathrm{D}_{n/\widehat{n}}$

(b) $\mathrm{D}_{n\mathrm{d}} = \langle \sigma_{\mathrm{h}} c(\pi/n), \sigma_{\mathrm{h}}\sigma_{\mathrm{v}} \rangle$

Multiplicity	Irreducible	$T^{\mu}(g)$		Symmetry Groups	
M	Representation, μ	$\sigma_{\mathrm{h}} c(\pi/n)$	$\sigma_{\mathrm{h}}\sigma_{\mathrm{v}}$	G^{μ}	Bifurcated Paths
1	$(+,+)_{\mathrm{D}_{n\mathrm{d}}}$	1	1	$\mathrm{D}_{n\mathrm{d}}$	No Bifurcation
	$(+,-)_{\mathrm{D}_{n\mathrm{d}}}$	1	−1	S_{2n}	S_{2n}
	$(-,+)_{\mathrm{D}_{n\mathrm{d}}}$	−1	1	D_{n}	D_{n}
	$(-,-)_{\mathrm{D}_{n\mathrm{d}}}$	−1	−1	$\mathrm{C}_{n\mathrm{v}}$	$\mathrm{C}_{n\mathrm{v}}$
2	$(j)_{\mathrm{D}_{n\mathrm{d}}}$	$-Q$	$-S$	$\mathrm{S}_{2n/\widehat{n}}$ ($\widehat{j}+\widehat{n}$: even)	$\mathrm{D}_{(n/\widehat{n})\mathrm{d}}$
	$1 \leq j \leq n-1$			$\mathrm{C}_{n/\widehat{n}}$ ($\widehat{j}+\widehat{n}$: odd)	$\mathrm{C}_{(n/\widehat{n})\mathrm{v}}$ or $\mathrm{D}_{n/\widehat{n}}$

(c) $\mathrm{C}_{n\mathrm{v}} = \langle c(2\pi/n), \sigma_{\mathrm{v}} \rangle$

Multiplicity	Irreducible	$T^{\mu}(g)$		Symmetry Groups	
M	Representation, μ	$c(2\pi/n)$	σ_{v}	G^{μ}	Bifurcated Paths
1	$(+,+)_{\mathrm{C}_{n\mathrm{v}}}$	1	1	$\mathrm{C}_{n\mathrm{v}}$	No Bifurcation
	$(+,-)_{\mathrm{C}_{n\mathrm{v}}}$	1	−1	C_{n}	C_{n}
1	$(-,+)_{\mathrm{C}_{n\mathrm{v}}}$	−1	1	$\mathrm{C}_{(n/2)\mathrm{v}}$	$\mathrm{C}_{(n/2)\mathrm{v}}$
(n: even)	$(-,-)_{\mathrm{C}_{n\mathrm{v}}}$	−1	−1	$\mathrm{C}_{(n/2)\mathrm{v}}$	$\mathrm{C}_{(n/2)\mathrm{v}}$
2	$(j)_{\mathrm{C}_{n\mathrm{v}}}$	Q^2	S	$\mathrm{C}_{(n/\widehat{n})}$	$\mathrm{C}_{(n/\widehat{n})\mathrm{v}}$
	$1 \leq j \leq \lfloor (n-1)/2 \rfloor$				

Table 13.4 Classification of critical points of systems equivariant to C_{nh}, D_n, S_{2n}, or C_n. $\widehat{n} = n/\gcd(n,j)$; $\widehat{j} = j/\gcd(n,j)$; $(*)$ denotes that the associated bifurcated path is nonexistent for a nonreciprocal system

(a) $C_{nh} = \langle c(2\pi/n), \sigma_h \rangle$

Multiplicity	Irreducible	$T^\mu(g)$		Symmetry Groups	
M	Representation, μ	$c(2\pi/n)$	σ_h	G^μ	Bifurcated Paths
1	$(+,+)_{C_{nh}}$	1	1	C_{nh}	No Bifurcation
	$(+,-)_{C_{nh}}$	1	-1	C_n	C_n
1	$(-,+)_{C_{nh}}$	-1	1	$C_{(n/2)h}$	$C_{(n/2)h}$
(n: even)	$(-,-)_{C_{nh}}$	-1	-1	S_n	S_n
2	$(j,+)_{C_{nh}}$	Q^2	I_2	$C_{(n/\widehat{n})h}$	$C_{(n/\widehat{n})h}$ $(*)$
	$(j,-)_{C_{nh}}$	Q^2	$-I_2$	$S_{2n/\widehat{n}}$ ($\widehat{n}$: even)	$S_{2n/\widehat{n}}$ $(*)$
				$C_{n/\widehat{n}}$ ($\widehat{n}$: odd)	$C_{n/\widehat{n}}$ $(*)$

(b) $D_n = \langle c(2\pi/n), \sigma_h\sigma_v \rangle$

Multiplicity	Irreducible	$T^\mu(g)$		Symmetry Groups	
M	Representation, μ	$c(2\pi/n)$	$\sigma_h\sigma_v$	G^μ	Bifurcated Paths
1	$(+,+)_{D_n}$	1	1	D_n	No Bifurcation
	$(+,-)_{D_n}$	1	-1	C_n	C_n
1	$(-,+)_{D_n}$	-1	1	$D_{n/2}$	$D_{n/2}$
(n: even)	$(-,-)_{D_n}$	-1	-1	$D_{n/2}$	$D_{n/2}$
2	$(j)_{D_n}$	Q^2	S	$C_{n/\widehat{n}}$	$D_{n/\widehat{n}}$

(c) $S_{2n} = \langle \sigma_h c(\pi/n) \rangle$

Multiplicity	Irreducible	$T^\mu(g)$	Symmetry Groups	
M	Representation, μ	$\sigma_h c(2\pi/n)$	G^μ	Bifurcated Paths
1	$(+)_{S_{2n}}$	1	S_{2n}	No Bifurcation
	$(-)_{S_{2n}}$	-1	C_n	C_n
2	$(j)_{S_{2n}}$	$-Q$	$S_{2n/\widehat{n}}$ ($\widehat{j}+\widehat{n}$: even)	$S_{2n/\widehat{n}}$ $(*)$
			$C_{n/\widehat{n}}$ ($\widehat{j}+\widehat{n}$: odd)	$C_{n/\widehat{n}}$ $(*)$

(d) $C_n = \langle c(2\pi/n) \rangle$

Multiplicity	Irreducible	$T^\mu(g)$	Symmetry Groups	
M	Representation, μ	$c(2\pi/n)$	G^μ	Bifurcated Paths
1	$(+)_{C_n}$	1	C_n	No Bifurcation
1 (n: even)	$(-)_{C_n}$	-1	$C_{n/2}$	$C_{n/2}$
2	$(j)_{C_n}$	Q^2	$C_{n/\widehat{n}}$	$C_{n/\widehat{n}}$ $(*)$

13.4.1 Bifurcation of $\mathrm{D}_{n\mathrm{h}}$-Equivariant System

The bifurcation of a $\mathrm{D}_{n\mathrm{h}}$-equivariant system is investigated.[11] Since $\mathrm{D}_{n\mathrm{h}} = \mathrm{D}_n \times \langle\sigma_\mathrm{h}\rangle$, the absolutely irreducible representations of $\mathrm{D}_{n\mathrm{h}}$ are obtained as the tensor products of those of D_n and $\langle\sigma_\mathrm{h}\rangle \simeq \mathrm{C_i}$. It turns out that these absolutely irreducible representations of $\mathrm{D}_{n\mathrm{h}}$ coincide with the real irreducible representations, which are one-dimensional or two-dimensional.

Denote by $(\nu_1,\nu_2,\nu_3)_{\mathrm{D}_{n\mathrm{h}}}$ the one-dimensional irreducible representation equivalent to the tensor product $(\nu_1,\nu_2)_{\mathrm{D}_n} \times (\nu_3)_{\langle\sigma_\mathrm{h}\rangle}$, where $\nu_1,\nu_2,\nu_3 \in \{+,-\}$. Also denote by $(j,\nu)_{\mathrm{D}_{n\mathrm{h}}}$ the two-dimensional irreducible representation equivalent to the tensor product $(j)_{\mathrm{D}_n} \times (\nu)_{\langle\sigma_\mathrm{h}\rangle}$, where $j = 1,2,\ldots,\lfloor(n-1)/2\rfloor$ and $\nu \in \{+,-\}$. Accordingly, we index the family of inequivalent irreducible representations of $\mathrm{D}_{n\mathrm{h}}$ over $\mathbb{R}$ by

$$R(\mathrm{D}_{n\mathrm{h}}) = \begin{cases} \{(\nu_1,\nu_2,+)_{\mathrm{D}_{n\mathrm{h}}} \mid \nu_1,\nu_2 = +,-\} & \\ \quad \cup \{(j,\nu)_{\mathrm{D}_{n\mathrm{h}}} \mid \nu = +,-;\ j = 1,\ldots,(n-1)/2\} & \text{if } n \text{ is odd,} \\ \{(\nu_1,\nu_2,\nu_3)_{\mathrm{D}_{n\mathrm{h}}} \mid \nu_1,\nu_2,\nu_3 = +,-\} & \\ \quad \cup \{(j,\nu)_{\mathrm{D}_{n\mathrm{h}}} \mid \nu = +,-;\ j = 1,\ldots,(n-2)/2\} & \text{if } n \text{ is even.} \end{cases} \tag{13.19}$$

The one-dimensional irreducible representations $(\nu_1,\nu_2,\nu_3)_{\mathrm{D}_{n\mathrm{h}}}$ and the two-dimensional irreducible representations $(j,\nu)_{\mathrm{D}_{n\mathrm{h}}}$ are defined by the representation matrices in Table 13.3(a).

Remark 13.3. Some conventional notations are used for the irreducible representations of $\mathrm{D}_{n\mathrm{h}}$ that are commonly employed in physics and chemistry (cf., Kettle, 1995 [116] and Kim, 1999 [117]). The correspondence of the present notation to a conventional notation is given below for $n = 4$. The conventional notation is also defined for the irreducible representations of the point group $\mathrm{D}_{n\mathrm{d}}$ treated in this appendix.

Present Notation	Conventional Notation
$(+,+,+)_{\mathrm{D_{4h}}}$	$\mathrm{A_{1g}}$
$(+,+,-)_{\mathrm{D_{4h}}}$	$\mathrm{A_{1u}}$
$(+,-,+)_{\mathrm{D_{4h}}}$	$\mathrm{A_{2g}}$
$(+,-,-)_{\mathrm{D_{4h}}}$	$\mathrm{A_{2u}}$
$(-,+,+)_{\mathrm{D_{4h}}}$	$\mathrm{B_{1g}}$
$(-,+,-)_{\mathrm{D_{4h}}}$	$\mathrm{B_{1u}}$
$(-,-,+)_{\mathrm{D_{4h}}}$	$\mathrm{B_{2g}}$
$(-,-,-)_{\mathrm{D_{4h}}}$	$\mathrm{B_{2u}}$
$(1,+)_{\mathrm{D_{4h}}}$	$\mathrm{E_u}$
$(1,-)_{\mathrm{D_{4h}}}$	$\mathrm{E_g}$

□

[11] We can treat a $\mathrm{D}_{\infty\mathrm{h}}$-equivariant system similarly by setting $n \to \infty$ in an appropriate way.

Simple Critical Point

For simple critical points, which are associated with one-dimensional irreducible representations, the analysis of the bifurcation equation is much the same as that for a D_n-equivariant system explained in §8.3.1 (see (8.20) in particular). A simple critical point is either a limit point or a pitchfork bifurcation point according to whether it is associated with the unit representation $(+,+,+)_{\mathrm{D}_{n\mathrm{h}}}$. The symmetry of the bifurcated path is represented by the subgroup G^{μ} in (7.22) for the associated one-dimensional irreducible representation $\mu = (\nu_1, \nu_2, \nu_3)_{\mathrm{D}_{n\mathrm{h}}}$. We have

$$
\begin{aligned}
G^{(+,+,+)_{\mathrm{D}_{n\mathrm{h}}}} &= \mathrm{D}_{n\mathrm{h}}, & G^{(+,+,-)_{\mathrm{D}_{n\mathrm{h}}}} &= \mathrm{C}_{n\mathrm{v}},\\
G^{(+,-,+)_{\mathrm{D}_{n\mathrm{h}}}} &= \mathrm{C}_{n\mathrm{h}}, & G^{(+,-,-)_{\mathrm{D}_{n\mathrm{h}}}} &= \mathrm{D}_{n},\\
G^{(-,+,+)_{\mathrm{D}_{n\mathrm{h}}}} &= \mathrm{D}_{(n/2)\mathrm{h}}, & G^{(-,+,-)_{\mathrm{D}_{n\mathrm{h}}}} &= \mathrm{D}_{(n/2)\mathrm{d}},\\
G^{(-,-,+)_{\mathrm{D}_{n\mathrm{h}}}} &= \mathrm{D}_{(n/2)\mathrm{h}}, & G^{(-,-,-)_{\mathrm{D}_{n\mathrm{h}}}} &= \mathrm{D}_{(n/2)\mathrm{d}}.
\end{aligned}
$$

Therefore, the group $\mathrm{D}_{n\mathrm{h}}$ is associated with the limit point, and the subgroups $\mathrm{C}_{n\mathrm{v}}$, $\mathrm{C}_{n\mathrm{h}}$, D_n, $\mathrm{D}_{(n/2)\mathrm{h}}$, and $\mathrm{D}_{(n/2)\mathrm{d}}$ are associated with simple pitchfork bifurcation points.

Remark 13.4. The group G^{μ} for $\mu = (-,+,-)_{\mathrm{D}_{n\mathrm{h}}}$, for example, can be determined as described below. The invariance to the actions σ_{v} and $\sigma_{\mathrm{h}} c(2\pi/n)$ is readily apparent in view of the definition of $T^{\mu}(g)$ in Table 13.3(a). The symmetry generated by these actions is expressed by the group:

$$G^{\mu} = \langle \sigma_{\mathrm{v}}, \sigma_{\mathrm{h}} c(2\pi/n) \rangle = \langle \sigma_{\mathrm{h}} c(2\pi/n) \sigma_{\mathrm{v}}, \sigma_{\mathrm{h}} c(2\pi/n) \rangle.$$

Making use of the indeterminacy of σ_{v} explained in Remark 13.1 in §13.2.2, we replace $c(2\pi/n)\sigma_{\mathrm{v}}$ by σ_{v} to obtain

$$G^{\mu} = \langle \sigma_{\mathrm{v}} \sigma_{\mathrm{h}}, \sigma_{\mathrm{h}} c(2\pi/n) \rangle = \mathrm{D}_{(n/2)\mathrm{d}},$$

where $\sigma_{\mathrm{h}} \sigma_{\mathrm{v}} = \sigma_{\mathrm{v}} \sigma_{\mathrm{h}}$ by (13.4). □

Double Critical Point

We now consider a double critical point and designate by $\mu = (j, \nu)_{\mathrm{D}_{n\mathrm{h}}}$ the associated two-dimensional irreducible representation of $\mathrm{D}_{n\mathrm{h}}$, where $1 \leq j \leq \lfloor (n-1)/2 \rfloor$ and $\nu \in \{+,-\}$. As in §8.5 and §8.6, we describe the bifurcation equation in complex coordinates $(z, \overline{z})$:

$$F(z, \overline{z}, \widetilde{f}) \approx \sum_{p=0} \sum_{q=0} A_{pq}(\widetilde{f}) z^p \overline{z}^q. \tag{13.20}$$

Recall from (8.43) that

$$A_{00}(0) = A_{10}(0) = A_{01}(0) = 0.$$

In §8.5, the action of $\mathrm{C}_{n\mathrm{v}}$ on $(z,\overline{z})$ is given by (8.49) with σ replaced by σ_{v}. On the other hand, we have an isomorphism $\mathrm{D}_{n\mathrm{h}} \simeq \mathrm{C}_{n\mathrm{v}} \times \langle \sigma_{\mathrm{h}} \rangle$. Consequently, the action of $\mathrm{D}_{n\mathrm{h}}$ is given by

$$\begin{aligned} &\sigma_{\mathrm{v}} \cdot z = \overline{z}, && \sigma_{\mathrm{v}} \cdot \overline{z} = z, \\ &c(2\pi/n) \cdot z = \omega z, && c(2\pi/n) \cdot \overline{z} = \overline{\omega}\,\overline{z}, \\ &\sigma_{\mathrm{h}} \cdot z = \alpha z, && \sigma_{\mathrm{h}} \cdot \overline{z} = \alpha \overline{z}, \end{aligned} \tag{13.21}$$

where

$$\omega = \exp(\mathrm{i}2\pi \widehat{j}/\widehat{n}), \qquad \alpha = \begin{cases} +1 & (\nu = +), \\ -1 & (\nu = -). \end{cases}$$

By (13.21) the equivariance of F to $\mathrm{D}_{n\mathrm{h}}$ is rewritten (cf., (8.53) and (8.54)) as

$$\begin{aligned} &A_{pq}(\widetilde{f}) \in \mathbb{R}, && p,q = 0,1,\ldots, \\ &A_{pq}(\widetilde{f}) = 0 && \text{unless } \; p-q-1 = m\check{n}, \qquad m \in \mathbb{Z}, \end{aligned}$$

where

$$\check{n} = \begin{cases} 2\widehat{n} & \text{if } \widehat{n} \text{ is odd and } \nu = -, \\ \widehat{n} & \text{if } \widehat{n} \text{ is even.} \end{cases}$$

Therefore, the bifurcation equation for a double point of a $\mathrm{D}_{n\mathrm{h}}$-equivariant system is fundamentally the same as that for a $\mathrm{D}_{\check{n}}$-equivariant system treated in Chapter 8.

The symmetry of the bifurcated solutions can be determined as follows. According to the argument in §8.6.1 it suffices to consider the symmetry

$$\Sigma(z) = \{ g \in \mathrm{D}_{n\mathrm{h}} \mid g \cdot z = z \}$$

for the solutions $z = z_0$, z_1 with $\arg z_0 = 0$ and $\arg z_1 = \pi/\check{n}$.

First, for $\mu = (j,+)_{\mathrm{D}_{n\mathrm{h}}}$, we have

$$\begin{aligned} \Sigma(z_0) &= \langle \sigma_{\mathrm{v}}, c(2\pi/(n/\widehat{n})), \sigma_{\mathrm{h}} \rangle = \mathrm{D}_{(n/\widehat{n})\mathrm{h}}, \\ \Sigma(z_1) &= \langle c(2\pi/n)^k \sigma_{\mathrm{v}}, c(2\pi/(n/\widehat{n})), \sigma_{\mathrm{h}} \rangle = \mathrm{D}_{(n/\widehat{n})\mathrm{h}}, \end{aligned}$$

where k is an appropriate integer and $c(2\pi/n)^k \sigma_{\mathrm{v}}$ is identified with σ_{v} in the expression of $\Sigma(z_1)$.

Next, we consider the case of $\mu = (j,-)_{\mathrm{D}_{n\mathrm{h}}}$. If $\widehat{n}$ is even, then we have

$$\begin{aligned} \Sigma(z_0) &= \langle \sigma_{\mathrm{v}}, c(2\pi/n)^{\widehat{n}}, \sigma_{\mathrm{h}} c(2\pi/n)^{\widehat{n}/2} \rangle \\ &= \langle c(\pi/(n/\widehat{n}))\sigma_{\mathrm{v}}\sigma_{\mathrm{h}}, \sigma_{\mathrm{h}} c(\pi/(n/\widehat{n})) \rangle. \end{aligned}$$

Making use of the indeterminacy of σ_{v} explained in Remark 13.1 in §13.2.2, we can replace $c(\pi/(n/\widehat{n}))\sigma_{\mathrm{v}}$ by σ_{v} in this expression to obtain

$$\Sigma(z_0) = \mathrm{D}_{(n/\widehat{n})\mathrm{d}}.$$

Similarly, we have $\Sigma(z_1) = \mathrm{D}_{(n/\widehat{n})\mathrm{d}}$. If $\widehat{n}$ is odd, then we have

$$\Sigma(z_0) = \langle \sigma_{\rm v}, c(2\pi/n)^{\widehat{n}} \rangle = \langle \sigma_{\rm v}, c(2\pi/(n/\widehat{n})) \rangle = {\rm C}_{(n/\widehat{n}){\rm v}},$$
$$\Sigma(z_1) = \langle c(2\pi/n)^k \sigma_{\rm v}\sigma_{\rm h}, c(2\pi/n)^{\widehat{n}} \rangle = {\rm D}_{(n/\widehat{n})}$$

for some k under the replacement of $c(2\pi/n)^k\sigma_{\rm v}$ by $\sigma_{\rm v}$.

To sum up, the symmetries of the bifurcated paths for $(j,+)_{{\rm D}_{n{\rm h}}}$ are labeled by ${\rm D}_{(n/\widehat{n}){\rm h}}$ and for $(j,-)_{{\rm D}_{n{\rm h}}}$ by

$$\begin{cases} {\rm D}_{(n/\widehat{n}){\rm d}} & \text{if } \widehat{n} \text{ is even,} \\ {\rm C}_{(n/\widehat{n}){\rm v}} \text{ or } {\rm D}_{n/\widehat{n}} & \text{if } \widehat{n} \text{ is odd.} \end{cases}$$

Remark 13.5. When n is odd, we have an isomorphism ${\rm D}_{n{\rm h}} \simeq {\rm D}_{2n} \simeq {\rm C}_{(2n){\rm v}}$ by (13.13) and (13.15), whereas the symmetry of the bifurcation solutions of a ${\rm C}_{(2n){\rm v}}$-equivariant system has been identified in (8.66). The symmetries $\Sigma(z_0)$ and $\Sigma(z_1)$ above can also be determined as the subgroups of ${\rm D}_{n{\rm h}}$ that correspond to the subgroups of ${\rm C}_{(2n){\rm v}}$ given by (8.66). See §13.4.2 for this procedure. □

13.4.2 Bifurcation of ${\rm D}_{n{\rm d}}$*-Equivariant System*

Since ${\rm D}_{n{\rm d}} \simeq {\rm D}_{2n}$ by (13.14), the bifurcation behavior of a ${\rm D}_{n{\rm d}}$-equivariant system can be elucidated through a translation of the corresponding results summarized in Table 8.1 in §8.3.

Let

$${\rm D}_N = \langle r, s \rangle = \{e, r, \ldots, r^{N-1}, s, sr, \ldots, sr^{N-1}\}$$

denote the (abstract) dihedral group introduced in (8.2) with the defining relations

$$r^i r^j = r^{i+j}, \qquad r^N = s^2 = (sr)^2 = e.$$

We put $N = 2n$. The isomorphism between ${\rm D}_{n{\rm d}}$ and ${\rm D}_N$ is established by the correspondences $\sigma_{\rm h} c(\pi/n) \leftrightarrow r$ and $\sigma_{\rm v}\sigma_{\rm h} \leftrightarrow s$.

The group ${\rm D}_{n{\rm d}}$ has four one-dimensional irreducible representations: $(+,+)_{{\rm D}_{n{\rm d}}}$, $(+,-)_{{\rm D}_{n{\rm d}}}$, $(-,+)_{{\rm D}_{n{\rm d}}}$, and $(-,-)_{{\rm D}_{n{\rm d}}}$; and $n-1$ two-dimensional irreducible representations: $(j)_{{\rm D}_{n{\rm d}}}$ for $j = 1,\ldots,n-1$; see Table 13.3(b) for their definitions. That is,

$$R({\rm D}_{n{\rm d}}) = \{(\nu_1,\nu_2)_{{\rm D}_{n{\rm d}}} \mid \nu_1,\nu_2 = +,-\} \cup \{(j)_{{\rm D}_{n{\rm d}}} \mid j = 1,\ldots,n-1\}. \tag{13.22}$$

The correspondence of the irreducible representations of ${\rm D}_{n{\rm d}}$ and ${\rm D}_N$ is given by

$$(\nu_1,\nu_2)_{{\rm D}_{n{\rm d}}} \mapsto (\nu_1,\nu_2)_{{\rm D}_N} \qquad (\nu_1,\nu_2 = +,-), \tag{13.23}$$
$$(j)_{{\rm D}_{n{\rm d}}} \mapsto (k_j)_{{\rm D}_N} \qquad (j = 1,\ldots,n-1), \tag{13.24}$$

where $k_j = n - j$.

Simple Critical Point

From the results for a D_N-equivariant system and the correspondence (13.23), it is apparent that a simple critical point is either a limit point or a pitchfork bifurcation point according to whether it is associated with the unit representation $(+,+)_{\mathrm{D}_{n\mathrm{d}}}$. The symmetry of the (bifurcation) path is represented by the subgroup G^{μ} in (7.22) for the associated one-dimensional irreducible representation $\mu = (\nu_1, \nu_2)_{\mathrm{D}_{n\mathrm{d}}}$. We have

$$G^{(+,+)_{\mathrm{D}_{n\mathrm{d}}}} = \mathrm{D}_{n\mathrm{d}}, \qquad G^{(+,-)_{\mathrm{D}_{n\mathrm{d}}}} = \mathrm{S}_{2n}, \qquad G^{(-,+)_{\mathrm{D}_{n\mathrm{d}}}} = \mathrm{D}_{n}, \qquad G^{(-,-)_{\mathrm{D}_{n\mathrm{d}}}} = \mathrm{C}_{n\mathrm{v}}.$$

Consequently, the group $\mathrm{D}_{n\mathrm{d}}$ is associated with the limit point; and the subgroups S_{2n}, D_n, and $\mathrm{C}_{n\mathrm{v}}$ are associated with simple pitchfork bifurcation points.

Double Critical Point

Recall from §8.3.1 that the bifurcation behavior of a D_N-equivariant system at a double critical point associated with the two-dimensional irreducible representation $(k)_{\mathrm{D}_N}$ is characterized by the index $\widehat{N} = N/\gcd(N,k)$. In particular, the symmetry of the bifurcation solutions is given by $\langle s, r^{\widehat{N}} \rangle$ or $\langle sr, r^{\widehat{N}} \rangle$. We are to translate this into a $\mathrm{D}_{n\mathrm{d}}$-equivariant system through the isomorphism between D_N and $\mathrm{D}_{n\mathrm{d}}$.

First, note that the index $\widehat{N} = N/\gcd(N,k)$ with $k = k_j = n - j$ is expressed by $\widehat{n} = n/\gcd(n, j)$ as

$$\widehat{N} = \begin{cases} 2\widehat{n} & \text{if } \widehat{n} + \widehat{j} \text{ is odd,} \\ \widehat{n} & \text{if } \widehat{n} + \widehat{j} \text{ is even.} \end{cases}$$

When $\widehat{n} + \widehat{j}$ is odd, $\langle s, r^{\widehat{N}} \rangle$ and $\langle sr, r^{\widehat{N}} \rangle$ correspond, respectively, to $\mathrm{D}_{n/\widehat{n}}$ and $\mathrm{C}_{(n/\widehat{n})\mathrm{v}}$. When $\widehat{n} + \widehat{j}$ is even, both $\widehat{n}$ and $\widehat{j}$ are odd, and both subgroups $\langle s, r^{\widehat{N}} \rangle$ and $\langle sr, r^{\widehat{N}} \rangle$ correspond to $\mathrm{D}_{(n/\widehat{n})\mathrm{d}}$. To sum up, the symmetries of the bifurcated paths are labeled by

$$\begin{cases} \mathrm{C}_{(n/\widehat{n})\mathrm{v}} \text{ or } \mathrm{D}_{n/\widehat{n}} & \text{if } \widehat{j} + \widehat{n} \text{ is odd,} \\ \mathrm{D}_{(n/\widehat{n})\mathrm{d}} & \text{if } \widehat{j} + \widehat{n} \text{ is even.} \end{cases}$$

Problems

13-1 Draw hierarchies of subgroups for $\mathrm{D}_{6\mathrm{h}}$ and $\mathrm{D}_{6\mathrm{d}}$ (cf., Fig. 13.7 in §13.2.2).

13-2 Show isomorphisms (13.10)–(13.17).

13-3 Following Remark 13.4 in §13.4.1, determine the subgroups G^{μ} associated with the one-dimensional irreducible representations $\mu = (\nu_1, \nu_2, \nu_3)_{\mathrm{D}_{n\mathrm{h}}}$ of $\mathrm{D}_{n\mathrm{h}}$.

13-4 Rewrite (13.21) to obtain the action of $\mathrm{D}_{n\mathrm{h}}$ on (w_1, w_2) defined by $z = w_1 + \mathrm{i}w_2$. Answer:

$$\sigma_{\rm v}: \begin{pmatrix} w_1 \\ w_2 \end{pmatrix} \mapsto \begin{pmatrix} w_2 \\ w_1 \end{pmatrix},$$

$$\sigma_{\rm h}: \begin{pmatrix} w_1 \\ w_2 \end{pmatrix} \mapsto \begin{pmatrix} \alpha w_1 \\ \alpha w_2 \end{pmatrix},$$

$$c(2\pi/n): \begin{pmatrix} w_1 \\ w_2 \end{pmatrix} \mapsto \begin{pmatrix} \cos(2\pi j/n) & -\sin(2\pi j/n) \\ \sin(2\pi j/n) & \cos(2\pi j/n) \end{pmatrix} \begin{pmatrix} w_1 \\ w_2 \end{pmatrix}.$$

13-5 Determine the symmetry of bifurcated paths for an S_{2n}-equivariant system by carrying out an analysis similar to that in §13.4.

Summary

- Initial and deformed configurations of cylindrical specimens have been labeled by the group $D_{\infty h}$ and its subgroups.
- The bifurcation rule of a $D_{\infty h}$-equivariant system has been derived.
- Occurrence of bifurcation in cylindrical soil specimens has been detected by the asymptotic and group-theoretic approaches.
- Recursive bifurcation and mode switching of the specimens have been observed.

Chapter 14
Echelon-Mode Formation

14.1 Introduction

The deformation pattern change of cylindrical soil specimens at the later stage has been discussed in Chapter 13. In this chapter, we move on to investigate pattern formation at the earlier stage with high spatial frequencies.[1]

Figure 14.1 displays: (a) a diamond pattern, (b) an oblique stripe pattern, and (c) an echelon mode of cylindrical sand and soil specimens observed at the earlier stage. In particular, the echelon mode[2] denotes a series of oblique parallel short wrinkles arranged with similar intervals, as depicted at the right of Fig. 14.1(c). Such patterns appear ubiquitously for materials with diverse shapes and deformation sizes, as depicted in Figs. 14.2 and 14.3 for rock and steel.

As we have studied up to this chapter, systems with the same symmetry inevitably follow the same rule of bifurcation, although the bifurcation mode that actually takes place varies with cases. This similarity, despite some variation, alludes to the assumption, or arguably the conclusion, that those materials are all governed by the same rule of symmetry-breaking bifurcation that is untangled by finding a pertinent symmetry group.

Diverse patterns in flows have been observed, for example, in the Couette–Taylor flow and in the Bénard convection.[3] Flow patterns are modeled as bifurcation phe-

---

[1] The contents of this chapter are mostly based on Ikeda, Murota, and Nakano, 1994 [95]; Murota, Ikeda, and Terada, 1999 [145]; and Ikeda et al., 2008 [104].

[2] The echelon mode is found for various materials: soils (e.g., Ikeda, Murota, and Nakano, 1994 [95]), rocks (e.g., Davis, 1984 [35]), and metals (e.g., Bai and Dodd, 1992 [10] and Poirier, 1985 [167]). The cross-checker pattern, which is interpreted as an echelon symmetry in this chapter, is found in metals (e.g., Voskamp and Hoolox, 1998 [205]) and in the zebra patterns on the ocean floors (e.g., Nicolas, 1995 [149]). A self-similar pattern model was introduced by Archambault et al., 1993 [6].

[3] In the *Couette–Taylor flow*, flows between two coaxial cylinders rotating with different angular velocities are investigated (cf., Taylor, 1923 [188]). For the Bénard convection, see, for example, Bénard, 1900 [13]; Chandrasekhar, 1961 [23]; and Koschmieder, 1966 [123], 1993 [125]. See also, for example, Drazin and Reid, 1981 [43] for hydrodynamic stability.

K. Ikeda and K. Murota, *Imperfect Bifurcation in Structures and Materials*, Applied Mathematical Sciences 149, DOI 10.1007/978-1-4419-7296-5_14,

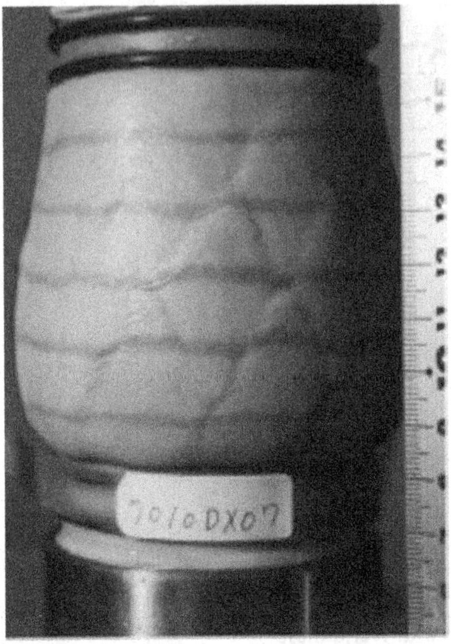

(a) Diamond pattern (sand)

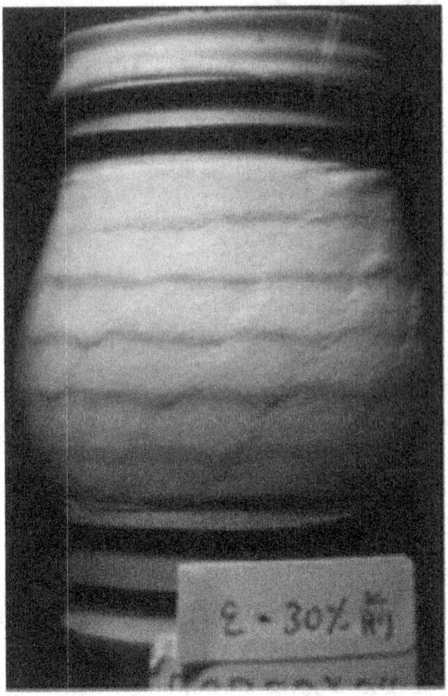

(b) Oblique stripe pattern (sand)

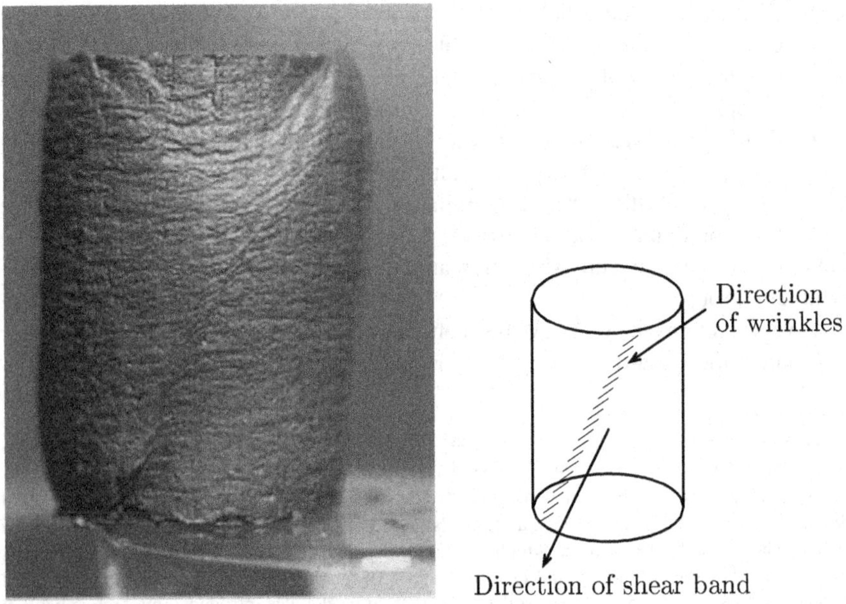

(c) Echelon mode (soil)

Fig. 14.1 Photographs of deformation patterns of cylindrical sand and soil specimens in triaxial compression tests. The horizontal black layers of sands in (a) and (b) are employed to display deformation from initial states. (c) is by Noda, 1994 [150].

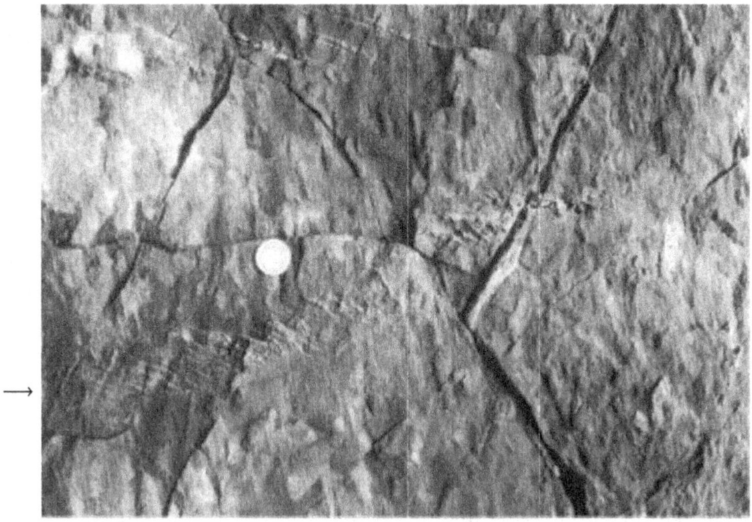

Fig. 14.2 Photograph of an echelon mode of a rock taken by M. Osada. The echelon cracks run diagonally from the arrowed part.

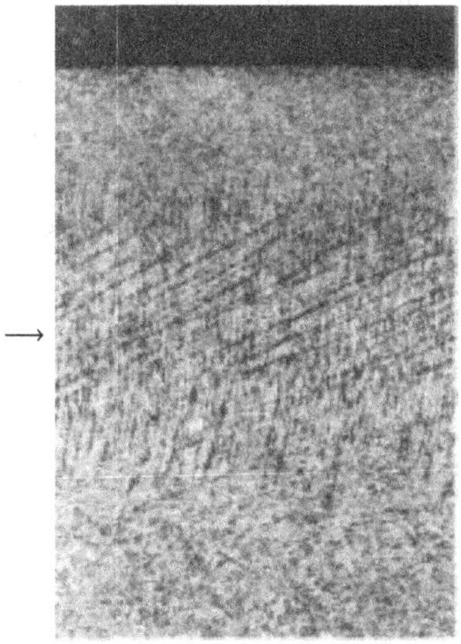

Fig. 14.3 Photograph of a cross-checker pattern of a steel specimen taken by N. Oguma: an optical micrograph of structural changes under the rolling track of a bearing steel specimen (maximum contact stress: 4.3 GPa and number of stress cycles: 90 million cycles). The cross-checkerlike cracks appear at the arrowed part.

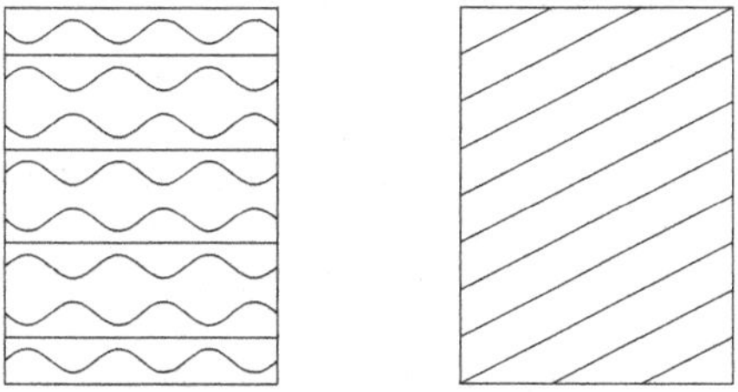

(a) Ribbon (b) Spiral

Fig. 14.4 Patterns in flows.

nomena of symmetric systems.[4] In particular, we can find an analogy between the pattern formation of the cylindrical soil specimens and that of the Couette–Taylor problem. Among diverse flow patterns acknowledged so far, the so-called ribbon in Fig. 14.4(a) resembles the diamond pattern and the spiral in Fig. 14.4(b) resembles the stripe pattern.

In the Couette–Taylor problem, it is commonplace and effective to model the symmetry of a cylindrical domain by group $\mathrm{SO}(2)\times\mathrm{O}(2)$, rather than by group $\mathrm{SO}(2)\times\mathbb{Z}_2$ that is geometrically more natural.[5] The use of the translational symmetry is the so-called *infinite-periodic-cylinder approximation*, which assumes that the cylindrical domain is to be periodically extended in the axial direction. The enlarged symmetry is more appropriate, in that the patterns in flows may be understood as a consequence of the local uniformity with only a secondary effect from the boundary conditions.

In the description of the symmetry of a uniform cylindrical domain, we consider the reflection symmetry with respect to a vertical plane containing the axis of a cylindrical domain. Such reflection is absent in the Couette–Taylor flow because it reverses the direction of the flow. To be more specific, we model the symmetry of a cylindrical domain by $\mathrm{O}(2)\times\mathrm{O}(2)$. In addition, we model the symmetry of a rectangular domain, such as steel in Fig. 14.3, also by $\mathrm{O}(2)\times\mathrm{O}(2)$; we employ the *infinite-periodic-domain approximation*, which assumes that the rectangular domain in the xy-plane has periodic boundaries and therefore is to be periodically extended in the x-direction and the y-direction.

By investigating the bifurcation behavior of an $\mathrm{O}(2)\times\mathrm{O}(2)$-equivariant system of equations, we show:

- A direct bifurcation point generically has multiplicity one, two, or four.
- Possible solutions branching from a bifurcation point of multiplicity four consist of the classical diamond pattern solution and a pair of oblique stripe pattern solutions.
- An echelon mode can appear as a secondary bifurcation from the oblique stripe pattern solutions.

Mathematically, the symmetry group representing the echelon mode in our modeling turns out to be a finite group, although $\mathrm{SO}(2)\times\mathrm{O}(2)$ and $\mathrm{O}(2)\times\mathrm{O}(2)$ are continuous groups. Furthermore, it is emphasized that the seemingly more "natural" $\mathrm{O}(2)\times\mathbb{Z}_2$-equivariance does not yield oblique stripe patterns, which produce echelon modes through secondary bifurcation.

The first half of this chapter is devoted to the mathematical analysis of the $\mathrm{O}(2)\times\mathrm{O}(2)$-equivariant system.

---

[4] See, for example, Schaeffer, 1980 [179]; Iooss, 1986 [106]; Bakker, 1991 [11]; Crawford and Knobloch, 1991 [32]; Chossat and Iooss, 1994 [27]; Chossat, 1994 [26]; Iooss and Adelmeyer, 1998 [107]; Moehlis and Knobloch, 2000 [140]; and Rabinovich, Ezersky, and Weidman, 2000 [170].

[5] The group $\mathrm{SO}(2)\times\mathrm{O}(2)$ acts as rotations about the axis of the cylindrical domain, translations in the axial direction and upside-down reflection, whereas group $\mathrm{SO}(2)\times\mathbb{Z}_2$ lacks the translational symmetry.

- Groups that express underlying symmetries of a cylindrical domain are introduced in §14.2.
- Subgroups that describe patterns with high spatial frequencies are presented in §14.3.
- A hierarchy of subgroups, which characterizes recursive bifurcation leading to an echelon mode, is introduced in §14.4.

The second half of this chapter is devoted to applications.

- Wrinkles, shear bands, and echelon modes observed on cylindrical soil specimens are investigated in §14.5.
- Diamond patterns, stripe patterns, and echelon modes are displayed by the numerical bifurcation analysis of an $O(2) \times O(2)$-symmetric rectangular plate in §14.6.
- Image simulations are conducted on steel and kaolin specimens to demonstrate the emergence of echelon patterns in §14.7.
- The emergence of a diamond pattern, prior to shear band formation, of a sand specimen is detected by experimental and numerical studies in §14.8.

Bifurcation rule for an $O(2) \times O(2)$-equivariant system is derived in §14.9, the appendix of this chapter.

14.2 Underlying Symmetries

We are primarily concerned with the deformation patterns of uniform (homogeneous and isotropic) material in a cylindrical domain (see Fig. 14.5):

$$\Omega = \{(x, y, z) \in \mathbb{R}^3 \mid x^2 + y^2 \leq R^2,\ -L/2 \leq z \leq L/2\},$$

where $R = R(f)$ is the radius and $L = L(f)$ is the length of the cylindrical domain, each of which might vary with a parameter f.

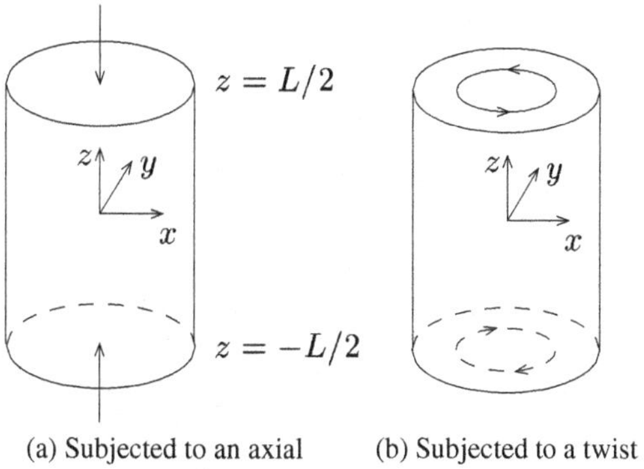

(a) Subjected to an axial compression (b) Subjected to a twist

Fig. 14.5 Cylindrical domain.

Geometrical Symmetry

The geometrical symmetry of the cylindrical domain is labeled by the group

$$\mathrm{D}_{\infty\mathrm{h}} = \langle \sigma_{\mathrm{v}}, \sigma_{\mathrm{h}}, c(\varphi) \rangle$$

of (13.3) in §13.2, where σ_{v} is the reflection with respect to a vertical plane, σ_{h} is the reflection with respect to a horizontal plane, and $c(\varphi)$ $(0 \leq \varphi < 2\pi)$ denotes the counterclockwise rotation about the z-axis at an angle of φ:

$$c(\varphi) : \theta \mapsto \theta + \varphi, \qquad \text{where} \quad \theta = \tan^{-1}(y/x). \tag{14.1}$$

We choose σ_{v} to be the reflection σ_y with respect to the xz-plane, defined as

$$\sigma_y : y \mapsto -y \qquad (\text{i.e., } \theta \mapsto -\theta), \tag{14.2}$$

and σ_{h} to be the reflection σ_z with respect to the xy-plane, defined as

$$\sigma_z : z \mapsto -z. \tag{14.3}$$

Then the symmetry group of the cylindrical domain is given by

$$G_0 = \langle \sigma_y, \sigma_z, c(\varphi) \rangle = \langle \sigma_y, c(\varphi), \sigma_z \rangle = \mathrm{C}_{\infty\mathrm{v}} \times \mathrm{C}_{1\mathrm{h}} \simeq \mathrm{O}(2) \times \mathbb{Z}_2, \tag{14.4}$$

which is expressed as the direct product of the group

$$\mathrm{C}_{\infty\mathrm{v}} = \langle \sigma_y, c(\varphi) \rangle = \{c(\varphi), \sigma_y c(\varphi) \mid 0 \leq \varphi < 2\pi\} \simeq \mathrm{O}(2) \tag{14.5}$$

for the axisymmetry and the other group

$$\mathrm{C}_{1\mathrm{h}} = \langle \sigma_z \rangle = \{e, \sigma_z\} \simeq \mathbb{Z}_2$$

for the *reflection symmetry* with respect to the xy-plane.

When the loading is compatible with the geometrical symmetry of the cylindrical domain (e.g., as in Fig. 14.5(a)), the governing equations for the deformation are equivariant to the group G_0 defined in (14.4) above.[6] Thus we have identified the group that naturally expresses the geometrical symmetry.

Underlying Symmetry

It is not difficult to see that the equations with geometrical symmetry in (14.4) do not admit, via direct bifurcation, solutions for the oblique stripe pattern and the echelon mode. In view of an experimental fact that such pattern and mode appear in the middle of the cylindrical domain, we may reasonably presume that the formation of these patterns is primarily governed by the local uniformity of the material with

[6] This is the case, for example, with the triaxial compression test on a cylindrical soil specimen.

only a secondary effect of the boundary conditions. To better express the local uniformity, at the sacrifice of the consistency with the boundary conditions, we employ the *infinite-periodic-cylinder approximation*: the cylindrical domain is periodically extended in the axial direction. Precisely stated, the cylindrical domain is assumed to have periodic boundary conditions on the top and bottom surfaces at $z = \pm L/2$. With this approximation the system of governing equations acquires an additional symmetry of the z-directional *translation* $t(l)$ for any length l ($0 \le l < L$) defined by

$$t(l) \,:\, z \mapsto z + l. \tag{14.6}$$

Using a group

$$\widetilde{\mathrm{C}}_{\infty\mathrm{v}} = \langle \sigma_z, t(l) \rangle = \{t(l), \sigma_z t(l) \mid 0 \le l < L\} \simeq \mathrm{O}(2), \tag{14.7}$$

we can replace the symmetry group G_0 by an enlarged group G defined by

$$G = \mathrm{C}_{\infty\mathrm{v}} \times \widetilde{\mathrm{C}}_{\infty\mathrm{v}} = \langle c(\varphi), t(l), \sigma_y, \sigma_z \rangle \simeq \mathrm{O}(2) \times \mathrm{O}(2). \tag{14.8}$$

In abstract terms, group G expresses the symmetry of a torus, with additional symmetry of reflections.

In what follows we consider the bifurcation behavior of an $\mathrm{O}(2)\times\mathrm{O}(2)$-equivariant system. It is elucidated that the echelon-mode formation can be viewed as a successive symmetry-breaking bifurcation.

To simplify the notation, we put

$$\xi = \frac{\theta}{2\pi}, \qquad \tilde{\xi} = \frac{z}{L} + \frac{1}{2}. \tag{14.9}$$

Then the circumferential surface of the cylindrical domain is described by the coordinates $(\xi, \tilde{\xi})$ with $0 \le \xi, \tilde{\xi} < 1$. We also put

$$\psi = \frac{\varphi}{2\pi}, \qquad \tilde{\psi} = \frac{l}{L}, \qquad \sigma = \sigma_y, \qquad \tilde{\sigma} = \sigma_z \tag{14.10}$$

to obtain $0 \le \psi, \tilde{\psi} < 1$,

$$r(\psi) = c(2\pi\psi) = c(\varphi), \qquad \tilde{r}(\tilde{\psi}) = t(L\tilde{\psi}) = t(l), \tag{14.11}$$

and

$$\mathrm{C}_{\infty\mathrm{v}} = \langle \sigma, r(\psi) \rangle, \qquad \widetilde{\mathrm{C}}_{\infty\mathrm{v}} = \langle \tilde{\sigma}, \tilde{r}(\tilde{\psi}) \rangle. \tag{14.12}$$

Note the following defining relations

$$\begin{aligned}
&\sigma\sigma = e, && r(1) = e, && r(\psi)\sigma r(\psi)\sigma = e,\\
&\tilde{\sigma}\tilde{\sigma} = e, && \tilde{r}(1) = e, && \tilde{r}(\tilde{\psi})\tilde{\sigma}\tilde{r}(\tilde{\psi})\tilde{\sigma} = e,\\
&\sigma\tilde{\sigma} = \tilde{\sigma}\sigma, && && r(\psi)\tilde{r}(\tilde{\psi}) = \tilde{r}(\tilde{\psi})r(\psi),\\
&r(\psi)\tilde{\sigma} = \tilde{\sigma}r(\psi), && && \sigma\tilde{r}(\tilde{\psi}) = \tilde{r}(\tilde{\psi})\sigma,
\end{aligned} \tag{14.13}$$

where e is the identity element. We extend the notation $r(\psi)$ for $\psi \in \mathbb{R}$ by defining $r(\psi) = r(\psi_0)$ using the fractional part $\psi_0 = \psi - \lfloor \psi \rfloor$ $(0 \leq \psi_0 < 1)$ of ψ; similarly for $\tilde{r}(\tilde{\psi})$.

Remark 14.1. When the cylindrical material is subjected to a twist (shear) about the z-axis, as in Fig. 14.5(b), the geometrical symmetry group is

$$G_0' = \langle c(\varphi), \sigma_y \sigma_z \rangle \simeq \mathrm{O}(2),$$

in which $\sigma_y \sigma_z$ represents the half-rotation about the x-axis. The infinite-periodic-cylinder approximation can be implemented by replacing G_0' by

$$G' = \mathrm{D}_{\infty\infty} = \langle \sigma_y \sigma_z, c(\varphi), t(l) \rangle = \langle \sigma\tilde{\sigma}, r(\psi), \tilde{r}(\tilde{\psi}) \rangle, \tag{14.14}$$

which is a subgroup of $G = \mathrm{C}_{\infty\mathrm{v}} \times \widetilde{\mathrm{C}}_{\infty\mathrm{v}}$. □

14.3 Subgroups for Patterns with High Spatial Frequencies

We introduce deformation patterns of three kinds with high spatial frequencies—diamond pattern, oblique stripe pattern, and echelon mode—and subgroups of $G = \mathrm{C}_{\infty\mathrm{v}} \times \widetilde{\mathrm{C}}_{\infty\mathrm{v}}$ that describe their symmetry. These patterns on expansion plans of the circumferential surface of the cylindrical domain are depicted in Fig. 14.6 and their contour views that resemble real behaviors are portrayed in Fig. 14.7. In Fig. 14.6, the symbols (•) denote the points with discrete translational symmetry, and the dashed lines show the periodic presence of diamond-shaped blocks.

Diamond Pattern

The diamond patterns, shown in Fig. 14.6(a) and Fig. 14.7(b), are made up of a set of $n \times \tilde{n}$ identical diamond-shaped blocks arranged regularly; furthermore, each block has upside-down, bilateral, and half-rotation symmetries. Such patterns can be labeled by a finite group

$$\mathrm{DI}_{n\tilde{n}} = \left\langle \sigma, \tilde{\sigma}, r\left(\frac{1}{2n}\right)\tilde{r}\left(\frac{1}{2\tilde{n}}\right), r\left(\frac{1}{2n}\right)\tilde{r}\left(\frac{-1}{2\tilde{n}}\right) \right\rangle, \tag{14.15}$$

where the elements $r(1/(2n))\tilde{r}(\pm 1/(2\tilde{n}))$ denote oblique translations at a length of $1/(2n)$ in the ξ-direction and at a length of $\pm 1/(2\tilde{n})$ in the $\tilde{\xi}$-direction. These translations move a diamond block to others and, therefore, represent the spatial symmetries of the diamond pattern.

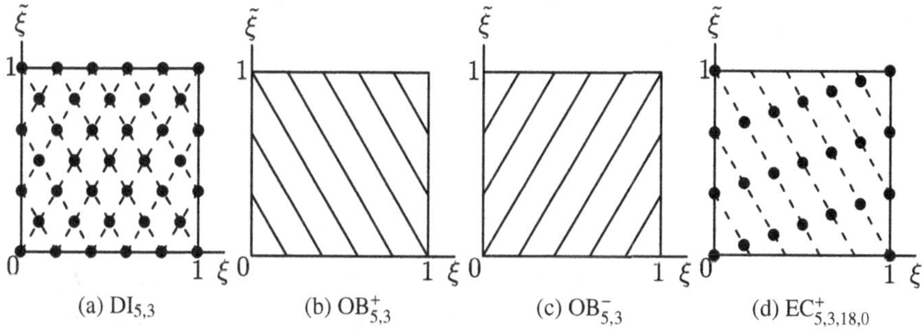

Fig. 14.6 Illustration of a series of patterns. •: point with discrete translational symmetry; ——: line with continuous translational symmetry.

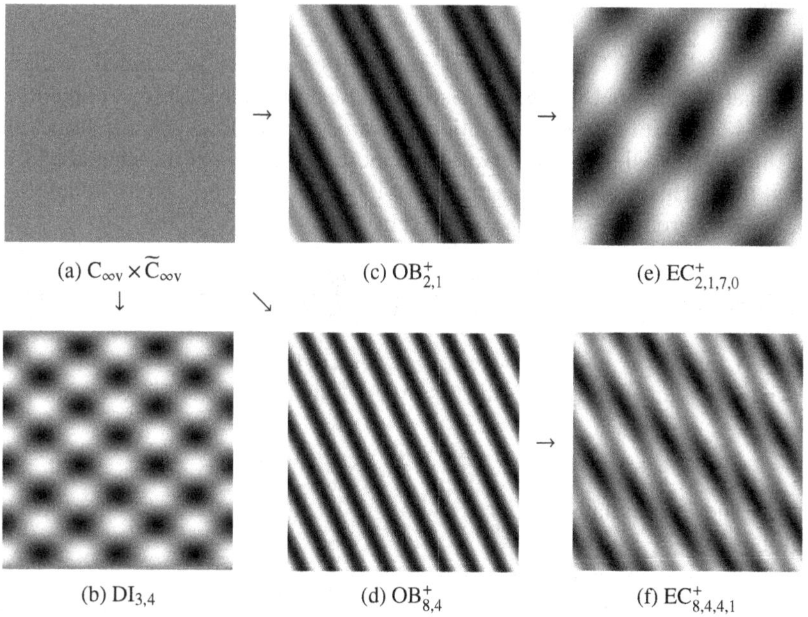

Fig. 14.7 Geometrical patterns expressed in terms of contour views. → : bifurcation.

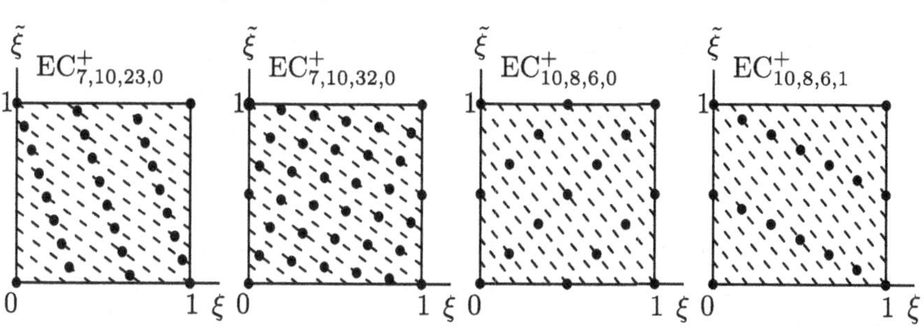

Fig. 14.8 $EC^{\pm}_{n\tilde{n}kl}$-symmetric modes. •: point with discrete translational symmetry.

Oblique Stripe Pattern

The symmetries of oblique stripe patterns, shown in Fig. 14.6(b) and (c) and Fig. 14.7(c) and (d), are described by groups

$$\mathrm{OB}^{+}_{n\tilde{n}} = \langle \sigma\tilde{\sigma}, \{r(\psi)\tilde{r}(\tilde{\psi}) \mid n\psi + \tilde{n}\tilde{\psi} \in \mathbb{Z}\}\rangle, \tag{14.16}$$

$$\mathrm{OB}^{-}_{n\tilde{n}} = \langle \sigma\tilde{\sigma}, \{r(\psi)\tilde{r}(\tilde{\psi}) \mid n\psi - \tilde{n}\tilde{\psi} \in \mathbb{Z}\}\rangle, \tag{14.17}$$

where $\mathbb{Z}$ is the set of integer numbers. The elements $\{r(\psi)\tilde{r}(\tilde{\psi}) \mid n\psi \pm \tilde{n}\tilde{\psi} = N\}$ (for a fixed integer N) indicate the invariance regarding arbitrary translation along an oblique straight line (shown, e.g., by the straight lines in Fig. 14.6(b) and (c)).

Echelon Mode

The patterns shown in Fig. 14.6(d) and Fig. 14.7(e) and (f) are interpreted as echelon modes that are expounded in §14.1 using several photographs. A series of points (•) in Fig. 14.6(d) with discrete translational symmetry resembles oblique stripes that have different directions and intervals from those of the other stripes denoted by the dashed lines. This is also the case in Fig. 14.7(e) and (f). This pattern is described by a finite subgroup of $\mathrm{C}_{\infty\mathrm{v}} \times \widetilde{\mathrm{C}}_{\infty\mathrm{v}}$ defined by

$$\mathrm{EC}^{+}_{n\tilde{n}kl} = \left\langle \sigma\tilde{\sigma}, r\left(-\frac{\tilde{n}}{dk}\right)\tilde{r}\left(\frac{n}{dk}\right), r\left(\frac{1}{d}\left(p - \frac{\tilde{n}l}{dk}\right)\right)\tilde{r}\left(\frac{1}{d}\left(\tilde{p} + \frac{nl}{dk}\right)\right)\right\rangle, \tag{14.18}$$

where n, $\tilde{n}$, and k are positive integers; $d = \gcd(n, \tilde{n})$ (= the greatest common divisor of n and $\tilde{n}$); l is an integer satisfying $0 \le l \le d-1$; and p and $\tilde{p}$ are integers such that

$$np + \tilde{n}\tilde{p} = d. \tag{14.19}$$

When n and $\tilde{n}$ are relatively prime (with $d = 1$ and $l = 0$), expression (14.18) is simplified to

$$\mathrm{EC}^{+}_{n\tilde{n}k0} = \left\langle \sigma\tilde{\sigma}, r\left(-\frac{\tilde{n}}{k}\right)\tilde{r}\left(\frac{n}{k}\right)\right\rangle,$$

since $r(p) = \tilde{r}(\tilde{p}) = e$ by (14.11). Similarly, we define

$$\mathrm{EC}^{-}_{n\tilde{n}kl} = \left\langle \sigma\tilde{\sigma}, r\left(\frac{\tilde{n}}{dk}\right)\tilde{r}\left(\frac{n}{dk}\right), r\left(\frac{1}{d}\left(p + \frac{\tilde{n}l}{dk}\right)\right)\tilde{r}\left(\frac{1}{d}\left(\tilde{p} + \frac{nl}{dk}\right)\right)\right\rangle, \tag{14.20}$$

where p and $\tilde{p}$ are integers such that

$$np - \tilde{n}\tilde{p} = d. \tag{14.21}$$

Note that $\mathrm{EC}^{+}_{n\tilde{n}kl}$ and $\mathrm{EC}^{-}_{n\tilde{n}kl}$ are finite subgroups (with $2dk$ elements) of $\mathrm{OB}^{+}_{n\tilde{n}}$ and $\mathrm{OB}^{-}_{n\tilde{n}}$, respectively. As presented in Fig. 14.8, $\mathrm{EC}^{\pm}_{n\tilde{n}kl}$-symmetric patterns vary drastically with the change of the values of k and l. The increase in k enhances the number dk of the points (•), whereas the change in l rearranges their location.

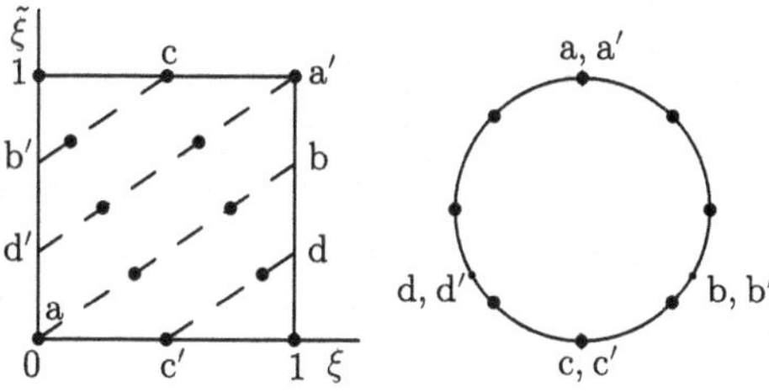

(a) $\mathrm{EC}^{-}_{2,3,8,0}$ $(d = 1)$

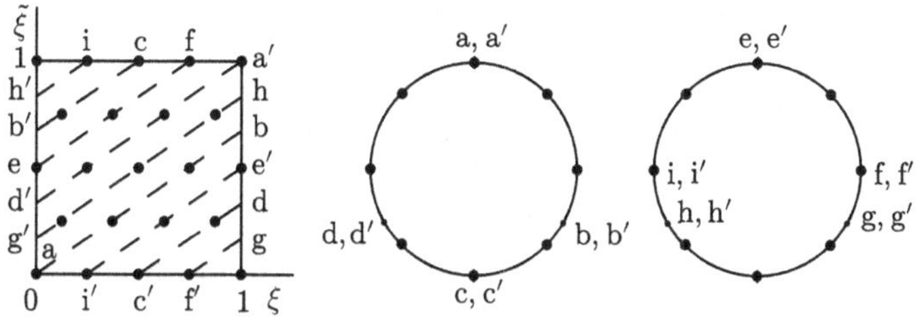

(b) $\mathrm{EC}^{-}_{4,6,8,0}$ $(d = 2)$

Fig. 14.9 Geometry of echelon symmetries. •: point with discrete translational symmetry.

The geometry of echelon symmetries is investigated further. First, for

$$\mathrm{EC}^{-}_{2,3,8,0} = \langle \sigma\tilde{\sigma}, r(3/8)\tilde{r}(1/4) \rangle$$

with $d = 1$, shown at the left of Fig. 14.9(a), we examine the dashed lines a–b, b′–c, c′–d, and d′–a′ that have $\mathrm{OB}^{-}_{2,3}$-symmetry as a whole. The dashed line a–b is connected continuously with b′–c inasmuch as points b and b′ are identified because of the periodic boundaries. Such is also the case for three pairs of points (c, c′), (d, d′), and (a, a′). Those lines, therefore, form a circle a–b–b′–c–c′–d–d′–a′ shown at the right of this figure. Then the points shown by (•), expressing the discrete translational symmetries, divide the circle into eight arcs of equal length. The integer $k = 8$ denotes the number of arcs.

Next, we investigate $\mathrm{EC}^{-}_{4,6,8,0}$ with $d = 2$ shown at the left of Fig. 14.9(b). A possible choice in (14.21) is $p = 2$, $\tilde{p} = 1$. Then,

$$\mathrm{EC}^{-}_{4,6,8,0} = \langle \sigma\tilde{\sigma}, r(3/8)\tilde{r}(1/4), \tilde{r}(1/2) \rangle.$$

For this case we can construct a pair of circles shown at the right of Fig. 14.9(b). The integer $d = 2$ denotes the number of circles. The second generator $\tilde{r}(1/2)$ maps point a on the first circle to point e on the second circle. Then the points shown by (•), expressing the discrete translational symmetries, divide each circle into $k = 8$ arcs of equal length. Note that $\mathrm{EC}^{-}_{4,6,8,0}$ consists of $2dk = 32$ elements.

Remark 14.2. Oblique stripe patterns without the symmetry of half-rotation $\sigma\tilde{\sigma}$ are labeled by the groups

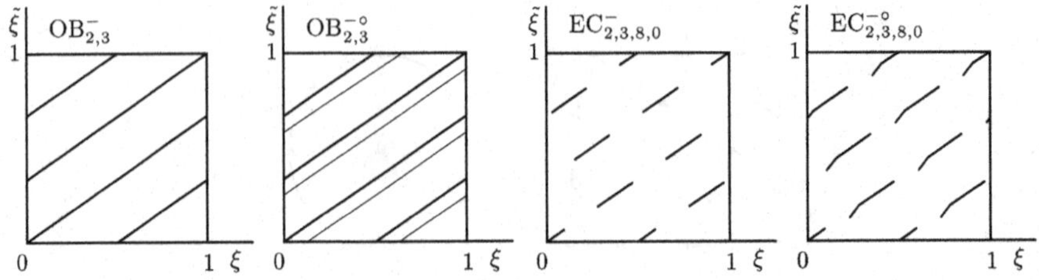

Fig. 14.10 Comparison of patterns with and without half-rotation symmetry $\sigma\tilde{\sigma}$.

$$\begin{aligned} \mathrm{OB}^{+\circ}_{n\tilde{n}} &= \langle \{ r(\psi)\tilde{r}(\tilde{\psi}) \mid n\psi + \tilde{n}\tilde{\psi} \in \mathbb{Z} \} \rangle, \\ \mathrm{OB}^{-\circ}_{n\tilde{n}} &= \langle \{ r(\psi)\tilde{r}(\tilde{\psi}) \mid n\psi - \tilde{n}\tilde{\psi} \in \mathbb{Z} \} \rangle. \end{aligned}$$

Patterns invariant to $\mathrm{OB}^{-}_{2,3}$ and $\mathrm{OB}^{-\circ}_{2,3}$ are compared at the left of Fig. 14.10.

Echelon modes without the half-rotation symmetry $\sigma\tilde{\sigma}$ are defined by

$$\mathrm{EC}^{+\circ}_{n\tilde{n}kl} = \left\langle r\left(-\frac{\tilde{n}}{dk}\right)\tilde{r}\left(\frac{n}{dk}\right), r\left(\frac{1}{d}\left(p - \frac{\tilde{n}l}{dk}\right)\right)\tilde{r}\left(\frac{1}{d}\left(\tilde{p} + \frac{nl}{dk}\right)\right)\right\rangle, \tag{14.22}$$

$$\mathrm{EC}^{-\circ}_{n\tilde{n}kl} = \left\langle r\left(\frac{\tilde{n}}{dk}\right)\tilde{r}\left(\frac{n}{dk}\right), r\left(\frac{1}{d}\left(p + \frac{\tilde{n}l}{dk}\right)\right)\tilde{r}\left(\frac{1}{d}\left(\tilde{p} + \frac{nl}{dk}\right)\right)\right\rangle. \tag{14.23}$$

Patterns invariant to $\mathrm{EC}^{-}_{2,3,8,0}$ and $\mathrm{EC}^{-\circ}_{2,3,8,0}$ are compared at the right of Fig. 14.10. □

14.4 Recursive Bifurcation Leading to Echelon Modes

The formation of echelon modes is explained through recursive bifurcation of a system equivariant to $\mathrm{C}_{\infty\mathrm{v}} \times \widetilde{\mathrm{C}}_{\infty\mathrm{v}} \simeq \mathrm{O}(2) \times \mathrm{O}(2)$. Echelon modes cannot appear directly from the $\mathrm{C}_{\infty\mathrm{v}} \times \widetilde{\mathrm{C}}_{\infty\mathrm{v}}$-equivariant system. Nonetheless a recursive bifurcation via an oblique stripe pattern is a possible mechanism to produce echelon modes, as clarified by the mathematical derivations of this recursive bifurcation in §14.9, the appendix of this chapter.

Direct Bifurcation

The direct bifurcation point of a $\mathrm{C}_{\infty\mathrm{v}} \times \widetilde{\mathrm{C}}_{\infty\mathrm{v}}$-equivariant system has the multiplicity $M = 1$, 2, or 4. We specifically address a *quadruple bifurcation point* with $M = 4$, at which bifurcated paths with the diamond pattern $\mathrm{DI}_{n\tilde{n}}$ and those with the oblique stripe patterns $\mathrm{OB}^{+}_{n\tilde{n}}$ and $\mathrm{OB}^{-}_{n\tilde{n}}$ branch simultaneously for some integers n and $\tilde{n}$ (cf., §14.9.1). That is, we encounter a bifurcation with symmetry breaking:

$$\mathrm{C}_{\infty\mathrm{v}} \times \widetilde{\mathrm{C}}_{\infty\mathrm{v}} \begin{array}{l} \nearrow \mathrm{DI}_{n\tilde{n}} \\ \rightarrow \mathrm{OB}^{+}_{n\tilde{n}} \\ \searrow \mathrm{OB}^{-}_{n\tilde{n}} \end{array}. \tag{14.24}$$

The main focus of this chapter is given to the recursive bifurcation leading to echelon modes via an oblique stripe pattern, in view of the experimental observation of oblique stripes for materials.[7] The direct bifurcation to the diamond pattern is briefly studied in §14.8 for a sand specimen, and bifurcation from the diamond pattern is left outside the scope of this book, although it is an interesting theoretical topic.

Recursive Bifurcation via Oblique Stripe Pattern

An echelon mode with $\mathrm{EC}^{\pm}_{n\tilde{n}kl}$-symmetry cannot appear directly from the uniform state with $\mathrm{C}_{\infty\mathrm{v}} \times \widetilde{\mathrm{C}}_{\infty\mathrm{v}}$-symmetry but, possibly, it is produced by a recursive bifurcation by way of the oblique stripe patterns with $\mathrm{OB}^{\pm}_{n\tilde{n}}$-symmetry. This recursive bifurcation reads as

$$\begin{array}{l} G = \mathrm{C}_{\infty\mathrm{v}} \times \widetilde{\mathrm{C}}_{\infty\mathrm{v}} \\ \quad \searrow (1) \qquad\qquad \searrow (4) \\ \qquad \mathrm{D}_{\infty\infty} \rightarrow (2) \rightarrow \mathrm{OB}^{\pm}_{n\tilde{n}} \rightarrow (2) \rightarrow \mathrm{EC}^{\pm}_{n\tilde{n}kl} \\ \qquad\qquad\qquad\qquad \searrow (1\,\mathrm{or}\,2) \\ \qquad\qquad\qquad\qquad\quad \mathrm{OB}^{\pm}_{m\tilde{m}} \rightarrow (2) \rightarrow \mathrm{EC}^{\pm}_{m\tilde{m}kl} \end{array} \tag{14.25}$$

where m and $\tilde{m}$, respectively, signify divisors of n and $\tilde{n}$ satisfying $n/m = \tilde{n}/\tilde{m} \geq 2$ and "$\searrow (M)$" or "$\rightarrow (M) \rightarrow$" denotes a bifurcation from a critical point of multiplicity M (= 1, 2, or 4).

Physical Scenario for Echelon Mode Formation

The following physical scenario for the emergence of echelon modes is suggested by the mathematical result described above. At an early stage of the deformation of a uniform material, oblique parallel stripes are formed through the direct bifurcation or the secondary bifurcation via a $\mathrm{D}_{\infty\infty}$-symmetric state (cf., (14.25)); that is,

$$\mathrm{C}_{\infty\mathrm{v}} \times \widetilde{\mathrm{C}}_{\infty\mathrm{v}} \rightarrow \mathrm{OB}^{\pm}_{n\tilde{n}} \qquad \text{or} \qquad \mathrm{C}_{\infty\mathrm{v}} \times \widetilde{\mathrm{C}}_{\infty\mathrm{v}} \rightarrow \mathrm{D}_{\infty\infty} \rightarrow \mathrm{OB}^{\pm}_{n\tilde{n}}.$$

Although these stripes may not be discernible at this stage, the material is weakened along the stripes. Then other patterns are formed by further bifurcation

$$\mathrm{OB}^{\pm}_{n\tilde{n}} \rightarrow \mathrm{OB}^{\pm}_{m\tilde{m}} \qquad \text{or} \qquad \mathrm{OB}^{\pm}_{n\tilde{n}} \rightarrow \mathrm{EC}^{\pm}_{n\tilde{n}kl}$$

---

[7] Oblique stripes are observed for soil specimens in §14.6 and steel and kaolin specimens in §14.7.

(cf., Fig. 14.7). In the first case ($\mathrm{OB}^{\pm}_{n\tilde{n}} \to \mathrm{OB}^{\pm}_{m\tilde{m}}$), the direction of the stripes (with the same physical property) does not change and their interval enlarges. In the second case ($\mathrm{OB}^{\pm}_{n\tilde{n}} \to \mathrm{EC}^{\pm}_{n\tilde{n}kl}$), on the other hand, the direction of the stripes[8] does change and their interval might enlarge or diminish.

There exist a number of possible $\mathrm{EC}^{\pm}_{n\tilde{n}kl}$-symmetric patterns that vary with physical and material properties. Naturally, different materials have different deformation patterns. However, the present argument implies that such a difference might be attributable merely to a difference in the values of k and l of the same family of $\mathrm{EC}^{\pm}_{n\tilde{n}kl}$-symmetric bifurcated solutions. For a deeper understanding of the phenomena, it is advised to observe echelon modes of materials of various kinds, which are quite diverse and complex, through a unified view of the recursive symmetry-breaking bifurcation like that depicted in (14.25).

Remark 14.3. Up to this point we have discussed possible bifurcation processes leading to echelon modes exclusively by way of $\mathrm{OB}^{\pm}_{n\tilde{n}}$- and $\mathrm{D}_{\infty\infty}$-symmetries, as summarized in (14.25). The main claim is that the echelon-mode is formed through a recursive bifurcation of a $\mathrm{C}_{\infty\mathrm{v}} \times \widetilde{\mathrm{C}}_{\infty\mathrm{v}}$-equivariant system. It is to be remarked that $\mathrm{EC}^{+\circ}_{n\tilde{n}kl}$ and $\mathrm{EC}^{-\circ}_{n\tilde{n}kl}$ introduced in (14.22) and (14.23) also represent the echelon mode. Possible subgroups leading to echelon modes represented by $\mathrm{EC}^{\pm\circ}_{n\tilde{n}kl}$ are not exhausted by $\mathrm{OB}^{\pm}_{n\tilde{n}}$ and $\mathrm{D}_{\infty\infty}$, but might at least include $\mathrm{OB}^{\pm\circ}_{n\tilde{n}}$ through a bifurcation process

$$\mathrm{C}_{\infty\mathrm{v}} \times \widetilde{\mathrm{C}}_{\infty\mathrm{v}} \to \mathrm{OB}^{\pm}_{n\tilde{n}} \to \mathrm{OB}^{\pm\circ}_{n\tilde{n}} \to \mathrm{EC}^{\pm\circ}_{n\tilde{n}kl}.$$

Figure 14.11 illustrates such possibilities. □

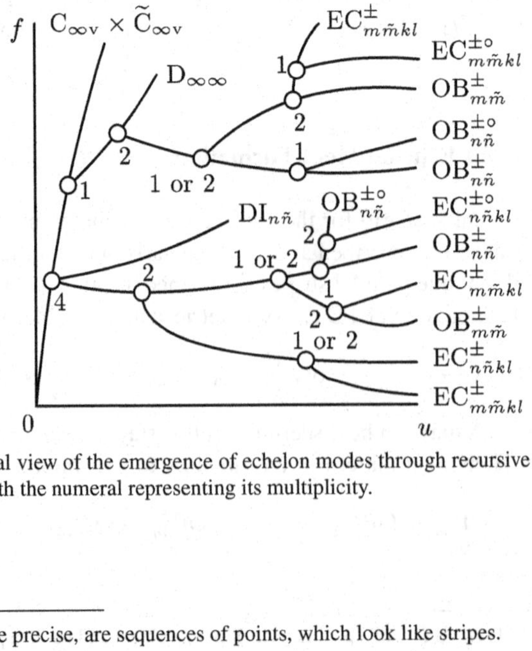

Fig. 14.11 General view of the emergence of echelon modes through recursive bifurcation. ∘: bifurcation point with the numeral representing its multiplicity.

[8] The stripes, to be precise, are sequences of points, which look like stripes.

14.5 Experiment on a Soil Specimen

The triaxial compression test on a soil specimen by Nakano, 1993 [147] is investigated as an example of echelon mode formation through a few distinct viewpoints, including: stress versus strain curves, deformation patterns, and symmetries.

Stress Versus Strain Curves

We observe the deviatoric stress σ_a versus axial strain ε_a curve in Fig. 14.12(a).

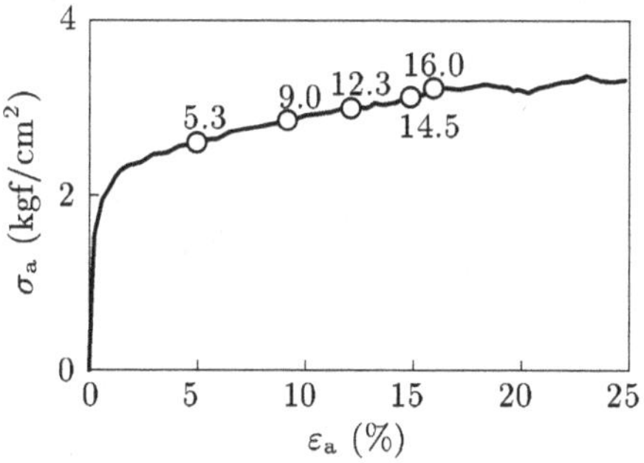

(a) Deviatoric stress σ_a versus axial strain ε_a curve

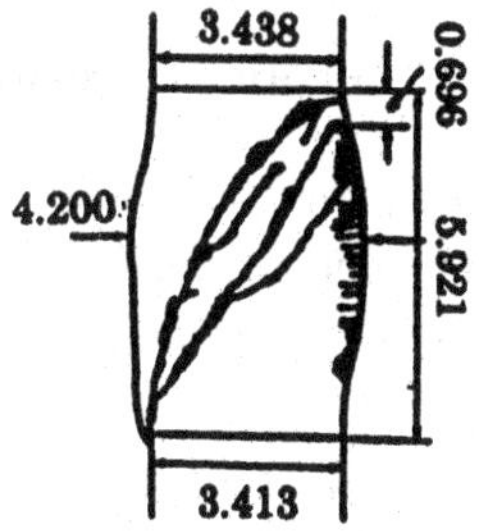

(c) Sketch at the final state taken from the rear

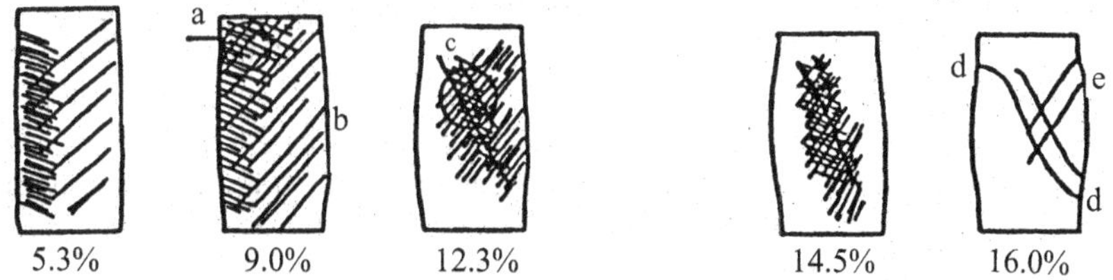

(b) Sketches of deformation patterns

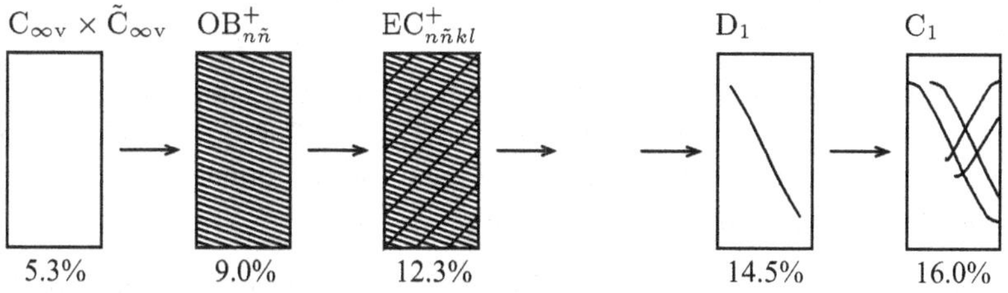

(d) Interpretation of deformation patterns

Fig. 14.12 Experiment on a cylindrical soil specimen (Nakano, 1993 [147]). The patterns at (○) in (a) are presented in (b) and (d).

- During the first stage of loading ($\varepsilon_a < 1.0\%$), the slope of the curve is steep and the specimen is nearly elastic.
- The slope is then reduced during $1.0\% < \varepsilon_a < 2.0\%$, and the specimen softens rapidly.
- The slope remains almost constant for $\varepsilon_a > 2.0\%$.
- The slope is slightly reduced at $\varepsilon_a = 20.0\%$ associated with the formation of a shear band.

In soil mechanics, it is commonplace to interpret the reduction of the slope ($1.0\% < \varepsilon_a < 2.0\%$) as the softening of soils, and the shear-band formation ($\varepsilon_a = 20.0\%$) as direct bifurcation. This interpretation, however, is to be denied in the succeeding argument based on the viewpoint of recursive bifurcation.

Deformation Patterns: Phenomenological Observation

The deformation patterns of the specimen are investigated from a phenomenological standpoint. Figure 14.12(b) shows the sketches of the deformation patterns, which are drawn at the points shown by (∘) on the stress versus strain curves in Fig. 14.12(a).

- The specimen almost retains its cylindrical shape until $\varepsilon_a = 5.0\%$.
- At $\varepsilon_a = 5.3\%$ its right side swells to lose the bilateral symmetry (σ_y-symmetry), and numerous oblique parallel wrinkles emerge on the surface of the specimen. A set of short parallel wrinkles at the left has a horizontal angle of 10°, whereas another set with a greater interval at the right has an angle of −35°.
- At $\varepsilon_a = 9.0\%$ a number of relatively long wrinkles show up in the upper-left part (zone a), and the right side (point b) swells significantly.
- At $\varepsilon_a = 12.3\%$ numerous wrinkles mutually intersect in the center (zone c) to form an echelonlike mesh. Some wrinkles observed earlier are not discernible at this stage.
- The localization of deformation progresses for $\varepsilon_a = 14.5\%$ and 16.0%.
- At $\varepsilon_a = 16.0\%$ most of the distinct wrinkles become indiscernible, but one of the wrinkles is further extended to reach point d. Figure 14.12(c) is a sketch of these wrinkles taken from the rear.

These shear bands, which comprise numerous wrinkles that had emerged earlier ($0 < \varepsilon_a < 9.0\%$) but became indiscernible later ($\varepsilon_a = 12.3\%$), look like echelon. The deformation of the specimen is indeed the recursive formation of wrinkles in different directions and with different intervals.

Remark 14.4. The formation of a shear band or a series of parallel shear bands is ascribed to a direct bifurcation in plastic bifurcation theory.[9] This is called the *shear-band mode bifurcation.* □

[9] See, for example, Hill and Hutchinson, 1975 [73] for theory and Vardoulakis, Goldscheider, and Gudehus, 1978 [201] for its application to soil.

Deformation Patterns: Symmetry

The deformation patterns of the specimen observed above are interpreted theoretically. Viewing the sketches shown in Fig. 14.12(b), one can point out the presence of five deformation patterns in Fig. 14.12(d):

- $C_{\infty v} \times \widetilde{C}_{\infty v}$-symmetric uniform deformation pattern,
- $OB^{+}_{n\tilde{n}}$-symmetric stripe pattern,
- $EC^{+}_{n\tilde{n}kl}$-symmetric transient pattern,
- D_1-symmetric pattern for the distinct wrinkle,[10]
- C_1-symmetric (asymmetric) pattern for the final stage of deformation.

In view of these patterns, one can presume the occurrence of recursive bifurcation, as portrayed in Fig. 14.13. Shear-band formation may be associated with secondary or further bifurcation, instead of direct bifurcation.

The emergence of the stripe pattern ($OB^{+}_{n\tilde{n}}$-symmetric mode) is the key phenomenon that indicates the occurrence of direct bifurcation. This pattern is already observed at $\varepsilon_a = 5.3\%$; therefore, the direct bifurcation

$$C_{\infty v} \times \widetilde{C}_{\infty v} \to OB^{+}_{n\tilde{n}}$$

should have taken place prior to $\varepsilon_a = 5.3\%$. It would be a rational hypothesis that the bifurcation, not the softening of the material, has triggered the sharp reduction of the slope of the σ_a versus ε_a curve ($1.0 < \varepsilon_a < 2.0\%$). Consequently, a doubt is cast upon the commonplace interpretation in soil mechanics to attribute such degradation to the softening of soils.

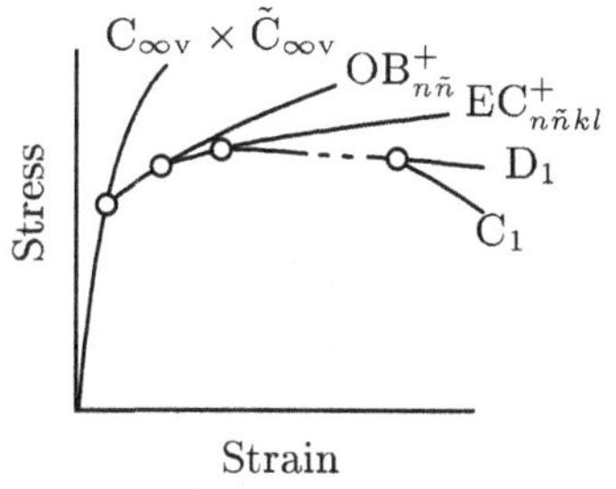

Fig. 14.13 General view of the recursive bifurcation of the soil specimen.

14.6 Rectangular Plate with Periodic Boundaries

We refer to a bifurcation analysis on a rectangular plate with periodic boundary conditions as another example[11] of an O(2)×O(2)-symmetric system.

[10] Here $D_1 = \langle \sigma_y \sigma_z \rangle$ denotes the half-rotation symmetry about the center of the specimen: the x-axis.

[11] This analysis is based on Murota, Ikeda, and Terada, 1999 [145].

14.6.1 Geometry and Potential Function

We consider the $L_x \times L_y$ rectangular plate, shown in Fig. 14.14, in the domain

$$\Omega = \{(x,y) \in \mathbb{R}^2 \mid -L_x/2 \le x \le L_x/2, \; -L_y/2 \le y \le L_y/2\}.$$

This plate is supported on uniformly distributed linear springs with the spring constant k, with periodic boundaries at four sides, subjected to the uniform uniaxial loading in the x-direction with the stress resultant of N. We consider only the z-directional displacement $u(x,y)$ of the plate.

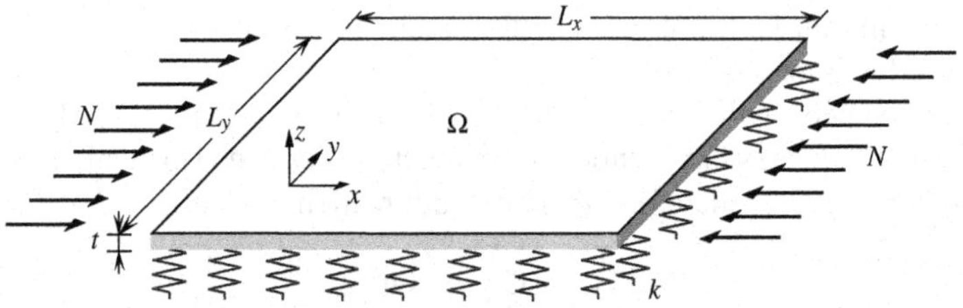

Fig. 14.14 Rectangular plate with four-side periodic boundaries.

The total potential energy for this plate reads as[12]

$$\begin{aligned}
U = {} & \frac{D}{2}\int\int_{\Omega}\left[\left(\frac{\partial^2 u}{\partial x^2}+\frac{\partial^2 u}{\partial y^2}\right)^2 + 2(1-\nu)\left\{\left(\frac{\partial^2 u}{\partial x \partial y}\right)^2 - \frac{\partial^2 u}{\partial x^2}\frac{\partial^2 u}{\partial y^2}\right\}\right] dx dy && \cdots \text{in-plane bending} \\
& + \frac{3D}{2t^2}\int\int_{\Omega}\left\{\left(\frac{\partial u}{\partial x}\right)^2 + \left(\frac{\partial u}{\partial y}\right)^2\right\}^2 dx dy && \cdots \text{membrane force} \\
& + \frac{k}{2}\int\int_{\Omega} u^2 dx dy && \cdots \text{linear springs} \\
& - \frac{2\pi^2 D}{L_y{}^2} f \int\int_{\Omega}\left(\frac{\partial u}{\partial x}\right)^2 dx dy && \cdots \text{external force,} \quad (14.26)
\end{aligned}$$

where E is Young's modulus; ν is Poisson's ratio; t is the plate thickness; $D = Et^3/[12(1-\nu^2)]$; and $f = L_y N/(4\pi^2 D)$ is a (nondimensional) loading parameter.

[12] See, for example, Timoshenko and Woinowsky–Krieger, 1959 [195] and Timoshenko and Gere, 1963 [194].

14.6.2 Underlying Symmetry

The domain Ω has the geometrical symmetry of the reflection σ_x with respect to the yz-plane and the reflection σ_y to the zx-plane.[13] Therefore, its symmetry is expressed by the direct product

$$G_0 = \mathrm{C}_{1\mathrm{v}} \times \widetilde{\mathrm{C}}_{1\mathrm{v}} = \langle \sigma_x, \sigma_y \rangle \simeq \mathbb{Z}_2 \times \mathbb{Z}_2, \tag{14.27}$$

where

$$\mathrm{C}_{1\mathrm{v}} = \langle \sigma_x \rangle = \{e, \sigma_x\} \simeq \mathbb{Z}_2, \qquad \widetilde{\mathrm{C}}_{1\mathrm{v}} = \langle \sigma_y \rangle = \{e, \sigma_y\} \simeq \mathbb{Z}_2.$$

Putting more emphasis on local uniformity than on global symmetry, we employ infinite-periodic-domain approximation: the domain is periodically extended in the x-direction and in the y-direction and is assumed to have periodic boundary conditions at $x = \pm L_x/2$ and at $y = \pm L_y/2$. With this approximation, the governing equation acquires an additional symmetry of the x-directional translation $t_x(l_x)$ for any length l_x ($0 \le l_x < L_x$) and the y-directional translation $t_y(l_y)$ for any length l_y ($0 \le l_y < L_y$) defined, respectively, by

$$t_x(l_x)\ :\ x \mapsto x + l_x, \qquad t_y(l_y)\ :\ y \mapsto y + l_y.$$

Accordingly, the group of equivariance G_0 is then replaced by an enlarged group

$$G = \mathrm{C}_{\infty\mathrm{v}} \times \widetilde{\mathrm{C}}_{\infty\mathrm{v}} = \langle \sigma_x, \sigma_y, t_x(l_x), t_y(l_y) \rangle \simeq \mathrm{O}(2) \times \mathrm{O}(2),$$

where

$$\begin{aligned}
\mathrm{C}_{\infty\mathrm{v}} &= \langle \sigma_x, t_x(l_x) \rangle = \{t_x(l_x), \sigma_x t_x(l_x) \mid 0 \le l_x < L_x\} \simeq \mathrm{O}(2),\\
\widetilde{\mathrm{C}}_{\infty\mathrm{v}} &= \langle \sigma_y, t_y(l_y) \rangle = \{t_y(l_y), \sigma_y t_y(l_y) \mid 0 \le l_y < L_y\} \simeq \mathrm{O}(2).
\end{aligned}$$

Therefore, all the mathematical arguments in this chapter for the group $\mathrm{O}(2) \times \mathrm{O}(2)$, developed mainly for the cylindrical domain with periodic boundaries, are also applicable to the rectangular domain with periodic boundaries.

To simplify the notation, we put

$$\xi = \frac{x}{L_x} + \frac{1}{2}, \qquad \tilde{\xi} = \frac{y}{L_y} + \frac{1}{2}, \qquad \psi = \frac{l_x}{L_x}, \qquad \tilde{\psi} = \frac{l_y}{L_y},$$
$$\sigma = \sigma_x, \qquad \tilde{\sigma} = \sigma_y, \qquad r(\psi) = t_x(L_x\psi) = t_x(l_x), \qquad \tilde{r}(\tilde{\psi}) = t_y(L_y\tilde{\psi}) = t_y(l_y).$$

Then we have $0 \le \xi, \tilde{\xi}, \psi, \tilde{\psi} < 1$; $\mathrm{C}_{\infty\mathrm{v}} = \langle \sigma, r(\psi) \rangle$ and $\widetilde{\mathrm{C}}_{\infty\mathrm{v}} = \langle \tilde{\sigma}, \tilde{r}(\tilde{\psi}) \rangle$ as in (14.12).

Remark 14.5. When a rectangular plate is subjected to a shear, as presented in Fig. 14.15(b), the geometrical symmetry group is

$$G_0' = \langle \sigma_x \sigma_y \rangle,$$

[13] Although this plate has an additional symmetry, the upside-down symmetry with respect to σ_z: $z \mapsto -z$, this symmetry is not treated here.

Fig. 14.15 Loading patterns for rectangular domains.

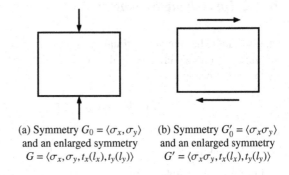

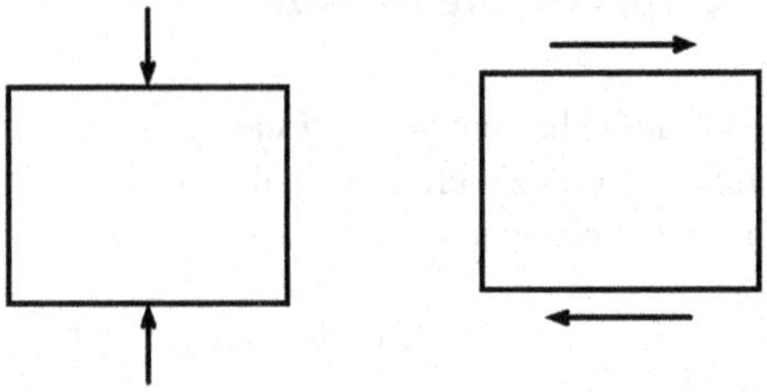

(a) Symmetry $G_0 = \langle \sigma_x, \sigma_y \rangle$ and an enlarged symmetry $G = \langle \sigma_x, \sigma_y, t_x(l_x), t_y(l_y) \rangle$

(b) Symmetry $G_0' = \langle \sigma_x \sigma_y \rangle$ and an enlarged symmetry $G' = \langle \sigma_x \sigma_y, t_x(l_x), t_y(l_y) \rangle$

in which $\sigma_x \sigma_y$ represents the in-plane half-rotation. The infinite-periodic-domain approximation can be implemented by replacing G_0' by

$$G' = \langle \sigma_x \sigma_y, t_x(l_x), t_y(l_y) \rangle,$$

which is isomorphic to $\mathrm{D}_{\infty\infty}$ introduced in (14.14). □

14.6.3 Numerical Analysis

To obtain a discretized form of the governing equation of the rectangular plate from (14.26), we expand the z-directional displacement $u(x,y)$ into the double Fourier series

$$u = \sum_{n=0}^{\infty} \sum_{\tilde{n}=0}^{\infty} [A_{n\tilde{n}} \sin(2\pi n \xi) \sin(2\pi \tilde{n} \tilde{\xi}) + B_{n\tilde{n}} \cos(2\pi n \xi) \cos(2\pi \tilde{n} \tilde{\xi}) \\ + C_{n\tilde{n}} \sin(2\pi n \xi) \cos(2\pi \tilde{n} \tilde{\xi}) + D_{n\tilde{n}} \cos(2\pi n \xi) \sin(2\pi \tilde{n} \tilde{\xi})], \quad (14.28)$$

where $A_{n\tilde{n}}$, $B_{n\tilde{n}}$, $C_{n\tilde{n}}$, and $D_{n\tilde{n}}$ are constants to be determined ($A_{0\tilde{n}} = A_{n0} = C_{0\tilde{n}} = D_{0\tilde{n}} = 0$ for $n, \tilde{n} = 0, 1, \ldots$).

Since the half-rotation $\sigma\tilde{\sigma}$ belongs to $\mathrm{DI}_{n\tilde{n}}$ of (14.15) and to $\mathrm{OB}^{\pm}_{n\tilde{n}}$ of (14.16) and (14.17), the solutions for the diamond pattern and for the oblique stripe pattern are invariant to $\sigma\tilde{\sigma}$. Hence, with a view to investigating such solutions, it is sufficient and convenient to employ a simplified series

$$u = \sum_{n=0}^{\infty} \sum_{\tilde{n}=0}^{\infty} [A_{n\tilde{n}} \sin(2\pi n \xi) \sin(2\pi \tilde{n} \tilde{\xi}) + B_{n\tilde{n}} \cos(2\pi n \xi) \cos(2\pi \tilde{n} \tilde{\xi})], \quad (14.29)$$

which includes only the terms invariant to the half-rotation $\sigma\tilde{\sigma}$.

For this reason, the bifurcation analysis of the rectangular plate is conducted using the simplified expansion of u in (14.29). Two different cases are compared:

- Periodic boundaries at four sides,
- Four-side simply supported boundaries.

Periodic Boundaries

The periodic boundaries correspond to a fictitious $O(2) \times O(2)$-symmetric system that exploits the translational symmetry arising from local uniformity.

We show in Fig. 14.16 the equilibrium paths computed with the periodic boundaries at the top and the associated schematic diagram at the bottom. Here, the normalized displacement u/t at $(\xi, \tilde{\xi}) = (0.62, 0.80)$ is used for the abscissa; the trivial

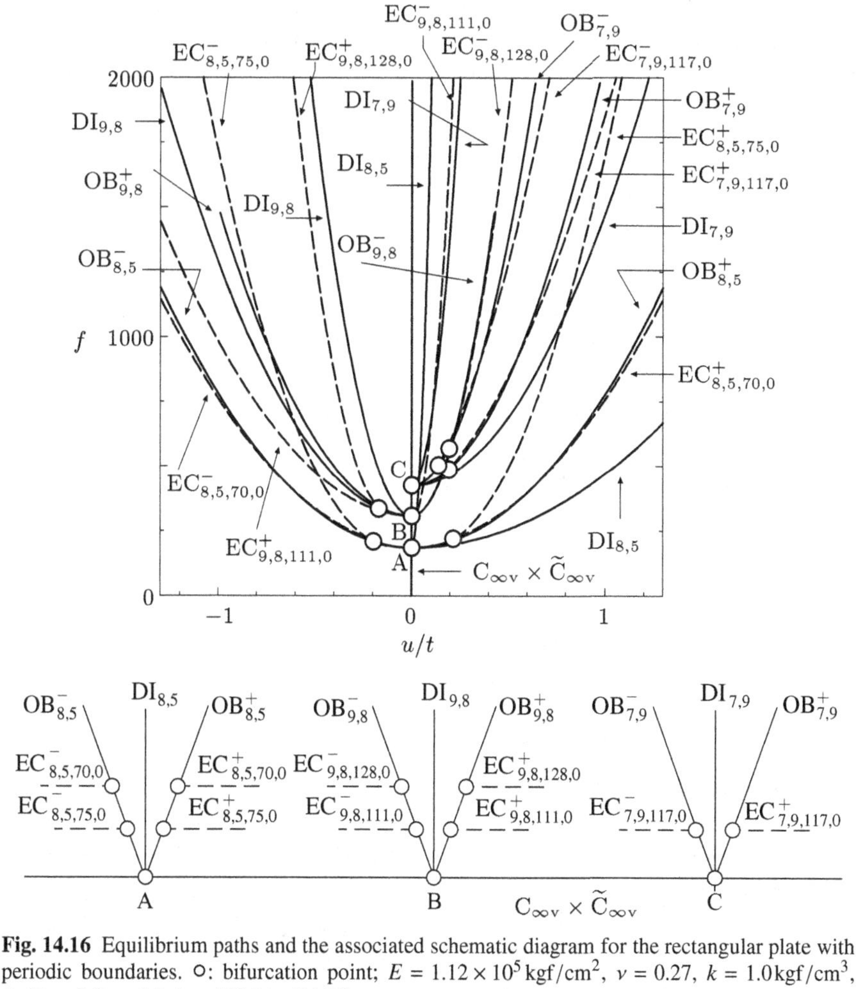

Fig. 14.16 Equilibrium paths and the associated schematic diagram for the rectangular plate with periodic boundaries. o: bifurcation point; $E = 1.12 \times 10^5\,\mathrm{kgf/cm^2}$, $\nu = 0.27$, $k = 1.0\mathrm{kgf/cm^3}$, $L_x/L_y = 1.0$, and $L_y/t = 500.0$ in (14.26).

solution with $C_{\infty v} \times \widetilde{C}_{\infty v}$-symmetry is given by $u/t = 0$; the dashed curves denote bifurcated paths with echelon symmetries labeled by $EC^{\pm}_{n\tilde{n}kl}$ for some integers n, $\tilde{n}$, k, and l; and the solid lines denote those labeled by other groups.

For the periodic boundaries, as is apparent from Fig. 14.16, three double bifurcation points A, B, and C exist on the $C_{\infty v} \times \widetilde{C}_{\infty v}$-symmetric trivial solution $u/t = 0$. A $DI_{n\tilde{n}}$-symmetric bifurcated path and a pair of $OB^{\pm}_{n\tilde{n}}$-symmetric paths branch simultaneously at each bifurcation point, which corresponds to $(n, \tilde{n}) = (8, 5)$, $(9, 8)$, or $(7, 9)$. For example, the bifurcation process

$$\begin{array}{l} C_{\infty v} \times \widetilde{C}_{\infty v} \rightarrow OB^{+}_{9,8} \rightarrow EC^{+}_{9,8,111,0} \\ \qquad\qquad\qquad\qquad\qquad\quad \searrow EC^{+}_{9,8,128,0} \end{array} \tag{14.30}$$

has led to $EC^{+}_{9,8,111,0}$- and $EC^{+}_{9,8,128,0}$-symmetric echelon modes via the $OB^{+}_{9,8}$-symmetric path, in agreement with the theoretical analysis in §14.9.2 in the appendix. Figure 14.17 presents some of the deformation patterns on the bifurcated paths.

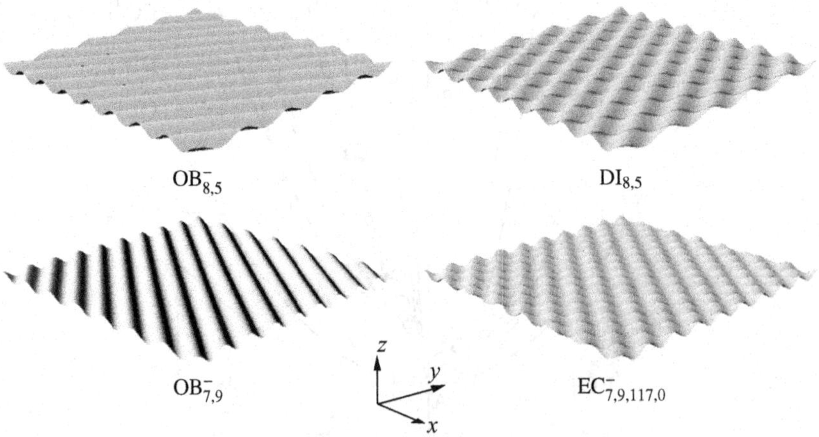

Fig. 14.17 Deformation patterns on bifurcated paths in Fig. 14.16.

Four-Side Simply Supported Boundaries

For the four-side simply supported boundaries, in which $u = 0$ on the boundaries, the Fourier series

$$u = \sum_{n=0}^{\infty} \sum_{\tilde{n}=0}^{\infty} A_{n\tilde{n}} \sin(\pi n \xi) \sin(\pi \tilde{n} \tilde{\xi}) \tag{14.31}$$

is to be employed, instead of (14.29). As shown in Fig. 14.18, three simple bifurcation points A, B, and C exist on the $C_{\infty v} \times \widetilde{C}_{\infty v}$-symmetric trivial solution $u/t = 0$.

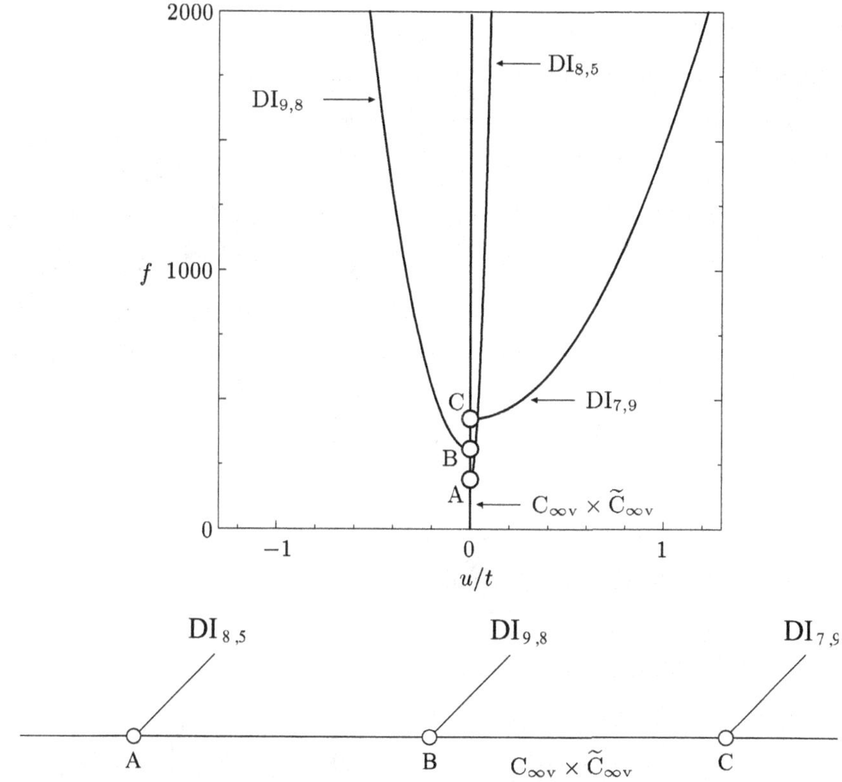

Fig. 14.18 Equilibrium paths and the associated schematic diagram for the rectangular plate with simply supported boundaries. ○: bifurcation point $E = 1.12 \times 10^5\,\mathrm{kgf/cm^2}$, $\nu = 0.27$, $k = 1.0\,\mathrm{kgf/cm^3}$, $L_x/L_y = 1.0$, and $L_y/t = 500.0$ in (14.26).

The solution paths with $\mathrm{DI}_{n\tilde{n}}$-symmetric diamond patterns $[(n,\tilde{n}) = (8,5), (9,8)$, and $(7,9)]$ branch at these bifurcation points. The bifurcation processes, such as (14.30), generating oblique stripe patterns and echelon modes, have been lost completely. This suffices to demonstrate the usefulness of the periodic boundaries in the simulation of these modes.

14.7 Image Simulations for Stripes on Kaolin and Steel

We conduct image simulations of the deformation patterns on kaolin and steel specimens with high spatial frequencies. The images of the deformation patterns of these specimens as depicted in Fig. 14.19 are processed in view of the rule of recursive bifurcation presented in §14.4 and the mathematical derivation for a $C_{\infty v} \times \tilde{C}_{\infty v}$-equivariant system in §14.9. As a result, we set forth the possible recursive bifurcation that produces those patterns.

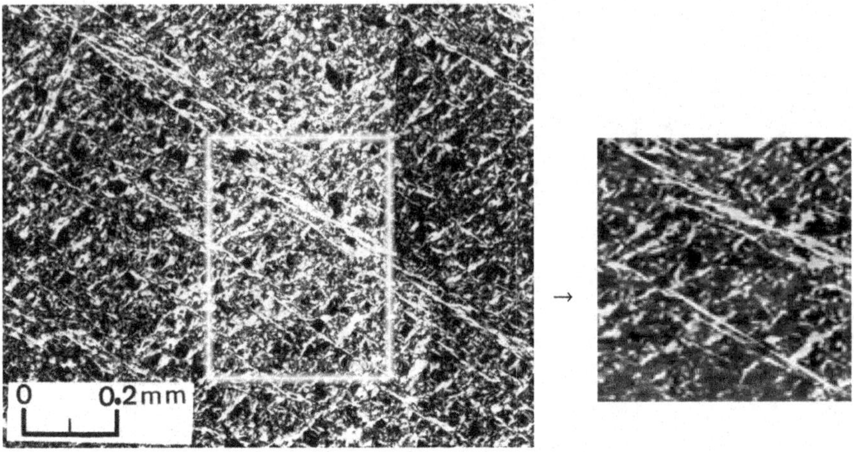

(a) Kaolin

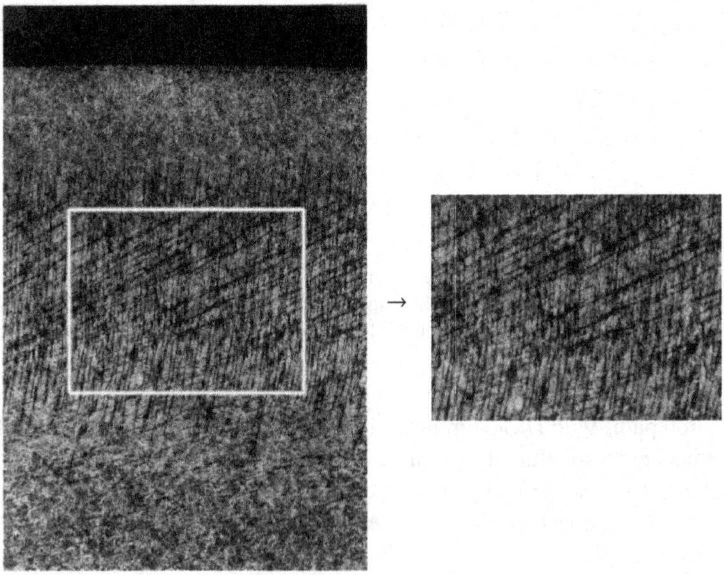

(b) Steel

Fig. 14.19 Photographs of kaolin and steel specimens processed by an image scanner at the left and the rectangular domains employed for the image analysis at the right (Ikeda et al., 2001 [84]). Photograph in (a) is by I. Sano and that in (b) by N. Oguma.

14.7.1 Image Simulation Procedure

We employ the rectangular domains shown at the right of Fig. 14.19 cut from the left of Fig. 14.19 for the image analysis for the kaolin and steel.[14] These domains have been chosen deliberately to satisfy the following characteristics.

- The domains are small enough, relative to the specimens, so as to reduce the influence of the boundaries.
- They are large enough, relative to the grains, to avoid the size effect.
- The right and left (respectively, the top and bottom) of the domains, to some extent, are to be connected periodically.

By virtue of these characteristics, the domains are sufficiently uniform and satisfy the periodic boundary conditions employed in §14.6.2.

The observed density u at the right of Fig. 14.19 for the kaolin and that for the steel are expanded into the double Fourier series

$$u(\xi,\tilde{\xi}) = \sum_{n=0}^{\infty}\sum_{\tilde{n}=0}^{\infty}[A_{n\tilde{n}}\cos 2\pi(n\xi+\tilde{n}\tilde{\xi}) + B_{n\tilde{n}}\cos 2\pi(n\xi-\tilde{n}\tilde{\xi}) + C_{n\tilde{n}}\sin 2\pi(n\xi+\tilde{n}\tilde{\xi}) + D_{n\tilde{n}}\sin 2\pi(n\xi-\tilde{n}\tilde{\xi})], \tag{14.32}$$

where, to make this representation unique, we put $A_{n0} = B_{n0}$ and $C_{n0} = D_{n0}$ for $n = 0,1,\ldots$; and $A_{0\tilde{n}} = B_{0\tilde{n}}$ and $C_{0\tilde{n}} = -D_{0\tilde{n}}$ for $\tilde{n} = 0,1,\ldots$; in particular, $C_{00} = D_{00} = 0$. The state prior to the bifurcation corresponds to the trivial solution u = constant. The Fourier series (14.32) is chosen to be compatible with the oblique stripe pattern. Specifically,

$$\cos 2\pi k(n\xi+\tilde{n}\tilde{\xi}), \qquad \cos 2\pi k(n\xi-\tilde{n}\tilde{\xi}), \qquad \sin 2\pi k(n\xi+\tilde{n}\tilde{\xi}), \qquad \sin 2\pi k(n\xi-\tilde{n}\tilde{\xi})$$

($k = 0,1,2,\ldots$) are invariant to $\mathrm{OB}^{+}_{n\tilde{n}}$, $\mathrm{OB}^{-}_{n\tilde{n}}$, $\mathrm{OB}^{+\circ}_{n\tilde{n}}$, and $\mathrm{OB}^{-\circ}_{n\tilde{n}}$, respectively.

We describe a method to reconstruct the deformation history of a specimen from the Fourier coefficients $A_{n\tilde{n}}$, $B_{n\tilde{n}}$, $C_{n\tilde{n}}$, and $D_{n\tilde{n}}$ with the theoretical knowledge of recursive bifurcation of a $\mathrm{C}_{\infty\mathrm{v}} \times \widetilde{\mathrm{C}}_{\infty\mathrm{v}}$-equivariant system. We denote by

$$G_1 \overset{\mu_1}{\rightarrow} G_2 \overset{\mu_2}{\rightarrow} G_3 \overset{\mu_3}{\rightarrow} \cdots \tag{14.33}$$

the associated chain of subgroups, where $G_1 = \mathrm{C}_{\infty\mathrm{v}} \times \widetilde{\mathrm{C}}_{\infty\mathrm{v}}$, and μ_i designates the irreducible representation of G_i associated with the bifurcation $G_i \rightarrow G_{i+1}$ for $i = 1,2,\ldots$. We estimate the bifurcation process (14.33) on the basis of the predominant Fourier coefficients and the symmetry of the Fourier terms mentioned above.

[14] A uniaxial compression test on kaolin clay and an endurance test on a steel ball bearing are conducted to obtain images of deformation patterns. The kaolin clay is suited for the visual observation of deformation patterns because of the geometrical characteristics of its grains that display optical anisotropy (Morgenstern and Tchalenko, 1967 [141]).

In view of the emergence of the Fourier terms, the postbifurcation stage of $G_i \to G_{i+1}$ is classified into two states: (1) activation of the Fourier terms related to the bifurcation mode and (2) activation of more terms by mode interference, as explained below.

Bifurcation Mode

The bifurcation $G_i \to G_{i+1}$ activates the bifurcation mode, that is, the Fourier terms in (14.32) for the associated irreducible representation μ_i. For instance, the change $\mathrm{OB}^+_{n\tilde{n}} \to \mathrm{EC}^+_{n\tilde{n}kl}$ of the symmetry is brought about by a bifurcation at a double point associated with a two-dimensional irreducible representation (labeled as (k,l) with $k \geq 1$ and $0 \leq l \leq d-1$ in §14.9.2). As derived in §14.9.4, the subspace of this two-dimensional irreducible representation is spanned by $\{\cos 2\pi(m\xi + \tilde{m}\tilde{\xi}), \sin 2\pi(m\xi + \tilde{m}\tilde{\xi})\}$ with the wave numbers $(m,\tilde{m})$ satisfying

$$k = \frac{|m\tilde{n} - \tilde{m}n|}{d}, \qquad l \equiv mp + \tilde{m}\tilde{p} \pmod d, \tag{14.34}$$

or by $\{\cos 2\pi(m\xi - \tilde{m}\tilde{\xi}), \sin 2\pi(m\xi - \tilde{m}\tilde{\xi})\}$ with the wave numbers $(m,\tilde{m})$ satisfying

$$k = \frac{|m\tilde{n} + \tilde{m}n|}{d}, \qquad l \equiv mp - \tilde{m}\tilde{p} \pmod d, \tag{14.35}$$

where $d = \gcd(n,\tilde{n})$ and $np + \tilde{n}\tilde{p} = d$, as defined in (14.19). It should be clear that $a \equiv b \pmod d$ means that $a - b$ is divisible by d.

Mode Interference

After the bifurcation, the solutions have G_{i+1}-symmetry. To reconstruct a solution with this symmetry, we collect those Fourier terms in (14.32) that are compatible with G_{i+1}. We denote by $u|G_{i+1}$ the partial sum of (14.32) consisting of those terms that are invariant to G_{i+1}. For instance, we have

$$u|\mathrm{OB}^+_{n\tilde{n}} = \sum_{k=0}^{\infty} A_{kn,k\tilde{n}} \cos 2\pi k(n\xi + \tilde{n}\tilde{\xi}). \tag{14.36}$$

The echelon pattern is simulated by

$$u|\mathrm{EC}^+_{n\tilde{n}kl} = \sum_{(m,\tilde{m})\in I_A} A_{m\tilde{m}} \cos 2\pi(m\xi + \tilde{m}\tilde{\xi}) + \sum_{(m,\tilde{m})\in I_B} B_{m\tilde{m}} \cos 2\pi(m\xi - \tilde{m}\tilde{\xi}), \tag{14.37}$$

where, as expounded in §14.9.4,

$$I_A = \{(m,\tilde{m}) \mid m = in + j(nl/d + k\tilde{p}) \geq 0,$$
$$\tilde{m} = i\tilde{n} + j(\tilde{n}l/d - kp) \geq 0,\ \ i,j \in \mathbb{Z}\}, \tag{14.38}$$
$$I_B = \{(m,\tilde{m}) \mid m = in + j(nl/d + k\tilde{p}) \geq 0,$$
$$\tilde{m} = -i\tilde{n} - j(\tilde{n}l/d - kp) \geq 0,\ \ i,j \in \mathbb{Z}\}. \tag{14.39}$$

It is emphasized that the bifurcation modes, which are activated by the bifurcation $G_i \to G_{i+1}$, are contained in the Fourier series $u|G_{i+1}$. The bifurcation mode is predominant in $u|G_{i+1}$ immediately after bifurcation, but the other Fourier terms with higher frequencies grow in magnitude thereafter. This postbifurcation growth of the other terms is called *mode interference*, which causes the localization of deformation.

14.7.2 Kaolin

We expand the kaolin image at the right of Fig. 14.19(a) into the Fourier series in (14.32). This image is the 128×128 pixel digitized data obtained by an image scanner from the domain shown at the left of Fig. 14.19(a). Figure 14.20 presents the density plot of the amplitudes of the Fourier terms, where a darker region has a larger amplitude. Although the distribution of the amplitudes appears to be random, it can offer us key information to untangle the underlying bifurcation mechanism, as described below.

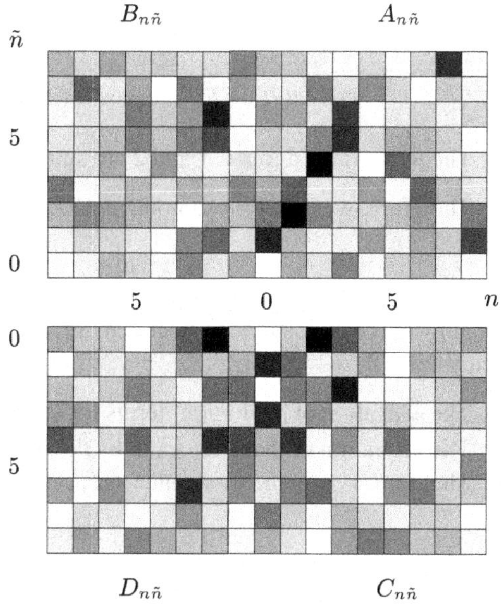

Fig. 14.20 Density plot of the amplitudes of the Fourier coefficients in (14.32) for kaolin specimen. The Fourier term with a darker region has a larger amplitude.

Stripe-Pattern Formation

The Fourier coefficient $A_{1,2}$ has the largest magnitude, and the coefficients $A_{2,4}$ and $A_{3,6}$ are also predominant. In view of (14.36), it would be rational to presume the presence of the recursive bifurcation process

$$\mathrm{C}_{\infty\mathrm{v}} \times \widetilde{\mathrm{C}}_{\infty\mathrm{v}} \to \mathrm{OB}^{+}_{3,6} \to \mathrm{OB}^{+}_{1,2} \tag{14.40}$$

to arrive at a series of stripe patterns labeled by $\mathrm{OB}^{+}_{3,6}$ and $\mathrm{OB}^{+}_{1,2}$ (see (14.25) for the hierarchy of the symmetry of the solution paths). The Fourier term $\cos 2\pi(3n+6\tilde{n})$ for $A_{3,6}$ is activated by the bifurcation $\mathrm{C}_{\infty\mathrm{v}} \times \widetilde{\mathrm{C}}_{\infty\mathrm{v}} \to \mathrm{OB}^{+}_{3,6}$, the term for $A_{1,2}$ is activated by the bifurcation $\mathrm{OB}^{+}_{3,6} \to \mathrm{OB}^{+}_{1,2}$, and the term for $A_{2,4}$ is by the mode interference after this bifurcation.

Echelon-Mode Formation

We consider the echelon-mode formation from the $\mathrm{OB}^{+}_{1,2}$-invariant path through the bifurcation

$$\mathrm{OB}^{+}_{1,2} \to \mathrm{EC}^{+}_{1,2,k,0} \tag{14.41}$$

for some k. For this case, with $n=1$, $\tilde{n}=2$, and $l=0$ $(d=1)$, we have $k=|m\tilde{n}-\tilde{m}n| = |2m-\tilde{m}|$ from (14.34) or $k=|m\tilde{n}+\tilde{m}n| = |2m+\tilde{m}|$ from (14.35).

We investigate the amplitudes of $A_{m\tilde{m}}$ and $B_{m\tilde{m}}$ using Fig. 14.20 to note the following issues.

- The most predominant terms are $A_{m\tilde{m}}$ with $(m,\tilde{m}) = (7,8)$, $(2,10)$, and $(5,4)$ with $k=|2m-\tilde{m}|=6$ associated with $\mathrm{EC}^{+}_{1,2,6,0}$-invariant Fourier terms

$$\cos 2\pi(7\xi+8\tilde{\xi}), \qquad \cos 2\pi(2\xi+10\tilde{\xi}), \qquad \cos 2\pi(5\xi+4\tilde{\xi}).$$

- The terms related to $k=1$ and 2 are sufficiently large.

Therefore, we can, for example, assume the presence of bifurcation process

$$\mathrm{OB}^{+}_{1,2} \to \mathrm{EC}^{+}_{1,2,6,0} \to \mathrm{EC}^{+}_{1,2,1,0}, \tag{14.42}$$

which yields echelon modes labeled by $\mathrm{EC}^{+}_{1,2,6,0}$ and $\mathrm{EC}^{+}_{1,2,1,0}$. We can see a possible sequence:

- The activation of the Fourier terms for $k=6$ by the bifurcation $\mathrm{OB}^{+}_{1,2} \to \mathrm{EC}^{+}_{1,2,6,0}$,
- The activation of those for $k=1$ by the bifurcation $\mathrm{EC}^{+}_{1,2,6,0} \to \mathrm{EC}^{+}_{1,2,1,0}$,
- That for $k=2$ by the mode interference after this bifurcation.

Loss of Half-Rotation Symmetry

We consider the bifurcations

$$\mathrm{OB}^{+}_{1,2} \to \mathrm{OB}^{+\circ}_{1,2}, \qquad \mathrm{EC}^{+}_{1,2,k,0} \to \mathrm{EC}^{+\circ}_{1,2,k,0}, \qquad k = 1,6, \tag{14.43}$$

which cause the loss of half-rotation symmetry $\sigma\tilde{\sigma}$ in association with the addition of sine terms ($\mathrm{EC}^{+\circ}_{1,2,1,0} = \mathrm{C}_1$).

For a reciprocal system, we can also assume the presence of the following bifurcation from an $\mathrm{OB}^{+\circ}_{1,2}$-invariant path,

$$\mathrm{OB}^{+\circ}_{1,2} \to \mathrm{EC}^{+\circ}_{1,2,k,0}.$$

In this case, the sets of wave numbers $(m,\tilde{m})$ corresponding to the number k are the same as those of the bifurcation (14.41), but the sine terms must be included in addition to the cosine terms. Therefore, we consider

$$u|\mathrm{OB}^{+\circ}_{n\tilde{n}} = \sum_{k=0}^{\infty} A_{kn,k\tilde{n}} \cos 2\pi k(n\xi + \tilde{n}\tilde{\xi}) + \sum_{k=1}^{\infty} C_{kn,k\tilde{n}} \sin 2\pi k(n\xi + \tilde{n}\tilde{\xi}). \tag{14.44}$$

Inasmuch as the amplitudes

$$\begin{cases} C_{0,1}, A_{0,1}, A_{3,5} & \text{for } k = 1, \\ C_{1,4}, C_{2,6}, A_{0,2} & \text{for } k = 2, \\ A_{7,8}, C_{3,0}, C_{5,4} & \text{for } k = 6 \end{cases}$$

are predominant, we can, for example, assume the presence of the bifurcation process

$$\mathrm{OB}^{+\circ}_{1,2} \dashrightarrow \mathrm{EC}^{+\circ}_{1,2,6,0} \dashrightarrow \mathrm{EC}^{+\circ}_{1,2,1,0} = \mathrm{C}_1. \tag{14.45}$$

Image Simulation

We have arrived at the hierarchical bifurcation shown in Fig. 14.21 as an assemblage of possible bifurcation processes in (14.40), (14.42), (14.43), and (14.45).

By assembling the Fourier terms related to the hierarchy of groups in Fig. 14.21, we predicted the progress of deformation of the kaolin specimen in Fig. 14.22, which shows a hierarchy of images from a uniform initial state to the final state shown at the right of Fig. 14.19, or Fig. 14.22(h). In Fig. 14.22, the transition ex-

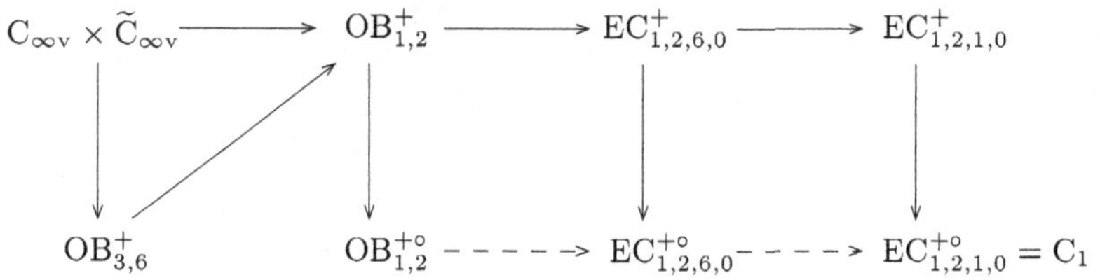

Fig. 14.21 Estimated possible bifurcation processes for the kaolin. The bifurcations indicated by the dashed arrows do not exist for a nonreciprocal system.

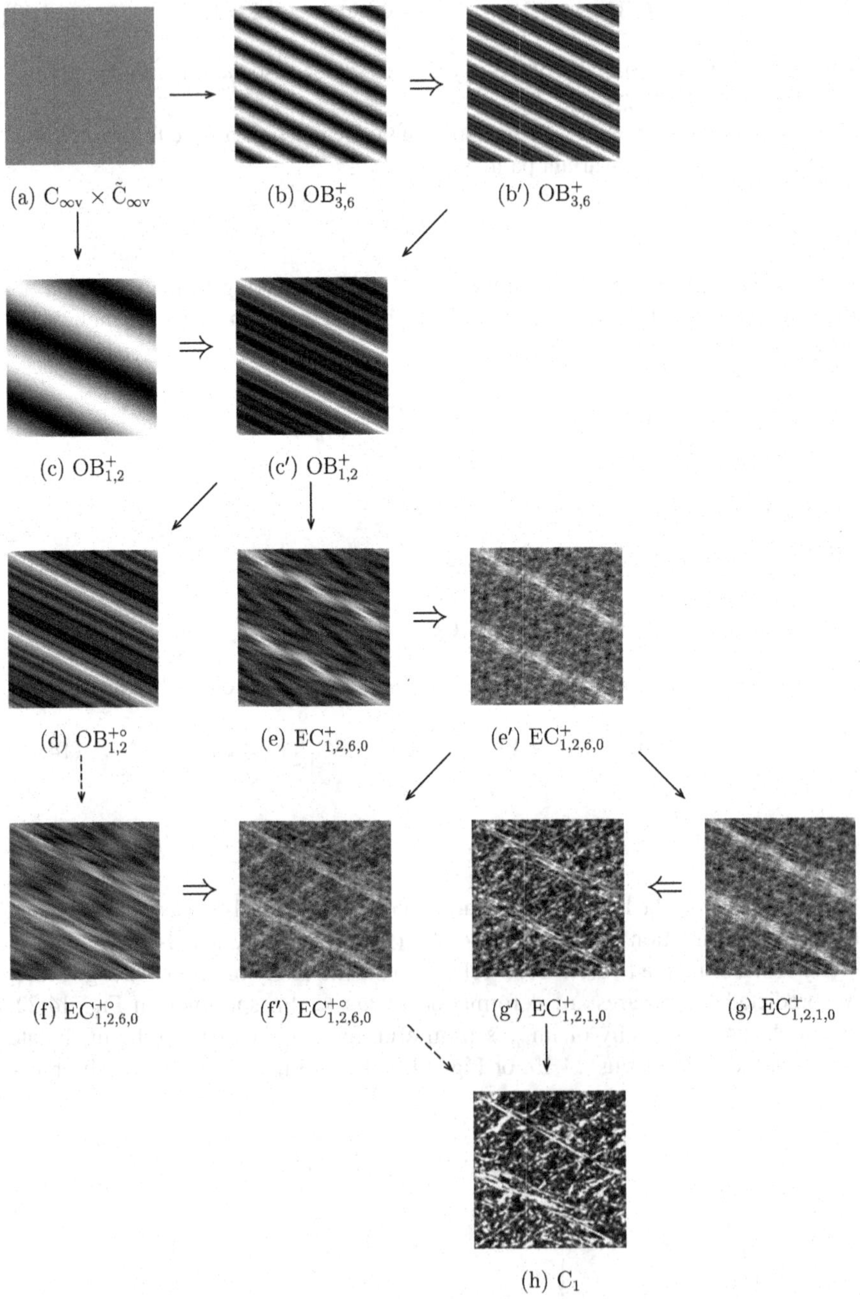

Fig. 14.22 Image simulation for a kaolin specimen: the progress of deformation expressed in terms of a hierarchy of images. (⟶): bifurcation; (⟹): mode interference; (– – →): bifurcation existent only for a reciprocal system.

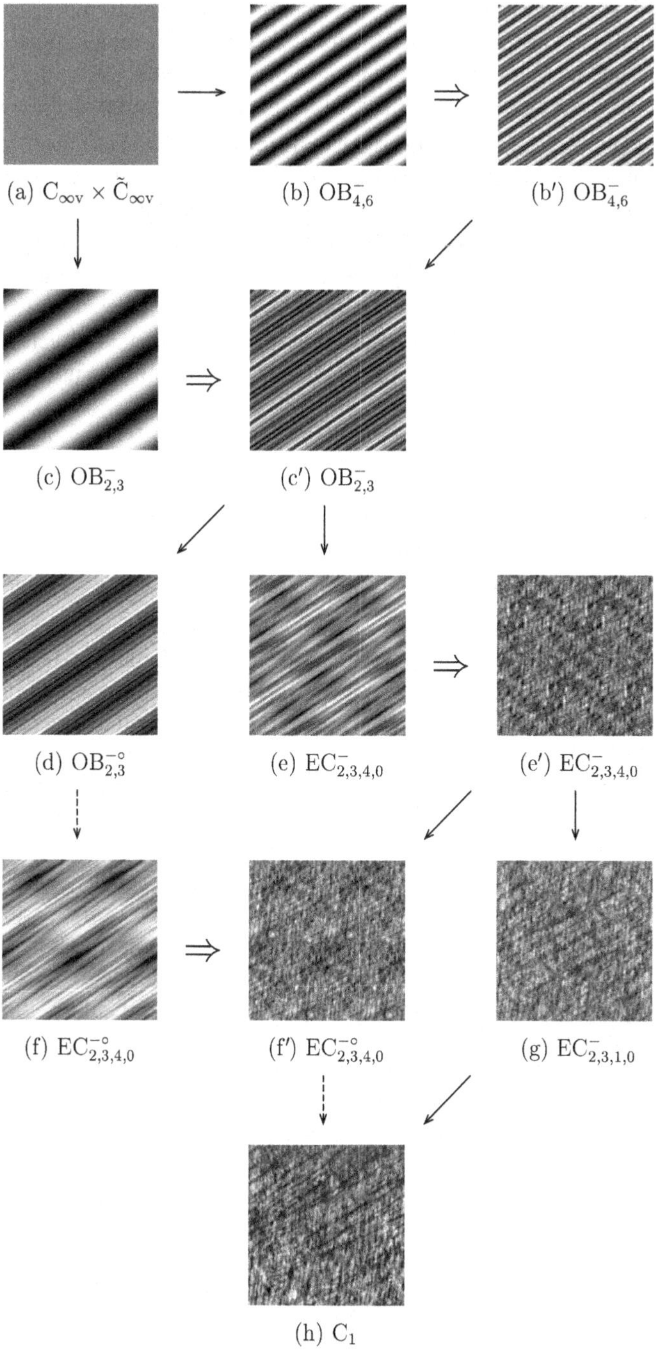

Fig. 14.23 Image simulation for a steel specimen: the progress of deformation expressed in terms of a hierarchy of images. ($\longrightarrow$): bifurcation; ($\Longrightarrow$): mode interference; ($--\rightarrow$): bifurcation existent only for a reciprocal system.

pressed by ($\longrightarrow$) means bifurcation, ($\Longrightarrow$) denotes mode interference, and ($\dashrightarrow$) is the bifurcation existent only for a reciprocal system. It is emphasized that some of these images might not actually take place in experiments.

Figure 14.22(c) corresponds to the $\mathrm{OB}^+_{1,2}$-invariant state immediately after bifurcation from the $\mathrm{C}_{\infty v} \times \widetilde{\mathrm{C}}_{\infty v}$-invariant uniform state. The state in Fig. 14.22(c) undergoes mode interference to arrive at the state in Fig. 14.22(c′). One can clearly see the emergence of a series of parallel stripe patterns labeled by $\mathrm{OB}^+_{3,6}$, $\mathrm{OB}^+_{1,2}$, and $\mathrm{OB}^{+\circ}_{1,2}$ in association with the bifurcation, and the localization along the stripes through the mode interference (Fig. 14.22(b), (b′), (c), (c′), (d)). Furthermore, at the onset of the bifurcation generating the echelon modes labeled by $\mathrm{EC}^+_{1,2,6,0}$ and $\mathrm{EC}^{+\circ}_{1,2,6,0}$, the stripes in a direction intersect with stripes in other directions to form discretized spatially periodic patterns: the echelon modes (cf., Fig. 14.22(e), (e′), (f), (f′)).

14.7.3 Steel

We address the steel specimen (image) at the right of Fig. 14.19(b), or Fig. 14.23(h). Figure 14.24 shows an estimated hierarchical bifurcation for the steel obtained by the same procedure as for the image simulation employed for kaolin. It is noteworthy that although different subgroups appear in the hierarchies for kaolin in Fig. 14.21 and for steel in Fig. 14.24, both of these hierarchies correspond to different branches of the same hierarchy in Fig. 14.11.

Figure 14.23 shows an image simulation of this hierarchical bifurcation expressed in terms of the progress of deformation from a uniform initial state in (a) to the final state in (h). It must be emphasized again that some of these images might not be observed in experiments. The mode interference denoted by ($\Longrightarrow$) in Fig. 14.23 is important in expressing the localization of deformation through the growth of the Fourier terms with higher frequencies. The images after mode interference express realistic localization of deformation.

The complexity of the echelon-mode formation notwithstanding, the recursive bifurcation rule, such as (14.25), presented in §14.4 has thus realized a successful

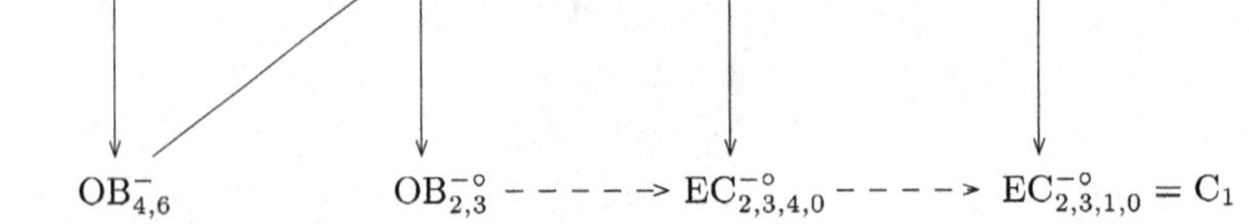

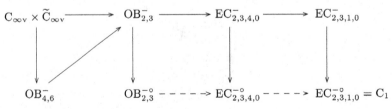

Fig. 14.24 Estimated possible bifurcation processes for the steel. The bifurcations indicated by the dashed arrows do not exist for a nonreciprocal system.

image simulation that offers an insight into the progress of deformation and is of great assistance in the study of damages on materials by bifurcations.

14.8 Diamond Pattern on Sand

The pattern formation of a sand specimen is investigated through experimentation, image simulation, and numerical bifurcation analysis.[15] By the image simulation of deformation of the sand specimen, we demonstrate that the direct bifurcation

$$\mathrm{C}_{\infty\mathrm{v}} \times \widetilde{\mathrm{C}}_{\infty\mathrm{v}} \to \mathrm{DI}_{n\tilde{n}}$$

occurs to break uniformity and engenders a diamond pattern. This bifurcation precedes the formation of echelonlike modes, which presents a sharp contrast with the kaolin and steel in §14.7 where oblique stripe patterns preceded the formation of echelon modes.

Remark 14.6. The formation of a diamondlike distributed deformation pattern is ascribed to a direct bifurcation in (soil) mechanics. This is called the *diffuse mode bifurcation*. □

14.8.1 Experiment

A characteristic shear band pattern is observed in an experiment.[16] Photographs are taken and numbered 1–8. The progression of localization of incremental strain fields between two neighboring photographs is obtained at the top row in Fig.14.25(a).

- During increments 1–4, the specimen displays the orientation of spatially distributed strain localization, which is weak and obscure.
- Two parallel oblique shear bands are observed during increments 3–5.
- During increments 5–8, some shear bands diminish gradually in favor of the emergence of two oblique shear bands in a different direction.

As a consequence, complex geometrical patterns are formed by parallel and crossing shear bands. The strain fields have thus been visualized. However, a bifurcation mode is not detected.

---

[15] This investigation is based on Ikeda et al., 2008 [104].

[16] A fine angular, siliceous sand (Hostun RF) specimen of 164.0 mm × 173.0 mm × 35.4 mm is tested by the plane strain compression apparatus in Desrues and Viggiani, 2004 [39]. The false relief stereophotogrammetry (FRS) method is used to digitize the displacement fields of the side of the specimen deforming under load. By FRS the deformation can be perceived directly as a fictitious relief by the well-known stereoscopic effect on successive pairs of photographs.

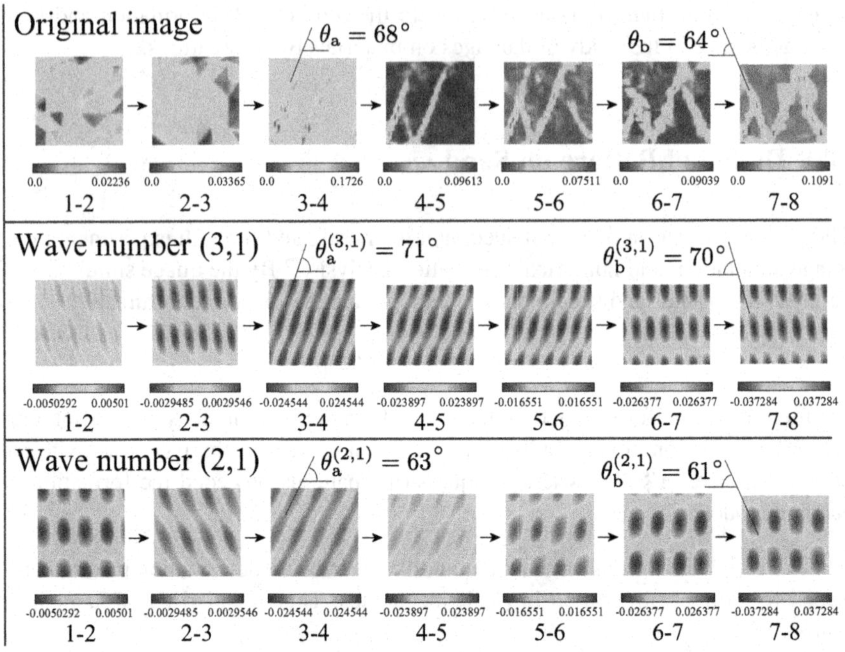

(a) Contour views of stereophotogrammetry-based incremental strain fields

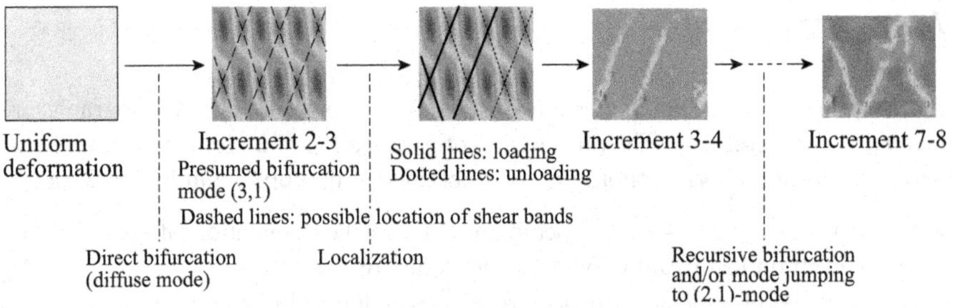

(b) Progress of diffuse-mode bifurcation, followed by localization, further bifurcation, and/or mode jumping

Fig. 14.25 Image simulation of incremental strain fields of patterned shear bands observed on Hostun sand rectangular parallelepiped specimens. In (a): top row, original images; second row, decomposed images for wave number (3,1); and third row, decomposed images for wave number (2,1). The strain means the deviatoric logarithmic strain.

To represent a bifurcation mode, the original incremental fields of the shear strain invariant (at the top row in Fig.14.25(a)) are expanded into the double Fourier series:[17]

$$\varepsilon(\xi,\tilde{\xi}) = \sum_{n=1}\sum_{\tilde{n}=1}(A_{n\tilde{n}} \sin 2\pi n\xi \cdot \cos 2\pi\tilde{n}\tilde{\xi} + B_{n\tilde{n}} \sin 2\pi n\xi \cdot \sin 2\pi\tilde{n}\tilde{\xi}$$
$$+ C_{n\tilde{n}} \cos 2\pi n\xi \cdot \cos 2\pi\tilde{n}\tilde{\xi} + D_{n\tilde{n}} \cos 2\pi n\xi \cdot \sin 2\pi\tilde{n}\tilde{\xi}),$$
$$0 \le \xi,\tilde{\xi} < 1, \qquad (14.46)$$

that can engender diamond and oblique stripe patterns

$$\begin{cases} C_1 \cos 2\pi(n\xi+\alpha)\cdot\cos 2\pi(\tilde{n}\tilde{\xi}+\beta): & \text{diamond pattern} \\ C_2 \cos 2\pi(n\xi \pm \tilde{n}\tilde{\xi}+\alpha): & \text{oblique stripe pattern} \end{cases}$$

for wave number $(n,\tilde{n})$ and for some constants C_1 and C_2, and $0 \le \alpha,\beta < 1$.

The history of the magnitude of the Fourier coefficient for each wave number is investigated as presented in Fig.14.26. Each wave number corresponds to a possible bifurcation mode (cf., (14.46)). Strong magnitudes are detected for two wave numbers: (3,1) and (2,1). The magnitude for mode (3,1) is predominant and increases sharply during increments 3–6. That for mode (2,1) increases stably during increments 6–8 and becomes predominant at the final state. Histories of the decomposed strain fields for (3,1) and (2,1) (at the second and third rows, respectively, in Fig.14.25(a)) display diamondlike and stripelike patterns. During increments 3–6, the inclination of shear bands in the original image is $\theta_{\rm a} = 68°$ and is close to $\theta_{\rm a}^{(3,1)} = 71°$ of localization patterns of (3,1). During increments 6–8, the inclination

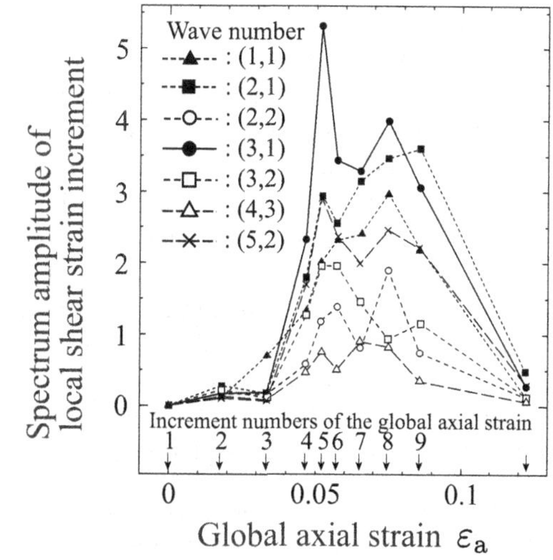

Fig. 14.26 History of intensity of decomposed strain fields between photographed points. Stereophotogrammetry-based incremental strain fields are expanded into the double Fourier series and further classified into distinct bifurcation modes. The intensity is defined as $(A_{n\tilde{n}}{}^2 + B_{n\tilde{n}}{}^2 + C_{n\tilde{n}}{}^2 + D_{n\tilde{n}}{}^2)^{1/2}$ using the coefficients in (14.46).

[17] The Fourier series (14.46) employed here for the sand differs from the Fourier series (14.32) for the kaolin and the steel. The series (14.32) is intended to detect oblique stripe patterns, whereas (14.46) is to detect diamond patterns, as well as the stripe patterns.

of shear bands is $\theta_b = 64°$ and is close to $\theta_b^{(2,1)} = 61°$ of localization patterns of (2,1). The change of inclination of shear bands, accordingly, can be explained as the change of the wave number of the predominant modes.

Based on this observation, we introduce an interpretation of the experimental behavior, as illustrated in Fig.14.25(b).

- During increment 2–3, the mode for wave number (3,1) is initiated possibly by a bifurcation
$$C_{\infty v} \times \widetilde{C}_{\infty v} \to DI_{3,1}$$
almost hidden behind the predominant uniform compressive deformation.
- Thereafter, the mode for (3,1) grows sharply to form localized deformation shown by the dashed lines that form a diamond pattern.
- During increment 3–4, among these possible locations of localization, loading progresses in some locations and develops into shear bands (shown by solid lines), whereas unloading progresses in other locations (shown by dotted lines).
- The shift of the predominant wave number from (3,1) to (2,1) observed herein renders geometrical patterns complex. This shift is ascribed to recursive bifurcation and/or mode jumping.[18]

The validity of this interpretation is assessed in the sequel, based on numerical simulation.

14.8.2 Numerical Simulation

Bifurcation analysis is conducted on the finite element model of a rectangular uniform domain as presented in Fig.14.27. Figure 14.28 shows computed equilibrium paths.[19] On the fundamental path, closely located bifurcation points are observable.[20] From the fundamental equilibrium path associated with uniform homogeneous deformation shown by the thick solid line, four bifurcated equilibrium paths (thin solid lines) branch at four closely located bifurcation points [i]–[iv] (∘). These bifurcated paths follow fairly well the experimental curve shown by the dashed line.

As shown by the postbifurcation progress of strains on the bifurcated paths at the left of Fig.14.29, the bifurcation modes at bifurcation points [i]–[iv] are spatially periodic, diamond-patternlike, diffuse modes with different but nearly equivalent wave numbers as shown at the left of Fig.14.29. They generate spatially periodic strain-localized locations that form echelonlike modes. Most of these locations undergo unloading; only a few of them undergo loading to engender shear bands with diverse and complex geometrical patterns as shown at the right of Fig.14.29.

[18] *Mode jumping* means a sudden and dynamic shift to a different wave number.

[19] Details of the numerical procedure used here are given in Ikeda, Yamakawa, and Tsutsumi, 2003 [105].

[20] Closely located simple bifurcation points appear extensively in the bifurcation of materials, and are called a *clustered bifurcation point* or a *point of accumulation* by Hill and Hutchinson, 1975 [73].

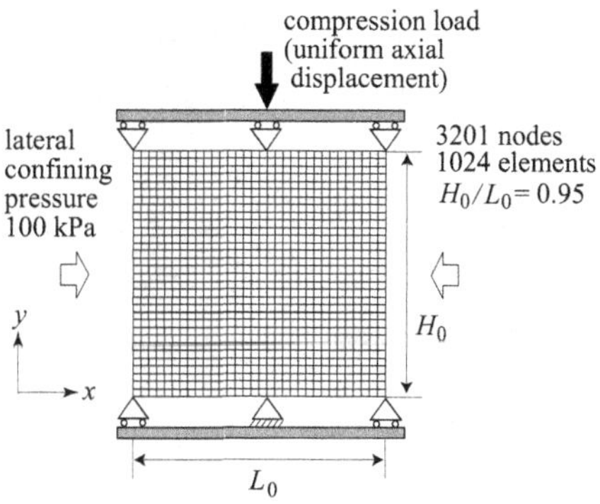

Fig. 14.27 Finite element model of a rectangular uniform domain that is divided into 32×32 elements (3201 nodes). The Drucker–Prager model is used with material properties: elastic bulk modulus, 12.50 MPa; elastic shear modulus, 5.77 MPa; critical stress ratio, 0.943; internal friction angle, 35.0°; and dilatancy angle, 10.0°.

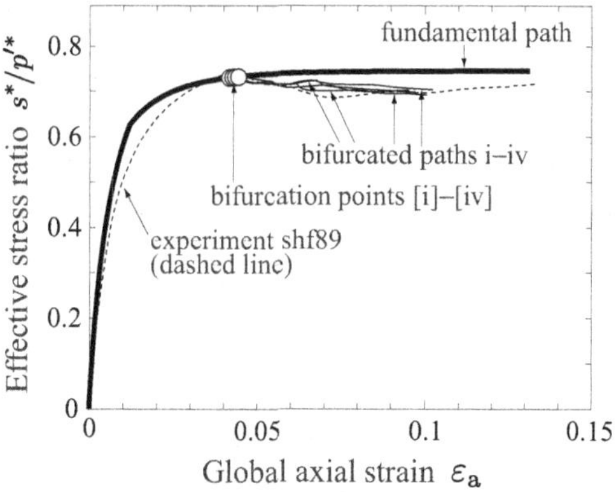

Fig. 14.28 Equilibrium paths obtained by elastoplastic finite-displacement numerical analysis. Thick solid line: fundamental (prebifurcation) path; thin solid lines: four bifurcated (postbifurcation) paths i–iv; dashed line: experimental curve; ○: bifurcation points [i]–[iv], which are located respectively at $\varepsilon_a = 0.0424$, 0.0430, 0.0432, and 0.0433.

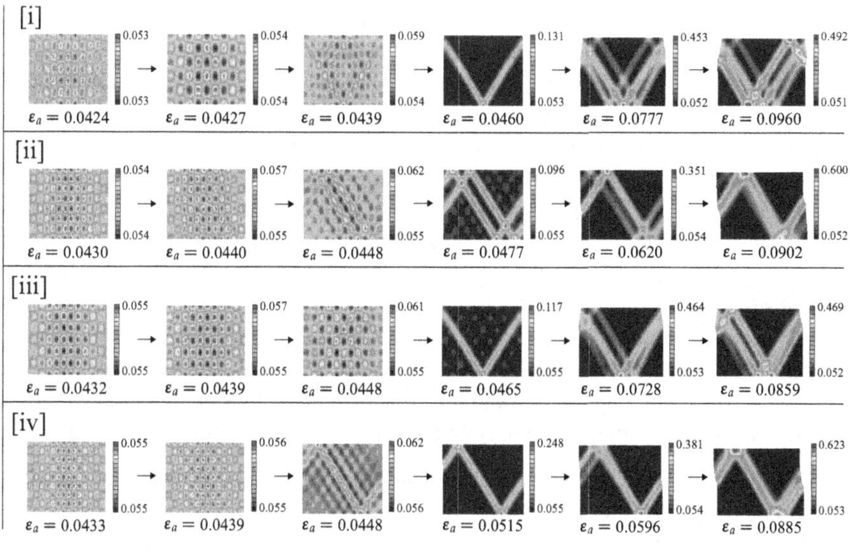

Fig. 14.29 Progress of the distribution of shear strains on bifurcated paths.

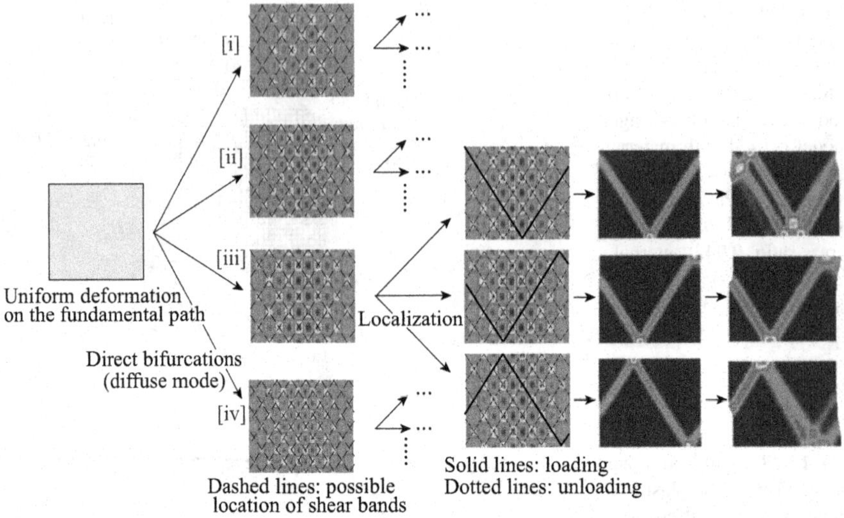

Fig. 14.30 Explosive increase of possible postbifurcation states through bifurcations and/or mode jumping.

Multiple possible states of equilibrium caused by bifurcations are observable. For example, three completely different solutions emanate from the same bifurcation point [iii] with slightly different numerical conditions as portrayed in Fig.14.30. The numerical equilibrium paths zigzag up and down,[21] which indicates possible further bifurcation and/or mode jumping. Such multiple states of equilibrium should be treated as an essential difficulty that is inherent in bifurcation phenomena. Because localized locations exist periodically, several possible locations have a similar likelihood to develop into shear bands. The locations that develop into shear bands change case by case, possibly because of imperfections arising from initial inhomogeneities. This mechanism to diversify the shear band patterns is designated as the "explosive increase of possible postbifurcation states." It resolves a paradox that has long puzzled us: The patterns of shear bands are quite diverse and every test appears to be unique.[22] The frustrating and surprising aspect is that experimental efforts to make specimens as homogeneous as possible end up with unpredictable diverse responses.

The occurrence of diffuse mode bifurcation, recursive bifurcation, and/or mode jumping, followed by localization into shear bands is the underlying mechanism forming geometrical patterns on sand (cf., Fig.14.25(b)). This understanding ends a long controversy of whether the initial bifurcation breaking uniformity is either:

- Shearband mode bifurcation that spontaneously engenders a shear band(s) without undergoing further bifurcation, or

[21] Such an up-and-down zigzag trend is also reported in Gajo, Bigoni, and Wood, 2004 [53].

[22] This statement is based on Desrues and Viggiani, 2004 [39].

- Diffuse mode bifurcation that causes distributed deformation initially and engenders a shear band(s) later.

Although shear band formation has heretofore attracted the most attention, this diffuse mode bifurcation is a catalyst that engenders geometrical patterns.

14.9 Appendix: Derivation of Bifurcation Rules

The bifurcation rule for systems equivariant to $\mathrm{C}_{\infty\mathrm{v}} \times \widetilde{\mathrm{C}}_{\infty\mathrm{v}}$ ($\simeq \mathrm{O}(2) \times \mathrm{O}(2)$) or its subgroups is derived in this appendix.

14.9.1 Bifurcation of O(2) × O(2)*-Equivariant System*

The direct bifurcation of a $\mathrm{C}_{\infty\mathrm{v}} \times \widetilde{\mathrm{C}}_{\infty\mathrm{v}} (\simeq \mathrm{O}(2) \times \mathrm{O}(2))$-equivariant system is investigated.

Irreducible Representations

We first enumerate all the irreducible representations of $\mathrm{C}_{\infty\mathrm{v}} \times \widetilde{\mathrm{C}}_{\infty\mathrm{v}}$. Since $\mathrm{C}_{\infty\mathrm{v}} \times \widetilde{\mathrm{C}}_{\infty\mathrm{v}}$ is a direct product of two groups $\mathrm{C}_{\infty\mathrm{v}}$ and $\widetilde{\mathrm{C}}_{\infty\mathrm{v}}$, each isomorphic to O(2), the family of the irreducible representations of $\mathrm{C}_{\infty\mathrm{v}} \times \widetilde{\mathrm{C}}_{\infty\mathrm{v}}$ is given by the family of the tensor products of two irreducible representations of O(2). The $\mathbb{R}$-irreducible representations of O(2) are absolutely irreducible.

We index the family of the inequivalent irreducible representations of $\mathrm{C}_{\infty\mathrm{v}} = \langle \sigma, r(\psi) \rangle$ by

$$R(\mathrm{C}_{\infty\mathrm{v}}) = \{(+)_{\mathrm{C}_{\infty\mathrm{v}}}, (-)_{\mathrm{C}_{\infty\mathrm{v}}}, (1)_{\mathrm{C}_{\infty\mathrm{v}}}, (2)_{\mathrm{C}_{\infty\mathrm{v}}}, \ldots\}. \tag{14.47}$$

Here $(+)_{\mathrm{C}_{\infty\mathrm{v}}}$ corresponds to the (one-dimensional) unit representation and $(-)_{\mathrm{C}_{\infty\mathrm{v}}}$ to the nonunit one-dimensional irreducible representation, being defined, respectively, by 1×1 representation matrices

$$T^{(+)_{\mathrm{C}_{\infty\mathrm{v}}}}(r(\psi)) = 1, \qquad T^{(+)_{\mathrm{C}_{\infty\mathrm{v}}}}(\sigma) = 1, \tag{14.48}$$

$$T^{(-)_{\mathrm{C}_{\infty\mathrm{v}}}}(r(\psi)) = 1, \qquad T^{(-)_{\mathrm{C}_{\infty\mathrm{v}}}}(\sigma) = -1. \tag{14.49}$$

The remaining irreducible representations, labeled $(1)_{\mathrm{C}_{\infty\mathrm{v}}}, (2)_{\mathrm{C}_{\infty\mathrm{v}}}, \ldots$ in (14.47), are two-dimensional, and are defined by

$$\begin{aligned} T^{(n)_{\mathrm{C}_{\infty\mathrm{v}}}}(r(\psi)) &= \begin{pmatrix} \cos(2\pi n\psi) & -\sin(2\pi n\psi) \\ \sin(2\pi n\psi) & \cos(2\pi n\psi) \end{pmatrix}, \\ T^{(n)_{\mathrm{C}_{\infty\mathrm{v}}}}(\sigma) &= \begin{pmatrix} 1 & 0 \\ 0 & -1 \end{pmatrix}, \qquad n = 1, 2, \ldots. \end{aligned} \tag{14.50}$$

With the apparent correspondence

$$r(\psi) \leftrightarrow \tilde{r}(\tilde{\psi}), \qquad \sigma \leftrightarrow \tilde{\sigma}$$

between $\mathrm{C}_{\infty\mathrm{v}}$ and $\widetilde{\mathrm{C}}_{\infty\mathrm{v}}$, we index the family of the inequivalent irreducible representations of $\widetilde{\mathrm{C}}_{\infty\mathrm{v}}$ by

$$R(\widetilde{\mathrm{C}}_{\infty\mathrm{v}}) = \{(+)_{\widetilde{\mathrm{C}}_{\infty\mathrm{v}}}, (-)_{\widetilde{\mathrm{C}}_{\infty\mathrm{v}}}, (1)_{\widetilde{\mathrm{C}}_{\infty\mathrm{v}}}, (2)_{\widetilde{\mathrm{C}}_{\infty\mathrm{v}}}, \ldots\}. \tag{14.51}$$

Then the family of the irreducible representations of $\mathrm{C}_{\infty\mathrm{v}} \times \widetilde{\mathrm{C}}_{\infty\mathrm{v}}$ is indexed by

$$R(\mathrm{C}_{\infty\mathrm{v}} \times \widetilde{\mathrm{C}}_{\infty\mathrm{v}}) = \{(k,\tilde{k}) \mid k,\tilde{k} = +,-,1,2,\ldots\}, \tag{14.52}$$

where $(k,\tilde{k}) = (k,\tilde{k})_{\mathrm{C}_{\infty\mathrm{v}} \times \widetilde{\mathrm{C}}_{\infty\mathrm{v}}}$ stands for the tensor product of the irreducible representation $(k)_{\mathrm{C}_{\infty\mathrm{v}}}$ of $\mathrm{C}_{\infty\mathrm{v}}$ and $(\tilde{k})_{\widetilde{\mathrm{C}}_{\infty\mathrm{v}}}$ of $\widetilde{\mathrm{C}}_{\infty\mathrm{v}}$. The dimension of the irreducible representation $(k,\tilde{k})$, being equal to the product of the dimensions of $(k)_{\mathrm{C}_{\infty\mathrm{v}}}$ and of $(\tilde{k})_{\widetilde{\mathrm{C}}_{\infty\mathrm{v}}}$, is equal to either 1, 2, or 4.

To categorize the bifurcated paths and group-theoretic bifurcation points of a $\mathrm{C}_{\infty\mathrm{v}} \times \widetilde{\mathrm{C}}_{\infty\mathrm{v}}$-equivariant system, we refer to the symmetry groups G^{μ} for the irreducible representations $\mu = (k,\tilde{k})$ (G^{μ} is defined by (7.22)); see Table 14.1 for their actual forms. The symmetry of bifurcated paths, which is to be determined by solving the bifurcation equations, is not necessarily the same as the symmetry G^{μ} of the kernel space of the Jacobian matrix (cf., (7.94)). We use the following notations:

Table 14.1 Classification of critical points of a $\mathrm{C}_{\infty\mathrm{v}} \times \widetilde{\mathrm{C}}_{\infty\mathrm{v}}$-equivariant system

Multiplicity	Irreducible	Symmetry Groups	
M	Representation, μ	G^{μ}	Bifurcated Paths
1	$(+,+)$	$\mathrm{C}_{\infty\mathrm{v}} \times \widetilde{\mathrm{C}}_{\infty\mathrm{v}}$	No Bifurcation
	$(+,-)$	$\mathrm{C}_{\infty\mathrm{v}} \times \widetilde{\mathrm{C}}_{\infty}$	$\mathrm{C}_{\infty\mathrm{v}} \times \widetilde{\mathrm{C}}_{\infty}$
	$(-,+)$	$\mathrm{C}_{\infty} \times \widetilde{\mathrm{C}}_{\infty\mathrm{v}}$	$\mathrm{C}_{\infty} \times \widetilde{\mathrm{C}}_{\infty\mathrm{v}}$
	$(-,-)$	$\mathrm{D}_{\infty\infty}$	$\mathrm{D}_{\infty\infty}$
2	$(+,\tilde{n})$	$\mathrm{C}_{\infty\mathrm{v}} \times \widetilde{\mathrm{C}}_{\tilde{n}}$	$\mathrm{C}_{\infty\mathrm{v}} \times \widetilde{\mathrm{C}}_{\tilde{n}\mathrm{v}}$
	$(-,\tilde{n})$	$\langle r(\psi), \sigma\tilde{r}(1/(2\tilde{n}))\rangle$	$\langle r(\psi), \sigma\tilde{r}(1/(2\tilde{n})), \tilde{\sigma}\rangle$
	$(n,+)$	$\mathrm{C}_{n} \times \widetilde{\mathrm{C}}_{\infty\mathrm{v}}$	$\mathrm{C}_{n\mathrm{v}} \times \widetilde{\mathrm{C}}_{\infty\mathrm{v}}$
	$(n,-)$	$\langle \tilde{r}(\tilde{\psi}), \tilde{\sigma}r(1/(2n))\rangle$	$\langle \sigma, \tilde{r}(\tilde{\psi}), \tilde{\sigma}r(1/(2n))\rangle$
4	$(n,\tilde{n})$	$\langle r(1/(2n))\tilde{r}(1/(2\tilde{n}))$, $r(1/(2n))\tilde{r}(-1/(2\tilde{n}))\rangle$	$\mathrm{DI}_{n\tilde{n}}$, $\mathrm{OB}^{\pm}_{n\tilde{n}}$

$$\mathrm{C}_\infty = \{r(\psi)\mid 0\le\psi<1\},\qquad \widetilde{\mathrm{C}}_\infty = \{\tilde{r}(\tilde{\psi})\mid 0\le\tilde{\psi}<1\},$$
$$\mathrm{C}_n = \langle r(1/n)\rangle,\qquad \widetilde{\mathrm{C}}_{\tilde{n}} = \langle \tilde{r}(1/\tilde{n})\rangle,$$
$$\mathrm{C}_{n\mathrm{v}} = \langle \sigma, r(1/n)\rangle,\qquad \widetilde{\mathrm{C}}_{\tilde{n}\mathrm{v}} = \langle \tilde{\sigma}, \tilde{r}(1/\tilde{n})\rangle,$$

and $\mathrm{C}_{\infty\mathrm{v}}$, $\widetilde{\mathrm{C}}_{\infty\mathrm{v}}$, $\mathrm{D}_{\infty\infty}$, $\mathrm{DI}_{n\tilde{n}}$, $\mathrm{OB}^+_{n\tilde{n}}$, and $\mathrm{OB}^-_{n\tilde{n}}$ are defined in (14.12)–(14.17). Recall that the multiplicity M of the critical (bifurcation) point is equal to the dimension of the associated irreducible representation.

The bifurcation equations at the critical points of multiplicity $M = 1, 2, 4$ are considered in the remainder of this section.

Simple Critical Point

A simple critical point (with $M = 1$) is associated with one of the four one-dimensional irreducible representations, $\mu = (k,\tilde{k})$ $(k,\tilde{k} = +,-)$. The critical point is, generically, a limit point if $\mu = (+,+)$, and it is a simple pitchfork bifurcation point otherwise. The symmetry of the bifurcated path agrees with the symmetry G^μ of the kernel space of the Jacobian matrix. As listed in Table 14.1, we obtain, from (14.48) and (14.49),

$$G^\mu = \begin{cases} \mathrm{C}_{\infty\mathrm{v}}\times\widetilde{\mathrm{C}}_\infty & \text{for } \mu=(+,-),\\ \mathrm{C}_\infty\times\widetilde{\mathrm{C}}_{\infty\mathrm{v}} & \text{for } \mu=(-,+),\\ \mathrm{D}_{\infty\infty} & \text{for } \mu=(-,-).\end{cases}$$

Double Bifurcation Point

We present the analysis of the bifurcation equation at a group-theoretic bifurcation point associated with the two-dimensional irreducible representation $(n,-)$. The other case of $(-,\tilde{n})$ can be treated similarly, and the cases of $(n,+)$ and $(+,\tilde{n})$ are immediate from the result for a $\mathrm{C}_{\infty\mathrm{v}}$-equivariant system included in §13.4. For the case of $(n,-)$, we make use of a slight variant of the standard technique for an O(2)-equivariant system found in the literature.[23]

With the use of a complex variable $z = w_1 + \mathrm{i}w_2$ to represent the two independent real variables w_1 and w_2, the bifurcation equation is expressed as

$$F(z,\bar{z},\widetilde{f}) = 0, \tag{14.53}$$

where we often omit the increment $\widetilde{f}$ of f in the subsequent derivation.

We can assume that the action of $\mathrm{C}_{\infty\mathrm{v}}\times\widetilde{\mathrm{C}}_{\infty\mathrm{v}}$ is defined by

[23] See, for example, Sattinger, 1983 [178] and Golubitsky, Stewart, and Schaeffer, 1988 [64].

$$
\begin{aligned}
r(\psi) &: z \mapsto \omega z, \\
\sigma &: z \mapsto \overline{z}, \\
\tilde{r}(\tilde{\psi}) &: z \mapsto z, \\
\tilde{\sigma} &: z \mapsto -z,
\end{aligned}
\tag{14.54}
$$

where $\omega = \exp(\mathrm{i}2\pi n\psi)$ and $\overline{\cdot}$ denotes the complex conjugate. The group $\mathrm{C}_{\infty\mathrm{v}} \times \widetilde{\mathrm{C}}_{\infty\mathrm{v}}$ is generated by the four elements $r(\psi)$, σ, $\tilde{r}(\tilde{\psi})$, and $\tilde{\sigma}$; therefore, the equivariance of the bifurcation equation (14.53) to the group $\mathrm{C}_{\infty\mathrm{v}} \times \widetilde{\mathrm{C}}_{\infty\mathrm{v}}$ is identical to the equivariance to the action of these four elements. Therefore, the equivariance condition can be written as

$$\omega F(z,\overline{z}) = F(\omega z, \overline{\omega}\overline{z}), \tag{14.55}$$

$$\overline{F(z,\overline{z})} = F(\overline{z}, z), \tag{14.56}$$

$$-F(z,\overline{z}) = F(-z,-\overline{z}). \tag{14.57}$$

We expand F as

$$F(z,\overline{z},\widetilde{f}) = \sum_{a=0}\sum_{b=0} A_{ab}(\widetilde{f}) z^a \overline{z}^b. \tag{14.58}$$

Substitution of (14.58) into (14.55) and (14.56) shows that the indices (a,b) for nonzero terms should satisfy

$$\exp[\mathrm{i}2\pi n\psi(a-b-1)] = 1, \tag{14.59}$$

$$A_{ab} = \overline{A_{ab}}, \tag{14.60}$$

where $A_{ab} = A_{ab}(\widetilde{f})$. Since (14.59) must hold for arbitrary (real) values of ψ, we have

$$a - b - 1 = 0. \tag{14.61}$$

With the use of (14.61), equation (14.58) can be rewritten as

$$F = z \sum_{a=0} A_a(\widetilde{f}) |z|^{2a}, \tag{14.62}$$

where $A_a(\widetilde{f}) = A_{a+1,a}(\widetilde{f})$ is real by (14.60). Then (14.57) is also satisfied. Since $(z,\widetilde{f}) = (0,0)$ corresponds to the critical point, we have $A_0(0) = 0$.

The solutions of (14.62) are either the trivial solution $z = 0$ or the bifurcated solution, which satisfies

$$\sum_{a=0} A_a(\widetilde{f}) |z|^{2a} = 0.$$

Hence the bifurcated solution, if any, takes the form $|z| = \Phi(\widetilde{f})$.

As the representative (up to conjugacy) of such solutions, we consider a solution with z real, because other solutions can be obtained from that solution by transformations with some elements of $\mathrm{C}_{\infty\mathrm{v}} \times \widetilde{\mathrm{C}}_{\infty\mathrm{v}}$. Using $z = \overline{z}$ in (14.54), we note that this solution is invariant to

$$\sigma, \qquad \tilde{\sigma}r\left(\frac{1}{2n}\right), \qquad \tilde{r}(\tilde{\psi}),$$

and, therefore, to the subgroup $\langle \sigma, \tilde{\sigma}r(1/(2n)), \tilde{r}(\tilde{\psi})\rangle$.

Quadruple Bifurcation Point

We solve the bifurcation equation at a group-theoretic quadruple bifurcation point of a $C_{\infty v} \times \widetilde{C}_{\infty v}$ ($\simeq O(2) \times O(2)$)-equivariant system by adapting the technique for an $O(2) \times SO(2)$-equivariant system.[24] We assume that this bifurcation point is associated with the four-dimensional irreducible representation indexed by $(n, \tilde{n})$ with n and $\tilde{n}$ being positive integers.

Instead of four real independent variables, w_1, w_2, w_3, and w_4, two complex variables $z_1 = w_1 + \mathrm{i}w_2$ and $z_2 = w_3 + \mathrm{i}w_4$ are used. Then the bifurcation equation takes the form

$$F_1(z_1, z_2, \widetilde{f}) = F_2(z_1, z_2, \widetilde{f}) = 0, \tag{14.63}$$

where $F_i(z_1, z_2, \widetilde{f})$ is an abbreviation for $F_i(z_1, z_2, \overline{z_2}, \overline{z_1}, \widetilde{f})$ (not for $F_i(z_1, z_2, \overline{z_1}, \overline{z_2}, \widetilde{f})$) for $i = 1, 2$, and the load increment $\widetilde{f}$ is often omitted in the derivation below.

The four-dimensional irreducible matrix representation of

$$C_{\infty v} \times \widetilde{C}_{\infty v} = \langle r(\psi), \sigma, \tilde{r}(\tilde{\psi}), \tilde{\sigma}\rangle$$

is given by

$$\widehat{T}(r(\psi)) = \begin{pmatrix} \omega & 0 & 0 & 0 \\ 0 & \omega & 0 & 0 \\ 0 & 0 & \overline{\omega} & 0 \\ 0 & 0 & 0 & \overline{\omega} \end{pmatrix}, \qquad \widehat{T}(\sigma) = \begin{pmatrix} 0 & 0 & 1 & 0 \\ 0 & 0 & 0 & 1 \\ 1 & 0 & 0 & 0 \\ 0 & 1 & 0 & 0 \end{pmatrix}, \tag{14.64}$$

$$\widehat{T}(\tilde{r}(\tilde{\psi})) = \begin{pmatrix} \zeta & 0 & 0 & 0 \\ 0 & \overline{\zeta} & 0 & 0 \\ 0 & 0 & \zeta & 0 \\ 0 & 0 & 0 & \overline{\zeta} \end{pmatrix}, \qquad \widehat{T}(\tilde{\sigma}) = \begin{pmatrix} 0 & 1 & 0 & 0 \\ 1 & 0 & 0 & 0 \\ 0 & 0 & 0 & 1 \\ 0 & 0 & 1 & 0 \end{pmatrix}, \tag{14.65}$$

where

$$\omega = \exp(\mathrm{i}2\pi n\psi), \qquad \zeta = \exp(\mathrm{i}2\pi\tilde{n}\tilde{\psi})$$

and $\overline{\cdot}$ means the complex conjugate; $\widehat{T}$ is the tensor product of the representation matrices of $(n)_{C_{\infty v}}$ and $(\widetilde{n})_{\widetilde{C}_{\infty v}}$. Assuming that $\widehat{T}$ represents the action of $C_{\infty v} \times \widetilde{C}_{\infty v}$ on $(z_1, z_2, \overline{z_2}, \overline{z_1})$, we obtain a succinct expression of this action:

$$\begin{aligned} r(\psi) &: z_1 \mapsto \omega z_1, & z_2 &\mapsto \omega z_2, \\ \sigma &: z_1 \mapsto \overline{z_2}, & z_2 &\mapsto \overline{z_1}, \\ \tilde{r}(\tilde{\psi}) &: z_1 \mapsto \zeta z_1, & z_2 &\mapsto \overline{\zeta} z_2, \\ \tilde{\sigma} &: z_1 \mapsto z_2, & z_2 &\mapsto z_1, \end{aligned} \tag{14.66}$$

---

[24] See, for example, Sattinger, 1983 [178]; Iooss, 1986 [106]; and Golubitsky, Stewart, and Schaeffer, 1988 [64].

where $r(\psi) : \overline{z_1} \mapsto \overline{\omega z_1}$, for example, is implied here. Indeed, this action is compatible with the relations in (14.13).

Since the group $C_{\infty v} \times \widetilde{C}_{\infty v}$ is generated by the four elements $r(\psi)$, σ, $\tilde{r}(\tilde{\psi})$, and $\tilde{\sigma}$, the equivariance of the bifurcation equation (14.63) to the group $C_{\infty v} \times \widetilde{C}_{\infty v}$ is identical to the equivariance to the action of these four elements. Therefore, the equivariance condition can be written as

$$\begin{aligned} \omega F_1(z_1,z_2) &= F_1(\omega z_1,\omega z_2), & \omega F_2(z_1,z_2) &= F_2(\omega z_1,\omega z_2),\\ \overline{F_2(z_1,z_2)} &= F_1(\overline{z_2},\overline{z_1}), & \overline{F_1(z_1,z_2)} &= F_2(\overline{z_2},\overline{z_1}),\\ \zeta F_1(z_1,z_2) &= F_1(\zeta z_1,\overline{\zeta} z_2), & \overline{\zeta} F_2(z_1,z_2) &= F_2(\zeta z_1,\overline{\zeta} z_2),\\ F_2(z_1,z_2) &= F_1(z_2,z_1), & F_1(z_1,z_2) &= F_2(z_2,z_1). \end{aligned}$$

From these relations we obtain a nonredundant set of equivariance conditions as

$$\omega F_1(z_1,z_2) = F_1(\omega z_1,\omega z_2), \tag{14.67}$$

$$\zeta F_1(z_1,z_2) = F_1(\zeta z_1,\overline{\zeta} z_2), \tag{14.68}$$

$$\overline{F_1(z_1,z_2)} = F_1(\overline{z_1},\overline{z_2}), \tag{14.69}$$

$$F_2(z_1,z_2) = F_1(z_2,z_1). \tag{14.70}$$

We expand F_1 as

$$F_1(z_1,z_2,\overline{z_2},\overline{z_1},\widetilde{f}) = \sum_{a=0}\sum_{b=0}\sum_{c=0}\sum_{d=0} A_{abcd}(\widetilde{f})\, {z_1}^a {z_2}^b\, \overline{z_1}^c\, \overline{z_2}^d. \tag{14.71}$$

Substitution of (14.71) into (14.67)–(14.69) shows that the indices (a,b,c,d) for nonzero terms should satisfy

$$\exp[\mathrm{i}2\pi n\psi(a+b-c-d-1)] = \exp[\mathrm{i}2\pi\tilde{n}\tilde{\psi}(a-b-c+d-1)] = 1, \tag{14.72}$$

$$A_{abcd} = \overline{A_{abcd}}, \tag{14.73}$$

where $A_{abcd} = A_{abcd}(\widetilde{f})$. Since (14.72) must hold for arbitrary (real) values of ψ and $\tilde{\psi}$, we have

$$a+b-c-d-1 = a-b-c+d-1 = 0;$$

that is,

$$a-c = 1, \qquad b = d. \tag{14.74}$$

With the use of (14.74), F_1 in (14.71) can be rewritten as

$$F_1 = z_1 \sum_{a=0}\sum_{b=0} \widehat{A}_{ab}(\widetilde{f})|z_1|^{2a}|z_2|^{2b}, \tag{14.75}$$

where $\widehat{A}_{ab}(\widetilde{f}) = A_{a+1,bab}(\widetilde{f})$ is real by (14.73). Substitution of (14.75) into (14.70) yields

$$F_2 = z_2 \sum_{a=0}\sum_{b=0} \widehat{A}_{ab}(\widetilde{f})|z_1|^{2b}|z_2|^{2a}. \tag{14.76}$$

Equations (14.75) and (14.76) give the concrete form of the bifurcation equation (14.63) as follows:

$$\begin{aligned} F_1 &= z_1 \sum_{a=0} \sum_{b=0} \widehat{A}_{ab}(\widetilde{f}) |z_1|^{2a} |z_2|^{2b} = 0, \\ F_2 &= z_2 \sum_{a=0} \sum_{b=0} \widehat{A}_{ab}(\widetilde{f}) |z_1|^{2b} |z_2|^{2a} = 0. \end{aligned} \tag{14.77}$$

Since $(z_1, z_2, \widetilde{f}) = (0,0,0)$ corresponds to the critical point, we have

$$\widehat{A}_{00}(0) = 0.$$

The solutions of (14.77) turn out to be:

(1) $z_1 = z_2 = 0$ (the trivial solution),
(2) $z_1 \neq 0$, $z_2 \neq 0$ (the diamond pattern solution),
(3) $z_1 \neq 0$, $z_2 = 0$ (a stripe pattern solution),
(4) $z_1 = 0$, $z_2 \neq 0$ (another stripe pattern solution).

The trivial solution (1) $z_1 = z_2 = 0$ corresponds to a $\mathrm{C}_{\infty\mathrm{v}} \times \widetilde{\mathrm{C}}_{\infty\mathrm{v}}$-symmetric fundamental path.

For the solution (2) $z_1 \neq 0$ and $z_2 \neq 0$, (14.77) becomes

$$\sum_{a=0} \sum_{b=0} \widehat{A}_{ab}(\widetilde{f}) |z_1|^{2a} |z_2|^{2b} = 0, \qquad \sum_{a=0} \sum_{b=0} \widehat{A}_{ab}(\widetilde{f}) |z_1|^{2b} |z_2|^{2a} = 0.$$

Generically, the solution to this system of equations is given by

$$|z_1| = |z_2| = \Phi(\widetilde{f}) \tag{14.78}$$

in a neighborhood of $(z_1, z_2, \widetilde{f}) = (0,0,0)$, where $\Phi = \Phi(\widetilde{f})$ is a solution to

$$\sum_{a=0} \sum_{b=0} \widehat{A}_{ab}(\widetilde{f}) \Phi^{2(a+b)} = 0. \tag{14.79}$$

Equation (14.78) shows that the bifurcated solutions form a two-dimensional sheet for each f.

As the representative (up to conjugacy) of such solutions, we consider a solution with $z_1 = z_2$ real. Then, as is seen from (14.66), this solution is invariant to

$$\sigma, \quad \tilde{\sigma}, \qquad \text{and} \qquad \{r(\psi)\tilde{r}(\tilde{\psi}) \mid n\psi \pm \tilde{n}\tilde{\psi} \in \mathbb{Z}\},$$

and, in turn, invariant to a subgroup

$$\left\langle \sigma, \tilde{\sigma}, r\left(\frac{1}{2n}\right)\tilde{r}\left(\frac{1}{2\tilde{n}}\right), r\left(\frac{1}{2n}\right)\tilde{r}\left(\frac{-1}{2\tilde{n}}\right) \right\rangle.$$

In (14.15) we have denoted this subgroup by $\mathrm{DI}_{n\tilde{n}}$. It is easy to verify (see §14.3) that the invariance to this subgroup characterizes a diamond pattern, as portrayed in Fig. 14.6(a) in §14.3.

For the solution (3) $z_1 \neq 0$ and $z_2 = 0$, equation $F_2 = 0$ is satisfied and $F_1/z_1 = 0$ yields

$$\sum_{a=0} \widehat{A}_{a0}(\widetilde{f})|z_1|^{2a} = 0, \tag{14.80}$$

which represents a sheet of bifurcated solutions, say, $|z_1| = \Phi(\widetilde{f})$. Among these solutions, we concentrate on a solution (z_1, z_2) with z_1 real (and $z_2 = 0$), because other solutions can be obtained from such a solution by means of transformations by some elements of $\mathrm{C}_{\infty\mathrm{v}} \times \widetilde{\mathrm{C}}_{\infty\mathrm{v}}$. It is observed from (14.66) that such a solution with z_1 real is invariant under the transformation by $\sigma\tilde{\sigma}$, and $r(\psi)\tilde{r}(\tilde{\psi})$ with $n\psi + \tilde{n}\tilde{\psi} \in \mathbb{Z}$, which generate $\mathrm{OB}^+_{n\tilde{n}}$ in (14.16). Thus we have arrived at an $\mathrm{OB}^+_{n\tilde{n}}$-symmetric oblique stripe pattern solution, as depicted in Fig. 14.6(b), with stripes parallel to a series of oblique straight lines $n\xi + \tilde{n}\tilde{\xi} \in \mathbb{Z}$.

The solution (4) $z_1 = 0$ and z_2 real represents another oblique stripe pattern solution invariant to $\mathrm{OB}^-_{n\tilde{n}}$ of (14.17), as illustrated in Fig. 14.6(c), with stripes parallel to another series of oblique straight lines $n\xi - \tilde{n}\tilde{\xi} \in \mathbb{Z}$.

14.9.2 Bifurcation of $\mathrm{OB}^{\pm}_{n\tilde{n}}$*-Equivariant System*

In this subsection we consider bifurcation of an $\mathrm{OB}^+_{n\tilde{n}}$-equivariant system. We recall (14.16):

$$\mathrm{OB}^+_{n\tilde{n}} = \langle \sigma\tilde{\sigma}, \{r(\psi)\tilde{r}(\tilde{\psi}) \mid n\psi + \tilde{n}\tilde{\psi} \in \mathbb{Z}\}\rangle, \qquad n \geq 1, \ \tilde{n} \geq 1.$$

Such a system plays a crucial role as an intermediate stage toward the echelon mode, as displayed schematically in (14.25). The companion case with $\mathrm{OB}^-_{n\tilde{n}}$-equivariance can of course be analyzed similarly.

Irreducible Representations

We need to know the irreducible representations before we can derive possible types of bifurcation equations. First, we consider

$$\mathrm{OB}^{+\circ}_{n\tilde{n}} = \{r(\psi)\tilde{r}(\tilde{\psi}) \mid n\psi + \tilde{n}\tilde{\psi} \in \mathbb{Z}\}, \tag{14.81}$$

which is an Abelian subgroup of index 2 of $\mathrm{OB}^+_{n\tilde{n}}$ (i.e., $|\mathrm{OB}^+_{n\tilde{n}}|/|\mathrm{OB}^{+\circ}_{n\tilde{n}}| = 2$). Let

$$d = \gcd(n, \tilde{n}) \tag{14.82}$$

denote the greatest common divisor of n and $\tilde{n}$ and then there exists a pair of integers p and $\tilde{p}$ such that

$$np + \tilde{n}\tilde{p} = d \geq 1. \tag{14.83}$$

Using these, we define

$$q(\phi) = r\left(\frac{\tilde{n}\phi}{d}\right)\tilde{r}\left(\frac{-n\phi}{d}\right), \qquad \rho = r\left(\frac{p}{d}\right)\tilde{r}\left(\frac{\tilde{p}}{d}\right) \tag{14.84}$$

to decompose $\mathrm{OB}_{n\tilde{n}}^{+\circ}$ into a direct product of two simpler groups:

$$\{q(\phi) \mid 0 \leq \phi < 1\} \simeq \mathrm{SO}(2), \qquad \{\rho^j \mid j = 0, 1, \ldots, d-1\} \simeq \mathrm{C}_d$$

as follows.

Lemma 14.1.

$$\mathrm{OB}_{n\tilde{n}}^{+\circ} = \{q(\phi) \mid 0 \leq \phi < 1\} \times \{\rho^j \mid j = 0, 1, \ldots, d-1\} \simeq \mathrm{SO}(2) \times \mathrm{C}_d. \tag{14.85}$$

Proof. Given (ϕ, j), we consider

$$\psi = \frac{\tilde{n}\phi + jp}{d}, \qquad \tilde{\psi} = \frac{-n\phi + j\tilde{p}}{d}. \tag{14.86}$$

Then, by (14.83), we have

$$n\psi + \tilde{n}\tilde{\psi} = n\left(\frac{\tilde{n}\phi + jp}{d}\right) + \tilde{n}\left(\frac{-n\phi + j\tilde{p}}{d}\right) = \frac{j(np + \tilde{n}\tilde{p})}{d} = j \in \mathbb{Z},$$

which shows that

$$q(\phi)\rho^j = r\left(\frac{\tilde{n}\phi + jp}{d}\right)\tilde{r}\left(\frac{-n\phi + j\tilde{p}}{d}\right) = r(\psi)\tilde{r}(\tilde{\psi})$$

belongs to $\mathrm{OB}_{n\tilde{n}}^{+\circ}$. Conversely, any $r(\psi)\tilde{r}(\tilde{\psi}) \in \mathrm{OB}_{n\tilde{n}}^{+\circ}$ is equal to $q(\phi)\rho^j$ for

$$\phi = \tilde{p}\psi - p\tilde{\psi}, \qquad j = n\psi + \tilde{n}\tilde{\psi}. \tag{14.87}$$

The one-to-one correspondence given by (14.86) and (14.87) establishes the identity in (14.85). □

The lemma presented above implies (cf., Remark 7.3 in §7.4.2) that the family of the $\mathbb{C}$-irreducible representations of $\mathrm{OB}_{n\tilde{n}}^{+\circ}$ coincides with the family of the tensor products of the $\mathbb{C}$-irreducible representations of the two groups SO(2) and C_d in (14.85). Hence the $\mathbb{C}$-irreducible representations of $\mathrm{OB}_{n\tilde{n}}^{+\circ}$ are one-dimensional and are given by

$$q(\phi)\rho^j \mapsto \exp(\mathrm{i}2\pi k\phi) \cdot \exp\left(\mathrm{i}2\pi\frac{jl}{d}\right) = \exp\left[\mathrm{i}2\pi\left(k\phi + \frac{jl}{d}\right)\right], \tag{14.88}$$

each indexed by a pair (k, l) with $k \in \mathbb{Z}$ and $l \in \{0, 1, \ldots, d-1\}$.

This in turn implies that the $\mathbb{R}$-irreducible representations of $\mathrm{OB}_{n\tilde{n}}^{+}$ are given as follows. They are either one-dimensional or two-dimensional.

The one-dimensional representations for $\mathrm{OB}_{n\tilde{n}}^{+} = \langle q(\phi), \rho, \sigma\tilde{\sigma}\rangle$ are induced from (14.88) for $\mathrm{OB}_{n\tilde{n}}^{+\circ} = \langle q(\phi), \rho\rangle$ with the following (k, l):

(i) $(k,l) = (0,0)$ for d odd; or
(ii) $(k,l) = (0,0)$, $(0,d/2)$ for d even.

When d is odd, $\mathrm{OB}^+_{n\tilde{n}}$ has two one-dimensional irreducible representations, comprising the unit representation, denoted as $(+,+)$, and the nonunit representation defined as

$$(-,+) : q(\phi) \mapsto 1, \quad \rho \mapsto 1, \quad \sigma\tilde{\sigma} \mapsto -1.$$

When d is even, there are two additional one-dimensional representations, defined, respectively, as

$$\begin{aligned}(+,-) &: q(\phi) \mapsto 1, \quad \rho \mapsto -1, \quad \sigma\tilde{\sigma} \mapsto 1,\\ (-,-) &: q(\phi) \mapsto 1, \quad \rho \mapsto -1, \quad \sigma\tilde{\sigma} \mapsto -1.\end{aligned}$$

The two-dimensional representations of $\mathrm{OB}^+_{n\tilde{n}}$ are induced from (14.88) with (k,l) such that

(i) $k \geq 1$ and $0 \leq l \leq d-1$; or
(ii) $k = 0$ and $1 \leq l \leq \lfloor (d-1)/2 \rfloor$.

In either case, we have

$$T^{(k,l)}(q(\phi)) = \begin{pmatrix} \cos(2\pi k\phi) & -\sin(2\pi k\phi) \\ \sin(2\pi k\phi) & \cos(2\pi k\phi) \end{pmatrix}, \tag{14.89}$$

$$T^{(k,l)}(\rho) = \begin{pmatrix} \cos(2\pi l/d) & -\sin(2\pi l/d) \\ \sin(2\pi l/d) & \cos(2\pi l/d) \end{pmatrix}, \tag{14.90}$$

$$T^{(k,l)}(\sigma\tilde{\sigma}) = \begin{pmatrix} 1 & 0 \\ 0 & -1 \end{pmatrix}. \tag{14.91}$$

To sum up (see Table 14.2), the family of the inequivalent $\mathbb{R}$-irreducible representations of $\mathrm{OB}^+_{n\tilde{n}}$ is given by

$$R(\mathrm{OB}^+_{n\tilde{n}}) = R_1(\mathrm{OB}^+_{n\tilde{n}}) \cup R_2(\mathrm{OB}^+_{n\tilde{n}}), \tag{14.92}$$

where

$$R_1(\mathrm{OB}^+_{n\tilde{n}}) = \begin{cases} \{(+,+),(-,+)\} & \text{for } d \text{ odd},\\ \{(+,+),(-,+),(+,-),(-,-)\} & \text{for } d \text{ even},\end{cases}$$
$$R_2(\mathrm{OB}^+_{n\tilde{n}}) = \{(k,l) \mid k \geq 1,\ 0 \leq l \leq d-1\} \cup \{(k,l) \mid k = 0,\ 1 \leq l \leq \lfloor (d-1)/2 \rfloor\}.$$

Bifurcation Analysis

With the exhaustive list (14.92) of irreducible representations we can proceed to the analyses of bifurcation equations. The solution path branching from a simple bifurcation point, with which one of the nonunit one-dimensional representations is associated, has the same symmetry as the kernel space (see Table 14.2).

Table 14.2 Classification of critical points of an $\mathrm{OB}^+_{n\tilde{n}}$-equivariant system. $d = \gcd(n,\tilde{n})$; $m = n \cdot \gcd(n,\tilde{n},l)/d$; $\tilde{m} = \tilde{n} \cdot \gcd(n,\tilde{n},l)/d$; $\mathrm{OB}^+_{n/2,\tilde{n}/2} = \langle q(\phi), \rho^2, \sigma\tilde{\sigma}\rangle$ in the entry for $(+,-)$

Multiplicity M	Irreducible Representation, μ	Symmetry Groups G^μ	Bifurcated Paths
1	$(+,+)$	$\mathrm{OB}^+_{n\tilde{n}}$	No Bifurcation
	$(-,+)$	$\mathrm{OB}^{+\circ}_{n\tilde{n}}$	$\mathrm{OB}^{+\circ}_{n\tilde{n}}$
	$(+,-)$ (d: even)	$\mathrm{OB}^+_{n/2,\tilde{n}/2}$	$\mathrm{OB}^+_{n/2,\tilde{n}/2}$
	$(-,-)$ (d: even)	$\langle q(\phi),\rho^2,\rho\sigma\tilde{\sigma}\rangle$	$\langle q(\phi),\rho^2,\rho\sigma\tilde{\sigma}\rangle$
2	(k,l) $(k \geq 1,\ 0 \leq l \leq d-1)$	$\mathrm{EC}^{+\circ}_{n\tilde{n}kl}$	$\mathrm{EC}^{+}_{n\tilde{n}kl}$
	$(0,l)$ $(1 \leq l \leq \lfloor (d-1)/2 \rfloor)$	$\mathrm{OB}^{+\circ}_{m\tilde{m}}$	$\mathrm{OB}^{+}_{m\tilde{m}}$

The analysis of the bifurcation at a group-theoretic double point is close to that in §14.9.1. Let (k,l) denote the two-dimensional irreducible representation associated with the double point.

As in (14.53) we consider the bifurcation equation $F(z,\overline{z},\widetilde{f}) = 0$ in the complex variable $z = w_1 + \mathrm{i}w_2$, where $F(z,\overline{z},\widetilde{f}) = 0$ is often abbreviated to $F(z,\overline{z}) = 0$. We can assume that the action of $\mathrm{OB}^+_{n\tilde{n}} = \langle q(\phi),\rho,\sigma\tilde{\sigma}\rangle$ is defined by

$$q(\phi) : z \mapsto \omega z, \qquad \rho : z \mapsto \zeta z, \qquad \sigma\tilde{\sigma} : z \mapsto \overline{z}, \tag{14.93}$$

where

$$\omega = \exp(\mathrm{i}2\pi k\phi), \qquad \zeta = \exp(\mathrm{i}2\pi l/d).$$

The equivariance condition can be written as

$$\omega F(z,\overline{z}) = F(\omega z, \overline{\omega}\,\overline{z}), \qquad \zeta F(z,\overline{z}) = F(\zeta z, \overline{\zeta}\,\overline{z}), \qquad \overline{F(z,\overline{z})} = F(\overline{z}, z). \tag{14.94}$$

These conditions prescribe the form of F again as (14.62):

$$F = z \sum_{a=0} A_a(\widetilde{f})\,|z|^{2a}, \tag{14.95}$$

where $A_a(\widetilde{f}) = A_{a+1,a}(\widetilde{f})$ is real and $A_0(0) = 0$. Hence the bifurcated solution, if any, takes the form $|z| = \Phi(\widetilde{f})$.

The symmetry of the bifurcated solution can be revealed as follows. As the representative of solutions $|z| = \Phi(\widetilde{f})$, we consider a solution with z real, because other solutions can be obtained from that solution by transformations with some elements of $\mathrm{OB}^+_{n\tilde{n}}$. The solution with z real is obviously invariant to the action of $\sigma\tilde{\sigma}$ by (14.93). The transformation

$$q(\phi)\rho^j : z \mapsto z \exp\left[\mathrm{i}2\pi\left(k\phi + \frac{jl}{d}\right)\right]$$

is an identity if and only if

$$k\phi + \frac{jl}{d} \in \mathbb{Z}. \tag{14.96}$$

If $k \geq 1$ (and $0 \leq l \leq d-1$), then (14.96) is equivalent to

$$\phi = \frac{1}{k}\left(-N - \frac{jl}{d}\right), \qquad N \in \mathbb{Z}.$$

Substituting this into (14.86) we obtain

$$(\psi, \tilde{\psi}) = N\left(-\frac{\tilde{n}}{dk}, \frac{n}{dk}\right) + j\left(\frac{1}{d}\left(p - \frac{\tilde{n}l}{dk}\right), \frac{1}{d}\left(\tilde{p} + \frac{nl}{dk}\right)\right),$$

which says that the vector $(\psi, \tilde{\psi})$ should lie on a lattice spanned by the two basis vectors

$$\left(-\frac{\tilde{n}}{dk}, \frac{n}{dk}\right) \qquad \text{and} \qquad \left(\frac{1}{d}\left(p - \frac{\tilde{n}l}{dk}\right), \frac{1}{d}\left(\tilde{p} + \frac{nl}{dk}\right)\right).$$

Thus we have shown that the bifurcated solution (with z real) is invariant to

$$\mathrm{EC}^{+}_{n\tilde{n}kl} = \left\langle \sigma\tilde{\sigma}, r\left(-\frac{\tilde{n}}{dk}\right)\tilde{r}\left(\frac{n}{dk}\right), r\left(\frac{1}{d}\left(p - \frac{\tilde{n}l}{dk}\right)\right)\tilde{r}\left(\frac{1}{d}\left(\tilde{p} + \frac{nl}{dk}\right)\right)\right\rangle$$

defined in (14.18), whereas the symmetry of the kernel space is described by

$$\mathrm{EC}^{+\circ}_{n\tilde{n}kl} = \left\langle r\left(-\frac{\tilde{n}}{dk}\right)\tilde{r}\left(\frac{n}{dk}\right), r\left(\frac{1}{d}\left(p - \frac{\tilde{n}l}{dk}\right)\right)\tilde{r}\left(\frac{1}{d}\left(\tilde{p} + \frac{nl}{dk}\right)\right)\right\rangle$$

defined in (14.22).

On the other hand, if $k = 0$ (and $1 \leq l \leq \lfloor (d-1)/2 \rfloor$), (14.96) is equivalent to the condition that j is a multiple of

$$j_0 = \frac{d}{\gcd(d, l)} = \frac{\gcd(n, \tilde{n})}{\gcd(n, \tilde{n}, l)},$$

where $d = \gcd(n, \tilde{n})$ by (14.82). Hence the bifurcated solution (with z real) is invariant to

$$\langle \sigma\tilde{\sigma}, q(\phi), \rho^{j_0} \rangle,$$

which agrees with $\mathrm{OB}^{+}_{m\tilde{m}}$ for $(m, \tilde{m})$ such that $\tilde{m}/m = \tilde{n}/n$ and $\gcd(m, \tilde{m}) = d/j_0$, that is, for

$$m = \frac{n}{j_0} = \frac{n \gcd(n, \tilde{n}, l)}{\gcd(n, \tilde{n})}, \qquad \tilde{m} = \frac{\tilde{n}}{j_0} = \frac{\tilde{n} \gcd(n, \tilde{n}, l)}{\gcd(n, \tilde{n})}.$$

On the other hand, the kernel space is invariant to a smaller subgroup $\mathrm{OB}^{+\circ}_{m\tilde{m}} \simeq \mathrm{SO}(2) \times \mathrm{C}_{\gcd(n,\tilde{n},l)}$.

Remark 14.7. The bifurcation of an $\mathrm{OB}^{+\circ}_{n\tilde{n}}$-equivariant system can be analyzed similarly. The $\mathbb{R}$-irreducible representations of $\mathrm{OB}^{+\circ}_{n\tilde{n}}$ are either one-dimensional or two-dimensional. One or two one-dimensional irreducible representations exist according to whether d is odd or even. The two-dimensional representations of $\mathrm{OB}^{+\circ}_{n\tilde{n}}$ are

Table 14.3 Classification of critical points of an $\mathrm{OB}^{+\circ}_{n\tilde{n}}$-equivariant system. $d = \gcd(n,\tilde{n})$, $m = n \cdot \gcd(n,\tilde{n},l)/d$, $\tilde{m} = \tilde{n} \cdot \gcd(n,\tilde{n},l)/d$); $(*)$ denotes that the associated bifurcated paths are nonexistent for a nonreciprocal system

Multiplicity	Irreducible	Symmetry Groups	
M	Representation, μ	G^{μ}	Bifurcated Paths
1	$+$	$\mathrm{OB}^{+\circ}_{n\tilde{n}}$	No Bifurcation
	$-$ (d: even)	$\mathrm{OB}^{+\circ}_{n/2,\tilde{n}/2}$	$\mathrm{OB}^{+\circ}_{n/2,\tilde{n}/2}$
2	(k,l) $(k \geq 1,\ 0 \leq l \leq d-1)$	$\mathrm{EC}^{+\circ}_{n\tilde{n}kl}$	$\mathrm{EC}^{+\circ}_{n\tilde{n}kl}$ $(*)$
	$(0,l)$ $(1 \leq l \leq \lfloor (d-1)/2 \rfloor)$	$\mathrm{OB}^{+\circ}_{m\tilde{m}}$	$\mathrm{OB}^{+\circ}_{m\tilde{m}}$ $(*)$

denoted as (k,l) with

$$k \geq 1,\ 0 \leq l \leq d-1; \qquad k = 0,\ \ 1 \leq l \leq \lfloor (d-1)/2 \rfloor$$

and are defined by (14.89) and (14.90).

The bifurcation equation at a group-theoretic double point takes the form of (14.95), where $A_a(\widetilde{f})$ is not necessarily real because of the lack of $\sigma\tilde{\sigma}$. The bifurcation equation has no nontrivial solution in general. However, for a reciprocal system, $A_a(\widetilde{f})$ is real by (8.58) and, in turn, the bifurcation equation has a nontrivial solution of the form $|z| = \Phi(\widetilde{f})$. The symmetry of the bifurcated solution is given in Table 14.3. □

14.9.3 Bifurcation of $\mathrm{D}_{\infty\infty}$*-Equivariant System*

We investigate the bifurcation of a system equivariant to the group

$$\mathrm{D}_{\infty\infty} = \langle \sigma\tilde{\sigma}, r(\psi), \tilde{r}(\tilde{\psi}) \rangle.$$

Such a system appears as a bifurcated path of a $\mathrm{C}_{\infty\mathrm{v}} \times \widetilde{\mathrm{C}}_{\infty\mathrm{v}}$-equivariant system (cf., §14.9.1), or as a system of equations describing the deformation under the shear (cf., Fig. 14.15(b) in §14.2).

Irreducible Representations

An irreducible representation of $\mathrm{D}_{\infty\infty}$ is either one-dimensional or two-dimensional. Therefore, the multiplicity of a group-theoretic bifurcation point is either one or two. Other than the unit representation denoted by $(+)$, there exists only one one-dimensional representation $(-)$, given by

$$T^{(-)}(r(\psi)) = 1, \qquad T^{(-)}(\tilde{r}(\tilde{\psi})) = 1, \qquad T^{(-)}(\sigma\tilde{\sigma}) = -1.$$

Two-dimensional representations are indexed by a pair $(n,\tilde{n}) \neq (0,0)$ of integers and are given by

$$T^{(n,\tilde{n})}(r(\psi)) = \begin{pmatrix} \cos(2\pi n\psi) & -\sin(2\pi n\psi) \\ \sin(2\pi n\psi) & \cos(2\pi n\psi) \end{pmatrix},$$
$$T^{(n,\tilde{n})}(\tilde{r}(\tilde{\psi})) = \begin{pmatrix} \cos(2\pi\tilde{n}\tilde{\psi}) & -\sin(2\pi\tilde{n}\tilde{\psi}) \\ \sin(2\pi\tilde{n}\tilde{\psi}) & \cos(2\pi\tilde{n}\tilde{\psi}) \end{pmatrix},$$
$$T^{(n,\tilde{n})}(\sigma\tilde{\sigma}) = \begin{pmatrix} 1 & 0 \\ 0 & -1 \end{pmatrix},$$

where it is noteworthy that the representations indexed by $(n,\tilde{n})$ and $(-n,-\tilde{n})$ are equivalent, and those indexed by $(n,\tilde{n})$ and $(n,-\tilde{n})$ are inequivalent.

Bifurcation Analysis

The bifurcated path from a simple bifurcation point for the nonunit one-dimensional representation has the symmetry described by the kernel of $T^{(-)}$, which is equal to

$$\langle r(\psi), \tilde{r}(\tilde{\psi})\rangle = \mathrm{C}_\infty \times \widetilde{\mathrm{C}}_\infty.$$

The analysis of the bifurcation at a group-theoretic double point is close to that in §14.9.1. Let $(n,\tilde{n}) \neq (0,0)$ denote the two-dimensional irreducible representation associated with the double point.

As in (14.53) we consider the bifurcation equation

$$F(z,\overline{z},\widetilde{f}) = 0$$

in the complex variable $z = w_1 + \mathrm{i}w_2$, where $F(z,\overline{z},\widetilde{f}) = 0$ is often abbreviated to $F(z,\overline{z}) = 0$. We may assume that the action of $\mathrm{D}_{\infty\infty}$ is defined by

$$r(\psi) : z \mapsto \omega z, \qquad \tilde{r}(\tilde{\psi}) : z \mapsto \zeta z, \qquad \sigma\tilde{\sigma} : z \mapsto \overline{z}, \tag{14.97}$$

where

$$\omega = \exp(\mathrm{i}2\pi n\psi), \qquad \zeta = \exp(\mathrm{i}2\pi\tilde{n}\tilde{\psi}).$$

The equivariance condition can be written as

$$\omega F(z,\overline{z}) = F(\omega z, \overline{\omega}\overline{z}), \qquad \zeta F(z,\overline{z}) = F(\zeta z, \overline{\zeta}\overline{z}), \qquad \overline{F(z,\overline{z})} = F(\overline{z}, z). \tag{14.98}$$

These conditions prescribe the form of F again as (14.62):

$$F = z\sum_{a=0}^{\infty} A_a(\widetilde{f})|z|^{2a},$$

Table 14.4 Classification of critical points of a $\mathrm{D}_{\infty\infty}$-equivariant system. n and $\tilde{n}$ are positive integers

Multiplicity M	Irreducible Representation, μ	Symmetry Groups G^{μ}	Symmetry Groups: Bifurcated Paths
1	$+$	$\mathrm{D}_{\infty\infty}$	No Bifurcation
	$-$	$\mathrm{C}_{\infty}\times\widetilde{\mathrm{C}}_{\infty}$	$\mathrm{C}_{\infty}\times\widetilde{\mathrm{C}}_{\infty}$
2	$(n,\tilde{n})$	$\mathrm{OB}^{+\circ}_{n\tilde{n}}$	$\mathrm{OB}^{+}_{n\tilde{n}}$
	$(n,-\tilde{n})$	$\mathrm{OB}^{-\circ}_{n\tilde{n}}$	$\mathrm{OB}^{-}_{n\tilde{n}}$
	$(n,0)$	$\langle r(1/n),\tilde{r}(\tilde{\psi})\rangle$	$\langle\sigma\tilde{\sigma},r(1/n),\tilde{r}(\tilde{\psi})\rangle$
	$(0,\tilde{n})$	$\langle r(\psi),\tilde{r}(1/\tilde{n})\rangle$	$\langle\sigma\tilde{\sigma},r(\psi),\tilde{r}(1/\tilde{n})\rangle$

where $A_a(\widetilde{f}) = A_{a+1,a}(\widetilde{f})$ is real and $A_0(0) = 0$. Hence the bifurcated solution, if any, takes the form $|z| = \Phi(\widetilde{f})$.

As the representative of such solutions, we consider a solution with z real, because other solutions can be obtained from that solution by transformations with some elements of $\mathrm{D}_{\infty\infty}$. Then the action (14.97) shows that this solution is invariant to the subgroup generated by

$$\sigma\tilde{\sigma} \qquad \text{and} \qquad \{r(\psi)\tilde{r}(\tilde{\psi}) \mid n\psi + \tilde{n}\tilde{\psi} \in \mathbb{Z}\}.$$

This subgroup is given as follows:

$$\begin{cases} \mathrm{OB}^{+}_{|n||\tilde{n}|} & \text{if } n\tilde{n} > 0, \\ \mathrm{OB}^{-}_{|n||\tilde{n}|} & \text{if } n\tilde{n} < 0, \\ \langle\sigma\tilde{\sigma}, r(1/n), \tilde{r}(\tilde{\psi})\rangle & \text{if } n \neq 0, \tilde{n} = 0, \\ \langle\sigma\tilde{\sigma}, r(\psi), \tilde{r}(1/\tilde{n})\rangle & \text{if } n = 0, \tilde{n} \neq 0. \end{cases}$$

The results are summarized in Table 14.4. Recall that $(n,\tilde{n})$ and $(-n,-\tilde{n})$ are equivalent, whereas $(n,\tilde{n})$ and $(n,-\tilde{n})$ are inequivalent.

14.9.4 Symmetry of Fourier Terms

To derive (14.34) and (14.35), which give the Fourier terms activated by bifurcation

$$\mathrm{OB}^{+}_{n\tilde{n}} \rightarrow \mathrm{EC}^{+}_{n\tilde{n}kl},$$

we first show that

$$\{\cos 2\pi(m\xi + \tilde{m}\tilde{\xi}), \sin 2\pi(m\xi + \tilde{m}\tilde{\xi})\}, \qquad (14.99)$$

where $m, \tilde{m} \geq 1$, spans a two-dimensional irreducible representation space of $\mathrm{OB}^+_{n\tilde{n}}$. Recall from §14.9.2 that the two-dimensional representations of $\mathrm{OB}^+_{n\tilde{n}}$ are indexed by (k,l) defined in (14.89)–(14.91).

By (14.84), as well as (14.1)–(14.3), (14.6), and (14.11), the actions of $q(\phi)$, ρ, and $\sigma\tilde{\sigma}$ on the two-dimensional space spanned by (14.99) can be written as (cf., (14.84))

$$q(\phi)\cdot\begin{pmatrix}\cos 2\pi(m\xi+\tilde{m}\tilde{\xi})\\ \sin 2\pi(m\xi+\tilde{m}\tilde{\xi})\end{pmatrix}=\begin{pmatrix}\cos 2\pi\dfrac{m\tilde{n}-\tilde{m}n}{d}\phi & -\sin 2\pi\dfrac{m\tilde{n}-\tilde{m}n}{d}\phi\\ \sin 2\pi\dfrac{m\tilde{n}-\tilde{m}n}{d}\phi & \cos 2\pi\dfrac{m\tilde{n}-\tilde{m}n}{d}\phi\end{pmatrix}\begin{pmatrix}\cos 2\pi(m\xi+\tilde{m}\tilde{\xi})\\ \sin 2\pi(m\xi+\tilde{m}\tilde{\xi})\end{pmatrix},$$

$$\rho\cdot\begin{pmatrix}\cos 2\pi(m\xi+\tilde{m}\tilde{\xi})\\ \sin 2\pi(m\xi+\tilde{m}\tilde{\xi})\end{pmatrix}=\begin{pmatrix}\cos 2\pi\dfrac{mp+\tilde{m}\tilde{p}}{d} & -\sin 2\pi\dfrac{mp+\tilde{m}\tilde{p}}{d}\\ \sin 2\pi\dfrac{mp+\tilde{m}\tilde{p}}{d} & \cos 2\pi\dfrac{mp+\tilde{m}\tilde{p}}{d}\end{pmatrix}\begin{pmatrix}\cos 2\pi(m\xi+\tilde{m}\tilde{\xi})\\ \sin 2\pi(m\xi+\tilde{m}\tilde{\xi})\end{pmatrix},$$

$$\sigma\tilde{\sigma}\cdot\begin{pmatrix}\cos 2\pi(m\xi+\tilde{m}\tilde{\xi})\\ \sin 2\pi(m\xi+\tilde{m}\tilde{\xi})\end{pmatrix}=\begin{pmatrix}1 & 0\\ 0 & -1\end{pmatrix}\begin{pmatrix}\cos 2\pi(m\xi+\tilde{m}\tilde{\xi})\\ \sin 2\pi(m\xi+\tilde{m}\tilde{\xi})\end{pmatrix}.$$

The comparison of these expressions with (14.89)–(14.91) shows that the Fourier terms (14.99) span a two-dimensional irreducible representation space. To be precise, define

$$k=\frac{|m\tilde{n}-\tilde{m}n|}{d},\qquad l\equiv mp+\tilde{m}\tilde{p}\pmod{d},$$

where $0 \leq l \leq d-1$, and $a \equiv b \pmod{d}$ means that $a-b$ is divisible by d; and if $k=0$, redefine l as $\min(l, d-l)$. If $k \geq 1$ or if $k=0$ and $1 \leq l \leq \lfloor (d-1)/2 \rfloor$, $\{\cos 2\pi(m\xi + \tilde{m}\tilde{\xi}), \sin 2\pi(m\xi+\tilde{m}\tilde{\xi})\}$ spans the irreducible representation space of (k,l). Expression (14.34) is thus derived.

Similarly, $\{\cos 2\pi(m\xi-\tilde{m}\tilde{\xi}), \sin 2\pi(m\xi-\tilde{m}\tilde{\xi})\}$ spans the irreducible representation space of (k,l) for

$$k=\frac{|m\tilde{n}+\tilde{m}n|}{d},\qquad l\equiv mp-\tilde{m}\tilde{p}\pmod{d},$$

where $k \geq 1$ and $0 \leq l \leq d-1$. Expression (14.35) then follows.

Next, we derive expression (14.37) of $u|\mathrm{EC}^+_{n\tilde{n}kl}$ by determining the Fourier terms invariant to the group $\mathrm{EC}^+_{n\tilde{n}kl}$ in (14.18). The invariance, with respect to $\sigma\tilde{\sigma}$, which changes ξ to $1-\xi$ and $\tilde{\xi}$ to $1-\tilde{\xi}$, imposes the disappearance of the sine terms in $u|\mathrm{EC}^+_{n\tilde{n}kl}$, whereas the cosine terms are apparently invariant to $\sigma\tilde{\sigma}$. By transforming $\cos 2\pi(m\xi+\tilde{m}\tilde{\xi})$ by the second element in (14.18), we obtain

$$\begin{aligned}r\left(-\frac{\tilde{n}}{dk}\right)\tilde{r}\left(\frac{n}{dk}\right)\cdot\cos 2\pi(m\xi+\tilde{m}\tilde{\xi}) &= \cos 2\pi\left\{\left(m\xi-\frac{m\tilde{n}}{dk}\right)+\left(\tilde{m}\tilde{\xi}+\frac{\tilde{m}n}{dk}\right)\right\}\\ &= \cos 2\pi\left\{\left(m\xi+\tilde{m}\tilde{\xi}\right)-\left(\frac{m\tilde{n}-\tilde{m}n}{dk}\right)\right\}.\end{aligned}$$

Hence the invariance condition for the second element is given by

$$m\tilde{n} - \tilde{m}n = jdk, \qquad j \in \mathbb{Z}. \tag{14.100}$$

By applying the third element to the term $\cos 2\pi(m\xi + \tilde{m}\tilde{\xi})$ and using (14.100), we obtain

$$\begin{aligned}
&r\left(\frac{1}{d}\left(p - \frac{\tilde{n}l}{dk}\right)\right)\tilde{r}\left(\frac{1}{d}\left(\tilde{p} + \frac{nl}{dk}\right)\right) \cdot \cos 2\pi(m\xi + \tilde{m}\tilde{\xi}) \\
&= \cos 2\pi\left\{m\left(\xi + \frac{p}{d} - \frac{\tilde{n}l}{d^2k}\right) + \tilde{m}\left(\tilde{\xi} + \frac{\tilde{p}}{d} + \frac{nl}{d^2k}\right)\right\} \\
&= \cos 2\pi\left\{(m\xi + \tilde{m}\tilde{\xi}) + \frac{mp + \tilde{m}\tilde{p}}{d} - \frac{l}{d}\frac{m\tilde{n} - \tilde{m}n}{dk}\right\} \\
&= \cos 2\pi\left\{(m\xi + \tilde{m}\tilde{\xi}) + \frac{mp + \tilde{m}\tilde{p} - lj}{d}\right\}.
\end{aligned}$$

Hence, the invariance condition added by the third element becomes

$$mp + \tilde{m}\tilde{p} - lj = id, \qquad i \in \mathbb{Z}. \tag{14.101}$$

The two equations (14.100) and (14.101) can be solved for $(m, \tilde{m})$ as

$$m = in + j(nl/d + k\tilde{p}), \qquad \tilde{m} = i\tilde{n} + j(\tilde{n}l/d - kp). \tag{14.102}$$

Thus, expression (14.38), except for nonnegativity constraints $m \geq 0$ and $\tilde{m} \geq 0$, has been derived.

Similar argument for $\cos 2\pi(m\xi - \tilde{m}\tilde{\xi})$ gives

$$m = in + j(nl/d + k\tilde{p}), \qquad \tilde{m} = -i\tilde{n} - j(\tilde{n}l/d - kp), \tag{14.103}$$

which shows (14.39), except for nonnegativity constraints.

To sum up, $u|\mathrm{EC}^+_{n\tilde{n}kl}$ consists of the Fourier terms $\cos 2\pi(m\xi + \tilde{m}\tilde{\xi})$ with $(m, \tilde{m})$ given by (14.102) and $\cos 2\pi(m\xi - \tilde{m}\tilde{\xi})$ with $(m, \tilde{m})$ given by (14.103). Therefore, (14.37) has been shown.

Problems

14-1 Show that $\mathrm{EC}^+_{n\tilde{n}kl}$ is composed of $2dk$ elements.

14-2 Show that $\mathrm{EC}^+_{n\tilde{n}kl}$ is a subgroup of $\mathrm{OB}^+_{n\tilde{n}}$.

14-3 Rewrite equation (14.54) to obtain the action of $\mathrm{C}_{\infty\mathrm{v}} \times \widetilde{\mathrm{C}}_{\infty\mathrm{v}}$ on (w_1, w_2) defined by $z = w_1 + \mathrm{i}w_2$.
Answer:

$$r(\psi): \begin{pmatrix} w_1 \\ w_2 \end{pmatrix} \mapsto \begin{pmatrix} \cos(2\pi n\psi) & -\sin(2\pi n\psi) \\ \sin(2\pi n\psi) & \cos(2\pi n\psi) \end{pmatrix} \begin{pmatrix} w_1 \\ w_2 \end{pmatrix},$$

$$\sigma: \begin{pmatrix} w_1 \\ w_2 \end{pmatrix} \mapsto \begin{pmatrix} 1 & 0 \\ 0 & -1 \end{pmatrix} \begin{pmatrix} w_1 \\ w_2 \end{pmatrix},$$

$$\tilde{r}(\tilde{\psi}): \begin{pmatrix} w_1 \\ w_2 \end{pmatrix} \mapsto \begin{pmatrix} 1 & 0 \\ 0 & 1 \end{pmatrix} \begin{pmatrix} w_1 \\ w_2 \end{pmatrix},$$

$$\tilde{\sigma}: \begin{pmatrix} w_1 \\ w_2 \end{pmatrix} \mapsto \begin{pmatrix} -1 & 0 \\ 0 & -1 \end{pmatrix} \begin{pmatrix} w_1 \\ w_2 \end{pmatrix}.$$

14-4 Show that the four-dimensional representation $\widehat{T}$ in (14.64) and (14.65) of $\mathrm{C}_{\infty\mathrm{v}} \times \widetilde{\mathrm{C}}_{\infty\mathrm{v}}$ in §14.9.1 is irreducible.

14-5 Rewrite equation (14.66) to obtain the action of $\mathrm{C}_{\infty\mathrm{v}} \times \widetilde{\mathrm{C}}_{\infty\mathrm{v}}$ on (w_1, w_2, w_3, w_4) defined by $z_1 = w_1 + \mathrm{i}w_2$ and $z_2 = w_3 + \mathrm{i}w_3$.

14-6 Draw a schematic view of $\mathrm{EC}^{+}_{9,8,19,0}$-invariant mode similar to that in Fig. 14.8 in §14.3.
Answer: See Fig. 7 of Ikeda, Murota, and Nakano, 1994 [95].

14-7 Obtain the double Fourier modes that are activated by a recursive bifurcation

$$\mathrm{C}_{\infty\mathrm{v}} \times \widetilde{\mathrm{C}}_{\infty\mathrm{v}} \to \mathrm{OB}^{+}_{6,8} \to \mathrm{OB}^{+}_{3,4} \to \mathrm{EC}^{+}_{3,4,6,0}.$$

14-8 Investigate the bifurcation of an O(2) × O(2) × O(2)-equivariant system.
Answer: See Tanaka, Saiki, and Ikeda, 2002 [187].

Summary

- The bifurcation rule of an O(2) × O(2)-equivariant system has been derived to reveal the mechanism of the possible emergence of an echelon mode.
- Echelon modes have been observed for soil specimens and the rectangular plate.
- Image simulation analyses have been conducted on kaolin, steel, and sand specimens to demonstrate the emergence of patterns with high spatial frequencies, such as diamond patterns, oblique stripe patterns, and echelon modes.

Chapter 15
Bifurcation of Steel Specimens

15.1 Introduction

The shear band formation of a steel specimen subjected to tension has been ascribed to the direct bifurcation from a uniform state in plastic bifurcation theory, and is simulated by numerical analyses.[1] However, direct bifurcation alone cannot account for the complexity of experimental deformation patterns. The recursive bifurcation that entails complex deformation patterns can be advanced as an underlying mechanism to create such complexity, as described in Chapters 13 and 14.

The aim of this chapter is to explain the complex deformation characteristics of rectangular parallelepiped steel specimens from the standpoint of recursive bifurcation.[2] We consider a rectangular parallelepiped domain with the periodic symmetry in one direction, say, the longitudinal direction. Group-theoretic bifurcation theory is applied to the group $O(2) \times \mathbb{Z}_2 \times \mathbb{Z}_2$ labeling the symmetry of this domain. In addition, we investigate the effect of cross-sectional shape on the deformation characteristics of the specimens. Another important aspect of the bifurcation phenomena of the specimens, hilltop bifurcation, is supplemented in §15.7.

This chapter is organized as follows.

- Symmetry of a rectangular parallelepiped domain is briefly described in §15.2.
- The rule of the recursive bifurcation for this symmetry is summarized in §15.3.
- An experimental study of the recursive bifurcation of steel specimens is presented in §15.4.
- Computational bifurcation analyses are conducted in §15.5.
- The bifurcation rule is derived in §15.6, the first appendix of this chapter.
- Hilltop bifurcation is studied in §15.7, the second appendix of this chapter.

---

[1] Plastic bifurcation theory was developed by Hill and Hutchinson, 1975 [73]. For the numerical analyses, see, for example, Tvergaard, Needleman, and Lo, 1981 [199]; Tvergaard and Needleman, 1984 [198]; and Petryk and Thermann, 1992 [165].

[2] The study of this chapter is based on Ikeda et al., 2001 [101].

K. Ikeda and K. Murota, *Imperfect Bifurcation in Structures and Materials*, Applied Mathematical Sciences 149, DOI 10.1007/978-1-4419-7296-5_15,

15.2 Symmetry of a Rectangular Parallelepiped Domain

We introduce the symmetry group and its subgroups of a rectangular parallelepiped domain (shown in Fig. 15.1):

$$\Omega = \left\{ (x,y,z) \in \mathbb{R}^3 \;\middle|\; -\frac{L_x}{2} \leq x \leq \frac{L_x}{2}, -\frac{L_y}{2} \leq y \leq \frac{L_y}{2}, -\frac{L_z}{2} \leq z \leq \frac{L_z}{2} \right\}.$$

This domain has a natural geometrical symmetry labeled by the group[3]

$$\begin{aligned} \mathrm{D_{2h}} &= \langle \sigma_x, \sigma_y, \sigma_z \rangle = \{e, \sigma_x, \sigma_y, \sigma_z, \sigma_x\sigma_y, \sigma_y\sigma_z, \sigma_z\sigma_x, \sigma_x\sigma_y\sigma_z\} \\ &= \langle \sigma_x \rangle \times \langle \sigma_y \rangle \times \langle \sigma_z \rangle \simeq \mathbb{Z}_2 \times \mathbb{Z}_2 \times \mathbb{Z}_2, \end{aligned}$$

where σ_x, σ_y, and σ_z, respectively, denote reflections with respect to the yz-plane, the zx-plane, and the xy-plane; $\sigma_x\sigma_y$, $\sigma_y\sigma_z$, and $\sigma_z\sigma_x$, respectively, denote half-rotations about the z-axis, the x-axis, and the y-axis; $\sigma_x\sigma_y\sigma_z$ is the inversion with respect to the origin; $\sigma_x\sigma_y = \sigma_y\sigma_x$, $\sigma_y\sigma_z = \sigma_z\sigma_y$, $\sigma_z\sigma_x = \sigma_x\sigma_z$, and ${\sigma_x}^2 = {\sigma_y}^2 = {\sigma_z}^2 = e$.

The proper subgroups of $\mathrm{D_{2h}}$ are

$$\left.\begin{array}{llll} \langle \sigma_x, \sigma_y \rangle, & \langle \sigma_y, \sigma_z \rangle, & \langle \sigma_z, \sigma_x \rangle, & \\ \langle \sigma_x, \sigma_y\sigma_z \rangle, & \langle \sigma_y, \sigma_z\sigma_x \rangle, & \langle \sigma_z, \sigma_x\sigma_y \rangle, & \langle \sigma_x\sigma_y, \sigma_y\sigma_z \rangle, \end{array}\right\} \quad (15.1)$$

$$\langle \sigma_x \rangle,\ \langle \sigma_y \rangle,\ \langle \sigma_z \rangle, \quad \langle \sigma_x\sigma_y \rangle,\ \langle \sigma_y\sigma_z \rangle,\ \langle \sigma_z\sigma_x \rangle, \quad \langle \sigma_x\sigma_y\sigma_z \rangle, \quad (15.2)$$

$$\langle e \rangle. \quad (15.3)$$

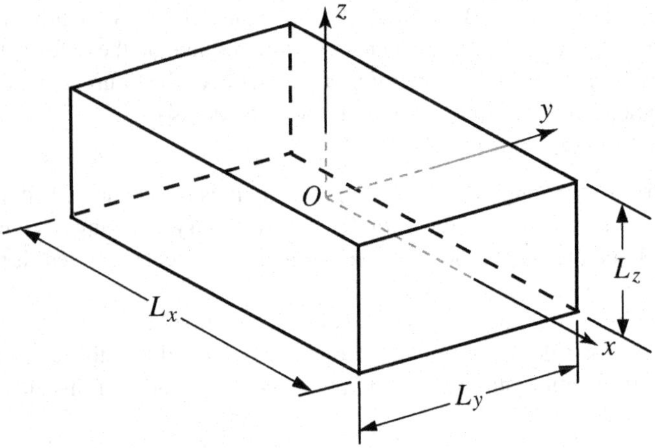

Fig. 15.1 Rectangular parallelepiped domain.

[3] The group $\mathrm{D_{2h}}$ was used in §13.2 to express a partial symmetry of the cylindrical domain. We have the correspondence of $\sigma_{\mathrm{v}} = \sigma_x$, $\sigma_{\mathrm{h}} = \sigma_z$, $c(\pi) = \sigma_x\sigma_y$, and $\sigma_{\mathrm{v}}c(\pi) = \sigma_y$.

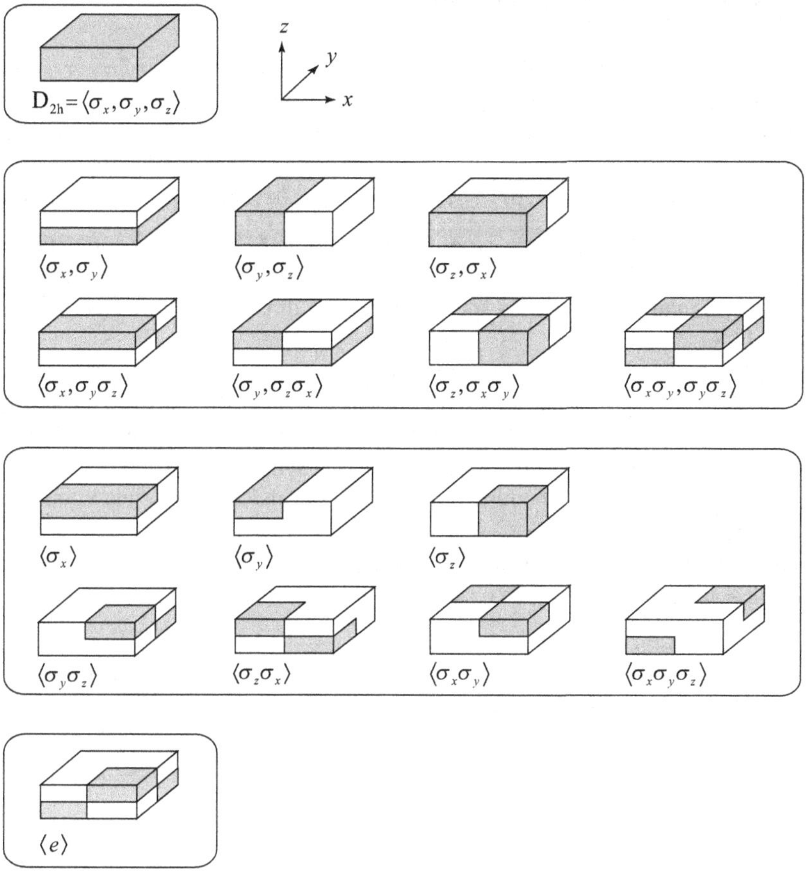

Fig. 15.2 Symmetries expressed by D_{2h} and its subgroups.

The symmetries expressed by D_{2h} and its subgroups are illustrated in Fig. 15.2.

Furthermore, because the rectangular parallelepiped steel specimens are long and uniform in the x-direction, we assume the *periodic symmetry* in this direction, that is, an additional symmetry of the x-directional translation $t_x(l_x)$ for any length l_x with $0 \le l_x < L_x$ defined by (cf., §14.6.2)

$$t_x(l_x) \; : \; x \mapsto x + l_x.$$

The rectangular parallelepiped domain, therefore, has the symmetry labeled by

$$G \equiv \langle \sigma_x, \sigma_y, \sigma_z, t_x(l_x) \rangle \simeq D_{\infty h} \times \mathbb{Z}_2 \simeq O(2) \times \mathbb{Z}_2 \times \mathbb{Z}_2. \tag{15.4}$$

Here $\langle \sigma_x, \sigma_y, \sigma_z, t_x(l_x) \rangle$ is a shorthand notation for the group generated by σ_x, σ_y, σ_z, and $t_x(l_x)$ with any $0 \le l_x < L_x$.

15.3 Recursive Bifurcation Rule

The rule of recursive bifurcation of a system equivariant to $G = \langle \sigma_x, \sigma_y, \sigma_z, t_x(l_x) \rangle$ is briefly introduced here, whereas the details are worked out in §15.6.

Among a series of possible bifurcations, we restrict ourselves to the bifurcation process

$$G = \langle \sigma_x, \sigma_y, \sigma_z, t_x(l_x) \rangle \rightarrow \mathrm{D}_{2\mathrm{h}} = \langle \sigma_x, \sigma_y, \sigma_z \rangle. \tag{15.5}$$

This restriction is in view of the experimental study in §15.4 and the computational study in §15.5, in which the necking with $\mathrm{D}_{2\mathrm{h}}$-symmetry appears around the center of the rectangular parallelepiped specimen through direct bifurcation, as illustrated in Fig. 15.3.

The bifurcation rule of a $\mathrm{D}_{2\mathrm{h}}$-symmetric system can be obtained as a part of the rule for the $\mathrm{D}_{\infty\mathrm{h}}$-symmetric system presented in Fig. 13.7 in §13.2. The direct bifurcation of this system has the symmetry labeled by one of the seven subgroups in (15.1), the secondary bifurcation engenders the symmetry labeled by one of the seven subgroups in (15.2), and further bifurcation engenders completely asymmetric states labeled by C_1.

By assembling the bifurcation process $G \rightarrow \mathrm{D}_{2\mathrm{h}}$ in (15.5), and the bifurcation rule below the group $\mathrm{D}_{2\mathrm{h}}$, we arrive at the rule in Fig. 15.4.

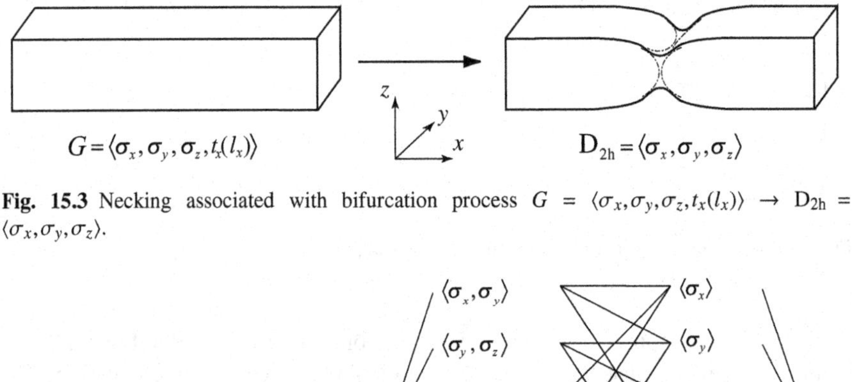

Fig. 15.3 Necking associated with bifurcation process $G = \langle \sigma_x, \sigma_y, \sigma_z, t_x(l_x) \rangle \rightarrow \mathrm{D}_{2\mathrm{h}} = \langle \sigma_x, \sigma_y, \sigma_z \rangle$.

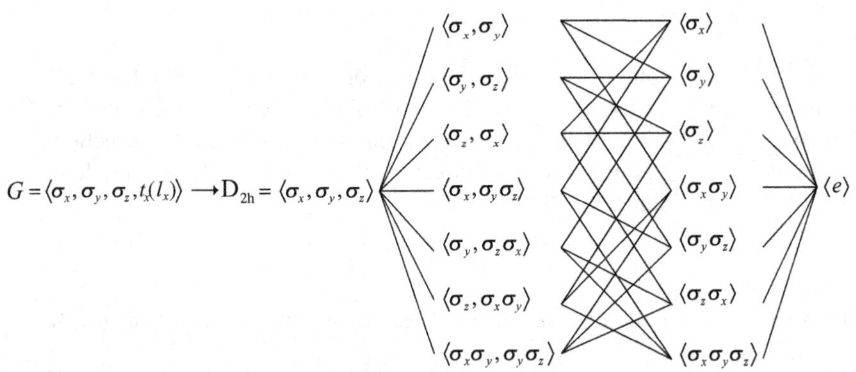

Fig. 15.4 Hierarchy of subgroups expressing the rule of bifurcation of a system invariant to $G = \langle \sigma_x, \sigma_y, \sigma_z, t_x(l_x) \rangle$. $G \rightarrow \mathrm{D}_{2\mathrm{h}}$ is chosen as the direct bifurcation from G among other possibilities (cf., Table 15.2 in §15.6).

Remark 15.1. In the Schoenflies notation, the subgroups $\langle\sigma_x,\sigma_y\sigma_z\rangle$ and $\langle\sigma_y,\sigma_z\sigma_x\rangle$ in (15.1) are identified as the same group D_{1d}. However, we distinguish these groups here because the rectangular parallelepiped domain under consideration is predominantly long in the x-direction and, therefore, the x-axis and the y-axis are not exchangeable. □

15.4 Experimental Study

With reference to the bifurcation rule presented in §15.3, we observe tensile instability behaviors of several rectangular parallelepiped steel specimens with different dimensions listed in Table 15.1 with the same material property. Specimen M is for calibration, and Specimens A, B, C, and D of four different width–thickness ratios are for observation of the effect of cross-sectional shape. A typical geometry of these steel specimens is depicted in Fig. 15.5. We put $L = L_x$, $W = L_y$, and $t = L_z$ in Fig. 15.1 in the sequel. A tensile load in the x-direction is applied to each specimen statically at constant and moderate temperature by a testing machine with a sufficient capacity.

The representative material properties for the specimens, which are used in §15.5 for a computational study, are determined based on a test on Specimen M. Fig-

Table 15.1 Dimensions of steel specimens

Specimen	Width W (mm)	Thickness t (mm)	Width–Thickness Ratio W/t	Length L (mm)
M	39.7	9.79	4.05	100
A	39.9	3.83	10.4	220
B	40.0	9.86	4.06	220
C	39.9	26.0	1.54	220
D	27.8	28.2	0.984	220

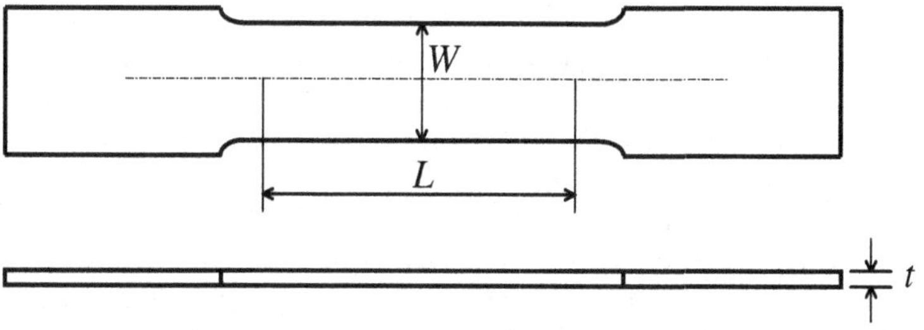

Fig. 15.5 Geometry of a steel specimen. L: length of the rectangular parallelepiped region; W: width of the region; t: thickness of the region.

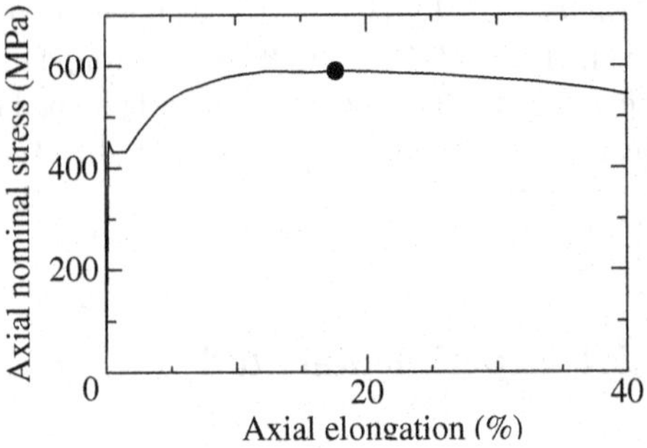

Fig. 15.6 Load versus elongation curve obtained in a tension test for Specimen M for calibration. ●: maximum load.

ure 15.6 shows the representative load versus elongation curve exhibiting several features:

- From the initial slope of the curve, Young's modulus is determined as 204 GPa.
- After a relatively small elastic deformation, plastic yielding occurs uniformly in the rectangular parallelepiped region of the specimen at the initial yield stress (429 MPa).
- As the specimen is further elongated, in association with the progress of plastic deformation, the load increases to the maximum load (591 MPa), thereafter a necking deformation occurs around the center of the specimen and is concentrated into the narrower region.
- Ultimate failure is encountered at the inherent straining limit of the material.

15.4.1 Effect of Cross-Sectional Shape

Specimens A, B, C, and D exhibited similar behaviors up to the maximum load, but the postpeak behaviors are substantially dependent on the cross-sectional shape to show the shape effect. This effect is investigated further in view of photographs in Figs. 15.7–15.10 taken during the tests. The states shown in (a), (b), and (c) in each photograph, respectively, correspond to the beginning of the nonuniform deformation, the appearance of prominent strain localization, and the failure mode.

First, we consider the thinnest Specimen A with a width–thickness ratio of $W/t = 10.4$. As portrayed in Fig. 15.7(a), diffuse necking appears around the center of this specimen. Because of strain localization triggered by the nonuniform deformation, two diagonal shear bands emerge in the middle part of the necking; see Fig. 15.7(b). Then the intense straining concentrates on one of the shear bands, and the other becomes less discernible. In association with the growth of this single distinct shear band, the specimen undergoes ductile failure as shown in Fig. 15.7(c).

Next, we investigate the deformation patterns of the thicker Specimens B, C, and D with width–thickness ratios of $W/t = 4.06$, 1.54, and 0.984, respectively. As portrayed in Figs. 15.8(a), 15.9(a), and 15.10(a), diffuse necking appears around the center; in contrast, the shear bands are not observed even in the state of fracturing,

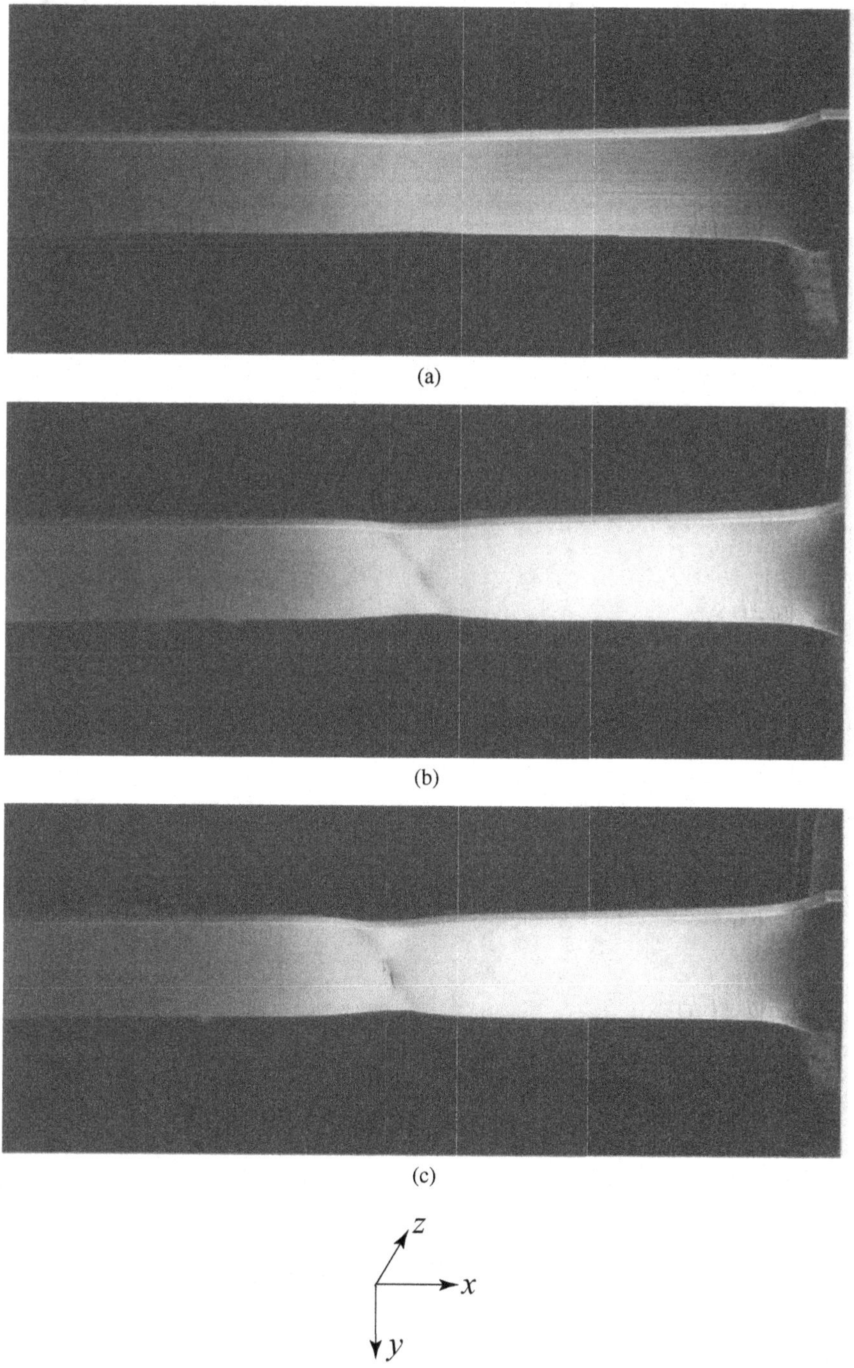

Fig. 15.7 Deformation pattern change of Specimen A ($W/t = 10.4$).

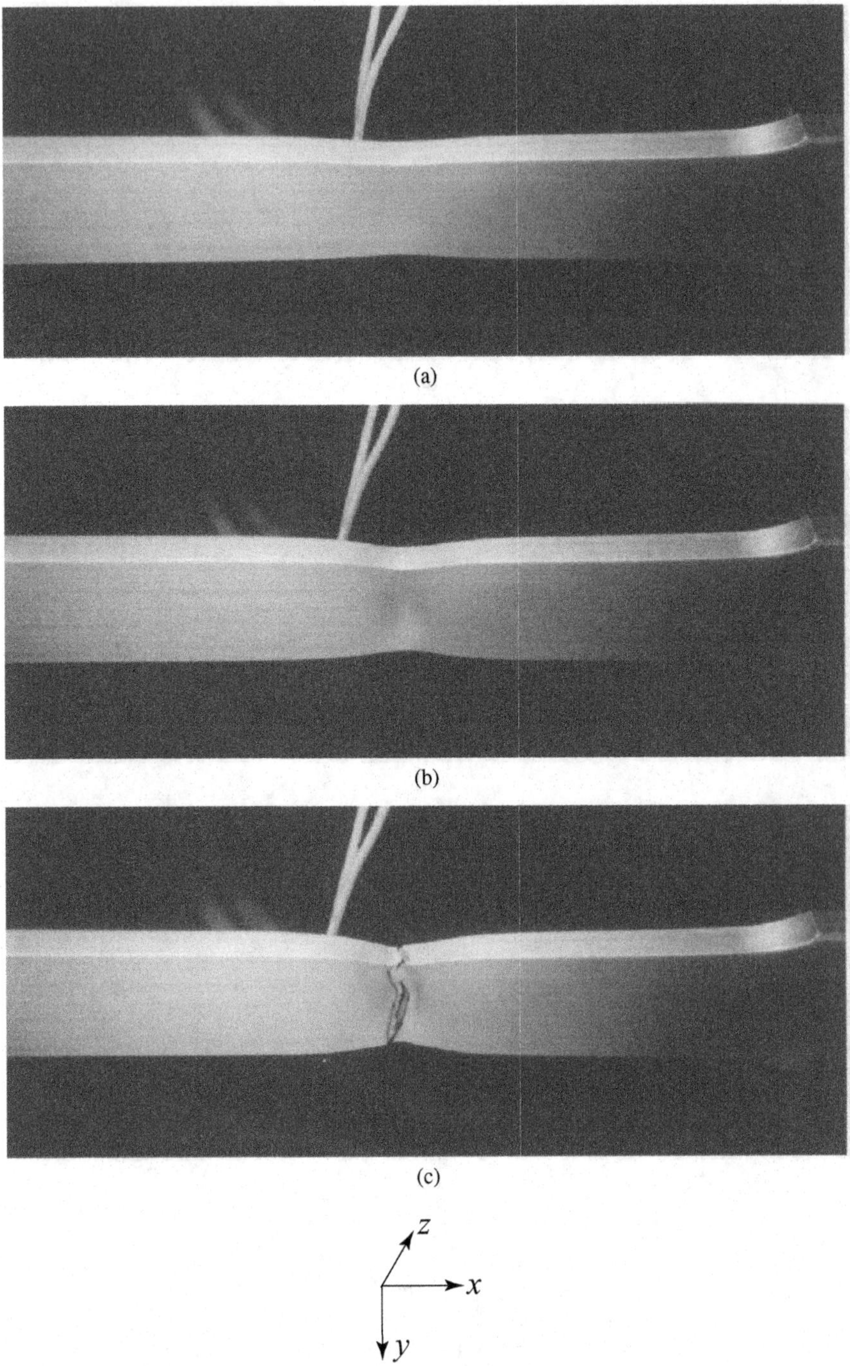

Fig. 15.8 Deformation pattern change of Specimen B ($W/t = 4.06$).

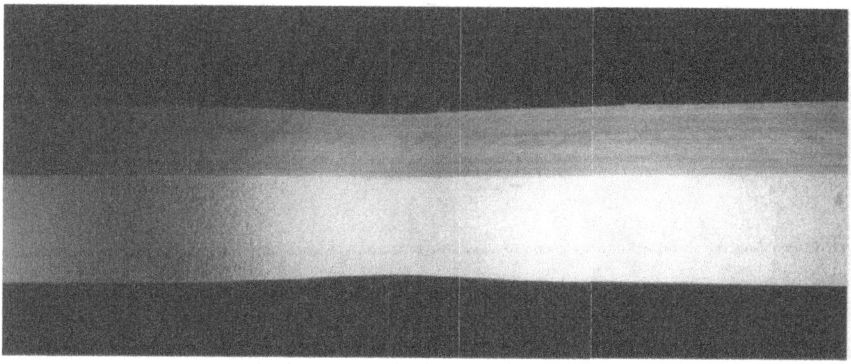

(a)

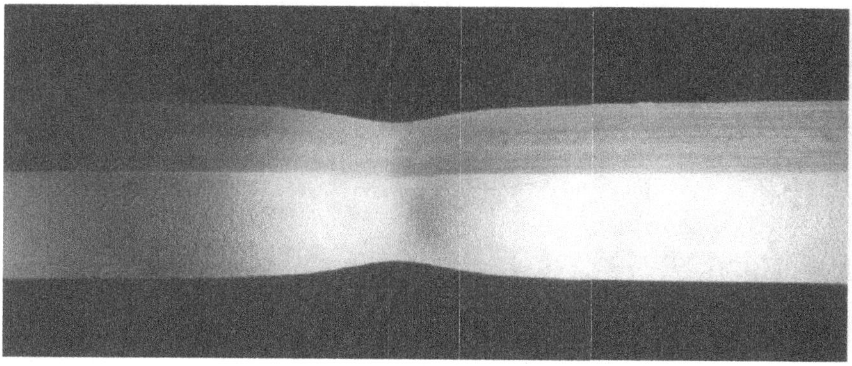

(b)

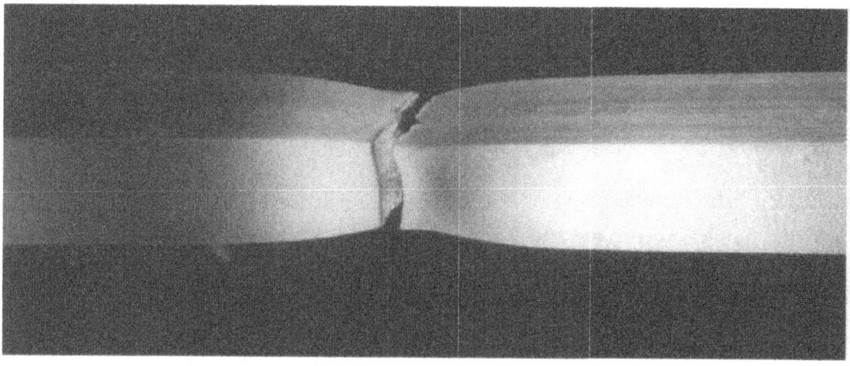

(c)

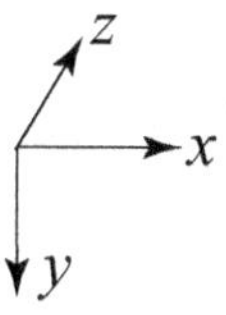

Fig. 15.9 Deformation pattern change of Specimen C ($W/t = 1.54$).

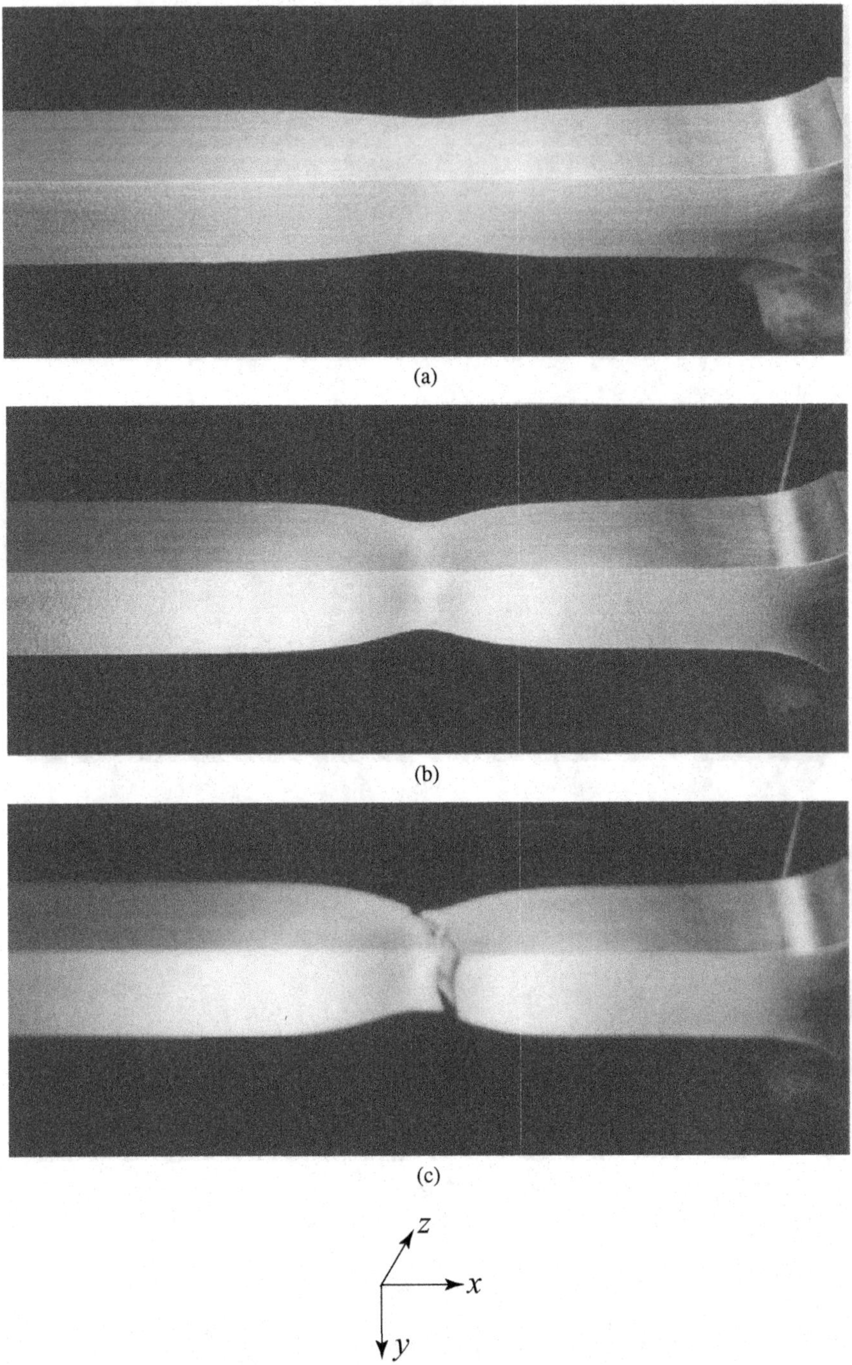

Fig. 15.10 Deformation pattern change of Specimen D ($W/t = 0.984$).

as portrayed in Figs. 15.8(b), 15.9(b), and 15.10(b). Instead, dimples are formed on all the side surfaces. The final states shown in Figs. 15.8(c), 15.9(c), and 15.10(c) display the occurrence of brittle failures. Notwithstanding the complexity of failure modes, they are all subjected to symmetry breaking.

As described, the width–thickness ratios of the specimens are influential on the resulting instability phenomena, especially at the final state. The transient pattern change leading to this state, which progresses rapidly and is therefore unclear, is discussed in the next subsection based on the theory of recursive bifurcation presented in §15.3.

15.4.2 Recursive Bifurcation

The experimental results presented above have displayed the complex instability phenomena triggered most likely by bifurcation. The rule of bifurcation presented in §15.3 is employed here to untangle this complexity and clarify the geometrical aspects of ultimate failures.

For all the specimens, the formation of the necking from the uniform state results from the direct bifurcation associated with

$$G = \langle \sigma_x, \sigma_y, \sigma_z, t_x(l_x) \rangle \rightarrow \mathrm{D}_{2\mathrm{h}} = \langle \sigma_x, \sigma_y, \sigma_z \rangle, \tag{15.6}$$

in which the symmetry of the uniform state is labeled by G and the necking by group $\mathrm{D}_{2\mathrm{h}}$.

First, the thinnest Specimen A ($W/t = 10.4$) is considered. As presented in Fig. 15.7, the diagonal shear bands are formed after the diffuse necking due to the intense localized straining. Such formation is characteristic from a physical standpoint, but is not a bifurcation in that both the state of necking and that of the diagonal shear bands have the same symmetry group $\mathrm{D}_{2\mathrm{h}}$. The secondary bifurcation takes place at the onset of the formation of a single distinct shear band, which is one of the diagonal shear bands. This secondary bifurcation is associated with a further reduction of symmetry described by

$$\mathrm{D}_{2\mathrm{h}} \rightarrow \langle \sigma_z, \sigma_x \sigma_y \rangle, \tag{15.7}$$

in which $\mathrm{D}_{2\mathrm{h}}$ denotes the symmetry of the diagonal shear bands and $\langle \sigma_z, \sigma_x \sigma_y \rangle$ indicates that of the single distinct shear band. Thereafter no bifurcation takes place even at the final failure state, because the symmetry of the specimen is labeled by the same group $\langle \sigma_z, \sigma_x \sigma_y \rangle$ throughout. The deformation pattern change and the loss of symmetry associated with the direct and secondary bifurcations (15.6) and (15.7) are illustrated in Fig. 15.11.

Next, the thicker Specimens B, C, and D ($W/t = 4.06$ to 0.984) are considered. These specimens undergo the direct bifurcation (15.6) to reach the state of diffuse necking. Then dimples are formed by localized straining; such formation is a nonbifurcation process. The specimens arrive at the final states shown in Figs. 15.8(c),

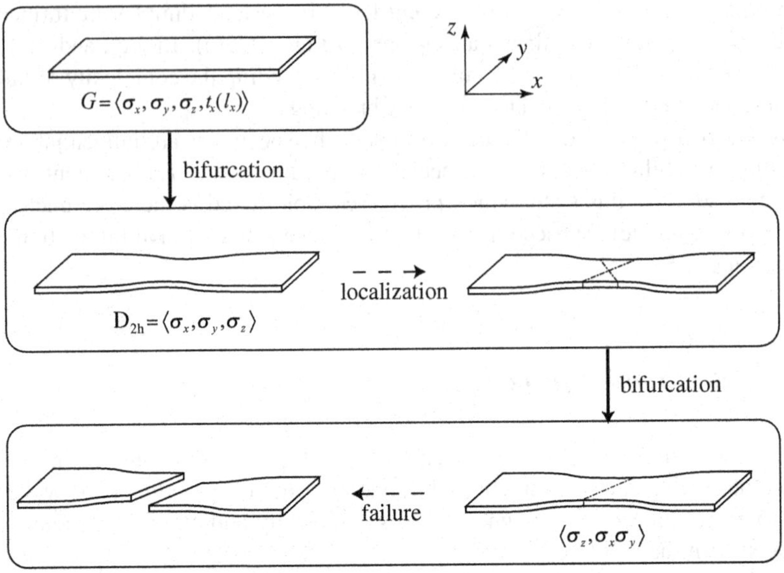

Fig. 15.11 Hierarchical deformation pattern change caused by recursive bifurcation of Specimen A ($W/t = 10.4$).

15.9(c), and 15.10(c). With reference to these states, we can advance the following possible secondary bifurcations:

$$
\begin{aligned}
\mathrm{D_{2h}} &\to \langle\sigma_y,\sigma_z\rangle && \text{for Specimens B and C,}\\
&\searrow \langle\sigma_y,\sigma_z\sigma_x\rangle && \text{for Specimen D.}
\end{aligned}
\qquad (15.8)
$$

These bifurcations, which take place just prior to their final failure, presumably cause the failure. The deformation pattern change and the loss of symmetry associated with the direct and secondary bifurcations (15.6) and (15.8) are illustrated in Figs. 15.12 and 15.13.

From (15.6)–(15.8), we can point out the presence of recursive bifurcation

$$
\begin{aligned}
&\qquad\qquad\qquad\qquad\quad \nearrow \langle\sigma_z,\sigma_x\sigma_y\rangle && \text{for Specimen A,}\\
G = \langle\sigma_x,\sigma_y,\sigma_z,t_x(l_x)\rangle &\to \mathrm{D_{2h}} \to \langle\sigma_y,\sigma_z\rangle && \text{for Specimens B and C,}\\
&\qquad\qquad\qquad\qquad\quad \searrow \langle\sigma_y,\sigma_z\sigma_x\rangle && \text{for Specimen D,}
\end{aligned}
\qquad (15.9)
$$

which is nothing but a part of the whole hierarchy in Fig. 15.4.

Remark 15.2. The introduction of group G, with an additional translational symmetry $t_x(l_x)$ to the natural geometrical symmetry $\mathrm{D_{2h}}$ in understanding the behaviors of steel specimens, may seem redundant, because the necking observed in experiments can be interpreted as a nonbifurcation process. However, as computational analyses on these specimens in §15.5 show, necking is indeed caused by bifurcation. Moreover, recall that enlarged symmetry has played a pivotal role in the understanding

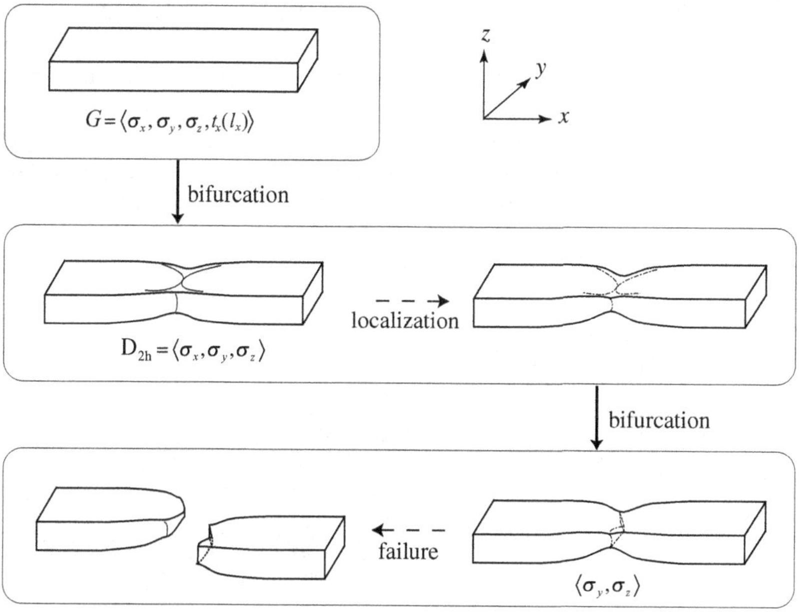

Fig. 15.12 Hierarchical deformation pattern change caused by recursive bifurcation of Specimens B ($W/t = 4.06$) and C ($W/t = 1.54$).

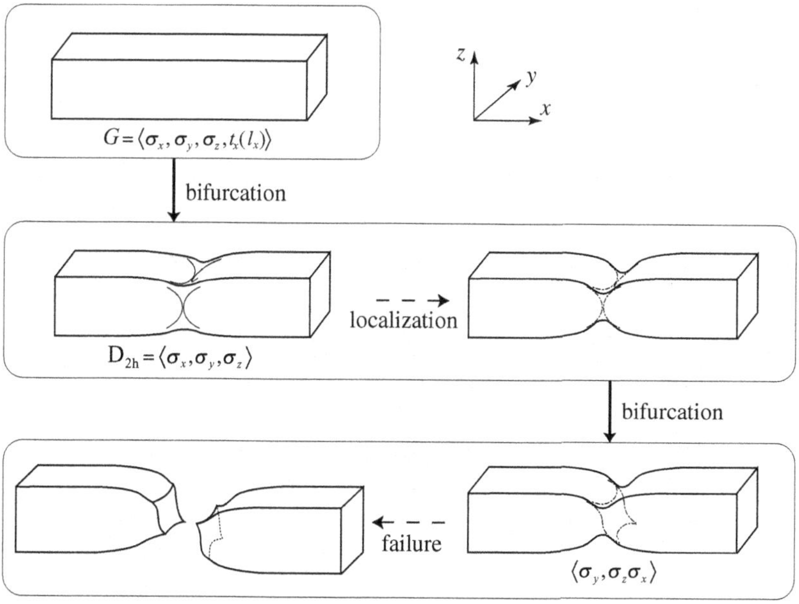

Fig. 15.13 Hierarchical deformation pattern change caused by recursive bifurcation of Specimen D ($W/t = 0.984$).

of patterns in the materials treated in Chapter 14. Patterns can be understood as a consequence of the local uniformity with only a secondary effect from the boundary conditions. A D_{2h}-symmetric state emerges through bifurcation from the G-symmetric state for the present case, although numerous patterns expressed by the symmetry groups in (15.12) are theoretically feasible and might possibly emerge for other cases. The enlarged symmetry expressed by G is also vital for the successful mathematical modeling of the bifurcation of steel specimens that undergo necking. □

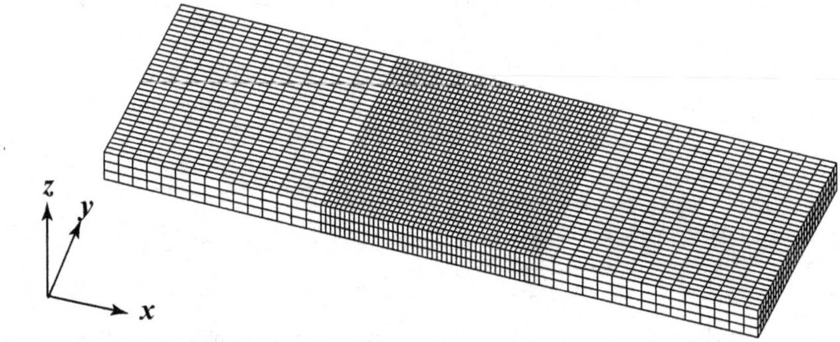

Fig. 15.14 Finite-element model of a rectangular parallelepiped specimen (6750 elements for Case A).

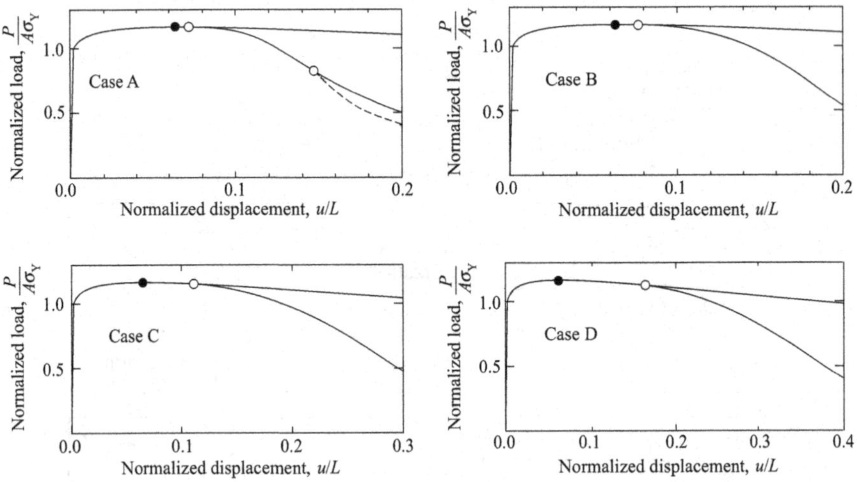

Fig. 15.15 Normalized load versus elongation curves of the specimens. ●: limit point; ○: bifurcation point; P: axial load; A: cross-sectional area; σ_Y: initial yield stress.

15.5 Computational Study

The recursive bifurcation of the rectangular parallelepiped steel specimens is also observed by numerical analyses for finite element models with classical finite-strain elastoplasticity.[4] The specimens with the same width–thickness ratios as those used in the experimental study in §15.4 are used, and are consistently labeled Cases A, B, C, and D. A typical finite-element mesh is portrayed in Fig. 15.14.

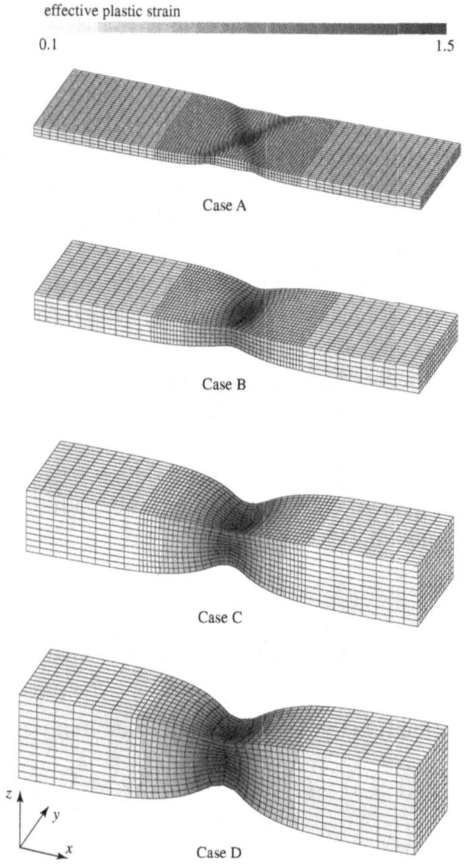

Fig. 15.16 Deformed configurations with the contour of effective plastic strain at load level $P/(A\sigma_Y) = 0.80$ (Case A), 0.70 (others).

[4] We set Young's modulus $E = 200$ GPa, Poisson's ratio $\nu = 0.3333$, and initial yield stress $\sigma_Y =$ 400 MPa. For plastic hardening, the following power law is assumed: $\bar{\sigma} = \sigma_Y(1 + e^p/e_Y)^{0.0625}$, where $e_Y = \sigma_Y/E = 1/500$ and e^p is the effective plastic strain. A tensile force is applied on the surfaces located at $x = \pm L/2$ and all the other surfaces are free from stress.

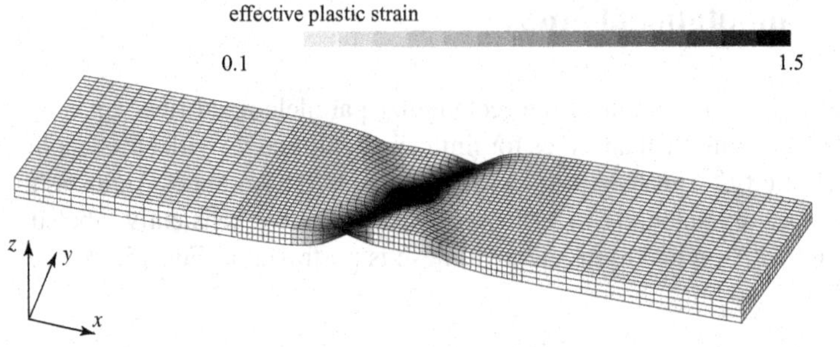

Fig. 15.17 Deformed configurations with the contour of effective plastic strain at load level $P/(A\sigma_Y) = 0.5$.

The load versus elongation curves are depicted in Fig. 15.15. The fundamental path exhibits one limit point shown as (•); the direct bifurcation point shown as (∘) is found after the limit point. Figure 15.16 shows the deformed configurations by contour plots of effective plastic strains. A localized shear band is formed for Case A; in contrast, the localized deformation is accumulated into the center of the rectangular parallelepiped domain for Cases B, C, and D. All of these deformation patterns remain in the course of the bifurcation process $G \to D_{2h}$ given by (15.6).

For Case A, the secondary bifurcation occurs, as depicted by the dotted line at the upper-left of Fig. 15.15. The corresponding deformed configuration shown in Fig. 15.17 agrees with the experimental result in §15.4 (cf., Fig. 15.7(c)). It, therefore, corresponds to the bifurcation process of $D_{2h} \to \langle \sigma_z, \sigma_x\sigma_y \rangle$ associated with (15.7). The successive pattern change due to recursive bifurcation does follow the bifurcation rule presented in §15.3.

15.6 Appendix: Derivation of Bifurcation Rule

The recursive bifurcation rule for a system equivariant to group

$$G = \langle \sigma_x, \sigma_y, \sigma_z, t_x(l_x) \rangle = \langle \sigma_x, t_x(l_x) \rangle \times \langle \sigma_y \rangle \times \langle \sigma_z \rangle \simeq O(2) \times \mathbb{Z}_2 \times \mathbb{Z}_2$$

in (15.4) is derived in this appendix. Noting that group $\langle \sigma_x, t_x(l_x) \rangle \times \langle \sigma_y \rangle$ is isomorphic to group $D_{\infty h}$ (cf., §13.4.1), we can index the family of inequivalent irreducible representations of $G = \langle \sigma_x, \sigma_y, \sigma_z, t_x(l_x) \rangle$ by

$$R(G) = \{(\nu_1, \nu_2, \nu_3),\ (n, \nu_2, \nu_3) \mid \nu_1, \nu_2, \nu_3 = +, -;\ n = 1, 2, \ldots\}. \qquad (15.10)$$

Here $(+,+,+)$ corresponds to the unit representation, which is associated with the limit point. Other one-dimensional representations

$$(-,+,+),\ (+,-,+),\ (-,-,+),\ (+,+,-),\ (-,+,-),\ (+,-,-),\ (-,-,-),$$

Table 15.2 Classification of critical points of a G-equivariant system. $G = \langle \sigma_x, \sigma_y, \sigma_z, t_x(l_x) \rangle$; $\sigma = \sigma_x t_x(l_x)$ for some l_x

	Irreducible	$T^\mu(g)$				Symmetry Groups	
M	Representation, μ	σ_x	σ_y	σ_z	$t_x(l_x)$	G^μ	Bifurcated Paths
1	$(+,+,+)$	1	1	1	1	G	No Bifurcation
	$(-,+,+)$	−1	1	1	1	$\langle \sigma_y, \sigma_z, t_x(l_x) \rangle$	$\langle \sigma_y, \sigma_z, t_x(l_x) \rangle$
	$(+,-,+)$	1	−1	1	1	$\langle \sigma_x, \sigma_z, t_x(l_x) \rangle$	$\langle \sigma_x, \sigma_z, t_x(l_x) \rangle$
	$(-,-,+)$	−1	−1	1	1	$\langle \sigma_z, \sigma_x\sigma_y, t_x(l_x) \rangle$	$\langle \sigma_z, \sigma_x\sigma_y, t_x(l_x) \rangle$
	$(+,+,-)$	1	1	−1	1	$\langle \sigma_x, \sigma_y, t_x(l_x) \rangle$	$\langle \sigma_x, \sigma_y, t_x(l_x) \rangle$
	$(-,+,-)$	−1	1	−1	1	$\langle \sigma_y, \sigma_z\sigma_x, t_x(l_x) \rangle$	$\langle \sigma_y, \sigma_z\sigma_x, t_x(l_x) \rangle$
	$(+,-,-)$	1	−1	−1	1	$\langle \sigma_x, \sigma_y\sigma_z, t_x(l_x) \rangle$	$\langle \sigma_x, \sigma_y\sigma_z, t_x(l_x) \rangle$
	$(-,-,-)$	−1	−1	−1	1	$\langle \sigma_x\sigma_y, \sigma_y\sigma_z, t_x(l_x) \rangle$	$\langle \sigma_x\sigma_y, \sigma_y\sigma_z, t_x(l_x) \rangle$
2	$(n,+,+)$	S	I_2	I_2	R	$\langle \sigma_y, \sigma_z, t_x(L_x/n) \rangle$	$\langle \sigma, \sigma_y, \sigma_z, t_x(L_x/n) \rangle$
	$(n,-,+)$	S	$-I_2$	I_2	R	$\langle \sigma_z, \sigma_y t_x(L_x/(2n)) \rangle$	$\langle \sigma, \sigma_z, \sigma_y t_x(L_x/(2n)) \rangle$
	$(n,+,-)$	S	I_2	$-I_2$	R	$\langle \sigma_y, \sigma_z t_x(L_x/(2n)) \rangle$	$\langle \sigma, \sigma_y, \sigma_z t_x(L_x/(2n)) \rangle$
	$(n,-,-)$	S	$-I_2$	$-I_2$	R	$\langle \sigma_y t_x(L_x/(2n)),$ $\sigma_z t_x(L_x/(2n)) \rangle$	$\langle \sigma, \sigma_y t_x(L_x/(2n)),$ $\sigma_z t_x(L_x/(2n)) \rangle$

defined by the one-dimensional representation matrices $T^\mu(\cdot)$ given in Table 15.2, are associated with simple bifurcation points.

The irreducible representations (n, ν_2, ν_3) in (15.10) are two-dimensional, and their actions are defined by the representation matrices listed in Table 15.2, where

$$I_2 = \begin{pmatrix} 1 & 0 \\ 0 & 1 \end{pmatrix}, \qquad R = \begin{pmatrix} \cos(n\varphi) & -\sin(n\varphi) \\ \sin(n\varphi) & \cos(n\varphi) \end{pmatrix}, \qquad S = \begin{pmatrix} 1 & 0 \\ 0 & -1 \end{pmatrix} \tag{15.11}$$

with the correspondence of

$$\varphi = 2\pi \frac{l_x}{L_x}, \qquad 0 \le \varphi < 2\pi, \;\; 0 \le l_x < L_x.$$

Through the analysis of the bifurcation equation, which is similar to that for a system equivariant to a dihedral group in §8.6, the symmetry of the bifurcating solutions associated with these two-dimensional irreducible representations is obtained as

$$\left\langle \sigma, \sigma_y, \sigma_z, t_x\left(\frac{L_x}{n}\right) \right\rangle, \quad \left\langle \sigma, \sigma_z, \sigma_y t_x\left(\frac{L_x}{2n}\right) \right\rangle, \quad \left\langle \sigma, \sigma_y, \sigma_z t_x\left(\frac{L_x}{2n}\right) \right\rangle,$$
$$\left\langle \sigma, \sigma_y t_x\left(\frac{L_x}{2n}\right), \sigma_z t_x\left(\frac{L_x}{2n}\right) \right\rangle, \qquad n = 1, 2, \ldots, \tag{15.12}$$

where $\sigma = \sigma_x t_x(l_x)$ for some l_x. The group D_{2h} (featured in §15.4.2) is obtained by setting $n = 1$ and $\sigma = \sigma_x$ in $\langle \sigma, \sigma_y, \sigma_z, t_x(L_x/n) \rangle$ in (15.12) and is therefore associated with $(1, +, +)$.

15.7 Appendix: Hilltop Bifurcation

In the mechanical instability of materials, such as steel, a nearly coincidental pair of critical points appears extensively.[5] In the pioneering work by Thompson and Schorrock, 1975 [193], such nearly coincidental pair of points are approximated by the *hilltop bifurcation point*,[6] at which a limit point and a simple pitchfork bifurcation point coincide exactly. They found that the hilltop bifurcation point follows the piecewise linear law, which is less severe than the two-thirds power law for a simple pitchfork bifurcation point.

We refer to the two-dimensional analysis[7] on rectangular steel specimens, for which hilltop bifurcation occurs. The same computational procedure taken for the parallelepiped steel specimens in §15.5 is employed. By imposing symmetry conditions, we analyze only one-fourth[8] of the whole domain shown at the top of Fig. 15.18. Two geometrical configurations of the specimens are employed: Case A with the aspect ratio $L/W = 2$ and Case B with $L/W = 10$ shown at the bottom of Fig. 15.18. Figure 15.19 presents the load versus elongation curves for Cases A

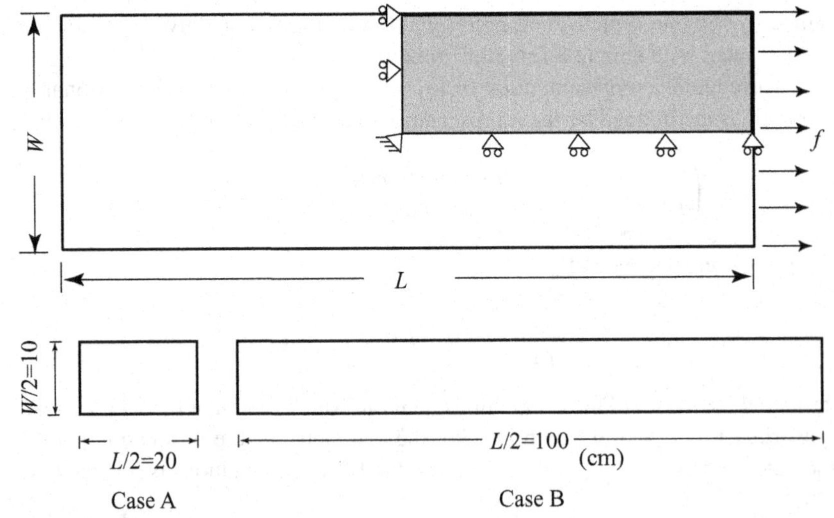

Fig. 15.18 Analysis domains for steel specimens.

[5] See, for a stressed atomic crystal lattice, for example, Thompson and Schorrock, 1975 [193] and Ikeda, Providéncia, and Hunt, 1993 [102]; and for steel specimens, for example, Needleman, 1972 [148] and Hutchinson and Miles, 1974 [81].

[6] A more account of hilltop bifurcation points, which are parametric critical points, can be found in Ikeda, Oide, and Terada, 2002 [100]; Ikeda, Ohsaki, and Kanno, 2005 [99]; and Ohsaki and Ikeda, 2006 [154], 2007 [155].

[7] This analysis is based on Okazawa et al., 2002 [157].

[8] The deformation is restricted to be D_{2h}-symmetric because the one-fourth of the whole domain is employed for the analysis. Therefore, $D_{\infty h} \rightarrow D_{2h}$ is the only bifurcation that can take place.

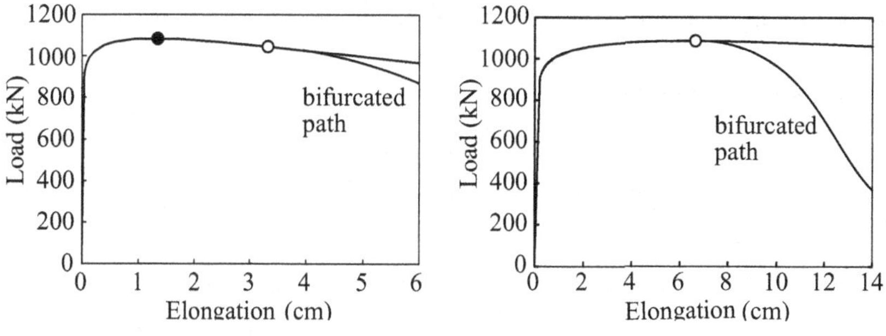

(a) Case A (b) Case B

Fig. 15.19 Load versus elongation curves for Cases A and B. ●: limit point; ○: bifurcation point.

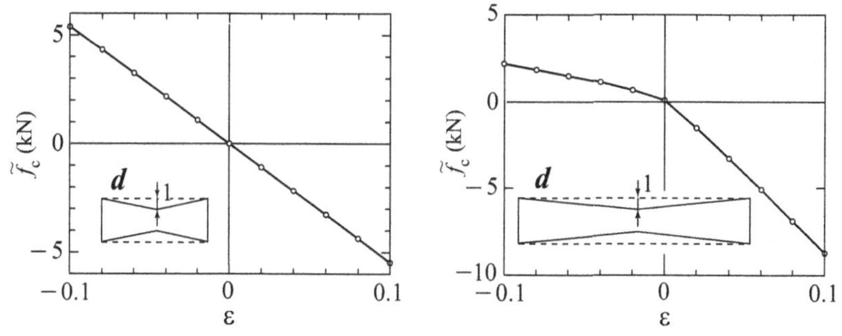

(a) Case A (b) Case B

Fig. 15.20 Imperfection sensitivity for Cases A and B.

and B. The strength of the specimens is governed by a limit point in Case A and by a hilltop bifurcation point in Case B. The steel specimens exhibit an interesting size effect that changes the type of critical points.

To investigate the imperfection sensitivity, we carry out the bifurcation analyses for Cases A and B with an imperfection pattern $\boldsymbol{d}$ shown in Fig. 15.20 but with different imperfection magnitudes $\varepsilon = -0.1$ to 0.1. The critical load increment $\widetilde{f_c}$ versus imperfection magnitude ε relation for Case A is shown in Fig. 15.20(a), which displays a linear law for a limit point consistent with (3.10) . The $\widetilde{f_c}$ versus ε relation for Case B is portrayed in Fig. 15.20(b), which displays a piecewise linear law for a hilltop bifurcation point consistent with the finding by Thompson and Schorrock, 1975 [193]. This shows the applicability of the imperfection sensitivity laws.

Problems

15-1 Verify that the proper subgroups of D_{2h} are given by (15.1)–(15.3).

15-2 Derive the symmetry of the bifurcated solutions in (15.12).

15-3 Show that a system of equations

$$F_1(w_1, w_2, f, \varepsilon) = 2w_1 w_2 + w_1 f + a\varepsilon = 0,$$
$$F_2(w_1, w_2, f, \varepsilon) = {w_1}^2 + {w_2}^2 + f + b\varepsilon = 0$$

has a double bifurcation point.

15-4 Derive the imperfection sensitivity law of the critical load f_c for the double bifurcation point in Problem 15-3.
Answer: $\widetilde{f_c} \approx -|a\varepsilon| - b\varepsilon$. See Ikeda, Oide, and Terada, 2002 [100] for details.

Summary

- The bifurcation rule of a rectangular parallelepiped steel specimen has been presented.
- Recursive bifurcation of steel specimens has been observed in experiment and computational analysis.

Chapter 16
Flower Patterns on Honeycomb Structures

16.1 Introduction

Honeycomb structures under compression display illuminative geometrical patterns.[1] As an example, Fig. 16.1(a) shows the so-called flower mode; a flowerlike pattern in (b), which is cut out from (a), comprises a regular hexagon and six identical cells surrounding this hexagon. Presented in (c) are its variants with different symmetries. In the numerical bifurcation analysis of the honeycomb structure to search for new patterns, it is pertinent to take advantage of group-theoretic analytical information.[2]

Such analytical information has been accumulated against different phenomena.

- An array of hexagonal cells is formed in a uniform fluid by Bénard convection.[3]
- A hexagonally periodic state is observed in the Faraday experiment on a surface wave (e.g., Kudrolli, Pier, and Gollub, 1998 [127]).

These indeed are totally different phenomena but they possess analogous patterns. The mechanism of bifurcation on a hexagonal lattice was investigated to reveal the existence of several patterns.[4] Planforms[5] associated with the 2×2 lattice, of the

[1] A series of characteristic deformation patterns of honeycomb structures subjected to uniaxial and biaxial in-plane compression were found during experiments (e.g., Gibson and Ashby, 1997 [60]). In particular, a flower mode was observed experimentally (e.g., Papka and Kyriakides, 1999 [161]), and was simulated successfully by finite-element analyses (e.g., Guo and Gibson, 1999 [66]).

[2] The theoretical and numerical analyses in this chapter are based on Saiki, Ikeda, and Murota, 2005 [175]. Corrections and revisions are made in this book to supplement deficiencies and to present more details.

[3] See, for example, Bénard, 1900 [13]; Chandrasekhar, 1961 [23]; and Koschmieder, 1966 [123], 1993 [125].

[4] These patterns, for example, are hexagons, antihexagons, rolls, regular triangle, and patchwork quilt. For related issues, see Buzano and Golubitsky, 1983 [21]; Golubitsky and Stewart, 2002 [63]; Melbourne, 1999 [136]; and Bressloff et al., 2001 [19]. Refer to Crawford, 1994 [31] for a study of a square lattice.

[5] *Planforms* mean the patterns associated with bifurcating solutions (cf., Golubitsky and Stewart, 2002 [63]).

K. Ikeda and K. Murota, *Imperfect Bifurcation in Structures and Materials*, Applied Mathematical Sciences 149, DOI 10.1007/978-1-4419-7296-5_16,

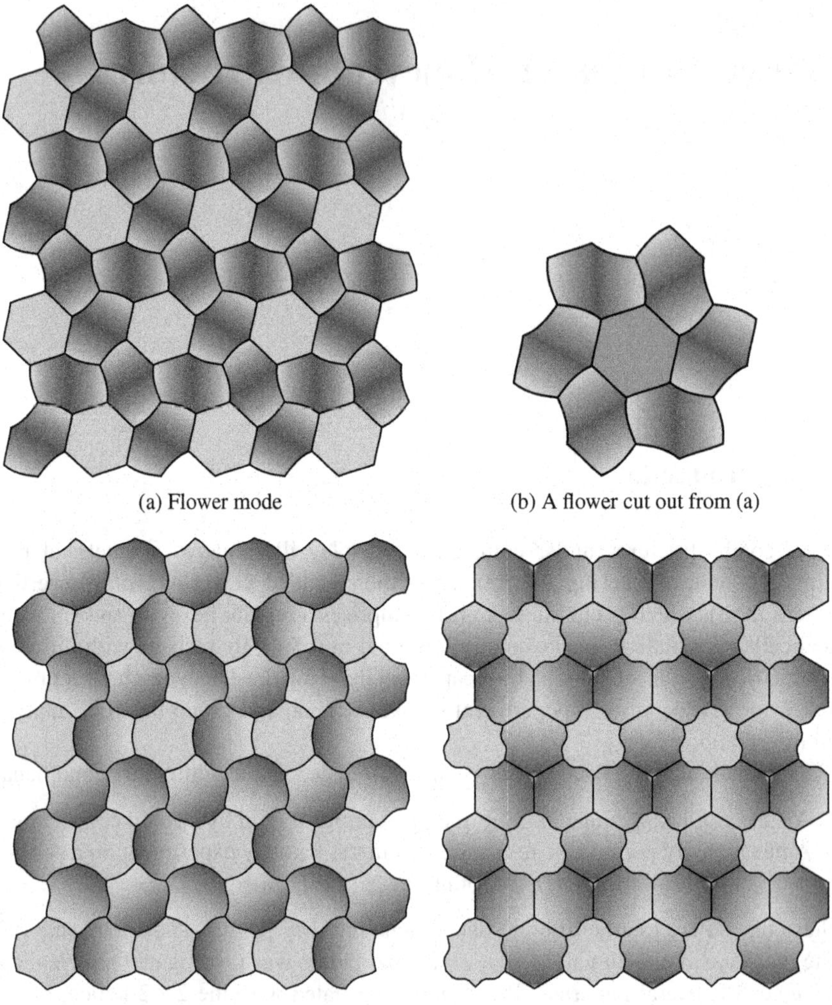

(a) Flower mode

(b) A flower cut out from (a)

(c) Variants of the flower pattern

Fig. 16.1 Planforms of flower patterns.

size next to the smallest, were studied[6] and those on all possible lattice sizes were also obtained.[7]

In this chapter bifurcated deformation patterns of a honeycomb structure are sought theoretically and numerically. As a sound engineering choice, 2×2 cells with periodic boundaries are employed as a representative volume element; an ex-

[6] See, for example, Kirchgässner, 1979 [118] and Judd and Silber, 2000 [110].

[7] See Dionne and Golubitsky, 1992 [42].

haustive list of possible bifurcating deformation patterns of 2×2 cells is obtained.[8] A numerical bifurcation analysis is conducted on an elastic in-plane honeycomb structure consisting of 2×2 cells to obtain a whole set of bifurcated paths. In the numerical bifurcation analysis, knowledge of the symmetries of the bifurcating solutions has turned out to be of great assistance in the tracing of the bifurcated paths at triple bifurcation points. Variants of flowerlike modes with different symmetries have been found and classified successfully on the basis of the list obtained herein.

This chapter is organized as follows.

- The symmetry group for 2×2 cells and its irreducible representations are presented in §16.2.
- The bifurcation of a honeycomb structure is investigated in §16.3.
- The bifurcation equation is derived in §16.4.
- The bifurcation equation is solved to determine bifurcated solutions in §16.5.
- Numerical bifurcation analysis of honeycomb cellular solids is conducted to find variants of flowerlike modes in §16.6.

16.2 Group Describing the Symmetry

A group describing the symmetry of a honeycomb structure is presented.

16.2.1 Symmetry of Representative Volume Element

We consider a regular hexagonal lattice in the xy-plane[9] portrayed in Fig. 16.2. This lattice has the following geometrical symmetries:

- $\overline{r}$: rotation about the z-axis at an angle of $\pi/3$,
- $\overline{s}$: reflection $y \mapsto -y$,
- $\overline{p_1}$: translation along the $\boldsymbol{\ell}_1$-axis that shifts a cell to another cell,
- $\overline{p_2}$: translation along the $\boldsymbol{\ell}_2$-axis that shifts a cell to another cell.

To simplify the numerical and theoretical bifurcation analysis of this lattice, the lattice with 2×2 cells shown in Fig. 16.2(c) is used as a *representative volume element* (RVE). More specifically, we choose as the RVE a parallelogram domain enclosed with $2\boldsymbol{\ell}_1$ and $2\boldsymbol{\ell}_2$ depicted in Fig. 16.2(a). A pattern of this RVE with 2×2 cells is repeated in the direction of the $\boldsymbol{\ell}_1$-axis and also in the direction of the $\boldsymbol{\ell}_2$-axis to arrive at a planform in an infinite domain.

---

8 Such a list is most coveted at this state of research where the patterns are found in an ad hoc manner.

9 Such a lattice is employed in the description of convection of fluids and nematic liquid crystals (cf., Peacock et al., 1999 [162]; Golubitsky and Stewart, 2002 [63]; and Chillingworth and Golubitsky, 2003 [25]).

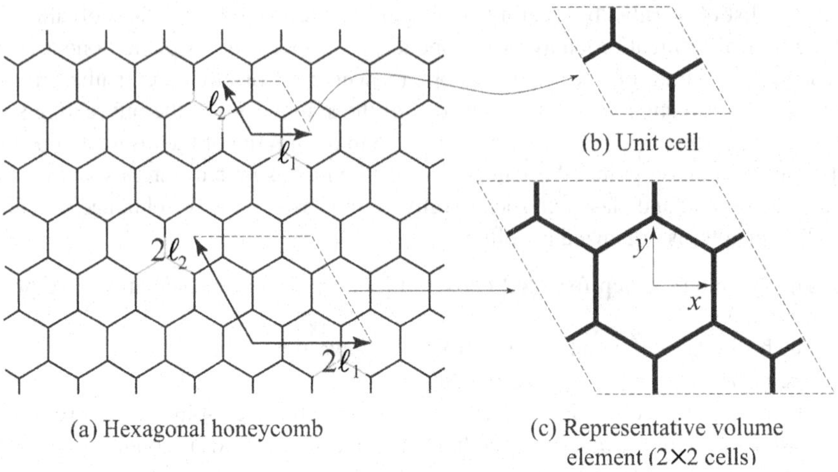

Fig. 16.2 Hexagonal honeycomb, its unit cell, and representative volume element (RVE).

The symmetry of the 2×2 cells is characterized by the invariance with respect to:

- r: rotation about the z-axis at an angle of $\pi/3$,
- s: reflection $y \mapsto -y$,
- p_1: *periodic translation* along the $\boldsymbol{\ell}_1$-axis (i.e., the x-axis),
- p_2: periodic translation along the $\boldsymbol{\ell}_2$-axis.

Consequently, the symmetry of the RVE is described by the group

$$G = \langle r, s, p_1, p_2 \rangle \tag{16.1}$$

with the fundamental relations given by

$$\begin{aligned} &r^6 = s^2 = (rs)^2 = p_1{}^2 = p_2{}^2 = e, \\ &rp_1 = p_1 p_2 r, \quad rp_2 = p_1 r, \quad sp_1 = p_1 s, \quad sp_2 = p_1 p_2 s, \quad p_2 p_1 = p_1 p_2. \end{aligned} \tag{16.2}$$

Note that each element of G can be represented uniquely in the form of

$$p_1{}^i p_2{}^j s^l r^m, \qquad i, j, l \in \{0, 1\};\ m \in \{0, 1, \ldots, 5\}. \tag{16.3}$$

Therefore, G consists of 48 elements.

The group G contains the dihedral group $\mathrm{D}_6 = \langle r, s \rangle$ and two-element groups $\mathbb{Z}_2 = \langle p_1 \rangle$ and $\tilde{\mathbb{Z}}_2 = \langle p_2 \rangle$ as its subgroups. Moreover, it has the structure of semidirect product (cf., Remark 16.1 below) of D_6 and $\mathbb{Z}_2 \times \tilde{\mathbb{Z}}_2$; that is,

$$G = \mathrm{D}_6 \dot{+} (\mathbb{Z}_2 \times \tilde{\mathbb{Z}}_2). \tag{16.4}$$

The expression of this kind is helpful in understanding geometrical patterns and is employed later from time to time.

Remark 16.1. In the standard terminology, a group G is said to be a *semidirect product* of its subgroups H and K, if H is a *normal subgroup* of G (i.e., $g^{-1}hg \in H$ for every $g \in G$), $H \cap K = \{e\}$, and each element $g \in G$ is represented as $g = hk$ with $h \in H$ and $k \in K$. In this case the representation $g = hk$ is unique. Our group G in (16.1) is the semidirect product of $H = \mathbb{Z}_2 \times \tilde{\mathbb{Z}}_2$ and $K = \mathrm{D}_6$ according to this definition and (16.3) gives the representation $g = hk$ with $h = {p_1}^i {p_2}^j \in H$ and $k = s^l r^m \in K$. The notation (16.4) follows the convention in the related literature of bifurcation analysis. □

Remark 16.2. It is easy to predict that $N_1 \times N_2$ cells with $N_1, N_2 > 2$ have different mathematical structures through the enlargement of the symmetry group from $\mathrm{D}_6 \dot{+} (\mathbb{Z}_2 \times \tilde{\mathbb{Z}}_2)$ to $\mathrm{D}_6 \dot{+} (\mathbb{Z}_{N_1} \times \tilde{\mathbb{Z}}_{N_2})$, where $\mathbb{Z}_N$ and $\tilde{\mathbb{Z}}_N$ for a positive integer N denote groups isomorphic to the cyclic group of order N. Such enlargement leads to the increase of possible bifurcating deformation patterns. Our choice of $N_1 = N_2 = 2$ is the simplest but nontrivial case that can engender the flower mode and its variants inherent in the symmetry $\mathrm{D}_6 \dot{+} (\mathbb{Z} \times \tilde{\mathbb{Z}}) = \langle \overline{r}, \overline{s}, \overline{p_1}, \overline{p_2} \rangle$ of the honeycomb structure. □

16.2.2 Irreducible Representations

The irreducible representation of the group $G = \langle r, s, p_1, p_2 \rangle = \mathrm{D}_6 \dot{+} (\mathbb{Z}_2 \times \tilde{\mathbb{Z}}_2)$ in (16.1) and (16.4) are presented. There exist four one-dimensional, two two-dimensional, and four three-dimensional irreducible representations over $\mathbb{R}$, which are all absolutely irreducible (cf., Problem 16-2).

The one-dimensional irreducible representations, labeled as $(+,+)$, $(+,-)$, $(-,+)$, and $(-,-)$, are given by

$$\begin{aligned}
&T^{(+,+)}(r) = 1, \quad && T^{(+,+)}(s) = 1, \quad && T^{(+,+)}(p_1) = 1, \quad && T^{(+,+)}(p_2) = 1,\\
&T^{(+,-)}(r) = 1, && T^{(+,-)}(s) = -1, && T^{(+,-)}(p_1) = 1, && T^{(+,-)}(p_2) = 1,\\
&T^{(-,+)}(r) = -1, && T^{(-,+)}(s) = 1, && T^{(-,+)}(p_1) = 1, && T^{(-,+)}(p_2) = 1,\\
&T^{(-,-)}(r) = -1, && T^{(-,-)}(s) = -1, && T^{(-,-)}(p_1) = 1, && T^{(-,-)}(p_2) = 1.
\end{aligned} \tag{16.5}$$

The two-dimensional irreducible representations, labeled as $(2, j)$ for $j = 1, 2$, are given by

$$\begin{aligned}
&T^{(2,j)}(r) = \begin{pmatrix} \cos(2\pi j/6) & -\sin(2\pi j/6) \\ \sin(2\pi j/6) & \cos(2\pi j/6) \end{pmatrix}, \qquad T^{(2,j)}(s) = \begin{pmatrix} 1 & 0 \\ 0 & -1 \end{pmatrix},\\
&T^{(2,j)}(p_1) = T^{(2,j)}(p_2) = \begin{pmatrix} 1 & 0 \\ 0 & 1 \end{pmatrix}, \qquad j = 1, 2.
\end{aligned} \tag{16.6}$$

The three-dimensional irreducible representations, labeled as $(3, j)$ for $j = 1, \ldots, 4$, are given by

$$
\begin{aligned}
&T^{(3,1)}(r) = \begin{pmatrix} 0\,1\,0 \\ 0\,0\,1 \\ 1\,0\,0 \end{pmatrix}, \qquad && T^{(3,1)}(s) = \begin{pmatrix} 1\,0\,0 \\ 0\,0\,1 \\ 0\,1\,0 \end{pmatrix}; \\
&T^{(3,2)}(r) = \begin{pmatrix} 0 & -1 & 0 \\ 0 & 0 & -1 \\ -1 & 0 & 0 \end{pmatrix}, \qquad && T^{(3,2)}(s) = \begin{pmatrix} 1\,0\,0 \\ 0\,0\,1 \\ 0\,1\,0 \end{pmatrix}; \\
&T^{(3,3)}(r) = \begin{pmatrix} 0\,1\,0 \\ 0\,0\,1 \\ 1\,0\,0 \end{pmatrix}, \qquad && T^{(3,3)}(s) = \begin{pmatrix} -1 & 0 & 0 \\ 0 & 0 & -1 \\ 0 & -1 & 0 \end{pmatrix}; \\
&T^{(3,4)}(r) = \begin{pmatrix} 0 & -1 & 0 \\ 0 & 0 & -1 \\ -1 & 0 & 0 \end{pmatrix}, \qquad && T^{(3,4)}(s) = \begin{pmatrix} -1 & 0 & 0 \\ 0 & 0 & -1 \\ 0 & -1 & 0 \end{pmatrix}; \\
&T^{(3,j)}(p_1) = \begin{pmatrix} 1 & 0 & 0 \\ 0 & -1 & 0 \\ 0 & 0 & -1 \end{pmatrix}, \qquad && T^{(3,j)}(p_2) = \begin{pmatrix} -1 & 0 & 0 \\ 0 & 1 & 0 \\ 0 & 0 & -1 \end{pmatrix}, \qquad j = 1,\dots,4.
\end{aligned}
\tag{16.7}
$$

These irreducible representations of $G = \langle r, s, p_1, p_2 \rangle$ are summarized in the form of a *character table* in Table 16.1, which shows trace($T^{\mu}(g)$), called the *character*, for

Table 16.1 Character table of $G = \langle r, s, p_1, p_2 \rangle$

	Class	C_1	C_2	C_3	C_4	C_5	C_6	C_7	C_8	C_9	C_{10}
	Size	1	8	8	1	6	6	3	3	6	6
	(+,+)	1	1	1	1	1	1	1	1	1	1
	(+,−)	1	1	1	1	−1	−1	1	1	−1	−1
	(−,+)	1	−1	1	−1	1	−1	1	−1	−1	1
	(−,−)	1	−1	1	−1	−1	1	1	−1	1	−1
μ	(2,1)	2	1	−1	−2	0	0	2	−2	0	0
	(2,2)	2	−1	−1	2	0	0	2	2	0	0
	(3,1)	3	0	0	3	1	1	−1	−1	−1	−1
	(3,2)	3	0	0	−3	1	−1	−1	1	1	−1
	(3,3)	3	0	0	3	−1	−1	−1	−1	1	1
	(3,4)	3	0	0	−3	−1	1	−1	1	−1	1

Class	Element
C_1	e
C_2	$r, r^5, p_1 r, p_1 r^5, p_2 r, p_2 r^5, p_1 p_2 r, p_1 p_2 r^5$
C_3	$r^2, r^4, p_1 r^2, p_1 r^4, p_2 r^2, p_2 r^4, p_1 p_2 r^2, p_1 p_2 r^4$
C_4	r^3
C_5	$s, sr^2, sr^4, p_1 s, p_2 sr^2, p_1 p_2 sr^4$
C_6	$sr, sr^3, sr^5, p_1 sr^3, p_2 sr^5, p_1 p_2 sr$
C_7	$p_1, p_2, p_1 p_2$
C_8	$p_1 r^3, p_2 r^3, p_1 p_2 r^3$
C_9	$p_1 sr, p_1 sr^5, p_2 sr, p_2 sr^3, p_1 p_2 sr^3, p_1 p_2 sr^5$
C_{10}	$p_1 sr^2, p_1 sr^4, p_2 s, p_2 sr^4, p_1 p_2 s, p_1 p_2 sr^2$

each $\mu \in R(G)$ and $g \in G$. Two elements g_1 and g_2 of G belong to the same *class* C_i if $g_2 = h^{-1} g_1 h$ for some $h \in G$; the character remains constant in a class since

$$\mathrm{trace}(T(h^{-1}gh)) = \mathrm{trace}(T(g)), \qquad g \in G$$

for any (not necessarily irreducible) representation T.

16.3 Bifurcation for Representative Volume Element

The direct bifurcations for the representative volume element of a honeycomb structure are investigated. In particular, an exhaustive list of symmetries of possible bifurcated solutions is presented.

As §16.2 shows, the RVE with 2×2 cells has the symmetry described by $G = \langle r, s, p_1, p_2 \rangle = \mathrm{D}_6 \dot{+} (\mathbb{Z}_2 \times \tilde{\mathbb{Z}}_2)$ in (16.1) and (16.4), and, accordingly, has group-theoretic simple, double, and triple critical points, associated with one-, two-, and three-dimensional irreducible representations μ of G.

16.3.1 Simple Critical Points

A simple critical point is associated with one of the four one-dimensional irreducible representations $\mu = (+,+), (+,-), (-,+)$, and $(-,-)$. We have

$$T^{\mu}(p_1) = 1, \qquad T^{\mu}(p_2) = 1$$

for $\mu = (+,+), (+,-), (-,+), (-,-)$ by (16.5); therefore, the equivariance condition for the bifurcation equation becomes identical with that for $\mathrm{D}_6 = \langle r, s \rangle$. The bifurcation rule for D_n presented in Table 8.1 in §8.3.1 is made consistent with the present case and is given in Table 16.2.

The simple critical point is, generically, a limit point if $\mu = (+,+)$; it is a pitchfork bifurcation point otherwise. The symmetry $\Sigma(\boldsymbol{u})$ (cf., (7.86)) of the solution $\boldsymbol{u}$ on the bifurcated path agrees with the symmetry G^{μ} of the kernel space of the Jacobian matrix (cf., (7.22) and (7.94)). As listed in Table 16.2, we have

$$\Sigma(\boldsymbol{u}) = G^{\mu} = \begin{cases} \langle r, p_1, p_2 \rangle = \mathrm{C}_6 \dot{+} (\mathbb{Z}_2 \times \tilde{\mathbb{Z}}_2) & \text{for } \mu = (+,-), \\ \langle r^2, s, p_1, p_2 \rangle = \mathrm{D}_3 \dot{+} (\mathbb{Z}_2 \times \tilde{\mathbb{Z}}_2) & \text{for } \mu = (-,+), \\ \langle r^2, sr, p_1, p_2 \rangle = \mathrm{D}_3^{2,6} \dot{+} (\mathbb{Z}_2 \times \tilde{\mathbb{Z}}_2) & \text{for } \mu = (-,-). \end{cases}$$

Figure 16.3 shows a representative volume element with 2×2 cells at the left. The upper-left is the representative volume element, and the lower-left is the element toned to clarify the spatial pattern. The element is assembled periodically to arrive at the planform shown at the right. This planform, associated with $\mu = (-,+)$, is invariant to $\langle \overline{r}^2, \overline{s}, \overline{p_1}, \overline{p_2} \rangle \simeq \mathrm{D}_3 \dot{+} (\mathbb{Z} \times \tilde{\mathbb{Z}})$; identical cells with local D_3-symmetry

Table 16.2 Classification of critical points and symmetries of bifurcated solutions

μ	G^{μ}		Symmetry $\Sigma(\boldsymbol{u})$ of Bifurcated Solutions $\boldsymbol{u}$
$(+,+)$	$\langle r,s,p_1,p_2\rangle$		$\langle r,s,p_1,p_2\rangle = \mathrm{D}_6\dot{+}(\mathbb{Z}_2\times\tilde{\mathbb{Z}}_2)$
$(+,-)$	$\langle r,p_1,p_2\rangle$		$\langle r,p_1,p_2\rangle = \mathrm{C}_6\dot{+}(\mathbb{Z}_2\times\tilde{\mathbb{Z}}_2)$
$(-,+)$	$\langle r^2,s,p_1,p_2\rangle$		$\langle r^2,s,p_1,p_2\rangle = \mathrm{D}_3\dot{+}(\mathbb{Z}_2\times\tilde{\mathbb{Z}}_2)$
$(-,-)$	$\langle r^2,sr,p_1,p_2\rangle$		$\langle r^2,sr,p_1,p_2\rangle = \mathrm{D}_3^{2,6}\dot{+}(\mathbb{Z}_2\times\tilde{\mathbb{Z}}_2)$
$(2,1)$	$\langle p_1,p_2\rangle$		$\langle sr^{k-1},p_1,p_2\rangle = \mathrm{D}_1^{k,6}\dot{+}(\mathbb{Z}_2\times\tilde{\mathbb{Z}}_2)$, $k=1,\dots,6$
$(2,2)$	$\langle r^3,p_1,p_2\rangle$		$\langle r^3,sr^{k-1},p_1,p_2\rangle = \mathrm{D}_2^{k,6}\dot{+}(\mathbb{Z}_2\times\tilde{\mathbb{Z}}_2)$, $k=1,2,3$
$(3,1)$	$\langle r^3\rangle \simeq \mathrm{C}_2$	Mode I	$\langle r^3,s,p_1\rangle$, $\langle r^3,sr^2,p_2\rangle$, $\langle r^3,sr,p_1p_2\rangle$
		Mode II	Nonexistent
		Flower	$\langle r,s\rangle$, $\langle p_1r,sr\rangle$, $\langle sr^2,p_1p_2r\rangle$, $\langle s,p_2r\rangle$
$(3,2)$	$\langle e\rangle = \mathrm{C}_1$	Mode I	$\langle p_2r^3,s,p_1\rangle$, $\langle p_1p_2r^3,sr^2,p_2\rangle$, $\langle p_1r^3,sr^4,p_1p_2\rangle$
		Mode II	$\langle p_1r^3,s\rangle$, $\langle p_1r^3,sr^3\rangle$, $\langle p_2r^3,sr^2\rangle$, $\langle p_2r^3,sr^5\rangle$, $\langle p_1p_2r^3,sr^4\rangle$, $\langle p_1p_2r^3,sr\rangle$
		Flower	$\langle r^2,s\rangle$, $\langle p_2r^2,sr^4\rangle$, $\langle p_1r^2,sr^2\rangle$, $\langle p_1p_2r^2,s\rangle$
$(3,3)$	$\langle r^3\rangle \simeq \mathrm{C}_2$	Mode I	$\langle r^3,p_2s\rangle$, $\langle r^3,p_1p_2sr^2\rangle$, $\langle r^3,p_1sr^4\rangle$
		Mode II	$\langle r^3,s\rangle$, $\langle r^3,sr^4\rangle$, $\langle r^3,sr^2\rangle$, $\langle r^3,p_1s\rangle$, $\langle r^3,p_1p_2sr^4\rangle$, $\langle r^3,p_2sr^2\rangle$
		Flower	$\langle r\rangle$, $\langle p_1r\rangle$, $\langle p_1p_2r\rangle$, $\langle p_2r\rangle$
$(3,4)$	$\langle e\rangle = \mathrm{C}_1$	Mode I	$\langle p_2r^3,sr^3,p_1\rangle$, $\langle p_1r^3,sr^5,p_2\rangle$, $\langle p_2r^3,sr,p_1p_2\rangle$
		Mode II	$\langle p_1r^3,sr^3\rangle$, $\langle p_1r^3,s\rangle$, $\langle p_2r^3,sr^5\rangle$, $\langle p_2r^3,sr^2\rangle$, $\langle p_1p_2r^3,sr\rangle$, $\langle p_1p_2r^3,sr^4\rangle$
		Flower	$\langle r^2,sr\rangle$, $\langle p_2r^2,sr\rangle$, $\langle p_1r^2,sr^5\rangle$, $\langle p_2r^4,sr^3\rangle$

$\mathrm{D}_6=\langle r,s\rangle$, $\mathrm{C}_6=\langle r\rangle$, $\mathbb{Z}_2=\langle p_1\rangle$, $\tilde{\mathbb{Z}}_2=\langle p_2\rangle$, $\mathrm{D}_m^{k,6}=\{r^{6i/m},sr^{k-1+(6i/m)}\mid i=0,1,\dots,m-1\}$

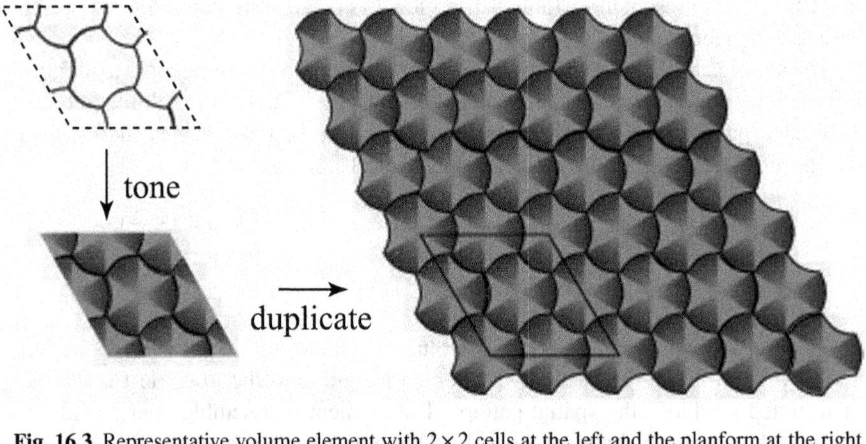

Fig. 16.3 Representative volume element with 2×2 cells at the left and the planform at the right associated with $\mu=(-,+)$. The planform has $\langle \overline{r^2},\overline{s},\overline{p_1},\overline{p_2}\rangle$-symmetry.

are repeated periodically by the translations $\overline{p_1}$ and $\overline{p_2}$. Consequently, at the simple bifurcation point, the local geometrical symmetry D_6 is broken, but the global *translational symmetry* is preserved.

16.3.2 Double Critical Points

A double critical point is associated with one of the two two-dimensional irreducible representations $\mu = (2,1)$ and $(2,2)$. Since we have

$$T^{(2,j)}(p_1) = T^{(2,j)}(p_2) = \begin{pmatrix} 1 & 0 \\ 0 & 1 \end{pmatrix}$$

for $j = 1,2$ by (16.6), the equivariance condition for the bifurcation equation becomes identical to that for $D_6 = \langle r, s \rangle$. The bifurcation rule for D_n presented in Table 8.1 in §8.3.1 is made consistent with the present case and is given in Table 16.2. The symmetry $\Sigma(\boldsymbol{u})$ of the solution $\boldsymbol{u}$ on the bifurcated path is larger than the symmetry G^{μ} of the kernel space of the Jacobian matrix, as listed in Table 16.2, and we have

$$\Sigma(\boldsymbol{u}) = \begin{cases} \langle sr^{k-1}, p_1, p_2 \rangle = D_1^{k,6} \dot{+} (\mathbb{Z}_2 \times \tilde{\mathbb{Z}}_2), & k = 1, \ldots, 6 \quad \text{for } \mu = (2,1), \\ \langle r^3, sr^{k-1}, p_1, p_2 \rangle = D_2^{k,6} \dot{+} (\mathbb{Z}_2 \times \tilde{\mathbb{Z}}_2), & k = 1,2,3 \quad \text{for } \mu = (2,2). \end{cases}$$

Consequently, at the double bifurcation point, the local geometrical symmetry D_6 is broken into $D_1^{k,6}$ $(k = 1, \ldots, 6)$ for $\mu = (2,1)$ and into $D_2^{k,6}$ $(k = 1,2,3)$ for $\mu = (2,2)$, but the global translational symmetry $\mathbb{Z}_2 \times \tilde{\mathbb{Z}}_2$ is preserved.

16.3.3 Triple Critical Points

A triple critical point is associated with one of the four three-dimensional irreducible representations $\mu = (3,1)$, $(3,2)$, $(3,3)$, and $(3,4)$. The symmetries of the bifurcated solutions at the triple points are determined by solving the bifurcation equation (7.52), which for this case reads as

$$F_i(w_1, w_2, w_3, f) = 0, \qquad i = 1,2,3, \tag{16.8}$$

where $\boldsymbol{w} = (w_1, w_2, w_3)^\top$ is a real vector and $(w_1, w_2, w_3, f) = (0,0,0,0)$ is assumed to correspond to the triple bifurcation point. Unlike the simple critical points in §16.3.1 and the double critical points in §16.3.2, the equivariance condition for the bifurcation equation cannot be reduced to that for $D_6 = \langle r, s \rangle$.

The symmetries of the solutions of the bifurcation equation (16.8) are obtained for each irreducible representation μ, as expounded in §16.4 and in §16.5, and the results are presented in Table 16.2. Both the local and the global translational sym-

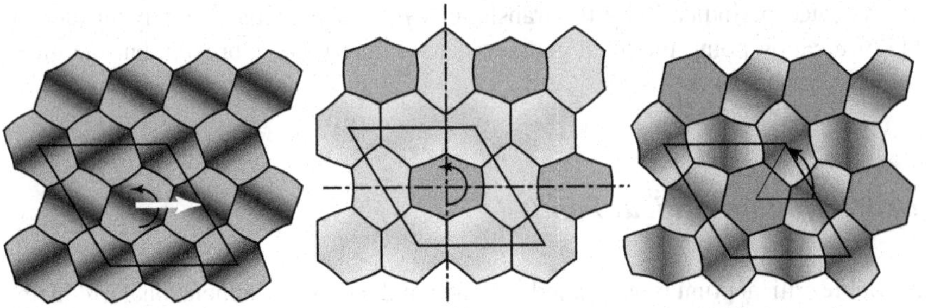

(a) Mode I
$\Sigma(\boldsymbol{u}) = \Sigma(\boldsymbol{w}) = \langle r^3, p_2 s\rangle$

(b) Mode II
$\Sigma(\boldsymbol{u}) = \Sigma(\boldsymbol{w}) = \langle r^3, s\rangle$

(c) Flower mode
$\Sigma(\boldsymbol{u}) = \Sigma(\boldsymbol{w}) = \langle r\rangle$

Fig. 16.4 Examples of deformation modes of bifurcated solutions for $\mu = (3,3)$. Arrow: translational symmetry; rotated arrow: rotation symmetry; dotted-dashed line: line of reflection symmetry.

metries are partially broken at the triple bifurcation point, whereas the global translational symmetry $\mathbb{Z}_2 \times \tilde{\mathbb{Z}}_2$ is preserved at the simple and double critical points.

There exist bifurcating patterns of three kinds:[10]

- Mode I—only one of w_i $(i = 1,2,3)$ is nonzero,
- Mode II—two of w_i $(i = 1,2,3)$ are nonzero,
- Flower mode—w_i $(i = 1,2,3)$ are all nonzero.

These three modes are explained for $\mu = (3,3)$ below.

- A mode I deformation pattern with $\langle r^3, p_2 s\rangle$-symmetry is portrayed in Fig. 16.4(a). This pattern has the local symmetry r^3 that rotates a cell at an angle of π, the symmetry with respect to the reflection s followed by oblique translation p_2, and the translational symmetry $p_1 = (p_2 s)^2$ that shifts a cell to another cell at the right, and so on.
- A mode II deformation pattern depicted in Fig. 16.4(b) is invariant to $\langle r^3, s\rangle \simeq \mathrm{D}_2$, and has the local D_2-symmetry, but has no translational symmetry.
- A flower mode presented in Fig. 16.4(c) is invariant to $\langle r\rangle \simeq \mathrm{C}_6$, and has the local C_6-symmetry.

16.4 Derivation of Bifurcation Equation

We derive the bifurcation equation (16.8):

$$F_1(w_1, w_2, w_3, f) = F_2(w_1, w_2, w_3, f) = F_3(w_1, w_2, w_3, f) = 0 \tag{16.9}$$

at a group-theoretic triple critical point associated with the three-dimensional irreducible representation $\mu = (3, j)$ $(j = 1,\dots,4)$ of the group

[10] For the names of these three patterns, see, for example, Ohno, Okumura, and Noguchi, 2002 [153].

$$G = \langle r, s, p_1, p_2 \rangle \simeq \mathrm{D}_6 \dot{+} (\mathbb{Z}_2 \times \tilde{\mathbb{Z}}_2)$$

in (16.1) and (16.4). It is assumed that $(w_1, w_2, w_3, f) = (0,0,0,0)$ corresponds to the triple bifurcation point.

16.4.1 A Representative Case

We choose $\mu = (3,3)$ as a representative case, before addressing all the cases $\mu = (3, j)$ $(j = 1, \ldots, 4)$. It is noteworthy that $\mu = (3,3)$ is associated with the flower mode hitherto obtained.[11]

For $\mu = (3,3)$, the action of G on $\boldsymbol{w} = (w_1, w_2, w_3)$ is given (cf., (16.7)) by

$$\begin{aligned}
p_1 &: w_1 \mapsto w_1, & w_2 &\mapsto -w_2, & w_3 &\mapsto -w_3, \\
p_2 &: w_1 \mapsto -w_1, & w_2 &\mapsto w_2, & w_3 &\mapsto -w_3, \\
r &: w_1 \mapsto w_2, & w_2 &\mapsto w_3, & w_3 &\mapsto w_1, \\
s &: w_1 \mapsto -w_1, & w_2 &\mapsto -w_3, & w_3 &\mapsto -w_2.
\end{aligned}$$

Since the group G is generated by the four elements p_1, p_2, s, and r, the equivariance of the bifurcation equation (16.9) to the group G is identical with its equivariance to the actions of these four elements. That is,

$$p_1: \quad F_1(w_1, w_2, w_3, f) = F_1(w_1, -w_2, -w_3, f), \tag{16.10}$$
$$-F_2(w_1, w_2, w_3, f) = F_2(w_1, -w_2, -w_3, f), \tag{16.11}$$
$$-F_3(w_1, w_2, w_3, f) = F_3(w_1, -w_2, -w_3, f); \tag{16.12}$$

$$p_2: \quad -F_1(w_1, w_2, w_3, f) = F_1(-w_1, w_2, -w_3, f), \tag{16.13}$$
$$F_2(w_1, w_2, w_3, f) = F_2(-w_1, w_2, -w_3, f), \tag{16.14}$$
$$-F_3(w_1, w_2, w_3, f) = F_3(-w_1, w_2, -w_3, f); \tag{16.15}$$

$$r: \quad F_2(w_1, w_2, w_3, f) = F_1(w_2, w_3, w_1, f), \tag{16.16}$$
$$F_3(w_1, w_2, w_3, f) = F_2(w_2, w_3, w_1, f), \tag{16.17}$$
$$F_1(w_1, w_2, w_3, f) = F_3(w_2, w_3, w_1, f); \tag{16.18}$$

$$s: \quad -F_1(w_1, w_2, w_3, f) = F_1(-w_1, -w_3, -w_2, f), \tag{16.19}$$
$$-F_3(w_1, w_2, w_3, f) = F_2(-w_1, -w_3, -w_2, f), \tag{16.20}$$
$$-F_2(w_1, w_2, w_3, f) = F_3(-w_1, -w_3, -w_2, f). \tag{16.21}$$

These equivariance conditions are equivalent to the following conditions:

[11] See, for example, Ohno, Okumura, and Noguchi, 2002 [153].

$$F_1(w_1,w_2,w_3,f) = F_1(w_1,-w_2,-w_3,f), \tag{16.22}$$
$$-F_1(w_1,w_2,w_3,f) = F_1(-w_1,w_2,-w_3,f), \tag{16.23}$$
$$-F_1(w_1,w_2,w_3,f) = F_1(-w_1,w_3,w_2,f), \tag{16.24}$$
$$F_2(w_1,w_2,w_3,f) = F_1(w_2,w_3,w_1,f), \tag{16.25}$$
$$F_3(w_1,w_2,w_3,f) = F_1(w_3,w_1,w_2,f). \tag{16.26}$$

We expand F_1 as

$$F_1(w_1,w_2,w_3,f) = \sum_{a=0}\sum_{b=0}\sum_{c=0} A_{abc}(f) {w_1}^a {w_2}^b {w_3}^c \tag{16.27}$$

and substitute (16.27) into the equivariance conditions (16.22)–(16.24) to obtain

$$(-1)^{b+c} = (-1)^{a+c-1} = 1, \tag{16.28}$$
$$A_{abc}(f) = (-1)^{a-1} A_{acb}(f) \tag{16.29}$$

for the nonzero terms in (16.27). By (16.28) we have $(a,b,c) = (\text{odd}, \text{even}, \text{even})$ or (even, odd, odd). Therefore, F_1 reduces to

$$\begin{aligned} F_1(w_1,w_2,w_3,f) = & \sum_{a:\text{odd}\geq 1}\ \sum_{b:\text{even}\geq 0}\ \sum_{c:\text{even}\geq 0} A_{abc}(f)\, {w_1}^a {w_2}^b {w_3}^c \\ & + \sum_{a:\text{even}\geq 0}\ \sum_{b:\text{odd}\geq 1}\ \sum_{c:\text{odd}\geq 1} A_{abc}(f)\, {w_1}^a {w_2}^b {w_3}^c \\ = & \; w_1 \sum_{a=0}\sum_{b=0}\sum_{c=0} A_{2a+1,2b,2c}(f)\, {w_1}^{2a} {w_2}^{2b} {w_3}^{2c} \\ & + w_2 w_3 \sum_{a=0}\sum_{b=0}\sum_{c=0} A_{2a,2b+1,2c+1}(f)\, {w_1}^{2a} {w_2}^{2b} {w_3}^{2c}. \end{aligned} \tag{16.30}$$

By the equivariance conditions (16.25) and (16.26), F_2 and F_3 are given, respectively, by

$$\begin{aligned} F_2(w_1,w_2,w_3,f) = & \; w_2 \sum_{a=0}\sum_{b=0}\sum_{c=0} A_{2a+1,2b,2c}(f)\, {w_2}^{2a} {w_3}^{2b} {w_1}^{2c} \\ & + w_3 w_1 \sum_{a=0}\sum_{b=0}\sum_{c=0} A_{2a,2b+1,2c+1}(f)\, {w_2}^{2a} {w_3}^{2b} {w_1}^{2c}, \end{aligned} \tag{16.31}$$

$$\begin{aligned} F_3(w_1,w_2,w_3,f) = & \; w_3 \sum_{a=0}\sum_{b=0}\sum_{c=0} A_{2a+1,2b,2c}(f)\, {w_3}^{2a} {w_1}^{2b} {w_2}^{2c} \\ & + w_1 w_2 \sum_{a=0}\sum_{b=0}\sum_{c=0} A_{2a,2b+1,2c+1}(f)\, {w_3}^{2a} {w_1}^{2b} {w_2}^{2c}. \end{aligned} \tag{16.32}$$

Example 16.1. The system of equations

$$F_1 = w_2 w_3({w_2}^2 - {w_3}^2), \qquad F_2 = w_3 w_1({w_3}^2 - {w_1}^2), \qquad F_3 = w_1 w_2({w_1}^2 - {w_2}^2)$$

satisfies the condition of equivariance. Note that the terms in these equations belong to the second series (even-degree part) on the right-hand side of (16.30)–(16.32). Accordingly, the second series needs to be taken into account in the bifurcation analysis. □

16.4.2 General Cases

We demonstrate the deriving of the bifurcation equation (16.9) for all cases $\mu = (3, j)$ $(j = 1,\dots,4)$.

We analyze the equivariance conditions, such as (16.10)–(16.21), on the bifurcation equation $F_i(w_1, w_2, w_3)$ $(i = 1,2,3)$ for each $\mu = (3, j)$ $(j = 1,\dots,4)$ to find that these conditions are equivalent to

$\mu = (3,1)$	$F_1(w_1,w_2,w_3,f) = F_1(w_1,-w_2,-w_3,f)$	A
	$-F_1(w_1,w_2,w_3,f) = F_1(-w_1,w_2,-w_3,f)$	A
	$F_1(w_1,w_2,w_3,f) = F_1(w_1,w_3,w_2,f)$	C$^+$
	$F_2(w_1,w_2,w_3,f) = F_1(w_2,w_3,w_1,f)$	D
	$F_3(w_1,w_2,w_3,f) = F_1(w_3,w_1,w_2,f)$	D
$\mu = (3,2), (3,4)$	$F_1(w_1,w_2,w_3,f) = F_1(w_1,-w_2,-w_3,f)$	A
	$-F_1(w_1,w_2,w_3,f) = F_1(-w_1,w_2,-w_3,f)$	A
	$-F_1(w_1,w_2,w_3,f) = F_1(-w_1,w_2,w_3,f)$	B
	$F_1(w_1,w_2,w_3,f) = F_1(w_1,w_3,w_2,f)$	C$^+$
	$F_2(w_1,w_2,w_3,f) = F_1(w_2,w_3,w_1,f)$	D
	$F_3(w_1,w_2,w_3,f) = F_1(w_3,w_1,w_2,f)$	D
$\mu = (3,3)$	$F_1(w_1,w_2,w_3,f) = F_1(w_1,-w_2,-w_3,f)$	A
	$-F_1(w_1,w_2,w_3,f) = F_1(-w_1,w_2,-w_3,f)$	A
	$-F_1(w_1,w_2,w_3,f) = F_1(-w_1,w_3,w_2,f)$	C$^-$
	$F_2(w_1,w_2,w_3,f) = F_1(w_2,w_3,w_1,f)$	D
	$F_3(w_1,w_2,w_3,f) = F_1(w_3,w_1,w_2,f)$	D

(16.33)

where the conditions for $\mu = (3,2)$ and $\mu = (3,4)$ turn out to be identical.

The conditions above are classified into the conditions for F_1:

$\mu = (3,1)$, (3,2), (3,3) (3,4)	Condition A $F_1(w_1,w_2,w_3) = F_1(w_1,-w_2,-w_3)$ $-F_1(w_1,w_2,w_3) = F_1(-w_1,w_2,-w_3)$
$\mu =$ (3,2), (3,4)	Condition B $-F_1(w_1,w_2,w_3) = F_1(-w_1,w_2,w_3)$
$\mu = (3,1)$, (3,2), (3,4)	Condition C$^+$ $F_1(w_1,w_2,w_3) = F_1(w_1,w_3,w_2)$
$\mu =$ (3,3)	Condition C$^-$ $-F_1(w_1,w_2,w_3) = F_1(-w_1,w_3,w_2)$

and the remaining conditions, Condition D:

$$F_2(w_1,w_2,w_3) = F_1(w_2,w_3,w_1), \qquad F_3(w_1,w_2,w_3) = F_1(w_3,w_1,w_2), \tag{16.34}$$

which give F_2 and F_3 in terms of F_1 in each case $(3,j)$ $(j = 1,\dots,4)$. The former conditions are translated to the coefficients of the expanded form:

$$F_1(w_1,w_2,w_3,f) = \sum_{a=0}\sum_{b=0}\sum_{c=0} A_{abc}(f) {w_1}^a {w_2}^b {w_3}^c \tag{16.35}$$

as follows.

First, for the nonzero terms in (16.35), Condition A gives

$$(-1)^{b+c} = (-1)^{a+c-1} = 1,$$

which means that $(a,b,c) = (\text{odd}, \text{even}, \text{even})$ or (even, odd, odd). Therefore, F_1 reduces to

$$\begin{aligned} F_1(w_1,w_2,w_3,f) &= \sum_{a:\text{odd}\ge 1}\ \sum_{b:\text{even}\ge 0}\ \sum_{c:\text{even}\ge 0} A_{abc}(f) {w_1}^a {w_2}^b {w_3}^c \\ &\quad + \sum_{a:\text{even}\ge 0}\ \sum_{b:\text{odd}\ge 1}\ \sum_{c:\text{odd}\ge 1} A_{abc}(f) {w_1}^a {w_2}^b {w_3}^c \\ &= w_1 \sum_{a=0}\sum_{b=0}\sum_{c=0} A_{2a+1,2b,2c}(f) {w_1}^{2a} {w_2}^{2b} {w_3}^{2c} \\ &\quad + w_2 w_3 \sum_{a=0}\sum_{b=0}\sum_{c=0} A_{2a,2b+1,2c+1}(f) {w_1}^{2a} {w_2}^{2b} {w_3}^{2c}. \end{aligned} \tag{16.36}$$

This form is common to all $\mu = (3,j)$ $(j = 1,\dots,4)$.

Next, Condition B, which is applicable to $\mu = (3,2)$ and $(3,4)$, brings the expression (16.36) to a simpler form

$$\begin{aligned} F_1(w_1,w_2,w_3,f) &= \sum_{a:\text{odd}\ge 1}\ \sum_{b:\text{even}\ge 0}\ \sum_{c:\text{even}\ge 0} A_{abc}(f) {w_1}^a {w_2}^b {w_3}^c \\ &= w_1 \sum_{a=0}\sum_{b=0}\sum_{c=0} A_{2a+1,2b,2c}(f) {w_1}^{2a} {w_2}^{2b} {w_3}^{2c}. \end{aligned} \tag{16.37}$$

Consequently, we have (16.37) for $\mu = (3,2)$ and $(3,4)$, whereas the expression (16.36) is applicable to $\mu = (3,1)$ and $(3,3)$.

Last, Conditions C^+ and C^- give, respectively,

$$A_{abc}(f) = A_{acb}(f) \qquad \text{for } \mu = (3,1),\ (3,2),\ (3,4), \tag{16.38}$$

$$A_{abc}(f) = (-1)^{a-1} A_{acb}(f) \qquad \text{for } \mu = (3,3). \tag{16.39}$$

The coefficients $A_{abc}(f)$ in (16.36) enjoy symmetry and/or antisymmetry prescribed above. The latter condition (16.39) implies, in particular, that

$$A_{0,2b+1,2b+1}(f) = 0, \qquad A_{0,2b+1,2c+1}(f) = -A_{0,2c+1,2b+1}(f), \qquad b,c = 0,1,2,\ldots, \tag{16.40}$$

which plays an important role for the existence of the mode II solution at the bifurcation point with $\mu = (3,3)$. It is also noted that (16.40) does not hold, generically, for $\mu = (3,1)$.

The expressions in (16.34) with the forms of F_1 in (16.36) or (16.37) give the forms of F_2 and F_3. For $\mu = (3,1)$ and $(3,3)$, we have

$$\begin{aligned} F_2(w_1,w_2,w_3,f) &= w_2 \sum_{a=0}\sum_{b=0}\sum_{c=0} A_{2a+1,2b,2c}(f)\, {w_2}^{2a}{w_3}^{2b}{w_1}^{2c} \\ &\quad + w_3 w_1 \sum_{a=0}\sum_{b=0}\sum_{c=0} A_{2a,2b+1,2c+1}(f)\, {w_2}^{2a}{w_3}^{2b}{w_1}^{2c}, \end{aligned} \tag{16.41}$$

$$\begin{aligned} F_3(w_1,w_2,w_3,f) &= w_3 \sum_{a=0}\sum_{b=0}\sum_{c=0} A_{2a+1,2b,2c}(f)\, {w_3}^{2a}{w_1}^{2b}{w_2}^{2c} \\ &\quad + w_1 w_2 \sum_{a=0}\sum_{b=0}\sum_{c=0} A_{2a,2b+1,2c+1}(f)\, {w_3}^{2a}{w_1}^{2b}{w_2}^{2c}. \end{aligned} \tag{16.42}$$

For $\mu = (3,2)$ and $(3,4)$, we have a simpler form

$$F_2(w_1,w_2,w_3,f) = w_2 \sum_{a=0}\sum_{b=0}\sum_{c=0} A_{2a+1,2b,2c}(f)\, {w_2}^{2a}{w_3}^{2b}{w_1}^{2c}, \tag{16.43}$$

$$F_3(w_1,w_2,w_3,f) = w_3 \sum_{a=0}\sum_{b=0}\sum_{c=0} A_{2a+1,2b,2c}(f)\, {w_3}^{2a}{w_1}^{2b}{w_2}^{2c}. \tag{16.44}$$

Because $(w_1,w_2,w_3,f) = (0,0,0,0)$ corresponds to the triple critical point and the Jacobian matrix

$$(\partial F_i/\partial w_j \mid i,j = 1,2,3)$$

at this point is equal to $A_{100}(0)I_3$ in either case, we have

$$A_{100}(0) = 0. \tag{16.45}$$

Therefore, we have

$$A_{100}(f) \approx Af \tag{16.46}$$

for some constant A, which is generically nonzero.

16.5 Solving of Bifurcation Equation

We solve the bifurcation equation (16.9) at a group-theoretic triple critical point associated with the three-dimensional irreducible representation $\mu = (3,j)$ $(j = 1,\ldots,4)$ of the group

$$G = \langle r,s,p_1,p_2\rangle \simeq \mathrm{D}_6 \dot{+} (\mathbb{Z}_2 \times \tilde{\mathbb{Z}}_2)$$

in (16.1) and (16.4).

The system of equations (16.9) with the equivariance (16.33) yields the following solutions of four types:

(1) $w_1 = w_2 = w_3 = 0$,
(2) Two of w_i $(i = 1,2,3)$ are zero and the remaining one is nonzero (e.g., $w_1 \neq 0$, $w_2 = w_3 = 0$),
(3) One of w_i $(i = 1,2,3)$ is zero and the others are nonzero (e.g., $w_1 = 0$, $w_2 \neq 0$, $w_3 \neq 0$),
(4) $w_1 \neq 0$, $w_2 \neq 0$, $w_3 \neq 0$.

The solutions (1)–(4) are designated, respectively, as (1) trivial, (2) mode I, (3) mode II, and (4) flower mode solutions. The symmetries of these bifurcated solutions are presented below.

16.5.1 *Irreducible Representation* (3,3)

For the irreducible representation $\mu = (3,3)$, from (16.36), (16.41), and (16.42), the system of bifurcation equation is expressed as

$$F_1(w_1,w_2,w_3,f) = w_1 \sum_{a=0}\sum_{b=0}\sum_{c=0} A_{2a+1,2b,2c}(f)\, {w_1}^{2a}{w_2}^{2b}{w_3}^{2c} + w_2w_3 \sum_{a=0}\sum_{b=0}\sum_{c=0} A_{2a,2b+1,2c+1}(f)\, {w_1}^{2a}{w_2}^{2b}{w_3}^{2c} = 0, \tag{16.47}$$

$$F_2(w_1,w_2,w_3,f) = w_2 \sum_{a=0}\sum_{b=0}\sum_{c=0} A_{2a+1,2b,2c}(f)\, {w_2}^{2a}{w_3}^{2b}{w_1}^{2c} + w_3w_1 \sum_{a=0}\sum_{b=0}\sum_{c=0} A_{2a,2b+1,2c+1}(f)\, {w_2}^{2a}{w_3}^{2b}{w_1}^{2c} = 0, \tag{16.48}$$

$$F_3(w_1,w_2,w_3,f) = w_3 \sum_{a=0}\sum_{b=0}\sum_{c=0} A_{2a+1,2b,2c}(f)\, {w_3}^{2a}{w_1}^{2b}{w_2}^{2c} + w_1w_2 \sum_{a=0}\sum_{b=0}\sum_{c=0} A_{2a,2b+1,2c+1}(f)\, {w_3}^{2a}{w_1}^{2b}{w_2}^{2c} = 0. \tag{16.49}$$

Trivial Solution

The trivial solution with $w_1 = w_2 = w_3 = 0$ is associated with the fundamental path.

Mode I Solution

For $w_1 \neq 0$ and $w_2 = w_3 = 0$, the equations $F_2 = F_3 = 0$ in (16.48) and (16.49) are satisfied, and $F_1 = 0$ in (16.47) becomes

$$\sum_{a=0} A_{2a+1,0,0}(f){w_1}^{2a} = 0. \tag{16.50}$$

This equation has two solutions of the form $w_1 = \pm\Phi_1(f) = \mathrm{O}(|f|^{1/2})$, since the leading part of the equation (16.50) is given as

$$Af + A_{300}(0){w_1}^2 = 0 \tag{16.51}$$

by (16.46).

The difference of two solutions $(w_1, w_2, w_3) = (\pm\Phi_1(f), 0, 0)$ is attributable solely to the definition of the sign of w_1. Similarly, when w_2 or w_3 is nonzero and the other two variables vanish, we have four solutions $(0, \pm\Phi_1(f), 0)$ and $(0, 0, \pm\Phi_1(f))$. Altogether, three bifurcated paths—six half-branches—exist, which are given by

$$(w_1, w_2, w_3) = (\pm\Phi_1(f), 0, 0),\ (0, \pm\Phi_1(f), 0),\ (0, 0, \pm\Phi_1(f)).$$

The difference of these solutions is attributable solely to the definition of w_1, w_2, and w_3, which, in principle, are exchangeable.

The symmetries of the solutions, which are depicted in Fig. 16.5, are expressed by

$$\Sigma(\boldsymbol{u}) = \Sigma(\boldsymbol{w}) = \begin{cases} \langle r^3, p_2 s\rangle & \text{for } \boldsymbol{w} = \pm(\Phi_1(f), 0, 0)^\top, \\ \langle r^3, p_1 p_2 s r^2\rangle & \text{for } \boldsymbol{w} = \pm(0, \Phi_1(f), 0)^\top, \\ \langle r^3, p_1 s r^4\rangle & \text{for } \boldsymbol{w} = \pm(0, 0, \Phi_1(f))^\top. \end{cases} \tag{16.52}$$

The three subgroups $\langle r^3, p_2 s\rangle$, $\langle r^3, p_1 p_2 s r^2\rangle$, and $\langle r^3, p_1 s r^4\rangle$ are pairwise conjugate in the sense of (7.99) since $p_2 s$, $p_1 p_2 s r^2$, and $p_1 s r^4$ all belong to the same class C_{10} in Table 16.1. Accordingly, the solutions associated with these three subgroups can be identified in numerical bifurcation analyses.

In (16.52), r^3 denotes a local half-rotation symmetry; $p_2 s$, $p_1 p_2 s r^2$, and $p_1 s r^4$ satisfying $p_1 = (p_2 s)^2$, $p_2 = (p_1 p_2 s r^2)^2$, and $p_1 p_2 = (p_1 s r^4)^2$ denote that a series of identical cells are arranged in the translational direction of p_1, p_2, and $p_1 p_2$; and $p_2 s$, $p_1 p_2 s r^2$, and $p_1 s r^4$ denote that a hexagonal cell after the local reflections s,

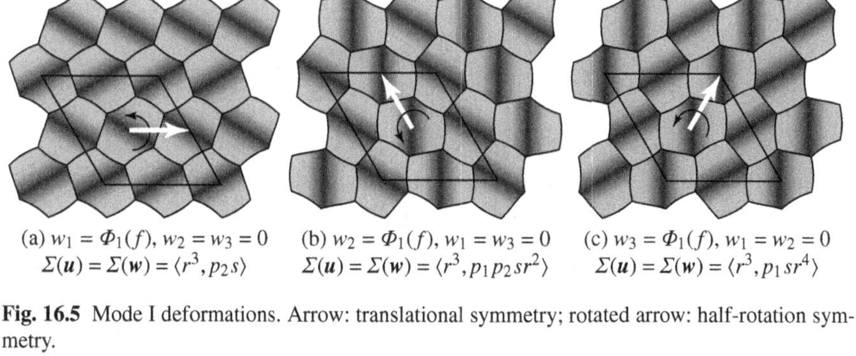

(a) $w_1 = \Phi_1(f)$, $w_2 = w_3 = 0$, $\Sigma(\boldsymbol{u}) = \Sigma(\boldsymbol{w}) = \langle r^3, p_2 s\rangle$
(b) $w_2 = \Phi_1(f)$, $w_1 = w_3 = 0$, $\Sigma(\boldsymbol{u}) = \Sigma(\boldsymbol{w}) = \langle r^3, p_1 p_2 s r^2\rangle$
(c) $w_3 = \Phi_1(f)$, $w_1 = w_2 = 0$, $\Sigma(\boldsymbol{u}) = \Sigma(\boldsymbol{w}) = \langle r^3, p_1 s r^4\rangle$

Fig. 16.5 Mode I deformations. Arrow: translational symmetry; rotated arrow: half-rotation symmetry.

sr^2, and sr^4 is identical with the neighboring cell in another translational direction of p_2, p_1p_2, and p_1. The difference of the three subgroups in (16.52) is ascribed to the difference in the direction of the translation as presented in Fig. 16.5. The solutions with the symmetries of (16.52), as a whole, are identified and called *mode I solution*.

Mode II Solution

For $w_1 = 0$, $w_2 \neq 0$, and $w_3 \neq 0$, the equations $F_i = 0$ $(i = 1,2,3)$ in (16.47)–(16.49) are reduced, respectively, to

$$\sum_{b=0}\sum_{c=0} A_{0,2b+1,2c+1}(f)\, {w_2}^{2b} {w_3}^{2c} = 0,$$
$$\sum_{a=0}\sum_{b=0} A_{2a+1,2b,0}(f)\, {w_2}^{2a} {w_3}^{2b} = 0,$$
$$\sum_{a=0}\sum_{c=0} A_{2a+1,0,2c}(f)\, {w_3}^{2a} {w_2}^{2c} = 0.$$

By (16.40), the above equations are rewritten as

$$\sum_{b>c\geq 0} A_{0,2b+1,2c+1}(f)\,({w_2}^{2b} {w_3}^{2c} - {w_3}^{2b} {w_2}^{2c}) = 0, \tag{16.53}$$
$$\sum_{a=0}\sum_{b=0} A_{2a+1,2b,0}(f)\, {w_2}^{2a} {w_3}^{2b} = 0, \tag{16.54}$$
$$\sum_{a=0}\sum_{b=0} A_{2a+1,2b,0}(f)\, {w_3}^{2a} {w_2}^{2b} = 0. \tag{16.55}$$

Since the leading term of (16.53) is given as

$$A_{031}(f)\,({w_2}^2 - {w_3}^2) = 0,$$

where $A_{031}(f) \neq 0$ generically, the solution must satisfy $|w_2| = |w_3|$. Then (16.54) and (16.55) are identical, and read as

$$\sum_{a=0}\sum_{b=0} A_{2a+1,2b,0}(f)\, {w_2}^{2(a+b)} = 0.$$

By (16.46), this can be solved for w_2 as $w_2 = \pm\Phi_2(f) = \mathrm{O}(|f|^{1/2})$.

Since the roles of w_1, w_2, and w_3 are interchangeable, six bifurcated paths—12 half-branches—exist, which are associated with

$$(w_1, w_2, w_3) = (0, \pm\Phi_2(f), \pm\Phi_2(f)),\ (\pm\Phi_2(f), 0, \pm\Phi_2(f)),\ (\pm\Phi_2(f), \pm\Phi_2(f), 0).$$

The symmetries of these solutions are depicted in Fig. 16.6, and are expressed by

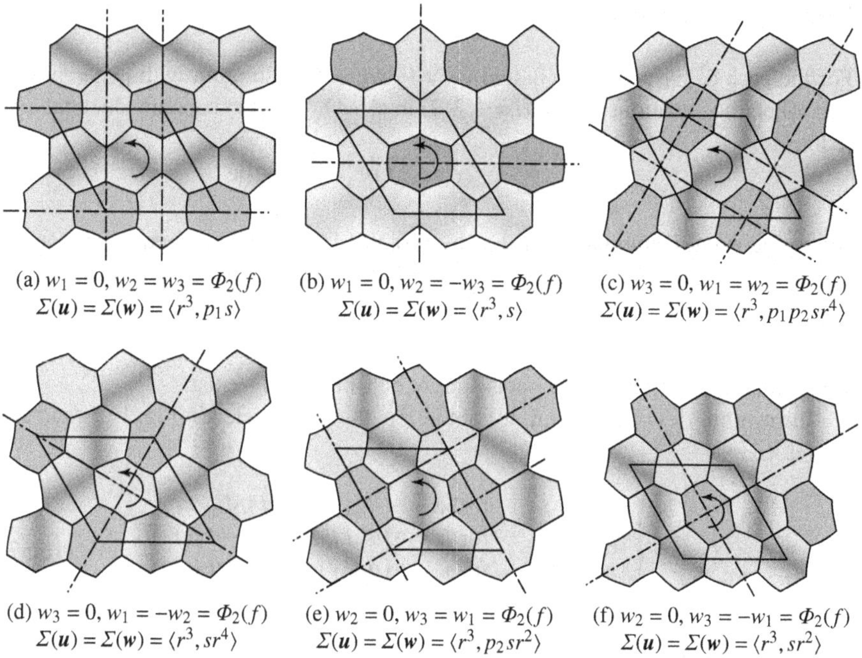

(a) $w_1 = 0, w_2 = w_3 = \Phi_2(f)$ $\Sigma(\boldsymbol{u}) = \Sigma(\boldsymbol{w}) = \langle r^3, p_1 s\rangle$

(b) $w_1 = 0, w_2 = -w_3 = \Phi_2(f)$ $\Sigma(\boldsymbol{u}) = \Sigma(\boldsymbol{w}) = \langle r^3, s\rangle$

(c) $w_3 = 0, w_1 = w_2 = \Phi_2(f)$ $\Sigma(\boldsymbol{u}) = \Sigma(\boldsymbol{w}) = \langle r^3, p_1 p_2 s r^4\rangle$

(d) $w_3 = 0, w_1 = -w_2 = \Phi_2(f)$ $\Sigma(\boldsymbol{u}) = \Sigma(\boldsymbol{w}) = \langle r^3, s r^4\rangle$

(e) $w_2 = 0, w_3 = w_1 = \Phi_2(f)$ $\Sigma(\boldsymbol{u}) = \Sigma(\boldsymbol{w}) = \langle r^3, p_2 s r^2\rangle$

(f) $w_2 = 0, w_3 = -w_1 = \Phi_2(f)$ $\Sigma(\boldsymbol{u}) = \Sigma(\boldsymbol{w}) = \langle r^3, s r^2\rangle$

Fig. 16.6 Mode II deformations. Dotted-dashed line: line of reflection symmetry; rotated arrow: half-rotation symmetry.

$$\Sigma(\boldsymbol{u}) = \Sigma(\boldsymbol{w}) = \begin{cases} \langle r^3, p_1 s\rangle & \text{for } \boldsymbol{w} = \pm(0, \Phi_2(f), \Phi_2(f))^\top, \\ \langle r^3, s\rangle & \text{for } \boldsymbol{w} = \pm(0, \Phi_2(f), -\Phi_2(f))^\top, \\ \langle r^3, p_1 p_2 s r^4\rangle & \text{for } \boldsymbol{w} = \pm(\Phi_2(f), \Phi_2(f), 0)^\top, \\ \langle r^3, s r^4\rangle & \text{for } \boldsymbol{w} = \pm(\Phi_2(f), -\Phi_2(f), 0)^\top, \\ \langle r^3, p_2 s r^2\rangle & \text{for } \boldsymbol{w} = \pm(\Phi_2(f), 0, \Phi_2(f))^\top, \\ \langle r^3, s r^2\rangle & \text{for } \boldsymbol{w} = \pm(\Phi_2(f), 0, -\Phi_2(f))^\top. \end{cases} \tag{16.56}$$

The six subgroups

$$\langle r^3, p_1 s\rangle, \quad \langle r^3, s\rangle, \quad \langle r^3, p_1 p_2 s r^4\rangle, \quad \langle r^3, s r^4\rangle, \quad \langle r^3, p_2 s r^2\rangle, \quad \langle r^3, s r^2\rangle$$

are all mutually conjugate since $p_1 s$, s, $p_1 p_2 s r^4$, $s r^4$, p_2, $p_2 s r^2$, and $s r^2$ all belong to the same class C_5 in Table 16.1.

The difference of these subgroups is ascribed to the difference in the line of reflection as presented in Fig. 16.6. The solutions with the symmetries of (16.56), as a whole, are identified and called *mode II solutions*.

Flower Mode Solution

For the case of $w_1 \neq 0$, $w_2 \neq 0$, and $w_3 \neq 0$, we seek a solution with $|w_1| = |w_2| = |w_3|$. Then the equations $F_i = 0$ $(i = 1,2,3)$ in (16.47)–(16.49) become identical and reduce to

$$\sum_{a=0}\sum_{b=0}\sum_{c=0} A_{2a+1,2b,2c}(f)\, w_1^{2(a+b+c)} = 0,$$

where the even-degree part vanishes by (16.39). By (16.46), this equation can be solved for w_1 as $w_1 = \pm\Phi_3(f)$ with $|w_1| = \mathrm{O}(|f|^{1/2})$. As a result, we have solutions with

$$(w_1, w_2, w_3) = (\pm\Phi_3(f), \pm\Phi_3(f), \pm\Phi_3(f)).$$

Consequently, four bifurcated paths—eight half-branches—exist.

The symmetries of the solutions on these paths are depicted in Fig. 16.7, and are expressed by

$$\Sigma(\boldsymbol{u}) = \Sigma(\boldsymbol{w}) = \begin{cases} \langle r \rangle & \text{for } \boldsymbol{w} = \pm(\Phi_3(f), \Phi_3(f), \Phi_3(f))^\top, \\ \langle p_1 r \rangle & \text{for } \boldsymbol{w} = \pm(\Phi_3(f), \Phi_3(f), -\Phi_3(f))^\top, \\ \langle p_1 p_2 r \rangle & \text{for } \boldsymbol{w} = \pm(\Phi_3(f), -\Phi_3(f), \Phi_3(f))^\top, \\ \langle p_2 r \rangle & \text{for } \boldsymbol{w} = \pm(-\Phi_3(f), \Phi_3(f), \Phi_3(f))^\top, \end{cases} \qquad (16.57)$$

as presented in Table 16.3.

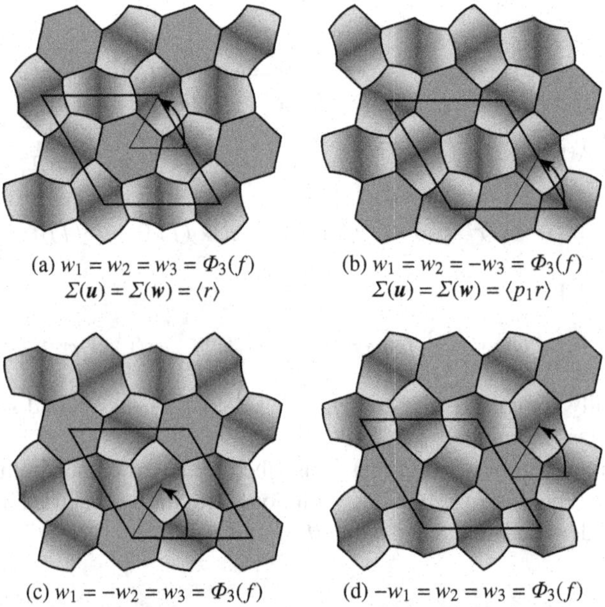

(a) $w_1 = w_2 = w_3 = \Phi_3(f)$ $\Sigma(\boldsymbol{u}) = \Sigma(\boldsymbol{w}) = \langle r \rangle$

(b) $w_1 = w_2 = -w_3 = \Phi_3(f)$ $\Sigma(\boldsymbol{u}) = \Sigma(\boldsymbol{w}) = \langle p_1 r \rangle$

(c) $w_1 = -w_2 = w_3 = \Phi_3(f)$ $\Sigma(\boldsymbol{u}) = \Sigma(\boldsymbol{w}) = \langle p_1 p_2 r \rangle$

(d) $-w_1 = w_2 = w_3 = \Phi_3(f)$ $\Sigma(\boldsymbol{u}) = \Sigma(\boldsymbol{w}) = \langle p_2 r \rangle$

Fig. 16.7 Flower mode deformations. Rotated arrow: rotation symmetry.

Table 16.3 Symmetries of the solutions at a triple bifurcation point associated with $\mu = (3,3)$

Type of Solutions	Symmetry Group	w_1, w_2, w_3
Trivial	$G = \langle r, s, p_1, p_2 \rangle$	$w_1 = w_2 = w_3 = 0$
Mode I	$\langle r^3, p_2 s \rangle$	$w_1 \neq 0$, $w_2 = w_3 = 0$
	$\langle r^3, p_1 p_2 s r^2 \rangle$	$w_2 \neq 0$, $w_1 = w_3 = 0$
	$\langle r^3, p_1 s r^4 \rangle$	$w_3 \neq 0$, $w_1 = w_2 = 0$
Mode II	$\langle r^3, p_1 s \rangle$	$w_1 = 0$, $w_2 = w_3$
	$\langle r^3, s \rangle$	$w_1 = 0$, $w_2 = -w_3$
	$\langle r^3, p_1 p_2 s r^4 \rangle$	$w_3 = 0$, $w_1 = w_2$
	$\langle r^3, s r^4 \rangle$	$w_3 = 0$, $w_1 = -w_2$
	$\langle r^3, p_2 s r^2 \rangle$	$w_2 = 0$, $w_3 = w_1$
	$\langle r^3, s r^2 \rangle$	$w_2 = 0$, $w_3 = -w_1$
Flower Mode	$\langle r \rangle$	$w_1 = w_2 = w_3$
	$\langle p_1 r \rangle$	$w_1 = w_2 = -w_3$
	$\langle p_1 p_2 r \rangle$	$w_1 = -w_2 = w_3$
	$\langle p_2 r \rangle$	$-w_1 = w_2 = w_3$

The four subgroups $\langle r \rangle$, $\langle p_1 r \rangle$, $\langle p_1 p_2 r \rangle$, and $\langle p_2 r \rangle$ are mutually conjugate since r, $p_1 r$, $p_1 p_2 r$, and $p_2 r$ all belong to the same class C_2 in Table 16.1. The difference of the four subgroups in (16.57) is ascribed to the difference in the location of the axis of $\pi/3$ rotation as presented in Fig. 16.7. The solutions with the symmetries of (16.57), as a whole, are identified and called *flower modes*.

Remark 16.3. A procedure is proposed here for the branch switching at the triple bifurcation point associated with $\mu = (3,3)$; the companion cases $\mu = (3,1)$, $(3,2)$ and $(3,4)$ can be treated similarly. Since the condition (7.84) is satisfied by (16.51), the direction of the solution path in the space of $\boldsymbol{u}$ is simply given by (7.85) with $M = 3$:

$$\sum_{i=1}^{3} \frac{\partial w_i}{\partial s}(0, \nu^0)\boldsymbol{\eta}_i,$$

lying in the subspace spanned by the right critical eigenvectors $\boldsymbol{\eta}_1$, $\boldsymbol{\eta}_2$, and $\boldsymbol{\eta}_3$ of the Jacobian matrix J_{c}^0. Accordingly, the bifurcated paths are to be sought in directions

$$\boldsymbol{\eta} = c_1\boldsymbol{\eta}_1 + c_2\boldsymbol{\eta}_2 + c_3\boldsymbol{\eta}_3$$

for some scaling constants c_1, c_2, and c_3. In the numerical bifurcation analysis, we choose $\boldsymbol{\eta}_1$, $\boldsymbol{\eta}_2$, and $\boldsymbol{\eta}_3$ in the directions of w_1, w_2, and w_3, respectively. This means that $\boldsymbol{\eta}_1$, $\boldsymbol{\eta}_2$, and $\boldsymbol{\eta}_3$ are chosen to represent the symmetries

$$\langle r^3, p_2 s \rangle, \qquad \langle r^3, p_1 p_2 s r^2 \rangle, \qquad \langle r^3, p_1 s r^4 \rangle, \tag{16.58}$$

respectively. Then the bifurcated paths are sought in the directions listed in Table 16.4. □

Table 16.4 Directions of bifurcated paths at a triple bifurcation point associated with $\mu = (3,3)$

η	Symmetry Group $\Sigma(\eta)$	Type of Solutions
η_1	$\langle r^3, p_2 s\rangle$	Mode I
η_2	$\langle r^3, p_1 p_2 s r^2\rangle$	
η_3	$\langle r^3, p_1 s r^4\rangle$	
$\eta_2+\eta_3$	$\langle r^3, p_1 s\rangle$	Mode II
$\eta_2-\eta_3$	$\langle r^3, s\rangle$	
$\eta_1+\eta_2$	$\langle r^3, p_1 p_2 s r\rangle$	
$\eta_1-\eta_2$	$\langle r^3, s r\rangle$	
$\eta_3+\eta_1$	$\langle r^3, p_2 s r^2\rangle$	
$\eta_3-\eta_1$	$\langle r^3, s r^2\rangle$	
$\eta_1+\eta_2+\eta_3$	$\langle r\rangle$	Flower mode
$\eta_1+\eta_2-\eta_3$	$\langle p_1 r\rangle$	
$\eta_1-\eta_2+\eta_3$	$\langle p_1 p_2 r\rangle$	
$-\eta_1+\eta_2+\eta_3$	$\langle p_2 r\rangle$	

16.5.2 *Irreducible Representation* $(3,1)$

For the irreducible representation $\mu = (3,1)$, the bifurcation equation takes the form in (16.47)–(16.49), the same as that for $\mu = (3,3)$.

Mode I Solution

For $w_1 \neq 0$, $w_2 = w_3 = 0$, the equations $F_2 = F_3 = 0$ in (16.48) and (16.49) are satisfied, and $F_1 = 0$ in (16.47) becomes

$$\sum_{a=0} A_{2a+1,0,0}(f) {w_1}^{2a} = 0, \tag{16.59}$$

which is of the same form as (16.50) for $\mu = (3,3)$. This equation has two solutions of the form $w_1 = \pm\Phi_1(f) = \mathrm{O}(|f|^{1/2})$. Similarly, when w_2 or w_3 is nonzero and the other two variables vanish, we have four solutions. Hence we have

$$(w_1, w_2, w_3) = (\pm\Phi_1(f), 0, 0),\ (0, \pm\Phi_1(f), 0),\ (0, 0, \pm\Phi_1(f)).$$

Altogether, three bifurcated paths—six half-branches—exist, which are invariant to

$$\langle r^3, s, p_1 \rangle, \qquad \langle r^3, sr^2, p_2 \rangle, \qquad \langle r^3, sr, p_1 p_2 \rangle,$$

as listed in Table 16.2.

Mode II Solution

If we set $w_1 = 0$, $w_2 \neq 0$, $w_3 \neq 0$ in the system of bifurcation equations in (16.47)–(16.49), then we have

$$\sum_{b=0}\sum_{c=0} A_{0,2b+1,2c+1}(f)\, w_2{}^{2b} w_3{}^{2c} = 0, \tag{16.60}$$

$$\sum_{a=0}\sum_{b=0} A_{2a+1,2b,0}(f)\, w_2{}^{2a} w_3{}^{2b} = 0, \tag{16.61}$$

$$\sum_{a=0}\sum_{c=0} A_{2a+1,0,2c}(f)\, w_3{}^{2a} w_2{}^{2c} = 0. \tag{16.62}$$

By (16.38), the leading part of (16.60) consisting of the terms for $(b,c) = (0,0)$, $(1,0)$, $(0,1)$ is

$$A_{011}(f) + A_{031}(f)(w_2{}^2 + w_3{}^2) = 0.$$

This equation has no solution inasmuch as $A_{011}(f) \approx A_{011}(0) \neq 0$ because of the absence of the condition (16.40), unlike for $\mu = (3,3)$. Therefore, the mode II solution is nonexistent.

Example 16.2. The system of equations

$$F_1 = w_2 w_3, \qquad F_2 = w_3 w_1, \qquad F_3 = w_1 w_2$$

satisfies the condition of equivariance, but has no mode II solution. In addition, this is a reciprocal system since

$$\frac{\partial F_1}{\partial w_2} = w_3 = \frac{\partial F_2}{\partial w_1}, \qquad \frac{\partial F_2}{\partial w_3} = w_1 = \frac{\partial F_3}{\partial w_2}, \qquad \frac{\partial F_3}{\partial w_1} = w_2 = \frac{\partial F_1}{\partial w_3}.$$

□

Flower Mode Solution

For the case of $w_1 \neq 0$, $w_2 \neq 0$, and $w_3 \neq 0$, we seek a solution with $|w_1| = |w_2| = |w_3|$. Then the system of bifurcation equations in (16.47)–(16.49) becomes identical and reads as

$$\sum_{a=0}\sum_{b=0}\sum_{c=0} A_{2a+1,2b,2c}(f)\, w_1{}^{2(a+b+c)} + \alpha |w_1| \sum_{a=0}\sum_{b=0}\sum_{c=0} A_{2a,2b+1,2c+1}(f)\, w_1{}^{2(a+b+c)} = 0, \tag{16.63}$$

where $\alpha = \text{sign}(w_1 w_2 w_3)$. By (16.46), the leading part of (16.63) is given by

$$Af + \alpha A_{011}(0)|w_1| = 0, \tag{16.64}$$

where $A_{011}(0) \neq 0$ (generically). Consequently, a solution of the form $w_1 = \mathrm{O}(f)$ to the equation (16.63) exists, which we set

$$w_1 = \begin{cases} \Phi_3(f) & \text{for } \alpha = 1, \\ \Phi_4(f) & \text{for } \alpha = -1. \end{cases}$$

It is noted that $\Phi_3(f) \approx -\Phi_4(f)$ and $\Phi_i(f) = \mathrm{O}(f)$ $(i = 3,4)$ when $|f|$ is small.

The flower mode solution, therefore, is existent. Four bifurcated paths—eight half-branches—exist, which are associated with

$$\begin{aligned}
&(w_1, w_2, w_3) \\
&= (\Phi_3(f), \Phi_3(f), \Phi_3(f)), \\
&\quad (-\Phi_3(f), -\Phi_3(f), \Phi_3(f)),\ (-\Phi_3(f), \Phi_3(f), -\Phi_3(f)),\ (\Phi_3(f), -\Phi_3(f), -\Phi_3(f)), \\
&\quad (-\Phi_4(f), -\Phi_4(f), -\Phi_4(f)), \\
&\quad (\Phi_4(f), \Phi_4(f), -\Phi_4(f)),\ (\Phi_4(f), -\Phi_4(f), \Phi_4(f)),\ (-\Phi_4(f), \Phi_4(f), \Phi_4(f)).
\end{aligned}$$

By starting with the assumption $|w_1| = |w_2| = |w_3|$, we have not excluded the possibility of solutions of other types with $w_1 w_2 w_3 \neq 0$.

16.5.3 *Irreducible Representation* (3, 2)

For the irreducible representation $\mu = (3,2)$, from (16.37), (16.43), and (16.44), the bifurcation equations are expressed as

$$F_1(w_1, w_2, w_3, f) = w_1 \sum_{a=0} \sum_{b=0} \sum_{c=0} A_{2a+1,2b,2c}(f)\, {w_1}^{2a} {w_2}^{2b} {w_3}^{2c} = 0, \tag{16.65}$$

$$F_2(w_1, w_2, w_3, f) = w_2 \sum_{a=0} \sum_{b=0} \sum_{c=0} A_{2a+1,2b,2c}(f)\, {w_2}^{2a} {w_3}^{2b} {w_1}^{2c} = 0, \tag{16.66}$$

$$F_3(w_1, w_2, w_3, f) = w_3 \sum_{a=0} \sum_{b=0} \sum_{c=0} A_{2a+1,2b,2c}(f)\, {w_3}^{2a} {w_1}^{2b} {w_2}^{2c} = 0. \tag{16.67}$$

Mode I Solution

For $w_1 \neq 0$, $w_2 = w_3 = 0$, the equations $F_2 = F_3 = 0$ in (16.66) and (16.67) are satisfied; moreover, $F_1 = 0$ in (16.65) reduces to

$$\sum_{a=0} A_{2a+1,0,0}(f) {w_1}^{2a} = 0. \tag{16.68}$$

A solution of the form $w_1 = \pm\Phi_1(f) = \mathrm{O}(|f|^{1/2})$ exists.

Consequently, the mode I solution is existent. Three bifurcated paths—six half-branches—exist, which are associated with

$$(w_1, w_2, w_3) = (\pm\Phi_1(f), 0, 0),\ (0, \pm\Phi_1(f), 0),\ (0, 0, \pm\Phi_1(f)).$$

Mode II Solution

For $w_1 = 0$, $w_2 \neq 0$, and $w_3 \neq 0$, $F_i = 0$ $(i = 1, 2, 3)$ in (16.65)–(16.67) become

$$\sum_{a=0}\sum_{b=0} A_{2a+1,2b,0}(f)\, {w_2}^{2a} {w_3}^{2b} = 0, \tag{16.69}$$

$$\sum_{a=0}\sum_{c=0} A_{2a+1,0,2c}(f)\, {w_3}^{2a} {w_2}^{2c} = 0. \tag{16.70}$$

For $|w_2| = |w_3|$, by (16.38), the two equations (16.69) and (16.70) reduce to a single equation

$$\sum_{a=0}\sum_{b=0} A_{2a+1,2b,0}(f)\, {w_2}^{2(a+b)} = 0.$$

The solution of this equation is existent by (16.46), and takes the form of $w_2 = \pm\Phi_2(f) = \mathrm{O}(|f|^{1/2})$.

Consequently, the mode II solution is existent. Six bifurcated paths—12 half-branches—exist, which are associated with

$$(w_1, w_2, w_3) = (0, \pm\Phi_2(f), \pm\Phi_2(f)),\ (\pm\Phi_2(f), 0, \pm\Phi_2(f)),\ (\pm\Phi_2(f), \pm\Phi_2(f), 0).$$

By starting with the assumption $|w_2| = |w_3|$, we have not excluded the possibility of solutions of other types with $w_2 w_3 \neq 0$.

Flower Mode Solution

For the case of $w_1 \neq 0$, $w_2 \neq 0$, and $w_3 \neq 0$, we seek a solution with $|w_1| = |w_2| = |w_3|$. Then (16.65)–(16.67) are identical and are expressed as

$$\sum_{a=0}\sum_{b=0}\sum_{c=0} A_{2a+1,2b,2c}(f)\, {w_1}^{2(a+b+c)} = 0,$$

the leading part of which reads as

$$Af + (A_{300}(0) + 2A_{120}(0))w_1^2 = 0$$

by (16.38) and (16.46). Therefore, a solution of the form $w_1 = \pm\Phi_3(f) = \mathrm{O}(|f|^{1/2})$ exists. Consequently, four bifurcated paths—eight half-branches—exist, which are associated with

$$(w_1, w_2, w_3) = (\pm\Phi_3(f), \pm\Phi_3(f), \pm\Phi_3(f)).$$

Again by starting with the assumption $|w_1| = |w_2| = |w_3|$ we have not excluded the possibility of solutions of other types with $w_1 w_2 w_3 \neq 0$.

16.5.4 Irreducible Representation (3,4)

The equivariance conditions on the bifurcation equation for $\mu = (3,4)$ are identical to those for $\mu = (3,2)$ by (16.33). Mode I, mode II, and flower mode solutions all exist for $\mu = (3,4)$, as is true for $\mu = (3,2)$. Yet the solutions for $\mu = (3,4)$ have different symmetries than those for $\mu = (3,2)$ (cf., Table 16.2).

16.6 Bifurcation Analysis of Honeycomb Cellular Solids

A series of characteristic geometrical patterns are obtained through numerical bifurcation analysis of a hexagonal honeycomb structure. These patterns are classified on the basis of the bifurcation rule presented in §16.3.

We consider the representative volume element (RVE) of honeycomb cellular solids with 2×2 cells with periodic boundary conditions: the RVE consists of lattice-like frames.[12] We conduct the bifurcation analysis of the RVE subjected to isotropic biaxial compression with a loading parameter f. The RVE has the fundamental path corresponding to the prebifurcation homogeneous state of deformation, and four bifurcation points A to D with multiplicity one or three exist on this path. These points are listed below.

Bifurcation Point	f	M	μ
A	0.0038	3	(3,3)
B	0.0100	1	(−,+)
C	0.0257	3	(3,2)
D	0.0941	3	(3,4)

M: multiplicity; μ: irreducible representation

[12] The so-called co-rotational formulation for large rotation and small strain problems, and the kinematic field of the Bernoulli–Euler beam are used to implement large rotations and small strains in the bifurcation analysis of RVE. The linear elastic relation between infinitesimal strain and stress in the co-rotated coordinates is employed. Cellular microstructures are assumed to be linearly elastic. The ratio of cell thickness to length is chosen as 0.109. More details are available in Saiki, Ikeda, and Murota, 2005 [175].

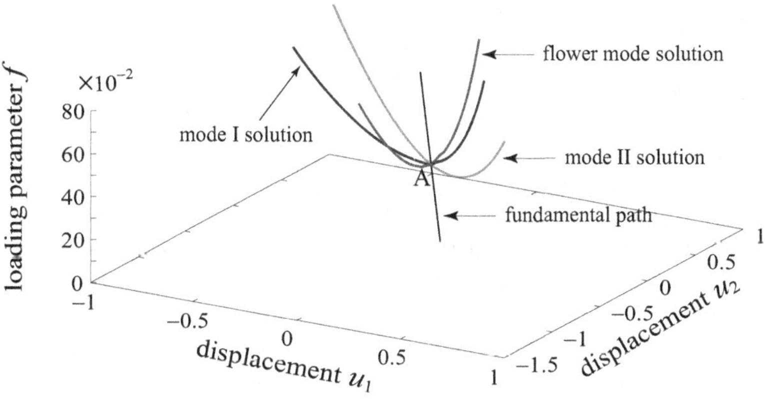

Fig. 16.8 Equilibrium paths in the neighborhood of the triple bifurcation point A.

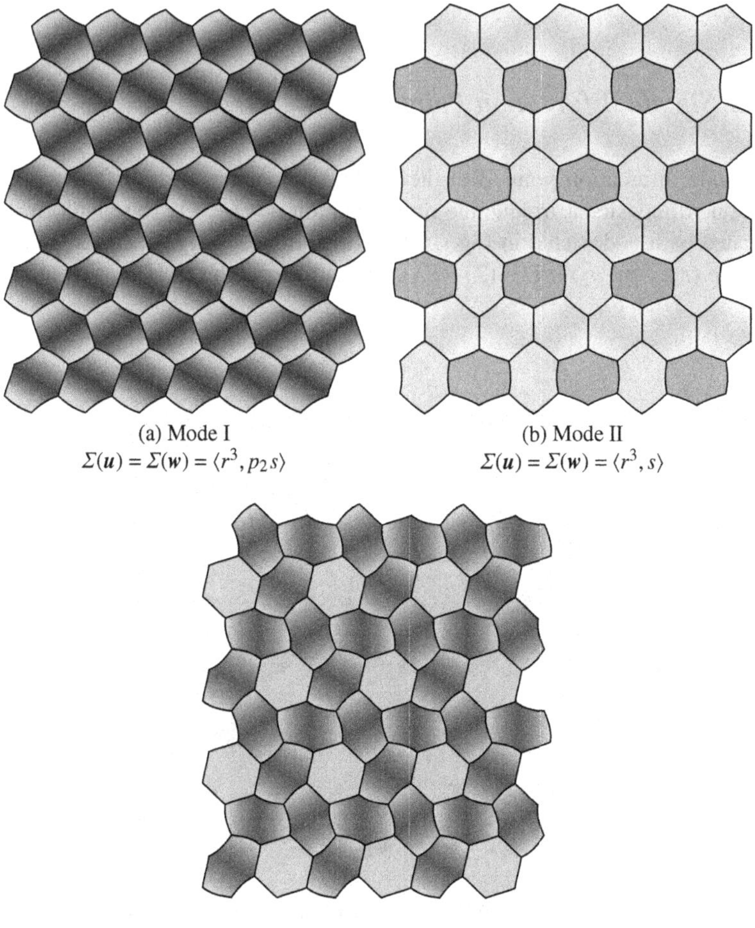

(a) Mode I
$\Sigma(\boldsymbol{u}) = \Sigma(\boldsymbol{w}) = \langle r^3, p_2 s\rangle$

(b) Mode II
$\Sigma(\boldsymbol{u}) = \Sigma(\boldsymbol{w}) = \langle r^3, s\rangle$

(c) Flower mode
$\Sigma(\boldsymbol{u}) = \Sigma(\boldsymbol{w}) = \langle r\rangle$

Fig. 16.9 Planforms for the triple bifurcation point A associated with $\mu = (3, 3)$.

16.6.1 Triple Bifurcation Point A

The bifurcation point A at $f = 0.0038$ is a triple bifurcation point corresponding to the irreducible representation $\mu = (3,3)$ that is examined in detail in §16.5.1. Three independent eigenvectors $\boldsymbol{\eta}_i$ $(i = 1,2,3)$ exist for the zero eigenvalue (repeated three times) of the Jacobian matrix J_{c}^0 at this bifurcation point. These three eigenvectors are chosen to have the symmetries in (16.58) and therefore are in the directions associated with mode I (cf., Remark 16.3 in §16.5.1). Furthermore, as listed in Table 16.4, appropriate linear combinations of the three eigenvectors give the directions for mode II and flower mode.

Bifurcated paths for the mode I, mode II, and flower mode branching at this bifurcation point are actually found as depicted in Fig. 16.8. Planforms that consist of planar assembly of RVEs are portrayed in Fig. 16.9, which are toned to emphasize geometrical patterns. Diverse patterns have therefore been engendered by the bifurcation.

16.6.2 Simple Bifurcation Point B

The simple bifurcation point B at the loading parameter $f = 0.0100$ corresponds to the one-dimensional irreducible representation $\mu = (-,+)$. The planform of the bifurcated path is depicted in Fig. 16.10, which displays a geometrical pattern invariant to $\langle r^2, s, p_1, p_2\rangle = \mathrm{D}_3 \dot{+} (\mathbb{Z}_2 \times \tilde{\mathbb{Z}}_2)$. The cells have the local D_3-symmetry that

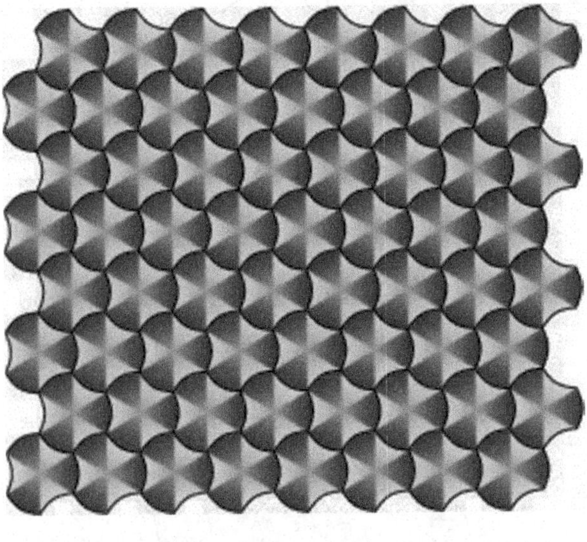

Fig. 16.10 Planform for the simple bifurcation point B associated with $\mu = (-,+)$ with the symmetry of $\Sigma(\boldsymbol{u}) = \Sigma(\boldsymbol{w}) = \langle r^2, s, p_1, p_2\rangle$.

are shifted one another by the translations p_1 and p_2. Consequently, the local geometrical symmetry D_6 is broken, but the global translational symmetry $\mathbb{Z}_2 \times \tilde{\mathbb{Z}}_2$ is preserved.

16.6.3 Triple Bifurcation Point C

The triple bifurcation point C at $f = 0.0257$ corresponds to the irreducible representation $\mu = (3,2)$. There are bifurcated patterns of three kinds—mode I, mode II, and flower mode. Their planforms, which are portrayed in Fig. 16.11, differ from the planforms for $\mu = (3,3)$ in Fig. 16.9 because of the absence of the half-rotation symmetry r^3. The planform in Fig. 16.11(c) therefore serves as a new variant of the standard flower mode in Fig. 16.9(c).

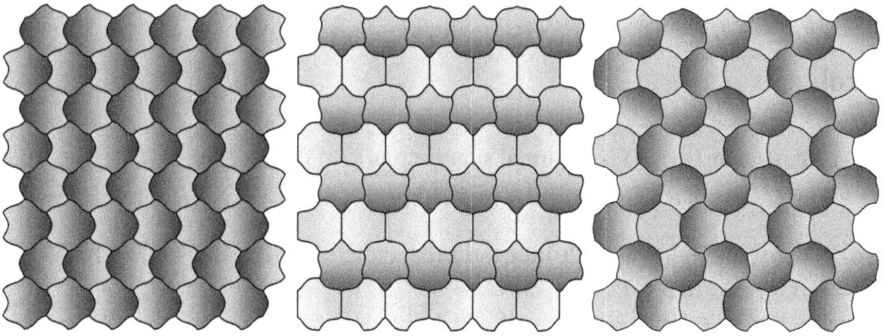

(a) Mode I $\Sigma(\boldsymbol{u}) = \Sigma(\boldsymbol{w}) = \langle p_2 r^3, s, p_1 \rangle$ (b) Mode II $\Sigma(\boldsymbol{u}) = \Sigma(\boldsymbol{w}) = \langle p_1 r^3, s \rangle$ (c) Flower mode $\Sigma(\boldsymbol{u}) = \Sigma(\boldsymbol{w}) = \langle r^2, s \rangle$

Fig. 16.11 Planforms for the triple bifurcation point C associated with $\mu = (3,2)$.

16.6.4 Triple Bifurcation Point D

Another variant of the flower mode is portrayed in Fig. 16.12, which corresponds to the bifurcation point D at the loading parameter $f = 0.0941$ associated with $\mu = (3,4)$. This flower mode differs from that for $\mu = (3,3)$ because of the symmetry with respect to $\pi/3$ rotation r^2 instead of that with respect to $\pi/6$ rotation and is also different from that for $\mu = (3,2)$ due to the presence of reflectional symmetries sr, sr^3, and sr^5, instead of s, sr^2, and sr^4.

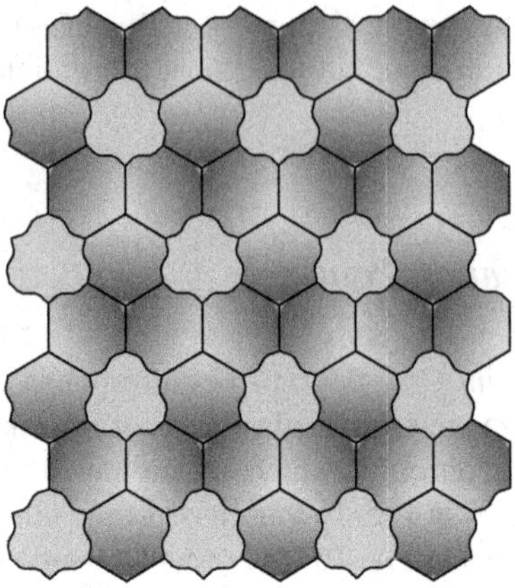

Fig. 16.12 Planform for the triple bifurcation point D associated with $\mu = (3,4)$: flower pattern with the symmetry of $\Sigma(\boldsymbol{u}) = \Sigma(\boldsymbol{w}) = \langle r^2, sr\rangle$.

Problems

16-1 Show that $\langle p_1, p_2\rangle$ is a normal subgroup of $G = \langle r, s, p_1, p_2\rangle$ in (16.1).
Answer: $r^{-1}p_1r = p_2$, $s^{-1}p_1s = p_1$, $r^{-1}p_2r = p_1p_2$, and $s^{-1}p_2s = p_1p_2$ by (16.2).

16-2 Verify the relation (7.20) for the irreducible representations (cf., §16.2.2) of $G = \langle r, s, p_1, p_2\rangle$ in (16.1).
Answer: $4\times 1^2 + 2\times 2^2 + 4\times 3^2 = 48 = |G|$.

16-3 Derive (16.22)–(16.26) from (16.10)–(16.21).

Summary

- Symmetries of bifurcating patterns of the 2×2 representative volume element have been obtained.
- Variants of flowerlike modes with different symmetries on a honeycomb structure have been found and classified.
- Illuminative geometrical patterns of these modes have been obtained by numerical bifurcation analysis.

References

1. Allgower, E., Böhmer, K., and Golubitsky M. eds. (1992) *Bifurcation and Symmetry.* International Series of Numerical Mathematics, Vol. 104. Birkhäuser, Basel.
2. Andersen, K.H., Abel, M., Krug, J., Ellegaard, C., Sondergaard, L.R., and Udesen, J. (2002) Pattern dynamics of vortex ripples in sand: Nonlinear modeling and experimental validation. *Phys. Rev. Lett.* **88**(23), 234302.
3. Antman, S.S. (1995) *Nonlinear Problems of Elasticity.* Applied Mathematical Sciences, Vol. 107. Springer-Verlag, New York; 2nd ed., 2005.
4. Arbocz, J. and Abramovich, H. (1979) *The Initial Imperfection Data Bank at the Delft University of Technology,* Part 1. Report LR-290. Department of Aerospace Engineering, Delft University of Technology.
5. Arbocz, J. and Hol, J.M.A.M. (1991) Collapse of axially compressed cylindrical shells with random imperfections. *AIAA J.* **29**(12), 2247–2256.
6. Archambault, G., Rouleau, A., Daigneault, R., and Flamand, R. (1993) Progressive failure of rock masses by a self-similar anastomosing process of rupture at all scales and its scale effect on their shear strength. In: *Scale Effects in Rock Masses,* Vol. 93, ed. A. Pinto da Cunha, pp. 133–141. Balkema, Rotterdam.
7. Asaoka, A. and Nakano, M. (1996) Deformation of triaxial clay specimen during undrained compression. Private communication.
8. Asaoka, A. and Noda, T. (1995) Imperfection-sensitive bifurcation of Cam-clay under plane strain compression with undrained boundaries. *Soils Foundations* **35**(1), 83–100.
9. Augusti, G., Barratta, A., and Casciati, F. (1984) *Probabilistic Methods in Structural Engineering.* Chapman and Hall, New York.
10. Bai, Y. and Dodd, B. (1992) *Adiabatic Shear Localization—Occurrence, Theories and Applications.* Pergamon Press, Oxford.
11. Bakker, P.G. (1991) *Bifurcations in Flow Patterns—Some Applications of the Qualitative Theory of Differential Equations in Fluid Dynamics.* Nonlinear Topics in the Mathematical Sciences, Vol. 2. Kluwer Academic, Dordrecht.
12. Bažant, Z.P. and Cedolin, L. (1991) *Stability of Structures.* Oxford University Press, New York.
13. Bénard, H. (1900) Les tourbillons cellulaires dans une nappe liquide. *Rev. Gén. Sci. Pure Appl.* **11**, 1261–1271, 1309–1328.
14. Ben-Haim, Y. and Elishakoff, I. (1990) *Convex Models of Uncertainty in Applied Mechanics.* Studies in Applied Mechanics, Vol. 25. Elsevier, Amsterdam.
15. Bolotin, V.V. (1958) *Statistical Methods in the Nonlinear Theory of Elastic Shells.* Izvestija Academii Nauk SSSR. Otdeleni Tekhnicheskikh Nauk, Vol. 3 (English translation, NASA TTF-85, 1962, 1–16).

16. Bolotin, V.V. (1969) *Statistical Methods in Structural Mechanics.* Holden-Day Series in Mathematical Physics, Vol. 7. Holden-Day, San Francisco.
17. Bolotin, V.V. (1984) *Random Vibrations of Elastic Systems.* Mechanics of Elastic Stability, Vol. 7. Martinus Nijhoff, The Hague.
18. Bossavit, A. (1986) Symmetry, groups, and boundary value problems—A progressive introduction to noncommutative harmonic analysis of partial differential equations in domains with geometrical symmetry. *Comput. Methods Appl. Mech. Engrg.* **56**, 167–215.
19. Bressloff, P.C., Cowan, J.D., Golubitsky, M., and Thomas, P.J. (2001) Scalar and pseudoscalar bifurcations: Pattern formation in the visual cortex. *Nonlinearity* **14**(4), 739–775.
20. Budiansky, B. (1974) Theory of buckling and post-buckling behavior of elastic structures. In: *Advances in Applied Mechanics,* Vol. 14, edited by C.-S. Yih, pp. 1–65. Academic Press, New York.
21. Buzano, E. and Golubitsky, M. (1983) Bifurcation on the hexagonal lattice and the planar Bénard problem. *Phil. Trans. Roy. Soc. London* **A 308**, 617–667.
22. Chadam, J., Golubitsky, M., Langford, W., and Wetton, B. eds. (1996) *Pattern Formation: Symmetry Methods and Applications.* Fields Institute Communications, Vol. 5. American Mathematical Society, Providence, RI.
23. Chandrasekhar, S. (1961) *Hydrodynamic and Hydromagnetic Stability.* Clarendon Press, London.
24. Chen, H.-C. and Sameh, A.H. (1989) A matrix decomposition method for orthotropic elasticity problems. *SIAM J. Matrix Anal. Appl.* **10**(1), 39–64.
25. Chillingworth, D.R.J. and Golubitsky, M. (2003) Planar pattern formation in bifurcation from a homeotropic nematic liquid crystal. In: *Bifurcation, Symmetry and Patterns*, ed. J. Buescu et al., Trends in Mathematics. Birkhäuser, Basel.
26. Chossat, P. (1994) *Dynamics, Bifurcation and Symmetry—New Trends and New Tools.* NATO Advanced Science Institutes Series C: Mathematical and Physical Sciences, Vol. 437. Kluwer Academic, Dordrecht.
27. Chossat, P. and Iooss, G. (1994) *The Couette–Taylor Problem.* Applied Mathematical Sciences, Vol. 102. Springer-Verlag, New York.
28. Chow, S. and Hale, J.K. (1982) *Methods of Bifurcation Theory.* Grundlehren der mathematischen Wissenschaften, Vol. 251. Springer-Verlag, New York.
29. Chow, S., Hale, J.K., and Mallet-Panet, J. (1975) Applications of generic bifurcations. I. *Arch. Rational Mech. Anal.* **59**, 159–188.
30. Chow, S., Hale, J.K., and Mallet-Panet, J. (1976) Applications of generic bifurcations. II. *Arch. Rational Mech. Anal.* **62**, 209–235.
31. Crawford, J.D. (1994) $D_4{+}T^2$ mode interactions and hidden rotational symmetry. *Nonlinearity* **7**(3), 697–739.
32. Crawford, J.D. and Knobloch, E. (1991) Symmetry and symmetry-breaking bifurcations in fluid dynamics. *Ann. Rev. Fluid Mech.* **23**, 341–387.
33. Crisfield, M.A. (1991) *Non-linear Finite Element Analysis of Solids and Structures*, Vol. 1. Wiley, Chichester.
34. Curtis, C.W. and Reiner, I. (1962) *Representation Theory of Finite Groups and Associative Algebras.* Wiley Classic Library, Vol. 45. Wiley (Interscience), New York.
35. Davis, G.H. (1984) *Structural Geology of Rocks & Regions.* Wiley, Singapore; 2nd ed., Davis, G.H. and Reynolds, S.J., 1996.
36. de Borst, R. (1988) Bifurcation in finite element models with non-associated flow law. *Internat. J. Numer. Analyt. Methods Geomech.* **12**, 99–116.
37. Dellnitz, M. and Werner, B. (1989) Computational methods for bifurcation problems with symmetries—With special attention to steady state and Hopf bifurcation points. *J. Comput. Appl. Math.* **26**, 97–123.
38. Desrues, J., Lanier, J., and Stutz, P. (1985) Localization of the deformation in tests on sand samples. *Engrg. Fracture Mech.* **21**(4), 909–921.

39. Desrues, J. and Viggiani, G. (2004) Strain localization in sand: An overview of the experimental results obtained in Grenoble using stereophotogrammetry. *Internat. J. Numer. Anal. Meth. Geomech.* **28**(4), 279–321.
40. Dinkevich, S. (1984) The spectral method of calculation of symmetric structures of finite size. *Trans. Canad. Soc. Mech. Engrg.* **8**(4), 185–194.
41. Dinkevich, S. (1991) Finite symmetric systems and their analysis. *Internat. J. Solids Structures* **27**(10), 1215–1253.
42. Dionne, B. and Golubitsky, M. (1992) Planforms in two and three dimensions. *Zeit. angew. Math. Phys.* **43**(1), 36–62.
43. Drazin, P.G. and Reid, W.H. (1981) *Hydrodynamic Stability*. Cambridge University Press, Cambridge, UK; 2nd ed., 2004.
44. Elishakoff, I. (1983) *Probabilistic Methods in the Theory of Structures*. Wiley Interscience, New York.
45. Elishakoff, I. and Arbocz, J. (1982) Reliability of axially compressed cylindrical shells with random axisymmetric imperfections. *Internat. J. Solids Structures* **18**(7), 563–585.
46. Elishakoff, I., Li, Y.W., and Starnes Jr., J.H. (2001) *Non-Classical Problems in the Theory of Elastic Stability*. Cambridge University Press, Cambridge, UK.
47. Elishakoff, I., Lin, Y.K., and Zhu, L.P. (1994) *Probabilistic and Convex Modelling of Acoustically Excited Structures*. Studies in Applied Mechanics, Vol. 39. Elsevier, Amsterdam.
48. Elishakoff, I., van Manen, S., Vermeulen, P.G., and Arbocz, J. (1987) First-order second-moment analysis of the buckling of shells with random initial imperfections. *AIAA J.* **25**(8), 1113–1117.
49. El Naschie, M.S. (1990) *Stress, Stability and Chaos in Structural Engineering: An Energy Approach*. McGraw-Hill, London.
50. Fujii, F. and Noguchi, H. (2002) The buckling mode extracted from the $LDL^{\top}$-decomposed large-order stiffness matrix. *Commun. Numer. Methods Engrg.* **18**(7), 459–467.
51. Fujii, H., Mimura, M., and Nishiura, Y. (1982) A picture of the global bifurcation diagram in ecological interacting and diffusing systems. *Physica D* **5**(1), 1–42.
52. Fujii, H. and Yamaguti, M. (1980) Structure of singularities and its numerical realization in nonlinear elasticity. *J. Math. Kyoto Univ.* **20**(3), 489–590.
53. Gajo, A., Bigoni, D., and Wood, D.M. (2004) Multiple shear band development and related instabilities in granular materials. *J. Mech. Phys. Solids* **52**(12), 2683–2724.
54. Galambos, J. (1978) *The Asymptotic Theory of Extreme Order Statistics*, Wiley, New York; 2nd ed., R.E. Krieger, Malabar, FL, 1987.
55. Gantmacher, F.R. (1959) *The Theory of Matrices*, Vol. I, Vol. II. Chelsea, New York.
56. Gatermann, K. (2000) *Computer Algebra Methods for Equivariant Dynamical Systems*. Lecture Notes in Mathematics, Vol. 1728. Springer-Verlag, Berlin.
57. Gatermann, K. and Hohmann, A. (1991) Symbolic exploitation of symmetry in numerical path-following. *Impact Comput. Sci. Engrg.* **3**, 330–365.
58. Gatermann, K. and Hosten, S. (2005) Computational algebra for bifurcation theory. *J. Symbolic Comput.* **40**, 1180–1207.
59. Gatermann, K. and Werner, B. (1994) Group theoretical mode interactions with different symmetries. *Internat. J. Bifurcation Chaos* **4**(1), 177–191.
60. Gibson, L.J. and Ashby, M.F. (1997) *Cellular Solids: Structure and Properties*, 2nd ed. Cambridge University Press, Cambridge, UK.
61. Godoy, L.A. (2000) *Theory of Elastic Stability—Analysis and Sensitivity*. Taylor & Francis, Philadelphia.
62. Golubitsky, M. and Schaeffer, D.G. (1985) *Singularities and Groups in Bifurcation Theory*, Vol. 1. Applied Mathematical Sciences, Vol. 51. Springer-Verlag, New York.
63. Golubitsky, M. and Stewart, I. (2002) *The Symmetry Perspective: From Equilibrium to Chaos in Phase Space and Physical Space*. Progress in Mathematics, Vol. 200. Birkhäuser, Basel.

64. Golubitsky, M., Stewart, I., and Schaeffer, D.G. (1988) *Singularities and Groups in Bifurcation Theory*, Vol. 2. Applied Mathematical Sciences, Vol. 69. Springer-Verlag, New York.
65. Govaerts, W.J.F. (2000) *Numerical Methods for Bifurcations of Dynamical Equilibria.* SIAM, Philadelphia.
66. Guo, X.E. and Gibson, L.J. (1999) Behavior of intact and damaged honeycomb: a finite element study. *Internat. J. Mech. Sci.* **41**(1), 85–105.
67. Hale, J. and Koçak, H. (1991) *Dynamics and Bifurcations.* Texts in Applied Mathematics, Vol. 3. Springer-Verlag, New York.
68. Hamermesh, M. (1962) *Group Theory and Its Application to Physical Problems.* Addison-Wesley Series in Physics. Addison-Wesley, Reading, MA.
69. Healey, T.J. (1985) Symmetry, bifurcation, and computational methods in nonlinear structural mechanics. Ph.D Thesis. University of Illinois.
70. Healey, T.J. (1988) A group theoretic approach to computational bifurcation problems with symmetry. *Comput. Methods Appl. Mech. Engrg.* **67**, 257–295.
71. Healey, T.J. and Treacy, J.A. (1991) Exact block diagonalization of large eigenvalue problems for structures with symmetry. *Internat. J. Numer. Methods Engrg.* **31**, 265–285.
72. Hill, R. (1958) A general theory of uniqueness and stability in elastic–plastic solids. *J. Mech. Phys. Solids* **16**, 236–240.
73. Hill, R. and Hutchinson, J.W. (1975) Bifurcation phenomena in the plane tension test. *J. Mech. Phys. Solids* **23**, 239–264.
74. Ho, D. (1974) Buckling load of non-linear systems with multiple eigenvalues. *Internat. J. Solids Structures* **10**, 1315–1330.
75. Hoyle, R. (2006) *Pattern Formation: An Introduction to Methods.* Cambridge Texts in Applied Mathematics. Cambridge University Press, Cambridge, UK.
76. Hui, D. (1986) Imperfection sensitivity of axially compressed laminated flat plate due to bending–stretching coupling. *Internat. J. Solids Structures* **22**(1), 13–22.
77. Hui, D. and Chen, Y.H. (1987) Imperfection-sensitivity of cylindrical panels under compression using Koiter's improved postbuckling theory. *Internat. J. Solids Structures* **23**(7), 969–982.
78. Hunt, G.W. (1977) Imperfection-sensitivity of semi-symmetric branching. *Proc. Roy. Soc. London Ser. A* **357**, 193–211.
79. Hunt, G.W., Bolt, H.M., and Thompson, J.M.T. (1989) Structural localization phenomena and the dynamical phase-space analogy. *Proc. Roy. Soc. London Ser. A* **425**, 245–267.
80. Hutchinson, J.W. and Koiter, W.T. (1970) Postbuckling theory. *Appl. Mech. Rev.* **23**(12), 1353–1366.
81. Hutchinson, J.W. and Miles, J.P. (1974) Bifurcation analysis of the onset of necking in an elastic/plastic cylinder under uniaxial tension. *J. Mech. Phys. Solids* **22**, 61–71.
82. Ikeda, K., Chida, T., and Yanagisawa, E. (1997) Imperfection sensitive strength variation of soil specimens. *J. Mech. Phys. Solids* **45**(2), 293–315.
83. Ikeda, K., Maruyama, K., Ishida, H., and Kagawa, S. (1997) Bifurcation in compressive behavior of concrete. *ACI Materials J.* **94**(6), 484–491.
84. Ikeda, K., Murakami, S., Saiki, I., Sano, I., and Oguma, N. (2001) Image simulation of uniform materials subjected to recursive bifurcation. *Internat. J. Engrg. Sci.* **39**(17), 1963–1999.
85. Ikeda, K. and Murota, K. (1990) Critical initial imperfection of structures. *Internat. J. Solids Structures* **26**(8), 865–886.
86. Ikeda, K. and Murota, K. (1990) Computation of critical initial imperfection of truss structures. *J. Engrg. Mech. Div. ASCE* **116**(10), 2101–2117.
87. Ikeda, K. and Murota, K. (1991) Bifurcation analysis of symmetric structures using block-diagonalization. *Comput. Methods Appl. Mech. Engrg.* **86**(2), 215–243.
88. Ikeda, K. and Murota, K. (1991) Random initial imperfections of structures. *Internat. J. Solids Structures* **28**(8), 1003–1021.

89. Ikeda, K. and Murota, K. (1993) Statistics of normally distributed initial imperfections. *Internat. J. Solids Structures* **30**(18), 2445–2467.
90. Ikeda, K. and Murota, K. (1996) Bifurcation as sources of uncertainty in soil shearing behavior. *Soils Foundations* **36**(1), 73–84.
91. Ikeda, K. and Murota, K. (1997) Recursive bifurcation as sources of complexity in soil shearing behavior. *Soils Foundations* **37**(3), 17–29.
92. Ikeda, K. and Murota, K. (1999) Systematic description of imperfect bifurcation behavior of symmetric systems. *Internat. J. Solids Structures* **36**, 1561–1596.
93. Ikeda, K., Murota, K., and Elishakoff, I. (1996) Reliability of structures subject to normally distributed initial imperfections. *Comput. & Structures* **59**(3), 463–469.
94. Ikeda, K., Murota, K., and Fujii, H. (1991) Bifurcation hierarchy of symmetric structures. *Internat. J. Solids Structures* **27**(12), 1551–1573.
95. Ikeda, K., Murota, K., and Nakano, M. (1994) Echelon modes in uniform materials. *Internat. J. Solids Structures* **31**(19), 2709–2733.
96. Ikeda, K., Murota, K., Yamakawa, Y., and Yanagisawa, E. (1997) Mode switching and recursive bifurcation in granular materials. *J. Mech. Phys. Solids* **45**(11–12), 1929–1953.
97. Ikeda, K., Murota, K., Yanagimoto, A., and Noguchi, H. (2007) Improvement of the scaled corrector method for bifurcation analysis using symmetry-exploiting block-diagonalization. *Comput. Methods Appl. Mech. Engrg.* **196**(9-12), 1648–1661.
98. Ikeda, K., Nishino, F., Hartono, W., and Torii, K. (1988) Bifurcation behavior of an axisymmetric elastic space truss. *Proc. JSCE, Structural Engrg./Earthquake Engrg.* **392**(I-9), 231–234.
99. Ikeda, K., Ohsaki, M., and Kanno, Y. (2005) Imperfection sensitivity of hilltop branching points of systems with dihedral group symmetry. *Internat. J. Nonlinear Mechanics* **40**(5), 755–774.
100. Ikeda, K., Oide, K., and Terada, K. (2002) Imperfection sensitive strength variation at hilltop bifurcation point. *Internat. J. Engrg. Sci.* **40**(7), 743–772.
101. Ikeda, K., Okazawa, S., Terada, K., Noguchi, H., and Usami, T. (2001) Recursive bifurcation of tensile steel specimens. *Internat. J. Engrg. Sci.* **39**(17), 1913–1934.
102. Ikeda, K., Providéncia, P., and Hunt, G.W. (1993) Multiple equilibria for unlinked and weakly-linked cellular forms. *Internat. J. Solids Structures* **30**(3), 371–384.
103. Ikeda, K., Sasaki, H., and Ichimura, T. (2006) Diffuse mode bifurcation of soil causing vortex-like shear investigated by group-theoretic image analysis. *J. Mech. Phys. Solids* **54**(2), 310–339.
104. Ikeda, K., Yamakawa, Y., Desrues, J., and Murota, K. (2008) Bifurcations to diversify geometrical patterns of shear bands on granular material. *Phys. Rev. Lett.* **100**(19) 198001.
105. Ikeda, K. Yamakawa, Y., and Tsutsumi, S. (2003) Simulation and interpretation of diffuse mode bifurcation of elastoplastic solids. *J. Mech. Phys. Solids* **51**(9), 1649–1673.
106. Iooss, G. (1986) Secondary bifurcations of Taylor vortices into wavy inflow or outflow boundaries. *J. Fluid Mech.* **173**, 273–288.
107. Iooss, G. and Adelmeyer, M. (1998) *Topics in Bifurcation Theory and Applications*, 2nd ed. Advanced Series Nonlinear Dynamics, Vol. 3. World Scientific, Singapore.
108. Iooss, G. and Joseph, D.D. (1990) *Elementary Stability and Bifurcation Theory*, 2nd ed. Undergraduate Texts in Mathematics. Springer-Verlag, New York.
109. Jacobson, N. (1989) *Basic Algebra I*, 2nd ed. Freeman, New York.
110. Judd, S.L. and Silber, M. (2000) Simple and superlattice turning patterns in reaction-diffusion systems: bifurcation, bistability, and parameter collapse. *Physica D* **136**(1-2), 45–65.
111. Keener, J.P. (1974) Perturbed bifurcation theory at multiple eigenvalues. *Arch. Rational Mech. Anal.* **56**, 348–366.
112. Keener, J.P. (1979) Secondary bifurcation and multiple eigenvalues. *SIAM J. Appl. Math.* **37**(2), 330–349.

113. Keener, J.P. and Keller, H.B. (1973) Perturbed bifurcation theory. *Arch. Rational Mech. Anal.* **50**, 159–175.
114. Keller, J.B. and Antman, S. eds. (1969) *Bifurcation Theory and Nonlinear Eigenvalue Problems.* Mathematics Lecture Note Series. Benjamin, New York.
115. Kendall, M. and Stuart, A. (1977) *The Advanced Theory of Statistics,* Vol. 1, 4th ed. Charles Griffin, London; 4th revised ed., Hodder Arnold, 1982.
116. Kettle, S.F.A. (1995) *Symmetry and Structure*, 2nd ed. Wiley, Chichester; 3rd ed., 2007.
117. Kim, S.K. (1999) *Group Theoretical Methods and Applications to Molecules and Crystals.* Cambridge University Press, Cambridge, UK.
118. Kirchgässner, K. (1979) Exotische Lösungen Bénardschen Problems, *Math. Meth. Appl. Sci.* **1**, 453–467.
119. Koiter, W.T. (1945) *On the Stability of Elastic Equilibrium.* Dissertation. Delft, Holland (English translation: NASA Technical Translation F10: 833, 1967).
120. Koiter, W.T. (1963) The effect of axisymmetric imperfections on the buckling of cylindrical shells under axial compression. *Proc. Roy. Netherlands Acad. Sci. Ser.* 13, **66**(5), 265–279.
121. Koiter, W.T. (1976) Current trends in the theory of buckling. In: *Buckling of Structures, Proceedings of the IUTAM Symposium at Cambridge*, pp. 1–16. Springer-Verlag, Berlin.
122. Kolymbas, D. (1981) Bifurcation analysis for sand samples with a non-linear constitutive equation. *Ingenieur-Archiv* **50**, 131–140.
123. Koschmieder, E.L. (1966) On convection of a uniformly heated plane. *Beitr. zur Phys. Atmosphäre* **39**, 1–11.
124. Koschmieder, E.L. (1974) Benard convection. *Adv. in Chem. Phys.* **26**, 177–188.
125. Koschmieder, E. L. (1993) *Bénard Cells and Taylor Vortices*. Cambridge Monograph on Mechanics and Applied Mathematics. Cambridge University Press, Cambridge, UK.
126. Krasnosel'skii, M.A. (1964) *Topological Methods in the Theory of Nonlinear Integral Equations.* International Series of Monographs on Pure and Applied Mathematics, Vol. 133. Pergamon Press, Oxford.
127. Kudrolli, A., Pier, B., and Gollub, J.P. (1998) Superlattice patterns in surface waves. *Physica D* **123**(1-4), 99–111.
128. Kuznetsov, Y.A. (1995) *Elements of Applied Bifurcation Theory.* Applied Mathematical Sciences, Vol. 112. Springer-Verlag, New York; 2nd ed., 1998; 3rd ed., 2004.
129. Lindberg, H.E. and Florence, A.L. (1987) *Dynamic Pulse Buckling.* Mechanics of Elastic Stability, Vol. 12. Martinus Nijhoff, Dordrecht.
130. Ludwig, W. and Falter, C. (1996) *Symmetries in Physics—Group Theory Applied to Physical Problems,* 2nd ed. Springer Series Solid-State Sciences, Vol. 64. Springer-Verlag, Berlin.
131. Maehara, T. and Murota, K. (2010) A numerical algorithm for block-diagonal decomposition of matrix $*$-algebras with general irreducible components, *Japan J. Indust. Appl. Math.* **27**, to appear.
132. Mainzer, K. (2005) *Symmetry and Complexity: The Spirit and Beauty of Nonlinear Science.* World Scientific Series on Nonlinear Science, Series A, Vol. 51. World Scientific, Singapore.
133. Marsden, J.E. and Hughes, T.J.R. (1983) *Mathematical Foundations of Elasticity.* Prentice-Hall Civil Engineering and Engineering Mechanics Series. Prentice-Hall, Englewood Cliffs, NJ.
134. Marsden, J.E. and Ratiu, S.R. (1994) *Introduction to Mechanics and Symmetry.* Texts in Applied Mathematics, Vol. 17. Springer-Verlag, New York; 2nd ed., 1999.
135. Matkowsky, B.J. and Reiss, E.L. (1977) Singular perturbations of bifurcations. *SIAM J. Appl. Math.* **33**(2), 230–255.
136. Melbourne, I. (1999) Steady-state bifurcation with Euclidean symmetry. *Trans. Amer. Math. Soc.* **351**(4), 1575–1603.

137. Melo, F., Umbanhowar, P.B., and Swinney, H.L. (1995) Hexagons, kinks, and disorder in oscillated granular layers. *Phys. Rev. Lett.* **75**(21), 3838–3841.
138. Miller, W., Jr. (1972) *Symmetry Groups and Their Applications.* Pure and Applied Mathematics, Vol. 50. Academic Press, New York.
139. Mitropolsky, Yu. A. and Lopatin A. K. (1988) *Nonlinear Mechanics, Groups and Symmetry.* Mathematics and Its Applications. Kluwer Academic, Dordrecht.
140. Moehlis, J. and Knobloch, E. (2000) Bursts in oscillatory systems with broken D_4 symmetry. *Physica D* **135**, 263–304.
141. Morgenstern, N.R. and Tchalenko, J.S. (1967) Microscopic structures in kaolin subjected to direct shear. *Géotechnique* **17**, 309–328.
142. Murota, K. and Ikeda, K. (1991) Computational use of group theory in bifurcation analysis of symmetric structures. *SIAM J. Sci. Statist. Comput.* **12**(2), 273–297.
143. Murota, K. and Ikeda, K. (1991) Critical imperfection of symmetric structures. *SIAM J. Appl. Math.* **51**(5) 1222–1254.
144. Murota, K. and Ikeda, K. (1992) On random imperfections for structures of regular-polygonal symmetry. *SIAM J. Appl. Math.* **52**(6), 1780–1803.
145. Murota, K., Ikeda, K., and Terada, K. (1999) Bifurcation mechanism underlying echelon mode formation. *Comp. Methods Appl. Mech. Engrg.* **170**(3–4), 423–448.
146. Murota, K., Kanno, Y., Kojima, M., and Kojima, S. (2010) A numerical algorithm for block-diagonal decomposition of matrix $*$-algebras with application to semidefinite programming. *Japan J. Indust. Appl. Math.* **27**, DOI 10.1007/s13160-010-0006-9.
147. Nakano, M. (1993) *Analysis of Undrained and Partially Drained Behavior of Soils and Application to Soft Clays under Embankment Loading.* Doctoral Thesis. Department of Civil Engineering, Nagoya University (in Japanese).
148. Needleman, A. (1972) A numerical study of necking in circular cylindrical bars. *J. Mech. Phys. Solids* **20**, 111–127.
149. Nicolas, A. (1995) *The Mid-Oceanic Ridges—Mountains below Sea Level.* Springer-Verlag, Berlin.
150. Noda, T. (1994) *Elasto-Plastic Behavior of Soil near/at Critical State and Soil–Water Coupled Finite Deformation Analysis.* Doctoral Thesis. Department of Civil Engineering, Nagoya University (in Japanese).
151. Noguchi, H. and Hisada, T. (1993) Sensitivity analysis in post-buckling problems of shell structures. *Comput. Struct.* **47**(4/5), 699–710.
152. Oden, J.T. and Ripperger, E.A. (1981) *Mechanics of Elastic Structures,* 2nd ed. Hemisphere, Washington, DC.
153. Ohno, N., Okumura, D., and Noguchi, H. (2002) Microscopic symmetric bifurcation condition of cellular solids based on a homogenization theory of finite deformation. *J. Mech. Phys. Solids* **50**(5), 1125–1153.
154. Ohsaki, M. and Ikeda, K. (2006) Imperfection sensitivity analysis of hill-top branching with many symmetric bifurcation points. *Internat. J. Solids Structures* **43**(16), 4704–4719.
155. Ohsaki, M. and Ikeda, K. (2007) *Stability and Optimization of Structures—Generalized Sensitivity Analysis.* Mechanical Engineering Series, Springer-Verlag, New York.
156. Okamoto, H. and Shoji, M. (2001) *The Mathematical Theory of Permanent Progressive Water–Waves.* Advanced Series in Nonlinear Dynamics. World Scientific, Singapore.
157. Okazawa, S., Oide, K., Ikeda, K., and Terada, K. (2002) Imperfection sensitivity and probabilistic variation of tensile strength of steel members. *Internat. J. Solids Structures,* **39**(6), 1651–1671.
158. Olver, P.J. (1986) *Applications of Lie Groups to Differential Equations.* Graduate Texts in Mathematics, Vol. 107. Springer-Verlag, New York; 2nd ed., 2000.
159. Olver, P.J. (1995) *Equivariance, Invariants, and Symmetry.* Cambridge University Press, Cambridge, UK.

160. Palassopoulos, G.V. (1989) Optimization of imperfection-sensitive structures. *J. Engrg. Mech. ASCE* **115**(8), 1663–1682.
161. Papka, S.D. and Kyriakides, S. (1999) Biaxial crushing of honeycombs—Part I: Experiments. *Internat. J. Solids Structures* **36**(29), 4367–4396.
162. Peacock, T., Mullin, T., and Binks, D. J. (1999) Bifurcation phenomena in flows of a nematic liquid crystal. *Internat. J. Bifurcation Chaos Appl. Sci. Engrg.* **9**(2), 427–441.
163. Peek, R. and Kheyrkhahan, M. (1993) Postbuckling behavior and imperfection sensitivity of elastic structures by the Lyapunov–Schmidt–Koiter approach. *Comput. Methods Appl. Mech. Engrg.* **108**, 261–279.
164. Peek, R. and Triantafyllidis, N. (1992) Worst shapes of imperfections for space trusses with many simultaneously buckling modes. *Internat. J. Solids Structures* **29**(19), 2385–2402.
165. Petryk, H. and Thermann, K. (1992) On discretized plasticity problems with bifurcations. *Internat. J. Solids Structures* **29**(6), 745–765.
166. Pignataro, M., Rizzi, N., and Luongo, A. (1991) *Stability, Bifurcation and Postcritical Behaviour of Elastic Structures.* Developments in Civil Engineering, Vol. 39. Elsevier, Amsterdam.
167. Poirier, J.-P. (1985) *Creep of Crystals: High-Temperature Deformation Processes in Metals, Ceramics and Minerals.* Cambridge Earth Science Series, Vol. 4. Cambridge University Press, Cambridge, UK.
168. Poston, T. and Stewart, I. (1978) *Catastrophe Theory and Its Applications.* Survey and Reference Works in Mathematics. Pitman, London.
169. Prince, E. (1994) *Mathematical Techniques in Crystallography and Materials Science.* Springer-Verlag, Berlin.
170. Rabinovich, M.I., Ezersky, A.B., and Weidman, P.D. (2000) *The Dynamics of Patterns.* World Scientific, Singapore.
171. Reiss, E.L. (1977) Imperfect bifurcation. In: *Applications of Bifurcation Theory,* ed. P.H. Rabinowitz, pp. 37–71. Academic Press, New York.
172. Roorda, J. (1965) *The Instability of Imperfect Elastic Structures.* Ph.D Dissertation. University of London.
173. Roorda, J. and Hansen, J.S. (1972) Random buckling behavior in axially loaded cylindrical shells with axisymmetric imperfections. *J. Spacecraft* **9**(2), 88–91.
174. Rosen, J. (1995) *Symmetry in Science.* Springer-Verlag, New York.
175. Saiki, I., Ikeda, K., and Murota, K. (2005) Flower patterns appearing on a honeycomb structure and their bifurcation mechanism. *Internat. J. Bifurcation Chaos Appl. Sci. Engrg.* **15**(2), 497–515.
176. Sattinger, D.H. (1979) *Group Theoretic Methods in Bifurcation Theory.* Lecture Notes in Mathematics, Vol. 762. Springer-Verlag, Berlin.
177. Sattinger, D.H. (1980) Bifurcation and symmetry breaking in applied mathematics. *Bull. Amer. Math. Soc.* **3**(2), 779–819.
178. Sattinger, D.H. (1983) *Branching in the Presence of Symmetry.* Regional Conference Series in Applied Mathematics, Vol. 40. SIAM, Philadelphia.
179. Schaeffer, D.G. (1980) Qualitative analysis of a model for boundary effects in the Taylor problem. *Math. Proc. Cambridge Philos. Soc.* **87**, 307–337.
180. Schofield, A.N. and Wroth, C.P. (1968) *Critical State Soil Mechanics.* McGraw-Hill, New York.
181. Serre, J.-P. (1977) *Linear Representations of Finite Groups.* Graduate Texts in Mathematics, Vol. 42. Springer-Verlag, New York.
182. Seydel, R. (1979) Numerical computation of branch points in ordinary differential equations. *Numer. Math.* **32**, 51–68.
183. Seydel, R. (1979) Numerical computation of branch points in nonlinear equations. *Numer. Math.* **33**, 339–352.
184. Seydel, R. (1994) *Practical Bifurcation and Stability Analysis,* 2nd ed. Interdisciplinary Applied Mathematics, Vol. 5. Springer-Verlag, New York; 3rd ed., 2009.

185. Singer, J., Abramovich, H., and Yaffe, R. (1978) *Initial Imperfection Measurements of Integrally Stringer-Stiffened Cylindrical Shells.* TAE Report 330. Department of Aeronautical Engineering, Technion, Israel Institute of Technology.
186. Stewart, I. and Golubitsky, M. (1992) *Fearful Symmetry—Is God a Geometer?* Blackwell, Oxford.
187. Tanaka, R., Saiki, I., and Ikeda, K. (2002) Group-theoretic bifurcation mechanism for pattern formation in three-dimensional uniform materials. *Internat. J. Bifurcation Chaos Appl. Sci. Engrg.* **12**(12), 2767–2797.
188. Taylor, G.I. (1923) The stability of a viscous fluid contained between two rotating cylinders. *Phil. Trans. Roy. Soc. London Ser. A* **223**, 289–343.
189. Terzaghi, K. and Peck, R.B. (1967) *Soil Mechanics in Engineering Practice,* 2nd ed. Wiley, New York; 3rd ed., Terzaghi, K., Peck, R.B., and Mesri, G., 1996.
190. Thompson, J.M.T. (1982) *Instabilities and Catastrophes in Science and Engineering.* Wiley Interscience, Chichester.
191. Thompson, J.M.T. and Hunt, G.W. (1973) *A General Theory of Elastic Stability.* Wiley, New York.
192. Thompson, J.M.T. and Hunt, G.W. (1984) *Elastic Instability Phenomena.* Wiley, Chichester.
193. Thompson, J.M.T. and Schorrock, P.A. (1975) Bifurcation instability of an atomic lattice. *J. Mech. Phys. Solids* **23**, 21–37.
194. Timoshenko, S.P. and Gere, J.M. (1963) *Theory of Elastic Stability.* McGraw-Hill, Tokyo.
195. Timoshenko, S. and Woinowsky–Krieger, S. (1959) *Theory of Plates and Shells,* 2nd ed. McGraw-Hill, New York.
196. Triantafyllidis, N. and Peek, R. (1992) On stability and the worst imperfection shape in solids with nearly simultaneous eigenmodes. *Internat. J. Solids Structures* **29**(18), 2281–2299.
197. Troger, H. and Steindl, A. (1991) *Nonlinear Stability and Bifurcation Theory: An Introduction for Engineers and Applied Scientists.* Springer-Verlag, Vienna.
198. Tvergaard, V. and Needleman, A. (1984) Analysis of the cup-cone fracture in a round tensile bar. *Acta Metall.* **32**(1), 157–169.
199. Tvergaard, V., Needleman, A., and Lo, K.K. (1981) Flow localization in the plane strain tensile test. *J. Mech. Phys. Solids* **29**(2), 115–142.
200. van der Waerden, B.L. (1955) *Algebra.* Springer-Verlag, New York.
201. Vardoulakis, I., Goldscheider, M., and Gudehus, G. (1978) Formation of shear bands in sand bodies as a bifurcation problem. *Internat. J. Numer. Anal. Methods Geomech.* **2**(2), 99–128.
202. Vardoulakis, I. and Sulem, J. (1995) *Bifurcation Analysis in Geomechanics.* Blackie Academic & Professional, Glasgow.
203. Venkataramani, S.C. and Ott, E. (1998) Spatiotemporal bifurcation phenomena with temporal period doubling: patterns in vibrated sand. *Phys. Rev. Lett.* **80**(16), 3495–3498.
204. von Kármán, T., Dunn, L.G., and Tsien, H.S. (1940) The influence of curvature on the buckling characteristics of structures. *J. Aero. Sci.* **7**(7), 276–289.
205. Voskamp, A.P. and Hoolox, G.E. (1998) Failsafe rating of ball bearing components. In: *Effect of Steel Manufacturing Processes on the Quality of Bearing Steels*, ed. J.J.C. Hoo, pp. 102–112, ASTM STP **987**.
206. Weibull, W. (1939) A statistical theory of the strength of materials. *Roy. Swedish Inst. Engrg. Res., Proc.* **151**.
207. Weibull, W. (1951) A statistical distribution function of wide applicability. *J. Appl. Mech.* **18**, 293–297.
208. Werner, B. and Spence, A. (1984) The computation of symmetry-breaking bifurcation points. *SIAM J. Numer. Anal.* **21**(2), 388–399.
209. Weyl, H. (1952) *Symmetry.* Princeton University Press, Princeton, NJ.

210. Wiggins, S. (1988) *Global Bifurcations and Chaos.* Applied Mathematical Sciences, Vol. 73. Springer-Verlag, New York.
211. Wolf, H., König, D., and Triantafyllidis, T. (2003) Experimental investigation of shear band patterns in granular material. *J. Struct. Geology* **25**(8), 1229–1240.
212. Wriggers, P. and Simo, J.C. (1990) A general procedure for the direct computation of turning and bifurcation points. *Internat. J. Numer. Methods Engrg.* **30**, 155–176.
213. Yamaki, N. (1984) *Elastic Stability of Circular Cylindrical Shells.* Applied Mathematics and Mechanics, Vol. 27. Elsevier, Amsterdam.
214. Ziegler, H. (1968) *Principles of Structural Stability,* GinnBlaisdel, Lexington; 2nd ed. Birkhäuser, Basel, 1977.
215. Zloković, G. (1989) *Group Theory and G-Vector Spaces in Structural Analysis.* Ellis Horwood, Chichester.

Index